T0320986

AN ILLUSTRATIVE INTRODUCTION TO MODERN ANALYSIS

AN ILLUSTRATIVE INTRODUCTION TO MODERN ANALYSIS

Nikos Katzourakis and **Eugen Vărvărucă**

CRC Press
Taylor & Francis Group
Boca Raton London New York

CRC Press is an imprint of the
Taylor & Francis Group, an **informa** business

A CHAPMAN & HALL BOOK

CRC Press
Taylor & Francis Group
6000 Broken Sound Parkway NW, Suite 300
Boca Raton, FL 33487-2742

© 2018 by Taylor & Francis Group, LLC
CRC Press is an imprint of Taylor & Francis Group, an Informa business

No claim to original U.S. Government works

Printed on acid-free paper
Version Date: 20171109

International Standard Book Number-13: 978-1-138-71827-2 (Hardback)

Library of Congress Cataloging-in-Publication Data

Names: Katzourakis, Nikos, author. | Varvaruca, Eugen, author.
Title: An Illustrative Introduction to Modern Analysis/Nikolaos Katzourakis and Eugen Varvaruca.
Description: Boca Raton : CRC Press, 2018. | Includes bibliographical references.
Identifiers: LCCN 2017032755 | ISBN 9781138718272 (hardback : alk. paper)
Subjects: LCSH: Mathematical analysis–Textbooks.
Classification: LCC QA300 .K38 2018 | DDC 515–dc23
LC record available at https://lccn.loc.gov/2017032755

Visit the Taylor & Francis Web site at
http://www.taylorandfrancis.com

and the CRC Press Web site at
http://www.crcpress.com

To our families

Contents

Preface

As the title suggests, this is a textbook on what we consider the cornerstones of modern Analysis, which we believe every fellow mathematician should learn during their academic studies. This does not mean that we have attempted to write an encyclopaedia of Analysis. On the contrary, the vastness of the field has forced us to make a drastic selection of topics, and to concentrate our efforts on giving a contemporary account of *the most fundamental notions and results*. Although the subject matter addressed here inherently involves a high level of abstraction, we have tried to present the relevant results in the *greatest possible simplicity*, prioritising clarity of presentation over (extreme) generality. The selection of topics is partly aligned with our personal taste—mostly inspired by applications to the theory of nonlinear Partial Differential Equations—but we nonetheless believe that the choices we have made will be considered appropriate by most of our colleagues.

Our textbook is primarily aimed at undergraduate and postgraduate level university students in the Mathematical Sciences, as well as students of other areas in which Mathematical Analysis constitutes an essential part of their background knowledge, for instance in Engineering and the Physical Sciences. We also hope that it might serve as a useful teaching resource for colleague mathematicians, as well as an accessible reference book for research students and academics of neighbouring research areas in which Analysis plays a crucial role, typically through the use of Differential Equations and of Mathematical Modelling.

Most definitely, there exist many classical and modern textbooks treating the topics we expound on here, of various levels of difficulty and addressed to a variety of audiences. *The central feature of our enterprise, which distinguishes it from other works* (reflected also on the title) is the attempt to *draw on the readers' geometric intuition* by promoting an *illustrative approach*, whenever possible, to the classical results of Analysis. To this end, our arguments are accompanied by almost one hundred (99) graphic images. *This, however, does not in any sense mean that we compromised on the mathematical rigour of our exposition*. Rather, the approach taken in this book is a reflection of our view that quite often mathematics is the endeavour to transform intuitive ideas into rigorous proofs.

Moreover, we give *lengthy but targeted motivations of "why we do what we do" as well as detailed comments on the "know-how"*, typically in the form of footnotes in order to avoid distracting the flow of the main text. Extensions towards more advanced themes which are not covered in this elementary exposition are also hinted. We aspire that *the outcome is an accessible and elementary textbook which does not contain just "dry" theorems and proofs.* Additionally, we have included numerous explicit examples and sometimes directions for further reading. Furthermore, our style of exposition is slightly informal and occasionally at variance from standard academic stereotypes, in an attempt to address our potential readers—the students!—in a more direct manner.

The material of this book is based on a series of lecture notes corresponding to courses and tutorials that we taught at the University of Reading in the UK during the academic years 2010–2017. Hence, it has been filtered through our academic experience regarding the ability of students to comprehend new analytical concepts. An extra feature of this work is that it includes almost two hundred (197) exercises of various levels of difficulty, varying from immediate applications of the theory to more advanced topics not covered in the main text, together with full solutions.

N.K. and E.V.,
Reading, UK
and Iaşi, Romania
September 2017

Acknowledgements. We are indebted to Tobias Kuna, Beatrice Pelloni and Titus Hilberdink for their underlying contribution to the material contained in this book. Some parts of our work draw heavily on earlier draft lecture note material and solved exercises they used for their teaching at the Department of Mathematics of the University of Reading during the years 2010–2017. We would also like to thank John Boras for his assistance in some aspects relevant to the preparation of a part of the source file of this book and Gregory Paschalides for the careful reading of an earlier version of this book and for his suggestions.

Chapter 1

Sets, mappings, countability and choice

1.1   Prerequisites for the reading of this book

Throughout this book we assume some (but not much!) familiarity on the reader's behalf with elementary first year undergraduate mathematics, which does not need to go deeper than knowledge of the main definitions and perhaps the statements of the main theorems. In particular, we assume throughout some knowledge of

- intuitive Set Theory (sets and operations on sets, the De Morgan's laws, mappings, functions),

- topological concepts on the Euclidean space (open, closed and compact sets, convergence of sequences),

- Real Analysis on the line (series, differentiation and Riemann integration, the fundamental theorems),

- elementary Linear Algebra (real and complex vector spaces, subspaces, matrices, linear mappings).

Given that the prerequisites of this book are totally standard and can be found in any undergraduate book of mathematics, we decided not to burden our exposition with a tremendous amount of repetition. When the necessity arises, we shall recall on the spot any important additional concept or result we need. Nonetheless, for the convenience of the reader and in order to facilitate independent reading, after some rudiments of set theory, we present some material on countability and cardinality of sets which may not comprise standard first year undergraduate mathematics material and are required in the sequel.

In terms of prerequisites not listed above, for Chapter 9 which concerns the Lebesgue functional spaces (Section 9.6), we shall additionally assume some familiarity with the elementary concepts of Vector Calculus on the Euclidean space, primarily multi-variable differentiation.

In this chapter we also present a more advanced concept which will underlie subsequent developments, that of the *Axiom of Choice*. As the nomenclature suggests, this is an axiom which tells us that *"we can choose some objects out of a collection of objects"*. Despite looking intuitively completely reasonable, this axiom is a delicate subject of some controversy in Mathematics because it *allows one to prove implicitly that a mathematical object exists without giving any clue of how to construct it explicitly.*

We close this introductory chapter by proving a statement that at first glance should look familiar from Linear Algebra: *every vector space has a basis.* However, as we shall discuss a little later, in the general case (without restriction on the dimension) this is not an algebraic triviality and the Axiom of Choice is required.

The material in Sections 1.3–1.5 is slightly more advanced than that in Section 1.2. However, it is not necessary for it to be read linearly after Section 1.2 and could instead be quickly scanned and then revisited later when the relevant concepts arise in the text.

1.2 A quick review of sets and mappings

We begin by recalling some standard terminology and notation. Mathematics may be regarded as the study of **sets** and their properties. A **set** is a fundamental object, that cannot be defined in terms of other (simpler) objects. Intuitively, however, a set is a collection of elements. Given a set X and an element x, if x belongs to X we use the notation $x \in X$, and if not we write $x \notin X$. Two sets X and Y are said to be **equal**, written $X = Y$, if they have exactly the same elements. Given two sets X and Y, we say that X is **included** in Y, and write $X \subseteq Y$, if every element of X is also an element of Y. Two sets X and Y are equal if and only if $X \subseteq Y$ and $Y \subseteq X$.

A set X may be specified either by indicating all its elements (this procedure is used mostly when X is a **finite** set), or by *identifying another set S and a property P* such that the elements of X are exactly those elements of S satisfying the given property P. In such a situation, we shall be using either of the standard notations

$$X = \big\{ x \in S : x \text{ satisfies } P \big\} \quad \text{or} \quad X = \big\{ x \in S \,\big|\, x \text{ satisfies } P \big\}$$

as appropriate, depending on the context and on the nature of the property P determining the elements of the set. Occasionally we may also use the terms "class", "family" and "collection" as being synonymous to "set", in order to avoid such cumbersome phrases as "set of sets".

We shall use the following standard notations for the usual sets of numbers: $\mathbb{N}$ for the set of natural numbers, $\mathbb{Z}$ for the set of integers, $\mathbb{Q}$ for the sets of rationals, $\mathbb{R}$ for the set of real numbers, and $\mathbb{C}$ for the set of complex numbers. We assume familiarity with the basic properties of these numbers.

For any set X, the collection of all subsets of X is called the **powerset** of X, and is denoted by $\mathcal{P}(X)$.

The familiar operations of **union** and **intersection** of two sets may be generalised to arbitrary **families of sets**. Suppose that, for a certain set I, we are given, for each $i \in I$, a subset A_i of a given set S. The set I is called an **index set** or **indexing set**, for the collection of sets denoted by $\{A_i\}_{i \in I} \subseteq \mathcal{P}(S)$. The **union**, and respectively, the **intersection**, of the family $\{A_i\}_{i \in I}$ are denoted by $\bigcup_{i \in I} A_i$, and respectively $\bigcap_{i \in I} A_i$, and are given by[1]

$$\bigcup_{i \in I} A_i := \Big\{ x \in S : \text{ there exists } i \in I \text{ such that } x \in A_i \Big\},$$

$$\bigcap_{i \in I} A_i := \Big\{ x \in S : x \in A_i \text{ for all } i \in I \Big\}.$$

We note that the standard notation ":=" will be our symbolisation for the definition of the quantity of the left hand side to be that of the right hand side, as we just used right above. Further, we will use the symbol "$\equiv$" to mean either that the objects on the left and right hand side are identified, or that they are identically equal. Given two subset A and B of a set S, the **(set) difference** of B and A is defined as

$$B \setminus A := \Big\{ x \in S : x \in B \text{ and } x \notin A \Big\}.$$

In particular, if $B = S$, the resulting set is called the **complement of A in S** and will occasionally be denoted by A^c, if the set S is clear from the context:

$$S \setminus A = \Big\{ x \in S : x \notin A \Big\}.$$

The following identities, known as the **De Morgan's Laws**, may then be easily verified: for any $\{A_i\}_{i \in I} \subseteq \mathcal{P}(S)$, we have

$$S \setminus \left(\bigcup_{i \in I} A_i \right) = \bigcap_{i \in I} (S \setminus A_i),$$

$$S \setminus \left(\bigcap_{i \in I} A_i \right) = \bigcup_{i \in I} (S \setminus A_i).$$

Given any sets X and Y, the **Cartesian product** of X and Y is the set, denoted by $X \times Y$, consisting of all **ordered pairs** (x, y) with $x \in X$ and $y \in Y$. If $X = Y$, then we may also write $X \times X \equiv X^2$. Cartesian products of a finite number of sets may be defined in a similar way. A **relation** between two sets X and Y is a subset of the Cartesian product $X \times Y$. A **map (or mapping) f from X to Y** is any relation between X and Y satisfying the following two properties:

---

[1] This is not as scary as it may appear at first glance. For example, let I be the open interval $(-1, 1)$ of $\mathbb{R}$ and, for each $i \in I$, let $A_i = (i-2, i+3)$, an open interval of $\mathbb{R}$. Then we have that $\bigcup_{i \in I} A_i = (-3, 4)$ and also that $\bigcap_{i \in I} A_i = (-1, 2)$.

(i) For any $x \in X$, there exists $y \in Y$ such that $(x, y) \in f$.

(ii) For any $x \in X$ and $y, y' \in Y$, if $(x, y), (x, y') \in f$, then $y = y'$.

If f is a mapping from X to Y then, for any element $x \in X$, the unique element $y \in Y$ such that $(x, y) \in f$ will be denoted by $f(x)$. Moreover, instead of writing $(x, f(x)) \in f$ and $f \subseteq X \times Y$, we shall use the standard symbolisations

$$x \mapsto f(x) \quad \text{and} \quad f : X \longrightarrow Y.$$

The set X is called the **domain** and the set Y is called the **target** (or **co-domain**) of the mapping f. If $Y = \mathbb{F}$, with $\mathbb{F}$ being either the field of real numbers $\mathbb{R}$ or the field of complex numbers $\mathbb{C}$, scalar-valued mappings $f : X \longrightarrow \mathbb{F}$ will be called (scalar) **functions**. Mappings $f : X \longrightarrow \mathbb{F}^d$ whose range is the d-dimensional Euclidean space may be occasionally be called **vector (or vector-valued) functions**. Let X and Y be sets and $f : X \longrightarrow Y$ a mapping. For any $A \subseteq X$, we define the **image of A through** f by

$$f(A) := \Big\{ y \in Y : \exists\, x \in X \text{ such that } y = f(x) \Big\}.$$

Since A can be any subset of X, the above formula shows that f induces the **(direct) image map**

$$f : \mathcal{P}(X) \longrightarrow \mathcal{P}(Y).$$

In particular, we define the **image** (or **range**) of a map $f : X \longrightarrow Y$ to be the subset $f(X)$ of Y. The notation $f|_A$ will symbolise the **restriction of f to the set** A, namely the map $f|_A : A \longrightarrow Y$ for which $f|_A(x) := f(x)$, for $x \in A$. For any $B \subseteq Y$, we define the **inverse image of B through** f by

$$f^{-1}(B) := \Big\{ x \in X : f(x) \in B \Big\}. \tag{$*$}$$

Since B can be any subset of Y, the above formula shows that f induces the so-called **inverse image map**

$$f^{-1} : \mathcal{P}(Y) \longrightarrow \mathcal{P}(X).$$

The inverse image map behaves nicely with respect to unions, intersections, inclusions and complements. Indeed, let $f : X \longrightarrow Y$ be a map between sets, I an index set, $A \in \mathcal{P}(X)$, $B, C \in \mathcal{P}(Y)$ and $\{B_i : i \in I\} \subseteq \mathcal{P}(Y)$. Then:

$$\left\{ \begin{aligned} &f^{-1}\left(\bigcap_{i \in I} B_i \right) = \bigcap_{i \in I} f^{-1}(B_i), \\[4pt] &f^{-1}\left(\bigcup_{i \in I} B_i \right) = \bigcup_{i \in I} f^{-1}(B_i), \\[4pt] &f^{-1}(Y \setminus B) = X \setminus (f^{-1}(B)), \\[4pt] &B \subseteq C \implies f^{-1}(B) \subseteq f^{-1}(C), \\[4pt] &(f|_A)^{-1}(B) = f^{-1}(B) \bigcap A. \end{aligned} \right. \tag{$\star$}$$

On the other hand, while the direct image map commutes with unions and behaves nicely under inclusions, in general does not commute with intersections. If $A, D \in \mathcal{P}(X)$, $B \in \mathcal{P}(Y)$ and $\{A_i : i \in I\} \subseteq \mathcal{P}(X)$, then

$$
\begin{cases}
f\left(\bigcup_{i \in I} A_i\right) = \bigcup_{i \in I} f(A_i), \\[2mm]
f\left(\bigcap_{i \in I} A_i\right) \subseteq \bigcap_{i \in I} f(A_i), \\[2mm]
A \subseteq D \implies f(A) \subseteq f(D), \\[2mm]
f(f^{-1}(B)) \subseteq B, \quad f^{-1}(f(A)) \supseteq A, \\[2mm]
f(A) \subseteq B \iff A \subseteq f^{-1}(B).
\end{cases}
\qquad (\star\star)
$$

However, for certain subsets $A, B \in \mathcal{P}(X)$ it may happen that

$$
f(A \cap B) \neq f(A) \cap f(B).
$$

For example, for $f = \sin$ and $X = Y = \mathbb{R}$, we have $\emptyset = f(\emptyset) = f(\{0\} \cap \{2\pi\}) \neq \{0\} = f(\{0\}) \cap f(\{2\pi\})$.

A mapping $f : X \longrightarrow Y$ is said to be **injective** if, for any $x', x'' \in X$ with $x' \neq x''$ we also have $f(x') \neq f(x'')$. If f is injective, then the **inverse map** f^{-1} can be defined on $f(X)$ by means of the equivalence

$$
f(x) = y \iff f^{-1}(y) = x
$$

and we have

$$
f^{-1} : f(X) \subseteq Y \longrightarrow X, \quad y \mapsto f^{-1}(y) = x. \qquad (\dagger)
$$

A mapping $f : X \longrightarrow Y$ is said to be **surjective** if

$$
f(X) = Y,
$$

that is, for every $y \in Y$ there exists $x \in X$, such that $f(x) = y$. If f is both injective and surjective, then the mapping is called **bijective** or, alternatively, **invertible**.

Note carefully that we have used the symbol f^{-1} to denote two different mappings: the **inverse map** $f^{-1} : f(X) \longrightarrow X$ in $(\dagger)$, which is defined only for injective maps $f : X \longrightarrow Y$, and the **inverse image map** f^{-1} in $(\ast)$, which is defined on sets (not at points), regardless of the invertibility or not of the mapping f (see Figure 1.1 below). However, in the event that $f : X \longrightarrow Y$ is invertible, we have that the inverse map and the inverse image coincide (up to the identification of singleton sets with points), in the sense that

$$
f(x) = y \iff f^{-1}(\{y\}) = \{x\}.
$$

Therefore, no confusion will arise by using the same symbol "f^{-1}" with these two different meanings.

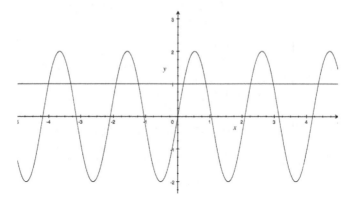

Figure 1.1: For any function $f : \mathbb{R} \longrightarrow \mathbb{R}$ which is neither strictly increasing or strictly decreasing, the inverse image function and the inverse function differ. For the example $f(x) = 2\sin(3x)$ above, we have that $f^{-1}(\{1\})$ is a sequence of points.

We close this section with a discussion regarding some concepts relevant to Real Analysis, about **infinity, extended arithmetics and suprema in the extended real line** $\mathbb{R}$. Let A be a subset of $\mathbb{R}$. Then, $\alpha \in \mathbb{R}$ is called the **supremum of** A if any of the next equivalent conditions hold:

(i) We have $x \le \alpha$ for all $x \in A$, and if b also satisfies $x \le b$ for all $x \in A$, then $\alpha \le b$.

(ii) We have $x \le \alpha$ for all $x \in A$ and for every $\varepsilon > 0$ there exists an $x \in A$ such that $\alpha - \varepsilon \le x \le \alpha$.

(iii) We have $x \le \alpha$ for all $x \in A$ and there exists a convergent sequence $(x_n)_1^\infty$ in A such that $\lim_{n \to \infty} x_n = \alpha$.

(iv) We have $x \le \alpha$ for all $x \in A$ and there exists a monotone increasing sequence $(x_n)_1^\infty$ in A such that $\lim_{n \to \infty} x_n = \alpha$.

If $\alpha \in A$, then α is the **maximum** of A. If A is not bounded from above we shall use the convention that $\sup A := +\infty$. Note that "$+\infty$" is not a number but instead an "additional point" to the elements of $\mathbb{R}$. We may then consider the **extended real line** and the **extended real half-line**

$$\overline{\mathbb{R}} := \mathbb{R} \cup \{-\infty, \infty\}, \qquad [0, \infty] := [0, \infty) \cup \{\infty\},$$

where "$-\infty$" is defined symmetrically. We shall also be using the conventions

$$\sup \emptyset = -\infty, \qquad \inf \emptyset = +\infty$$

which quantify the fact that "every arbitrarily small number is an upper bound of the empty set" and "every arbitrarily large number is a lower bound of the empty set". If $A \subseteq \overline{\mathbb{R}}$, the following statement are equivalent:

(i) For any $b \in \mathbb{R}$, there exists an element $x \in A$ with $b < x$.

(ii) There exists a sequence $(x_n)_1^\infty$ in A which diverges to $+\infty$.

(iii) if $\infty \notin A$ then there exists a strictly monotone increasing sequence $(x_n)_1^\infty$ in A which diverges to $+\infty$.

Note that $\sup A = \infty$ if either $\infty \in A$ or if A is an unbounded set. An equivalent formulation of the completeness of $\mathbb{R}$ is that any monotone sequence in $\overline{\mathbb{R}}$ converges. Further, if $(a_n)_1^\infty \subseteq [0, \infty]$ is a sequence of *non-negative* extended real numbers, then the series $\sum_{n=1}^\infty a_n$ converges in $[0, \infty]$.

Finally, if $X \neq \emptyset$ is **any set** and $f : X \longrightarrow [-\infty, \infty]$ is a (extended) real valued function, we will use either of the notations below as equivalent, to denote the **supremum of the image set** $f(X)$, namely the supremum of $f(x)$, when x is running through the set X:

$$\sup f(X) \equiv \sup_{x \in X} f(x) \equiv \sup \big\{ f(x) \, : \, x \in X \big\}.$$

Then, $\sup f(X)$ is an extended real number in $[-\infty, \infty]$. Similar remarks apply to the infimum $\inf f(X)$ as well, in a symmetric fashion.

1.3 Cardinality and countability of sets

In this section we would like to identify an appropriate way to compare any two sets in terms of their "size". For **finite** sets this is easy, since to any such set we can assign the **number of elements** of the set, which is a natural number. Then, any two such numbers can be compared using the standard ordering relation "$\leq$" on $\mathbb{N}$. For instance, if we assume that different letters are used to denote different objects, then the set $\{a, b, c\}$ contains the same number of elements as the set $\{u, v, w\}$, while containing fewer elements than the set $\{x, y, z, t\}$. Note that the "nature" of the elements is irrelevant to the above discussion because we only care about the "size" and the "relative size" of sets.

On the other hand, while it is reasonable to declare that any **infinite** set has more elements than any finite set, given any two **infinite** sets we would like to be able to say more about their relative size than merely the fact that they are both infinite. This will be achieved by means of the concept of **cardinality**, which is an extension to the class of all sets of the concept of **number of elements** valid for finite sets. In what follows we define this concept rigorously, and use it in particular to compare the relative size of the usual sets of numbers $\mathbb{N}, \mathbb{Z}, \mathbb{Q}, \mathbb{R}$.

To this aim, we first observe that two finite sets X and Y have the same number of elements if and only if there exists a **bijective mapping** $f :$ $X \longrightarrow Y$. Indeed, the act of counting the elements of a given finite set may be regarded itself as the process of exhibiting a bijection between the set in question and a set of the form $\{1, 2, \ldots, n\}$, for some $n \in \mathbb{N}$. Namely, we identify a first element, then a second element, and so on. We also observe that a finite set X has fewer elements than some other finite set Y if and only if there exists an **injective mapping** $f : X \longrightarrow Y$. This can be seen for example by exhibiting a bijection between each of X and Y and a set of the form $\{1, 2, \ldots, n\}$, for some $n \in \mathbb{N}$.[2] The most important point to note, however, it that these alternative ways of comparing the size of any two finite sets, namely the existence of a bijective (or an injective) mapping between them, are at our disposal irrespective of the finiteness of the sets in question! This leads to the following definitions.

Definition 1.1. Two sets X and Y are said to have **the same cardinality** if there exists a bijective mapping $f : X \longrightarrow Y$. In such a situation, we write

$$\mathrm{card}(X) = \mathrm{card}(Y).$$

Definition 1.2. A set X is said to have **smaller cardinality** than a set Y if there exists an injective mapping $f : X \longrightarrow Y$. In such a situation, we write

$$\mathrm{card}(X) \leq \mathrm{card}(Y).$$

Note carefully that **we have not defined what the cardinality of a set X is as a mathematical object!** While it is possible to give a rigorous meaning to $\mathrm{card}(X)$, we refrain from doing that since such an endeavour would take us too far afield, and is ultimately not very important for this book. Intuitively, however, one should think of $\mathrm{card}(X)$ as "something that all the sets that can be put in a bijection with X have in common" (for finite sets, this role is played by the number of elements of X). As a consequence, the symbols "$\leq$" and "$=$" should also not be taken to have any independent meaning either, and their usage here should not be confused with other usages of these symbols (e.g., the inequality of cardinals may have nothing to with inequalities among real numbers).

In order to be considered a meaningful means of comparison of "sizes" of sets, cardinality needs to order the sets, in the sense of satisfying the typical properties of a partial ordering, which we recall right next.

---

[2]In fact, two finite sets may well be compared in terms of their size without knowing precisely how many elements each of them has, but instead using directly the existence of injective (or bijective) mappings between them. An illustrative example of such a situation arising in practice is the following. Suppose that a large number of students enter a large lecture room, and we would like to know whether the number of seats in the room is greater or smaller than the number of students. Rather than counting both sets and comparing the resulting numbers, we can just ask each student to take a seat. Then there are three possible situations: all students are seated and there are some empty seats left, all seats are occupied and there are some students left standing, and all students occupy all seats with no seats left empty and no students standing. The occurrence of the first situation obviously means that the number of students is less than the number of seats; in such a case, the mapping that assigns to each student its seat is injective from the set of students to the set of seats. A similar discussion can be made in each of the other two situations.

Definition 1.3. Let X be a non-empty set. A **(partial) ordering on** X is a subset $\prec$ of $X \times X$ with the properties that, for every $x, y, z \in X$, we have that:

- $x \prec x$ (reflexivity);

- if $x \prec y$ and $y \prec x$ then $x = y$ (antisymmetry);

- if $x \prec y$ and $y \prec z$ then $x \prec z$ (transitivity).

A **(partially) ordered set** is a pair $(X, \prec)$, where X is a set and $\prec$ is a (partial) order relation on X. (Note that we are using here a standard notational convention to write $x \prec y$ instead of $(x, y) \in \prec$.)

Returning to the notion of cardinality, note that $\mathrm{card}(X) \leq \mathrm{card}(X)$. Next, it is an easy exercise on the composition of maps to confirm that if $\mathrm{card}(X) \leq \mathrm{card}(Y)$ and $\mathrm{card}(Y) \leq \mathrm{card}(Z)$, then it follows that $\mathrm{card}(X) \leq \mathrm{card}(Z)$. The following theorem completes the ordering properties.

Theorem 1.4 (Schröder and Bernstein). *Let X, Y be two sets. If $\mathrm{card}(X) \leq \mathrm{card}(Y)$ and $\mathrm{card}(Y) \leq \mathrm{card}(X)$, then $\mathrm{card}(X) = \mathrm{card}(Y)$.*

We refrain from giving the non-trivial proof of this result, which goes deeper into the realm of set theory and exceeds the scope of this elementary book. The proof is non-trivial because the assumption of the theorem means only that there exist injective maps $f_1 : X \longrightarrow Y$ and $f_2 : Y \longrightarrow X$, but since these need not be inverses of each other, a construction of a bijective map $f : X \longrightarrow Y$ needs to be carried out.

We now examine the cardinalities of some of the usual sets of numbers. Since the set of natural numbers $\mathbb{N}$ is the simplest infinite set, it is appropriate to study the cardinality of any set in relation to the cardinality of $\mathbb{N}$.

Definition 1.5. A set X is called **countable** if there exists an injective function $f : X \longrightarrow \mathbb{N}$. Otherwise, it is called **uncountable**.

Equivalently, by Lemma 1.6 that follows, a set X is **countable** if and only if either $X = \emptyset$ or there exists a surjective function $g : \mathbb{N} \longrightarrow X$.

A countable set which is not finite is called **countably infinite**.

The next simple result will allow us to define countability by utilising either injective of surjective mappings.

Lemma 1.6. *Let $X \neq \emptyset$ be any set. Then, there exists a injective map $f : X \longrightarrow Y$ if and only if there exists a surjective map $g : Y \longrightarrow X$.*

Proof. Assume we are given an injective $f : X \longrightarrow Y$. Choose an $x_0 \in X$ and define
$$g(y) := \begin{cases} f^{-1}(y), & \text{for } y \in f(X), \\ x_0, & \text{for } y \in Y \setminus f(x), \end{cases}$$
which is surjective, because for each $x \in X$ there exists $y_x := f(x)$ such that

$g(y_x) = x$. Conversely, suppose that we are given a surjective map $g : Y \longrightarrow X$. Then, for each $x \in X$ **one can choose**[3] an element $y_x \in Y$ satisfying $g(y_x) = x$. By setting $f(x) := y_x$, the map f defined in this way is injective: indeed, for $x' \neq x''$, there exist $y', y'' \in Y$ such that $g(y') = x'$, $g(y'') = x''$. Since g is a well-defined map, we have $y' \neq y''$.   □

Remark 1.7. Note that X may well be the empty set $\emptyset$, which by convention is considered countable. Clearly, any finite set is countable and $\mathbb{N}$ itself is countable (take f or g to be the identity $\mathbb{N} \longrightarrow \mathbb{N}$). Further, $\mathbb{Z}$ is countable, which is obvious if one writes $\mathbb{Z}$ as $\{0, 1, -1, 2, -2, ...\}$. Note the remarkable fact that

$$\mathbb{N} \subseteq \mathbb{Z}, \ \mathbb{N} \neq \mathbb{Z}, \ \text{but} \ \mathrm{card}(\mathbb{N}) = \mathrm{card}(\mathbb{Z}) \ !!!$$

Strict (proper) countable subsets appear only in infinite sets.

Intuitively, the function g in the definition can be understood as "tagging the elements of X with numbers", whilst, we can interpret f as "laying the elements of X in an infinite chain of numbered boxes". The following lemmas list some properties of countability.

Lemma 1.8. *Let X, Y be sets and $h : X \longrightarrow Y$ a map.*

(i) *Any subset of a countable set is countable.*

(ii) *If X is countable, then $h(X)$ is countable.*

(iii) *If h is injective and $h(X)$ is countable, then X is countable.*

Proof. It is left as an (easy) exercise on the definition!   □

Lemma 1.9. *Let X, Y be countable sets. Then the Cartesian product $X \times Y$ is countable. Moreover, if $X_1, \ldots, X_N$ are countable sets, where $N \in \mathbb{N}$, then the Cartesian product $X_1 \times X_2 \times ... \times X_N$ is also countable.*

In particular, the (finite) Cartesian product $\mathbb{N} \times \mathbb{N} \times ... \times \mathbb{N}$ is countable.

Proof. It suffices to discuss only the case of $\mathbb{N} \times \mathbb{N}$ (why?). One can easily verify that the mapping given by $(m, n) \mapsto 2^n 3^m$ is an injective mapping from $\mathbb{N} \times \mathbb{N}$ to $\mathbb{N}$. We conclude by invoking the definition of countability.   □

---

[3]But can we really choose such a point? And why? Can you explain it? Note that we (have to) do simultaneously infinitely many choices of elements from non-empty sets. Why don't we just attempt to construct somehow the element y_x we select? At this point we are "hiding under the carpet" for the sake of simplicity an essential point which will concern us largely in the next section, that of *infinite choices*. We refrain from discussing this topic any further and we keep this footnote as "food for thought" until we revisit this more subtle topic in Sections 1.3-1.4, clarifying therein how and why this choice is indeed possible. We merely confine ourselves to noting that *no such issues arise when we have the particular set $X = \mathbb{N}$, which is the case of interest here, instead of a general set*. In the case of $\mathbb{N}$, selecting an element of any non-empty subset is straightforward since any such subset of $\mathbb{N}$ has a smallest element, and thus, an unambiguous selection of an element of that set can be made.

Lemma 1.10 (Countable unions of countable sets). *Let $X_1, X_2, \ldots$ be a countable family of countable sets. Then, the union*

$$\bigcup_{i=1}^{\infty} X_i$$

is also a countable set.

Proof. By assumption, for each i there exists a surjective map $f_i : \mathbb{N} \longrightarrow X_i$. We define the map

$$g \; : \; \mathbb{N} \times \mathbb{N} \longrightarrow \bigcup_{i=1}^{\infty} X_i, \quad (m, n) \mapsto f_n(m).$$

Then it follows that g is surjective (why?). □

Lemma 1.11. *The set of rationals $\mathbb{Q}$ is countable.*

Proof. Let $\mathbb{Q}_+ = \{q \in \mathbb{Q} : q > 0\}$ denote the positive rationals; we may define $\mathbb{Q}_-$ analogously. Hence $\mathbb{Q} = \mathbb{Q}_+ \cup \{0\} \cup \mathbb{Q}_-$. We note that $\mathbb{Q}_+$ is countable because the following map is surjective:

$$g \; : \; \mathbb{N} \times \mathbb{N} \longrightarrow \mathbb{Q}_+, \quad (m, n) \mapsto \frac{m}{n}.$$

A similar argument can be used to show that $\mathbb{Q}_-$ is countable. Then by Lemma 1.10 $\mathbb{Q}$ is countable, being a finite union of countable sets. □

Are all infinite sets countable? Actually no, as the next lemma asserts.

Lemma 1.12. *The following statements hold:*

(i) *The unit interval $(0, 1) \subseteq \mathbb{R}$ is uncountable.*

(ii) *The set of real numbers $\mathbb{R}$ is uncountable.*

(iii) *Any interval (open or otherwise) in $\mathbb{R}$ is uncountable.*

Informally, the meaning of the above lemma is that uncountable sets have "many more elements" than countable sets, even if an uncountable set is bounded (e.g., $[0, 1]$) and the countable set is unbounded (e.g., $\mathbb{N}$).

Proof. (i) Let $g : \mathbb{N} \longrightarrow (0, 1)$ be any map. Let us write

$$g(n) \; = \; 0.c_1(n)\, c_2(n) \ldots$$

for a decimal expansion of $g(n)$ for $n \in \mathbb{N}$, where $c_i(n) \in \{0, 1, \ldots, 9\}$. We shall construct a number $x \in (0, 1)$ such that $x \in (0, 1) \setminus g(\mathbb{N})$. Then, it will follow that g cannot be surjective and hence $(0, 1)$ cannot be countable. To this end, we define the number x via a decimal expansion $0.a_1 a_2\ldots$ of it. Indeed, if $c_n(n) \neq 1$, define $a_n := 1$ and if $c_n(n) = 1$ define $a_n := 0$. Hence $x \neq g(n)$ for any $n \in \mathbb{N}$ because the n-th decimal of x is different from to the n-th decimal of $g(n)$.

(ii) By Lemma 1.8, $\mathbb{R}$ is uncountable because it contains an uncountable set.

(iii) This follows from the fact that for any $a < b$, the function

$$f : (0,1) \longrightarrow (a,b), \quad t \mapsto f(t) := (1-t)a + tb$$

is bijective (complete the details as an exercise!). $\qquad\square$

The next result says that if we remove a countable set (i.e. a "small" one) from an uncountable set (i.e. a "large" one), it remains uncountable.

Lemma 1.13. *If X is an uncountable set and $Y \subseteq X$ is a countable set, then $X \setminus Y$ is a uncountable set.*

Proof. It is left as an easy exercise for the reader. $\qquad\square$

As we have seen, in the universe of infinite sets there exist at least two cardinalities (intuitively, "numbers of elements"), that of $\mathbb{N}$ and $\mathbb{R}$. The natural question now arises of whether there exist any more cardinalities.

It is obvious that for any set X we have

$$\mathrm{card}(X) \leq \mathrm{card}(\mathcal{P}(X)).$$

(If it is not obvious, then it is an exercise on the use of the definition with the aid of the map $X \ni x \mapsto \{x\} \in \mathcal{P}(X)$. Note that this map is *not* the identity, since it sends the element x to the singleton set $\{x\}$ which contains just one element, x.) We now show that the previous inequality is actually strict:

Lemma 1.14. *For any set X, it holds that*

$$\mathrm{card}(X) \neq \mathrm{card}(\mathcal{P}(X)).$$

Proof. We argue by contradiction. Let

$$f : X \longrightarrow \mathcal{P}(X), \quad X \ni x \mapsto f(x) \subseteq X$$

be a surjective (set-valued!) map. We define the set

$$A := \{x \in X : x \notin f(x)\}.$$

By the surjectivity of f, there exists an element $a \in X$ such that $f(a) = A$. By the definition of A, we have that $a \in A$ if and only if $a \notin f(a) = A$. This contradiction establishes the claim. $\qquad\square$

By iteration of the powerset operation we can create countably many different cardinalities:

$$\mathcal{P}(\mathbb{N}), \ \mathcal{P}(\mathcal{P}(\mathbb{N})), \ \mathcal{P}(\mathcal{P}(\mathcal{P}(\mathbb{N}))), \dots.$$

In the lemma below (whose proof is omitted since it is outside of the scope of this book) we explain how the real line $\mathbb{R}$ emerges from the countable set $\mathbb{N}$:

Lemma 1.15. *It holds that* $\mathrm{card}(\mathcal{P}(\mathbb{N})) = \mathrm{card}(\mathbb{R})$.

In addition, as we show next, the cardinality of $\mathbb{N}$ is the smallest possible among all infinite sets:

Lemma 1.16. *Each infinite set X contains a countably infinite subset. Hence, for each infinite set X it holds that* $\mathrm{card}(\mathbb{N}) \leq \mathrm{card}(X)$.

Proof. Choose $x_1 \in X$. Then choose some $x_2 \in X \setminus \{x_1\}$ and repeat the process recursively (the details are left as an exercise for the reader!). $\qquad\square$

Remark 1.17. It follows that the smallest infinite cardinality is $\mathrm{card}(\mathbb{N})$ (known as $\aleph_0$). The question of whether the next larger cardinality is $\mathrm{card}(\mathbb{R})$ is called the *continuum hypothesis*. Even though at first sight it seems that the question of whether there exists an infinite subset of $\mathbb{R}$ that cannot be put into a bijection either with $\mathbb{N}$ or with $\mathbb{R}$ *ought to have an answer* (either affirmative, or negative), it has actually been proven that this hypothesis cannot be settled either way in the realm of our normal axiomatic framework (that is, without adding extra axioms to the "usual" mathematics we work in)!

1.4   The Axiom of Choice and Zorn's Lemma

Now we delve into a discussion of a more subtle topic, which can be omitted on the first reading; our discussion can be consulted when the concepts arise later in the exposition.

Suppose that we have finitely-many non-empty sets $A_1, \ldots, A_n$. Then, one can construct a *selection mapping*, namely, a map which selects for each of the sets A_i an element belonging to it:

$$f \; : \; \{1, \ldots, n\} \longrightarrow \bigcup_{i=1}^{n} A_i, \quad i \mapsto f(i) \in A_i.$$

This can be done in an algorithmic fashion in a number of n steps: select any element $a_1 \in A_1$ and set $f(1) := a_1$. Continue by selecting an element $a_2 \in A_2$ and set $f(2) := a_2$, etc. However, the situation becomes substantially trickier when one has to make **infinitely many** (and, in particular, **uncountably many**) **choices**. Quite surprisingly, the analogous statement cannot be proved from the "usual" principles of Mathematics! Accordingly, it has to be required as a new axiom:

Axiom 1.18 (Axiom of Choice). *For any index set I and any family $\{A_i\}_{i \in I}$ of non-empty sets, there exists a selection mapping:*

$$f \; : \; I \longrightarrow \bigcup_{i \in I} A_i, \quad i \mapsto f(i) \in A_i.$$

To be somewhat more precise, the issue here is that not every argument or construction that is intuitively reasonable can be accepted as legitimate in Mathematics, but has to be built up, making use only of well specified principles of Logic, from a collection of carefully chosen statements that are accepted as axioms. The need of a rigorous axiomatisation of Mathematics, and in particular of Set Theory, arose at the beginning of the 20th century with the discovery of **paradoxes**, seemingly innocuous basic statements that are, however, self-contradictory. The most famous paradox is probably that of Russell, which involves "the set of all sets that are not elements of themselves" or, formally:

$$S = \Big\{A \,:\, A \notin A\Big\}.$$

It is immediate to verify that $S \in S$ if and only $S \notin S$, which means that we have a statement that is true if and only if it is false, a situation that contradicts the basic principles of Logic, and is therefore utterly unacceptable!!! Where could such a big problem come from in such a simple situation? After careful thought, the root of the problem has been identified by mathematicians in the lack of precision of the "definition" of a set. The resolution of this paradox (and of others) has been found in the identification of a suitable, now standard, collection of "reasonable" axioms for Set Theory, called the **Zermelo–Fraenkel Axioms**. It is a consequence of the axioms that not all mathematical objects that could naively be thought as "sets" are actually sets, and, in particular, the above formula that has been used to define S is not an admissible way to define a set!

The situation occurring in the Axiom of Choice can be regarded in a similar vein. Even though, for each $i \in I$, it is possible to choose a value $f(i) \in A_i$, it is considered questionable by some mathematicians whether this is an admissible way to define a function, for which the values at all all points should be prescribed "simultaneously" (as opposed to "one at a time")! Moreover, it has been proved that the Axiom of Choice is *independent of the ZF Axioms*, in the sense that it can neither be proved nor disproved from them, the situation being somewhat similar with that for the *Parallel Postulate* in Euclidean Geometry.

On the other hand, the Axiom of Choice is needed in Analysis in only a small number of occasions. The reason is that, although for *general sets* without any additional structure one has no other means to make uncountably many selections of elements, it usually turns out that the sets of interest have more structure, a fact which enables selections of elements via explicit constructions.

The most frequent way the Axiom of Choice is usually invoked in Mathematics is through a statement equivalent to it, which has the nomenclature "lemma", mostly for historical reasons. In order to state it, we need some further terminology.

Definition 1.19. Let $(X, \prec)$ be a partially ordered set. A subset A of X

is said to be **totally ordered** (or **linearly ordered**, or **chain**) if, for any $x, y \in A$, either $x \prec y$ or $y \prec x$.

Definition 1.20. Let $(X, \prec)$ be a partially ordered set. Given any subset A of X, an element M of X is said to be an **upper bound** for A if $x \prec M$ for every $x \in A$.

Definition 1.21. Let $(X, \prec)$ be a partially ordered set. An element M of X is said to be **maximal** if there exists no element $M' \in X$ such that $M \prec M'$ and $M' \neq M$.

The following statement is equivalent to the Axiom of Choice, in the sense that if one assumes the validity of the Axiom of Choice, then one can prove the validity of the Zorn Lemma, and the other way around.

Lemma 1.22 (Zorn's Lemma). *Let $(X, \prec)$ be a partially ordered set with the property that any totally ordered subset of X has an upper bound. Then X contains at least one maximal element.*

As we already said, it has been shown that, under the ZF Axioms, the Axiom of Choice and Zorn's Lemma are equivalent statements. Most mathematicians, including the authors of this book, are happy to accept the Axiom of Choice or, equivalently, the Zorn Lemma, as a valid axiom for Set Theory and therefore for Mathematics, without raising philosophical objections to its non-constructive nature. There also exist further mathematical statements equivalent to the Axiom of Choice, which we refrain from stating. We will see, however, one of them in the next section.

1.5 Literally every vector space has a basis

Consider a vector space over the field $\mathbb{F}$, where $\mathbb{F}$ is either $\mathbb{R}$ of $\mathbb{C}$. We know from Linear Algebra that if the space has **finite** dimension, then one can exhibit the existence of a basis (and in fact of infinitely many) in a constructive manner, by (finite) induction. In fact, one calls "dimension" the cardinality of any basis of the space.

If however the space **does not have finite dimension** (when it is, as we say, **infinite-dimensional**), namely if no basis of finitely-many vectors exists, then, exhibiting "some kind of basis" becomes a real problem which cannot be addressed by a recursive construction!

Typically, the usual algebraic notion of basis of an infinite-dimensional vector space has to be adapted accordingly and is called **Hamel basis**, given below. This reduces precisely to the notion of basis known from Linear Algebra if the space has finite dimension. If the space is infinite-dimensional, then the following definition is in effect.

Definition 1.23. Let X be a vector space over the field $\mathbb{F}$ (where $\mathbb{F}$ is either $\mathbb{R}$ of $\mathbb{C}$). A set of vectors $E := \{e_i : i \in I\}$, where I is some index set, is called a **Hamel basis of** X when for every $x \in X$, there exist:

• an integer $N \in \mathbb{N}$,

• a finite number of scalars $\lambda_1, \ldots, \lambda_N \in \mathbb{F}$, and,

• a finite collection of vectors $e_{i_1}, \ldots, e_{i_N} \in E$,

such that:

$$x = \sum_{k=1}^{N} \lambda_k \, e_{i_k}.$$

Note that the integer N, the coefficients $\lambda_1, \ldots, \lambda_N$ and the vectors $e_{i_1}, \ldots, e_{i_N}$ from the set E in the above expansion **depend on the vector** x.

Note that, in the absence of any analytic structure on X allowing one to "take limits" (as e.g. we shall see in later chapters), **a Hamel basis is the unique notion of basis** for a general vector space with no additional structure! In particular, in such a setting there is no way to write a putative "infinite sum" that one may hope could represent an expansion of a vector with respect to a "basis"! In order for such infinite sums to make sense, it is necessary to have a way to be able to **"take limits"** in the space (and therefore "do Analysis").

Establishing that a Hamel basis of a vector space exists (without the Linear-Algebraic hypothesis that the space is finite-dimensional) is a non-trivial matter, that can only be settled in the **non-constructive**, though perspicuous, matter indicated below. The proof is based on the Axiom of Choice, through an application of Zorn's lemma.

Theorem 1.24 (Existence of a Hamel basis). *Every vector space over the field $\mathbb{F}$ (where $\mathbb{F}$ is either $\mathbb{R}$ of $\mathbb{C}$) possesses a Hamel basis.*[4]

Proof. Let X be a vector space over $\mathbb{F}$. Without loss of generality, we may assume that the space has infinite dimension. Let $\mathfrak{L}$ symbolise the set of all linearly independent subsets of X:

$$\mathfrak{L} := \left\{ L \in \mathcal{P}(X) \, \middle| \, \forall v_1, \ldots v_n \in L : \sum_{i=1}^{n} \lambda_i v_i = 0 \implies \lambda_1 = \cdots = \lambda_n = 0 \right\}.$$

The set $\mathfrak{L}$ is nonempty and can be partially ordered by the usual inclusion "$\subseteq$", that is, by the ordering obtained by taking $A \prec B$ *if and only if* $A \subseteq B$. Let $\mathfrak{L}'$ be a subset of $\mathfrak{L}$ that is totally ordered by the "$\subseteq$", and let $\bigcup \mathfrak{L}'$ be the union of all the elements of $\mathfrak{L}'$.

Since $(\mathfrak{L}', \subseteq)$ is totally ordered, the union of all of its elements $\bigcup \mathfrak{L}'$ is an element of $\mathfrak{L}$. Therefore $\bigcup \mathfrak{L}'$ is an upper bound for $\mathfrak{L}'$ in $(\mathfrak{L}, \subseteq)$, namely, it is an element of $\mathfrak{L}$ which includes every element of $\mathfrak{L}'$.

---

[4]The proof can perhaps be omitted on the first reading of the book.

Since $\mathfrak{L}$ is nonempty and every totally ordered subset of $(\mathfrak{L}, \subseteq)$ has an upper bound, Zorn's Lemma 1.22 yields that $\mathfrak{L}$ has a maximal element, say $L_{\max} \in \mathfrak{L}$.

It remains to show that $L_{\max}$ is a Hamel basis of X. Since $L_{\max} \in \mathfrak{L}$, it is a linearly independent subset of X. If for the sake of contradiction we suppose that $L_{\max}$ does not span X, there would exist some vector $x \in X \setminus \{0\}$ which cannot be expressed as a linear combination of elements of $L_{\max}$. As a consequence, we would have that $L_{\max} \bigcup \{x\}$ is also linearly independent and $L_{\max} \subsetneq L_{\max} \bigcup \{x\}$, a fact which constitutes a contradiction to the maximality of $L_{\max}$. Hence, $L_{\max}$ is a Hamel basis of the vector space X. $\qquad \square$

We close this chapter by noting that the statement that every vector space admits a Hamel basis is in fact one of those statements whose validity is equivalent to that of the Axiom of Choice.

1.6 Exercises

Exercise 1.1. Prove all the properties $(\star)$-$(\star\star)$ that the image map and the inverse image map satisfy.

Exercise 1.2. Show that:

1. Any subset of a countable set is countable.

2. If $h : X \longrightarrow Y$ is an injective map between sets and $h(X)$ is countable, then X is countable.

Exercise 1.3. Let us denote by $\{0,1\}^{\mathbb{N}}$ the set of all maps from $\mathbb{N} \longrightarrow \{0,1\}$, that is the set of all sequences which takes only values 0 or 1. Show that $\{0,1\}^{\mathbb{N}}$ is uncountable.

Exercise 1.4. Let X be a set. When is the powerset $\mathcal{P}(X)$ of X, that is the set of all subsets of X, countable?

Exercise 1.5. Show that, in the real vector space

$$C([0,1], \mathbb{R}) = \left\{ f : [0,1] \longrightarrow \mathbb{R} \mid f \text{ is continuous on } [0,1] \right\},$$

the family $\{f_k : k \in \mathbb{N} \cup \{0\}\}$ given, for each $k \in \mathbb{N} \cup \{0\}$, by

$$f_k(x) := x^k \quad \text{for all } x \in [0,1],$$

is a linearly independent set. Deduce that $C([0,1], \mathbb{R})$ is an infinite-dimensional vector space.

Chapter 2

Metric spaces and normed spaces

2.1 Abstracting the concepts of Analysis

Mathematical Analysis is based on the concept of limit. Convergence of sequences, continuity, differentiability and integrability of functions, are all based on the consideration of certain limiting processes. You are undoubtedly familiar with the notions of convergent sequences in $\mathbb{R}$ and with the continuity of real-valued functions defined on subsets of $\mathbb{R}$.

In case you have already seen the extension of the aforementioned concepts to the Euclidean space $\mathbb{R}^m$, $m \geq 1$, you must have certainly noticed a similarity in the definitions, in that they are essentially the same, except for the modulus of the difference of two numbers being replaced by the Euclidean distance between the two points. This is no coincidence. Beyond the formal technicalities, these concepts are devised to capture essentially the same underlying ideas. For example, convergence of a sequence means that, as the index of the sequence grows large, the terms of the sequences get as close we wish to a certain object, called limit.

Hence on a second (more mature) inspection of the analytical notions on $\mathbb{R}$ and $\mathbb{R}^m$, we realise that the common underlying concept is *proximity*, measured by the "distance" between the objects in question. Moreover, it is often the case in mathematics that one is interested in the (possible) "convergence" of sequences of objects of a different nature than that of the elements of $\mathbb{R}^m$. For example, you have seen already that one can study the so-called uniform convergence of sequences of functions.

It is very pleasant that all these and many other instances of convergence of sequences (and similarly, of continuity of functions) may be usefully studied *axiomatically* in an abstract framework, that of **metric spaces**. *The importance of such an enterprise which will justify our endeavours is that we shall develop a general theory, preserving the main ideas we learn in Real Analysis, but having a **very wide applicability** to seemingly unrelated contexts, way beyond $\mathbb{R}$!*

A metric space is an (arbitrary) set X in which to any two elements (points) $x, y \in X$ we associate a nonnegative real number, the **distance** $d(x, y)$ between them. Hence, the distance d is a real-valued function with non-negative values on the Cartesian product

$$d \; : \; X \times X \longrightarrow [0, \infty), \quad (x, y) \mapsto d(x, y).$$

However, not any function on the Cartesian product $X \times X$ qualifies to be called distance! It has to satisfy certain *reasonable conditions*. The first one, already assumed, is that

$$d(x, y) \geq 0.$$

Namely, *distances between the points* $x, y \in X$ *are "lengths"* and therefore have to be non-negative numbers. In addition, *the distance of x from y has to be the same as the distance of y from x*. Hence, this results in requiring the symmetry condition

$$d(x, y) \; = \; d(y, x) \quad \text{for any } x, y \in X.$$

Further, the distance of any $x \in X$ from itself has to be zero, that is

$$d(x, x) \; = \; 0.$$

More importantly, the following natural converse needs to hold: *if the distance between any two points $x, y \in X$ vanishes, the points have to coincide*:

$$d(x, y) \; = \; 0 \quad \Longrightarrow \quad x = y.$$

The final property, called the **triangle inequality**, requires the following natural property: suppose we wish to "travel" from a point x to a point y in the set X in the shortest possible route. Then, by travelling through a third point z in the shortest possible route, this cannot decrease the total distance we travel (see Figure 2.1). In symbols, this can be expressed as

$$d(x, y) \; \leq \; d(x, z) + d(z, y) \quad \text{for any } x, y, z \in X.$$

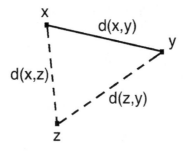

Figure 2.1: A depiction of the triangle inequality in $\mathbb{R}^2$ with respect to the usual distance function given by the Euclidean metric.

It turns out that these axioms alone suffice to develop an abstract analytical theory of great beauty and elegance, that constitutes the foundation of nearly all modern developments in Analysis. In fact, we believe it is not an overstatement to say that **most of Mathematical Analysis is a consequence of the triangle inequality!**

The principal purpose of this chapter is to develop an axiomatic generalisation of Real Analysis to the more abstract realm of metric spaces. A closely associated theme will be the study of a privileged class of metric spaces, called **normed spaces**. These are (abstract) vector spaces endowed with a metric having special properties which are "tied" to the linear structure of the space and resemble those of the Euclidean norm. Along the way, we shall make connexions to more familiar contexts and also present important examples which led the mathematicians of previous generations to develop this abstract theory.

2.2 Metrics on sets and norms on vector spaces

As indicated in the previous introductory section, we shall work in an axiomatic abstract framework, in which we define and study concepts of Analysis that you are already familiar with from the particular case of the set $\mathbb{R}$ of real numbers, and possibly that of the Euclidean space $\mathbb{R}^m$, $m \geq 1$.

The first fundamental concept, that we now introduce more formally, is that of a **metric space**.

Definition 2.1. Let X be a set. A **metric** (or **distance**) on X is a function $d : X \times X \longrightarrow \mathbb{R}$ with the following properties:

$(M1)$ $d(x,y) \geq 0$ for all $x, y \in X$, and $d(x,y) = 0$ if and only if $x = y$;

$(M2)$ $d(x,y) = d(y,x)$ for all $x, y \in X$;

$(M3)$ $d(x,z) \leq d(x,y) + d(y,z)$ for all $x, y, z \in X$.

A **metric space** is a pair (X, d), where X is a set and d is a metric on X. The property $(M3)$ is called the **triangle inequality**.

The first example of a metric is the one familiar to you from your course of Analysis (or Calculus in Several Variables).

Example 2.2. In $\mathbb{R}^m$, $m \in \mathbb{N}$, the function $d_2 : \mathbb{R}^m \times \mathbb{R}^m \longrightarrow \mathbb{R}$ given, for any $x = (x_1, \ldots, x_m)$ and $y = (y_1, \ldots, y_m)$, by

$$d_2(x,y) := \left(\sum_{k=1}^{m} (x_k - y_k)^2 \right)^{1/2},$$

is obviously a metric, called the **Euclidean metric, standard metric,** or

usual metric. An analogous metric may be defined on $\mathbb{C}^m$, namely $d_2 :$ $\mathbb{C}^m \times \mathbb{C}^m \longrightarrow \mathbb{R}$ given, for any $x = (x_1, \ldots, x_m)$ and $y = (y_1, \ldots, y_m)$, by

$$d_2(x, y) := \left(\sum_{k=1}^m |x_k - y_k|^2 \right)^{1/2}.$$

The next example is a very simple one, in which the distance between any two different elements of a set is defined to be equal to 1. This example is not important *per se*, nor for its occurrence in applications, but rather because it can usefully serve as a test case for various conjectures one may make about general metric spaces.

Example 2.3. Let X be a non-empty set. Define a metric d_0 on X by

$$d_0(x, y) := \begin{cases} 1, & \text{if } x \neq y, \\ 0, & \text{if } x = y. \end{cases}$$

This is called the **discrete metric** on X. Axioms (M1) and (M2) are immediate. For (M3), if $x = z$ the required property is obvious, whilst if $x \neq z$ then, for any $y \in X$, we have either $x \neq y$ or $y \neq z$ (or both), and hence $d_0(x, z) = 1 \leq d_0(x, y) + d_0(y, z)$.

The metric spaces that one most often encounters in applications have, in addition to the metric structure, an additional **vector space** structure, that is, two operations are defined on the space, one called **addition** and the other called **multiplication by scalars**, that have the familiar properties studied in Linear Algebra. We recall from Chapter 1 that throughout what follows, $\mathbb{F}$ symbolises either the real number field $\mathbb{R}$ or the complex number field $\mathbb{C}$ and the elements of $\mathbb{F}$ will be called scalars.

We consider below the situation when a vector spaces is endowed with a **norm**, a notion which generalises the usual notion of **length** of a vector in Euclidean spaces. As we shall immediately see, the norm gives rise in a very natural way to a distance, and thus to a metric space structure on the vector space in question.

Definition 2.4. Let X be a vector space over $\mathbb{F}$. A **norm** on X is a function $\|\cdot\| : X \to \mathbb{R}$ with the following properties:

($N1$) $\|x\| \geq 0$ for all $x \in X$, and $\|x\| = 0$ if and only if $x = 0$;

($N2$) $\|\lambda x\| = |\lambda| \|x\|$ for all $x \in X$ and $\lambda \in \mathbb{F}$;

($N3$) $\|x + y\| \leq \|x\| + \|y\|$ for all $x, y \in X$ (the **triangle inequality**).

A **normed space** (or **normed vector space**, or **normed linear space**) is a pair $(X, \|\cdot\|)$, where X is a vector space over $\mathbb{F}$ and $\|\cdot\|$ is a norm on X.

A normed space may always be regarded as a metric space, in a way that is made precise by the following result.

Proposition 2.5. *Let* $(X, \| \cdot \|)$ *be a normed space. Let* $d : X \times X \longrightarrow \mathbb{R}$ *be given by*

$$d(x, y) := \|x - y\|, \qquad \text{for all } x, y \in X.$$

Then d *is a metric on the normed space* X.

Definition 2.6. Let $(X, \| \cdot \|)$ be a normed space. The metric d of Proposition 2.5 is called the **metric induced by the norm** (or **metric associated to the norm**) of the space.

Remark 2.7 (Metric structure of a normed space). The above construction of a metric from a norm on a vector space is so important that, whenever a normed space will be under consideration, it will **automatically be regarded as metric space too**, with the **metric being induced by the norm** as described in Definition 2.6 above.

Proof. Let $x, y, z \in X$, arbitrary. To check $(M1)$, note that directly from $(N1)$ we have

$$d(x, y) = \|x - y\| \geq 0,$$

and

$$d(x, y) = \|x - y\| = 0 \iff x - y = 0 \iff x = y,$$

hence $(M1)$ indeed holds. To check $(M2)$, note that by $(N2)$ we have

$$d(x, y) = \|x - y\| = \|(-1)(y - x)\| = |-1| \|y - x\| = d(y, x),$$

so that $(M2)$ holds. To check $(M3)$, note that by $(N3)$ we have

$$
\begin{aligned}
d(x, z) = \|x - z\| &= \|(x - y) + (y - z)\| \\
&\leq \|x - y\| + \|y - z\| \\
&= d(x, y) + d(y, z),
\end{aligned}
$$

so that $(M3)$ indeed holds. Therefore, d is indeed a metric on X. $\qquad \square$

Note that the usual distance on the Euclidean space $\mathbb{F}^m$ is actually induced by the norm (called **Euclidean norm**)

$$\|x\|_2 := \left(\sum_{k=1}^{m} |x_k|^2 \right)^{1/2} \qquad \text{for all } x = (x_1, \dots, x_m) \in \mathbb{F}^m.$$

In addition, on $\mathbb{F}^m$ one may consider several other norms, such as the following:

Example 2.8. Let $\| \cdot \|_1, \| \cdot \|_\infty : \mathbb{F}^m \longrightarrow \mathbb{R}$ be given by

$$\|x\|_1 := \sum_{k=1}^{m} |x_k| \quad \text{for all } x = (x_1, \dots, x_m) \in \mathbb{F}^m,$$

$$\|x\|_\infty := \max_{k \in \{1, \dots, m\}} |x_k| \quad \text{for all } x = (x_1, \dots, x_m) \in \mathbb{F}^m.$$

Then $\| \cdot \|_1, \| \cdot \|_\infty$ are norms on $\mathbb{F}^m$ (exercise!) which induce the metrics d_1 and d_∞ given, for any $x = (x_1, \ldots, x_m)$ and $y = (y_1, \ldots, y_m)$, by

$$d_1(x, y) := \sum_{k=1}^{m} |x_k - y_k|,$$

$$d_\infty(x, y) := \max_{k \in \{1, \ldots, m\}} |x_k - y_k|.$$

Many (if not all) of **the spaces of interest in Analysis are spaces whose elements are functions and mappings.** The general setting is described right below.

Definition 2.9. Let S be a non-empty set and Y be a vector space over $\mathbb{F}$. Let

$$F(S, Y) := \{f : S \longrightarrow Y\}.$$

Then, for any $f, g \in F(S, Y)$ and any $\lambda \in \mathbb{F}$, we may **define** $f + g$ and λf as elements of $F(S, Y)$ by[1]

$$\begin{cases} (f + g)(x) := f(x) + g(x) & \text{for all } x \in S, \\ (\lambda f)(x) := \lambda f(x) & \text{for all } x \in S. \end{cases}$$

Then, with the above defined operations, it can be shown that $F(S, Y)$ is a vector space called the **linear space of all vector functions from S to Y.**

In order to confirm that $F(S, Y)$ is a vector space, one has to check (this is rather tedious, but unfortunately unavoidable) that all the axioms of a vector space are satisfied. For illustration, let us verify that, for every $f, g, h \in F(S, Y)$,

$$(f + g) + h = f + (g + h).$$

Indeed, for every $x \in S$, using the associativity of the addition in Y, we get

$$\begin{aligned} ((f + g) + h)(x) &= (f + g)(x) + h(x) \\ &= ((f(x) + g(x)) + h(x) \\ &= f(x) + ((g(x) + h(x)) \\ &= f(x) + (g + h)(x) \\ &= ((f + (g + h))(x). \end{aligned}$$

The fact that all the other axioms of a vector space are satisfied by $F(S, Y)$ is left as an exercise for the reader.

Many of the vector spaces that we shall encounter are either of the form $F(S, Y)$ for suitable S and Y, or are vector subspaces of such spaces. Below is such an occurrence:

---

[1] Note that in the left-hand side of the above equalities we are defining vector addition and multiplication by scalars for elements of $F(S, Y)$, where these operations were previously undefined, whilst in the right-hand side we are using the known operations of the vector space Y. Namely, $F(S, Y)$ inherits a linear structure from the linear structure of Y.

Example 2.10. Consider the vector space of sequences over $\mathbb{F}$, given by

$$\mathbf{s} := \Big\{ x = (x_1, x_2, \dots) \ : \ x_k \in \mathbb{F} \text{ for all } k \in \mathbb{N} \Big\}.$$

The algebraic operations making $\mathbf{s}$ a vector space are given by

$$\left\{ \begin{array}{rl} x + y := & (x_1 + y_1, x_2 + y_2, \dots), \\ \lambda x := & (\lambda x_1, \lambda x_2, \dots). \end{array} \right.$$

for all $x = (x_1, x_2, \dots)$, $y = (y_1, y_2, \dots) \in \mathbf{s}$ and $\lambda \in \mathbb{F}$. In other words, $\mathbf{s}$ is the space of all sequences of scalars, and we have defined the vector addition and the multiplication by scalars in the usual way. However, recall that a sequence is merely a mapping $x : \mathbb{N} \longrightarrow \mathbb{F}$, which is characterised by the ordered list of all its values $(x(1), x(2), \dots)$, and we have written x_k instead of $x(k)$ for $k \in \mathbb{N}$. Thus we have the following equality of sets

$$\mathbf{s} = F(\mathbb{N}, \mathbb{F}).$$

Moreover, the operations defined on $\mathbf{s}$ are defined in the same way as those on $F(\mathbb{N}, \mathbb{F})$. Therefore, $\mathbf{s}$ is a vector space.

The space $\mathbf{s}$ is "too large" to be able to define any useful norms on it. Rather, we shall consider various vector subspaces of $\mathbf{s}$, on which appropriate norms may be defined, such as in the following representative examples.

Definition 2.11.
• Let ℓ^∞ be the set of sequences in $\mathbb{F}$ given by

$$\ell^\infty := \Big\{ x = (x_1, x_2, \dots) \ : \ x_k \in \mathbb{F} \text{ for all } k \in \mathbb{N}, \ \{x_k : k \in \mathbb{N}\} \text{ bounded in } \mathbb{F} \Big\}.$$

Then ℓ^∞ is a vector subspace of $\mathbf{s}$. In fact, $\ell^\infty = B(\mathbb{N}, \mathbb{F})$. Moreover, the function given on ℓ^∞ by

$$\|x\|_\infty := \sup_{k \in \mathbb{N}} |x_k| \quad \text{for all } x = (x_1, x_2, \dots) \in \ell^\infty$$

is a norm on ℓ^∞.
• Let ℓ^1 be the set of sequences in $\mathbb{F}$ given by

$$\ell^1 := \Big\{ x = (x_1, x_2, \dots) \ : \ x_k \in \mathbb{F} \text{ for all } k \in \mathbb{N}, \ \sum_{k=1}^{\infty} |x_k| < \infty \Big\}.$$

Then ℓ^1 is also a vector subspace of $\mathbf{s}$, and the function given on ℓ^1 by

$$\|x\|_1 := \sum_{k=1}^{\infty} |x_k| \quad \text{for all } x = (x_1, x_2, \dots) \in \ell^1$$

is a norm on ℓ^1.

It follows that both pairs $(\ell^\infty, \|\cdot\|_\infty)$ and $(\ell^1, \|\cdot\|_1)$ are normed spaces. Further vector subspaces of **s**, endowed with appropriate norms, will be considered later on. Let us now indicate a different example of a normed space.

Definition 2.12. Let S be a non-empty set. We set

$$B(S, \mathbb{F}) := \left\{ f : S \longrightarrow \mathbb{F} \mid f \text{ is bounded on } S \right\}.$$

Then, $B(S, \mathbb{F})$ is called **the space of (scalar) bounded functions on** $\mathbb{F}$. The function

$$\|\cdot\|_\infty : \quad B(S, \mathbb{F}) \longrightarrow \mathbb{R}$$

given by

$$\|f\|_\infty := \sup_{x \in S} |f(x)| \quad \text{for all } f \in B(S, \mathbb{F}),$$

is a norm, which is called the **supremum norm** or **uniform norm**. The metric d_∞ induced by the supremum norm is given, for any $f, g \in B(S, \mathbb{F})$, by

$$d_\infty(f, g) := \sup_{x \in S} |f(x) - g(x)|,$$

and is called the **supremum metric** or **uniform metric**.

It is easy to check that $B(S, \mathbb{F})$ is a vector subspace of $F(S, \mathbb{F})$. Let us check only $(N3)$, for the sake of illustration. Let $f, g \in B(S, \mathbb{F})$, arbitrary. For any $y \in S$, one has

$$
\begin{aligned}
|f(y) + g(y)| &\leq |f(y)| + |g(y)| \\
&\leq \sup_{x \in S} |f(x)| + \sup_{x \in S} |g(x)| \\
&= \|f\|_\infty + \|g\|_\infty.
\end{aligned}
$$

Since this is true for all $y \in S$, it follows that

$$\|f + g\|_\infty \leq \|f\|_\infty + \|g\|_\infty,$$

as required.

We conclude this section with some remarks regarding metrics and norms.

Remark 2.13.
• **A generic metric space** (X, d) **need not have any linear structure at all, so it does not even make sense to talk about norms on** X. While many important examples of metric spaces are in fact normed spaces, this is, however, not always the case. For instance, a proper (strict) subset A of the Euclidean space $\mathbb{F}^m$ endowed with the distance induced by the norm (whichever norm we wish to put on $\mathbb{F}^m$) is **not** a normed space anymore, because A is not a vector space!

• **Even if** X **is a vector space, not all metrics on a normed space**

can arise from a norm. For example, the discrete metric d_0 cannot possibly arise from any norm since it only takes the values 0 and 1, whilst by $(N2)$, the metric induced by a norm would have to take all values in $[0, \infty)$. Indeed, for any $t > 0$, there exists a vector $x \in X \setminus \{0\}$ such that $\|x\| = t$. This follows by setting $x := ty/\|y\|$, for any vector $y \in X \setminus \{0\}$.

The above observations suggest that a metric structure on a set is (much) more general than that of a norm structure, even if the set is a normed space.

2.3 Bounded sets, convergence and continuity

In this section we aim at abstracting well-known concepts of Real Analysis to the metric setting. The idea is to revisit in each case the Euclidean definitions and isolate the essential idea which depends only on the underlying (Euclidean) distance.

We begin by introducing the notion of a bounded set in a metric space. The idea is a direct extension of boundedness stemming from our Euclidean intuition that a set is bounded set if and only if it is contained into some ball.

Definition 2.14. Let (X, d) be a metric space. A subset A of X is said to be **bounded** if there exist $x_0 \in X$ and $M \geq 0$ such that

$$d(x, x_0) \leq M \quad \text{for all } x \in A.$$

Remark 2.15. If A is a bounded set in a metric space (X, d), then the role of x_0 in the above definition can be played by any point $\tilde{x}_0 \in X$. Indeed, let $x_0 \in X$ and $M \geq 0$ be such that

$$d(x, x_0) \leq M \quad \text{for all } x \in A,$$

and let $\tilde{x}_0$ be an arbitrary point in X. Then, for any $x \in A$,

$$d(x, \tilde{x}_0) \leq d(x, x_0) + d(x_0, \tilde{x}_0) \leq M + d(x_0, \tilde{x}_0) =: \tilde{M}.$$

Thus the boundedness condition for A is satisfied with respect to $\tilde{x}_0 \in X$, provided $\tilde{M}$ is defined as above.

The above remark motivates the fact that, in a normed vector space, the point x_0 in the above definition of a bounded set can be (and usually is) chosen to be the origin (see Figure 2.2).

Definition 2.16. Let $(X, \|\cdot\|)$ be a normed space. A subset A of X is said to be **bounded** if there exists $M \geq 0$ such that

$$\|x\| \leq M \quad \text{for all } x \in A.$$

We now generalise the definition of sequences to that of **sequences in a metric space**, or, more generally, **sequences in a given set**.

Definition 2.17. Let X be a non-empty set. A **sequence** in X is a mapping $x : \mathbb{N} \longrightarrow X$ sending each integer n to an element $x(n) \in X$.

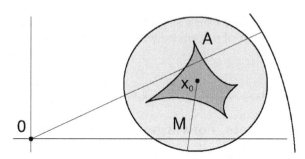

Figure 2.2: A depiction of a bounded set A in $\mathbb{R}^2$. Any ball which contains A centred at some x_0 satisfies the definition of boundedness, but in the Euclidean case we have the prominent point $x = 0$. The latter is not available in an abstract set!

As it is customary, we will be using the notation x_n instead of $x(n)$ for the nth term of the sequence, and the notation $(x_n)_{n=1}^\infty$ (or merely $(x_n)_1^\infty$ if the index set is clear) for the sequence itself. The notation $\{x_n : n \in \mathbb{N}\}$ will be used mostly to denote the set of all terms of the sequence.

Definition 2.18. Let (X, d) be a metric space. A sequence $(x_n)_1^\infty$ in X is said to be **bounded** if the set $\{x_n : n \in \mathbb{N}\}$ is bounded, meaning that there exists $x_0 \in X$ and $M \geq 0$ such that

$$d(x_n, x_0) \leq M \quad \text{for all } n \in \mathbb{N}.$$

Our next goal is to define convergence of a sequence in a metric space. Recall that in Real Analysis the definition of a convergent sequence is meant to express in a precise mathematical way the intuitive idea that the terms of the sequence get as close as one wishes to a certain real number (the limit) if n is sufficiently large. Note that closeness of any two real numbers can be quantified by means of the value of the distance between them, which is the modulus of the absolute value of their difference (see Figure 2.3).

If, however, we are given a sequence in a metric space (X, d) and an element which we would like it to play the role of a limit for the sequence, our definition of convergence, given below, will capture the same intuitive idea. This is possible since there is available a way to quantify how "close" any two elements of X are, namely by means of the distance d between them.

Definition 2.19. Let (X, d) be a metric space, let $(x_n)_1^\infty$ be a sequence in X and let $x_0 \in X$. We say that $(x_n)_1^\infty$ **converges to x_0 in (X, d)** as $n \to \infty$, and write $x_n \longrightarrow x_0$ in (X, d) as $n \to \infty$, if for any $\varepsilon > 0$ there exists $N \in \mathbb{N}$ such that

$$d(x_n, x_0) < \varepsilon \quad \text{for all } n \geq N.$$

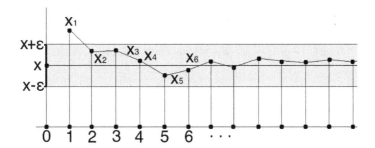

Figure 2.3: An illustration of a convergent sequence in $\mathbb{R}^2$. The idea of convergence is purely "metrical" in nature: for any $\varepsilon > 0$, there is an index $N \in \mathbb{N}$ such that for all subsequent terms, the distance of x_n from the limit x is at most ε. Equivalently, x_n lies in $(x - \varepsilon, x + \varepsilon)$ for all $n \geq N$.

Remark 2.20. It is sometimes useful to express the convergence condition of a sequence $(x_n)_1^\infty$ to a point x_0 in X in the following equivalent way: *for any $\varepsilon > 0$ there exists $N \in \mathbb{N}$ such that*

$$d(x_n, x_0) \leq \varepsilon \quad \text{for all } n \geq N.$$

The reader is advised to verify this equivalence. **We shall use without further comment this alternative form of the convergence condition whenever it is more convenient.**

Remark 2.21. If (X, d) is a metric space and $(x_n)_1^\infty \subseteq X$, $x_0 \in X$, then

$$\left(d(x_n, x_0) \right)_{n=1}^\infty$$

is a sequence of real numbers. It is immediate from the definitions that $x_n \longrightarrow x_0$ in (X, d) if and only if the sequence of real numbers $(d(x_n, x_0))_{n=1}^\infty$ converges to 0 in $\mathbb{R}$, in the sense familiar from Real Analysis.

We now examine more concrete forms taken by the convergence condition in certain metric spaces.

Example 2.22. In any discrete metric space (X, d_0), a sequence converges if and only if it is eventually constant.

Example 2.23. Consider the normed space $(\mathbb{F}^m, \| \cdot \|_2)$. For a sequence in $\mathbb{F}^m$ we shall usually use the notation $(x^{(n)})_{n=1}^\infty$, meaning that for each $n \in \mathbb{N}$, $x^{(n)}$ is an element of $\mathbb{F}^m$, namely it has the form

$$x^{(n)} = \left(x_1^{(n)}, x_2^{(n)}, \ldots, x_m^{(n)} \right).$$

It is known from analysis in several variables that, for any $x^{(0)} \in \mathbb{F}^m$ with

$x^{(0)} = (x_1^{(0)}, x_2^{(0)}, \ldots, x_m^{(0)})$, and for any sequence $(x^{(n)})_1^\infty$ in $\mathbb{F}^m$, we have that $x^{(n)} \longrightarrow x^{(0)}$ in $(\mathbb{F}^m, \| \cdot \|_2)$ if and only if

$$x_k^{(n)} \longrightarrow x_k^{(0)} \text{ in } (\mathbb{F}, |\cdot|) \text{ for each } k \in \{1, \ldots, m\},$$

as $n \to \infty$.

Example 2.24. In the space $B(S, \mathbb{F})$ of scalar bounded functions, convergence with respect to the supremum norm coincides with uniform convergence on S, familiar from Real Analysis (at least in the case of $S = [\alpha, \beta]$). Indeed, let $(f_n)_1^\infty \subseteq B(S, \mathbb{F})$, $f_0 \in B(S, \mathbb{F})$. Then claimed result follows from the following chain of equivalences:

$$f_n \longrightarrow f_0 \text{ uniformly on } S \text{ as } n \to \infty$$

$$\Longleftrightarrow \forall \varepsilon > 0 \, \exists \, N \in \mathbb{N} : \, \forall n \geq N, \, \left|f_n(x) - f_0(x)\right| \leq \varepsilon, \, \forall x \in S$$

$$\Longleftrightarrow \forall \varepsilon > 0, \, \exists \, N \in \mathbb{N} : \, \forall n \geq N, \, \sup_{x \in S} |f_n(x) - f_0(x)| \leq \varepsilon$$

$$\Longleftrightarrow \forall \varepsilon > 0, \, \exists \, N \in \mathbb{N} : \, \forall n \geq N, \, \|f_n - f_0\|_\infty \leq \varepsilon$$

$$\Longleftrightarrow f_n \longrightarrow f_0 \text{ in } (B(S, \mathbb{F}), \| \cdot \|_\infty), \text{ as } n \to \infty.$$

As expected, the next result confirms that for any convergent sequence, its limit is unique, if it exists.

Proposition 2.25 (Uniqueness of limits). *Suppose that (X, d) is a metric space, $(x_n)_1^\infty \subseteq X$, $x_0 \in X$ and $\tilde{x}_0 \in X$. If*

$$x_n \longrightarrow x_0 \quad and \quad x_n \longrightarrow \tilde{x}_0 \text{ in } (X, d), \text{ as } n \to \infty,$$

then $x_0 = \tilde{x}_0$.

Proof. Suppose that $x_n \longrightarrow x_0$ and $x_n \longrightarrow \tilde{x}_0$ in (X, d). This means, by an earlier remark, that $d(x_n, x_0) \longrightarrow 0$ and $d(x_n, \tilde{x}_0) \longrightarrow 0$ in $\mathbb{R}$ as $n \to \infty$. It follows that

$$0 \leq d(x_0, \tilde{x}_0) \leq d(x_0, x_n) + d(x_n, \tilde{x}_0) \longrightarrow 0.$$

By the "Sandwich (or Squeeze) Theorem"[2] in Real Analysis, it follows that $d(x_0, \tilde{x}_0) = 0$, which implies that $x_0 = \tilde{x}_0$. $\qquad \square$

The connection between the convergence of a sequence and its boundedness is given by the following result.

---

[2]We recall that the "Sandwich (or Squeeze) Theorem" asserts the following fact regarding the convergence of sequences. Suppose that $(a_n)_1^\infty$, $(b_n)_1^\infty$, $(c_n)_1^\infty$ are sequences in $\mathbb{R}$ such that $a_n \leq b_n \leq c_n$ for all $n \in \mathbb{N}$ and let $l \in [-\infty, \infty]$. Then, if $l \in \mathbb{R}$ and $a_n \longrightarrow l$ and $c_n \longrightarrow l$ as $n \to \infty$, it follows that $b_n \longrightarrow l$ as $n \to \infty$ as well. If $l = \infty$, it suffices to have only that $a_n \longrightarrow \infty$ and if $l = -\infty$, it suffices to have only that $c_n \longrightarrow -\infty$.

Proposition 2.26 (Convergence implies boundedness). *Let (X, d) be a metric space. Then every convergent sequence in (X, d) is bounded.*

Proof. Let $(x_n)_1^\infty$ be an arbitrary sequence in X such that $x_n \longrightarrow x_0$ in (X, d). Taking $\varepsilon := 1$ in the definition of convergence, we obtain the existence of $N \in \mathbb{N}$ such that $d(x_n, x_0) < 1$ for all $n \geq N$. This implies that

$$d(x_n, x_0) \leq \max \left\{ d(x_1, x_0), \ldots, d(x_{N-1}, x_0), 1 \right\} \quad \text{for all } n \in \mathbb{N},$$

which shows that $(x_n)_1^\infty$ is bounded in X. $\qquad \square$

The next result shows that, in a normed space, convergent sequences behave nicely with respect to the algebraic operations.

Proposition 2.27. *Let $(X, \|\cdot\|)$ be a normed space. Suppose that $x_0, y_0 \in X$, $\lambda_0 \in \mathbb{F}$ and also that $(x_n)_1^\infty$, $(y_n)_1^\infty \subseteq X$ and $(\lambda_n)_1^\infty \subseteq \mathbb{F}$. Then the following properties hold true:*

(i) *If $x_n \longrightarrow x_0$ and $y_n \longrightarrow y_0$ in $(X, \|\cdot\|)$ as $n \to \infty$, then*

$$x_n + y_n \longrightarrow x_0 + y_0 \quad \text{in } (X, \|\cdot\|) \text{ as } n \to \infty.$$

(ii) *If $x_n \longrightarrow x_0$ in $(X, \|\cdot\|)$ and $\lambda_n \longrightarrow \lambda_0$ in $(\mathbb{F}, |\cdot|)$ as $n \to \infty$, then*

$$\lambda_n x_n \longrightarrow \lambda_0 x_0 \quad \text{in } (X, \|\cdot\|) \text{ as } n \to \infty.$$

(iii) *If $x_n \longrightarrow x_0$ in $(X, \|\cdot\|)$ as $n \to \infty$, then $\|x_n\| \longrightarrow \|x_0\|$ in $(\mathbb{R}, |\cdot|)$ as $n \to \infty$.*

Proof. (i) The claimed result is a consequence of

$$\left\| (x_n + y_n) - (x_0 + y_0) \right\| \leq \|x_n - x_0\| + \|y_n - y_0\| \longrightarrow 0 + 0 = 0.$$

(ii) Since $(\lambda_n)_1^\infty$ is a convergent sequence in $\mathbb{F}$, it is a bounded sequence, namely there exists $M \geq 0$ such that $|\lambda_n| \leq M$ for all $n \in \mathbb{N}$. This implies that

$$\begin{aligned}
\|\lambda_n x_n - \lambda_0 x_0\| &= \left\| (\lambda_n x_n - \lambda_n x_0) + (\lambda_n x_0 - \lambda_0 x_0) \right\| \\
&\leq \left\| \lambda_n(x_n - x_0) \right\| + \left\| (\lambda_n - \lambda_0) x_0 \right\| \\
&\leq |\lambda_n| \, \|x_n - x_0\| + |\lambda_n - \lambda_0| \, \|x_0\| \\
&\leq M\|x_n - x_0\| + |\lambda_n - \lambda_0| \, \|x_0\|,
\end{aligned}$$

as $n \to \infty$, from where the claimed result follows.

(iii) We prove first the following inequality, which is also of independent interest:

$$\left| \|x\| - \|y\| \right| \leq \|x - y\| \quad \text{for all } x, y \in X.$$

Indeed, using the triangle inequality we get that

$$\|x\| = \big\|(x-y)+y\big\| \le \|x-y\| + \|y\|,$$
$$\|y\| = \big\|(y-x)+x\big\| \le \|x-y\| + \|x\|.$$

Hence

$$\max\Big\{\|x\|-\|y\|, \|y\|-\|x\|\Big\} \le \|x-y\|,$$

as claimed.

Now, using this inequality, we obtain that

$$0 \le \Big|\|x_n\|-\|x_0\|\Big| \le \|x_n - x_0\| \longrightarrow 0,$$

as $n \to \infty$, from where the claimed result follows.   □

In what follows, we consider mappings between metric spaces. Let (X, d) and (Y, ρ) be metric spaces, $x_0 \in X$, and $f : X \longrightarrow Y$ be a mapping. We define what it means for f to be **continuous** at x_0. As in the familiar case from Real Analysis, the notion of continuity expresses mathematically in a precise way the intuitive idea that the values of f get as close as one wishes to $f(x_0)$ at all points sufficiently close to x_0 (Figure 2.4).

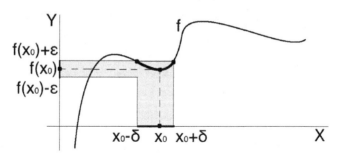

Figure 2.4: The geometric idea of continuity for a mapping $f : X \longrightarrow Y$ with $X = Y = \mathbb{R}$ at x_0. For any arbitrarily small $\varepsilon > 0$, there is a small $\delta > 0$ such that, when $|x - x_0| < \delta$, we have $|f(x) - f(x_0)| < \varepsilon$.

Note that closeness is measured in two possibly different ways in the spaces X and Y, namely in X we use the metric d whilst in Y we use the metric ρ.

Definition 2.28. Let (X, d) and (Y, ρ) be metric spaces, $f : X \longrightarrow Y$ a map, and $x_0 \in X$. The map f is said to be **continuous at** x_0 if for any $\varepsilon > 0$ there exists $\delta > 0$ such that $\rho(f(x), f(x_0)) < \varepsilon$ for all $x \in X$ with the property that $d(x, x_0) < \delta$. The mapping f is said to be **continuous on** X if it is continuous at all points $x_0 \in X$.

It turns out that, as in the familiar case of real functions of a real variable, continuity of mappings between metric spaces can be characterised by means of convergent sequences.

Theorem 2.29 (Sequential characterisation of continuity). *Let (X, d) and (Y, ρ) be metric spaces, $f : (X, d) \longrightarrow (Y, \rho)$ a mapping and $x_0 \in X$. Then, the next statements are equivalent:*

(i) *The map f is continuous at x_0.*

(ii) *For any sequence $(x_n)_1^\infty \subseteq X$ for which $x_n \longrightarrow x_0$ in (X, d) as $n \to \infty$, we have*

$$f(x_n) \longrightarrow f(x_0) \text{ in } (Y, \rho) \quad \text{as } n \to \infty.$$

Proof. (i) $\implies$ (ii): Suppose first that f is continuous at x_0. Let $(x_n)_1^\infty \subseteq X$ be any sequence with $x_n \longrightarrow x_0$ in (X, d) as $n \to \infty$. Fix $\varepsilon > 0$, arbitrary. Since f is continuous at x_0, there exists $\delta > 0$ such that $\rho(f(x), f(x_0)) < \varepsilon$ for all $x \in X$ with the property that $d(x, x_0) < \delta$. Since $d(x_n, x_0) \longrightarrow 0$ as $n \to \infty$, there exists $N \in \mathbb{N}$ such that $d(x_n, x_0) < \delta$ for all $n \geq N$. It follows that

$$\rho(f(x_n), f(x_0)) < \varepsilon, \quad \text{for all } n \geq N.$$

Since $\varepsilon > 0$ was arbitrary, we conclude that $\rho(f(x_n), f(x_0)) \longrightarrow 0$ in $\mathbb{R}$ as $n \to \infty$, which means that $f(x_n) \longrightarrow f(x_0)$ in (Y, ρ) as $n \to \infty$.

(ii) $\implies$ (i): Suppose now that for any sequence $(x_n)_1^\infty \subseteq X$ converging to x_0 in (X, d), $f(x_n)$ converges to $f(x_0)$ in (Y, ρ) as $n \to \infty$. Suppose for a contradiction that f is *not* continuous at x_0. Then there exists $\varepsilon_0 > 0$ such that, for all $\delta > 0$, there exists $x \in X$ such that

$$d(x, x_0) < \delta \quad \text{and} \quad \rho(f(x), f(x_0)) \geq \varepsilon_0.$$

For each $n \in \mathbb{N}$, let us take $\delta = 1/n$. It follows that, for all $n \in \mathbb{N}$, there exists $x_n \in X$ such that

$$d(x_n, x_0) < \frac{1}{n} \quad \text{and} \quad \rho(f(x_n), f(x_0)) \geq \varepsilon_0.$$

By the "Sandwich (or Squeeze) Theorem" in Real Analysis, the relation $d(x_n, x_0) < \frac{1}{n}$ for all $n \in \mathbb{N}$ implies that $d(x_n, x_0) \longrightarrow 0$ in $\mathbb{R}$, and hence that $x_n \longrightarrow x_0$ in (X, d) as $n \to \infty$.

On the other hand, the relation $\rho(f(x_n), f(x_0)) \geq \varepsilon_0$ for all $n \in \mathbb{N}$ shows that $(\rho(f(x_n), f(x_0)))_1^\infty$ does not converge to 0 in $\mathbb{R}$, and hence $(f(x_n))_1^\infty$ does not converge to $f(x_0)$ in (Y, ρ) as $n \to \infty$, contradicting our original assumption. Hence f must be continuous at x_0, as required. $\qquad \square$

2.4 Balls, open sets and closed sets

In what follows we develop a more geometric way of characterising convergence of sequences and continuity of functions in metric spaces. To that aim, we now introduce and study several types of subsets of a metric space which are analogous to the corresponding notions in Euclidean spaces.

Definition 2.30. Let (X, d) be a metric space. Given any $x_0 \in X$ and $r > 0$, we define the following sets:

(i) the **open ball of centre x_0 and radius r**:

$$\mathbb{B}_r(x_0) = \{x \in X \ : \ d(x_0, x) < r\};$$

(ii) the **closed ball of centre x_0 and radius r**:

$$\mathbb{D}_r(x_0) = \{x \in X \ : \ d(x_0, x) \leq r\};$$

(iii) the **sphere of centre x_0 and radius r**

$$\mathbb{S}_r(x_0) = \{x \in X \ : \ d(x_0, x) = r\}.$$

Sometimes we shall use such notation as $\mathbb{B}_r^X(x_0)$ or $\mathbb{B}_r^d(x_0)$ to emphasise the dependence of the ball on the set X or on the metric d.

Example 2.31. Let us consider, on the vector space $\mathbb{R}^2$, the norms $\| \cdot \|_1$, $\| \cdot \|_2$ and $\| \cdot \|_\infty$, which induce the metrics d_1, d_2 and d_∞. The balls centred at the origin with respect to these metrics are given by (see Figure 2.5)

$$\mathbb{B}_r^{d_1}\left((x_1^0, x_2^0)\right) = \left\{(x_1, x_2) \ : \ |x_1 - x_1^0| + |x_2 - x_2^0| < r\right\},$$

$$\mathbb{B}_r^{d_2}\left((x_1^0, x_2^0)\right) = \left\{(x_1, x_2) \ : \ (x_1 - x_1^0)^2 + (x_2 - x_2^0)^2 < r^2\right\},$$

$$\mathbb{B}_r^{d_\infty}\left((x_1^0, x_2^0)\right) = \left\{(x_1, x_2) \ : \ \max\left\{|x_1 - x_1^0|, |x_2 - x_2^0|\right\} < r\right\}.$$

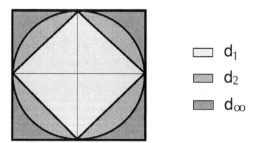

Figure 2.5: A depiction of the balls on $\mathbb{R}^2$ for the different metrics d_1, d_2 and d_∞.

In the study of metric spaces it is often useful to draw schematically pictures of balls in the same way as in the Euclidean plane. However, as we have just seen even in the plane the balls look entirely different from the Euclidean ones if we consider metrics different from the Euclidean ones. Thus one should bear in mind that our proofs, whilst informed by geometric intuition, must be based on the axioms of a metric space solely.

Example 2.32. Let (X, d_0) be a discrete metric space. Then, for any $x_0 \in X$, we have that $\mathbb{B}_r(x_0) = \{x_0\}$ if $0 < r \leq 1$, whilst $\mathbb{B}_r(x_0) = X$ if $r > 1$.

The following notion will play a most important role in all that follows.

Definition 2.33. Let (X, d) be a metric space. A subset A of X is said to be **open** in (X, d) if for any $x \in A$ there exists $r > 0$ (depending on x) such that $\mathbb{B}_r(x) \subseteq A$.

Roughly speaking, we think of a set as being open if, for any given point in the set, all nearby points (namely all points contained in a little open ball centred at that point) are also in the set. An equivalent way of think of open sets is as those which "do not contain any part of their boundary" (see Figure 2.6).

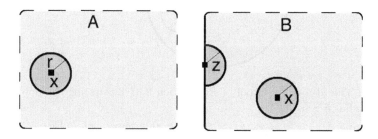

Figure 2.6: An illustration of an open set A and a non-open set B on the plane $\mathbb{R}^2$ with respect to the Euclidean distance d_2. The set B contains a "boundary part" of its lateral left side. For A, for any $x \in A$ there is a ball of radius r centred at x inside A. However, for B, for any point z on its "boundary", there is no ball centred at z which is contained inside B!

Definition 2.34. Let (X, d) be a metric space. A subset A of X is said to be **closed** in (X, d) if its complement, $X \setminus A$, is open.

Remark 2.35. The concepts of open and closed sets are not mutually exclusive! A give subset of a metric space might be both open and closed. For instance, for **any** metric space (X, d), the subsets $\emptyset$ and X are both open and closed (such sets are sometimes called "**clopen**" by some authors[3]. However, $\emptyset$ and X are not the only examples. It can be shown that when X can be written as $X = A \bigcup B$ and A, B are disjoint, namely $A \bigcap B = \emptyset$, then the sets A, B are "clopen".

Example 2.36. In $\mathbb{R}$ with the usual Euclidean metric given by the absolute value of the difference of numbers, any interval of the form (a, b) is an open set, whilst intervals of the form $[a, b)$ are not open sets.

---

[3]The pseudo-word "clopen" does not really exist in the English language, it is a mathematical hybrid of the terms clo-sed and o-pen.

The next result shows that the terminology **open ball**, that we have used prior to defining the notion of an **open set**, is in fact appropriate.

Proposition 2.37. *Let (X, d) be a metric space. Then every open ball $\mathbb{B}_r(x_0)$, where $x_0 \in X$ and $r > 0$, is an open set in (X, d).*

Proof. Let x be an arbitrary element of $\mathbb{B}_r(x_0)$, so that $d(x_0, x) < r$. We must show that there exists $\delta > 0$ such that $\mathbb{B}_\delta(x) \subseteq \mathbb{B}_r(x_0)$ (see Figure 2.7 for the intuitive idea of the construction).

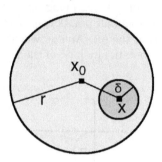

Figure 2.7: The idea of the proof of Proposition 2.37 illustrating the appropriate choice of radius $\delta > 0$.

Let us define $\delta := r - d(x_0, x) > 0$. With this choice of radius δ, for any $y \in \mathbb{B}_\delta(x)$ we have

$$d(x_0, y) \leq d(x_0, x) + d(x, y) < d(x_0, x) + \delta = r,$$

which shows that $y \in \mathbb{B}_r(x_0)$. This means that the inclusion $\mathbb{B}_\delta(x) \subseteq \mathbb{B}_r(x_0)$ holds, as required. □

The open sets in a metric space have the following important properties.

Proposition 2.38. *Let (X, d) be a metric space. Then the following hold:*

(i) *$\emptyset$ and X are open sets.*

(ii) *Let $\{U_i\}_{i \in I}$ be an arbitrary (possibly infinite) family of open subsets of X, where I is any index set. Then $\bigcup_{i \in I} U_i$ is an open set.*

(iii) *Let $U_1, \ldots, U_n$ be a finite family of open subsets of X, where $n \in \mathbb{N}$. Then $\bigcap_{i=1}^{n} U_i$ is an open set.*

Remark 2.39. The intersection of an infinite family of open sets need not be open in general. Indeed, in $\mathbb{R}$ with the usual metric, each of the intervals $(-1/n, 1/n)$ is open, but their intersection for all $n \in \mathbb{N}$ is not, since

$$\bigcap_{n=1}^{\infty} \left(-\frac{1}{n}, \frac{1}{n} \right) = \{0\}$$

and $\{0\}$ is not an open subset of $\mathbb{R}$.

Proof. (i) is immediate from the definition.

(ii) Let $x \in \bigcup_{i \in I} U_i$, arbitrary. Then $x \in U_{i_0}$ for some $i_0 \in I$. Since U_{i_0} is open, there exists $r > 0$ such that $\mathbb{B}_r(x) \subseteq U_{i_0}$. Therefore, we have

$$\mathbb{B}_r(x) \subseteq \bigcup_{i \in I} U_i.$$

Hence the set $\bigcup_{i \in I} U_i$ is open.

(iii) If $\bigcap_{i=1}^{n} U_i = \emptyset$, then the claimed result holds trivially. Otherwise, let $x \in \bigcap_{i=1}^{n} U_i$, arbitrary. Then $x \in U_i$ for all $i \in \{1, \dots, n\}$. Since each U_i is open, there exists $r_i > 0$ such that $\mathbb{B}_{r_i}(x) \subseteq U_i$ for each $i \in \{1, \dots, n\}$. Set $r := \min\{r_1, \dots, r_n\}$. Then $r > 0$ and

$$\mathbb{B}_r(x) \subseteq \mathbb{B}_{r_i}(x) \subseteq U_i$$

for each $i \in \{1, \dots, n\}$ and hence $\mathbb{B}_r(x) \subseteq \bigcap_{i=1}^{n} U_i$. Therefore, $\bigcap_{i=1}^{n} U_i$ is open. $\qquad\square$

The following result indicates another close connection between the open balls and the open sets of a metric space.

Proposition 2.40. *A subset of a metric space (X, d) is open if and only if it can be written as the union of open balls.*

Proof. Suppose first that we have

$$A = \bigcup_{i \in I} \mathbb{B}_{r_i}(x_i),$$

for some index set I, where $x_i \in X$ and $r_i > 0$ for all $i \in I$. Since, as we have proved, $\mathbb{B}_{r_i}(x_i)$ is an open set, it follows A is open, being a union of open sets.

Conversely, suppose that A is open. Then for each $x \in A$, there exists $r_x > 0$ such that $\mathbb{B}_{r_x}(x) \subseteq A$. This implies that $\bigcup_{x \in A} \mathbb{B}_{r_x}(x) \subseteq A$. On the other hand, for each $x \in A$ one has that $x \in \mathbb{B}_{r_x}(x)$, and this implies that $A \subseteq \bigcup_{x \in A} \mathbb{B}_{r_x}(x)$. In conclusion, we see that

$$A = \bigcup_{x \in A} \mathbb{B}_{r_x}(x),$$

and therefore A is a a union of open balls (see Figure 2.8). $\qquad\square$

Remark 2.41 (Reformulations of convergence and continuity). Let (X, d) be a metric space. Observe that the definition of the convergence of a sequence $(x_n)_1^{\infty}$ to x_0 may be restated using open balls as follows: for any $\varepsilon > 0$ there exists $N \in \mathbb{N}$ such that $x_n \in \mathbb{B}_{\varepsilon}(x_0)$ for all $n \geq N$. Similarly, if (X, d) and (Y, ρ) are metric spaces, the definition of continuity at a point $x_0 \in X$ for a map $f : X \longrightarrow Y$ may be restated using open balls as follows: for any $\varepsilon > 0$ there exists $\delta > 0$ such that (recall Figure 2.4 as well)

$$f\big(\mathbb{B}_\delta^X(x_0)\big) \subseteq \mathbb{B}_\varepsilon^Y\big(f(x_0)\big).$$

Note that, in view of the formulas $(\star)$ and $(\star\star)$ of Chapter 1 that the inverse image map and the image map satisfy, the above inclusion can be reformulated equivalently as

$$\mathbb{B}_\delta^X(x_0) \subseteq f^{-1}\big(\mathbb{B}_\varepsilon^Y\big(f(x_0)\big)\big).$$

These comments will turn out to be very important in subsequent generalisations of convergence and continuity.

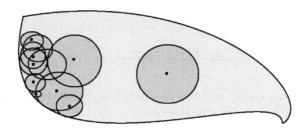

Figure 2.8: An illustration of the idea of the claim of Proposition 2.40. If the set is open, we can write it as the union of open balls centred at all of its points.

Because of the close connection between open sets and open balls (every open ball is an open set, and every open set contains an open ball centred at any of its points), one may easily obtain characterisations of convergence of sequences, and of continuity of functions, using open sets only, whilst not making use explicitly of the distance d. Such matters will be important in Chapter 4, in which we shall study **topological spaces**, i.e. spaces in which there exists a notion of **open set** which need not be associated with any metric. The formulation of these characterisations is simplified by the use of the following concept.

Definition 2.42. Let (X, d) be a metric space and $x_0 \in X$. A subset U of X is said to be a **neighbourhood** of x_0 if U is an open set in (X, d) that contains x_0.

The announced new characterisations of convergence and continuity are given below. Their proofs are left as easy exercises for the reader.

Proposition 2.43 (A "topological reformulation" of convergence). *Let (X, d) be a metric space, $(x_n)_1^\infty \subseteq X$ a sequence and $x_0 \in X$. Then $x_n \longrightarrow x_0$ in (X, d) as $n \to \infty$ if and only if for every neighbourhood U of x_0 there exists $N \in \mathbb{N}$ such that $x_n \in U$ for all $n \geq N$.*

Also, we have:

Proposition 2.44 (A "topological reformulation" of continuity). *Let (X, d) and (Y, ρ) be metric spaces, $f : X \longrightarrow Y$ a mapping and $x_0 \in X$. Then f is continuous at x_0 if and only if for any neighbourhood V of $f(x_0)$ there exists a neighbourhood U of x_0 such that $f(U) \subseteq V$.*

We note that, in view of the formulas $(\star)$ and $(\star\star)$ of Chapter 1, the inclusion $f(U) \subseteq V$ is equivalent to $U \subseteq f^{-1}(V)$. The next result gives an elegant characterisation of the global continuity of a map between metric spaces, either in terms of open sets, or of closed sets.

Proposition 2.45 (Topological criteria for global continuity). *Let (X, d) and (Y, ρ) be metric spaces and $f : X \longrightarrow Y$ a mapping between them. Then the following statements are equivalent:*

(i) *f is continuous on X;*

(ii) *$f^{-1}(V)$ is an open set in (X, d) for any open set V in (Y, ρ);*

(iii) *$f^{-1}(F)$ is a closed set in (X, d) for any closed set F in (Y, ρ).*

Proof of Proposition 2.45. (i) $\Longrightarrow$ (ii): Suppose first that (i) holds. Let V be an arbitrary open set in (Y, ρ). Let $x_0 \in f^{-1}(V)$, arbitrary. Then $f(x_0) \in V$ and, since V is open, there exists $\varepsilon > 0$ such that $\mathbb{B}_\varepsilon(f(x_0)) \subseteq V$. On the other hand, since f is continuous at x_0, there exists $\delta > 0$ such that

$$f(\mathbb{B}_\delta(x_0)) \subseteq \mathbb{B}_\varepsilon(f(x_0)).$$

This implies that $f(\mathbb{B}_\delta(x_0)) \subseteq V$, which means that $\mathbb{B}_\delta(x_0) \subseteq f^{-1}(V)$. Since x_0 was arbitrary, it follows that $f^{-1}(V)$ is open in (X, d).

(ii) $\Longrightarrow$ (i): Suppose now that (ii) holds. Let $x_0 \in X$, arbitrary. We aim to show that f is continuous at x_0. Fix $\varepsilon > 0$, arbitrary. Then $\mathbb{B}_\varepsilon(f(x_0))$ is open in (Y, ρ). Hence, by assumption, $f^{-1}(\mathbb{B}_\varepsilon(f(x_0)))$ is open in (X, d). However, since $x_0 \in f^{-1}(\mathbb{B}_\varepsilon(f(x_0)))$, it follows that there exists $\delta > 0$ such that

$$\mathbb{B}_\delta(x_0) \subseteq f^{-1}(\mathbb{B}_\varepsilon(f(x_0))).$$

This is equivalent to $f(\mathbb{B}_\delta(x_0)) \subseteq \mathbb{B}_\varepsilon(f(x_0))$. Since $\varepsilon > 0$ was arbitrary, this shows that f is continuous at x_0. Hence f is continuous on X.

(ii) $\Longrightarrow$ (iii): Suppose again that (ii) holds. Let $F \subseteq Y$ be any closed subset of Y. Then $V := Y \setminus F$ is open and hence, by assumption, $f^{-1}(V)$ is open in X. But then

$$f^{-1}(F) = f^{-1}(Y \setminus V) = X \setminus f^{-1}(V),$$

so that $f^{-1}(F)$ is closed in X.

(iii) $\Longrightarrow$ (ii): Suppose now that (iii) holds. Let V be an arbitrary open subset of Y. Let $F := Y \setminus V$, so that F is closed in Y. By assumption, $f^{-1}(F)$ is closed in X. But then

$$f^{-1}(V) = f^{-1}(Y \setminus F) = X \setminus f^{-1}(F),$$

so that $f^{-1}(V)$ is open in X. $\qquad\square$

Remark 2.46. Let $f : (X,d) \longrightarrow (Y,\rho)$ be a continuous mapping on X. Then:

(i) It need not be true that $f(U)$ is an open set in Y for every open set U in X. For example, take the continuous function $f(x) = |x|$ ($f : \mathbb{R} \longrightarrow \mathbb{R}$, usual metrics). Then $\mathbb{R}$ is open, but the image set $f(\mathbb{R}) = [0, \infty)$ is not open (see Figure 2.9).

(ii) It is not necessarily true that the inverse image of an open ball is an open ball (although it is an open set). For example, take the continuous function $f(x) = |x|$ ($f : \mathbb{R} \longrightarrow \mathbb{R}$, usual metrics). Then $(1,2)$ is an open ball in $\mathbb{R}$, but the inverse image set equals

$$f^{-1}((1,2)) = (-2,-1) \bigcup (1,2),$$

which is not an open ball (Figure 2.9).

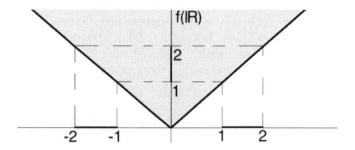

Figure 2.9: An illustration of the examples in Remark 2.46.

2.5 Metric topologies and equivalent metrics

As we have seen in the preceding results, convergence of sequences and continuity of functions in metric spaces can be described purely in terms of open sets. These facts show the importance of the notion of an open set. We therefore give the collection of all the open sets of a metric space a special name.

Definition 2.47. Let (X,d) be a metric space. The collection of all open subsets of X, denoted by $\mathcal{T}_d$, namely

$$\mathcal{T}_d = \{U \subseteq X : U \text{ is open in } (X,d)\},$$

is called the **metric topology** of (X,d), or **the topology of X induced by the metric d**.

We now study the situation when several metrics are defined on a set X, or when several norms are defined on a vector space X. We have already encountered such situations for the space $\mathbb{F}^m$, on which we have defined the norms $\|\cdot\|_1, \|\cdot\|_2$ and $\|\cdot\|_\infty$, with respective induced metrics d_1, d_2, and d_∞. As illustrated in Example 2.31, the open balls with respect to these metrics are quite different geometrically and thus, at first sight, it may appear that the open sets with respect to these metrics should also be different. As we shall see, this is actually not true! In fact, the open sets with respect to these metrics are the same. On the other hand, there exist also situations where different metrics on the same set induce indeed different topologies.

To study such issues systematically, it will be convenient to introduce the following terminology.

Definition 2.48. Let X be a set.

(i) Two metrics d and ρ on X are said to be **topologically equivalent** if they induce the same topology on X, namely if

$$\mathcal{T}_d = \mathcal{T}_\rho$$

or, in other words, if the metrics d, ρ give rise to the same open sets.

(ii) Two metrics d and ρ on X are said to be **Lipschitz equivalent** if there exist constants $m, M > 0$ such that

$$md(x,y) \leq \rho(x,y) \leq Md(x,y) \quad \text{for all } x, y \in X.$$

(iii) Suppose in addition that X is a vector space. Two norms $\|\cdot\|$ and $\|\|\cdot\|\|$ on X are said to be **equivalent** if there exist constants $m, M > 0$ such that

$$m\|x\| \leq \|\|x\|\| \leq M\|x\| \quad \text{for all } x \in X.$$

Remark 2.49. It is immediate that the topological equivalence of metrics, the Lipschitz equivalence of metrics, as well as the equivalence of norms, as defined above, are indeed **equivalence relations** in the sense of set theory (that is: each of them is reflexive, symmetric and transitive).

Remark 2.50. Given two norms $\|\cdot\|$ and $\|\|\cdot\|\|$ on X, let d and ρ be their respective induced metrics. It is immediate to check that the norms $\|\cdot\|$ and $\|\|\cdot\|\|$ are equivalent if and only if the metrics d and ρ are Lipschitz equivalent.

The main result connecting these notions is the following.

Proposition 2.51. *Let X be a set. If two metrics on X are Lipschitz equivalent, then they are topologically equivalent.*

Proof. Let d and ρ be two Lipschitz equivalent metrics on X, and let $m, M > 0$ be such that

$$md(x,y) \leq \rho(x,y) \leq Md(x,y) \quad \text{for all } x, y \in X.$$

Let A be any open set in (X, d). Let x_0 be an arbitrary point of A. It follows that there exists $r > 0$ such that $\mathbb{B}_r^d(x_0) \subseteq A$. We now claim that $\mathbb{B}_{mr}^e(x_0) \subseteq \mathbb{B}_r^d(x_0)$. This is indeed true since, for any $x \in \mathbb{B}_{mr}^e(x_0)$ we have

$$d(x, x_0) \leq \frac{1}{m}\rho(x, x_0) < \frac{1}{m}mr = r,$$

so that $x \in \mathbb{B}_r^d(x_0)$. This implies that $\mathbb{B}_{mr}^\rho(x_0) \subseteq A$. Since $x_0 \in A$ was arbitrary, it follows that A is an open set in (X, ρ). In a similar way one can prove that any open set in (X, ρ) is also open in (X, d). Therefore, the metrics d and ρ are topologically equivalent. $\qquad\square$

Remark 2.52 (Topological equivalence vs. shape of balls). Let us now return to the example of the vector space $\mathbb{F}^m$, with the three norms $\|\cdot\|_1$, $\|\cdot\|_2$ and $\|\cdot\|_\infty$. The following inequalities, which may be easily proved,

$$\|z\|_\infty \leq \|z\|_2 \leq \|z\|_1 \leq m\|z\|_\infty \quad \text{for all } z \in \mathbb{F}^m,$$

show that the norms $\|\cdot\|_1$, $\|\cdot\|_2$ and $\|\cdot\|_\infty$ are equivalent on $\mathbb{F}^m$. It follows that the metrics d_1, d_2 and d_∞ are Lipschitz equivalent, and therefore topologically equivalent. Thus, even though the open balls with respect to the three metrics are different (see Figure 2.5), the open sets are the same! Since, as we have seen, convergence of sequences and continuity of functions can be expressed in terms of the family of the open sets only (and without explicit reference to the actual metric), it follows that, for the purposes of studying convergence and continuity on $\mathbb{F}^m$ it does not matter which of the three metrics one is using. In particular, the characterisation of convergent sequences given in Example 2.8 for $(\mathbb{F}^m, \|\cdot\|_2)$ applies equally well to $(\mathbb{F}^m, \|\cdot\|_1)$ and $(\mathbb{F}^m, \|\cdot\|_\infty)$.

2.6   Metric and normed subspaces

Suppose that a metric space is given, and we are particularly interested in a subset of it. Is there an easy way to regard the subset itself as a metric space? This is indeed so, since we can just restrict to the subset the metric of the original space!

Definition 2.53. Let (X, d) be a metric space and Y be a subset of X. Then the restriction of d to $Y \times Y$ (which we continue to denote by d) is a metric on Y, which is called the **induced metric on** Y, or **subspace metric of** Y. In this way, (Y, d) is a metric space, which is called a **metric subspace** of (X, d).

The same philosophy applies to normed spaces as well.

Definition 2.54. Let $(X, \| \cdot \|)$ be a normed vector space, and let Y be a vector subspace of X. Then the restriction of $\| \cdot \|$ to Y (which we continue to denote by $\| \cdot \|$) is a norm on Y. The normed space $(Y, \| \cdot \|)$ is called a **normed subspace** of X.

It is important to note that, if $(X, \|\cdot\|)$ is a normed space, and Y is a subset of X, then Y is **NOT** a normed subspace of X unless it is vector subspace. Nevertheless, since X is a metric space (with the metric d induced by the norm), then Y is a metric subspace of X (with the metric induced by d).

Convergence of a sequence of elements in a subspace is related to convergence in the original space in the following way.

Proposition 2.55. *Let (X, d) be a metric space and Y be a subspace of X.*

(i) *If $(y_n)_1^\infty \subseteq Y, y_0 \in Y$ and $y_n \longrightarrow y_0$ in (Y, d) as $n \to \infty$, then $y_n \longrightarrow y_0$ in (X, d) as $n \to \infty$.*

(ii) *If $(y_n)_1^\infty \subseteq Y$, $x_0 \in X$ and $y_n \longrightarrow x_0$ in (X, d) as $n \to \infty$, then $y_n \longrightarrow x_0$ in (Y, d) as $n \to \infty$ if and only if $x_0 \in Y$.*

The proof of this result is left as an easy exercise for the reader.

As an illustrative example, consider the subspace $(0, 1]$ of $\mathbb{R}$ with the standard metric. Then $1/n \in (0, 1]$ for all $n \in \mathbb{N}$, but $(1/n)_1^\infty$ does not converge in $(0, 1]$ since there exists no $\alpha \in (0, 1]$ such that $1/n \longrightarrow \alpha$ (whilst $1/n \longrightarrow 0$ in $\mathbb{R}$, note that $0 \notin (0, 1]$). Thus, the limit point 0 is outside the subspace. (As we shall see later, this cannot happen if the subspace is closed.)

The open balls of a subspace are related to the open balls of the original space in the following simple way.

Proposition 2.56. *Let (X, d) be a metric space and Y be a subset of X. Then, for any $y_0 \in Y$ and $r > 0$,*

$$\mathbb{B}_r^Y(y_0) = \mathbb{B}_r^X(y_0) \bigcap Y.$$

Namely, the balls of the metric subspace can be characterised as those sets which are obtained by intersecting an open ball of the bulk space with the subspace.

Example 2.57. Let us take $X = \mathbb{R}^2$ with the usual metric and consider the subspace $Y := \big((-1, 1] \bigcup \{3\}\big) \times \mathbb{R}$. Then (see Figure 2.10)

$$\mathbb{B}_1^Y((-1, 0)) = \mathbb{B}_1((-1, 0)) \bigcap \{(x, y) \in \mathbb{R}^2 : x > -1\},$$

$$\mathbb{B}_1^Y((1, 0)) = \mathbb{B}_1^Y((1, 0)) \bigcap \{(x, y) \in \mathbb{R}^2 : x \leq 1\},$$

$$\mathbb{B}_1^Y((3, 0)) = \{3\} \times (-1, 1).$$

Let (X, d) be a metric space and let $Y \subseteq X$. As a metric space with the induced metric, Y has its own open sets, which may not necessarily be open in X. For example, Y itself need not be open in (X, d), though Y is always open in (Y, d). The connection between the open sets of Y and those of X is given by the following result.

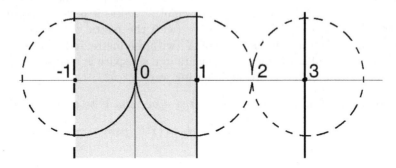

Figure 2.10: An illustration of the balls relative to the subspace Y of the plane $\mathbb{R}^2$ given by $Y := \big((-1, 1] \bigcup \{3\}\big) \times \mathbb{R}$ in Example 2.57.

Proposition 2.58. *Let (X, d) be a metric space and Y be a subset of X. Then a subset U of Y is open in (Y, d) if and only if there exists an open set V in (X, d) such that $U = V \bigcap Y$.*

Proof. Suppose first that $U = V \bigcap Y$, for some open set V in (X, d). We want to show that U is open in (Y, d). Let $y \in U$, arbitrary. Then $y \in V$ and, since V is open in (X, d), there exists $r > 0$ such that $\mathbb{B}_r^X(y) \subseteq V$. But then, by the previous proposition,

$$\mathbb{B}_r^Y(y) = \mathbb{B}_r^X(y) \bigcap Y \subseteq V \bigcap Y = U.$$

Since $y \in U$ was arbitrary, it follows that U is open in (Y, d).

Now, let $U \subseteq Y$ be an open set in (Y, d). Then, by Proposition 2.40,

$$U = \bigcup_{i \in I} \mathbb{B}_{r_i}^Y(y_i),$$

for some family $\{\mathbb{B}_{r_i}^Y(y_i)\}_{i \in I}$ of open balls in (Y, d), where I is an index set, $y_i \in Y$ and $r_i > 0$ for each $i \in I$. But, by Proposition 2.56, we have that $\mathbb{B}_{r_i}^Y(y_i) = \mathbb{B}_{r_i}^X(y_i) \bigcap Y$. Thus

$$U = \bigcup_{i \in I} \left(\mathbb{B}_{r_i}^X(y_i) \bigcap Y \right) = \left(\bigcup_{i \in I} \mathbb{B}_{r_i}^X(y_i) \right) \bigcap Y.$$

By setting $V := \bigcup_{i \in I} \mathbb{B}_{r_i}^X(y_i)$, it follows that the set V is open in (X, d) and $U = V \bigcap Y$, as required. $\qquad \square$

Example 2.59. Consider the set $Y = [1,2] \cup (3,4]$ with the metric induced when regarded as a subspace of $\mathbb{R}$ with the usual metric. Then each of $[1,2]$, $(1,2]$, $(3,4)$ and $(3,4]$ is an open subset of Y. This is indeed so because $[1,2] = Y \cap (0,3)$, $(1,2] = Y \cap (1,3)$, $(3,4) = Y \cap (3,4)$ and $(3,4] = Y \cap (3,5)$. However, only $(3,4)$ is open in $\mathbb{R}$! This shows that "openness" is a property relative to the (sub)space we consider.

2.7 Interior, closure and boundary of a set

Let (X,d) be a metric space and A a subset of X. In general, A need not be either open or closed. It turns out to be useful to associate to every A an open and a closed subset of X that are closely related to A. These are, roughly speaking, the "largest open set contained in A" and the "smallest closed set containing A".

Definition 2.60. Let (X,d) be a metric space and A be a subset of X.

(i) A point $x \in A$ is said to be an **interior point** of A if there exists $r > 0$ such that $\mathbb{B}_r(x) \subseteq A$. The set of all interior points of A is called the **interior of** A, and is denoted by A°:

$$A^\circ := \left\{ x \in A \,\middle|\, \exists\, r > 0 : \mathbb{B}_r(x) \subseteq A \right\}.$$

Note that $A^\circ \subseteq A$.

(ii) A point $x \in X$ is said to be a **closure point** of A if, for every $r > 0$, one has that $\mathbb{B}_r(x) \cap A \neq \emptyset$. The set of all closure points of A is called the **closure of** A, and is denoted by $\overline{A}$:

$$\overline{A} := \left\{ x \in X \,\middle|\, \forall\, r > 0 : \mathbb{B}_r(x) \cap A \neq \emptyset \right\}.$$

Note that in general $\overline{A} \not\subseteq A$.

(iii) The **boundary of** A, denoted by ∂A, is defined to be the set

$$\partial A := \overline{A} \setminus A^\circ.$$

The elements of the boundary of A are called **boundary points** of A.

Example 2.61. (i) In $\mathbb{R}$ with the usual metric: $(a,b]^\circ = (a,b)$, $\overline{(a,b]} = [a,b]$ and

$$\partial(a,b) = \partial[a,b] = \partial[a,b) = \partial(a,b] = \{a,b\}.$$

Also, the boundary of the set of rationals is $\mathbb{R}$ itself:

$$\partial\mathbb{Q} = \overline{\mathbb{Q}} \setminus \mathbb{Q}^\circ = \mathbb{R} \setminus \emptyset = \mathbb{R}.$$

(ii) In $\mathbb{R}^2$ with the usual metric, we have

$$\partial \mathbb{B}_r(0) = \mathbb{D}_r(0) \setminus \mathbb{B}_r(0) = \mathbb{S}_r(0) = \left\{(x, y) \in \mathbb{R}^2 \ : \ x^2 + y^2 = r^2\right\}.$$

The most important properties of the interior and the closure of a set in a metric space are summarised in the following result.

Proposition 2.62. *Let (X, d) be a metric space and A a subset of X. Then:*

(i) $A^\circ \subseteq A \subseteq \overline{A}$;

(ii) A° *is an open set;*

(iii) $\overline{A}$ *is a closed set;*

(iv) A *is open if and only if $A = A^\circ$;*

(v) A *is closed if and only if $A = \overline{A}$.*

Proof. (i) The inclusion $A^\circ \subseteq A$ is trivial. We now show that $A \subseteq \overline{A}$ by proving that the complements of these sets satisfy the reverse inclusion. Let x be an arbitrary element of $X \setminus \overline{A}$. Then there exists $r > 0$ such that $\mathbb{B}_r(x) \cap A = \emptyset$, which means that $\mathbb{B}_r(x) \subseteq X \setminus A$. This implies, in particular, that $x \in X \setminus A$. We have thus proved that $X \setminus \overline{A} \subseteq X \setminus A$, a fact which is equivalent to $A \subseteq \overline{A}$.

(ii) Let x be an arbitrary element of A°. Then there exists $r > 0$ such that $\mathbb{B}_r(x) \subseteq A$. Since $\mathbb{B}_r(x)$ is an open set, for every $y \in \mathbb{B}_r(x)$ there exists $\delta > 0$ such that $\mathbb{B}_\delta(y) \subseteq \mathbb{B}_r(x) \subseteq A$. It follows that all such points y belong to A°, and therefore $\mathbb{B}_r(x) \subseteq A^\circ$. This shows that A° is an open set.

(iii) We now show that $\overline{A}$ is a closed set by proving that its complement is open. Let x be an arbitrary element of $X \setminus \overline{A}$. Then there exists $r > 0$ such that $\mathbb{B}_r(x) \cap A = \emptyset$, which means that $\mathbb{B}_r(x) \subseteq X \setminus A$. Since $\mathbb{B}_r(x)$ is an open set, for every $y \in \mathbb{B}_r(x)$ there exists $\delta > 0$ such that $\mathbb{B}_\delta(y) \subseteq \mathbb{B}_r(x) \subseteq X \setminus A$, a fact which implies that $\mathbb{B}_\delta(y) \cap A = \emptyset$. It follows that all such points y belong to $X \setminus \overline{A}$, and therefore $\mathbb{B}_r(x) \subseteq X \setminus \overline{A}$. This shows that $X \setminus \overline{A}$ is an open set, and therefore $\overline{A}$ is a closed set.

(iv) Suppose first that A is open. It is immediate from the definition of A° that in this case $A \subseteq A^\circ$. Then, taking also into account (i), it follows that $A = A^\circ$. If we now assume that $A = A^\circ$, the fact that A is open follows from (ii).

(v) Suppose first that A is closed. Then $X \setminus A$ is open. It follows that, for any $x \in X \setminus A$ there exists $r > 0$ such that $\mathbb{B}_r(x) \subseteq X \setminus A$, so that $x \notin \overline{A}$. This means that $X \setminus A \subseteq X \setminus \overline{A}$, a fact which is equivalent to $\overline{A} \subseteq A$. Then, taking also into account (i), it follows that $A = \overline{A}$. If we now assume that $A = \overline{A}$, the fact that A is closed follows from (iii). $\qquad\square$

Further properties of the interior and closure will be studied in Chapter 4 in the more general setting of topological spaces. In what follows we shall merely indicate a few properties which rely explicitly on the metric structure of the space.

The next result shows that the closure of a set in a metric space can be characterised by means of convergent sequences.

Proposition 2.63 (Sequential characterisation of the closure). *Let (X, d) be a metric space and A be a subset of X. Then, for any $x \in X$, $x \in \overline{A}$ if and only if there exists a sequence $(x_n)_1^\infty \subseteq A$ such that $x_n \longrightarrow x$ in (X, d) as $n \to \infty$. This equivalence can be rephrased in the following way:*

$$\overline{A} = \left\{ x \in X \mid \exists\, (x_n)_1^\infty \subseteq A \; : \; x_n \longrightarrow x \text{ in } (X, d) \text{ as } n \to \infty \right\}.$$

Proof. ($\Longrightarrow$): Suppose first that $x \in \overline{A}$. By taking, for each $n \in \mathbb{N}$, $r := 1/n$ in Definition 2.60, we obtain that $\mathbb{B}_{1/n}(x) \cap A \neq \emptyset$. Hence we can choose $x_n \in \mathbb{B}_{1/n}(x) \cap A$, for each $n \in \mathbb{N}$. Then $(x_n)_1^\infty \subseteq A$ and

$$0 \leq d(x_n, x) < \frac{1}{n} \quad \text{for all } n \in \mathbb{N},$$

which implies, by the Squeeze Theorem, that $x_n \longrightarrow x$ in (X, d) as $n \to \infty$.

($\Longleftarrow$): Conversely, suppose that there exists $(x_n)_1^\infty \subseteq A$ such that $x_n \longrightarrow x$ in (X, d) as $n \to \infty$. We prove that $x \in \overline{A}$. Indeed, let $r > 0$ be arbitrary. Then, by the definition of convergence, there exists $N \in \mathbb{N}$ such that $x_n \in \mathbb{B}_r(x)$ for all $n \geq N$. In particular, $x_N \in \mathbb{B}_r(x) \cap A$, so that $\mathbb{B}_r(x) \cap A \neq \emptyset$. This shows that $x \in \overline{A}$. $\square$

We may now easily extend the concept of density, familiar from the Real Analysis, to metric spaces.

Definition 2.64. A subset A of a metric space (X, d) is said to be **dense** in (X, d) if the closure of A equals the space itself: $\overline{A} = X$.

Remark 2.65. We recall from Definition 2.60 that the closure of a subset A of a metric space (X, d) is the set of points $x \in X$ for which $\mathbb{B}_r(x) \cap A \neq \emptyset$ for all $r > 0$. Thus, a subset A of X is dense in (X, d) if and only if

$$\mathbb{B}_r(x) \cap A \neq \emptyset \quad \text{for all } x \in X \text{ and for all } r > 0.$$

Further, by the sequential characterisation of the closure of a set (Proposition 2.63), one can easily recast the density condition in terms of sequences.

Let (X, d) be a metric space. We shall now study the relation between the closure of open balls (of certain centres and radii) and the closed balls (of the same centres and radii). These two classes of sets are **not** always the same! To this end, consider the next example:

Example 2.66. Let (X, d_0) be any discrete metric space. Then, the reader may easily check directly from the definitions that, for any $x \in X$, we have

$$\overline{\mathbb{B}_1(x)} = \mathbb{B}_1(x) = \{x\}.$$

Namely, the closure of the open ball of radius 1 centred at x coincides with the singleton set $\{x\}$. On the other hand, the closed ball of radius 1 centred at x is the whole space: $\mathbb{D}_1(x) = X$.

However, such pathologies are excluded in normed vector spaces, as the next result attests. Namely, in normed spaces, closed balls are the same as closures of (open) balls. In the general case of metric spaces, we have just an one-sided inclusion.

Proposition 2.67 (Closed balls vs. closures of balls in metric and normed spaces).
(i) *Let (X, d) be a metric space. Then for any $x_0 \in X$ and $r > 0$, we have that*

$$\overline{\mathbb{B}_r(x_0)} \subseteq \mathbb{D}_r(x_0).$$

(ii) *Let $(X, \|\cdot\|)$ be a normed space. Then for any $x_0 \in X$ and $r > 0$, we have that*

$$\overline{\mathbb{B}_r(x_0)} = \mathbb{D}_r(x_0).$$

Proof. (i) Let $x \in \overline{\mathbb{B}_r(x_0)}$ be an arbitrary point. Then, by Proposition 2.63, there exists $(x_n)_1^\infty \subseteq \mathbb{B}_r(x_0)$ such that $x_n \longrightarrow x$. Since $d(x_n, x_0) < r$ for all $n \in \mathbb{N}$ and also $x_n \longrightarrow x$ in (X, d) as $n \to \infty$, this yields

$$d(x, x_0) = \lim_{n \to \infty} d(x_n, x_0).$$

Hence, it follows that $d(x, x_0) \leq r$, and hence $x \in \mathbb{D}_r(x_0)$. This proves that indeed $\overline{\mathbb{B}_r(x_0)} \subseteq \mathbb{D}_r(x_0)$.

(ii) It is enough to prove that $\mathbb{D}_r(x_0) \subseteq \overline{\mathbb{B}_r(x_0)}$. In view of Proposition 2.63, it is sufficient to prove that, for any $x \in \mathbb{D}_r(x_0)$, there exists a sequence $(x_n)_1^\infty \subseteq \mathbb{B}_r(x_0)$ such that $x_n \longrightarrow x$ in $(X, \|\cdot\|)$ as $n \to \infty$. Let $x \in \mathbb{D}_r(x_0)$ be an arbitrary point. It follows that $\|x - x_0\| \leq r$. For each $n \in \mathbb{N}$, let us consider the following convex combination of x and x_0:

$$x_n := \frac{1}{n} x_0 + \frac{n-1}{n} x.$$

Then, we have

$$\|x_n - x_0\| = \frac{n-1}{n} \|x - x_0\| \leq \frac{n-1}{n} r < r,$$

which implies that $x_n \in \mathbb{B}_r(x_0)$. On the other hand,

$$\|x_n - x\| = \frac{1}{n} \|x_0 - x\| \longrightarrow 0 \quad \text{as } n \to \infty.$$

This shows that $x \in \overline{\mathbb{B}_r(x_0)}$, as required, therefore completing the proof. $\square$

2.8 Some concrete examples of normed spaces

In this section we present some further classes of normed vector spaces which are of paramount importance throughout Analysis. These spaces consist of either sequences of numbers (scalars), or scalar-valued functions. Since every sequence can be seen as a function with domain $\mathbb{N}$, it is fair to say that essentially all examples are of *"functional spaces"*, namely spaces whose elements are functions.

2.8.1 The Euclidean space

Let us first consider the familiar vector space $\mathbb{F}^m$, for some $m \in \mathbb{N}$. The elements of $\mathbb{F}^m$ have the form $x = (x_1, \ldots, x_m)$. We have already seen several examples of norms on $\mathbb{F}^m$, namely $\|\cdot\|_2, \|\cdot\|_1, \|\cdot\|_\infty$. These norms are in fact part of a larger family of norms on $\mathbb{F}^m$, that we now exhibit.

Definition 2.68. For each $p \in (1, \infty)$, we define the function

$$\|\cdot\|_p \ : \ \mathbb{F}^n \longrightarrow [0, \infty)$$

by setting

$$\|x\|_p := \left(|x_1|^p + \ldots + |x_m|^p \right)^{1/p} \quad \text{for } x = (x_1, x_2, \ldots, x_m) \in \mathbb{F}^m.$$

One can show that, for every *fixed* $x \in \mathbb{F}^m$, the norms $\|\cdot\|_1$ and $\|\cdot\|_\infty$ can be approximated by $\|\cdot\|_p$ as p tends to the extreme values $p = 1, \infty$:

$$\|x\|_1 = \lim_{p \to 1} \|x\|_p,$$

$$\|x\|_\infty = \lim_{p \to \infty} \|x\|_p.$$

We leave the verification of the details as an exercise for the reader. Let us now show what the reader might have already suspected, namely:

Proposition 2.69. *For each $p \in (1, \infty)$, the function $\|\cdot\|_p$ is a norm on $\mathbb{F}^m$.*

For the proof we need the next fact which is an elementary inequality for non-negative numbers, but comprises an essential tool in our investigations of functional spaces.

Lemma 2.70 (Young inequality). *For any $a, b \geq 0$, $p, q \in (1, \infty)$ such that*

$$\frac{1}{p} + \frac{1}{q} = 1,$$

we have the estimate

$$ab \leq \frac{a^p}{p} + \frac{b^q}{q}.$$

Proof. Since "ln" is a concave and strictly increasing function on $(0, \infty)$, for any $a, b > 0$ we have that

$$\ln\left(\frac{1}{p}a^p + \frac{1}{q}b^q\right) \geq \frac{1}{p}\ln(a^p) + \frac{1}{q}\ln(b^q) = \ln(ab).$$

The conclusion ensues by the properties of the logarithm. $\square$

Definition 2.71. The exponent q in Lemma 2.70 is called the **conjugate exponent of p and is frequently denoted by p'**, that is

$$p' := \frac{p}{p-1}.$$

Note that $(p')' = p$, and hence the exponent p is the conjugate exponent of p'. Therefore, we may also say that p, p' **are conjugate to each other.** In particular, we have $2' = 2$.

We now deduce from Young's Inequality another important inequality:

Lemma 2.72 (Hölder's Inequality). *If $p, q \in (1, \infty)$ are conjugate exponents, then for all $z, w \in \mathbb{F}^m$ with $z = (z_1, \dots, z_m)$ and $w = (w_1, \dots, w_m)$ it holds that*

$$\sum_{k=1}^{m} |z_k|\,|w_k| \leq \left(\sum_{k=1}^{m} |z_k|^p\right)^{1/p} \left(\sum_{k=1}^{m} |w_k|^q\right)^{1/q}.$$

Remark 2.73. Hölder's Inequality may be written in short as

$$\sum_{k=1}^{m} |z_k|\,|w_k| \leq \|z\|_p \|w\|_q$$

and by defining the **dot product on the Euclidean space $\mathbb{F}^m$** as

$$z \cdot \overline{w} := \sum_{k=1}^{m} z_k \, \overline{w}_k$$

(where the bar denotes complex conjugation[4] when $\mathbb{F} = \mathbb{C}$), it follows that

$$\left|z \cdot \overline{w}\right| \leq \|z\|_p \|w\|_q.$$

Note that, when written in this form, the inequality can be easily seen to hold also for $p = 1$ and $q = \infty$ or $q = 1$ and $p = \infty$. Hence, $\{1, \infty\}$ *can be regarded as conjugate exponents to each other.*

---

[4]When expanded, the complex dot product on $\mathbb{C}^m$ takes the form

$$z \cdot \overline{w} = \sum_{k=1}^{m} \big(\text{Re}(z_k) + i\text{Im}(z_k)\big)\big(\text{Re}(w_k) - i\text{Im}(w_k)\big).$$

On the subspace $\mathbb{R}^m$ of $\mathbb{C}^m = \mathbb{R}^m + i\mathbb{R}^m$ whereon the imaginary parts vanish, it reduces to the standard real dot product.

Proof. Let us set $\alpha := \|z\|_p$, $\beta := \|w\|_q$. It suffices to prove Hölder's Inequality in the case when $\alpha \neq 0$, $\beta \neq 0$, since in the other case it is trivially satisfied. Note that the inequality can be rearranged as

$$\sum_{k=1}^{m} \frac{|z_k|}{\alpha} \frac{|w_k|}{\beta} \leq 1.$$

Using Young's Inequality for each term in the left-hand side of the above, we get

$$\sum_{k=1}^{m} \frac{|z_k|}{\alpha} \frac{|w_k|}{\beta} \leq \frac{1}{p}\left(\frac{\sum_{k=1}^{m}|z_k|^p}{\alpha^p}\right) + \frac{1}{q}\left(\frac{\sum_{k=1}^{m}|w_k|^q}{\beta^q}\right) = \frac{1}{p} + \frac{1}{q} = 1,$$

since we have

$$\alpha^p = \sum_{k=1}^{m}|z_k|^p, \qquad \beta^p = \sum_{k=1}^{m}|w_k|^p.$$

This completes the proof of Hölder's Inequality. $\qquad\square$

By combining Lemma 2.70, Remark 2.73 and Lemma 2.72, we obtain the next consequence which will be needed in the sequel:

Corollary 2.74 (Young inequality for vectors). *For any $z, w \in \mathbb{F}^m$ and any conjugate exponents $p, q \in (1, \infty)$ (Definition 2.71), it holds that*

$$|z \cdot \overline{w}| \leq \frac{\|z\|_2^p}{p} + \frac{\|w\|_2^q}{q}.$$

We are now ready to establish the main result.

Proof of Proposition 2.69. Since the first two axioms of a norm are straightforward to check, it suffices to check the triangle inequality (N3). For any $p \in (1, \infty)$, the inequality

$$\|x + y\|_p \leq \|x\|_p + \|y\|_p,$$

for any $x = (x_1, \ldots, x_m)$, $y = (y_1, \ldots, y_m)$, is written explicitly as

$$\left(\sum_{k=1}^{m}|x_k + y_k|^p\right)^{1/p} \leq \left(\sum_{k=1}^{m}|x_k|^p\right)^{1/p} + \left(\sum_{k=1}^{m}|y_k|^p\right)^{1/p}. \qquad (\square)$$

To prove it, it suffices to do this in the case when $\|x + y\|_p \neq 0$, since in the other case it is trivially satisfied. By using Hölder's Inequality and the fact that p, q are conjugate exponents, we have

$$\sum_{k=1}^{m}|x_k + y_k|^p \leq \sum_{k=1}^{m}|x_k|\,|x_k + y_k|^{p-1} + \sum_{k=1}^{m}|y_k|\,|x_k + y_k|^{p-1}$$

and hence

$$
\sum_{k=1}^{m} |x_k + y_k|^p \leq \left(\sum_{k=1}^{m} |x_k|^p \right)^{1/p} \left(\sum_{k=1}^{m} |x_k + y_k|^{(p-1)q} \right)^{1/q}
$$

$$
+ \left(\sum_{k=1}^{m} |y_k|^p \right)^{1/p} \left(\sum_{k=1}^{m} |x_k + y_k|^{(p-1)q} \right)^{1/q}
$$

$$
\leq \left(\sum_{k=1}^{m} |x_k + y_k|^p \right)^{1/q} \left[\left(\sum_{k=1}^{m} |x_k|^p \right)^{1/p} + \left(\sum_{k=1}^{m} |y_k|^p \right)^{1/p} \right].
$$

Rearranging this, using the fact that $\|x + y\|_p \neq 0$, and again the fact that p, q are conjugate exponents, we obtain Minkowski's Inequality. This completes the proof. $\qquad\square$

Definition 2.75.

(i) The inequality ($\square$) is called **Minkowski's Inequality**.

(ii) The special case when $p = 2$ (and thus also $q = 2$) in Hölder's Inequality is also known as the **Cauchy-Schwarz Inequality**:

$$
\sum_{k=1}^{m} |z_k w_k| \leq \left(\sum_{k=1}^{m} |z_k|^2 \right)^{1/2} \left(\sum_{k=1}^{m} |w_k|^2 \right)^{1/2}.
$$

2.8.2   The class of ℓ^p spaces

We now examine some spaces of sequences (vector subspaces of the space we earlier denoted by **s**) which complement the spaces ℓ^1 and ℓ^∞ which have been defined earlier.

Definition 2.76. For each $p \in (1, \infty)$, we consider the set, which is a subset of the space **s** of all sequences of scalars, given by

$$
\ell^p := \left\{ x = (x_1, x_2, \ldots) \, : \, x_k \in \mathbb{F} \text{ for all } k \in \mathbb{N}, \ \sum_{k=1}^{\infty} |x_k|^p < \infty \right\},
$$

and on ℓ^p we consider the non-negative function

$$
\|x\|_p := \left(\sum_{k=1}^{\infty} |x_k|^p \right)^{1/p} \quad \text{for all } x = (x_1, x_2, \ldots) \in \ell^p.
$$

Proposition 2.77. *For any $p \in (1, \infty)$, $(\ell^p, \| \cdot \|_p)$ is a normed vector space.*

Proof. The fact that ℓ^p is a vector space will be obtained by showing that ℓ^p is a vector subspace of the vector space **s**. It is clear that, for any $p \in (1, \infty)$,

$\lambda x \in \ell^p$ for any $x \in \ell^p$ and $\lambda \in \mathbb{F}$. We now consider any $x = (x_1, x_2, \ldots)$, $y = (y_1, y_2, \ldots) \in \ell^p$ and show that $x + y \in \ell^p$ and the triangle inequality holds:

$$\|x + y\|_p \leq \|x\|_p + \|y\|_p.$$

We start with Minkowski's Inequality for finite $m \in \mathbb{N}$:

$$\left(\sum_{k=1}^{m} |x_k + y_k|^p \right)^{1/p} \leq \left(\sum_{k=1}^{m} |x_k|^p \right)^{1/p} + \left(\sum_{k=1}^{m} |y_k|^p \right)^{1/p}.$$

Since $x, y \in \ell^p$, the right hand side of the above estimate is bounded by the respective inequality arising by taking supremum[5] over all $m \in \mathbb{N}$:

$$\left(\sum_{k=1}^{m} |x_k + y_k|^p \right)^{1/p} \leq \left(\sum_{k=1}^{\infty} |x_k|^p \right)^{1/p} + \left(\sum_{k=1}^{\infty} |y_k|^p \right)^{1/p}.$$

The validity of the above estimate for all $m \in \mathbb{N}$ implies that $x + y \in \ell^p$ and by letting $m \to \infty$ (or equivalently, by taking supremum over all $m \in \mathbb{N}$) we infer that

$$\left(\sum_{k=1}^{\infty} |x_k + y_k|^p \right)^{1/p} \leq \left(\sum_{k=1}^{\infty} |x_k|^p \right)^{1/p} + \left(\sum_{k=1}^{\infty} |y_k|^p \right)^{1/p}. \qquad (\Box\Box)$$

The above is exactly the triangle inequality for $\| \cdot \|_p$ that we also wanted to prove. Since the first two conditions in the definition of a norm are clearly true, it follows that $\| \cdot \|_p$ is indeed a norm on ℓ^p. $\qquad \Box$

We now isolate in a definition the "infinite-dimensional" analogues of two inequalities we saw earlier.

Definition 2.78.
(i) The inequality ($\Box\Box$) is known as **Minkowski's Inequality for sequences/series**.

(ii) The following inequality is known as **Hölder's Inequality for sequences/series**:

$$\sum_{k=1}^{\infty} |z_k| \, |w_k| \leq \left(\sum_{k=1}^{\infty} |z_k|^p \right)^{1/p} \left(\sum_{k=1}^{\infty} |w_k|^q \right)^{1/q},$$

and it holds true for any conjugate exponents $p, q \in (1, \infty)$ and any $z = (z_1, \ldots, z_k, \ldots) \in \ell^p$ and $w = (w_1, \ldots, w_k, \ldots) \in \ell^q$.

---

[5]Note that if $(a_m)_1^\infty \subseteq \mathbb{R}$ is an increasing sequence of numbers, then

$$\sup_{n \in \mathbb{N}} a_m = \lim_{n \to \infty} a_m.$$

In our case, the sequences $m \mapsto \left(\sum_{k=1}^{m} |x_k|^p \right)^{1/p}$ and $m \mapsto \left(\sum_{k=1}^{m} |x_k|^p \right)^{1/p}$ are increasing.

The convergence of the series in the left-hand side of Hölder's Inequality is a consequence of the convergence of the two series in the right-hand side. Its proof is an immediate consequence Hölder's Inequality for finite $m \in \mathbb{N}$ (Lemma 2.72) by letting $m \to \infty$. When written is short as

$$\sum_{k=1}^{\infty} |z_k|\,|w_k| \le \|z\|_p \|w\|_q,$$

Hölder's Inequality is easily seen to hold true for $p = 1$ and $q = \infty$ as well.

2.8.3 Spaces of continuous functions

We now examine a family of normed spaces whose elements are continuous functions. This class of spaces in a sense complements the space of continuous functions with the uniform norm[6] we discussed earlier.

Definition 2.79. Let $a, b \in \mathbb{R}$ with $a < b$. We consider the vector space

$$C([a,b], \mathbb{F}) := \Big\{ f : [a,b] \longrightarrow \mathbb{F} \,\Big|\, f \text{ is continuous on } [a,b] \Big\}.$$

For any $p \in [1, \infty)$, let us define

$$\|f\|_p := \left(\int_a^b |f(x)|^p \, \mathrm{d}x \right)^{1/p} \qquad \text{for any } f \in C([a,b], \mathbb{F}).$$

(Recall from Real Analysis that any continuous function on $[a,b]$ is Riemann integrable.)

Proposition 2.80. *For any* $p \in [1, \infty)$, *the pair* $\big(C([a,b], \mathbb{F}), \|\cdot\|_p \big)$ *is a normed vector space.*

Proof. The first two axioms of a norm are easy to check. Let us remark only that the fact that $\|f\|_p = 0$ implies that $f = 0$ uses the continuity of f, a similar conclusion not being valid if f was merely Riemann integrable. Let us now check the triangle inequality, namely that for any $f, g \in C([a,b], \mathbb{F})$ it holds that

$$\|f + g\|_p \le \|f\|_p + \|g\|_p.$$

Written explicitly,

$$\left(\int_a^b |f(x) + g(x)|^p \, \mathrm{d}x \right)^{1/p} \le \left(\int_a^b |f(x)|^p \, \mathrm{d}x \right)^{1/p} + \left(\int_a^b |g(x)|^p \, \mathrm{d}x \right)^{1/p},$$

---

[6]This subsection is intended to be a "taster", making our first contact with a celebrated class of spaces known as the "*L^p spaces*". These spaces will be studied very systematically in subsequent chapters and will provide reference examples for many interesting phenomena. However, we temporarily only consider the subspace of continuous functions of the L^p spaces and we defer studying the general spaces, the reason being that we first need to learn a new theory of integration which is more general than Riemann's approach!

this is known as **Minkowski's Inequality for integrals** and we shall study it in subsequent chapters in much greater generality.

The case $p = 1$ is straightforward, so let us consider in what follows the case $p \in (1, \infty)$. Let $q \in (1, \infty)$ be the conjugate exponent of p. We prove first **Hölder's Inequality for integrals**: for any $u, v \in C([a, b], \mathbb{F})$,

$$\int_a^b |u(x)v(x)| \, dx \leq \left(\int_a^b |u(x)|^p \, dx \right)^{1/p} \left(\int_a^b |v(x)|^q \, dx \right)^{1/q},$$

which in short form is

$$\|uv\|_1 \leq \|u\|_p \|v\|_q.$$

(The above is easily seen to be valid also if $p = 1$ and $q = \infty$.) We prove Hölder's Inequality only in the case when $\|u\|_p \neq 0$ and $\|v\|_q \neq 0$, the proof in the other case being trivial. Let us set for convenience

$$\alpha := \|u\|_p, \quad \beta := \|v\|_q.$$

Then, our inequality takes the form

$$\int_a^b \frac{|u(x)v(x)|}{\alpha\beta} \, dx \leq 1,$$

which we need to prove. For any $x \in [a, b]$, we use Young's Inequality to write

$$\frac{|u(x)|}{\alpha} \frac{|v(x)|}{\beta} \leq \frac{1}{p} \frac{|u(x)|^p}{\alpha^p} + \frac{1}{q} \frac{|v(x)|^q}{\beta^q}.$$

Integrating this inequality on $[a, b]$, we obtain

$$\int_a^b \frac{|u(x)v(x)|}{\alpha\beta} \, dx \leq \frac{1}{p} \left(\frac{1}{\alpha^p} \int_a^b |u(x)|^p \, dx \right) + \frac{1}{q} \left(\frac{1}{\beta^q} \int_a^b |v(x)|^q \, dx \right)$$

$$= \frac{1}{p} + \frac{1}{q}$$

$$= 1,$$

where we have used the definition of α and β. This establishes Hölder's Inequality for integrals.

We now prove Minkowski's Inequality in the case when $\|f + g\|_p \neq 0$, the proof in the other case being trivial. Using Hölder's Inequality and the fact that p and q are conjugate exponents, we obtain

$$\int_a^b |f(x) + g(x)|^p \, dx \leq \int_a^b |f(x)||f(x) + g(x)|^{p-1} \, dx$$

$$+ \int_a^b |g(x)||f(x) + g(x)|^{p-1} \, dx$$

and hence

$$\int_a^b |f(x) + g(x)|^p \, dx \leq \left(\int_a^b |f(x)|^p \, dx \right)^{1/p} \left(\int_a^b |f(x) + g(x)|^{(p-1)q} \, dx \right)^{1/q}$$

$$+ \left(\int_a^b |g(x)|^p \, dx \right)^{1/p} \left(\int_a^b |f(x) + g(x)|^{(p-1)q} \, dx \right)^{1/q}.$$

Rearranging this, using the fact that $\|f + g\|_q \neq 0$, and again the fact that p, q are conjugate exponents, we obtain Minkowski's Inequality. This completes the proof of the fact that $\| \cdot \|_p$ is a norm on $C([a, b], \mathbb{F})$. $\qquad \square$

Let us close this section by recording a consequence of the Hölder inequality of integrals, whose proof consists of repeating the arguments given above but through an application of Corollary 2.74 instead of Lemma 2.70:

Corollary 2.81 (Hölder inequality of integrals for mappings). *For any continuous maps $u, v \in C([a, b], \mathbb{F}^m)$, we have the estimate*

$$\int_a^b |u(x) \cdot \overline{v}(x)| \, dx \leq \left(\int_a^b (\|u(x)\|_2)^p \, dx \right)^{1/p} \left(\int_a^b (\|v(x)\|_2)^q \, dx \right)^{1/q}.$$

2.9   Separable metric spaces and the ℓ^p class

In this section we introduce an important property regarding metric spaces which will play a very significant role in subsequent considerations. This property is, in a certain broad sense, a "smallness condition" for a metric space different from that of the smallness of their diameter. If this property is satisfied, the space usually possesses much nicer properties overall.

The motivational example for the introduction of this property is the real line $\mathbb{R}$ and more generally the Euclidean space $\mathbb{R}^m$ for any $m \in \mathbb{N}$. Indeed, some of the pleasant properties of $\mathbb{R}^m$ arise as a consequence that $\mathbb{R}^m$ **possesses dense sequences in it**, one of them being of course the rationals $\mathbb{Q}^m$. Namely, for any vector $x \in \mathbb{R}^m$, there exists a sequence of vectors $(x_n)_1^\infty \subseteq \mathbb{Q}^m$ with rational coefficients such that $x_n \longrightarrow x$ as $n \to \infty$.

As we alluded to above, the property of metric spaces we are interested in refers to those contain dense sequences. This is given formally in the next definition.

Definition 2.82. A metric set (X, d) is said to be **separable** if it contains a countable dense subset.

In general, verifying whether a metric or normed space is separable is not an easy task. Nonetheless, in the particular class of ℓ^p spaces we introduced earlier, one can show that the space ℓ^p is separable when $p \in [1, \infty)$, but ℓ^∞ is **not** separable. Accordingly, we have the next two results.

Proposition 2.83. *For any $p \in [1, \infty)$, the normed space $(\ell^p, \| \cdot \|_p)$ is separable.*

Proof. Fix $p \in [1, \infty)$. We need to find a countable dense subset A of ℓ^p. Let us consider, for any $n \in \mathbb{N}$, the vector $e_n := (0, \ldots, 1, 0, \ldots)$, whose nth entry is 1 and for which all other entries are 0. We first claim that any $x = (x_1, x_2, \ldots) \in \ell^p$ can be represented as the sum of the following convergent series

$$x = \sum_{k=1}^{\infty} x_k e_k \quad \text{in } (\ell^p, \| \cdot \|_p).$$

This is indeed so since

$$\left\| x - \sum_{k=1}^{n} x_k e_k \right\|_p^p = \sum_{k=n+1}^{\infty} |x_k|^p \longrightarrow 0 \quad \text{as } n \to \infty. \tag{$*$}$$

Let $\mathbb{Q}_{\mathbb{F}}$ denote the set $\mathbb{Q}$ if $\mathbb{F} = \mathbb{R}$ and the set $\mathbb{Q} + i\mathbb{Q}$ if $\mathbb{F} = \mathbb{C}$. It is obvious that $\mathbb{Q}_{\mathbb{F}}$ is dense in $(\mathbb{F}, | \cdot |)$, as a consequence of the density of $\mathbb{Q}$ in $\mathbb{R}$. Let us now consider the set

$$A := \bigcup_{m=1}^{\infty} \left\{ \sum_{k=1}^{m} q_k e_k : q_k \in \mathbb{Q}_{\mathbb{F}} \text{ for all } k \in \{1, \ldots, m\} \right\}.$$

Then A is countable, being a countable union of countable sets. We now show that A is dense in $(\ell^p, \| \cdot \|_p)$. To this end, fix $x \in X$ and $r > 0$, both arbitrary. We need to prove that there exists $a \in A$ such that $\|x - a\|_p < r$. It is a consequence of $(*)$ that there exists $m \in \mathbb{N}$ such that

$$\left\| x - \sum_{k=1}^{m} x_k e_k \right\|_p < \frac{r}{2}.$$

Let us set $a := \sum_{k=1}^{m} q_k e_k$, where $q_k \in \mathbb{Q}_{\mathbb{F}}$ are to be suitably chosen later. In any case, we have $a \in A$. Then

$$\left\| \left(\sum_{k=1}^{m} x_k e_k \right) - a \right\|_p \leq \sum_{k=1}^{m} |x_k - q_k| \|e_k\|_p < \frac{r}{2},$$

provided $q_k \in \mathbb{Q}_{\mathbb{F}}$ is chosen so that $|x_k - q_k| < r/(2m)$ for all $k \in \{1, \ldots, m\}$, which is possible since $\mathbb{Q}_{\mathbb{F}}$ is dense in $\mathbb{F}$. With this choice of $a \in A$, it follows from the triangle inequality that

$$\|x - a\|_p \leq \left\| x - \sum_{k=1}^{m} x_k e_k \right\|_p + \left\| \left(\sum_{k=1}^{m} x_k e_k \right) - a \right\|_p < \frac{r}{2} + \frac{r}{2} = r.$$

This shows that the countable set A is dense in $(\ell^p, \|\cdot\|_p)$, and hence $(\ell^p, \|\cdot\|_p)$ is separable. $\qquad\square$

On the other hand, the situation is radically different for $p = \infty$:

Proposition 2.84. *The normed space $(\ell^\infty, \|\cdot\|_\infty)$ is not separable.*

Proof. The idea is to find an *uncountable* collection of open balls in $(\ell^\infty, \|\cdot\|_\infty)$ that are pairwise disjoint. Then, any dense set A in $(\ell^\infty, \|\cdot\|_\infty)$ must contain at least one element from each of these balls. These elements will all be distinct, due to the fact that the balls are disjoint, and therefore A would necessarily be uncountable. This argument shows that $(\ell^\infty, \|\cdot\|_\infty)$ cannot contain any countable dense subset, and therefore is not separable.

The construction of an *uncountable* collection of open balls that are pairwise disjoint can be carried out as follows. Let $f : \mathcal{P}(\mathbb{N}) \longrightarrow \ell^\infty$ be given by

$$S \mapsto f(S) := \left(x_1^{(S)}, x_2^{(S)}, \ldots\right) \quad \text{for all } S \in \mathcal{P}(\mathbb{N}),$$

where

$$x_k^{(S)} := \begin{cases} 1, & \text{if } k \in S, \\ 0, & \text{if } k \notin S. \end{cases}$$

We define the subset of ℓ^∞ given by

$$M := f\big(\mathcal{P}(\mathbb{N})\big) = \left\{x^{(S)} : S \in \mathcal{P}(\mathbb{N})\right\}.$$

It is easy to check that f is a bijection from $\mathcal{P}(N)$ to M, and therefore (recalling the results of Chapter 1),

$$\mathrm{card}(M) = \mathrm{card}\big(\mathcal{P}(\mathbb{N})\big),$$

which shows, in view of Lemma 1.14, that M is uncountable. Now observe that, for any $S, T \in \mathcal{P}(\mathbb{N})$ with $S \neq T$, there exists $k_0 \in \mathbb{N}$ (in particular, k_0 belongs to the symmetric difference set $S\Delta T$) such that $x_{k_0}^{(S)} \neq x_{k_0}^{(T)}$, and therefore $\left|x_{k_0}^{(S)} - x_{k_0}^{(T)}\right| = 1$, which implies that

$$\left\|x^{(S)} - x^{(T)}\right\|_\infty = 1.$$

It is easy to see that, in turn, the above relation implies that

$$\mathbb{B}_{1/2}\big(x^{(S)}\big) \bigcap \mathbb{B}_{1/2}\big(x^{(T)}\big) = \emptyset.$$

We have thus found an uncountable collection

$$\left\{\mathbb{B}_{1/2}(x) : x \in M\right\}$$

of open balls in $(\ell^\infty, \|\cdot\|_\infty)$ which are pairwise disjoint. In view of the comments at the beginning of the proof, this implies that $(\ell^\infty, \|\cdot\|_\infty)$ is not separable. $\qquad\square$

2.10 Exercises

Exercise 2.1. Let $\| \cdot \|_1, \| \cdot \|_2, \| \cdot \|_\infty : \mathbb{F}^m \longrightarrow \mathbb{R}$ be given, for every $z = (z_1, \ldots, z_m) \in \mathbb{F}^m$, by

$$\|z\|_1 = \sum_{k=1}^{m} |z_k|, \quad \|z\|_2 = \left(\sum_{k=1}^{m} |z_k|^2 \right)^{1/2}, \quad \|z\|_\infty = \max_{k \in \{1,\ldots,m\}} |z_k|.$$

Show that each of $\| \cdot \|_1$, $\| \cdot \|_2$ and $\| \cdot \|_\infty$ is a norm on the vector space $\mathbb{F}^m$.

Exercise 2.2. Let (X, d) be a metric space.

(i) Show that for every x, y, a, b in X, we have

$$\big| d(x, y) - d(a, b) \big| \leq d(x, a) + d(y, b).$$

(ii) Let $(x_n)_1^\infty$ and $(y_n)_1^\infty$ be sequences in X such that $x_n \longrightarrow x_0$ and $y_n \longrightarrow y_0$ in (X, d) as $n \to \infty$, where $x_0, y_0 \in X$. Show that

$$d(x_n, y_n) \longrightarrow d(x_0, y_0) \quad \text{in } \mathbb{R} \text{ as } n \to \infty.$$

[Hint: You may find Part (i) useful.]

Exercise 2.3. Let $(X, \| \cdot \|)$ be a normed space, $(x_n)_1^\infty$ a sequence in X, and $x_0 \in X$ be such that $x_n \longrightarrow x_0$ in $(X, \| \cdot \|)$. For any $n \in \mathbb{N}$, let

$$y_n := \frac{1}{n} \big(x_1 + x_2 + \cdots + x_n \big).$$

Show that $y_n \longrightarrow x_0$ in $(X, \| \cdot \|)$ as $n \to \infty$.

Exercise 2.4. Let (X, d) be a metric space. Show that any finite union of bounded sets in X is bounded.

Exercise 2.5. Let (X, d) be a metric space.

(i) Let C be a bounded subset of X. Show that the set of real numbers $\{ d(x, y) : x, y \in C \}$ is bounded. We may thus define the *diameter* of C as

$$\text{diam}(C) := \sup \{ d(x, y) : x, y \in C \}.$$

(ii) Show that if A, B are bounded subsets of X with $A \cap B \neq \emptyset$ then

$$\text{diam}(A \cup B) \leq \text{diam} A + \text{diam} B.$$

(iii) Suppose $X = \mathbb{R}$, with the usual metric. Give an example of bounded sets A, B with $A \cap B = \emptyset$ such that

$$\text{diam}(A \cup B) > \text{diam} A + \text{diam} B.$$

Exercise 2.6. Let (X, d) be a metric space and let $a \in X$ be a given point. Let $f : X \longrightarrow \mathbb{R}$ be given by

$$f(x) = d(x, a) \quad \text{for all } x \in X.$$

Show that f is continuous on X, where $\mathbb{R}$ is considered as a metric space with the standard metric.

Exercise 2.7. Let (X, d) be a metric space and A be a non-empty subset of X. For each $x \in X$, let us define

$$d(x, A) := \inf\{d(x, a) : a \in A\}.$$

(i) Show that, for any $x, y \in X$ and for any $b \in A$,

$$d(x, A) \leq d(x, b) \leq d(x, y) + d(y, b).$$

(ii) Show that, for any $x, y \in X$,

$$d(x, A) \leq d(x, y) + d(y, A).$$

(iii) Deduce that, for any $x, y \in X$,

$$|d(x, A) - d(y, A)| \leq d(x, y).$$

(iv) Show that the function $g : X \longrightarrow \mathbb{R}$ given by

$$g(x) = d(x, A) \quad \text{for all } x \in X$$

is continuous from (X, d) to $\mathbb{R}$ with the standard metric.

Exercise 2.8. Let (X, d) be a metric space and $f : X \longrightarrow X$ be a continuous mapping on X. Let $g : X \longrightarrow \mathbb{R}$ be given by

$$g(x) = d(x, f(x)) \quad \text{for all } x \in X.$$

Show that g is a continuous function on X, where $\mathbb{R}$ is considered as a metric space with the usual metric.

Exercise 2.9. Let (X, d) be a metric space, $x_0, y_0 \in X$ and $r > 0$. Show that if $y_0 \in \mathbb{B}_{r/2}(x_0)$, then $\mathbb{B}_{r/2}(y_0) \subseteq \mathbb{B}_r(x_0)$.

Exercise 2.10. (i) Let (X, d) be a metric space, and let $x_0, y_0 \in X$ with $x_0 \neq y_0$. Show that there exists $r > 0$ such that $\mathbb{B}_r(x_0) \cap \mathbb{B}_r(y_0) = \emptyset$.

(ii) Let (X, d) be a metric space containing more than one point. Show that it is impossible that the only open subsets of X are $\emptyset$ and X.

Exercise 2.11. A set A in a (real or complex) vector space X is said to be *convex* if

$$(1 - \lambda)x + \lambda y \in A \quad \text{for all } x, y \in A \text{ and for all } \lambda \in [0, 1].$$

Let $(X, \|\cdot\|)$ be a normed vector space. Show that, for any $x_0 \in X$ and $r > 0$, the closed ball

$$\mathbb{D}_r(x_0) := \{x \in X : \|x - x_0\| \leq r\}$$

is a convex set.

Exercise 2.12. Let (X, d) be a metric space. Show that any finite subset of X is closed.

Exercise 2.13. Show that in any discrete metric space every open ball is either a single point or the whole space.

Exercise 2.14. Let (X, d_X) and (Y, d_Y) be metric spaces, and consider their Cartesian product set $X \times Y$. Let $d_2 : (X \times Y) \times (X \times Y) \longrightarrow \mathbb{R}$ be given, for any $(x_1, y_1), (x_2, y_2) \in X \times Y$, by

$$d_2\Big((x_1, y_1), (x_2, y_2)\Big) = \Big((d_X(x_1, x_2))^2 + (d_Y(y_1, y_2))^2\Big)^{1/2}.$$

(i) Show that d_2 is a metric on $X \times Y$.

(ii) Show that, if U is an open set in (X, d_X) and V is an open set in (Y, d_Y), then $U \times V$ is an open set in $(X \times Y, d_2)$.

(iii) In the case when $X = Y = \mathbb{R}$, and each of d_X and d_Y is the usual metric on $\mathbb{R}$, give an example of an open set in the metric space $(\mathbb{R} \times \mathbb{R}, d_2)$ which is *not* of the form $U \times V$ for any U and V open sets in $\mathbb{R}$ with the usual metric.

Exercise 2.15. Let (X, d) be a metric space

(i) Let $(x_n)_1^\infty \subseteq X$, $x_0 \in X$. Show that $x_n \longrightarrow x_0$ in (X, d) if and only if

$$\forall U \text{ open in } X, U \ni x_0 \exists N \in \mathbb{N} : \forall n \in \mathbb{N}, n \geq N \Longrightarrow x_n \in U.$$

(ii) Let (Y, ρ) be a metric space, $f : X \longrightarrow Y$ and $x_0 \in X$. Show that f is continuous at x_0 if and only if

$$\forall V \text{ open in } Y, V \ni f(x_0), \exists U \text{ open in } X \text{ with } U \ni x_0 : f(U) \subseteq V.$$

Exercise 2.16. Let (X, d) be a metric space, and $f, g : X \longrightarrow \mathbb{R}$ be continuous functions on X, where $\mathbb{R}$ is considered as a metric space with the standard metric.

(i) Show that if $g(x_0) \neq 0$ for some $x_0 \in X$, then there exists $\delta_0 > 0$ such that $g(x) \neq 0$ for all $x \in \mathbb{B}_{\delta_0}(x_0)$.

(ii) Prove, using the definition of continuity, that if $g(x) \neq 0$ for all $x \in X$, then the function $f/g : X \longrightarrow \mathbb{R}$ is continuous on X.

Exercise 2.17. Let (X, d) be a metric space and $r > 0$.

(i) Let $a \in X$. Show that $\{x \in X : d(x, a) > r\}$ is open in (X, d).

(ii) Let A be a non-empty subset of X. Show that $\{x \in X : d(x, A) < r\}$ and $\{x \in X : d(x, A) > r\}$ are open in (X, d). (Recall that, for any $x \in X$, $d(x, A) := \inf\{d(x, a) : a \in A\}$.)

[Hint: It may be helpful to write some of the sets in question as the inverse images of open sets in $\mathbb{R}$ through some continuous functions $g : (X, d) \longrightarrow \mathbb{R}$.]

Exercise 2.18. Let

$$X = \Big\{ p : [0, 1] \longrightarrow \mathbb{R} : p \text{ is a polynomial with real coefficients} \Big\}.$$

Show by a simple argument that X is a real vector space. Let $\| \cdot \|_\infty, \| \cdot \|_1 : X \longrightarrow \mathbb{R}$ be given by

$$\|p\|_\infty := \max_{x \in [0,1]} |p(x)| \quad \text{for all } p \in X,$$

$$\|p\|_1 := \int_0^1 |p(x)| \, dx \quad \text{for all } p \in X.$$

Show that $\| \cdot \|_\infty$ and $\| \cdot \|_1$ are norms on X which are not equivalent.

Exercise 2.19. Let (X, d) be a metric space. Define $\rho : X \times X \longrightarrow \mathbb{R}$ by

$$\rho(x, y) = \min\{1, d(x, y)\} \quad \text{for all } x, y \in X.$$

Show that :

(i) ρ is a metric on X.

(ii) The metrics d and ρ are topologically equivalent.

(iii) If $X = \mathbb{R}$ and d is the standard metric, then the metrics d and ρ are not Lipschitz equivalent.

Exercise 2.20. Let (X, d) be a metric space. Let $\rho : X \times X \longrightarrow \mathbb{R}$ be given by

$$\rho(x, y) = \frac{d(x, y)}{1 + d(x, y)} \quad \text{for all } x, y \in X.$$

Show that :

(i) ρ is a metric on X.

(ii) The metrics d and ρ are topologically equivalent.

(iii) If $X = \mathbb{R}$ and d is the usual metric, then the metrics d and ρ are not Lipschitz equivalent.

Exercise 2.21. Let

$$\mathbf{s} := \left\{ x = (x_1, x_2, \dots) : x_k \in \mathbb{F} \text{ for all } k \in \mathbb{N} \right\}.$$

Let $d : \mathbf{s} \times \mathbf{s} \longrightarrow \mathbb{R}$ be given by

$$d(x, y) := \sum_{k=1}^{\infty} \frac{1}{2^k} \frac{|x_k - y_k|}{1 + |x_k - y_k|},$$

for all $x = (x_1, \dots, x_k, \dots), y = (y_1, y_2, \dots) \in \mathbf{s}$.

(i) Show that d is well-defined (in the sense that the series in the above definition is convergent) and that d is a metric on $\mathbf{s}$.

(ii) Let $(x^{(n)})_1^{\infty}$ be a sequence in $\mathbf{s}$, where

$$x^{(n)} = (x_1^{(n)}, \dots, x_k^{(n)}, \dots), \quad n \in \mathbb{N}.$$

Show that $(x^{(n)})_1^{\infty}$ converges to $x^{(0)} = (x_1^{(0)}, \dots, x_k^{(0)}, \dots)$ in the metric space $(\mathbf{s}, d)$ if and only if $(x_k^{(n)})_1^{\infty}$ converges to $x_k^{(0)}$ in $(\mathbb{F}, |\cdot|)$ for each $k \in \mathbb{N}$.

Exercise 2.22. Consider $Y = [0, 1) \cup [2, 3] \cup \{4\} \cup [5, \infty)$ as a metric subspace of $\mathbb{R}$ with the standard metric.

(i) Find the following open balls of Y:

$$\mathbb{B}_{1/2}(0); \quad \mathbb{B}_2(2); \quad \mathbb{B}_1(4); \quad \mathbb{B}_5(4); \quad \mathbb{B}_3(6).$$

(ii) Which of the following sets are open in Y? Justify your answer.

$$[0, 1); \quad (2, 3); \quad (0, 1) \cup \{4\}; \quad [5, 6]; \quad [5, 6).$$

Exercise 2.23. Let (X, d) be a metric space and A be a subset of X. Recall that, for every $x \in X$, we have defined

$$d(x, A) := \inf\{d(x, a) : a \in A\}.$$

Show that, for every $x \in X$, the following is true:

$$x \in \overline{A} \quad \text{if and only if} \quad d(x, A) = 0.$$

Exercise 2.24. Let A be a bounded subset of $\mathbb{R}$ with the usual metric. Let $\alpha := \inf A$ and $\beta := \sup A$. Show that $\alpha, \beta \in \overline{A}$.

Exercise 2.25. Let (X, d) be a metric space, and A, B be non-empty subsets of X. Let

$$d(A, B) = \inf\{d(a, b) : a \in A, b \in B\}.$$

Suppose that $\overline{A} \cap \overline{B} \neq \emptyset$. Show that $d(A, B) = 0$.

Exercise 2.26. Let (X, d) be a metric space. Show that, if A is a bounded subset of X, then $\overline{A}$ is also bounded and, moreover, $\operatorname{diam} \overline{A} = \operatorname{diam} A$. (Recall that, for any bounded set C, the diameter of C is defined by

$$\operatorname{diam}(C) := \sup\{d(x, y) : x, y \in C\}.)$$

Exercise 2.27. For any $p \in [1, \infty)$, let $\|\cdot\|_p : \mathbb{F}^m \longrightarrow \mathbb{R}$ be given by

$$\|x\|_p := \left(\sum_{k=1}^{m} |x_k|^p\right)^{1/p} \qquad \text{for all } x = (x_1, \ldots, x_m) \in \mathbb{F}^m.$$

Let also

$$\|x\|_\infty := \max_{k \in \{1, \ldots, m\}} |x_k| \qquad \text{for all } x = (x_1, \ldots, x_m) \in \mathbb{F}^m.$$

Prove that $\|\cdot\|_p$ and $\|\cdot\|_q$ are equivalent norms for any $p, q \in [1, \infty]$.

Exercise 2.28. By using Hölder's Inequality for Integrals, or otherwise, prove that, for any $a, b \in \mathbb{R}$ with $a < b$, and any α, β with $1 \leq \alpha < \beta < \infty$, there exists a constant M, depending only on a, b, α and β, such that

$$\left(\int_a^b |f(x)|^\alpha \, dx\right)^{1/\alpha} \leq M \left(\int_a^b |f(x)|^\beta \, dx\right)^{1/\beta} \qquad \text{for all } f \in C([a, b], \mathbb{F}).$$

Exercise 2.29. Let (X, d) be a separable metric space and Y be a subset of X. Show that the metric subspace (Y, d) is itself separable.

[Hint: A countable dense subset of X need not intersect Y at all. To construct a countable dense subset of Y, consider a collection of elements of Y belonging to balls of certain radii centred at points of the original countable dense subset of X.]

Chapter 3

Completeness and applications

3.1 The significance of completeness

Our discussion in Chapter 2 seems to suggest that, in order to prove that a sequence $(x_n)_1^\infty$ in a metric space (X, d) is convergent, it is necessary to know in advance the element which it converges to (its limit). On the other hand, one is often faced with the emergence of sequences which asymptotically behave *like convergent*, but their limit might either be difficult to exhibit explicitly, or even may not eventually be convergence because their putative "limit" might not be contained in the space! For instance, a bounded increasing sequence of rational numbers in $\mathbb{Q}$ may or may not converge to a rational number, although it certainly converges to a real number in $\mathbb{R}$.

This phenomenon suggests that it would be beneficial to have a method to determine whether a sequence converges by means of a condition that depends only on the terms of the sequence itself, and not on the a priori knowledge of a potential limit. It is likely that you have encountered in the study of sequences of real numbers the concept of a **Cauchy sequence** which is equivalent in that setting to that of a convergent sequence. We now define Cauchy sequences in general metric spaces. As in the case of $\mathbb{R}$, a Cauchy sequence in a metric space is one in which, roughly speaking, the terms of the sequence get as close to each other as one wishes, provided that we consider only indices which are sufficiently large.

However, unlike the case of sequences of real numbers, it turns out not to be true in general that every Cauchy sequence is convergent. It is important therefore to study carefully the relation between Cauchy sequences and convergent sequences, and to identify instances of metric spaces in which the two concepts are equivalent. The desirable property of a subset of a metric space that all Cauchy sequences are convergent is called, as we shall see right next, **completeness**.

This chapter is devoted to the study of metric spaces and normed vector spaces in relation to the property of completeness (or its failure). To this end, we examine more closely those spaces and subspaces which are complete, establishing along the way several important results valid on spaces which possess this favourable property of completeness.

3.2   Complete metric spaces and Banach spaces

Definition 3.1. Let (X, d) be a metric space. A sequence $(x_n)_1^\infty$ in X is said to be a **Cauchy sequence** in (X, d) if for all $\varepsilon > 0$ there exists $N \in \mathbb{N}$ such that

$$d(x_n, x_m) < \varepsilon \quad \text{for all } n, m \geq N.$$

We commence our study of Cauchy sequences with the next simple result.

Proposition 3.2. *Let (X, d) be a metric space. Then every convergent sequence in (X, d) is a Cauchy sequence.*

Proof. Let $(x_n)_1^\infty \subseteq X$ be such that $x_n \longrightarrow x_0$ in (X, d) as $n \to \infty$. Fix $\varepsilon > 0$, arbitrary. Since $x_n \longrightarrow x_0$ as $n \to \infty$, there exists $N \in \mathbb{N}$ such that $d(x_n, x_0) < \varepsilon/2$ for all $n \geq N$. But then, for every $n, m \geq N$, we have that

$$d(x_n, x_m) \leq d(x_n, x_0) + d(x_m, x_0) < \frac{\varepsilon}{2} + \frac{\varepsilon}{2} = \varepsilon.$$

Since $\varepsilon > 0$ was arbitrary, it follows that $(x_n)_1^\infty$ is indeed a Cauchy sequence. $\qquad \square$

To study the validity of the converse of this result, it is useful to recall the proof of the fact that every Cauchy sequence of real numbers is convergent. In that setting, the proof is carried out in three steps:

(i) The given Cauchy sequence is necessarily bounded.

(ii) Due to the fact that it is bounded, the sequence has a convergent subsequence.

(iii) Due to the fact that it has a convergent subsequence, the Cauchy sequence is convergent.

It turns out that the arguments in steps (i) and (ii) can be carried out in any metric space (X, d). On the other hand, *step (ii) is not valid in general.* An investigation of the circumstances of its validity will be carried out in Chapter 5, where the concept of **(sequential) compactness** will be studied. For the time being, let us prove the statements corresponding to steps (i) and (iii). To this end, we need to define the notion of a **subsequence** of a given sequence in an arbitrary set. This is just a direct generalisation of the Euclidean notion.

Definition 3.3. Let X be a set and $(x_n)_1^\infty$ a sequence in X. A **subsequence** of $(x_n)_1^\infty$ is any sequence of the form $(x_{\phi(k)})_{k=1}^\infty$, where $\phi : \mathbb{N} \longrightarrow \mathbb{N}$ is a strictly increasing function. One usually writes n_k instead of $\phi(k)$, where

$$n_1 < n_2 < \cdots < n_k < \cdots$$

and then the subsequence is symbolised as $(x_{n_k})_{k=1}^\infty$ or, in brief, $(x_{n_k})_1^\infty$.

We then have the following result, as claimed above.

Proposition 3.4. *Let (X, d) be a metric space.*

(i) *Every Cauchy sequence in (X, d) is bounded.*

(ii) *If a Cauchy sequence in (X, d) has a convergent subsequence, then the sequence itself is convergent to the same limit as the convergent subsequence.*

Proof. (i) Let $(x_n)_1^\infty$ be a Cauchy sequence in (X, d). Taking $\varepsilon := 1$ in the definition, we get that there exists $N \in \mathbb{N}$ such that $d(x_n, x_m) < 1$ for all $n, m \geq N$. In particular, $d(x_n, x_N) < 1$ for all $n \geq N$. Hence

$$d(x_n, x_N) \leq \max\Big\{ d(x_1, x_N), \ldots, d(x_{N-1}, x_N), 1 \Big\} \quad \text{for all } n \in \mathbb{N}.$$

This shows that $(x_n)_1^\infty$ is bounded.

(ii) Suppose that $(x_n)_1^\infty$ is a Cauchy sequence, and $(x_{n_k})_1^\infty$ is a convergent subsequence with limit x_0. Fix $\varepsilon > 0$, arbitrary. Since $x_{n_k} \longrightarrow x_0$ as $k \to \infty$, there exists $K \in \mathbb{N}$ such that

$$d(x_{n_k}, x_0) < \varepsilon/2 \quad \text{for all } k \geq K.$$

On the other hand, since $(x_n)_1^\infty$ is Cauchy, there exists $N \in \mathbb{N}$ such that

$$d(x_n, x_m) < \varepsilon/2 \quad \text{for all } n, m \geq N.$$

Then, for any $n \geq N$, we may choose some $k \geq K$ such that $n_k \geq N$ (this is possible since $n_k \geq k$), to obtain that

$$d(x_n, x_0) \leq d(x_n, x_{n_k}) + d(x_{n_k}, x_0) < \frac{\varepsilon}{2} + \frac{\varepsilon}{2} = \varepsilon.$$

Since $\varepsilon > 0$ was arbitrary, it follows that $x_n \longrightarrow x_0$ in (X, d) as $n \to \infty$, as required. $\qquad \square$

Remark 3.5. It may well happen in certain metric spaces that there exist Cauchy sequences which are not convergent. Consider, for example, the metric subspace $Y := (0, 1]$ of $\mathbb{R}$ with the usual metric. It can be easily checked that the sequence $(1/n)_1^\infty \subseteq Y$ is a Cauchy sequence in Y, but does not converge in Y. Further, for $Y := \mathbb{Q}$ again considered as a metric subspace of $\mathbb{R}$, we have that the sequence of fractions $(a_n)_1^\infty \subseteq \mathbb{Q}$ given by

$$a_1 := 2.7, \quad a_2 := 2.71, \quad a_3 := 2.718, \quad a_4 := 2.7182, \quad a_5 := 2.71828, \quad \ldots$$

is Cauchy in $\mathbb{Q}$ but converges to the irrational number $e \in \mathbb{R} \setminus \mathbb{Q}$.

It turns out to be useful to give a special name to those metric spaces in which every Cauchy sequence is convergent.

Definition 3.6. A metric space (X, d) is said to be **complete** if every Cauchy sequence in (X, d) is convergent. A subset Y of a metric space (X, d) is said to be **complete** if the metric subspace (Y, d) is complete.

Moreover, the complete normed vector spaces are so important in Analysis that they are also given a special name.

Definition 3.7. A complete normed space $(X, \|\cdot\|)$ is called a **Banach space**.

As we already know, the normed vector space $(\mathbb{F}^m, \|\cdot\|_p)$ is a Banach space for any $p \in [1, \infty]$ (because it is complete!). In this book we shall study further classes of Banach spaces quite systematically. In particular, we shall later establish that, for any $p \in [1, \infty]$, the normed vector space $(\ell^p, \|\cdot\|_p)$ is also a Banach space.

The next result refers to the completeness of the subspaces of a given metric space. In particular, it provides a way of exhibiting metric spaces that are not complete.

Proposition 3.8. *Let (X, d) be a metric space and Y a subset of X.*

  (i) *If (Y, d) is complete, then Y is closed in X.*

  (ii) *If (X, d) is complete and Y is closed in X, then (Y, d) is complete.*

Proof. (i) Let $x \in \overline{Y}$, arbitrary. Then there exists a sequence $(y_n)_1^\infty \subseteq Y$ such that $y_n \longrightarrow x$ in (X, d). It follows that $(y_n)_1^\infty$ is a Cauchy sequence in (X, d), and hence in (Y, d) also. Since (Y, d) is complete, there exists $y \in Y$ such that $y_n \longrightarrow y$ in (Y, d) as $n \to \infty$, and hence in (X, d) as well. But the uniqueness of limits in (X, d) shows that $x = y$, which means that $x \in Y$. We have therefore proved that

$$\overline{Y} \subseteq Y,$$

which shows that Y is closed in X, because we always have the reverse inclusion $Y \subseteq \overline{Y}$.

(ii) Let $(y_n)_1^\infty \subseteq Y$ be an arbitrary Cauchy sequence in (Y, d). Then $(y_n)_1^\infty$ is a Cauchy sequence in (X, d) also, and, since (X, d) is complete, it follows that there exists $x \in X$ such that $x_n \longrightarrow x$ in (X, d) as $n \to \infty$. Then $x \in \overline{Y}$, and the assumption that Y is closed in X now shows that $x \in Y$. Therefore, (Y, d) is a complete metric space. □

When restricted to the case of normed spaces, the above result takes the following particular form.

Proposition 3.9. *Let $(X, \|\cdot\|)$ be a normed space and Y a normed (vector) subspace of X.*

  (i) *If $(Y, \|\cdot\|)$ is a Banach space, then Y is closed in X.*

  (ii) *If $(X, \|\cdot\|)$ is a Banach space and Y is closed in X, then $(Y, \|\cdot\|)$ is a Banach space.*

3.3 Cantor's intersection and Baire's Category Theorems

In this section we establish two relatively simple but extremely important results which hold true on complete metric spaces. Both of them are of "existential" flavour, namely they guarantee the existence of some object. The first one below (whose version for $X = \mathbb{R}$ might already be familiar from Real Analysis) says that if we have a decreasing sequence of sets $(F_n)_1^\infty$ whose diameters "shrink to zero", then in the limit as $n \to \infty$ we obtain a single point. We first record for the sake of completeness the next notion.

Definition 3.10. Let (X, d) be a metric space. The **diameter** of the space X is defined as the (extended) number in $[0, \infty]$ given by (see Figure 3.1)

$$\operatorname{diam}(X) := \sup \left\{ d(x, y) : x, y \in X \right\}.$$

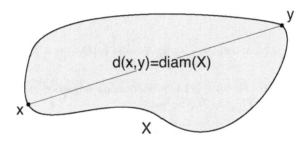

Figure 3.1: Intuitively, the diameter of a set is the largest possible distance between any two of its points. Note that the "sup" may not be realised by a pair of points, for instance if the set is open (e.g. what is the diameter of $(-1, 1) \subseteq \mathbb{R}$?).

Theorem 3.11 (Cantor's intersection theorem). *Let (X, d) be a complete metric space. Let $\{F_n\}_1^\infty$ be a decreasing sequence of closed sets in X*

$$F_1 \supseteq F_2 \supseteq \cdots \supseteq F_n \supseteq \cdots$$

for which their diameters tend to zero (see Figure 3.2):

$$\operatorname{diam}(F_n) \longrightarrow 0 \quad as \quad n \to \infty.$$

Then the intersection $\bigcap_{n=1}^\infty F_n$ is a singleton set:

$$\exists \, x \in X : \quad \bigcap_{n=1}^\infty F_n = \{x\}.$$

Remark 3.12. The converse of the above statement is also true, but we shall not need it for our present purposes. Also, all three assumptions of the result are necessary:

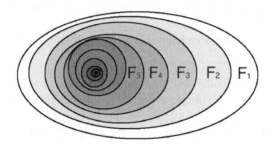

Figure 3.2: An illustration of a decreasing sequence of closed sets on $\mathbb{R}^2$ which "shrink" to a point, as in the statement of Theorem 3.11.

(i) Consider the metric space $\mathbb{Q}$ with the usual distance, the sequence of fractions $(a_n)_1^\infty \subseteq \mathbb{Q}$ given by

$$a_1 := 2.7, \quad a_2 := 2.71, \quad a_3 := 2.718, \quad a_4 := 2.7182, \quad a_5 := 2.71828, \quad \ldots$$

and the sets

$$F_n := \left[a_n - \frac{1}{\sqrt{n}}, \, a_n + \frac{1}{\sqrt{n}} \right] \bigcap \mathbb{Q}, \quad n \in \mathbb{N}.$$

Then, the sequence $\{F_n : n \in \mathbb{N}\}$ is a decreasing sequence of relatively closed sets (in $\mathbb{Q}$) and $\operatorname{diam}(F_n) = 2/\sqrt{n}$, but

$$\mathbb{Q} \bigcap \left(\bigcap_{n=1}^{\infty} F_n \right) = \emptyset.$$

(ii) Consider the metric space $\mathbb{R}$ with the usual distance and the sequence

$$F_n := \left[-1 - \frac{1}{n}, \, 1 + \frac{1}{n} \right], \quad n \in \mathbb{N}.$$

Then, the sequence $\{F_n : n \in \mathbb{N}\}$ is a decreasing sequence of relatively closed sets, but

$$\bigcap_{n=1}^{\infty} F_n = [-1, 1].$$

(iii) Consider again the metric space $\mathbb{R}$ with the usual distance and the sequence $F_n := (0, 1/n)$, $n \in \mathbb{N}$. Then, the sequence $\{F_n : n \in \mathbb{N}\}$ is a decreasing and $\operatorname{diam}(F_n) = 1/n$, but

$$\bigcap_{n=1}^{\infty} F_n = \emptyset.$$

Proof. Let $\{F_n\}_1^\infty$ be a sequence of sets with the above properties. We first prove that $\bigcap_{n=1}^\infty F_n$ has at most one element. Indeed, let $x_0, \tilde{x}_0 \in \bigcap_{n=1}^\infty F_n$. Then $x_0, \tilde{x}_0 \in F_n$ for all $n \in \mathbb{N}$ and hence

$$d(x_0, \tilde{x}_0) \leq \text{diam}(F_n)$$

for any $n \in \mathbb{N}$. Therefore, $x_0 = \tilde{x}_0$ and uniqueness ensues.

We now prove that $\bigcap_{n=1}^\infty F_n$ has at least one element. For any $n \in \mathbb{N}$, let us pick some element $x_n \in F_n$. We prove that $(x_n)_1^\infty$ is a Cauchy sequence. Indeed, for any $\varepsilon > 0$, let us choose $N \in \mathbb{N}$ such that $\text{diam}(F_N) < \varepsilon$. Then, for any $n, m \geq N$, we have that $x_n \in F_n \subseteq F_N$ and $x_m \in F_m \subseteq F_N$, so that

$$d(x_n, x_m) \leq \text{diam}(F_N) < \varepsilon.$$

Hence $(x_n)_1^\infty$ is a Cauchy sequence. Since (X, d) is complete, it follows that there exists $x_0 \in X$ such that $x_n \longrightarrow x_0$ as $n \to \infty$. We claim that $x_0 \in F_k$ for all $k \in \mathbb{N}$. Indeed, for any $k \in \mathbb{N}$, we have that

$$x_k, \ x_{k+1}, \ x_{k+2}, \ \ldots \ \in F_k$$

and since F_k is closed and $x_n \longrightarrow x_0$ as $n \to \infty$, it follows that $x_0 \in F_k$. Hence $x_0 \in \bigcap_{k \in \mathbb{N}} F_k$. This completes the proof. $\qquad\square$

Next we come to a further very important result which comes by two different, but closely related, versions. We have already seen instance where an intersection of infinitely-many open sets might not be open, and in fact may even be empty (recall e.g. Remark 3.12(ii) above). However, if the open sets are also *dense*, then, the intersection is always non-empty.[1]

Theorem 3.13 (Baire's Theorem–version 1). *Let (X, d) be a complete metric space. Let $\{V_n : n \in \mathbb{N}\}$ be a sequence of dense open sets in X, i.e., such that $\overline{V_n} = X$, for all $n \in \mathbb{N}$. Then, their intersection $\bigcap_{n=1}^\infty V_n$ is dense in X (and in particular not empty):*

$$\overline{\bigcap_{n=1}^\infty V_n} = X.$$

Example 3.14. Note that the above intersection of open sets may well fail to be open. For instance, consider $\mathbb{R}$ as a metric space with the usual structure and let $\{q_j : j \in \mathbb{N}\} = \mathbb{Q}$ be an enumeration of the rationals. We consider the sequence of subsets of $\mathbb{R}$:

$$V_n := \bigcup_{j=1}^\infty \left(q_j - \frac{1}{n \, 2^j}, \, q_j + \frac{1}{n \, 2^j} \right), \quad n \in \mathbb{N}.$$

---

[1]We recall that a set $D \subseteq X$ is dense when its closure equals the entire space: $\overline{D} = X$. Equivalently, every $x \in X$ can be approximated by a sequence $(x_n)_1^\infty \subseteq D$.

Then, each V_n is open as a union of open intervals and is dense because it contains a dense set, $\mathbb{Q}$ itself. However, the intersection of all V_n equals

$$\bigcap_{n=1}^{\infty} V_n = \mathbb{Q}$$

and therefore it is dense but not open.

Proof. In order to prove that $\bigcap_1^{\infty} V_n$ is dense in X, we shall show that

$$\mathbb{B}_r(x_0) \bigcap_{n=1}^{\infty} V_n \neq \emptyset,$$

for any $x_0 \in X$ and $r > 0$. Fix $x_0 \in X$ and $r > 0$ both arbitrary. In the sequel we shall use repeatedly the following:

Observation. *If V is an open dense set and A is open, there exists $z \in X$ and $s > 0$ such that $\mathbb{D}_s(z) \subseteq V \bigcap A$.*

First, since V_1 is open and dense, and $\mathbb{B}_r(x_0)$ is open, there exist $x_1 \in X$ and $r_1 > 0$ such that

$$\mathbb{D}_{r_1}(x_1) \subseteq V_1 \bigcap \mathbb{B}_r(x_0).$$

Then, since V_2 is open and dense, and $\mathbb{B}_{r_1}(x_1)$ is open, there exist $x_2 \in X$ and $r_2 > 0$ such that

$$\mathbb{D}_{r_2}(x_2) \subseteq V_2 \bigcap \mathbb{B}_{r_1}(x_1).$$

Continuing in this way, using the fact for each $n \in \mathbb{N}$, V_n is open and dense and $\mathbb{B}_{r_{n-1}}(x_{n-1})$ is open, we obtain $x_n \in X$ and $r_n > 0$ such that

$$\mathbb{D}_{r_n}(x_n) \subseteq V_n \bigcap \mathbb{B}_{r_{n-1}}(x_{n-1}).$$

Moreover, at each step of this construction, we may choose r_n such that $0 < r_n \leq 1/n$. Note that

$$\mathbb{D}_{r_1}(x_1) \supseteq \mathbb{D}_{r_2}(x_2) \supseteq \cdots \supseteq \mathbb{D}_{r_n}(x_n) \supseteq \cdots$$

and

$$\operatorname{diam}\left(\mathbb{D}_{r_n}(x_n)\right) \longrightarrow 0 \text{ as } n \to \infty,$$

so that, by Theorem 3.11, we have that there exists $x \in X$ such that

$$\bigcap_{n=1}^{\infty} \mathbb{D}_{r_n}(x_n) = \{x\}.$$

Since $\mathbb{D}_{r_1}(x_1) \subseteq \mathbb{B}_r(x_0)$ and $\mathbb{D}_{r_n}(x_n) \subseteq V_n$ for all $n \in \mathbb{N}$, it follows that

$$x \in \mathbb{B}_r(x_0) \bigcap_{n=1}^{\infty} V_n.$$

Therefore, the set $\bigcap_{n=1}^{\infty} V_n$ is indeed dense in X. $\qquad\square$

The following version of Baire's theorem is also valid and actually it is equivalent to Theorem 3.13 above.

Theorem 3.15 (Baire's Theorem–version 2). *Let (X, d) be a complete metric space and suppose that X can be written as a countable union of closed subsets $\{F_n : n \in \mathbb{N}\}$ of it:*

$$X = \bigcup_{n=1}^{\infty} F_n.$$

Then, at least one of these closed sets has non-empty interior. Namely, there exists $N \in \mathbb{N}$ such that:

$$(F_N)^{\circ} \neq \emptyset.$$

In other words, at least one set F_N contains an open ball $\mathbb{B}_r(x_0)$ for some $x_0 \in X$ and $r > 0$.

Proof. Suppose for the sake of contradiction that $F_n^{\circ} = \emptyset$ for all $n \in \mathbb{N}$. We set

$$V_n := X \setminus F_n,$$

for every $n \in \mathbb{N}$. Then each V_n is open. Moreover, each V_n is also dense, since

$$\overline{V_n} = \overline{X \setminus F_n} = X \setminus F_n^{\circ} = X.$$

Recall now that by assumption we have

$$X = \bigcup_{n=1}^{\infty} F_n,$$

which by taking complements yields

$$\emptyset = X \setminus \left(\bigcup_{n=1}^{\infty} F_n \right) = \bigcap_{n=1}^{\infty} (X \setminus F_n) = \bigcap_{n=1}^{\infty} V_n.$$

The above equality contradicts the conclusion of Theorem 3.13 and this establishes the desired result. $\square$

In Chapter 11 we shall see some important applications of Cantor's and Baire's theorems to the theory of Banach spaces.

3.4 Series in normed vector spaces

We are now in a good position to introduce the concept of a **convergent series** in a normed space and to identify a sufficient condition for the convergence of such a series. Both the given definition and the convergence result are extensions of facts the reader has already seen in Real Analysis.

Definition 3.16. Let $(X, \|\cdot\|)$ be a normed space, and $(x_k)_1^\infty$ be a sequence in X. We say that the **series** $\sum_{k=1}^\infty x_k$ **converges** in $(X, \|\cdot\|)$ if the sequence of partial sums $(s_n)_1^\infty$, defined by

$$s_n := \sum_{k=1}^n x_k \quad \text{for any } n \in \mathbb{N},$$

converges in $(X, \|\cdot\|)$. In this situation, if $s := \lim_{n \to \infty} s_n$, then we shall write

$$\sum_{k=1}^\infty x_k \equiv s$$

and say that s is the value[2] (or sum) of the series $\sum_{k=1}^\infty x_k$. Finally, we shall say that the series $\sum_{k=1}^\infty x_k$ is **absolutely convergent** in X if the series of non-negative numbers $\sum_{k=1}^\infty \|x_k\|$ converges in $[0, \infty)$.

Note that $\sum_{k=1}^\infty x_k$ is an element of the Banach space X and convergence of the sequence $s_n = \sum_{k=1}^n x_k$ to its limit means

$$\left\| \sum_{k=1}^\infty x_k - \sum_{k=1}^n x_k \right\| \longrightarrow 0, \quad \text{as } n \to \infty.$$

We will now show that the modifier "absolutely" describes a generally stronger notion of convergence.

Proposition 3.17. *Let $(X, \|\cdot\|)$ be a Banach space. Then any absolutely convergent series in X is convergent.*

Proof. Suppose that $\sum_{k=1}^\infty x_k$ is absolutely convergent. This means that $\sum_{k=1}^\infty \|x_k\|$ converges. Let, for any $n \in \mathbb{N}$,

$$s_n := \sum_{k=1}^n x_k, \qquad t_n := \sum_{k=1}^n \|x_k\|.$$

We now show that $(s_n)_1^\infty$ is a Cauchy sequence in $(X, \|\cdot\|)$. Indeed, for any $n, m \in \mathbb{N}$ with $n > m$, we have

$$\|s_n - s_m\| = \left\| \sum_{k=m+1}^n x_k \right\| \leq \sum_{k=m+1}^n \|x_k\| = t_n - t_m.$$

---

[2]Note that, in the above definition, we are using in an essential way the linear structure of X. It does not make any sense to speak of (convergence of) series in metric spaces without a linear structure! In addition, we cannot talk about series and of their convergence in a mere vector space! We need an additional metric or at least (as we shall see later) "topological structure"! In (mere) Algebraic structures only finite sums make sense, nothing can converge, nothing is "infinite" and in general everything is static and "nothing moves"!

Since $(t_n)_1^\infty$ is, by assumption, convergent in $\mathbb{R}$ and therefore a Cauchy sequence, the above implies that $(s_n)_1^\infty$ is a Cauchy sequence in $(X, \|\cdot\|)$. Since $(X, \|\cdot\|)$ it a Banach space, it follows that $(s_n)_1^\infty$ converges, which means that the series $\sum_{k=1}^\infty x_k$ converges in $(X, \|\cdot\|)$, as required. $\qquad\square$

Remark 3.18. It may be proven that a certain converse of the above proposition is also true. Namely, if the normed space $(X, \|\cdot\|)$ has the property that any absolutely convergent series in X is convergent, then $(X, \|\cdot\|)$ is a Banach space. Therefore, completeness of normed vector spaces can be characterised by the absolute convergence of series in the space.

3.5 Are our favourite normed spaces actually complete?

In this section we take up the task of studying whether the examples of normed spaces we have exhibited so far are complete or not.

3.5.1 Completeness of the ℓ^p spaces

We begin with the normed spaces $(\ell^p, \|\cdot\|_p)$, where $1 \le p \le \infty$, which have been introduced in Definitions 2.76 and 2.11.

Theorem 3.19 (The ℓ^p spaces are Banach spaces). *For each $p \in [1, \infty]$, the normed space $(\ell^p, \|\cdot\|_p)$ is a Banach space.*

Proof. We give here the proof only in the case $p \in [1, \infty)$. The result in the case $p = \infty$ will be a mere consequence of Theorem 3.20, that follows since $(\ell^\infty, \|\cdot\|_\infty)$ coincides with $(B(\mathbb{N}, \mathbb{F}), \|\cdot\|_\infty)$.

Let $(x^{(n)})_1^\infty$ be a Cauchy sequence in $(\ell^p, \|\cdot\|_p)$, which by definition means that

$$\text{for any } \varepsilon > 0, \ \exists N \in \mathbb{N} \text{ such that } \left\|x^{(n)} - x^{(m)}\right\|_p \le \varepsilon \text{ for all } n, m \ge N.$$

By Proposition 3.4(i), it follows that $(x^{(n)})_1^\infty$ is bounded in $(\ell^p, \|\cdot\|_p)$, meaning that there exists $M \ge 0$ such that

$$\|x^{(n)}\|_p \le M \quad \text{for all } n \in \mathbb{N}.$$

Writing, for every $n \in \mathbb{N}$, $x^{(n)} = (x_1^{(n)}, \dots, x_k^{(n)}, \dots)$, the Cauchy condition takes the form

$$\text{for any } \varepsilon > 0, \ \exists N \in \mathbb{N} \text{ such that } \sum_{k=1}^\infty \left|x_k^{(n)} - x_k^{(m)}\right|^p \le \varepsilon^p \text{ for all } n, m \ge N.$$

Fix $k \in \mathbb{N}$, arbitrary. The above condition implies that

for any $\varepsilon > 0$, $\exists N \in \mathbb{N}$ such that $\left| x_k^{(n)} - x_k^{(m)} \right| \leq \varepsilon$ for all $n, m \geq N$.

This means exactly that $(x_k^{(n)})_{n=1}^{\infty}$ is a Cauchy sequence in $(\mathbb{F}, |\cdot|)$. Since $\mathbb{F}$ is complete, the next limit exists:

$$\lim_{n \to \infty} x_k^{(n)} =: x_k^{(0)}$$

and in fact this is the case for every $k \in \mathbb{N}$. Let us now use the symbolisation $x^{(0)} := (x_1^{(0)}, \ldots, x_k^{(0)}, \ldots)$. When written explicitly, the boundedness condition implies that, for any $K \in \mathbb{N}$,

$$\sum_{k=1}^{K} |x_k^{(n)}|^p \leq M^p \quad \text{for all } n \in \mathbb{N}.$$

For any fixed $K \in \mathbb{N}$, passing to the limit as $n \to \infty$ in the above[3], we obtain that

$$\sum_{k=1}^{K} |x_k^{(0)}|^p \leq M^p.$$

The fact that this is true for all $K \in \mathbb{N}$ allows us to infer (by taking supremum over all $K \in \mathbb{N}$, or equivalently the limit as $K \to \infty$) that $x^{(0)} \in \ell^p$ and also $\|x^{(0)}\|_p \leq M$.

We now go back to the Cauchy condition. Fix $\varepsilon > 0$, arbitrary, and let N be given by the Cauchy condition. It follows that, for every $K \in \mathbb{N}$, and for every $n, m \geq N$, we have that

$$\sum_{k=1}^{K} |x_k^{(n)} - x_k^{(m)}|^p \leq \varepsilon^p.$$

For any fixed $n \geq N$ and $K \in \mathbb{N}$, arbitrary, we pass to the limit as $m \to \infty$ in the above relation (again, it is crucial that we are dealing with a finite sum, rather than a series), therefore obtaining that

$$\sum_{k=1}^{K} |x_k^{(n)} - x_k^{(0)}|^p \leq \varepsilon^p.$$

For any fixed $n \geq N$, the validity of this for all $K \in \mathbb{N}$ shows that indeed $\|x^{(n)} - x^{(0)}\|_p \leq \varepsilon$, for any $n \geq N$. Since $\varepsilon > 0$ was arbitrary, this finally establishes that $x^{(n)} \longrightarrow x^{(0)}$ in $(\ell^p, \|\cdot\|_p)$ as $n \to \infty$. $\qquad\square$

We note that Theorem 3.19 will be generalised significantly in Chapter 9, when we shall discuss a much broader class of spaces containing the class of ℓ^p spaces as a particular case.

---

[3]Note that in this argument it is crucial that we are dealing with a finite sum, rather than a series.

3.5.2 Spaces of bounded and continuous functions

For any set X, the normed space $(B(X, \mathbb{F}), \| \cdot \|_\infty)$ has been introduced in Definition 2.12. When (X, d) is a metric space, one may also consider the set

$$C_b(X, \mathbb{F}) := \left\{ f : X \longrightarrow \mathbb{F} \mid f \text{ is continuous and bounded on } X \right\}.$$

Then it is easy to check that $C_b(X, \mathbb{F})$ is a vector subspace of $B(X, \mathbb{F})$. The latter allows us to infer that $(C_b(X, \mathbb{F}), \| \cdot \|_\infty)$ is a normed subspace of $(B(X, \mathbb{F}), \| \cdot \|_\infty)$. We may now establish the following important result regarding the completeness of these spaces.

Theorem 3.20.

(i) *For any set X, the space $(B(X, \mathbb{F}), \| \cdot \|_\infty)$ is a Banach space.*

(ii) *For any metric space (X, d), the space $(C_b(X, \mathbb{F}), \| \cdot \|_\infty)$ is a Banach space.*

Proof. (i) We only need to prove the completeness of $(B(X, \mathbb{F}), \| \cdot \|_\infty)$. Let $(f_n)_1^\infty$ be a Cauchy sequence in $(B(X, \mathbb{F}), \| \cdot \|_\infty)$, meaning that

for any $\varepsilon > 0$ $\exists N \in \mathbb{N}$ such that $\| f_n - f_m \|_\infty \leq \varepsilon$ for all $n, m \geq N$.

We need to show that there exists $f_0 \in B(X, \mathbb{F})$ such that $f_n \longrightarrow f_0$ in $(B(X, \mathbb{F}), \| \cdot \|_\infty)$ as $n \to \infty$. Note first that, by Proposition 3.4(i), the sequence $(f_n)_1^\infty$ is bounded in $(B(X, \mathbb{F}), \| \cdot \|_\infty)$, meaning that there exists $M \geq 0$ such that $\| f_n \|_\infty \leq M$ for all $n \in \mathbb{N}$. The latter can be written explicitly as

$$|f_n(x)| \leq M \quad \text{for all } x \in X \text{ and for all } n \in \mathbb{N}.$$

We now write the Cauchy condition explicitly:

for any $\varepsilon > 0$ $\exists N \in \mathbb{N}$ such that $\left| f_n(x) - f_m(x) \right| \leq \varepsilon$

for all $x \in X$ and all $n, m \geq N$. Fix $x \in X$, arbitrary. The above condition implies that $(f_n(x))_1^\infty$ is a Cauchy sequence in $(\mathbb{F}, | \cdot |)$. Since $\mathbb{F}$ is complete, the next limit exists:

$$\lim_{n \to \infty} f_n(x) \in \mathbb{F}.$$

Since this is true for any $x \in X$, we may define a function $f_0 : X \longrightarrow \mathbb{F}$ by

$$f_0(x) := \lim_{n \to \infty} f_n(x) \quad \text{for all } x \in X.$$

Namely, f_0 is the function which equals the *pointwise* limit of the sequence of functions $(f_n)_1^\infty$, but nothing has thus far established regarding the desired uniform convergence. This is established right next.

For each $x \in X$, letting $n \to \infty$ in the condition

$$|f_n(x)| \leq M \quad \text{for all } n \in \mathbb{N},$$

we obtain that

$$|f_0(x)| \leq M \quad \text{for all } n \in \mathbb{N}.$$

Thus, $f_0 \in B(X, \mathbb{F})$. We now get back to the Cauchy condition. Fix $\varepsilon > 0$, and let $N \in \mathbb{N}$ be given by the Cauchy condition. Then, for any $x \in X$ and $n, m \geq N$, we have

$$|f_n(x) - f_m(x)| \leq \varepsilon.$$

For any fixed $x \in X$ and any $n \geq N$, arbitrary, we pass to the limit in the above relation as $m \to \infty$, obtaining

$$|f_n(x) - f_0(x)| \leq \varepsilon.$$

For any fixed $n \geq N$, arbitrary, since the above is true for all $x \in X$, we deduce that

$$\|f_n - f_0\|_\infty \leq \varepsilon.$$

Since the above is true for any $\varepsilon > 0$, it follows that $f_n \longrightarrow f_0$ as $n \to \infty$ in $(B(X, \mathbb{F}), \| \cdot \|_\infty)$. In conclusion, we have proved that $(B(X, \mathbb{F}), \| \cdot \|_\infty)$ is a Banach space.

(ii) In view of part (i) and Proposition 3.9, to show that $(C_b(X, \mathbb{F}), \| \cdot \|_\infty)$ is complete, it suffices to show that it is closed in $B(X, \mathbb{F})$. Thus, let

$$f_0 \in \overline{C_b(X, \mathbb{F})}$$

be an arbitrary element, where $\overline{C_b(X, \mathbb{F})}$ symbolises the closure of $C_b(X, \mathbb{F})$ in $(B(X, \mathbb{F}), \| \cdot \|_\infty)$. Then there exists a sequence $(f_n)_1^\infty \subseteq C_b(X, \mathbb{F})$ such that

$$f_n \longrightarrow f_0 \quad \text{in } (B(X, \mathbb{F}), \| \cdot \|_\infty) \quad \text{as } n \to \infty.$$

We need to show that $f_0 \in C_b(X, \mathbb{F})$.[4]

Fix $x_0 \in X$ and $\varepsilon > 0$, both arbitrary and recall that we have $f_n \longrightarrow f_0$ as $n \to \infty$ in $(B(X, \mathbb{F}), \| \cdot \|_\infty)$. Therefore, we may select some $N \in \mathbb{N}$ such that $\|f_N - f_0\|_\infty < \varepsilon/3$. Since f_N is continuous and bounded, there exists an open neighbourhood V of x_0 such that

$$\left| f_N(x) - f_N(x_0) \right| < \frac{\varepsilon}{3} \quad \text{for all } x \in V.$$

It follows that, for any $x \in V$,

$$\begin{aligned}
\left| f_0(x) - f_0(x_0) \right| &\leq \left| f_0(x) - f_N(x) \right| + \left| f_N(x) - f_N(x_0) \right| + \left| f_N(x_0) - f_0(x_0) \right| \\
&\leq 2\|f_N - f_0\|_\infty + \left| f_N(x) - f_N(x_0) \right| \\
&< \frac{2\varepsilon}{3} + \frac{\varepsilon}{3} \\
&= \varepsilon.
\end{aligned}$$

---

[4]The argument below is the result known from Real Analysis that "the uniform limit of a sequence of continuous functions is itself continuous", but we shall give the proof anyway for the sake of completeness.

Hence f_0 is continuous at x_0. The arbitrariness of x_0 shows that f is indeed continuous on X, as required. We have thus proved that

$$\overline{C_b(X, \mathbb{F})} \subseteq C_b(X, \mathbb{F}),$$

which means that $C_b(X, \mathbb{F})$ is closed in $\big(B(X, \mathbb{F}), \|\cdot\|_\infty\big)$, as required. $\qquad\square$

It will be important for later applications to note here that one may also consider similar spaces of functions with values in $\mathbb{F}^m$ rather than in $\mathbb{F}$. On $\mathbb{F}^m$ one may use several different norms, but we shall restrict attention to the particular choices

$$|y|_\infty := \max_{k \in \{1, \dots, m\}} |y_k|, \quad |y|_2 := \left(\sum_{k=1}^m |y_k|^2 \right)^{1/2},$$

for $y = (y_1, \dots, y_m) \in \mathbb{F}^m$. The most popular choice is that of the Euclidean norm on $\mathbb{F}^m$.[5] Then one can consider the following spaces: for any set X, the space

$$B(X, \mathbb{F}^m) := \Big\{ f : X \longrightarrow \mathbb{F}^m \ \Big| \ f \text{ is bounded on } X \Big\},$$

and for any metric space (X, d), the space

$$C_b(X, \mathbb{F}^m) := \Big\{ f : X \longrightarrow \mathbb{F}^m \ \Big| \ f \text{ is continuous and bounded on } X \Big\},$$

where, since the norms $|\cdot|_\infty$ and $|\cdot|_2$ are equivalent on $\mathbb{F}^m$, both continuity and boundedness of functions with values in $\mathbb{F}^m$ have the same meaning irrespective of whether we consider $|\cdot|_\infty$ or $|\cdot|_2$ as a norm on $\mathbb{F}^m$. We may thus consider on the vector space $B(X, \mathbb{F}^m)$ (as well as on its vector subspace $C_b(X, \mathbb{F}^m)$ whenever (X, d) is a metric space) the norms

$$\|f\|_{\infty,(\infty)} := \sup_{x \in X} |f(x)|_\infty \quad \text{for all } f \in B(X, \mathbb{F}^m),$$

and

$$\|f\|_{\infty,(2)} := \sup_{x \in X} |f(x)|_2 \quad \text{for all } f \in B(X, \mathbb{F}^m).$$

We then have the following result, whose proof is nearly identical to that of the preceding result, except for obvious changes in notation. Therefore, we leave the details as a simple exercise for the reader.

Theorem 3.21.

(i) *For any set X, the normed vector spaces $\big(B(X, \mathbb{F}^m), \|\cdot\|_{\infty,(\infty)}\big)$ and $\big(B(X, \mathbb{F}^m), \|\cdot\|_{\infty,(2)}\big)$ are Banach spaces.*

---

[5]The choice of norms on $\mathbb{F}^m$ plays little role since, as we shall see later in Chapter 5, all of them are equivalent and thus give rise to equivalent norms on the functional spaces.

(ii) *For any metric space (X, d), the normed vector spaces $\big(C_b(X, \mathbb{F}^m), \| \cdot \|_{\infty,(\infty)}\big)$ and $\big(C_b(X, \mathbb{F}^m), \| \cdot \|_{\infty,(2)}\big)$ are Banach spaces.*

We now indicate a family of examples of normed spaces which are not Banach spaces. For any $a, b \in \mathbb{R}$ with $a < b$ and any $p \in [1, \infty)$, the normed space $(C([a, b], \mathbb{F}), \| \cdot \|_p)$ has been introduced in Definition 2.79.

Proposition 3.22. *For any $p \in [1, \infty)$, the normed space $\big(C([a, b], \mathbb{F}), \| \cdot \|_p\big)$ is not a Banach space.*[6]

Proof. We construct a Cauchy sequence in $(C([a, b], \mathbb{F}), \| \cdot \|_p)$ which does not converge in $(C([a, b], \mathbb{F}), \| \cdot \|_p)$. We do this only in the case $[a, b] = [-1, 1]$, the general case being treated by a similar construction. For each $n \in \mathbb{N}$, let $f_n : [-1, 1] \longrightarrow \mathbb{F}$ be given by (see Figure 3.3)

$$f_n(x) := \begin{cases} 0, & \text{for} \quad -1 \le x \le 0, \\ nx, & \text{for} \quad 0 < x < \dfrac{1}{n}, \\ 1, & \text{for} \quad \dfrac{1}{n} \le x \le 1. \end{cases}$$

One checks easily that all functions f_n are continuous on $[-1, 1]$. Fix $N \in \mathbb{N}$ and let $n, m \ge N$. Then $f_n(x) = f_m(x)$ for all $x \in [-1, 0] \bigcup [1/N, 1]$, whilst for any $x \in [0, 1/N]$, we have

$$\big| f_n(x) - f_m(x) \big| \le 1.$$

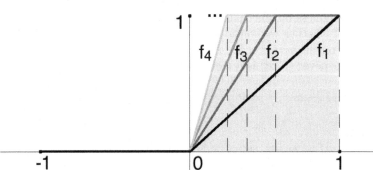

Figure 3.3: An illustration of the graphs of the functions f_n. As $n \to \infty$, a jump is created at $x = 0$ and the continuous functions convergence pointwise on $[-1, 1] \setminus \{0\}$ to a discontinuous function! Remember that we need uniform convergence for the limit of continuous functions to be continuous.

---

[6]In later chapters we shall see that the introduction of a "generalised" integral for discontinuous functions, called the Lebesgue integral, allows us to complete the normed vector space $(C([a, b], \mathbb{F}), \| \cdot \|_p)$ by enlarging it to a space of less regular functions which are limits of Cauchy sequences under the $\| \cdot \|_p$-norm.

This allows us to estimate

$$\|f_n - f_m\|_p^p = \int_{-1}^{1} |f_n(x) - f_m(x)|^p \, \mathrm{d}x$$

$$= \int_{0}^{1/N} |f_n(x) - f_m(x)|^p \, \mathrm{d}x$$

$$\leq \frac{1}{N},$$

and therefore

$$\|f_n - f_m\|_p \leq \frac{1}{N^{1/p}} \quad \text{for all} \ \ n, m \geq N.$$

The above bound clearly implies that $(f_n)_1^\infty$ is a Cauchy sequence in the space $\big(C([-1,1], \mathbb{F}), \|\cdot\|_p\big)$.

It remains to prove that $(f_n)_1^\infty$ does not converge in $\big(C([-1,1], \mathbb{F}), \|\cdot\|_p\big)$ as $n \to \infty$. Suppose for the sake of contradiction that there existed $f_0 \in C([-1,1], \mathbb{F})$ such that $f_n \longrightarrow f_0$ in $\big(C([-1,1], \mathbb{F}), \|\cdot\|_p\big)$ as $n \to \infty$. This implies that

$$\int_{-1}^{0} |f_n(x) - f_0(x)|^p \, \mathrm{d}x \longrightarrow 0 \quad \text{as} \ n \to \infty,$$

and that, for every $\alpha \in (0,1]$,

$$\int_{\alpha}^{1} |f_n(x) - f_0(x)|^p \, \mathrm{d}x \longrightarrow 0 \quad \text{as} \ n \to \infty.$$

Note that $f_n(x) = 0$ for all $x \in [-1, 0]$, for all $n \in \mathbb{N}$, which implies that

$$\int_{-1}^{0} |f_0(x)|^p \, \mathrm{d}x = 0.$$

Since f_0 is continuous, it follows that $f_0(x) = 0$ for all $x \in [-1, 0]$. On the other hand, fix $\alpha \in (0, 1]$, arbitrary. For every $n \geq 1/\alpha$, we have that $f_n(x) = 1$ for all $x \in [\alpha, 1]$. This implies that

$$\int_{\alpha}^{1} |1 - f_0(x)|^p \, \mathrm{d}x = 0.$$

Since f_0 is continuous, it follows that $f_0(x) = 1$ for all $x \in [\alpha, 1]$. However, since $\alpha \in (0, 1]$ was arbitrary, we deduce that $f_0(x) = 1$ for all $x \in (0, 1]$. Combined with the fact that $f_0(x) = 0$ for all $x \in [-1, 0]$, this shows that f_0 has different one-sided limits (from the left and from the right) at $x_0 = 0$, a fact which contradicts the continuity of f_0 at 0. The conclusion is that $(f_n)_1^\infty$ does not converge in $\big(C([-1,1], \mathbb{F}), \|\cdot\|_p\big)$, and therefore $\big(C([-1,1], \mathbb{F}), \|\cdot\|_p\big)$ is not a Banach space. $\qquad\square$

3.6 The Banach fixed-point Theorem

In this section we establish a **very important** result which can be stated and proved in the generality of metric spaces and it can be seen as a fundamental consequence of completeness. The idea and the setting is as follows. Suppose that (X, d) is a complete metric space and we are given a mapping $f : X \longrightarrow X$ from X to itself. Then, if f *strictly contracts (decreases) the distances on X*, that is for any pair of points $x, y \in X$, we have that $d(f(x), f(y))$ is strictly smaller than $d(x, y)$, and this in addition happens in a uniform fashion for all $x, y \in X$, then *the equation*

$$f(x) = x$$

has a unique solution $\tilde{x} \in X$! Obviously, this a very important assertion because it gives existence of a unique object which might be difficult or even impossible to be constructed otherwise. In addition, the method of proof gives a algorithmic method allowing to approximate it! The latter fact is especially significant to Numerical Analysts who are interested in computer simulations.

The relevant result, known as the **contraction mapping principle** (or the **Banach fixed point theorem**), has numerous important applications in Analysis, particularly in the theory of Differential Equations, both Ordinary and Partial. For instance, it is an essential ingredient in the proof of the Inverse and the Implicit Function Theorem of Analysis in several variables.

One standard application in the realm of Ordinary Differential Equations will be given in the next section. We begin with the formal definition of "contractions".

Definition 3.23. Let (X, d) be metric space, $f : X \longrightarrow X$ a mapping and $\tilde{x}$ a point in X.

(i) f is said to be a **contraction** on X if there exists $\alpha \in [0, 1)$ such that

$$d(f(x), f(y)) \leq \alpha \, d(x, y) \quad \text{for all } x, y \in X.$$

(ii) The point $\bar{x} \in X$ in the metric space is called a **fixed point** of f if it satisfies the equation

$$f(\tilde{x}) = \tilde{x}.$$

It can be easily seen that any contraction on X is a continuous mapping. The main result of this section is the following.

Theorem 3.24 (Banach's Fixed-Point Theorem, or the Contraction Mapping Principle).
Let (X, d) be a complete metric space with $X \neq \emptyset$ and $f : X \longrightarrow X$ a contraction on X. Then, there exist a unique fixed point $\tilde{x} \in X$, namely a solution to the equation

$$f(\tilde{x}) = \tilde{x}.$$

In addition, $\tilde{x}$ can be approximated in (X, d) by the sequence of iterations

$$f(x_0), \; f(f(x_0)), \; f(f(f(x_0))), \; \ldots$$

where x_0 is an arbitrary point of $x \in X$.

Proof. Let $x_0 \in X$, arbitrary. We define the sequence

$$x_n := f(x_{n-1}), \quad n \in \mathbb{N}.$$

In this way we construct a sequence $(x_n)_1^\infty$. The plan is to show that it converges and if we symbolise as $\tilde{x}$ the limit of $(x_n)_1^\infty$ in (X, d), we shall then show that $f(\tilde{x}) = \tilde{x}$ and that $\tilde{x}$ is the unique such element.

We begin by showing that $(x_n)_1^\infty$ is a Cauchy sequence in (X, d). Observe that, for every $j \in \mathbb{N}$,

$$d(x_{j+1}, x_j) = d\big(f(x_j), f(x_{j-1})\big) \leq \alpha \, d(x_j, x_{j-1}). \qquad (\bullet)$$

Repeated applications of the above estimate give for every $k \in \mathbb{N}$ that

$$\begin{aligned}
d\big(x_{k+1}, x_k\big) &\leq \alpha \, d\big(x_k, x_{k-1}\big) \\
&\leq \alpha^2 \, d\big(x_{k-1}, x_{k-2}\big) \\
&\leq \cdots \\
&\leq \alpha^k \, d(x_1, x_0).
\end{aligned}$$

Let $N \in \mathbb{N}$, to be chosen more specifically later on, and let us consider arbitrary $n, m \geq N$. The goal is to estimate $d(x_m, x_n)$. This is nontrivial only when $m \neq n$ and hence there is no loss of generality to assume $m > n$. Repeated applications of the triangle inequality lead to

$$d(x_m, x_n) \leq d(x_m, x_{m-1}) + d(x_{m-1}, x_{m-2}) + \cdots + d(x_{n+1}, x_n).$$

By changing the order the order of summation and utilising the estimate $(\bullet)$ and that $0 \leq \alpha < 1$, we infer

$$\begin{aligned}
d(x_m, x_n) &\leq \Big(\alpha^n + \cdots + \alpha^{m-1}\Big) d(x_1, x_0) \\
&\leq \alpha^n \frac{1 - \alpha^{m-n}}{1 - \alpha} \, d(x_1, x_0) \\
&\leq \frac{\alpha^N}{1 - \alpha} \, d(x_1, x_0).
\end{aligned}$$

Now we may show convergence. Given any $\varepsilon > 0$, we choose N so large that

$$\frac{\alpha^N}{1 - \alpha} \, d(x_1, x_0) < \varepsilon.$$

Then, for all $n, m \geq N$ we have $d(x_n, x_m) < \varepsilon$. Hence $(x_n)_1^\infty$ is a Cauchy sequence. Since (X, d) is complete, the next limit exists:

$$\lim_{n \to \infty} x_n =: \tilde{x} \in X.$$

By using the continuity of f, we obtain that

$$f(x_n) \longrightarrow f(\tilde{x}) \ \text{ in } (X, d), \ \text{ as } n \to \infty.$$

On the other hand, we have

$$f(x_n) = x_{n+1} \longrightarrow \tilde{x} \ \text{ in } (X, d), \text{ as } n \to \infty.$$

The uniqueness of limits allows us to infer that $f(\tilde{x}) = \tilde{x}$. This establishes the existence part of the theorem and the desired approximation claim.

We finally show that $\tilde{x}$ is the unique fixed point. Suppose for the sake of contradiction that there existed $\tilde{x}_1$ and $\tilde{x}_2$ such that $f(\tilde{x}_1) = \tilde{x}_1$ and $f(\tilde{x}_2) = \tilde{x}_2$. Then

$$d(\tilde{x}_1, \tilde{x}_2) = d\big(f(\tilde{x}_1), f(\tilde{x}_2)\big) \leq \alpha \, d(\tilde{x}_1, \tilde{x}_2).$$

Since $0 \leq \alpha < 1$, the above estimate yields that $d(\tilde{x}_1, \tilde{x}_2) = 0$, so that $\tilde{x}_1 = \tilde{x}_2$. This completes the uniqueness assertion and the result ensues. $\qquad \square$

3.7   An application to Differential Equations

In this section we digress from the previous developments made in the context of abstract metric and/or normed spaces and we shall consider instead a more concrete application of the Contraction Mapping Principle to the theory of Ordinary Differential Equations. In fact, we shall establish a result of so-called local well posedness for the initial value problem. The jargon in the last sentence is explained right below:

Let $I \subseteq \mathbb{R}$ be an open interval, $m \in \mathbb{N}$, $\Omega \subseteq \mathbb{R}^m$ an open set and

$$\mathbf{f} \ : \ I \times \Omega \longrightarrow \mathbb{R}^m$$

a given continuous mapping on $I \times \Omega$. Here $\mathbb{R}^m$ is considered as a normed space with respect to any norm that is equivalent to the Euclidean norm, whilst $I \times \Omega$ is considered as a metric subspace of the normed space $\mathbb{R}^{m+1}$ (with any norm).

Given $t_0 \in I$ and $\mathbf{x}_0 \in \Omega$, we are interested in the existence of local solutions to the so-called **initial-value problem** (or **Cauchy problem**) for the **system of first-order Ordinary Differential Equations**

$$\begin{cases} \dot{\mathbf{x}}(t) = \mathbf{f}(t, \mathbf{x}(t)), \ \ t \neq t_0, \\ \mathbf{x}(t_0) = \mathbf{x}_0. \end{cases} \tag{IVP}$$

This means that we are seeking a continuously differentiable mapping

$$\mathbf{x} \ : \ J \longrightarrow \Omega \subseteq \mathbb{R}^m,$$

where $J \subseteq I$ is an interval containing t_0, such that the above relations hold for $t \in J$. The considerations below are simplified slightly if one chooses to work in $\mathbb{R}^m$ with the norm $|\cdot|_\infty$ given by

$$|\mathbf{y}|_\infty = \max_{k \in \{1, \dots, m\}} |y_k| \quad \text{for all } \mathbf{y} = (y_1, \dots, y_m) \in \mathbb{R}^m.$$

The next result is one of the cornerstones of the theory of Ordinary Differential Equations in Euclidean spaces.

Theorem 3.25 (Picard's Theorem). *Let $\mathbf{f} : I \times \Omega \longrightarrow \mathbb{R}^m$ be a given continuous mapping, $t_0 \in I$, $\mathbf{x}_0 \in \Omega$, $\alpha > 0$ and $R > 0$ be such that[7]*

$$[t_0 - \alpha, t_0 + \alpha] \subseteq I, \quad \mathbb{D}_R(\mathbf{x}_0) \subseteq \Omega.$$

Let $M > 0$ be such that

$$|\mathbf{f}(t, \mathbf{x})|_\infty \leq M$$

for all $(t, \mathbf{x}) \in [t_0 - \alpha, t_0 + \alpha] \times \mathbb{D}_R(\mathbf{x}_0)$. Suppose in addition that there exists $L > 0$ such that

$$\left| \mathbf{f}(t, \mathbf{y}) - \mathbf{f}(t, \mathbf{z}) \right|_\infty \leq L \left| \mathbf{y} - \mathbf{z} \right|_\infty$$

for all $(t, \mathbf{y}), (t, \mathbf{z}) \in [t_0 - \alpha, t_0 + \alpha] \times \mathbb{D}_R(\mathbf{x}_0)$. Let $\delta > 0$ be such that

$$\delta < \min \left\{ \alpha, \frac{R}{M}, \frac{1}{L} \right\}.$$

Then (IVP) has a unique solution C^1 solution $\mathbf{x} : [t_0 - \delta, t_0 + \delta] \longrightarrow \mathbb{D}_R(\mathbf{x}_0)$.

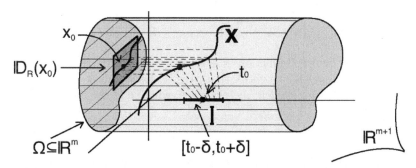

Figure 3.4: An illustration of the setting of the existence result for Ordinary Differential Equations for $m = 2$.

Proof. We first note that, by the Fundamental Theorem of Calculus, the problem (IVP) for a continuously differentiable solution

$$\mathbf{x} : \quad [t_0 - \delta, t_0 + \delta] \longrightarrow \mathbb{D}_R(\mathbf{x}_0)$$

---

[7]Here $\mathbb{D}_R(\mathbf{x}_0)$ denotes the closed ball with centre $\mathbf{x}_0$ and radius R in the $|\cdot|_\infty$ norm.

may be written in an equivalent way as the integral equation

$$\mathbf{x}(t) = \mathbf{x}_0 + \int_{t_0}^{t} \mathbf{f}(s, \mathbf{x}(s)) \, ds. \tag{IE}$$

Indeed, (IE) is straightforwardly obtained from (IVP) by integration, whilst (IVP) can be obtained from (IE) by differentiation. The equation (IE) may be considered in the class of *continuous* mappings

$$\mathbf{x} \; : \; [t_0 - \delta, t_0 + \delta] \longrightarrow \mathbb{D}_R(\mathbf{x}_0).$$

It can easily be seen that any continuous solution $\mathbf{x}$ of (IE) is necessarily continuously differentiable. Let us denote in what follows, for convenience, $J := [t_0 - \delta, t_0 + \delta]$. The form of equation (IE) suggests introducing the space

$$C_b(J, \mathbb{R}^m) := \Big\{ \mathbf{u} : J \longrightarrow \mathbb{R}^m \ \Big| \ \mathbf{u} \text{ is continuous (and bounded) on } J \Big\},$$

where $\mathbb{R}^m$ is considered as a normed space with the $| \cdot |_\infty$ norm. The spaces $C_b(J, \mathbb{R}^m)$ is considered endowed with the norm

$$\|u\|_{\infty,(\infty)} := \sup_{t \in J} |\mathbf{u}(t)|_\infty \quad \text{for all } \mathbf{u} \in C_b(J, \mathbb{R}^m),$$

and, as noted in Theorem 3.21(ii), $\big(C(J, \mathbb{R}^m), \|\cdot\|_{\infty,(\infty)}\big)$ is a Banach space. For simplicity of the notation, in what follows we shall utilise the symbolisation

$$(\mathcal{X}, \| \cdot \|) \equiv \big(C(J, \mathbb{R}^m), \| \cdot \|_{\infty,(\infty)}\big).$$

Let $\mathcal{A}$ be the subset of $\mathcal{X}$ be given by

$$\mathcal{A} := \Big\{ \mathbf{x} \in \mathcal{X} \ : \ |\mathbf{x}(t) - \mathbf{x}_0|_\infty \leq R \ \text{ for all } t \in J \Big\},$$

so that $\mathcal{A}$ is the set of all functions $\mathbf{x} \in \mathcal{X}$ with the property that $\mathbf{x}(t) \in \mathbb{D}_R(\mathbf{x}_0)$ for all $t \in J$. It is easy to check that $\mathcal{A}$ is a closed subset of $(\mathcal{X}, \| \cdot \|)$, and therefore $(\mathcal{A}, d)$ is a *complete* metric space, where d is the metric induced by $\| \cdot \|$. We consider now the operator $T : \mathcal{A} \longrightarrow \mathcal{X}$ given by the right-hand side of (IE), namely

$$(T[\mathbf{x}])(t) := \mathbf{x}_0 + \int_{t_0}^{t} \mathbf{f}(s, \mathbf{x}(s)) \, ds \quad \text{for all } t \in J.$$

Note that (IE) takes the form

$$\mathbf{x} = T[\mathbf{x}] \quad \text{where } \mathbf{x} \in \mathcal{A},$$

that is, we need to prove the existence of a unique fixed point in $\mathcal{A}$ of the function T. This will follow by an application of the Banach Fixed-Point Theorem once we prove that

$$T(\mathcal{A}) \subseteq \mathcal{A},$$

so that T may be regarded as a mapping from $\mathcal{A}$ into itself (rather than from $\mathcal{A}$ to $\mathcal{X}$), and that T is a contraction on $\mathcal{A}$. Note first that, for any $\mathbf{x} \in \mathcal{A}$ we have that, for any $t \in J$,

$$
\begin{aligned}
\left|(T[\mathbf{x}])(t) - \mathbf{x}_0\right|_\infty &= \left| \int_{t_0}^t \mathbf{f}(s, \mathbf{x}(s))\, ds \right|_\infty \\
&\leq \left| \int_{t_0}^t \left| \mathbf{f}(s, \mathbf{x}(s)) \right|_\infty ds \right| \\
&\leq M |t - t_0| \\
&\leq M\delta \\
&< R,
\end{aligned}
$$

since $\delta < R/M$. The fact that this is true for all $t \in J$ shows that $T[\mathbf{x}] \in \mathcal{A}$. Therefore, T is a mapping from $\mathcal{A}$ into itself:

$$
T : \quad \mathcal{A} \longrightarrow \mathcal{A}.
$$

We now check that T is a contraction on $\mathcal{A}$. Let $\mathbf{y}, \mathbf{z} \in \mathcal{A}$, arbitrary. For any $t \in J$, we have that

$$
\begin{aligned}
\left|(T[\mathbf{y}])(t) - (T[\mathbf{z}])(t)\right|_\infty &= \left| \int_{t_0}^t \mathbf{f}(s, \mathbf{y}(s))\, ds - \int_{t_0}^t \mathbf{f}(s, \mathbf{z}(s))\, ds \right|_\infty \\
&\leq \left| \int_{t_0}^t \left| \mathbf{f}(s, \mathbf{y}(s)) - \mathbf{f}(s, \mathbf{z}(s)) \right|_\infty ds \right| \\
&\leq L \int_{t_0}^t \left| \mathbf{y}(s) - \mathbf{z}(s) \right|_\infty ds \\
&\leq L\,|t - t_0|\,\|\mathbf{y} - \mathbf{z}\| \\
&\leq L\delta\,\|\mathbf{y} - \mathbf{z}\|.
\end{aligned}
$$

The fact that the above is true for all $t \in J$ implies that

$$
\left\|T[\mathbf{y}] - T[\mathbf{z}]\right\| \leq L\delta\,\|\mathbf{y} - \mathbf{z}\|.
$$

Since $\mathbf{y}, \mathbf{z} \in \mathcal{A}$ were arbitrary, and $L\delta < 1$, this means that T is a contraction on $\mathcal{A}$. Since $\mathcal{A}$ is complete, the Banach Fixed-Point Theorem yields that the equation $T[\mathbf{x}] = \mathbf{x}$ has a unique solution $\mathbf{x} \in \mathcal{A}$. This proves the claimed result. $\qquad\square$

3.8   Exercises

Exercise 3.1. Show that any discrete metric space (X, d_0) is complete.

Exercise 3.2. Show that $(\mathbb{Z}, d)$ is a complete metric space, where $\mathbb{Z}$ is the set of all integers, and

$$d(m, n) = |m - n| \quad \text{for all } m, n \in \mathbb{Z}.$$

Exercise 3.3. Which of the following subspaces of either $\mathbb{R}$ or $\mathbb{R}^2$ with the usual metric are complete?

$$A := \{1/n : n \in \mathbb{N}\} \cup \{0\}, \ \ B := \mathbb{Q} \cap [0, 1], \ \ C := \left\{(x, y) \in \mathbb{R}^2 : x > 0, y \geq 1/x\right\}.$$

Justify your answers.

Exercise 3.4. Let $\rho : \mathbb{R} \times \mathbb{R} \longrightarrow \mathbb{R}$ be given by

$$\rho(x, y) = |\arctan x - \arctan y| \quad \text{for all } x, y \in \mathbb{R}.$$

Show that:

(i) ρ is a metric on $\mathbb{R}$;

(ii) The metric space (X, ρ) is not complete.

Exercise 3.5. Let $\mathbb{N}$ be the set of natural numbers and $d : \mathbb{N} \times \mathbb{N} \longrightarrow \mathbb{R}$ be given by

$$\rho(m, n) = \left|\frac{1}{m} - \frac{1}{n}\right| \quad \text{for all } m, n \in \mathbb{N}.$$

Show that:

(i) ρ is a metric on $\mathbb{N}$;

(ii) The metric space $(\mathbb{N}, \rho)$ is not complete.

Exercise 3.6. Let (X, d) be a metric space and let Y, Z be complete metric subspaces of X. Show that the metric subspace $Y \cup Z$ is complete.

Exercise 3.7. Let d and ρ be two Lipschitz equivalent metrics on a set X. Show that the metric space (X, d) is complete if and only if the metric space (X, ρ) is complete.

Exercise 3.8. Show that the set

$$Y := \left\{f \in C([-1, 1], \mathbb{R}) \ : \ f(-1) = f(1)\right\}$$

is a vector subspace of $C([-1, 1], \mathbb{R})$. Determine whether $(Y, \|\cdot\|_\infty)$ is a Banach space or not, carefully justifying your answer (where $\|\cdot\|_\infty$ is defined in the usual way on $C([-1, 1], \mathbb{R})$, and therefore also on Y).

Exercise 3.9. Consider the set

$$\mathbf{c_0} := \Big\{ x = (x_1, x_2, \ldots) \ : \ x_k \in \mathbb{F} \ \forall k \in \mathbb{N}, \ x_k \longrightarrow 0 \text{ in } \mathbb{F} \text{ as } k \to \infty \Big\},$$

with

$$\|x\|_\infty := \sup_{k \in \mathbb{N}} |x_k| \quad \text{for all } x = (x_1, x_2, \ldots) \in \mathbf{c_0}.$$

Show that $(\mathbf{c_0}, \|\cdot\|_\infty)$ is a Banach space.

[Hint: You may use without proof the fact that $(\ell^\infty, \|\cdot\|_\infty)$ is a Banach space.]

Exercise 3.10. Consider the set

$$\mathbf{c} := \Big\{ x = (x_1, x_2, \ldots) \ : \ x_k \in \mathbb{F} \ \forall k \in \mathbb{N}, \ \lim_{k \to \infty} x_k \text{ exists} \Big\}.$$

(i) Show that $\mathbf{c}$ is a vector space.

(ii) Let $\|\cdot\|_\infty : \mathbf{c} \longrightarrow \mathbb{R}$ be given by

$$\|x\|_\infty := \sup_{k \in \mathbb{N}} |x_k| \quad \text{for all } x = (x_1, x_2, \ldots) \in \mathbf{c}.$$

Show that $(\mathbf{c}, \|\cdot\|_\infty)$ is a Banach space.

[Hint: You may use without proof the fact that $(\ell^\infty, \|\cdot\|_\infty)$ is a Banach space.]

Exercise 3.11. Consider the space

$$Y := \Big\{ x = (x_1, x_2, \ldots) \in \ell^\infty \ : \ \exists K = K(x) \text{ such that } x_k = 0 \ \forall k > K \Big\}.$$

Show that $(Y, \|\cdot\|_\infty)$ is a normed space which is not a Banach space.

Exercise 3.12. Consider the space Y of Exercise 3.11. Give an example, in the normed space $(Y, \|\cdot\|_\infty)$, of a series which is absolutely convergent but not convergent. Justify your answer carefully.

Exercise 3.13. Let $(X, \|\cdot\|)$ be a normed space in which every absolutely convergent series is convergent. Show that $(X, \|\cdot\|)$ is a Banach space.

[Hint: Consider an arbitrary Cauchy sequence in X, construct from it an absolutely convergent series, and use the convergence of the series to deduce the convergence of a subsequence of the original Cauchy sequence.]

Exercise 3.14. Let (X, d) be a metric space, and let $(x_n)_1^\infty$ and $(y_n)_1^\infty$ be Cauchy sequences in X. Show that the sequence of real numbers $(d(x_n, y_n))_1^\infty$ converges in $\mathbb{R}$ with the usual metric.

Exercise 3.15. Let (X, d) be a metric space, and Y be a dense subset of X (meaning that $\overline{Y} = X$). Suppose that every Cauchy sequence in Y converges to a point in X. Show that (X, d) is complete.

[Hint: For any Cauchy sequence $(x_n)_1^\infty$ in X, it may be useful to consider a sequence $(y_n)_1^\infty$ in Y such that $d(x_n, y_n) < \frac{1}{n}$ for all $n \in \mathbb{N}$.]

Exercise 3.16. Let $a, b \in \mathbb{R}$ with $a < b$, and let $f : [a, b] \longrightarrow [a, b]$ be a function which is continuous on $[a, b]$, differentiable on (a, b), and with the property that there exists $\alpha \in (0, 1)$ such that $|f'(z)| \leq \alpha$ for all $z \in (a, b)$. Show that the equation $f(x) = x$ has a unique solution in $[a, b]$.

Chapter 4

Topological spaces and continuity

4.1 From metric to topological spaces and beyond

In Chapters 2 and 3 we introduced and studied the rudiments of *the theory of metric spaces*. As we have already expounded, metric spaces form a comprehensive framework which allows the generalisation of a substantial part of the concepts of Analysis, like convergence, limits and continuity, from Euclidean spaces to arbitrary sets having a *metric structure* that enables one to define "proximity" via distances.

Although this approach might seem sufficiently abstract and general, unfortunately, there still exist numerous limiting processes of great analytic interest which cannot be described by any metric structure! Such important instances will be presented in subsequent chapters of this book and will concern us in great depth and detail.

Therefore, we have to generalise the basic analytic notions even further, to a more general framework which does not depend on the existence of any underlying metric. It turns out that one can rely solely on a notion of "proximity" via a concept of open neighbourhoods, introduced axiomatically. This is the main goal of the present chapter, in which we present the axiomatic abstract framework of **topological spaces**.

The axioms defining a **topology** and the concepts associated to it (such as continuity), were developed as the result of an arduous and lengthy endeavour during the 19th and early 20th century to isolate and abstract the underlying ideas related to limiting processes. These concepts are ubiquitous in mathematics, so the definition of topology is designed to be applicable to a tremendous variety of different contexts. The toll is that the axioms may at first glance seem unmotivated, but nonetheless they have been carefully crafted to capture effectively the functionality of limiting processes in the most general possible setting.

However, nowadays the branch of Topology has expanded tremendously and is primarily concerned with the study various kinds of *"shapes of spaces"*, with a view towards perhaps classifying them (i.e. understand how many such exist), or understanding certain invariants or other aspects of their structure, like for instance how they behave under *continuous deformations* (see e.g.

Figure 4.1). These more advanced topics will not concern us in this elementary book.

The structure of this chapter parallels roughly that of Chapter 2, in that we study to what extent the properties proved there for metric spaces are valid in the setting of topological spaces as well, whilst also indicating a few additional properties that are specific to topological spaces.

Figure 4.1: An illustration of the continuous deformation of a coffee mug to a doughnut! Mathematically, a continuous deformation is called "homotopy". In a sense, the doughnut and the mug shapes are equivalent from the viewpoint of topology.

4.2   The endgame of abstraction of Analysis

In Chapter 2 we have seen that, in metric spaces, the notions of convergence of sequences and of continuity of mappings do not really depend on the precise way the distances between any two objects are measured, but rather on the open sets induced by those distances.

In particular:

- In Proposition 2.43 we proved that, for any metric space (X, d) and any $(x_n)_1^\infty \subseteq X$, $x_0 \in X$, we have $x_n \longrightarrow x_0$ as $n \to \infty$ if and only if for every open set $U \subseteq X$ with $U \ni x_0$, there exists $N \in \mathbb{N}$ such that $x_n \in U$ for all $n \geq N$.

- In Proposition 2.45 we proved that, for any metric spaces (X, d) and (Y, ρ) and any mapping $f : X \longrightarrow Y$ between them, the mapping f is continuous if and only if for any open set $V \subseteq Y$, the set $f^{-1}(V)$ is open in X.

Hence, *convergence and continuity can be formulated without any explicit mention to the distance*, but merely to the open sets the postulated distances generate on the spaces!

In addition, we have seen that the class of open sets on a metric space remains the same if we use an equivalent metric. For instance, for any $p \in [1, \infty]$, the metric d_p defines on the space $\mathbb{F}^m$ the same open sets, although the shape of the respective balls for these metrics may well be different (recall for instance Figure 2.5 in Chapter 2).

The above considerations suggest that **distance functions are auxiliary to the definition of convergence and continuity**, since we merely use a distance function to produce neighbourhoods and open sets. Hence, a more general theory of convergence and continuity could perhaps exist even in the absence of any distance function, provided we have at our disposal a certain family of sets, called **open sets**, that enjoy similar properties to those of the open sets induced by a metric distance.

But what should these axiomatic properties be, allowing for a "theory of convergence" to work?

To find them, we should have as our starting point those *properties of open sets of metric space which are independent of the metric structure per se*. The idea has already arisen in Proposition 2.38. Therein we proved that the open sets on a metric space (X, d) satisfy the following set of properties:

(i) The sets $\emptyset$ and X are always open.

(ii) If $\{U_i\}_{i \in I}$ is any family of open sets, then $\bigcup_{i \in I} U_i$ is open.

(iii) If $U_1, \ldots, U_n$ are open sets, then $\bigcap_{i=1}^n U_i$ is open.

Quite surprisingly, *the properties* (i)-(iii) *above can be taken as the axiomatic properties of the definition of a class of subsets on any set X* which we shall shortly call **topology** and its elements **open sets**. This is exactly what is achieved by the theory of **topological spaces**, which is based on this axiomatisation for the notion of open sets.

The definition of a topological space relies solely upon set theory and is the most general concept of a mathematical space allowing for analytical notions such as convergence and continuity to make sense. Topological spaces are a central unifying notion and appear in virtually every branch of modern mathematical Analysis.

4.3   Open sets and closed sets

Now we delve into our rigorous study of **topological spaces**, namely of sets in which a family of subsets is defined which possesses exactly the properties (i)-(iii) of Proposition 2.38, as we explained above. We stress that even though members of the family of will be called **open sets**, they are axiomatically defined and **we do not assume the existence of any metric on X that induces them!**

Our first main concept is therefore the following.

Definition 4.1. Let X be a set. A **topology** on X is a collection $\mathcal{T}$ of subsets of X satisfying:

(T1) $\emptyset \in \mathcal{T}$, $X \in \mathcal{T}$.

(T2) For any index set I, if $\{U_i\}_{i \in I}$ is any family of sets such that $U_i \in \mathcal{T}$ for all $i \in I$, then $\bigcup_{i \in I} U_i \in \mathcal{T}$.

(T3) For any $n \in \mathbb{N}$, if $U_1, \ldots, U_n$ are such that $U_i \in \mathcal{T}$ for all $i \in \{1, \ldots, n\}$, then $\bigcap_{i=1}^{n} U_i \in \mathcal{T}$.

A **topological space** is a pair $(X, \mathcal{T})$, where X is a set and $\mathcal{T}$ is a topology on X. The elements of $\mathcal{T}$ are called the **open sets** of the topology $\mathcal{T}$, or of the topological space $(X, \mathcal{T})$.

One of the most important classes of topological spaces is that provided by the **metric spaces**. Indeed, for any metric space (X, d), Proposition 2.38 exhibits that the **metric topology** $\mathcal{T}_d$ (introduced in Definition 2.47) is indeed a topology on X in the sense of Definition 4.1.

Throughout what follows, we shall employ the convention that **any metric space (X, d) will always be understood to be a topological space with the metric topology $\mathcal{T}_d$.**

It is natural to ask whether there exist topologies that do not correspond to any possible metric structures on the set in question. As we shall see shortly, the answer is positive. This justifies the following terminology.

Definition 4.2. A topological space $(X, \mathcal{T})$ is said to be **metrisable** if there exists a metric d on X such that $\mathcal{T} = \mathcal{T}_d$.

Note that, if a topological space is metrisable, then the same topology may be induced by more than one metrics: for example, on $\mathbb{F}^m$ the topology $\mathcal{T}_{d_2}$ is also induced by the metrics d_1 and d_∞. Right below we provide more examples of topologies on sets.

Example 4.3 (Discrete topology). For any set X, recall the definition of the Powerset and let us take

$$\mathcal{T} := \mathcal{P}(X).$$

Thus, in this topology, every subset of X is declared to be an open set. It is immediate to verify that $(X, \mathcal{P}(X))$ satisfies $(T1) - (T3)$ of Definition 4.1. This topology is metrisable, being induced by the discrete metric d_0 on X. It is also "maximal", in the sense that it is the largest topology one can define on a set X.

Example 4.4 (Indiscrete topology). For any set X, let us take

$$\mathcal{T} := \{\emptyset, X\}.$$

In this topology, only $\emptyset$ and X are declared to be open sets. It is immediate to verify that $(X, \{\emptyset, X\})$ satisfies $(T1) - (T3)$ of Definition 4.1.

This topology is **not metrisable** if X contains at least two points. Indeed, in any metric space there exists an open ball which is neither empty nor the entire space, and thus the topology induced by that metric cannot coincide with the indiscrete topology. More precisely, let (X, d) be a metric space. If X contains at least two elements x, y with $x \neq y$ then, if we define $r := d(x, y)$, we have that $x \in \mathbb{B}_r(x)$ and $y \notin \mathbb{B}_r(x)$, so that $\mathbb{B}_r(x)$ is an open set which is neither empty nor equal to X. Hence, the topology $\{\emptyset, X\}$ cannot arise from *any* metric d. This topology is also "minimal", in the sense that it is the topology with the fewest possible elements one can define on a set X.

Example 4.5 (Finite complement topology). For any set X, let

$$\mathcal{T} := \{\emptyset\} \bigcup \{U \subseteq X \,:\, X \setminus U \text{ is finite}\}.$$

Thus, a subset U of X is declared to be open if either it is empty or $X \setminus U$ is a finite set. One can easily check that this collection of subsets satisfies indeed $(T1) - (T3)$ of Definition 4.1.

Example 4.6. Consider a set with three elements $X = \{a, b, c\}$, where $a \neq b \neq c \neq a$ and take $\mathcal{T} := \{\emptyset, \{a\}, \{a, b, c\}\}$. Then, it can be verified that $\mathcal{T}$ is a topology on X.[1]

Let us point out that most examples of non-metrisable topological spaces which are of genuine significance in Analysis are too complicated to be presented right now. These non-metrisable topologies will be defined at the end of this chapter and will concern us largely in Chapter 12 and onwards.

The following notion of **neighbourhood** is an appropriate extension to that used in the metric space setting.

Definition 4.7. Let $(X, \mathcal{T})$ be a topological space and $x_0 \in X$. A subset U of X is said to be a **neighbourhood** of x_0 if U is an open set in (X, τ) that contains x_0.

The following characterisation of open sets in topological spaces turns out to be very useful.

Proposition 4.8. *Let $(X, \mathcal{T})$ be a topological space. A subset U of X is open in $(X, \mathcal{T})$ if and only if it includes a neighbourhood of any of its points. The latter statement means that for any $x \in X$, there exists $U_x \in \mathcal{T}$ with $U_x \ni x$ such that $U_x \subseteq U$.*

Proof. Suppose first that $U \in \mathcal{T}$. Then for any $x \in U$, we may choose $U_x := U$, which obviously has all the required properties.

Conversely, suppose that $U \subseteq X$ has the property that for any $x \in X$ there exists $U_x \in \mathcal{T}$ with $U_x \ni x$ such that $U_x \subseteq U$. Then one may easily check that U can be written as

$$U = \bigcup_{x \in U} U_x$$

---

[1]Unlike the previous examples, this particular one has no special significance in other contexts. It is provided merely to illustrate the fact that is very easy to construct topologies on spaces!

and therefore $U \in \mathcal{T}$ as a consequence of axiom $(T2)$.   □

Now we introduce the notion of closed sets in a topological space as those which are complements of open sets.

Definition 4.9. A subset A of a topological space $(X, \mathcal{T})$ is said to be **closed** in $(X, \mathcal{T})$ if its complement, $X \setminus A$, is open in $(X, \mathcal{T})$.

Closed sets in a topological space have the following properties which in a sense are complementary to those of the open sets.

Proposition 4.10. *Let $(X, \mathcal{T})$ be a topological space. Then the following hold:*

$(\widetilde{T1})$ *$\emptyset$ and X are closed.*

$(\widetilde{T2})$ *For any index set I, if $\{F_i\}_{i \in I}$ is any family of closed subsets of X, then $\bigcap_{i \in I} F_i$ is closed.*

$(\widetilde{T3})$ *For any $n \in \mathbb{N}$, if $F_1, \ldots, F_n$ are closed subsets of X, then $\bigcup_{i=1}^{n} F_i$ is closed.*

Proof. We only prove $(\widetilde{T2})$, the other claims being proved in a similar way. Let $\{F_i\}_{i \in I}$ be a family of closed subsets of X. Then $\{X \setminus F_i\}_{i \in I}$ is a family of open sets. By $(T2)$, $\bigcup_{i \in I} (X \setminus F_i)$ is open. However, by the De Morgan laws it follows that

$$\bigcup_{i \in I} (X \setminus F_i) = X \setminus \bigcap_{i \in I} F_i,$$

and hence $\bigcap_{i \in I} F_i$ is closed.   □

We now revisit the topologies of the examples we considered earlier and discuss their corresponding closed sets:

Example 4.11.

(i) In any metric space (X, d), any finite subset of X is closed.

(ii) In any discrete topological space, all subsets are closed.

(iii) In any indiscrete topological space, only $\emptyset$ and X are closed.

(iv) In the Finite Complement Topology, the closed sets are exactly the finite sets and the space itself. This is so because $U \subseteq X$ is open in this topology if and only if $U = \emptyset$ or $X \setminus U$ is finite.

The next example demonstrates a fundamental fact already noticed in the setting of metric spaces. Namely, that in $(\widetilde{T3})$, an infinite union of closed sets need not be closed.[2]

---

[2]By taking complement, one easily sees that, in regard to the respective hypothesis $(T3)$, an arbitrary intersection of open sets need not be open.

Remark 4.12. Let us consider $\mathbb{R}$ with its usual metric topology. Then, the subsets $[1/n, 1]$ are closed for all $n \in \mathbb{N}$. However, their union equals

$$\bigcup_{n=1}^{\infty} \left[\frac{1}{n}, 1 \right] = (0, 1]$$

and therefore the countable union is not a closed set.

The subsets of a topological space which have the form of **countable unions of closed sets** or of **countable intersections of open sets** (despite not being either closed or open in general), appear very frequently and have been given the next particular respective names: F_σ-**set** and G_δ-**set**.

4.4 Topological subspaces

In this brief section we are concerned with the issue of assigning a natural topology to a subset of a topological space that would rightly allows us to call it a "subspace". To this end, recall that for a subset Y of a metric space (X, d), the open sets of the metric subspace (Y, d) are given by $V \cap Y$, where V is open in X. This suggests a natural generalisation to topological spaces.

Definition 4.13. Let $(X, \mathcal{T})$ be a topological space, and Y be a subset of X. We consider the class of subsets

$$\mathcal{T}|_Y := \left\{ V \cap Y : V \in \mathcal{T} \right\}.$$

Then, $\mathcal{T}|_Y$ is a topology on Y (as one can easily check), called the **induced topology on Y** or the **subspace topology of Y**. The pair $(Y, \mathcal{T}|_Y)$ is called a **topological subspace** of $(X, \mathcal{T})$.

The closed sets of a topological subspace have a very simple characterisation which allows one to express them as intersections of the subspace with closed sets of the original space.

Proposition 4.14. *Let $(X, \mathcal{T})$ be a topological space and Y be a subset of X. Then the closed sets of Y in the subspace topology are exactly those of the form $F \cap Y$, where F is closed in X.*

Proof. Suppose first that F is an arbitrary closed subset of X. Then $X \setminus F$ is open in X, so that $(X \setminus F) \cap Y$ is open in Y. But then, note that

$$Y \setminus (F \cap Y) = Y \setminus F = (X \setminus F) \cap Y,$$

which shows that $Y \setminus (F \cap Y)$ is open in Y. Hence $F \cap Y$ is a closed subset of Y in the subspace topology.

Suppose now that G is an arbitrary closed subset of Y in the subspace topology. Then $Y \setminus G$ is open in Y and hence it has the form $V \cap Y$, for some V open in X. But then, note that

$$G = Y \setminus (Y \setminus G) = Y \setminus (V \cap Y) = Y \setminus V = (X \setminus V) \cap Y,$$

which shows that $G = F \cap Y$, where $F := X \setminus V$ is closed in X, as claimed. □

4.5   Convergence and continuity

In this section we discuss the notion of convergence of sequences in general topological spaces, as well as the concept of continuity of a mapping between two topological spaces together with their main properties. Both these concepts are defined in a natural way, by taking as our definition a property which is satisfied in metric spaces, see Propositions 2.43 and 2.45 recalled at the beginning of this chapter.

We begin with the notion of sequential convergence in topological spaces.

Definition 4.15. Let $(X, \mathcal{T})$ be a topological space. We say that a sequence $(x_n)_1^\infty \subseteq X$ **converges** to $x_0 \in X$ in $(X, \mathcal{T})$ and write

$$x_n \longrightarrow x_0 \text{ in } (X, \mathcal{T}) \text{ as } n \to \infty,$$

if for any $U \in \mathcal{T}$ with $U \ni x_0$, there exists $N = N(U) \in \mathbb{N}$ such that $x_n \in U$ for all $n \geq N$.

Although the notion of Definition 4.15 seems natural, there are many subtleties associated with the (very abstract) structure of topological spaces which are not present in the metric space realm. Below we present a rather striking one: unlike the case of metric spaces, *in topological spaces the limit of a convergent sequence need not always be unique!*

Example 4.16 (Non-uniqueness of limits). Let $\mathcal{T} := \{\emptyset, X\}$ be the indiscrete topology on a given set X. In this space *every sequence converges to every point in the space!* To see this, let $(x_n)_1^\infty \subseteq X$ and $x_0 \in X$ be arbitrary. Let U be open and containing x_0. Then $U = X$ (since $U \neq \emptyset$). But then $x_n \in U$ for all $n \in \mathbb{N}$. Hence indeed $x_n \longrightarrow x_0$ in $(X, \mathcal{T})$ as $n \to \infty$.

Nonetheless, uniqueness of the limit of a convergent sequence is guaranteed if one introduces an additional (mild) hypothesis as follows.

Definition 4.17. A topological space $(X, \mathcal{T})$ is said to be a **Hausdorff space** if for every $x, y \in X$ with $x \neq y$ there exist disjoint open sets U_x, U_y such that $x \in U_x$, $y \in U_y$, and $U_x \cap U_y = \emptyset$ (see Figure 4.2). This property is called the **Hausdorff axiom**, or **Hausdorff property**.

The alternative terminology "$\mathcal{T}_2$" is a common topological synonym for the Hausdorff property and *should not be confused with the property* (T2) *of Definition 4.1.*

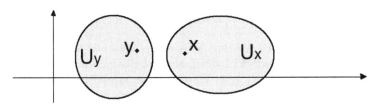

Figure 4.2: An illustration of the Hausdorff property on $\mathbb{R}^2$.

As it might have been already suspected, every metric space (X, d) is a Hausdorff space when considered with the metric topology:

Lemma 4.18. *Let* (X, d) *be a metric space. Then, the topological space* $(X, \mathcal{T}_d)$ *satisfies the Hausdorff property.*

Proof. Given any two distinct points x, y in X, let $r \in \mathbb{R}$ be such that $r \in (0, d(x, y)/2)$. Then $\mathbb{B}_r(x)$ and $\mathbb{B}_r(y)$ are open sets with $\mathbb{B}_r(x) \ni x$ and $\mathbb{B}_r(y) \ni y$. Also, they are by construction disjoint: $\mathbb{B}_r(x) \cap \mathbb{B}_r(y) = \emptyset$. □

The Hausdorff axiom is sufficient to ensure the uniqueness of limits of convergent sequences.

Proposition 4.19 (Uniqueness of limits for Hausdorff spaces). *Let* $(X, \mathcal{T})$ *be a Hausdorff space and* $(x_n)_1^\infty$ *a sequence in* X. *If* $x_n \longrightarrow x_0$ *and* $x_n \longrightarrow \tilde{x}_0$ *in* $(X, \mathcal{T})$ *as* $n \to \infty$ *for some* $x_0, \tilde{x}_0 \in X$, *then* $x_0 = \tilde{x}_0$.

Proof. Suppose for a contradiction that $x_0 \neq \tilde{x}_0$. Since $(X, \mathcal{T})$ is by assumption a Hausdorff space, there exist disjoint open sets U, V with $x_0 \in U$ and $\tilde{x}_0 \in V$. Since $x_n \longrightarrow x_0$ in $(X, \mathcal{T})$ as $n \to \infty$, it follows that there exists $N_1 \in N$ such that $x_n \in U$ for all $n \geq N_1$. Similarly, since $x_n \longrightarrow \tilde{x}_0$ in $(X, \mathcal{T})$ as $n \to \infty$, it follows that there exists $N_2 \in \mathbb{N}$ such that $x_n \in V$ for all $n \geq N_2$. We deduce that

$$x_n \in U \cap V, \text{ for all } n \geq \max\{N_1, N_2\},$$

which contradicts the fact that $U \cap V = \emptyset$. This contradiction yields that $x_0 = \tilde{x}_0$. □

Even though the majority of topological spaces of interest do satisfy the Hausdorff axiom, its validity is not relevant for most of the results that follow. **Unless explicitly stated otherwise, we shall NOT assume that the topological spaces under consideration satisfy the Hausdorff axiom.** For instance, it can be checked that the indiscrete topology is non-Hausdorff. Right next is a simple more interesting example on $\mathbb{R}^2$ showing how things might be in a non-Hausdorff topology:

Example 4.20 (A non-Hausdorff topology). On $\mathbb{R}^2$, consider the class of sets given by the empty set together with $\mathbb{R}^2$ itself and all the half-spaces emanating from any vertical line with direction towards $+\infty$ along the horizontal axis:

$$\mathcal{T} := \{\emptyset\} \bigcup \left\{ (a, \infty) \times \mathbb{R} : a \in \mathbb{R} \cup \{-\infty\} \right\}.$$

Then, it is very easy to see that $\mathcal{T}$ is a topology and satisfies $(T1) - (T3)$. However, $\mathcal{T}$ is not Hausdorff! Indeed, any two distinct points $x = (x_1, x_2), y = (y_1, y_2) \in \mathbb{R}^2$ with, say, $x_1 < y_1$ cannot be separated by two open sets. The only open neighbourhoods U_x, U_y of x, y which are not $\mathbb{R}^2$ itself are sets of the form

$$U_x = (x_1 - \varepsilon, \infty) \times \mathbb{R} , \quad U_y = (y_1 - \delta, \infty) \times \mathbb{R}, \quad \text{for some } \varepsilon, \delta > 0$$

and evidently $U_x \cap U_y \neq \emptyset$, regardless of how small we may choose $\varepsilon, \delta > 0$ since U_x, U_y always overlap on the half-space $(y_1, \infty) \times \mathbb{R} \neq \emptyset$ (see Figure 4.3). In fact, $U_y \subseteq U_x$!

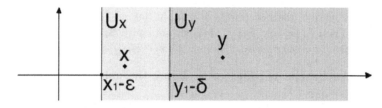

Figure 4.3: An illustration of the non-Hausdorff topology on $\mathbb{R}^2$ of Example 4.20.

Now we turn our attention to the concept of continuity of a mapping between two topological spaces. This is defined again by axiomatising the distance-free restatement of continuity in metric spaces, as given in Proposition 2.44.

Definition 4.21. Let $(X, \mathcal{T})$ and $(Y, \mathcal{S})$ be topological spaces, $f : X \longrightarrow Y$ a mapping between them and $x_0 \in X$.

(i) The mapping f is said to be **continuous at** x_0 if for any open $V \in \mathcal{S}$ with $V \ni f(x_0)$, there exists an open $U \in \mathcal{T}$ with $U \ni x_0$ such that

$$f^{-1}(V) \supseteq U.$$

(ii) The mapping f is said to be **continuous on** X if it is continuous at every point $x_0 \in X$.

Remark 4.22. We note that, in view of the identities $(\star)$ and $(\star\star)$ of Chapter 1 that the image map and the inverse image map satisfy, the condition $f^{-1}(V) \supseteq U$ of Definition 4.21(i) above can equivalently be restated as

$$f(U) \subseteq V.$$

As in the case of metric spaces, global continuity on X of a mapping $f : X \longrightarrow Y$ admits a characterisation using either open sets, or closed sets.

Proposition 4.23. *Let* $(X, \mathcal{T})$, $(Y, \mathcal{S})$ *be topological spaces and* $f : X \longrightarrow Y$ *a mapping between them. Then the following statements are equivalent:*

(i) *f is continuous on X;*

(ii) *$f^{-1}(V)$ is an open set in $(X, \mathcal{T})$ for any open set V in $(Y, \mathcal{S})$;*

(iii) *$f^{-1}(F)$ is a closed set in $(X, \mathcal{T})$ for any closed set F in $(Y, \mathcal{S})$.*

Proof. (i) $\Longrightarrow$ (ii): Suppose first that (i) holds. Let V be an arbitrary open set in $(Y, \mathcal{S})$. Let $x_0 \in f^{-1}(V)$ be arbitrary. Then, since f is continuous at x_0, there exists $U_{x_0} \in \mathcal{T}$ with $U_{x_0} \ni x_0$ such that $f(U_{x_0}) \subseteq V$. The last condition is equivalent to $U_{x_0} \subseteq f^{-1}(V)$. The fact that $f^{-1}(V)$ is open in X now follows from Proposition 4.8.

(ii) $\Longrightarrow$ (i): Suppose now that (ii) holds. Let $x_0 \in X$, arbitrary. We prove that f is continuous at x_0. Let $V \in \mathcal{S}$ with $V \ni f(x_0)$, arbitrary. Let $U := f^{-1}(V)$. Then $U \in \mathcal{T}$ by assumption, and also, obviously, $U \ni x_0$ and $f(U) \subseteq V$. It follows that f is continuous at x_0. Since $x_0 \in X$ was arbitrary, it follows that f is continuous on X, as required.

The proof of the equivalence of the statements (ii) and (iii) is identical with the corresponding one given in Proposition 2.45. □

Remark 4.24. We would like to underline that the continuity of a mapping $f : X \longrightarrow Y$ between two topological spaces is a property depending on the topologies that the spaces are equipped. In particular, *a mapping $f : X \longrightarrow Y$ between two sets X, Y might be continuous for certain topologies $\mathcal{T}_X, \mathcal{T}_Y$ on X, Y, but not for other topologies*, although as a mapping between sets, f remains unchanged! The next example demonstrates this in a rather striking fashion:

Example 4.25 (Dependence of continuity on the topologies)**.** Let $X \neq \emptyset$ be a non-empty set and consider the identity mapping from X to itself:

$$\text{id} : \quad X \longrightarrow X, \quad \text{id}(x) := x.$$

Then, for any topology $\mathcal{T}$ on X, id is continuous when considered as a mapping

$$\text{id} : \quad (X, \mathcal{T}) \longrightarrow (X, \mathcal{T}),$$

namely when both the target space and the domain space are endowed with the same topology. This is immediate since for any open set $U \in \mathcal{T}$, the set $(\text{id})^{-1}(U) = U \in \mathcal{T}$ is also open.

However, the identity map id is (generally) **discontinuous** when the target space and the domain space are equipped with different topologies. This is always the case when the topology of the domain space has less open sets than the topology of the target space. For instance, when

$$\text{id} \ : \ (X, \mathcal{T}_{\text{indiscrete}}) \longrightarrow (X, \mathcal{T}_{\text{discrete}}).$$

Indeed, recall that $\mathcal{T}_{\text{discrete}} = \mathcal{P}(X)$, whilst $\mathcal{T}_{\text{indiscrete}} = \{\emptyset, X\}$. Hence, for any open set $U \in \mathcal{P}(X)$ which is neither the empty set nor X itself, the set $(\text{id})^{-1}(U) = U$ is not open for $\mathcal{T}_{\text{indiscrete}}$, since its only elements are $\emptyset, X$!

We have seen that in metric spaces, continuity of a function may be characterised using convergent sequences. An analogous results is **not** valid in full generality in topological spaces, but we can prove a closely related result if one makes the following extra assumption.[3]

Definition 4.26. A topological space $(X, \mathcal{T})$ is said to satisfy **the first axiom of countability** if, for every $x_0 \in X$, there exists a sequence $(U_n)_1^{\infty} \subseteq \mathcal{T}$ such that $U_n \ni x_0$ for every $n \in \mathbb{N}$,

$$U_1 \supseteq U_2 \supseteq \cdots \supseteq U_n \supseteq \cdots ,$$

and for every $U \in \mathcal{T}$ with $U \ni x_0$ there exists $n \in \mathbb{N}$ such that $U_n \subseteq U$.

Remark 4.27 (Metric spaces are first countable). Every metric space (X, d), considered as a topological space with the metric topology, satisfies the first axiom of countability. Indeed, to see this consider for any point $x_0 \in X$ the sequence of open balls $(U_n)_1^{\infty}$ with $U_n := \mathbb{B}_{1/n}(x_0)$. Then, we have that $U_1 \supseteq U_2 \supseteq \cdots \supseteq \{x_0\}$ and for any open set $U \subseteq X$ with $x_0 \in U$, there exists a large enough $n \in \mathbb{N}$ such that $U_n \subseteq U$.

The result attests that continuity of mapping can be characterised in terms if sequential convergence for those topological spaces which satisfy the first countability axiom.

Theorem 4.28 (Sequential characterisation of continuity). *Let $(X, \mathcal{T})$, $(Y, \mathcal{S})$ be topological spaces. Let also $f : X \longrightarrow Y$ be a mapping and $x_0 \in X$. Consider the following statements:*

(i) *The function f is continuous at x_0.*

(ii) *For any sequence $(x_n)_1^{\infty} \subseteq X$ for which $x_n \longrightarrow x_0$ in $(X, \mathcal{T})$ as $n \to \infty$, we have*

$$f(x_n) \longrightarrow f(x_0) \text{ in } (Y, \mathcal{S}) \text{ as } n \to \infty.$$

---

[3]Since *the standard notion of sequences is not really adequate for the description of convergence in non-metrisable topological spaces,* to this end more advanced tools have been developed which generalise it. One of them, that of so-called **nets**, is obtained by replacing the countable index set of $(x_n)_1^{\infty}$ by a more general ordered set I, that is $(x_i)_{i \in I}$. Except for some subtleties of the type that subnets of nets are not subsets in the way that subsequences are subsets of sequences, nets formally perform in a similar fashion to that of sequences. However, in the present quite elementary approach to Analysis, every possible effort has been made to avoid the use of nets because they frequently cause difficulties to the less experienced folks.

Then, (i) *implies* (ii). *If, in addition,* $(X, \mathcal{T})$ *satisfies the first axiom of countability, then* (ii) *implies* (i).

Note that *only the domain space* X (and not the target space Y) is assumed to satisfy the countability hypothesis.

Proof. (i) $\Longrightarrow$ (ii): Suppose that (i) holds. Let $(x_n)_1^\infty \subseteq X$ be an arbitrary sequence such that $x_n \longrightarrow x_0$ in $(X, \mathcal{T})$. We aim at showing that

$$f(x_n) \longrightarrow f(x_0) \text{ in } (Y, \mathcal{S}) \text{ as } n \to \infty.$$

Let $V \in \mathcal{S}$ with $V \ni f(x_0)$, arbitrary. Since f is continuous at x_0, there exists $U \in \mathcal{T}$ with $U \ni x_0$ such that $f(U) \subseteq V$. Since $x_n \longrightarrow x_0$ in $(X, \mathcal{T})$ as $n \to \infty$, there exists $N \in \mathbb{N}$ such that $x_n \in U$ for all $n \geq N$. But then $f(x_n) \in f(U) \subseteq V$ for all $n \geq N$. This shows $f(x_n) \longrightarrow f(x_0)$ in $(Y, \mathcal{S})$ as $n \to \infty$.

(ii) $\Longrightarrow$ (i): Let $(X, \mathcal{T})$ satisfy the first axiom of countability, and suppose that (ii) holds. We suppose for the sake of contradiction that f is not continuous at x_0. Then, there exists $V_0 \in \mathcal{S}$ with $V_0 \ni f(x_0)$, such that for every $U \in \mathcal{T}$ with $U \ni x_0$, $f(U)$ is not a subset of V. In particular, by taking $U := U_n$ for each $n \in \mathbb{N}$ (where $(U_n)_1^\infty \subseteq \mathcal{T}$ is the sequence of open sets in the definition of the first axiom of countability), we obtain the existence of $x_n \in U_n$ such that $f(x_n) \notin V_0$. The fact that $f(x_n) \notin V_0$ for any $n \in \mathbb{N}$ implies that

$$f(x_n) \nrightarrow f(x_0) \text{ in } (Y, \mathcal{S}) \text{ as } n \to \infty.$$

On the other hand, we claim that $x_n \longrightarrow x_0$ in $(X, \mathcal{T})$. Indeed, let $U \in \mathcal{T}$ with $U \ni x_0$, arbitrary. Let $N \in \mathbb{N}$ be such that $U_N \subseteq U$. Then, for any $n \geq N$, we have that $x_n \in U_n \subseteq U_N \subseteq U$, so that $x_n \in U$. This shows that

$$x_n \longrightarrow x_0 \text{ in } (X, \mathcal{T}) \text{ as } n \to \infty.$$

We have thus obtained a contradiction to (ii). As a consequence, it follows that f is indeed continuous at x_0. $\square$

The next two results show that continuous mappings behave as expected with respect to compositions and, in the event that our target space is a normed vector space, the algebraic operations of addition and scalar multiplication. We begin with the case of compositions.

Proposition 4.29 (Composition of continuous mappings). *Let* $(X, \mathcal{T})$, $(Y, \mathcal{S})$ *and* $(Z, \mathcal{R})$ *be topological spaces. Let also* $f : X \longrightarrow Y$ *and* $g : Y \longrightarrow Z$ *be mappings. If* f *is continuous at the point* $x_0 \in X$ *and* g *is continuous at the point* $f(x_0) \in Y$, *then* $g \circ f : X \longrightarrow Z$ *is continuous at* x_0.

Proof. Let $W \in \mathcal{R}$ be an arbitrary set such that $W \ni g(f(x_0))$. Since g is continuous at $f(x_0)$, there exist $V \in \mathcal{S}$ with $V \ni f(x_0)$ such that $g(V) \subseteq W$. Also, since f is continuous at x_0, there exists $U \in \mathcal{T}$ with $U \ni x_0$ such that $f(U) \subseteq V$. Hence, for every $x \in U$, we have that $f(x) \in V$ and $g(f(x)) \in W$. Hence, $(g \circ f)(U) \subseteq W$. Therefore, $g \circ f$ is continuous at x_0. $\square$

The significance of the above result lies in that it shows that continuity is indeed a "topological" property,[4] independent of the possible presence of additional other structures (like metrics, norms, etc.) that possibly induce the topologies.

In relation to the above remarks, the following topological concepts are very important.

Definition 4.30. Let $(X, \mathcal{T})$, $(Y, \mathcal{S})$ be topological spaces and let also $f : X \longrightarrow Y$ be a continuous and injective (i.e., "$1 - 1$") mapping between them. Consider the image set $f(X) \subseteq Y$ as a topological subspace of Y with the induced topology.

(i) If the inverse mapping $f^{-1} : f(X) \longrightarrow X$ is also continuous, then f is called an **embedding** (or a **topological embedding**) of X into Y.

(ii) If f is surjective (i.e., onto) and the inverse mapping $f^{-1} : Y \longrightarrow X$ is also continuous, then f is called an **homeomorphism** between X and Y. In such an event, the topological spaces X, Y are called **homeomorphic** to each other.

Remark 4.31 (Identifying homeomorphic spaces).

(i) It is immediate from the definition above that if a mapping $f : X \longrightarrow Y$ is an homeomorphism, then its inverse $f^{-1} : Y \longrightarrow X$ is also an homeomorphism. Therefore, homeomorphisms allow us to identify topological spaces. Namely, two spaces which are **homeomorphic** are indistinguishable from the viewpoint of topology, although they might be very different otherwise![5] Hence, **the homeomorphisms are the "isomorphisms" of topology,** namely, the topology-preserving mappings.

(ii) Embeddings allow to identify spaces with sets contained in other spaces. This might be particularly useful if we need to realise a complicated space as a subspace of a more familiar one.

Example 4.32. Let $\mathbb{R}, \mathbb{R}^2$ be considered with their usual topologies.

(i) Consider the function $f : \mathbb{R} \longrightarrow \mathbb{R}$ given by (see Figure 4.4)

$$f(x) := \frac{2}{\pi} \arctan(x), \quad x \in \mathbb{R}.$$

---

[4]In Algebra, the fact that continuity is preserved under compositions is referred to as that continuous mappings are "morphism" between topological spaces. The terminology 'morphism" means that it "preserves the structure" of the space. For topological spaces, this structure is the topology. The seemingly odd synthetical "morphi-" is a transliteration of the Greek word "μορφή" which means "form" or "shape".

[5]At this point is might be useful to recall a similar occurrence in Linear Algebra. For example, the vector space $\mathbb{R}^2$ is linearly isomorphic to the vector space $V := \{(t \cos + s \sin) : t, s, \in \mathbb{R}\}$ and hence indistinguishable from the viewpoint of their linear structure, even though they contain of a different nature! $\mathbb{R}^2$ contains pairs of numbers, whilst V contains maps!

Then, f is an embedding of $\mathbb{R}$ into $\mathbb{R}$ with $f(\mathbb{R}) = (-1, 1)$, when $(-1, 1)$ is considered with the subspace topology. In addition, it is an homeomorphism of $\mathbb{R}$ with the interval $(-1, 1)$.

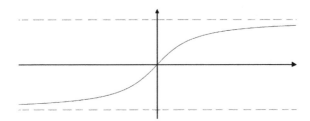

Figure 4.4: The graph of the function f embedding $\mathbb{R}$ into $(-1, 1)$.

(ii) Consider the mapping $F : \mathbb{R} \longrightarrow \mathbb{R}^2$ given by

$$F(x) := \big(\cos(2\pi x), \sin(2\pi x) \big), \quad x \in \mathbb{R}.$$

Then, the following are true (see Figure 4.5):

(a) The restriction $F|_{(0,1)}$ is an embedding of the interval $(0, 1)$ into $\mathbb{R}^2$ and its image $F((0, 1))$ equals to the punctured circle $\partial \mathbb{B}_1(0) \setminus \{(1, 0)\}$ of $\mathbb{R}^2$, that is

$$F((0, 1)) = \Big\{ (s, t) \in \mathbb{R}^2 : s^2 + t^2 = 1, \ (s, t) \neq (1, 0) \Big\}.$$

In addition, F is an homeomorphism of $(0, 1)$ (with the induced topology from $\mathbb{R}$) with $F((0, 1))$ (with the induced topology from $\mathbb{R}^2$).

(b) The restriction $F|_{(0,1]}$ is **not** an embedding of $(0, 1]$ into the circle $F((0, 1]) = \partial \mathbb{B}_1(0)$, although the map $F|_{(0,1]}$ is continuous and bijective, because the inverse is not continuous.

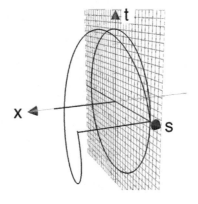

Figure 4.5: The graph of the mapping F embedding $(0, 1)$ into the circle.

Next we discuss algebraic properties of continuous mappings, in the case that the target space has the structure of either a normed vector space or of the field of scalars ($\mathbb{R}$ or $\mathbb{C}$).

Proposition 4.33 (Algebra of Continuous Functions). *Let $(X, \mathcal{T})$ be a topological space and $(Y, \|\cdot\|)$ a normed space. Let $f, g : X \longrightarrow Y$ be mappings and let also $\alpha \in \mathbb{F}$ and $x_0 \in X$. Then the following statements hold:*

(i) *If f and g are continuous at x_0, then $f + g$ is continuous at x_0.*

(ii) *If f is continuous at x_0, then αf is continuous at x_0.*

(iii) *If $Y = \mathbb{F}$ and both f and g are continuous at x_0, then fg is continuous at x_0.*

(iv) *If $Y = \mathbb{F}$, g is continuous at x_0 and $g(x_0) \neq 0$, then there exists $U \in \mathcal{T}$ with $U \ni x_0$ such that $g(x) \neq 0$ for all $x \in U$, and $1/g : U \longrightarrow \mathbb{F}$ is continuous at x_0.*

Proof. We only prove (i), the proof of the other parts being entirely similar. Let V be an arbitrary open set in Y with $V \ni (f(x_0) + g(x_0))$. Then there exists $\varepsilon > 0$ such that $\mathbb{B}_\varepsilon(f(x_0) + g(x_0)) \subseteq V$. We shall show that there exists $U \in \mathcal{T}$ with $U \ni x_0$ such that

$$(f + g)(U) \subseteq \mathbb{B}_\varepsilon(f(x_0) + g(x_0)) \subseteq V.$$

Since f is continuous at x_0, there exists $U_1 \in \mathcal{T}$ with $U \ni x_0$ such that

$$\|f(x) - f(x_0)\| < \frac{\varepsilon}{2} \quad \text{for all } x \in U_1.$$

Similarly, since g is continuous at x_0, there exists $U_2 \in \mathcal{T}$ with $U_2 \ni x_0$ such that

$$\|g(x) - g(x_0)\| < \frac{\varepsilon}{2} \quad \text{for all } x \in U_2.$$

Let $U := U_1 \bigcap U_2$. Then $U \in \mathcal{T}$ by $(T3)$, and $U \ni x_0$. Also, for all $x \in U$ we have

$$\left\|\big(f(x) + g(x)\big) - \big(f(x_0) + g(x_0)\big)\right\| \leq \|f(x) - f(x_0)\| + \|g(x) - g(x_0)\|$$
$$< \frac{\varepsilon}{2} + \frac{\varepsilon}{2} = \varepsilon.$$

This means that

$$(f + g)(U) \subseteq \mathbb{B}_\varepsilon(f(x_0) + g(x_0)) \subseteq V,$$

as required. Thus, $f + g$ is continuous at x_0. $\qquad\square$

4.6 Interior, closure and boundary of a set

We now indicate the appropriate generalisations to the setting of topological spaces of the notions of **interior**, **closure** and **boundary**.

Definition 4.34. Let $(X, \mathcal{T})$ be a topological space and A be a subset of X.

(i) A point $x \in X$ is said to be an **interior point** of A if there exists $U \in \mathcal{T}$ with $U \ni x$ such that $U \subseteq A$. The set of all interior points of A is called the **interior of** A, and is denoted by A°:

$$A^\circ := \left\{ x \in A \,\middle|\, \exists\, U \in \mathcal{T} \text{ with } U \ni x \,:\, U \subseteq A \right\}.$$

(ii) A point $x \in X$ is said to be a **closure point** of A if, for every $U \in \mathcal{T}$ with $U \ni x$, one has that $U \cap A \neq \emptyset$. The set of all closure points of A is called the **closure of** A, and is denoted by $\overline{A}$:

$$\overline{A} := \left\{ x \in X \,\middle|\, \forall\, U \in \mathcal{T} \text{ with } U \ni x \,:\, U \cap A \neq \emptyset \right\}.$$

(iii) The **boundary of** A, denoted by ∂A, is defined to be the set

$$\partial A := \overline{A} \setminus A^\circ.$$

The elements of the boundary of A are called **boundary points** of A.

In the case when the topology $\mathcal{T}$ on X is induced by a metric d, one has to verify that the notions of **interior point**, **interior of a set**, **closure point**, **closure of a set** as defined in above agree with those for metric spaces as given in Definition 2.60. This is indeed the case and we leave the verification of this fact as an exercise for the reader. It is obvious for the definitions that, for every subset A of a topological space $(X, \mathcal{T})$, one has the inclusions

$$A^\circ \subseteq A \subseteq \overline{A}.$$

The interior and the closure of a set have the following basic properties.

Proposition 4.35 (Properties of interior and closure). *Let $(X, \mathcal{T})$ be a topological space and A be a subset of X. Then the following hold:*

(i) A° *is open and for any open set $V \subseteq X$ we have that*

$$V \subseteq A \implies V \subseteq A^\circ.$$

(ii) *The interior can be represented as*

$$A^\circ = \bigcup \left\{ V : V \subseteq A, V \text{ open} \right\}.$$

(iii) *A is open if and only if $A = A°$.*

(iv) $X \setminus \overline{A} = (X \setminus A)°$.

(v) $X \setminus A° = \overline{X \setminus A}$.

(vi) *The closure can be represented as*

$$\overline{A} = \bigcap \Big\{ F : F \supseteq A, \, F \text{ closed} \Big\}.$$

(vii) $\overline{A}$ *is closed and for any closed set $F \subseteq X$ we have that*

$$A \subseteq F \implies \overline{A} \subseteq F.$$

(viii) *A is closed if and only if $A = \overline{A}$.*

Remark 4.36. In view of (i)-(ii) and (iv)-(v), it is appropriate to refer informally to $A°$ as the **largest open set contained in** A and to $\overline{A}$ as the **smallest closed set containing** A.

Proof. (i) Let $x \in A°$, arbitrary. By the definition of $A°$, there exists $U_x \in \mathcal{T}$ with $U_x \ni x$ such that $U_x \subseteq A$. But then, for every $y \in U_x$ one has that $U_x \ni y$ and $U_x \in \mathcal{T}$, so that $y \in A°$. This shows that $U_x \subseteq A°$. The fact that $A°$ is open now follows from Proposition 4.8. Let V be an arbitrary open set with $V \subseteq A$. Then, for all $x \in V$, one has that $V \ni x$ and $V \in \mathcal{T}$, and therefore $x \in A°$. This shows that $V \subseteq A°$.

(ii) Since $A°$ is open and $A° \subseteq A$, it follows that $A°$ is contained in $\bigcup \{V : V \subseteq A, V \text{ open}\}$. On the other hand, by (i), every V open with $V \subseteq A$ is contained in $A°$, and hence the union of all such V is contained in $A°$. The conclusion follows.

(iii) is an immediate consequence of (i) and (ii).

(iv) Rephrasing the definition of $\overline{A}$, we have that a point $x \in X$ is in $X \setminus \overline{A}$ if and only if there exists $U \in \mathcal{T}$ with $U \ni x$ such that $U \subseteq X \setminus A$. But this is exactly the definition of $(X \setminus A)°$.

(v) By applying (iv) with $X \setminus A$ in the role of A, we obtain the conclusion.

(vi) By using (iv) and (ii) and De Morgan's Laws, we obtain that

$$\overline{A} = X \setminus (X \setminus A)°$$
$$= X \setminus \bigcup \Big\{ V : V \subseteq X \setminus A, V \text{ open} \Big\}$$
$$= \bigcap \Big\{ X \setminus V : V \subseteq X \setminus A, V \text{ open} \Big\}.$$

But for any V open with $V \subseteq X \setminus A$, the set $X \setminus V$ is closed and $X \setminus V \supseteq A$, and as V ranges over the family of all open subsets of X, $X \setminus V$ ranges over the family of all closed subsets of X. Thus, by denoting $F := X \setminus V$, where V is open in X, we obtain the required result.

(vii) Since any intersection of closed sets is closed, (vi) yields that $\overline{A}$ is closed. Moreover, (vi) also shows that the closure $\overline{A}$ is a subset of any closed set which contains A.

(viii) is an immediate consequence of (vii) and (vi). $\qquad\square$

Remark 4.37. Note that, for any subset A of a topological space $(X, \mathcal{T})$, one has that

$$\partial A = \overline{A} \cap \overline{X \setminus A}.$$

The interior and the closure of a set also have the following properties in relation to the operations of union and intersection.

Proposition 4.38. *Let $(X, \mathcal{T})$ be a topological space and A, B be any subsets of X. Then*

(i) *$A \subseteq B$ implies $A^\circ \subseteq B^\circ$ and $\overline{A} \subseteq \overline{B}$;*

(ii) *$(A \cap B)^\circ = A^\circ \cap B^\circ$ and $\overline{A \cup B} = \overline{A} \cup \overline{B}$;*

(iii) *$(A^\circ)^\circ = A^\circ$ and $\overline{\overline{A}} = \overline{A}$.*

Proof. We leave the verification of the details as an exercise for the reader. $\quad\square$

Similarly to the case of metric spaces, one may generalise the (topological) notion of density of a subset in an abstract topological space.

Definition 4.39. A subset A of a topological space $(X, \mathcal{T})$ is said to be **dense** in X if $\overline{A} = X$.

Example 4.40. Let X be an *infinite* set, and $\mathcal{T}$ be the Finite Complement Topology on X. Then any infinite subset of X is dense in X, while no finite subset of X is dense. Indeed, recall that the closed sets in (X, τ) are, apart from X itself, exactly the finite ones. If A is a finite subset of X, then A is closed, and hence $\overline{A} = A \neq X$, so that A is not dense. On the other hand, if A is infinite, then $\overline{A}$ is a closed set which includes A, and therefore is also infinite. It follows that necessarily $\overline{A} = X$, and hence A is dense in X.

A characterisation of the closure using sequential convergence as in Proposition 2.63 is not valid in general topological spaces. One can give, however, the following related result in the interesting case that the spaces are first countable.

Proposition 4.41 (Characterising closure via sequential convergence)**.** *Let $(X, \mathcal{T})$ be a topological space, A a subset of X and $x \in X$.*

(i) *If there exists a sequence $(x_n)_1^\infty \subseteq A$ such that $x_n \longrightarrow x$ in $(X, \mathcal{T})$ as $n \to \infty$, then $x \in \overline{A}$.*

(ii) *If $(X, \mathcal{T})$ satisfies the first axiom of countability and $x \in \overline{A}$, then there exists a sequence $(x_n)_1^\infty \subseteq A$ such that $x_n \longrightarrow x$ in $(X, \mathcal{T})$ as $n \to \infty$.*

Hence, for first countable topological spaces, we have the next representation of the closure which parallels that of metric spaces:

$$\overline{A} = \left\{ x \in X \,\middle|\, \exists \, (x_n)_1^\infty \subseteq A \; : \; x_n \longrightarrow x \text{ in } (X, \mathcal{T}) \text{ as } n \to \infty \right\}.$$

Proof. (i) Let $V \in \mathcal{T}$ with $V \ni x$, arbitrary. Since $x_n \longrightarrow x$ in $(X, \mathcal{T})$ as $n \to \infty$, there exists $N \in \mathbb{N}$ such that $x_n \in V$ for all $n \geq N$. In particular, we have that $x_N \in V \cap A$. This shows that $x \in \overline{A}$.

(ii) Suppose that $(X, \mathcal{T})$ satisfies the first axiom of countability, let $x \in \overline{A}$, and let $(U_n)_1^\infty \subseteq \mathcal{T}$ be the sequence associated to x by the first axiom of countability. Then, since $x \in \overline{A}$, it follows that for every $n \in \mathbb{N}$ there exists $x_n \in U_n \cap A$. Note that $(x_n)_1^\infty \subseteq A$. We claim that also $x_n \longrightarrow x$ in $(X, \mathcal{T})$ as $n \to \infty$. Indeed, let $V \in \mathcal{T}$ with $V \ni x_0$, arbitrary. Let $N \in \mathbb{N}$ be such that $U_N \subseteq V$. Then, for any $n \geq N$, we have that $x_n \in U_n \subseteq U_N \subseteq V$, so that $x_n \in V$. This shows that $x_n \longrightarrow x$ in $(X, \mathcal{T})$ as $n \to \infty$, as required. $\square$

4.7 Subbases, weak and product topology

In this final section[6] we turn our attention to the following natural questions regarding topological spaces:

(Q1) Suppose we are given a non-empty set X and a family of subsets $\mathcal{E} \subseteq \mathcal{P}(X)$ (which may not be a topology). Is there a way to define a topology $\mathcal{T}$ on X that contains $\mathcal{E}$ and is *as small as possible* (in terms of set inclusion)?

(Q2) Suppose we are given a mapping $f : X \longrightarrow Y$ between topological spaces $(X, \mathcal{T}_X), (Y, \mathcal{T}_Y)$. Is there a way to check if f is continuous by checking that the sets $f^{-1}(V)$ are open for *as small a collection as possible* of elements $V \in \mathcal{T}_Y$?

(Q3) Suppose we are given a non-empty set X, a topological space $(Y, \mathcal{T}_Y)$ and a mapping $f : X \longrightarrow Y$. Is there a way to define a topology $\mathcal{T}_X$ on X that is *as small as possible* and that will make f continuous?

---

[6]The reading of this subsection can be postponed until before the beginning of Chapter 12. Notwithstanding, reading it at this point in time is highly recommended since, in a sense, it justifies the developments earlier in this chapter as a standalone theory beyond a mere abstraction of the concept of metric spaces.

(Q4) Suppose we are given two topological spaces $(X, \mathcal{T}_X), (Y, \mathcal{T}_Y)$. Is there a natural way to define a topology on the Cartesian product set $X \times Y$ that will make it a topological space in a way that respects $\mathcal{T}_X, \mathcal{T}_Y$?

As we shall see, the same simple underlying ideas and constructions allow us to answer all these questions **affirmatively**, and providing these answers is the content of this section. To this end, we begin with some terminology.

Definition 4.42. Let $X \neq \emptyset$ be a set.

(i) Let $\mathcal{T}', \mathcal{T}''$ be two topologies on the set X. We say that the topology $\mathcal{T}'$ is **weaker** that the topology $\mathcal{T}''$ if

$$\mathcal{T}' \subseteq \mathcal{T}''.$$

In such an event, one might also say that $\mathcal{T}''$ is **stronger** that $\mathcal{T}'$.

(ii) Let $\mathcal{E} \subseteq \mathcal{P}(X)$ a family of subsets of X and $\mathcal{T}$ a topology on X.

We say that $\mathcal{E}$ is **a subbasis for the topology** $\mathcal{T}$ (or that $\mathcal{T}$ is **the topology generated by** $\mathcal{E}$) when $\mathcal{T}$ is the weakest topology on X containing the family $\mathcal{E}$. Namely, when $\mathcal{E} \subseteq \mathcal{T}$ and the next implication holds:

$$\mathcal{E} \subseteq \mathcal{S} \text{ and } \mathcal{S} \text{ topology on } X \implies \mathcal{T} \subseteq \mathcal{S}.$$

In such an event, we shall symbolise the topology generated by the family of sets $\mathcal{E}$ as $\tau(\mathcal{E})$.

Remark 4.43. (i) Clearly, the indiscrete topology $\{\emptyset, X\}$ is the weakest possible topology on a set X, whilst the discrete topology $\mathcal{P}(X)$ is the strongest possible. Any other topology $\mathcal{T}$ one might define, lies in between these two extremes:

$$\{\emptyset, X\} \subseteq \mathcal{T} \subseteq \mathcal{P}(X).$$

(ii) Our reference to "the" topology generated by the family of sets $\mathcal{E}$ symbolised as $\tau(\mathcal{E})$ in Definition 4.42(ii) alludes to that $\tau(\mathcal{E})$ is unique (and exists). This is indeed the case, as we shall see right next.

The result below answers affirmatively our first question (Q1) posed at the beginning of the section:

Lemma 4.44 (Generation of topology by a subbasis, I)**.** *Let $X \neq \emptyset$ be a set and $\mathcal{E} \subseteq \mathcal{P}(X)$ a family of subsets of X. Then, there exists a unique topology $\tau(\mathcal{E})$ generated by $\mathcal{E}$, which is represented as the intersection of all topologies containing $\mathcal{E}$:*

$$\tau(\mathcal{E}) = \bigcap \left\{ \mathcal{T} : \mathcal{T} \text{ topology on } X, \mathcal{T} \supseteq \mathcal{E} \right\}.$$

Proof. First note that there exists at least one topology containing $\mathcal{E}$, namely the discrete topology $\mathcal{P}(X)$. In order to conclude, it suffices to note that the intersection of (arbitrarily many) topologies is a topology itself. Indeed, let $\{\mathcal{T}_i : i \in I\}$ be a family of topologies on X. We wish to show that

$$\mathcal{T} := \bigcap_{i \in I} \mathcal{T}_i$$

is also a topology on X and satisfies $(T1) - (T3)$.

$(T1)$: It is obvious that $\mathcal{T}$ contains $\emptyset, X$ since $\emptyset, X \in \mathcal{T}_i$ for all $i \in I$.

$(T2)$: If a family of sets $\{U_a : a \in A\}$ belongs to $\mathcal{T}$, then each of the sets U_a belongs to all topologies $\mathcal{T}_i$ for all $i \in I$. Hence, their union $\bigcup_{a \in A} U_a$ belongs to $\mathcal{T}_i$ for all $i \in I$. Thus, it belongs to the intersection $\bigcap_{i \in I} \mathcal{T}_i$.

$(T3)$: If $U_1, \ldots, U_n$ belong to $\mathcal{T}$, then each of these belongs to each of the $\mathcal{T}_i$ for any $i \in I$. Since each $\mathcal{T}_i$ satisfies $(T3)$, we have that $U_1 \cap \cdots \cap U_n \in \mathcal{T}_i$ for all $i \in I$, and hence $U_1 \cap \cdots \cap U_n \in \mathcal{T}$, establishing $(T3)$. $\square$

Lemma 4.44 above provides a representation formula for the topology generated by a family of sets as the intersection of all possible topologies containing that family, but gives no information as to what the elements of that topology actually are. The next result clarifies the picture:

Proposition 4.45 (Generation of topology by a subbasis, II). *Let $X \neq \emptyset$ be a set and $\mathcal{E} \subseteq \mathcal{P}(X)$ a family of subsets of X. Then, the topology $\tau(\mathcal{E})$ generated by $\mathcal{E}$ consists of $\emptyset$, X and arbitrary unions of finite intersections of elements of the family $\mathcal{E}$:*

$$\tau(\mathcal{E}) = \left\{ \emptyset, \ X, \ \bigcup_{i \in I} \left(\bigcap_{j=1}^{n} E_{ij} \right) \ : \ E_{ij} \in \mathcal{E}, \text{ for all } i \in I \text{ and } j = 1, \ldots, n \right\}.$$

In the above formula, both $n \in \mathbb{N}$ and the index set I are arbitrary.

We note that the operations of union and intersection above **cannot be interchanged**. The proof is based on the following two seemingly technical lemmas which have elementary proofs.

Lemma 4.46. *Let $(X, \mathcal{T})$ be a topological space and $\mathcal{N} \subseteq \mathcal{P}(X)$ a family of subsets of X. Then, the next two assertions are equivalent:*

(i) *Every $U \in \mathcal{T} \setminus \{\emptyset\}$ can be written as a union of members of the family $\mathcal{N}$.*

(ii) *We have that $\mathcal{N} \subseteq \mathcal{T}$ and $\mathcal{N}$ can be written as a union of families parameterised by the points of X*

$$\mathcal{N} = \bigcup_{x \in X} \mathcal{N}_x$$

such that, for any $x \in X$, the following hold true:

(a) $x \in N$, *for all $N \in \mathcal{N}_x$.*

(b) *If $U \in \mathcal{T}$ and $x \in U$, then there exists $N \in \mathcal{N}_x$ with $N \subseteq U$.*

The class of sets $\mathcal{N}$ above has a special name:

Definition 4.47. Let $(X, \mathcal{T})$ be a topological space and $\mathcal{N} \subseteq \mathcal{P}(X)$ a family of subsets of X. If either of the statements (i) or (ii) of Lemma 4.46 hold true for $\mathcal{N}$, then $\mathcal{N}$ is called a **neighbourhood basis for the topology $\mathcal{T}$ of X**, and $\mathcal{N}_x$ is called a **neighbourhood basis at x** for any $x \in X$.

Proof of Lemma 4.46. (i) $\Longrightarrow$ (ii): If every non-empty $U \in \mathcal{T}$ is a union of elements of sets from $\mathcal{N}$, then the class of sets $\mathcal{N}_x := \{N \in \mathcal{N} : N \ni x\}$ satisfies (a)-(b) of (ii) for any $x \in X$.

(ii) $\Longrightarrow$ (i): If (ii) holds, then for any $U \in \mathcal{T}$ with $U \ni x$ there exists $N_x \in \mathcal{N}_x$ with $x \in N_x \subseteq U$. Therefore, $U = \bigcup_{x \in U} N_x$ and (i) ensues. $\qquad\square$

Lemma 4.48. *Let X be a set and $\mathcal{N} \subseteq \mathcal{P}(X)$ a family of subsets of X. Then, the next two assertions are equivalent:*

(i) *$\mathcal{N}$ is a neighbourhood basis for a topology on X.*

(ii) *The following hold true:*

    (a) *For any $x \in X$, there exists $N \in \mathcal{N}$ with $x \in N$.*

    (b) *If $U, V \in \mathcal{N}$ and $x \in U \cap V$, then there exists $W \in \mathcal{N}$ with $x \in W \subseteq U \cap V$.*

Proof of Lemma 4.48. Since (i) $\Longrightarrow$ (ii) is obvious, it suffices to establish that (ii) $\Longrightarrow$ (i). To this end, we set

$$\mathcal{T} := \left\{ U \in \mathcal{P}(X) \,\middle|\, \forall\, x \in U, \, \exists\, V \in \mathcal{N} : x \in V \subseteq U \right\}.$$

Then, by (a) we have that $X \in \mathcal{T}$ and also $\emptyset \in \mathcal{T}$ trivially. Further, $\mathcal{T}$ is closed under arbitrary unions. If $U_1, U_2 \in \mathcal{T}$ and $x \in U_1 \cap U_2$, there exist $V_1, V_2 \in \mathcal{N}$ with $x \in V_1 \subseteq U_1$ and $x \in V_2 \subseteq U_2$. Hence, by (b) there exists $W \in \mathcal{N}$ such that $x \in W \subseteq V_1 \cap V_2$. Therefore, $U_1 \cap U_2 \in \mathcal{T}$. By induction, $\mathcal{T}$ is closed under finite intersections. Hence, $\mathcal{T}$ is a topology on X for which $\mathcal{N}$ is the neighbourhood basis. $\qquad\square$

Now we return to the proof of Proposition 4.45.

Proof. The family of finite intersections of sets from $\mathcal{E}$ together with X satisfies the conditions (ii)(a)-(ii)(b) of Lemma 4.48, therefore they form the neighbourhood basis for a topology on X. By Lemma 4.46, it follows that the family of all arbitrary unions of such intersections together with $\emptyset$ is a topology on X which is contained in $\tau(\mathcal{E})$. Since $\tau(\mathcal{E})$ is the weakest such topology, it is equal to it. $\qquad\square$

Now we are ready to answer the second question (Q2) posed at the beginning of this section.

Proposition 4.49 (Criterion for continuity). *Let $(X, \mathcal{T})$, $(Y, \mathcal{S})$ be topological spaces and $f : X \longrightarrow Y$ a mapping between them. Suppose that $\mathcal{S}$ is generated by a neighbourhood subbasis $\mathcal{E}$, i.e., $\mathcal{S} = \tau(\mathcal{E})$. Then, the following are equivalent:*

(i) *The mapping f is continuous.*

(ii) *The set $f^{-1}(V)$ belongs to $\mathcal{T}$, for all $V \in \mathcal{E}$.*

Hence, in view of Proposition 4.49, it suffices to check that f inverts only the sets on a subbasis of a topology of the target space to open sets in the domain space.

Proof. The equivalence is an immediate consequence of Proposition 4.45 and that the inverse image map commutes with unions and intersections (see formulas $(\star)$-$(\star\star)$ in Chapter 1). $\qquad\square$

We now conclude by answering the questions (Q3) and (Q4) posed at the beginning of the section. It turns out that the answer of (Q4) follows from the answer of (Q3) via the next **very important general construction**, which will be the starting point of our investigations in Chapter 12.

Lemma 4.50. *Let $X \neq \emptyset$ be a set and we have a family of mappings*

$$\mathcal{F} := \Big\{ f_i \, : \, X \longrightarrow Y_i \, \Big| \, i \in I \Big\}$$

*valued in the topological spaces $\{(Y_i, \mathcal{T}_i) : i \in I\}$. Then, there **exists a unique weakest topology on X which makes all the mappings in $\mathcal{F}$ continuous**.*

Proof. Consider the family of subsets $\mathcal{E} \subseteq \mathcal{P}(X)$:

$$\mathcal{E} := \Big\{ f_i^{-1}(V_i) \, : \, V_i \in \mathcal{T}_i, \, i \in I \Big\}.$$

Then, the topology $\tau(\mathcal{E})$ generated by $\mathcal{E}$ has the property of being the topology with the fewest possible open sets which makes all $f \in \mathcal{F}$ continuous. The existence and uniqueness claims follow directly from Proposition 4.44, whilst the continuity of all $f \in \mathcal{F}$ follows from Proposition 4.49. $\qquad\square$

The topology above has a celebrated name:

Definition 4.51. Suppose that $X \neq \emptyset$ is a set and we are given a family of mappings

$$\mathcal{F} := \big\{ f_i \, : \, X \longrightarrow Y_i \, \big| \, i \in I \big\}$$

whose targets are the topological spaces $\{(Y_i, \mathcal{T}_i) : i \in I\}$.

Then, the topology on X generated by $\{f_i^{-1}(V_i) : V_i \in \mathcal{T}_i, i \in I\}$ is called the **weak topology of X generated by the family $\mathcal{F}$**, is symbolised by $\mathcal{T}^w$ and is the topology whose subbasis consists of all sets of the form $f_i^{-1}(V_i)$, when V_i runs through the open sets of Y_i for all $i \in I$:

$$\mathcal{T}^w := \tau\left(\left\{f_i^{-1}(V_i) : V_i \in \mathcal{T}_i,\ i \in I\right\}\right).$$

Remark 4.52. From Proposition 4.45 we have that **the weak topology $\mathcal{T}^w$ on X generated by $\mathcal{F}$ is the collection of all arbitrary unions of finite intersections of elements of the form $f_i^{-1}(V_i)$, where $i \in I$ and $V_i \in \mathcal{T}_i$.**

The particular choice of $\mathcal{F} = \{f : X \longrightarrow Y\}$ for a single topological space answers explicitly (Q3). Finally, we answer (Q4) in the form of a definition.

Definition 4.53. Let $\{(X_i, \mathcal{T}_i) : i \in I\}$ be a family of topological spaces.

- The **product space** $\prod_{i \in I} X_i$ is defined as[7]

$$\prod_{i \in I} X_i := \left\{ f : I \longrightarrow \bigcup_{i \in I} X_i \ \middle|\ f(i) \in X_i,\ i \in I \right\}.$$

- The **product topology** on $X := \prod_{i \in I} X_i$ is the weak topology generated by the projection mappings $\pi_i : X \longrightarrow X_i$, $\pi_i(f) = f(i)$:

$$\mathcal{T}^{\mathrm{prod}} := \tau\left(\left\{\pi_i^{-1}(U_i) : U_i \in \mathcal{T}_i,\ i \in I\right\}\right).$$

It can be seen that if I is finite, then $\prod_{i \in I} X_i$ reduces to the usual Cartesian product space $\prod_{i=1}^n X_i$ and one might write its elements in the more standard way as $(f_1, \ldots, f_n)$ (by identifying f with its image), instead of $\{f(i) : i \in I\}$. The case of $I = \{1, 2\}$ gives an explicit answer to our question (Q4).

4.8 Exercises

Exercise 4.1. Let $(X, \mathcal{T})$ be a topological space. Let U be open in $(X, \mathcal{T})$ and F be closed in $(X, \mathcal{T})$. Show that $U \setminus F$ is open in $(X, \mathcal{T})$ and $F \setminus U$ is closed in $(X, \mathcal{T})$.

Exercise 4.2. Let $\mathbb{N} := \{1, 2, 3, \ldots\}$ be the set of natural numbers, and let $\mathcal{T} := \{\emptyset, \mathbb{N}, A_1, A_2, A_3, \ldots, A_k, \ldots\}$, where $A_k = \{1, 2, \ldots, k\}$ for each $k \in \mathbb{N}$. Show that $\mathcal{T}$ is a topology on $\mathbb{N}$.

---

[7]It is perhaps worth noting that the Cartesian product of a collection of non-empty sets is non-empty as a consequence of the Axiom of Choice which is tacitly invoked here.

Exercise 4.3. Let X be a set. Let $\mathcal{T} := \{U \subseteq X : X \setminus U \text{ is finite}\} \cup \{\emptyset\}$. Show that $\mathcal{T}$ is a topology on X.

Exercise 4.4. Let X be an infinite set, and let $\mathcal{T}$ be the finite complement topology on X (i.e., $U \in \mathcal{T}$ if and only if either $U = \emptyset$ or $X \setminus U$ is finite). Let $(x_n)_1^\infty$ be a sequence in X such that $x_i \neq x_j$ for all $i, j \in \mathbb{N}$ with $i \neq j$. Show that, for any $x_0 \in X$, the sequence $(x_n)_1^\infty$ converges to x_0 in $(X, \mathcal{T})$.

Exercise 4.5. Let X be an infinite set, and $\mathcal{T}$ be the finite complement topology on X. Show that:

(i) If U and V are non-empty open sets in $(X, \mathcal{T})$, then $U \cap V$ is non-empty.

(ii) If $f : (X, \mathcal{T}) \longrightarrow \mathbb{R}$ is continuous on X, where $\mathbb{R}$ is considered with the usual topology, then f is constant.

Exercise 4.6. Let X be an infinite set, and $\mathcal{T}$ be the finite complement topology on X. Show that $(X, \mathcal{T})$ is not a Hausdorff space.

Exercise 4.7. Let $(X, \mathcal{T})$ be a Hausdorff space, and let x_1, x_2, x_3 be distinct points in X. Show that there exists $U_1, U_2, U_3 \in \mathcal{T}$ such that $U_k \ni x_k$ for each $k \in \{1, 2, 3\}$, and $U_i \cap U_j = \emptyset$ for all $i, j \in \{1, 2, 3\}$ with $i \neq j$.

Exercise 4.8. Let $(X, \mathcal{T})$ be a Hausdorff topological space. Show that any finite subset of X is closed.

Exercise 4.9. Let $(X, \mathcal{T})$ be a topological space, and $f, g : X \longrightarrow \mathbb{R}$ be continuous functions on X, where $\mathbb{R}$ is considered as a topological space with the standard topology. Show that if $g(x) \neq 0$ for all $x \in X$, then the function $f/g : X \longrightarrow \mathbb{R}$ is continuous on X.

Exercise 4.10. Let $(X, \mathcal{T})$ be a topological space and A, B be subsets of X. Show that

(i) $A \subseteq B$ implies $A^\circ \subseteq B^\circ$ and $\overline{A} \subseteq \overline{B}$;

(ii) $(A \cap B)^\circ = A^\circ \cap B^\circ$ and $\overline{A \cup B} = \overline{A} \cup \overline{B}$;

(iii) $(A^\circ)^\circ = A^\circ$ and $\overline{\overline{A}} = \overline{A}$.

Exercise 4.11. Let $(X, \mathcal{T})$ be a topological space, and A, B be subsets of X. Show that

$$A^\circ \cup B^\circ \subseteq (A \cup B)^\circ \qquad \text{and} \qquad \overline{A \cap B} \subseteq \overline{A} \cap \overline{B}.$$

If $X = \mathbb{R}$ and $\mathcal{T}$ is the usual topology, give an example of sets A and B for which

$$A^\circ \cup B^\circ \neq (A \cup B)^\circ,$$

and an example for which

$$\overline{A \cap B} \neq \overline{A} \cap \overline{B}.$$

Exercise 4.12. Let $(X, \mathcal{T})$ be a topological space, and A be a subset of X. Show that

(i) A is closed in X if and only if $\partial A \subseteq A$;

(ii) $\partial A = \emptyset$ if and only if A is both open and closed in $(X, \mathcal{T})$.

Exercise 4.13. Let $(X, \mathcal{T})$ and $(Y, \mathcal{S})$ be topological spaces, and $f : X \longrightarrow Y$ be a function. Show that f is continuous on X if and only if $f^{-1}(A^\circ) \subseteq (f^{-1}(A))^\circ$ for any subset A of Y. (A° means the interior of A in $(Y, \mathcal{S})$, whilst $(f^{-1}(A))^\circ$ means the interior of $f^{-1}(A)$ in $(X, \mathcal{T})$.)

Exercise 4.14. Let $(X, \mathcal{T})$ and $(Y, \mathcal{S})$ be topological spaces, and $f : X \longrightarrow Y$ be a function. Show that f is continuous on X if and only if $f(\overline{A}) \subseteq \overline{f(A)}$ for every subset A of X. ($\overline{A}$ means the closure of A in $(X, \mathcal{T})$, whilst $\overline{f(A)}$ means the closure of $f(A)$ in $(Y, \mathcal{S})$.)

Exercise 4.15. Let $(X, \mathcal{T})$ and $(Y, \mathcal{S})$ be topological spaces, where $(Y, \mathcal{S})$ is a Hausdorff space, and let $f, g : X \longrightarrow Y$ be continuous functions on X.

(i) Show that the set $\{x \in X : f(x) = g(x)\}$ is closed in X.

(ii) Deduce that if $f(x) = g(x)$ for all $x \in A$, where A is a dense subset of X (meaning that $\overline{A} = X$), then $f = g$ on X.

Chapter 5

Compactness and sequential compactness

5.1 Compactness beyond Euclidean spaces

The Euclidean space $\mathbb{F}^m$ has the very pleasant property, expressed by the **Bolzano–Weierstrass Theorem**, that every bounded sequence has a convergent subsequence:

$$\left.\begin{array}{l} (x_n)_1^\infty \subseteq \mathbb{F}^m \; : \\[2mm] \sup_{n \in \mathbb{N}} \|x_n\| \leq C \end{array}\right\} \quad \Longrightarrow \quad \exists \; (x_{n_k})_1^\infty \text{ and } x \in \mathbb{F}^m \; : \; x_{n_k} \longrightarrow x \text{ as } k \to \infty.$$

This property enables us to demonstrate that certain objects exist without constructing them explicitly. Namely, it **allows us to establish existence of objects obtained via a limiting process that might be difficult or even impossible to be obtained otherwise**. A notable example is the set of points at which a real-valued function on a closed bounded subset of $\mathbb{F}^m$ attains its minimum and its maximum value. Unfortunately, this property is generally *false* if $\mathbb{F}^m$ is replaced by a metric space (X, d), let alone by a general topological space. Below is such a simple occurrence:

Example 5.1. Let X be an infinite set (e.g. $\mathbb{N}$) considered as a metric space with the discrete metric d_0. Let $(x_n)_1^\infty \subseteq X$ be any sequence such that $x_i \neq x_j$ for all $i \neq j$. Then $(x_n)_1^\infty$ is bounded, but has no convergent subsequence. This happens because the convergent sequences of a discrete metric space are exactly those which are eventually constant, which is not the case here for any subsequence of $(x_n)_1^\infty$.

Motivated by the Euclidean Bolzano–Weierstrass property and its significance, mathematicians were interested in understanding and studying the class of topological spaces which possess this property, namely those privileged spaces in which *any sequences has a convergent subsequence*, thus coining the term **sequentially compact** topological spaces. A primary goal in this chapter is to study these favourable spaces more closely.

Roughly speaking, a sequentially compact space is one which is, "eventually confined" in the sense that no sequence in it can wander around infinitely, but rather some part of it (a subsequence) must necessarily eventually "accumulate" and settle into a certain point. Under the light of this new terminology, the Bolzano–Weierstrass theorem expresses the fact that *closed and bounded subsets of $\mathbb{F}^m$ are sequentially compact*. Thus, investigating and determining which spaces are sequentially compact is a naturally emerging question.

A seemingly unrelated concept, but which turns out to be closely connected to sequential compactness is described by the **Heine–Borel Theorem**, which was originally observed in the real line $\mathbb{R}$:

If a closed bounded interval $[a, b]$ can be covered by a family of open intervals (or, more generally, open sets), then there exists a sub-selection of finitely many of these sets which still cover $[a, b]$.

Here *cover (or covering) means a collection of sets whose union contains the given set*. This property regarding open coverings holds more generally for closed and bounded sets in $\mathbb{F}^m$ and, like sequential compactness, is also useful in many other contexts as well, particularly in terms of establishing existential results when a constructive method is not available. The sets which satisfy the Heine-Borel property are called **compact**.

Compactness in this form of "open coverings" is one of the most subtle and hard-to-motivate notions in Analysis, whilst its usefulness is also not immediately apparent. Roughly speaking, a compact set is "confined" in the sense that no matter how it were covered by a family of open sets (possibly a "huge" number of them), then finitely many of them would actually sufficient to cover it, and thus the space could not possibly be spread out very widely. On $\mathbb{F}^m$, the compact sets are precisely the closed and bounded ones. For instance, neither $(0, 1)$ nor $\mathbb{R}$ are compact. Their respective coverings by intervals of the form

$$\big\{(0, \varepsilon) : 0 < \varepsilon < 1\big\}, \quad \big\{(-N, N) : N \in \mathbb{N}\big\},$$

have no finite sub-selections which still cover either $(0, 1)$ or $\mathbb{R}$, respectively.

Motivated by the Heine-Borel property which is purely topological and involves only open sets, the concept of the so-called **compact topological spaces** emerged, that is of those spaces for which every open covering has a finite sub-covering. This important property of compactness for a topological space, if satisfied, is frequently used to **pass from local to global properties** of the space. That is, suppose that one can prove that a certain nice property holds in an open neighbourhood of each point; then these neighbourhoods provide a (typically, infinite) cover of the space. The extraction of a finite subcover out of it becomes an essential tool for demonstrating that the property holds over the entire space.

Remarkably and contrary perhaps to first appearances, compactness is equivalent to sequential compactness in the context of *metric* spaces. This equivalence breaks down when going to more generalised space frameworks and understanding when these concepts coincide is an important issue. A further principal goal in this chapter is to understand and determine for which spaces compactness is true and under what additional conditions. We are confident that the readers will come to appreciate for themselves the usefulness of the compactness concepts and their fundamental consequences after seeing its numerous applications in the present and subsequent chapters, encompassing the pragmatic view that "the proof of the pudding is in the eating".

5.2 The class of sequentially compact sets

We commence our study of topological spaces possessing the "Bolzano-Weierstrass property" by designating a name for them.[1]

Definition 5.2. A topological space $(X, \mathcal{T})$ is said to be **sequentially compact** if every sequence contained in X has a subsequence which is convergent in $(X, \mathcal{T})$:

$$(x_n)_1^\infty \subseteq X \implies \exists \, (x_{n_k})_1^\infty \text{ and } x \in X : x_{n_k} \longrightarrow x \text{ in } \mathcal{T} \text{ as } k \to \infty.$$

A subset $Y \subseteq X$ of a topological space $(X, \mathcal{T})$ is said to be **sequentially compact** if the topological subspace $(Y, \mathcal{T}|_Y)$ is sequentially compact when endowed with the subspace topology.

To place now already know results involving sequential compactness into this new context, we recall without proof from Analysis in Several Variables the following results, the second of which is expressed using the newly introduced terminology.

Theorem 5.3 (Bolzano-Weierstrass Theorem in $\mathbb{F}^m$). *Every bounded sequence in $(\mathbb{F}^m, \| \cdot \|)$ has a convergent subsequence.*

Theorem 5.4. *A subset Y of $(\mathbb{F}^m, \| \cdot \|_2)$ is sequentially compact if and only if it is closed and bounded.*

We now address the question of how sequential compactness is related to closedness and boundedness for general metric spaces. We shall see that, contrary to the case of $\mathbb{F}^m$ of Theorem 5.4, one needs to *strengthen the notion of boundedness* appropriately to achieve equivalence.

Theorem 5.5. *Let (X, d) be a metric space. Then every sequentially compact subset of X is closed and bounded.*

---

[1] We note that (perhaps to the readers' disappointment!) in the sequel we shall be concerned primarily with sequential compactness in metric spaces, rather than in general topological spaces. The reason is that, as we have already explained, if a topological space fails to satisfy the first axiom of countability, sequences are not really appropriate to characterise convergence in them and more advanced concepts (like "nets") are required.

Proof. Let Y be a sequentially compact subset of the metric space (X, d).

We first prove that Y is closed. Let $x \in \overline{Y}$ be an arbitrary point. Then, there exists a sequence $(y_n)_1^\infty \subseteq Y$ such that $y_n \longrightarrow x$ in (X, d) as $n \to \infty$. Since Y is sequentially compact, there exists $y \in Y$ and a subsequence $(y_{n_k})_1^\infty$ of $(y_n)_1^\infty$ such that $y_{n_k} \longrightarrow y$ in (Y, d) as $k \to \infty$, and hence in (X, d) as well. However, the fact that $y_n \longrightarrow x$ in (X, d) implies that also $y_{n_k} \longrightarrow x$ in (X, d) as $k \to \infty$. By the uniqueness of limits in X, it follows that $x = y$, and therefore $x \in Y$. This shows that $\overline{Y} \subseteq Y$, which means that the subspace Y is closed.

We now prove that Y is bounded. Suppose for the sake of contradiction that Y is not bounded. Fix some point $z_0 \in Y$. Since Y is not bounded, it follows that for any $n \in \mathbb{N}$, there exists a point $y_n \in Y$ such that $d(y_n, z_0) > n$. In this way, we construct a sequence $(y_n)_1^\infty \subseteq Y$. Since Y is sequentially compact, there exist $y_0 \in Y$ and a subsequence $(y_{n_k})_1^\infty$ of $(y_n)_1^\infty$ such that $y_{n_k} \longrightarrow y_0$ in (Y, d). Note that, by the triangle inequality

$$n_k < d(y_{n_k}, z_0) \leq d(y_{n_k}, y_0) + d(y_0, z_0) \quad \text{for all } k \subset \mathbb{N}.$$

This inequality, however, cannot possibly hold for all k sufficiently large, since, as $k \to \infty$, the left-hand side becomes arbitrarily large, whilst the right-hand side remains bounded. The contradiction just obtained shows that Y must be bounded. □

The converse of Theorem 5.5 is **not** true in general. However, a strengthened version of the notion of boundedness, called **total boundedness**, turns out to be the appropriate notion in this generalised context.

Definition 5.6. Let (X, d) be a metric space. A subset Y of X is said to be **totally bounded** if for every $\varepsilon > 0$ there exist $n = n \in \mathbb{N}$ and $y_1, \ldots, y_n \in Y$ (all depending on ε) such that

$$Y \subseteq \bigcup_{k=1}^n \mathbb{B}_\varepsilon(y_k).$$

Remark 5.7. We note that in the above definition the balls are understood to be with respect to X, but in fact it is immaterial whether the above open balls are considered as balls in the space X or as balls in the subspace Y.

The next simple fact claims that total boundedness implies boundedness:

Lemma 5.8. *Let (X, d) be a metric space. Then, any totally bounded subset is bounded.*

Proof. Let $A \subseteq X$ be totally bounded and choose $\varepsilon = 1$ and respective points $a_1, \ldots, a_n \in A$ in Definition 5.6 such that

$$A \subseteq \bigcup_{k=1}^n \mathbb{B}_1(a_k).$$

Since each of these balls is bounded, and there are only a finite number of them, it is easy to check that their union is bounded, and hence A is also bounded. □

The converse of Lemma 5.6 is *not true* in general, as the example right next confirms. Hence, for general metric spaces, boundedness is strictly weaker that total boundedness.

Example 5.9. Let X be an infinite set (e.g. $\mathbb{N}$) endowed with the discrete metric d_0. Then, X is bounded (being contained in all balls of radius 2) but not totally bounded since it is not contained in any finite union of open balls of radius 1.

The next result attests that the centres of the balls may be taken to be in X if desired (rather than in Y as the definition of total boundedness requires) without altering the notion.

Proposition 5.10. *Let (X, d) be a metric space, and Y be a subset of X. Then the following statements are equivalent:*

(i) *Y is totally bounded;*

(ii) *for every $\varepsilon > 0$ there exist $n \in \mathbb{N}$ and $x_1, \ldots, x_n \in X$ such that*

$$Y \subseteq \bigcup_{k=1}^{n} \mathbb{B}_{\varepsilon}^{X}(x_k).$$

Proof. It is obvious that (i) implies (ii). Conversely, suppose that (ii) holds. Fix $\varepsilon > 0$, arbitrary, and use $\varepsilon/2$ instead of ε in (ii). It follows that there exist $n \in \mathbb{N}$ and $x_1, \ldots, x_n \in X$ such that

$$Y \subseteq \bigcup_{k=1}^{n} \mathbb{B}_{\varepsilon/2}^{X}(x_k).$$

We now assume with no loss of generality that

$$\mathbb{B}_{\varepsilon/2}^{X}(x_k) \bigcap Y \neq \emptyset \quad \text{for each } k \in \{1, \ldots, n\}.$$

Indeed, if this were not the case, then those balls which do not intersect Y could be discarded. Thus, let us pick some $y_k \in \mathbb{B}_{\varepsilon/2}^{X}(x_k) \bigcap Y$, for each $k \in \{1, \ldots, n\}$. Then $\mathbb{B}_{\varepsilon/2}^{X}(x_k) \subseteq \mathbb{B}_{\varepsilon}^{X}(y_k)$ for each $k \in \{1, \ldots, n\}$, and hence

$$Y \subseteq \bigcup_{k=1}^{n} \mathbb{B}_{\varepsilon}^{X}(y_k).$$

Since $\varepsilon > 0$ was arbitrary, this shows that Y is totally bounded. □

The end goal of the present section is to establish a connection between sequential compactness, completeness, and total boundedness in metric spaces, which is Theorem 5.12 that follows. To this end, we first need the next result of independent interest which characterises totally bounded sets in terms of the existence of Cauchy subsequences to any sequence in the set.

Theorem 5.11 (Characterisation of total boundedness via sequences). *A metric space (X, d) is totally bounded if and only if every sequence in X has a Cauchy subsequence.*

Proof. ($\Longleftarrow$): We argue by contraposition, proving that if X is not totally bounded, then there exists a sequence $(x_n)_1^\infty$ in X which has no Cauchy subsequence. Let us assume therefore that X is not totally bounded. Then there exists $\varepsilon_0 > 0$ such that X cannot be written as a finite union of open balls of radius ε_0. Pick some $x_1 \in X$. Since $X \neq \mathbb{B}_{\varepsilon_0}(x_1)$, one can pick $x_2 \in X \setminus \mathbb{B}_{\varepsilon_0}(x_1)$. Since

$$X \neq \mathbb{B}_{\varepsilon_0}(x_1) \bigcup \mathbb{B}_{\varepsilon_0}(x_2),$$

one can pick

$$x_3 \in X \setminus \left(\mathbb{B}_{\varepsilon_0}(x_1) \bigcup \mathbb{B}_{\varepsilon_0}(x_2) \right).$$

Suppose that, for some $n \in \mathbb{N}$, $x_1, \ldots, x_{n-1}$ have been constructed. Since

$$X \neq \bigcup_{i=1}^{n-1} \mathbb{B}_{\varepsilon_0}(x_i),$$

one can pick a point

$$x_n \in X \setminus \bigcup_{i=1}^{n-1} \mathbb{B}_{\varepsilon_0}(x_i)$$

which, therefore, satisfies

$$d(x_n, x_i) \geq \varepsilon_0 \quad \text{for all } i \in \{1, \ldots, n-1\}.$$

The sequence $(x_n)_1^\infty$ constructed in this way has the property that

$$d(x_i, x_j) \geq \varepsilon_0 \quad \text{for all } i \neq j.$$

Now for every subsequence $(x_{n_k})_{k=1}^\infty$ of $(x_n)_1^\infty$, we still have

$$d(x_{n_i}, x_{n_j}) \geq \varepsilon_0 \quad \text{for all } i \neq j.$$

This shows that $(x_{n_k})_{k=1}^\infty$ is not a Cauchy sequence. This proves the claimed result.

($\Longrightarrow$): Let (X, d) be a totally bounded metric space. This means that, for any $\varepsilon > 0$, X can be written as a finite union of open balls of radius ε.

Let $(x_n)_1^\infty$ be an arbitrary sequence in X. We shall prove that $(x_n)_1^\infty$ has a Cauchy subsequence. To this aim, we apply the total boundedness condition with $\varepsilon := 1/k$, for each $k \in \mathbb{N}$.

For $k = 1$, X is a finite union of open balls of radius 1. Since $\mathbb{N}$ is infinite, there exist $a_1 \in X$ and an infinite subset S_1 of $\mathbb{N}$ such that

$$\{x_n : n \in S_1\} \subseteq \mathbb{B}_1(a_1).$$

This implies that

$$d(x_n, x_m) < 2 \quad \text{for all } n, m \in S_1.$$

For $k = 2$, X is a finite union of open balls of radius $1/2$. Since S_1 is infinite, there exist $a_2 \in X$ and an infinite subset S_2 of S_1 such that

$$\{x_n : n \in S_2\} \subseteq \mathbb{B}_{1/2}(a_2).$$

This implies that

$$d(x_n, x_m) < 2/2 \quad \text{for all } n, m \in S_2.$$

Suppose now that, for some $k \in \mathbb{N}$, the sets $S_1, \ldots, S_{k-1}$ have been constructed. Since X is a finite union of open balls of radius $1/k$, and S_{k-1} is infinite, there exist $a_k \in X$ and an infinite subset S_k of S_{k-1} such that

$$\{x_n : n \in S_k\} \subseteq \mathbb{B}_{1/k}(a_k).$$

This implies that

$$d(x_n, x_m) < 2/k \quad \text{for all } n, m \in S_k.$$

In this way we therefore construct infinite sets

$$\mathbb{N} \supseteq S_1 \supseteq S_2 \supseteq \cdots \supseteq S_k \supseteq \cdots, \qquad (*)$$

such that, for each $k \in \mathbb{N}$,

$$d(x_n, x_m) < 2/k \quad \text{for all } n, m \in S_k.$$

We now consider each set S_k as an ordered set, namely we consider its elements arranged in increasing order. Let n_1 be the first element of S_1, n_2 be the second element of S_2, and so on, n_k be the k-th element of the ordered set S_k, for each $k \in \mathbb{N}$. A moment's reflection shows that, because of $(*)$, we have $n_1 < \cdots < n_k < \cdots$ We now consider the subsequence $(x_{n_k})_1^\infty$ of $(x_n)_1^\infty$. Note that, for each $k \in \mathbb{N}$,

$$n_k \in S_k \subseteq S_{k-1} \subseteq \cdots \subseteq S_1.$$

Let $\varepsilon > 0$ be arbitrary and pick some $N \in \mathbb{N}$ such that $2/N < \varepsilon$. Then, for every $i, j \in \mathbb{N}$ with $i, j \geq N$, we have that $n_i \in S_i \subseteq S_N$ and $n_j \in S_j \subseteq S_N$, and therefore

$$d(x_{n_i}, x_{n_j}) < 2/N < \varepsilon.$$

Since $\varepsilon > 0$ was arbitrary, this shows that the subsequence $(x_{n_k})_1^\infty$ is a Cauchy sequence, as required. $\qquad \square$

We may now state and prove the principal result of this section.

Theorem 5.12 (Sequentially compact $\Leftrightarrow$ complete and totally bounded). *A metric space (X, d) is sequentially compact if and only if it is complete and totally bounded.*

Proof. ($\Longrightarrow$): Let (X, d) be a sequentially compact metric space. Let $(x_n)_1^\infty$ be an arbitrary sequence in X. Then, by assumption, $(x_n)_1^\infty$ has a convergent subsequence, which is obviously also a Cauchy subsequence. Therefore, by Theorem 5.11, X is totally bounded. Now let $(x_n)_1^\infty$ be an arbitrary Cauchy sequence. Then, by assumption, $(x_n)_1^\infty$ has a convergent subsequence. Hence, by Proposition 3.4(ii), $(x_n)_1^\infty$ converges. Therefore, (X, d) is complete.

($\Longleftarrow$): Let (X, d) be a metric space which is totally bounded and complete. Let $(x_n)_1^\infty$ be an arbitrary sequence in X. Since X is totally bounded, then, by Theorem 5.11, $(x_n)_1^\infty$ has a Cauchy subsequence and, moreover, since X is complete, that Cauchy subsequence is convergent. Therefore, (X, d) is sequentially compact. $\square$

5.3   The class of compact sets

In this section we commence our journey of compactness in topological and metric spaces. The principal result of this section is to establish that compactness (as defined below and alluded to in Section 5.1) is equivalent to its sequential counterpart in the context of *metric* spaces. As as side result, we shall also give a proof to the Heine-Borel theorem regarding compact sets on $\mathbb{R}$. We begin with some terminology.

Definition 5.13. Let X be a set and A a subset of X. A family $\mathcal{U}_I = \{U_i\}_{i \in I}$ of subsets of X, where I is an index set, is said to be a **cover (or covering)** of A if

$$A \subseteq \bigcup_{i \in I} U_i.$$

Given any cover $\mathcal{U}_I = \{U_i\}_{i \in I}$ of A, a **subcover (or subcovering)** of $\mathcal{U}_I$ is a family $\mathcal{U}_J = \{U_i\}_{i \in J}$, where $J \subseteq I$, which is still a cover of A.

Said differently, a subcover is a sub-selection of sets which compose the original covering and remain a cover as well. For example, $\mathcal{U}_I = \{[1, 3], [2, 4], [3, 5]\}$ is a cover of $[1, 5] \subseteq \mathbb{R}$ and $\mathcal{U}_J = \{[1, 3], [3, 5]\}$ is a subcover of it still covering $[1, 5]$. According to the following definition, a cover is called **open** if it consists of open sets.

Definition 5.14. Let $(X, \mathcal{T})$ be a topological space and A a subset of X. A cover $\mathcal{U}_I = \{U_i\}_{i \in I}$ of A is said to be an **open cover** if

$$U_i \in \mathcal{T} \quad \text{for all } i \in I.$$

We are now able to officially define our notion of compactness, as discussed in Section 5.1.

Definition 5.15.

- A topological space $(X, \mathcal{T})$ is said to be **compact** if every open cover of X has a finite subcover.

- A subset Y of X is said to be **compact** if the topological subspace $(Y, \mathcal{T}|_Y)$ is compact when endowed with the induced subspace topology.

Remark 5.16. According to Definition 5.15, a topological space $(X, \mathcal{T})$ is compact when for any $\mathcal{U}_I = \{U_i\}_{i \in I} \subseteq \mathcal{T}$ such that

$$X = \bigcup_{i \in I} U_i,$$

there exists a **finite** set $J \subseteq I$, such that

$$X = \bigcup_{j \in J} U_j.$$

Similarly, a subset Y of X is compact when for any $\mathcal{V}_I = \{V_i\}_{i \in I} \subseteq \mathcal{T}|_Y$ such that $Y = \bigcup_{i \in I} V_i$ there exists a finite set $J \subseteq I$ such that $Y = \bigcup_{i \in J} V_i$.

The compactness of a subset a topological space admits the following equivalent characterisation which involves covers of Y with **open subsets of X**, rather than open in the subspace Y.[2]

Proposition 5.17. *Let $(X, \mathcal{T})$ be a topological space and Y be a subset of X. Then the following statements are equivalent:*

(i) *Y is compact;*

(ii) *every open cover of Y (i.e., cover by open subsets of X) has a finite subcover, i.e., for any $\mathcal{U}_I = \{U_i\}_{i \in I} \subseteq \mathcal{T}$ such that*

$$Y \subseteq \bigcup_{i \in I} U_i,$$

*there exists a **finite** set $J \subseteq I$ such that $Y \subseteq \bigcup_{j \in J} U_j$.*

Proof. (i) $\implies$ (ii): Let Y be a compact subset of X. Let $\{U_i\}_{i \in I} \subseteq \mathcal{T}$ be such that $Y \subseteq \bigcup_{i \in I} U_i$. Then

$$Y = \bigcup_{i \in I} (U_i \cap Y) \quad \text{and} \quad U_i \cap Y \in \mathcal{T}|_Y \text{ for all } i \in I.$$

---

[2]We draw to the reader's attention the fact that this property is in some textbooks given as the definition of compactness. It is not so important which property one considers as definition, but rather, to be aware that compactness can be expressed in two equivalent ways.

By utilising the compactness of Y, we get that there exists $J \subseteq I$, J finite, such that $Y = \bigcup_{i \in J}(U_i \cap Y)$, which yields that (ii) holds because

$$Y \subseteq \bigcup_{i \in J} U_i.$$

(ii) $\Longrightarrow$ (i): Suppose that (ii) holds. Let $\{V_i\}_{i \in I} \subseteq \mathcal{T}|_Y$ be such that $Y = \bigcup_{i \in I} V_i$. By the definition of the subspace topology $\mathcal{T}|_Y$, for each $i \in I$ there exists $U_i \in \mathcal{T}$ such that $V_i = U_i \cap Y$. Then we have that $Y \subseteq \bigcup_{i \in I} U_i$. By (*), there exists $J \subseteq I$, J finite, such that $Y \subseteq \bigcup_{i \in J} U_i$. But then

$$Y = \bigcup_{i \in J}(U_i \cap Y) = \bigcup_{i \in J} V_i.$$

This shows that Y is compact. $\qquad\square$

Let us now state, in the newly introduced terminology, and prove the result that historically lies at the origins of the notion of compactness.

Theorem 5.18 (Heine–Borel Theorem). *For any $a, b \in \mathbb{R}$ with $a \leq b$, the interval $[a, b]$ is a compact subset of $\mathbb{R}$ with the standard metric.*

Proof. Suppose for the sake of contradiction that $[a, b]$ is not compact. Let $\{U_i\}_{i \in I}$ be an open cover of $[a, b]$ which does not have any finite subcover. Then at least one of the intervals

$$\left[a, \frac{a+b}{2}\right] \quad \text{and} \quad \left[\frac{a+b}{2}, b\right]$$

cannot be covered by any finite subfamily of $\{U_i\}_{i \in I}$. Let us call this interval $[a_1, b_1]$ (if both intervals have this property, choose the leftmost one). Now repeat this process inductively. At step n, we define $[a_n, b_n]$ as either

$$\left[a_{n-1}, \frac{a_{n-1} + b_{n-1}}{2}\right] \quad \text{or} \quad \left[\frac{a_{n-1} + b_{n-1}}{2}, b_{n-1}\right],$$

so that $[a_n, b_n]$ cannot be covered by any finite subfamily of $\{U_i\}_{i \in I}$. In this way we construct sequences $(a_n)_1^\infty$ and $(b_n)_1^\infty$ such that

$$a \leq a_1 \leq a_2 \leq \dots \leq a_n \leq \dots \leq b_n \leq \dots \leq b_2 \leq b_1 \leq b,$$

and with $b_n - a_n = (b-a)/2^n$. Since $(a_n)_1^\infty$ is increasing and bounded above, whilst $(b_n)_1^\infty$ is decreasing and bounded below, the previous relations imply that there exists $\alpha \in [a, b]$ with $a_n \longrightarrow \alpha$ and $b_n \longrightarrow \alpha$ as $n \to \infty$. Since $\alpha \in [a, b] \subseteq \bigcup_{i \in I} U_i$, there exists $i_0 \in I$ such that $\alpha \in U_{i_0}$. Since U_{i_0} is open, there exists $r > 0$ such that $(\alpha - r, \alpha + r) \subseteq U_{i_0}$. Since $a_n \longrightarrow \alpha$ and $b_n \longrightarrow \alpha$ as $n \to \infty$, there exists $N \in \mathbb{N}$ such that

$$[a_N, b_N] \subseteq (\alpha - r, \alpha + r) \subseteq U_{i_0}.$$

This contradicts the fact that $[a_N, b_N]$ cannot be covered by any finite subfamily of $\{U_i\}_{i \in I}$. In conclusion, $[a, b]$ is compact. $\qquad\square$

For the sake of comparison, let us recall from Section 5.1 that the interval $(0, 1)$ and the line $\mathbb{R}$ are not compact because the respective covers $\{(0, \varepsilon) : 0 < \varepsilon < 1\}$ and $\{(-n, n) : n \in \mathbb{N}\}$ have no finite subcovers.

Although seemingly quite different, the notions of sequential compactness and compactness are equivalent in metric spaces, as the very important Theorem 5.21 that follows asserts. The proof of Theorem 5.21 is based on the next two lemmas, the first of which is as follows.

Lemma 5.19. *Let (X, d) be a metric space, $(x_n)_1^\infty$ a sequence in X and $x_0 \in X$. Then the following statements are equivalent:*

(i) *$(x_n)_1^\infty$ has a subsequence which converges to x_0;*

(ii) *for any $\varepsilon > 0$, the set $A_\varepsilon := \{n \in \mathbb{N} : x_n \in \mathbb{B}_\varepsilon(x_0)\}$ is infinite.*

Proof. (i) $\Longrightarrow$ (ii): Suppose first that (i) holds, and let $x_{n_k} \longrightarrow x_0$ in (X, d) as $k \to \infty$, where $(x_{n_k})_1^\infty$ is a subsequence of $(x_n)_1^\infty$. Then, for each $\varepsilon > 0$, there exists $K \in \mathbb{N}$ such that $x_{n_k} \in \mathbb{B}_\varepsilon(x_0)$ for all $k \geq K$. This clearly implies (ii).

(ii) $\Longrightarrow$ (i) Suppose now that (ii) holds. We apply the respective statement in the case $\varepsilon := 1/k$, for each $k \in \mathbb{N}$. For $k = 1$, pick $n_1 \in \mathbb{N}$ such that $x_{n_1} \in \mathbb{B}_1(x_0)$. For $k = 2$, pick $n_2 \in \mathbb{N}$ with $n_2 > n_1$ such that $x_{n_2} \in \mathbb{B}_{1/2}(x_0)$. Suppose now that, for some $k \in \mathbb{N}$, we have chosen $n_1, \ldots, n_{k-1}$. Since $A_{1/n}$ is infinite, it is possible to pick $n_k > n_{k-1}$ such that $x_{n_k} \in \mathbb{B}_{1/k}(x_0)$. In this way, we construct a subsequence of indices $n_1 < n_2 < \cdots < n_k < \cdots$ such that $x_{n_k} \in \mathbb{B}_{1/k}(x_0)$ for all $k \in \mathbb{N}$. It follows that $(x_{n_k})_1^\infty$ is a subsequence of $(x_n)_1^\infty$, and the fact that $d(x_{n_k}, x_0) < 1/k$ implies, by the Squeeze Theorem, that $x_{n_k} \longrightarrow x_0$ in (X, d) as $k \to \infty$, as required. $\qquad\square$

Our next auxiliary result concerns a subtle property of open coverings of sequentially compact metric spaces.

Lemma 5.20 (Lebesgue number of a covering). *Let (X, d) be a sequentially compact metric space and let $\{U_i\}_{i \in I}$ be an open cover of X. Then there exists $\delta > 0$ such that for every $x \in X$ there exists $i \in I$ such that $\mathbb{B}_\delta(x) \subseteq U_i$.*

Any radius $\delta > 0$ satisfying the property of Lemma 5.20 with respect to a cover $\{U_i\}_{i \in I}$ is called a **Lebesgue number** of the cover (see Figure 5.1).

Note that the **key point** in Lemma 5.20 is that the value of the radius is **independent** of the position of the centre; otherwise, the statement would be trivial.

Proof. Suppose for a contradiction that the claimed property fails. In particular, for $\delta := 1/n$, where $n \in \mathbb{N}$, we infer that there exists $x_n \in X$ with

$$\mathbb{B}_{\frac{1}{n}}(x_n) \not\subseteq U_i \quad \text{for any } i \in I. \tag{$\dagger$}$$

Consider the sequence $(x_n)_1^\infty$. Since (X, d) is sequentially compact, there exist

$x_0 \in X$ and a subsequence $(x_{n_k})_1^\infty$ of $(x_n)_1^\infty$ such that $x_{n_k} \longrightarrow x_0$ in (X,d) as $k \to \infty$. Since

$$x_0 \in X = \bigcup_{i \in I} U_i,$$

there exists $i_0 \in I$ such that $x_0 \in U_{i_0}$, and since $U_{i_0} \in \mathcal{T}$, there exists $r > 0$ such that $\mathbb{B}_r(x_0) \subseteq U_{i_0}$. Now, for any $k \in \mathbb{N}$,

$$\mathbb{B}_{\frac{1}{n_k}}(x_{n_k}) \subseteq \mathbb{B}_{\frac{1}{n_k}+d(x_{n_k},x_0)}(x_0).$$

Since $1/n_k \longrightarrow 0$ and $d(x_{n_k},x_0) \longrightarrow 0$ as $k \to \infty$, there exists $K \in \mathbb{N}$ such that $1/n_K + d(x_{n_K},x_0) < r$. Therefore,

$$\mathbb{B}_{\frac{1}{n_K}}(x_{n_K}) \subseteq \mathbb{B}_{\frac{1}{n_K}+d(x_{n_K},x_0)}(x_0) \subseteq \mathbb{B}_r(x_0) \subseteq U_{i_0},$$

a fact which contradicts (†). The lemma ensues. $\qquad\square$

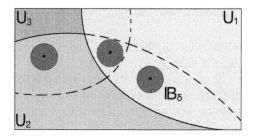

Figure 5.1: The idea of the Lebesgue number in Lemma 5.20 is that there exists a $\delta > 0$ small enough such that if one "slides" a small ball of this radius δ around any point, it will always be included entirely in one of the sets of the covering. In this picture we demonstrate this in the case of a cover of a (compact) rectangle in $\mathbb{R}^2$ by three (relatively) open sets $\{U_1, U_2, U_3\}$.

We may now come to the main result of this section.

Theorem 5.21 (Compact $\Leftrightarrow$ Sequentially compact). *A metric space (X,d) is compact if and only if it is sequentially compact.*

Note that, while the notion of compactness is expressed using open sets, the notion of sequential compactness involves convergent (sub)sequences. The underlying idea in the proof of Theorem 5.21 is the fact that both of these notions (i.e, "open set" and "convergent sequence") can be expressed by means of the notion of an "open ball".

Proof of Theorem 5.21. ($\Longrightarrow$): Let (X,d) be a compact metric space. Suppose for a contradiction that (X,d) is not sequentially compact. Let $(x_n)_1^\infty$ be a sequence in X which does not have any convergent subsequence. By Lemma 5.19, this implies that for any $x \in X$ there exists some $\varepsilon_x > 0$ such that

$$\{n \in \mathbb{N} : x_n \in \mathbb{B}_{\varepsilon_x}(x)\}$$

is finite. For any $x \in X$, we shall use the symbolisation $U_x := \mathbb{B}_{\varepsilon_x}(x)$. Clearly, each U_x is open and $X = \bigcup_{x \in X} U_x$. Since X is compact, it follows that there exist $p \in \mathbb{N}$ and $y_1, \ldots, y_p \in X$ such that

$$X = U_{y_1} \bigcup \cdots \bigcup U_{y_p}.$$

The above implies that

$$\mathbb{N} = \{n \in \mathbb{N} : x_n \in X\} = \bigcup_{j=1}^{p} \{n \in \mathbb{N} : x_n \in U_{y_j}\}.$$

This is impossible, however, since $\mathbb{N}$ is infinite, while the set in the right-hand side of the above is finite, being a finite union of finite sets (recall that, by construction, the set $\{n \in \mathbb{N} : x_n \in U_{y_j}\}$ is finite for each $j \in \{1, \ldots, p\}$). Therefore, (X, d) is sequentially compact.

($\Longleftarrow$): Let (X, d) be a sequentially compact metric space. We now prove that (X, d) is compact. To this aim, let $\{U_i\}_{i \in I}$ be an arbitrary open cover of X. Let $\delta > 0$ be a Lebesgue number of the open cover $\{U_i\}_{i \in I}$, whose existence is ensured by Lemma 5.20. This means that, for any $x \in X$ there exists $i \in I$ such that $\mathbb{B}_\delta(x) \subseteq U_i$.

Now we use the fact, proved in Theorem 5.12, that any sequentially compact metric space is totally bounded. Thus, by using in the total boundedness condition the chosen value δ (a Lebesgue number of the cover $\{U_i\}_{i \in I}$), we get that there exist $p \in \mathbb{N}$ and $y_1, \ldots, y_p \in X$ such that

$$X = \mathbb{B}_\delta(y_1) \bigcup \cdots \bigcup \mathbb{B}_\delta(y_p).$$

As a consequence of the definition of δ, we obtain that for each $j \in \{1, \ldots, p\}$ there exists $i_j \in I$ such that $\mathbb{B}_\delta(y_j) \subseteq U_{i_j}$. It now follows that

$$X = U_{i_1} \bigcup \cdots \bigcup U_{i_p},$$

which means that the open cover $\{U_i\}_{i \in I}$ has a finite subcover. Therefore, (X, d) is compact. □

It is worth mentioning that, for non-metrisable topological spaces, **neither of the statements "$(X, \mathcal{T})$ is compact" and "$(X, \mathcal{T})$ is sequentially compact" implies the other**. We refrain from presenting the details of the relevant (pathological counter)examples which are relatively complicated and would take us too far afield.[3]

---

[3]Let us merely mention without any details that the uncountable Cartesian product $[0, 1]^{\mathbb{R}}$ is compact but not sequentially compact and $[0, \omega_1)$ where ω_1 is the first uncountable ordinal is sequentially compact but not compact. The reader is invited to look for instance at the (more advanced) book [40] for details.

5.4 Compactness and finite dimensionality in normed spaces

In this section we revisit the class of normed vector spaces and study them in relation to their compactness properties. We shall see that a insuperable dichotomy emerges:

• In the case of **finite-dimensional** normed vector spaces we establish the important result that *any two norms are equivalent*. This means there is only one norm-induced topology on such a space.[4] A consequence of this fact is that the completeness and compactness properties of any finite-dimensional normed space are essentially the same as those of $(\mathbb{F}^m, \|\cdot\|_2)$. When combining this result with Theorems 5.3-5.4, we infer that the **closed balls of any finite-dimensional normed vector space are compact.**

• While the norm-induced topologies in finite dimensions are fully understood, in infinite dimensions one of the most important tool in the constructions of certain objects (i.e., the possibility of extracting convergent subsequences out of any bounded sequence) becomes unavailable. Accordingly, we demonstrate the striking result that **the closed unit balls of infinite-dimensional normed vector spaces are NOT compact**, neither sequentially nor otherwise. Thus, one has to identify different approaches to achieve similar aims.[5]

We begin with the case of finite dimensions. Accordingly, the first result of this section is the following.

Theorem 5.22 (Norm equivalence in finite-dimensional vector spaces). *Let X be a finite-dimensional vector space. Then any two norms on X are equivalent.*

Proof. Let $\{e_1, \ldots, e_n\}$ be a (Hamel) basis for X. Then any $x \in X$ can be written in a unique way as

$$x = \sum_{k=1}^{n} \lambda_k \, e_k, \qquad (*)$$

where $\lambda_1, \ldots, \lambda_n \in \mathbb{F}$. Let us define, for any such $x \in X$,

$$\|x\|_2 := \left(\sum_{k=1}^{n} |\lambda_k|^2 \right)^{1/2}.$$

---

[4]Note carefully that this has *not* yet been showed in the text. Instead, what we have previously seen is that *all the possible* $\|\cdot\|_p$*-norms induce the same topology on* $\mathbb{F}^m$. We advise the reader to have a quick look back at Chapter 2 and in particular at Remark 2.52.

[5]To this end and for our curious readers, we mention that our subsequent Chapters 12-13 contain developments exactly towards the direction of somehow "ascertaining compactness for a weaker topology", because of the failure of compactness of closed and bounded sets in infinite dimensions (with respect to the natural norm topology).

Then one can easily check that $\| \cdot \|_2$ is a norm on X. To prove the required result, it suffices to show that any norm $\| \cdot \|$ on X is equivalent to $\| \cdot \|_2$. To this end, let us define the constant (which depends only on the selected basis)

$$M := \left(\sum_{k=1}^{n} \|e_k\|^2 \right)^{1/2}.$$

Then, for any $\lambda_1, \ldots, \lambda_n \in \mathbb{F}$, the axioms of a norm, the Cauchy–Schwarz Inequality and $(*)$ imply that

$$
\begin{aligned}
\|x\| &= \|\lambda_1 e_1 + \cdots + \lambda_n e_n\| \\
&\leq |\lambda_1| \|e_1\| + \cdots + |\lambda_n| \|e_n\| \\
&\leq \left(\sum_{k=1}^{n} \|e_k\|^2 \right)^{1/2} \left(\sum_{k=1}^{n} |\lambda_k|^2 \right)^{1/2}
\end{aligned}
$$

which yields

$$
\begin{aligned}
\|x\| &\leq M \|\lambda_1 e_1 + \cdots + \lambda_n e_n\|_2 \\
&= M \|x\|_2.
\end{aligned}
$$

Thus we have proved that there exists $M > 0$ such that

$$\|x\| \leq M \|x\|_2 \quad \text{for all } x \in X.$$

We now show that there exists $m > 0$ such that

$$\|x\| \geq m \|x\|_2 \quad \text{for all } x \in X. \tag{$\dagger$}$$

To this aim, we claim that it suffices to show there exists $m > 0$ such that

$$\|z\| \geq m \quad \text{for all } z \in X \text{ with } \|z\|_2 = 1. \tag{$\dagger\dagger$}$$

Indeed, $(\dagger)$ is trivially true $x = 0$. If $x \in X \setminus \{0\}$, we rewrite $(\dagger)$ as

$$\left\| \frac{x}{\|x\|_2} \right\| \geq m \quad \text{for all } x \in X$$

and the equivalence between $(\dagger)$ and $(\dagger\dagger)$ is obvious through the substitution $z := x/\|x\|_2$, since $\|z\|_2 = 1$ if for any $x \in X \setminus \{0\}$.

It remains to prove the validity of $(\dagger\dagger)$. To that aim, we study the function $z \mapsto \|z\|$ on the set $\{z : \|z\|_2 = 1\}$. In view of the representation $(*)$, we therefore consider the function $F : \mathbb{F}^n \longrightarrow \mathbb{R}$ given by

$$F(\underline{\lambda}) := \|\lambda_1 e_1 + \cdots + \lambda_n e_n\|,$$

for any $\underline{\lambda} = (\lambda_1, \ldots, \lambda_n) \in \mathbb{F}^n$. In particular, we shall study the behaviour of F on the unit sphere $\mathbb{S}_1(0)$ in $(\mathbb{F}^n, \|\cdot\|_2)$.[6] The goal is to show that F is a continuous function on a compact set and therefore attains its infimum which also is positive. Note first that F is continuous on $\mathbb{F}^n$, since for any $\underline{\alpha} = (\alpha_1, \ldots, \alpha_n), \underline{\beta} = (\beta_1, \ldots, \beta_n) \in \mathbb{F}^n$, we have

$$
\begin{aligned}
\left| F(\underline{\alpha}) - F(\underline{\beta}) \right| &= \left| \|\alpha_1 e_1 + \ldots + \alpha_n e_n\| - \|\beta_1 e_1 + \ldots + \beta_n e_n\| \right| \\
&\leq \left\| (\alpha_1 - \beta_1)e_1 + \cdots + (\alpha_n - \beta_n)e_n \right\| \\
&\leq M \|\underline{\alpha} - \underline{\beta}\|_2,
\end{aligned}
$$

where M has been defined earlier in the proof, and we have used again the Cauchy–Schwarz Inequality in the last step. Since f is continuous on the (sequentially) compact set $\mathbb{S}_1(0)$ in $\mathbb{F}^n$, by the Weierstrass Theorem 5.30, there exists $\underline{\mu} = (\mu_1, \ldots, \mu_n) \in \mathbb{S}_1(0)$ such that

$$
F(\underline{\mu}) \leq F(\underline{\lambda}) \quad \text{for all } \underline{\lambda} \in \mathbb{S}_1(0).
$$

Let $m := F(\underline{\mu})$. Clearly $m \geq 0$, and we now claim that $m > 0$ (see Figure 5.2). Indeed, if we suppose for a contradiction that $m = 0$, it follows that

$$
\|\mu_1 e_1 + \cdots + \mu_n e_n\| = 0,
$$

which implies, in view of the fact that $\{e_1, \ldots, e_n\}$ is a basis, that

$$
\mu_1 = \cdots = \mu_n = 0,
$$

contradicting the fact that $\underline{\mu} \in \mathbb{S}_1(0)$.

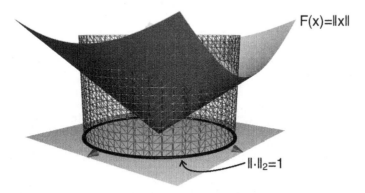

Figure 5.2: In illustration of the idea of the proof of the equivalence between the Euclidean norm $\|\cdot\|_2$ and an arbitrary norm $\|\cdot\|$ on $\mathbb{R}^2$. The function $F(x) = \|x\|$ attains its minimum and its maximum of the unit circle $\{\|\cdot\|_2 = 1\}$ which is a compact set not containing the origin.

---

[6]Note that we are using $\|\cdot\|_2$ with two different meanings: one as a norm on X, and one as the Euclidean norm on $\mathbb{F}^n$; it should be clear, however, from the context which one of them is used at any instance.

We have thus proved the existence of $m > 0$ such that

$$\left\| \lambda_1 e_1 + \cdots + \lambda_n e_n \right\| \geq m \quad \text{for all } (\lambda_1, .., \lambda_n) \text{ with } \left(\sum_{k=1}^{n} |\lambda_k|^2 \right)^{1/2} = 1,$$

which is merely another way of writing (††) in terms of the expansion of z with respect to the given basis. The theorem ensues. $\qquad\square$

The next result shows that all finite-dimensional normed spaces have entirely similar properties to the familiar ones for Euclidean spaces.

Theorem 5.23. *Let $(X, \| \cdot \|)$ be a finite-dimensional normed vector space. Then the following hold:*

(i) $(X, \| \cdot \|)$ *is Banach space.*

(ii) *Any bounded sequence in $(X, \| \cdot \|)$ has a convergent subsequence.*

(iii) *Any closed bounded subset of $(X, \| \cdot \|)$ is (sequentially) compact.*

Roughly speaking, the idea of the proof is that, if we define $\| \cdot \|_2$ as in the proof of Theorem 5.22, then $(X, \| \cdot \|)$ will have similar properties to $(X, \| \cdot \|_2)$ because of the equivalence of the norms, whilst $(X, \| \cdot \|_2)$ will have similar properties to $(\mathbb{F}^m, \| \cdot \|_2)$ because of the similar way in which the norms are defined. We now make these arguments precise.

Proof. Let $\{e_1, \ldots, e_m\}$ be a (Hamel) basis for X. Then, for any $x \in X$, there exists a unique $\underline{\lambda} = (\lambda_1, \ldots, \lambda_m) \in \mathbb{F}^m$ such that

$$x = \lambda_1 e_1 + \ldots + \lambda_m e_m.$$

This correspondence $x \mapsto \underline{\lambda}$ is a (linear) bijection between X and $\mathbb{F}^m$. For any $x \in X$ as above, we define

$$\|x\|_2 := \left(\sum_{k=1}^{m} |\lambda_k|^2 \right)^{1/2}.$$

Then $\| \cdot \|_2$ is a norm on X, and $\|x\|_2 = \|\underline{\lambda}\|_2$, where $\| \cdot \|_2$ also denotes the standard Euclidean norm on $\mathbb{F}^m$. Recall also that, by Theorem 5.22, $\| \cdot \|$ and $\| \cdot \|_2$ are equivalent norms on X.

(i) Let $(x_n)_1^\infty$ be an arbitrary Cauchy sequence in $(X, \| \cdot \|)$. Then $(x_n)_1^\infty$ is also a Cauchy sequence in $(X, \| \cdot \|_2)$. Let $(\underline{\lambda}^{(n)})_1^\infty$ be the corresponding sequence in $\mathbb{F}^m$. Since

$$\left\| \underline{\lambda}^{(n)} - \underline{\lambda}^{(p)} \right\|_2 = \|x_n - x_p\|_2$$

for all $n, p \in \mathbb{N}$, it follows that $(\underline{\lambda}^{(n)})_1^\infty$ is a Cauchy sequence in $(\mathbb{F}^m, \| \cdot \|_2)$, where $m = \dim X$. Since $(\mathbb{F}^m, \|\cdot\|_2)$ is complete, there exists $\underline{\lambda}^{(0)} \in \mathbb{F}^m$ such that $\underline{\lambda}^{(n)} \longrightarrow \underline{\lambda}^{(0)}$ in $(\mathbb{F}^m, \|\cdot\|_2)$ as $n \to \infty$. Let

$$x_0 := \lambda_1^{(0)} e_1 + \cdots + \lambda_m^{(0)} e_m$$

where $\{\lambda_1^{(0)}, \ldots, \lambda_m^{(0)}\}$ are the components of the vector $\underline{\lambda}^{(0)}$. Since

$$\|x_n - x_0\|_2 = \|\underline{\lambda}^{(n)} - \underline{\lambda}^{(0)}\|_2$$

for all $n \in \mathbb{N}$, it follows that $x_n \longrightarrow x_0$ in $(X, \|\cdot\|_2)$. In turn, this implies that $x_n \longrightarrow x_0$ in $(X, \|\cdot\|)$ as $n \to \infty$. Therefore, $(X, \|\cdot\|)$ is a Banach space.

(ii) Let $(x_n)_1^\infty$ be an arbitrary bounded sequence in $(X, \|\cdot\|)$. Then $(x_n)_1^\infty$ is also a bounded sequence in $(X, \|\cdot\|_2)$. It follows that the corresponding sequence $(\underline{\lambda}^{(n)})_1^\infty$ is bounded in $(\mathbb{F}^m, \|\cdot\|_2)$. Therefore, by the Bolzano–Weierstrass Theorem 5.3, there exists a subsequence $(\underline{\lambda}^{(n_k)})_1^\infty$ which converges in the space $(\mathbb{F}^m, \|\cdot\|_2)$ to some $\underline{\lambda}^{(0)} \in \mathbb{F}^m$ as $k \to \infty$. Let

$$x_0 := \lambda_1^{(0)} e_1 + \ldots + \lambda_m^{(0)} e_m,$$

where $\{\lambda_1^{(0)}, \ldots, \lambda_m^{(0)}\}$ are the components of the vector $\underline{\lambda}^{(0)}$. Since

$$\|x_{n_k} - x_0\|_2 = \|\underline{\lambda}^{(n_k)} - \underline{\lambda}^{(0)}\|_2,$$

for all $n \in \mathbb{N}$, it follows that $x_{n_k} \longrightarrow x$ in $(X, \|\cdot\|_2)$ as $k \to \infty$. In turn, this yields that $x_{n_k} \longrightarrow x_0$ in $(X, \|\cdot\|)$ as $k \to \infty$, as required.

(iii) Let Y be any closed bounded subset of $(X, \|\cdot\|)$. Let $(x_n)_1^\infty$ be an arbitrary sequence in Y. Since Y is bounded, by part (ii) there exist $x_0 \in X$ and a subsequence $(x_{n_k})_1^\infty$ such that $x_{n_k} \longrightarrow x_0$ in $(X, \|\cdot\|)$ as $k \to \infty$. Since Y is closed, it follows that $x_0 \in Y$. This establishes that Y is (sequentially) compact. □

The next result is an immediate consequence of part (i) of the preceding theorem, combined with Proposition 3.9.

Theorem 5.24 (Finite-dimensional subspaces are closed). *Let $(X, \|\cdot\|)$ be a normed space, and Y be a finite-dimensional vector subspace of X. Then Y is closed in $(X, \|\cdot\|)$.*

We have seen earlier that in any finite-dimensional normed space any closed bounded set is (sequentially) compact, and thus in particular the closed unit ball is compact. The next result, which is very important, implies that the compactness of the closed unit ball characterises finite-dimensional normed spaces.[7]

---

[7]At this point it is useful to recall that, in view of Proposition 2.67, closed balls $\mathbb{D}_r(x)$ and closures of open balls $\overline{\mathbb{B}}_r(x)$ coincide in normed vector spaces. Further, the respective spheres $\mathbb{S}_r(x)$ coincide with the boundaries $\partial \mathbb{B}_r(x) = \partial \mathbb{D}_r(x)$ of either.

Theorem 5.25 (The failure of compactness in infinite dimensions).
Let $(X, \|\cdot\|)$ be an infinite-dimensional normed vector space. Then the closed unit ball $\overline{\mathbb{B}}_1(0)$ is **not** *(sequentially) compact in $(X, \|\cdot\|)$.*

To prove Theorem 5.25 it is necessary to exhibit a sequence in $\overline{\mathbb{B}}_1(0)$ which has no convergent subsequence, having at our disposal no further information other than the infinite dimensionality of X. Since X may or may not be complete, it seems reasonable that our objective should be to construct a sequence with no Cauchy subsequence.

Aiming to define such a non-Cauchy sequence recursively, we would like that at every step the newly constructed term lies "far enough" from all the preceding terms. The key ingredient to achieve such a construction is the following result, which establishes that one can find on the unit sphere elements which are almost as far way from a given vector subspace as possible.

Theorem 5.26 (Riesz's Lemma). *Let $(X, \|\cdot\|)$ be a normed space and Y be a closed vector subspace of X such that $Y \neq X$. Then for every $\alpha \in (0,1)$ (as close to 1 as desired, see Figure 5.3), there exists $z_\alpha \in \partial\mathbb{B}_1(0)$ such that*

$$\|z_\alpha - y\| \geq \alpha \quad \text{for all } y \in Y. \tag{$\diamond$}$$

Proof of Theorem 5.26. Let us define, for any $x \in X$,

$$d(x, Y) := \inf\left\{\|x - y\| : y \in Y\right\}.$$

Note that $(\diamond)$ can then be reformulated as $d(z_\alpha, Y) \geq \alpha$. Observe also that, for every $z \in \partial\mathbb{B}_1(0)$, we have that $d(z, Y) \leq 1$.

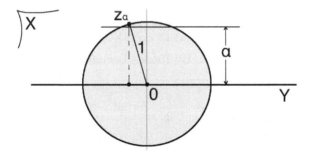

Figure 5.3: In illustration of the idea of the Riesz lemma for $X = \mathbb{R}^2$ and $Y = \mathbb{R} \times \{0\}$. In this case it is geometrically obvious that the conclusion is true, but the claim holds true in infinite dimensions as well, wherein the situation becomes far less clear!

We prove first that for every $x \in X \setminus Y$ we have that $d(x, Y) > 0$. Indeed, suppose for the sake of contradiction that this is not so, namely $d(x, Y) = 0$. Then for every $n \in \mathbb{N}$ there exists $y_n \in Y$ such that $\|x - y_n\| < 1/n$. This implies that $y_n \longrightarrow x$ in $(X, \|\cdot\|)$ as $n \to \infty$ and hence, since $(y_n)_1^\infty \subseteq Y$ and Y is closed, it follows that $x \in Y$, a contradiction.

Fix now $\alpha \in (0,1)$, as close to 1 as desired. Choose $x_0 \in X \setminus Y$ and let us use the symbolisation $d_0 := d(x_0, Y) > 0$. We are going to choose a unit vector z_α of the form

$$z_\alpha := \frac{x_0 - w}{\|x_0 - w\|},$$

for some $w \in Y$ to be suitably chosen.[8] Then, for any $y \in Y$, we have that

$$\begin{aligned}
\|z_\alpha - y\| &= \left\| \frac{x_0 - w}{\|x_0 - w\|} - y \right\| \\
&= \frac{\|x_0 - (w + \|x_0 - w\|y)\|}{\|x_0 - w\|} \\
&\geq \frac{d_0}{\|x_0 - w\|}.
\end{aligned}$$

The required conclusion is then satisfied, provided that $w \in Y$ can be chosen such that

$$\|x_0 - w\| \leq \frac{d_0}{\alpha}.$$

In view of the definition of d_0 and of the fact that $0 < \alpha < 1$, such a choice for $w \in Y$ is indeed possible. This completes the proof. $\qquad\square$

Proof of Theorem 5.25. The proof is based on repeated applications of Riesz's Lemma. Let $x_1 \in X$ with $\|x_1\| = 1$. Let us define $X_1 := \mathrm{span}[\{x_1\}]$. Then X_1 is 1-dimensional and by Theorem 5.24 it is a closed subspace of X. By Riesz's Lemma, there exists $x_2 \in X$ with $\|x_2\| = 1$ such that

$$\|x_2 - y\| \geq \frac{1}{2} \quad \text{for all } y \in X_1.$$

Let now

$$X_2 := \mathrm{span}[\{x_1, x_2\}] \subseteq X.$$

By Theorem 5.24, X_2 is closed. By Riesz's Lemma, there exists $x_3 \in X$ with $\|x_3\| = 1$ such that

$$\|x_3 - y\| \geq \frac{1}{2} \quad \text{for all } y \in X_2.$$

Suppose that, for some $n \in \mathbb{N}$, $x_1, \ldots, x_{n-1}$ have been constructed. Let us define the (finite-dimensional) space

$$X_{n-1} := \mathrm{span}[\{x_1, \ldots, x_{n-1}\}] \subseteq X.$$

By Theorem 5.24, X_{n-1} is closed. By Riesz's Lemma, there exists $x_n \in X$ with $\|x_n\| = 1$ such that

$$\|x_n - y\| \geq \frac{1}{2} \quad \text{for all } y \in X_{n-1}.$$

---

[8]Note that such a choice for z_α is not as special as it may first seem. Indeed, apart from the normalising factor, we are just seeking z_α as the sum between a vector contained in Y and a vector contained in $X \setminus Y$.

In particular, for every $k \in \{1, \ldots, n-1\}$, we have $\|x_n - x_k\| \geq 1/2$. In this way we construct a sequence $(x_n)_1^\infty \subseteq \partial \mathbb{B}_1(0) \subseteq \overline{\mathbb{B}}_1(0)$ such that

$$\|x_i - x_j\| \geq \frac{1}{2} \quad \text{for all } i, j \in \mathbb{N} \text{ with } i \neq j.$$

This implies that, for any subsequence $(x_{n_k})_1^\infty$ of $(x_n)_1^\infty$, we also have that

$$\|x_{n_i} - x_{n_j}\| \geq \frac{1}{2} \quad \text{for all } i, j \in \mathbb{N} \text{ with } i \neq j.$$

This yields that $(x_{n_k})_1^\infty$ is not a Cauchy sequence, and therefore it cannot be convergent. Therefore, $\overline{\mathbb{B}}_1(0)$ is not sequentially compact. $\quad\square$

Remark 5.27 (All balls lack compactness)**.** Note that the normalisation of the unit ball with radius equal to one in Theorem 5.25 is merely for convenience, since a simpl translation and dilation argument shows that, as a consequence, all balls lack compactness because

$$\overline{\mathbb{B}}_r(x) = x + r\,\overline{\mathbb{B}}_1(0) = x + r\{x \in X : \|x\| \leq 1\}.$$

The proof actually shows that the sequence $(x_n)_1^\infty \subset \overline{\mathbb{B}}_1(0)$ constructed above has no Cauchy subsequence, yielding the next corollary:

Corollary 5.28 (Failure of total boundedness in infinite dimensions)**.** *Let* $(X, \|\cdot\|)$ *be an infinite-dimensional normed vector space. Then, the closed unit ball* $\overline{\mathbb{B}}_1(0)$ *of the space is not totally bounded.*

Proof. It is an immediate consequence of Theorem 5.11. $\quad\square$

Finally, since every open ball contains a closed ball of smaller radius, it follows from Remark 5.27 that no open ball and in general no bounded open set is totally bounded.

5.5 Continuous and uniformly continuous maps

In this section we revisit continuous (real-valued) functions as well general maps defined on topological and metric spaces, now in relation to compactness and/or sequential compactness of the underlying spaces. We also introduce and study a stronger version of continuity, called **uniform continuity**, a special case of which is already known from Real Analysis.

We begin with the following two theorems, the second of which is merely a particular case of the first, which are natural extensions to the abstract topological realm of the Weierstrass Theorem from Real Analysis.

Theorem 5.29 (Weierstrass Theorem for general maps). *Let $(X, \mathcal{T})$ and $(Y, \mathcal{S})$ be topological spaces and $f : X \longrightarrow Y$ a continuous mapping. Then, for every compact subset $K \subseteq X$, $f(K)$ is a compact subset of Y.*

Proof. Let K be a compact subset of X. Let $\{V_i\}_{i \in I} \subseteq \mathcal{S}$ be an arbitrary open cover of $f(K)$, meaning that $f(K) \subseteq \bigcup_{i \in I} V_i$. By formulas $(\star)$-$(\star\star)$ of Chapter 1, the inverse image commutes with unions. Hence, it follows that

$$K \subseteq f^{-1}\left(\bigcup_{i \in I} V_i \right) = \bigcup_{i \in I} f^{-1}(V_i).$$

Since f is continuous on X and $V_i \in \mathcal{S}$ for all $i \in I$, it follows that $f^{-1}(V_i) \in \mathcal{T}$ for all $i \in I$. Hence $\{f^{-1}(V_i)\}_{i \in I}$ is an open cover of K and hence, since K is compact, there exist $J \subseteq I$, J finite, such that

$$K \subseteq \bigcup_{i \in J} f^{-1}(V_i) = f^{-1}\left(\bigcup_{i \in J} V_i \right).$$

This implies that $f(K) \subseteq \bigcup_{i \in J} V_i$. Therefore, $f(K)$ is compact. $\qquad\square$

We invite the reader to notice the simplicity of the argument in the proof above. Once the "compactness framework" has been set up, the idea is a direct application of the definition of compactness combined with the fact the inverse image commutes with unions.

In particular, for $Y = \mathbb{R}$ one might prove the next specialised result which, by utilising the global ordering structure of $\mathbb{R}$, provides the additional information that the function *attains* its supremum and its infimum.

Theorem 5.30 (Weierstrass Theorem for real functions). *Let $(X, \mathcal{T})$ be a topological space and $f : X \longrightarrow \mathbb{R}$ a continuous function, where $\mathbb{R}$ has the usual topology. Then, for every compact subset $K \subseteq X$, $f(K)$ is a compact subset of $\mathbb{R}$ and there exist $x_m, x_M \in K$ such that*

$$f(x_m) \leq f(x) \leq f(x_M) \quad \text{for all } x \in K.$$

Proof. Let K be an arbitrary compact subset of X. By Theorem 5.29, $f(K)$ is a compact subset of $\mathbb{R}$. Since $\mathbb{R}$ is a metric space, it follows that $f(K)$ is sequentially compact and, moreover, $f(K)$ is bounded and closed. Since $f(K)$ is bounded, one can define $\alpha := \inf f(K)$ and $\beta := \sup f(K)$, where $\alpha, \beta \in \mathbb{R}$. By the properties of suprema and infima we have that $\alpha, \beta \in \overline{f(K)}$. Since the set $f(K)$ is closed, this yields $\alpha, \beta \in f(K)$. Thus, there exist $x_m, x_M \in K$ such that $\alpha = f(x_m)$, $\beta = f(x_M)$. But then the definition of α and β implies that $f(K)$ is contained in the interval $[f(x_m), f(x_M)]$ (see Figure 5.4). $\qquad\square$

We now introduce, for functions between metric spaces, a seemingly stronger version of the notion of continuity, in that the distance between two values of a function can be made as small as one wishes, provided that the two points in the domain are sufficiently close, irrespective of their position in the domain space.

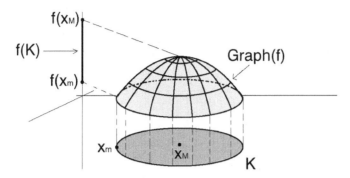

Figure 5.4: An illustration of a function $f : K \longrightarrow \mathbb{R}$ with $K \subseteq \mathbb{R}^2$ is compact (a closed disc). In this particular example, the point x_m where the infimum is attained is on the boundary ∂K and the point x_M where the supremum is attained is in the interior K°. The image $f(K)$ is always contained in the interval $[f(x_m), f(x_M)]$.

Such functions are called **uniformly continuous**. This is a notion that the readers have most likely encountered in the special case of real-valued functions of one variable. We then prove that any continuous function on a compact metric space is uniformly continuous there. This result has important applications in Analysis.[9]

Definition 5.31. Let (X, d) and (Y, ρ) be metric spaces. A mapping $f : X \longrightarrow Y$ is called **uniformly continuous** on X if, for any $\varepsilon > 0$, there exists[10] $\delta = \delta(\varepsilon) > 0$ such that:

for any $x, y \in X$ with $d(x, y) < \delta$, we have $\rho\big(f(x), f(y)\big) < \varepsilon$.

Remark 5.32. Recall that the continuity of f on X can be expressed as: for any $x \in X$ and for any $\varepsilon > 0$ there exists $\delta = \delta(x, \varepsilon) > 0$ such that:

for any $y \in X$ with $d(x, y) < \delta$, we have $\rho\big(f(x), f(y)\big) < \varepsilon$.

The difference between the two definitions is that, in the case of continuity, δ **generally depends on both** ε **and** x (and we symbolised this dependence explicitly as $\delta(x, \varepsilon)$ for easier memorability), whilst in the case of uniformly continuity it is required that δ **depends only on** ε.

Clearly, any uniformly continuous map $f : X \longrightarrow Y$ is continuous on X. However, the converse is not true in general:

---

[9]For example, uniform continuity is the key ingredient in the proof that any continuous function on a closed bounded interval $[a, b]$ is Riemann integrable.

[10]As we have occasionally done in the case of sequences, we explicitly symbolise the number δ as $\delta(\varepsilon)$ to emphasise its dependence on the choice of ε.

Example 5.33. The function $f : (0, \infty) \longrightarrow \mathbb{R}$ given by $f(x) = 1/x$ for $x > 0$ is continuous on $(0, \infty)$ but not uniformly continuous. Indeed, by defining the sequences $(x_n)_1^\infty$ and $(y_n)_1^\infty$ as

$$x_n := 1/n , \quad y_n := 1/(n+1) \quad \text{for all } n \in \mathbb{N},$$

we have that (see Figure 5.5)

$$|x_n - y_n| = \frac{1}{n(n+1)} \longrightarrow 0 \text{ as } n \to \infty$$

but $\big|f(x_n) - f(y_n)\big| = 1$, for all $n \in \mathbb{N}$.

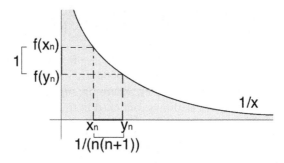

Figure 5.5: The uniform continuity of the function $f(x) = 1/x$ fails because we cannot choose uniformly the $\delta(\varepsilon)$ irrespectively of the point. For $\varepsilon = 1$, we have $\delta(\varepsilon) = 1/(n(n+1))$ which "shrinks" as $n \to \infty$.

Quite remarkably, if the metric space (X, d) is compact then the converse is also true as well and continuity implies uniform continuity.

Theorem 5.34. *Let (X, d) be a (sequentially) compact metric space, (Y, ρ) be a metric space and $f : X \longrightarrow Y$ be a continuous mapping on X. Then f is uniformly continuous on X.*

We give two different proof of this result, one using the sequential compactness of X and one using compactness. The first is probably simpler, while the second illustrates the general principle that compactness allows us to pass from local properties to global ones.

First proof of Theorem 5.34 (using sequential compactness). Suppose for the sake of contradiction that f is not uniformly continuous on X. Then there would exist $\varepsilon_0 > 0$ such that for all $\delta > 0$ there would exist (δ-dependent) $x_\delta, y_\delta \in X$ with

$$d(x_\delta, y_\delta) < \delta \quad \text{and} \quad \rho\big(f(x_\delta), f(y_\delta)\big) \geq \varepsilon_0.$$

In particular, for $\delta := 1/n$, where $n \in \mathbb{N}$, there exist (n-dependent) points $x_n, y_n \in X$ with

$$d(x_n, y_n) < 1/n \quad \text{and} \quad \rho\big(f(x_n), f(y_n)\big) \geq \varepsilon_0.$$

Since X is sequentially compact, there exists a subsequence $(x_{n_k})_1^\infty$ of $(x_n)_1^\infty$ and there exists $x_0 \in X$ such that $x_{n_k} \longrightarrow x_0$ in (X, d), the last relation being equivalent to $d(x_{n_k}, x_0) \longrightarrow 0$ as $k \to \infty$. But then, by the triangle inequality,

$$
\begin{aligned}
d(y_{n_k}, x_0) &\leq d(y_{n_k}, x_{n_k}) + d(x_{n_k}, x_0) \\
&\leq \frac{1}{n_k} + d(x_{n_k}, x_0) \longrightarrow 0 \quad \text{as } n \to \infty,
\end{aligned}
$$

and therefore, by the Squeeze Theorem we infer that $d(y_{n_k}, x_0) \longrightarrow 0$, that is, $y_{n_k} \longrightarrow x_0$ in (X, d) as $k \to \infty$.

Since f is continuous at x_0, and $x_{n_k} \longrightarrow x_0$, $y_{n_k} \longrightarrow x_0$ in (X, d) as $k \to \infty$, the sequential characterisation of continuity shows that $f(x_{n_k}) \longrightarrow f(x_0)$ and $f(y_{n_k}) \longrightarrow f(x_0)$ in (Y, ρ) as $k \to \infty$. This means that $\rho\big(f(x_{n_k}), f(x_0)\big) \longrightarrow 0$ and $\rho\big(f(y_{n_k}), f(x_0)\big) \longrightarrow 0$ as $k \to \infty$. But then,

$$
\begin{aligned}
\rho\big(f(x_{n_k}), f(y_{n_k})\big) &\leq \rho\big(f(x_{n_k}), f(x_0)\big) + \rho\big(f(y_{n_k}), f(x_0)\big) \\
&\longrightarrow 0 + 0 = 0 \quad \text{as } k \to \infty,
\end{aligned}
$$

and hence, by the Squeeze Theorem, $\rho\big(f(x_{n_k}), f(y_{n_k})\big) \longrightarrow 0$. This, however, contradicts the fact that $\rho\big(f(x_{n_k}), f(y_{n_k})\big) \geq \varepsilon_0$ for all $k \in \mathbb{N}$. In conclusion, f is uniformly continuous on X. □

Second proof of Theorem 5.34 (using compactness). Fix $\varepsilon > 0$, arbitrary. Then, for every $x \in X$, the continuity of f at x shows that there exists $\delta(x, \varepsilon) > 0$ such that

$$
\rho\big(f(x), f(y)\big) < \varepsilon/2 \quad \text{for all } y \in X \text{ with } d(x, y) < \delta(x, \varepsilon).
$$

Since

$$
X = \bigcup_{x \in X} \mathbb{B}_{\delta(x, \varepsilon)/2}(x)
$$

and X is compact, there exist $n \in \mathbb{N}$ and $x_1, \ldots, x_n \in X$ such that

$$
X = \bigcup_{k=1}^n \mathbb{B}_{\delta(x_k, \varepsilon)/2}(x_k).
$$

We define

$$
\delta(\varepsilon) := \min \left\{ \frac{\delta(x_1, \varepsilon)}{2}, \ldots, \frac{\delta(x_k, \varepsilon)}{2} \right\}
$$

and note that, as the notation suggests, that δ depends on ε only. Let $x, y \in X$ arbitrary with $d(x, y) < \delta$. Then there exists $k \in \{1, \ldots, n\}$ such that $x \in \mathbb{B}_{\delta(x_k, \varepsilon)/2}(x_k)$, which means that $d(x, x_k) < \delta(x_k, \varepsilon)/2$. But then, by the definition of δ,

$$
d(y, x_k) \leq d(y, x) + d(x, x_k) < \delta + \frac{\delta(x_k, \varepsilon)}{2} \leq \delta(x_k, \varepsilon).
$$

Therefore, since $d(x, x_k) < \delta(x_k, \varepsilon)$ and $d(y, x_k) < \delta(x_k, \varepsilon)$, it follows that

$$\rho\big(f(x), f(y)\big) \leq \rho\big(f(x), f(x_k)\big) + \rho\big(f(y), f(x_k)\big) < \frac{\varepsilon}{2} + \frac{\varepsilon}{2} = \varepsilon.$$

Hence f is indeed uniformly continuous. $\qquad\qquad\qquad\qquad\qquad\qquad\Box$

5.6   Compact sets in the space of continuous functions

As we saw earlier, in any infinite-dimensional normed space the closed unit ball is not (sequentially) compact. A consequence is that, unlike the finite-dimensional case, in infinite-dimensional normed spaces $(X, \| \cdot \|)$ closedness and boundedness of a set are **not enough** to ensure its compactness:

$$\left.\begin{array}{l} (x_n)_1^\infty \subseteq X : \\[4pt] \sup_{n \in \mathbb{N}} \|x_n\| \leq C \end{array}\right\} \;\;\not\Rightarrow\;\; \exists\, (x_{n_k})_1^\infty \text{ and } x \in X : x_{n_k} \longrightarrow x \text{ as } k \to \infty.$$

Therefore, it is of interest to find characterisations of compactness for subsets of specific normed spaces which are fundamental in applications throughout Analysis. This means that **we wish to find additional conditions, on top of boundedness, to be imposed so that the validity of the implication above is guaranteed.** A particular space of interest is the space of real continuous functions over a topological space. The purpose of the present and final section for this chapter is to establish a *celebrated criterion which guarantees compactness of families of continuous functions*, named after the mathematicians Cesare Arzelà and Giulio Ascoli.

Let $(X, \mathcal{T})$ be a compact topological space, and note that, by the Weierstrass Theorem 5.3, any continuous real-valued function on X is bounded, whilst its supremum and infimum are realised.[11] We may thus consider the normed space $\big(C(X), \| \cdot \|_\infty\big)$, where

$$C(X) \equiv C(X, \mathbb{R}) := \Big\{ f : X \longrightarrow \mathbb{R} \,\Big|\, f \text{ is continuous on } X \Big\},$$

and

$$\|f\|_\infty := \sup_{x \in X} |f(x)| \quad \text{for all } f \in C(X).$$

which is, by identical arguments to those in the proof of Theorem 3.20 (that deals with the particular case when X has a metric structure), a Banach space.[12]

---

[11] Therefore, since the supremum is always attained on compact spaces, we may safely write "max" in the place of "sup" in the definition of $\| \cdot \|_\infty$, if so desired.

[12] As it is standard, if the target space of the mappings is $\mathbb{R}$, then it can be safely suppressed and we can abbreviate the space of real-valued functions $C(X, \mathbb{R})$ to $C(X)$.

The main result of this section is a necessary and sufficient condition for a subset of $\big(C(X), \|\cdot\|_\infty\big)$ to be compact. For this, we need to introduce the notion of **equicontinuity**.

Definition 5.35. Let $(X, \mathcal{T})$ be a topological space (not necessarily compact). A family of continuous functions $\mathcal{F} \subseteq C(X)$ is said to be **equicontinuous at** x_0, where $x_0 \in X$, if for any $\varepsilon > 0$ there exists an open set $V \in \mathcal{T}$ with $V \ni x_0$ such that, for any $x \in V$, we have that

$$\big|f(x) - f(x_0)\big| < \varepsilon \quad \text{for all } f \in \mathcal{F}.$$

The family $\mathcal{F}$ is said to be **equicontinuous on** X if it is equicontinuous at every $x_0 \in X$.

The idea of equicontinuity is that one can choose the neighbourhood V **uniformly** for all functions $f \in \mathcal{F}$. The particular case of metric spaces is very illustrative:

Remark 5.36 (Equicontinuity of functions on metric spaces). Let (X, d) be a metric space. A family $\mathcal{F} \subseteq C(X)$ is **equicontinuous at** x_0, where $x_0 \in X$, if for any $\varepsilon > 0$ there exists $\delta = \delta(\varepsilon, x_0) > 0$ such that

$$d(x, x_0) < \delta \text{ for } x \in X \quad \text{implies} \quad \big|f(x) - f(x_0)\big| < \varepsilon, \quad \text{for all } f \in \mathcal{F}.$$

Hence, the point is that δ depends on ε and on the point x_0, but it is **uniform** for all functions $f \in \mathcal{F}$.

We remark that the notion of equicontinuity also makes sense when the target is not $\mathbb{R}$ but a general metric space. Let us now examine some examples to gain some intuition on this notion.

Example 5.37. Any **finite** family $\mathcal{F}$ contained in $C(X)$, where $(X, \mathcal{T})$ is a topological space, is necessarily equicontinuous on X. Indeed, suppose that $\mathcal{F} = \{f_1, \ldots, f_n\}$, for some $n \in \mathbb{N}$. Let $x_0 \in X$, arbitrary, and fix $\varepsilon > 0$, also arbitrary. Since, for each $k \in \{1, \ldots, n\}$, the function f_k is continuous at x_0, we have that there exists $V_k \in \mathcal{T}$ with $V_k \ni x_0$ such that

$$\big|f_k(x) - f_k(x_0)\big| < \varepsilon \quad \text{for all } x \in V_k.$$

Defining $V := \bigcap_{k=1}^n V_k$, we clearly have (by $(T2)$ of the definition of topologies) that $V \in \mathcal{T}$ and $V \ni x_0$, whilst for any $x \in V$ we have

$$\big|f(x) - f(x_0)\big| < \varepsilon \quad \text{for all } f \in \mathcal{F} = \{f_1, \ldots, f_n\}.$$

The arbitrariness of $\varepsilon > 0$ and $x_0 \in X$ shows that $\mathcal{F}$ is equicontinuous on X.

Example 5.38. Let, for any $n \in \mathbb{N}$, $f_n : [0, 1] \to \mathbb{R}$ be given by

$$f_n(x) := x^n \quad \text{for all } x \in [0, 1].$$

Then the family $\{f_n : n \in \mathbb{N}\} \subseteq C([0,1])$, is equicontinuous at 0 (see Figure 5.6(a)). Indeed, for any $\varepsilon \in (0,1)$, let us choose $\delta := \varepsilon$, and thus $V := [0,\delta)$, an open set in $[0,1]$, with $0 \in V$. Then, for any $x \in V$,

$$|f_n(x)| = |x|^n \le |x| < \delta = \varepsilon \quad \text{for all } n \in \mathbb{N}.$$

Example 5.39. Let, for any $n \in \mathbb{N}$, $f_n : [0,1] \to \mathbb{R}$ be given by

$$g_n(x) := x^{1/n} \quad \text{for all } x \in [0,1].$$

Then the family $\{g_n : n \in \mathbb{N}\} \subseteq C([0,1])$, is not equicontinuous at 0 (see Figure 5.6(b)).

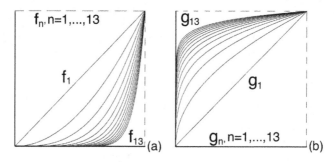

Figure 5.6: In illustration of the first 13 terms of the sequences of functions $f_n(x) = x^n$ and $g_n(x) = x^{1/n}$, $n \in \mathbb{N}$, in (a) and (b) respectively. Equicontinuity fails in the case of the $\{g_n : n \in \mathbb{N}\}$ because as n increases, the sequence tends to "break" at the origin, creating a discontinuity ($g_n(0) = 0$ for all n, while $g_n(x) \longrightarrow 1$ as $n \to \infty$ for any fixed $x \in (0,1)$). That is, the g_n become "less and less continuous" at the origin, whilst the f_n become "more and more" continuous.

Indeed, suppose for the sake of contradiction that it is. Then, in particular, for $\varepsilon := 1/2$ there would exist an open set V containing 0, and therefore containing an interval $[0,\delta)$, for some $\delta \in (0,1)$, such that, for any $x \in V$,

$$|g_n(x)| < 1/2 \quad \text{for all } n \in \mathbb{N}.$$

This condition would imply, in particular, that

$$(\delta/2)^{1/n} < 1/2 \quad \text{for all } n \in \mathbb{N},$$

or, equivalently, $\delta < 2(1/2)^n$ for all $n \in \mathbb{N}$, which is obviously not possible for any $\delta > 0$. In conclusion, the family $\{g_n : n \in \mathbb{N}\}$ is not equicontinuous at 0.

The following result offers a full characterisation of the compact sets in the space $\big(C(X), \|\cdot\|_\infty\big)$.

Theorem 5.40 (Arzelà–Ascoli Theorem). *Let $(X, \mathcal{T})$ be a compact topological space and $\mathcal{F} \subseteq C(X)$ a set of continuous functions on X. Then, the following are equivalent:*

(i) $\mathcal{F}$ *is (sequentially) compact in* $\left(C(X), \|\cdot\|_\infty\right)$.

(ii) $\mathcal{F}$ *is closed, bounded, and equicontinuous on* X.

Proof. (i) $\Longrightarrow$ (ii): Suppose first that a subset $\mathcal{F}$ of $C(X)$ is (sequentially) compact in $\left(C(X), \|\cdot\|_\infty\right)$. By Theorem 5.5, it follows that $\mathcal{F}$ is necessarily closed and bounded. It remains to show that $\mathcal{F}$ is equicontinuous on X.

Fix $x_0 \in X$ and $\varepsilon > 0$, both arbitrary. Since $\mathcal{F}$ is sequentially compact, Theorem 5.12 yields that it is also totally bounded. As a consequence, there exist $n \in \mathbb{N}$ and $f_1, \ldots, f_n \in \mathcal{F} \subseteq C(X)$ such that

$$\mathcal{F} \subseteq \mathbb{B}^\infty_{\varepsilon/3}(f_1) \bigcup \cdots \bigcup \mathbb{B}^\infty_{\varepsilon/3}(f_n), \qquad (\diamondsuit)$$

where the notation $\mathbb{B}^\infty_r(f)$ is used for the open ball of centre f and radius r in $\left(C(X), \|\cdot\|_\infty\right)$. Since $f_1, \ldots, f_n$ are continuous at x_0, the argument given in Example 5.37 shows that there exists $V \in \mathcal{T}$ with $V \ni x_0$ such that, for any $x \in V$, we have

$$\left|f_k(x) - f_k(x_0)\right| < \varepsilon \quad \text{for all } k \in \{1, \ldots, n\}.$$

Consider now an arbitrary $f \in \mathcal{F}$. Let $k \in \{1, \ldots, n\}$ be selected as determined by $(\diamondsuit)$, namely, any $f \in \mathcal{F}$ belongs to at least one of the balls $\mathbb{B}^\infty_{\varepsilon/3}(f_1), \ldots,$ $\mathbb{B}^\infty_{\varepsilon/3}(f_n)$ and hence there exists at least one index k so that $\|f - f_k\|_\infty < \varepsilon/3$. It follows that, for any $x \in V$, we have

$$\begin{aligned}
\left|f(x) - f(x_0)\right| &\leq \left|f(x) - f_k(x)\right| + \left|f_k(x) - f_k(x_0)\right| + \left|f_k(x_0) - f(x_0)\right| \\
&\leq 2\|f_k - f\|_\infty + \left|f_k(x) - f_k(x_0)\right| \\
&< \varepsilon,
\end{aligned}$$

noting also that V does not depend on $f \in \mathcal{F}$. It follows that $\mathcal{F}$ is indeed equicontinuous on X, as required.

(ii) $\Longrightarrow$ (i): Suppose now that $\mathcal{F}$ is closed, bounded, and equicontinuous on X. Since $\left(C(X), \|\cdot\|_\infty\right)$ is a Banach space, and $\mathcal{F}$ is a closed subset, it follows that the metric subspace $(\mathcal{F}, d_\infty)$ is complete, where d_∞ is the metric on $C(X)$ induced by the norm $\|\cdot\|_\infty$. To show that $\mathcal{F}$ is (sequentially) compact, it remains to prove that $\mathcal{F}$ is totally bounded.

Fix an arbitrary $\varepsilon > 0$. We need to prove that there exist $n \in \mathbb{N}$ and $f_1, \ldots, f_n \in \mathcal{F}$ such that

$$\mathcal{F} \subseteq \mathbb{B}^\infty_\varepsilon(f_1) \bigcup \cdots \bigcup \mathbb{B}^\infty_\varepsilon(f_n), \qquad (\diamondsuit\diamondsuit)$$

Since $\mathcal{F}$ is equicontinuous on X, it follows that for every $x \in X$ there exists $V_x \in \mathcal{T}$ with $V_x \ni x$ such that, for any $z \in V_x$,

$$\left|f(z) - f(x)\right| < \frac{\varepsilon}{3} \quad \text{for all } f \in \mathcal{F}. \qquad (\dagger)$$

Note that we may write X as the union of open sets

$$X = \bigcup_{x \in X} V_x,$$

and since $(X, \mathcal{T})$ is compact, it follows that there exist $m \in \mathbb{N}$ and $x_1, \ldots, x_m \in X$ such that X has a finite subcover of such sets

$$X = \bigcup_{k=1}^{m} V_{x_k}.$$

We now consider on $\mathbb{R}^m$ the norm $|\cdot|_\infty$ given by

$$|y|_\infty := \max_{k \in \{1, \ldots, m\}} |y_k| \quad \text{for all } y = (y_1, \ldots, y_m) \in \mathbb{R}^m.$$

We define the mapping $\pi : \mathcal{F} \longrightarrow \mathbb{R}^m$ by setting

$$\pi(f) := (f(x_1), \ldots, f(x_m)) \quad \text{for all } f \in \mathcal{F}.$$

The boundedness of $\mathcal{F}$ in $(C(X), \|\cdot\|_\infty)$ ensures that $\pi(\mathcal{F})$ is a bounded subset of $(\mathbb{R}^m, |\cdot|_\infty)$, and therefore totally bounded.[13] This means that there exist $n \in \mathbb{N}$ and $f_1, \ldots, f_n \in \mathcal{F}$ such that

$$\pi(\mathcal{F}) \subseteq \mathbb{B}_{\varepsilon/3}(\pi(f_1)) \bigcup \cdots \bigcup \mathbb{B}_{\varepsilon/3}(\pi(f_n)),$$

the above open balls being taken in the space $(\mathbb{R}^m, |\cdot|_\infty)$.

We now claim that, for the above choice of $n \in \mathbb{N}$ and $f_1, \ldots, f_n \in \mathcal{F}$, relation $(\Diamond\Diamond)$ is satisfied. Indeed, let $f \in \mathcal{F}$, arbitrary. Let $i \in \{1, \ldots, n\}$ be such that

$$\left|\pi(f) - \pi(f_i)\right|_\infty < \frac{\varepsilon}{3}. \tag{‡}$$

We shall prove that

$$\|f - f_i\|_\infty < \varepsilon.$$

To this end, let $x_f \in X$ be a point realising the supremum in the norm $\|f - f_i\|_\infty$, whose existence is ensured by the Weierstrass Theorem 5.30:

$$\|f - f_i\|_\infty = \sup_{x \in X} \left|f(x) - f_i(x)\right| = \left|f(x_f) - f_i(x_f)\right|.$$

Let $k \in \{1, \ldots, m\}$ be such that $x_f \in V_{x_k}$. We then have

$$
\begin{aligned}
\|f - f_i\|_\infty &= \left|f(x_f) - f_i(x_f)\right| \\
&\leq \left|f(x_f) - f(x_k)\right| + \left|f(x_k) - f_i(x_k)\right| + \left|f_i(x_k) - f_i(x_f)\right| \\
&\leq \left|\pi(f) - \pi(f_i)\right|_\infty + \left|f(x_f) - f(x_k)\right| + \left|f_i(x_f) - f_i(x_k)\right| \\
&< \varepsilon,
\end{aligned}
$$

where we have used (‡) to estimate the first term and (†) combined with $x_f \in V_{x_k}$ to estimate the remaining two terms. This proves the relation $(\Diamond\Diamond)$. In conclusion, $\mathcal{F}$ is totally bounded and the result ensues. □

---

[13]This last assertion of total boundedness depends in an essential way on the fact that the image of π lies in the finite-dimensional normed space $(\mathbb{R}^m, |\cdot|_\infty)$.

5.7 Exercises

Exercise 5.1. Which of the following subsets of $\mathbb{R}$ or $\mathbb{R}^2$, considered as metric spaces in the usual way, are sequentially compact? Justify your answers.

(i) $A = (1,5]$; $B = [-2, \infty)$; $C = \mathbb{Q} \cap [0,1]$;

(ii) $D = \{(x,y) \in \mathbb{R}^2 : (x-1)^2 + (y+2)^2 = 4\}$;

$E = \{(x^4 + y^5, x^5 + y^4) : x^2 + y^2 = 1\}$;

$F = \{(x,y) \in \mathbb{R}^2 : x \sin y \leq 1\}$.

Exercise 5.2. Let (X,d) be a sequentially compact metric space, and $(F_n)_1^\infty$ be a sequence of non-empty closed sets in X such that

$$F_1 \supseteq F_2 \supseteq \cdots F_n \supseteq F_{n+1} \supseteq \cdots$$

Show that $\bigcap_1^\infty F_n \neq \emptyset$.
[Hint: Consider, for each $n \in \mathbb{N}$, a point $x_n \in F_n$, and examine the behaviour of the sequence $(x_n)_1^\infty$.]

Exercise 5.3. Let (X,d) be a metric space. For any A, B non-empty subsets of X, we define

$$d(A,B) := \inf\{d(a,b) : a \in A, b \in B\}.$$

In what follows, suppose that A, B are such that $A \cap B = \emptyset$.

(i) If A is sequentially compact and B is closed, prove that $d(A,B) > 0$.

(ii) If A and B are sequentially compact, prove that there exist $a_0 \in A$ and $b_0 \in B$ such that $d(A,B) = d(a_0, b_0)$.

(iii) In the metric space $\mathbb{R}$ with the usual metric consider the sets $A = \mathbb{N}$ and $B = \{n - \frac{1}{n} : n \in \mathbb{N}\}$. Show that A and B are closed, $A \cap B = \emptyset$, and $d(A,B) = 0$.

Exercise 5.4. Let $(X, \mathcal{T})$ and $(Y, \mathcal{S})$ be topological spaces and $f : X \longrightarrow Y$ a continuous map on X. Show that $f(K)$ is a sequentially compact subset of Y for every sequentially compact subset K of X.

Exercise 5.5. Let (X,d) be a sequentially compact metric space. Let also $f : X \longrightarrow X$ be a map such that

$$d(f(x), f(y)) < d(x,y) \quad \text{for all } x, y \in X \text{ with } x \neq y.$$

Show that there exists a unique point $x_0 \in X$ such that $f(x_0) = x_0$.
[Hint: Show that $\inf\{d(x, f(x)) : x \in X\}$ is attained, at a point x_0 say, and deduce that $f(x_0) = x_0$.]

Exercise 5.6. Let $(X, \mathcal{T})$ be a Hausdorff topological space. Show that for every compact subset K of X and for every $x \in X \setminus K$ there exist open sets U and V such that $x \in U$, $K \subseteq V$, and $U \cap V = \emptyset$.
[Hint: Use the Hausdorff condition for x and for each $y \in K$, and then try to use the compactness of K.]

Exercise 5.7. Let $(X, \mathcal{T})$ be a topological space and Y be a subset of X. Show that:

 (i) If $(X, \mathcal{T})$ is a Hausdorff space and Y is compact, then Y is closed in X.

 [Hint: Use the preceding question.]

 (ii) If $(X, \mathcal{T})$ is compact and Y is closed in X, then Y is compact.

Exercise 5.8. Let X be a set, and $\mathcal{T}$ be the finite complement topology on X. Show that the topological space $(X, \mathcal{T})$ is compact.

Exercise 5.9. Show that a topological space $(X, \mathcal{T})$ is compact if and only if the following property holds:

$$\text{for any family } \{F_i\}_{i \in I} \text{ of closed sets with } \bigcap_{i \in I} F_i = \emptyset,$$

$$\text{there exists a finite subset } J \text{ of } I \text{ such that } \bigcap_{i \in J} F_j = \emptyset.$$

Exercise 5.10. Let K be a compact subset of a topological space $(X, \mathcal{T})$. Let $\{U_i\}_{i \in \mathbb{N}}$ be a family of open subsets of X such that

$$U_1 \subseteq U_2 \subseteq \cdots \subseteq U_i \subseteq U_{i+1} \subseteq \cdots \qquad (*)$$

and $K \subseteq \bigcup_{i=1}^{\infty} U_i$. Show that there exists $i_0 \in \mathbb{N}$ such that $K \subseteq U_{i_0}$.

Exercise 5.11. Let $(X, \mathcal{T})$ be a topological space and let Y, Z be compact topological subspaces of X. Show that the topological subspace $Y \cup Z$ is compact.

Exercise 5.12. Give an (explicit) example of a sequence contained in the closed unit ball $\mathbb{D}_1(0)$ of the normed space $(\ell^\infty, \|\cdot\|_\infty)$ which does not have any convergent subsequence.

Exercise 5.13. Let $(X, \|\cdot\|)$ be a normed space, and Y a finite-dimensional vector subspace of X, such that $Y \neq X$. Show that there exists $z \in X$ with $\|z\| = 1$ such that
$$\|z - y\| \geq 1 \quad \text{for all } y \in Y.$$

[Hint: You may try to follow the proof of Riesz's Lemma, and identify the reason why in the present situation one can get a stronger result.]

Chapter 6

The Lebesgue measure on the Euclidean space

6.1 Transcending lengths, areas and volumes

We wish to study the concepts of length, area, volume, as well as higher dimensional analogues and abstractions of these ideas quite systematically. In $1, 2$ or 3 dimensions, we already know how to use the Riemann integral to calculate the area or the volume[1] of certain regions of $\mathbb{R}^2$ and $\mathbb{R}^3$. For example, we know how to calculate the area under the graph of "nice" (differentiable or at least continuous) functions (see Figure 6.1).

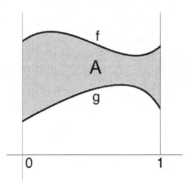

Figure 6.1: The grey area A between the graphs of the "nice" functions f, g can be calculated by the integral $A = \int_0^1 (f(x) - g(x)) \mathrm{d}x$.

However, measure theory is *not* an application of Calculus. The most important new point of our study is that we shall develop tools of Analysis which will allow the study of areas and volumes for *"wild" non-smooth objects*.

But why do that?

---

[1]There is no need to panic if you haven't seen any Vector Calculus yet, you can just ignore the comment about volumes in 3 (and higher) dimensions! No knowledge of higher dimensional Riemann integrals is needed in the sequel. This comment (and any future ones) is given for the benefit of those readers who have already seen double integrals and calculations of volumes of bodies.

The reason is that there exist many shapes of interest which are very wild and that cannot be studied with only the mathematical tools of classical Calculus! As an example, consider the function

$$f : \mathbb{R} \longrightarrow \mathbb{R}, \quad f(x) := \begin{cases} 1, & x \text{ irrational}, \\ 0, & x \text{ rational}. \end{cases}$$

This function is so irregular that we cannot even draw its graph, let alone calculate the classical area under its graph given that its Riemann integral does not exist! For such sets and functions...

We cannot define a classical notion of area because they are very "wild", although it is easily seen that they "do take up some space" and hence we should be able to somehow define a notion of area/volume for them.

These mathematical problems have reflections and applications to the real-world as well: interesting examples of wild sets occur even in nature; consider for instance the snowflakes, or certain types of so-called "fractal-shape" broccoli (which you can eat, but not easily define their volume, see Figures 6.2-6.3)!

Figure 6.2: The Romanesco Broccoli is a so-called "fractal" set.

Figure 6.3: The snowflake is also a so-called "fractal" set.

Further, the chapter is not restricted to the consideration of volumes or areas. These are only the easiest and most straightforward examples of what will be called **measures**, that is, volume-like mappings which behave much like the volume. Examples of measures arise e.g. in physics: mass distributions in gravitational problems, charge distributions in electromagnetism, fluxes, vortex distributions and more.

Finally, there exists a further example of such great importance that itself alone could provide enough justification for the theory we are about to develop. To this end, let $P(A)$ denote the probability of an event A to occur. Given any two events A and B, the event $A \cup B$ is the event which can be described as " either A happens or B happens". In the case when A and B are mutually exclusive, a situation that can be formulated mathematically as $A \cap B = \emptyset$, we have the relation $P(A \cup B) = P(A) + P(B)$. You probably saw a formula like this in any elementary probability classes. In other words, a probability, is merely a "measure" in our sense and behaves much like the notion of volume.

So, as mathematicians, we treat this large class of different problems "all in one go" and in a smart way:

Instead of treating every example separately, we develop a unified general mathematical theory of "measures" and "integrals" which applies to all instances we may encounter.

(Hence we do the hard work just once, not every time for each example over and over again as in the other sciences!) In order to develop a general theory which will allow to define and study general "measures" and "integrals" we need to...

...isolate the common underlying features that all the examples possess.

To begin with, what features do length, area and volume have in common? Firstly, they all measure sizes, and so these quantities are *non-negative numbers*. For example, there is no subset of $\mathbb{R}$, smooth or otherwise, with length -3! However, zero and "infinity" are admissible values: we all unanimously agree that for any number $a \in \mathbb{R}$ the set $\{a\}$ has zero length, whilst $\mathbb{R}$ itself has infinite length.

Secondly, *length, area and volume must be preserved when we divide the object under consideration into disjoint pieces.* For example, if we cut a surface into pieces the overall area should be equal to the sum of the individual areas of the pieces.

To formalise this in a mathematical manner, let $A \subseteq \mathbb{R}^d$ be a set. If $d = 1$ then A is a subset of $\mathbb{R}$ and we denote by $\mathcal{L}^1(A)$ the length of A. If $d = 2$, A is a subset of $\mathbb{R}^2$ and we denote by $\mathcal{L}^2(A)$ the area of A. In general, we denote by $\mathcal{L}^d(A)$ the d-dimensional volume of A. For clarity let us fix $d = 2$, since the d-dimensional case is completely analogous. So let $A \subseteq \mathbb{R}^2$, assumed to be chopped it into pieces in such a way that the pieces do not overlap and, taken together, cover A exactly (Figure 6.4).

However, chopping our shape up in finitely many pieces is not enough if we want to compute the area from simpler shapes like rectangles (e.g. in $\mathbb{R}^3$ you cannot split a ball into finitely may cubes! See Figure 6.5 for $n = 2$).

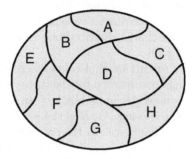

Figure 6.4: The area of the union set, that is $\mathcal{L}^2(A \cup B \cup ... \cup H)$, must be the same as the sum of the areas of the pieces, that is $\mathcal{L}^2(A) + \mathcal{L}^2(B) + ... + \mathcal{L}^2(H)$.

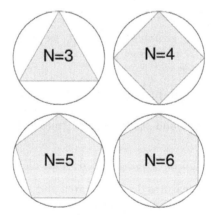

Figure 6.5: The area of the disc can be approximated (either from the inside or from the outside) by the area of solid polygons with N-many vertices, as $N \to \infty$.

In conclusion, by putting these requirements together, for any sequence of sets $A_1, A_2, ... \subseteq A$ such that

$$A = \bigcup_{i=1}^{\infty} A_i, \qquad A_i \cap A_j = \emptyset \text{ for } i \neq j,$$

the "measure" of the set should satisfy

$$\mathcal{L}^2 \left(\bigcup_{i=1}^{\infty} A_i \right) = \sum_{i=1}^{\infty} \mathcal{L}^2(A_i).$$

It is not difficult to see that all examples of "measures" above (probabilities, masses, etc.) satisfy these properties.

Note also that the above properties have nothing to do with the specific structure of $\mathbb{R}^2$ and slight variants of them make sense for general sets. In the next chapter we shall indeed develop general tools and concepts of measures on abstract sets. You are probably now wondering why on earth we need to go to an even greater level of abstraction. The answer is that even simple applications require to "measure" on some space different from the Euclidean space $\mathbb{R}^d$ (imagine for instance the problem of measuring the area of a portion of a curved or doughnut-like surface).

We will see later that the above countable unions cannot be relaxed to arbitrary unions, if we are to retain the preservation of volume (additivity) property. A further problem arising that we shall encounter later is that even in this generality **not all sets can be measured**! That is, there are *very bizarre* sets even in $\mathbb{R}$ at which we cannot associate any meaningful notion of length. We defer this discussion to later sections once we understand some basics first. However, we shall derive general properties of measures and show that all reasonable sets are measurable.

Finally, we should emphasize that the aim of this chapter (as well as of the next two) is not that of computing the area or the volume for as many bodies as possible. Instead, we shall discover relations between objects in the form of theorems. Actually, *essentially we shall never need to compute any particular volume*. This chapter will provide us with a toolbox of tremendous potential. For the sake of comparison, recall that you almost never needed to compute a (Riemann) integral from its very definition: any complicated integral is computed using properties (namely, theorems) like integration by parts, the chain rule or the connection between integration and differentiation. In all of modern Analysis (in particular in Functional Analysis, Probability Theory, Harmonic Analysis, Partial Differential Equations, Calculus of Variations), only the Lebesgue integral is used. Actually, measure theory plays for modern Analysis the role that Calculus used to play for classical Analysis until about 100 years ago.

6.2   The pathway to measuring sets on $\mathbb{R}^d$

Herein we shall define for any set in $\mathbb{R}^d$ its area, volume, or a higher dimensional analogue of these notions, starting from rectangles and cuboids. As we already said, we shall discover that we cannot associate a measure to all subsets! Only sufficiently regular subsets will have a notion of measure. When the concepts will be formalised, these will be called the *measurable sets*.

In particular, herein we consider the fundamental measure on $\mathbb{R}^d$ associated to the volume which is called the (classical) *Lebesgue measure on $\mathbb{R}^d$*. If $d = 1$, then "volume in $\mathbb{R}$" is typically called length, whilst if $d = 2$ then "volume in $\mathbb{R}^2$" is typically called area (for $d \geq 3$ we just call it volume). We recall that the plan is to associate a notion of volume (measure) to as many sets as possible in a way that, if the set is smooth (e.g. the graph of a nice function) then it coincides with the usual volume we can calculate via Riemann integrals. General *measures* on arbitrary sets (which may have nothing to do with volume whatsoever) will be studied in the next chapter(s).

The plan. This construction of the Lebesgue measure on $\mathbb{R}^d$ is organised in the following step-wise fashion:

- Step 1. We first define the Lebesgue measure on the so-called **elementary sets** of $\mathbb{R}^d$. These sets are very simple, just unions of boxes, and their Lebesgue measure is just their usual volume.

- Step 2. We extend the above notion to **all** subsets of $\mathbb{R}^d$ by approximation and we call the result the **outer Lebesgue measure**.

- Step 3. As the terminology of step 2 suggests, the previous step does not yield the desired Lebesgue measure. The problem is that **there exist bizarre sets which CANNOT be measured and hence have no notion of volume!** By **restricting** the outer Lebesgue measure to a (very large) class of sets of $\mathbb{R}^d$, the **measurable sets**, we obtain the desired notion of measure (volume) for these sets.

6.3   Elementary sets and content

We begin by defining our cornerstone objects.

Definition 6.1. (i) A set $B \subseteq \mathbb{R}^d$ is called a **box in** $\mathbb{R}^d$, if either $B = \emptyset$ or it has the Cartesian product form

$$B = I_1 \times \cdots \times I_d = \prod_{j=1}^{d} I_j,$$

where, for each $j \in \{1, ..., d\}$, I_j is a (possibly degenerate) interval, open, closed or half-open/half-closed. This means that there exist real numbers $a_j \leq b_j$ such that each I_j equals one of the following sets: $[a_j, b_j]$, (a_j, b_j), $(a_j, b_j]$ or $[a_j, b_j)$.

Note that if $a_j = b_j$ for some index j, then we have the singleton set $I_j = \{a_j\} (= \{b_j\})$. The empty set $\emptyset$ is also considered as a (trivial) box.

(ii) The (**d-dimensional**) **volume** (or **Jordan content**) of a box $B \subseteq \mathbb{R}^d$ is defined as

$$\mathcal{V}^d(B) := \begin{cases} \displaystyle\prod_{j=1}^{d} \left(\sup I_j - \inf I_j \right), & \text{if } B = \displaystyle\prod_{j=1}^{d} I_j, \\ 0, & \text{if } B = \emptyset. \end{cases}$$

The above definition means that if $B \neq \emptyset$ and each I_j has either of the forms $[a_j, b_j]$, (a_j, b_j), $(a_j, b_j]$ or $[a_j, b_j)$, then

$$\mathcal{V}^d(B) = (b_1 - a_1) \cdots (b_d - a_d)$$

because $\sup I_j = b_j$ and $\inf I_j = a_j$.

If for any of the indices $j \in \{1, \ldots, d\}$ we have $a_j = b_j$ (degenerate box), the necessarily the volume of the box vanishes: $\mathcal{V}^d(B) = 0$.

(iii) A set $E \subseteq \mathbb{R}^d$ is called a d-dimensional **elementary set** if it can be represented as a **finite** union of (perhaps overlapping) boxes (see Figure 6.6):

$$E = \bigcup_{i=1}^{m} B_i, \quad \{B_1, \ldots, B_m\} \text{ boxes in } \mathbb{R}^d.$$

The **class of all elementary sets of** $\mathbb{R}^d$ will be symbolised as $\mathcal{E}(\mathbb{R}^d)$:

$$\mathcal{E}(\mathbb{R}^d) := \left\{ E \in \mathcal{P}(\mathbb{R}^d) \mid E \text{ is an elementary set} \right\}.$$

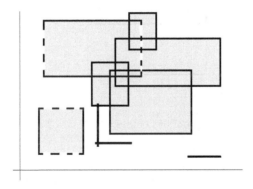

Figure 6.6: An illustration of an elementary set on the plane.

Remark 6.2. (i) If $d = 1$, any 1-dimensional box is an interval in $\mathbb{R}$, any 2-dimensional box is a rectangle in $\mathbb{R}^2$ and any 3-dimensional box is a cuboid (rectangular parallelepiped) in $\mathbb{R}^3$. However, "lower-dimensional" degenerate boxes are not excluded. For instance, $\{1\}$ is a box in $\mathbb{R}$ and $[0, 1] \times [1, 2) \times \{3\}$ is a box in $\mathbb{R}^3$.

(ii) The **representation of an elementary set** as a union of boxes **is not unique**. For example, the box $[0, 1] \subseteq \mathbb{R}$ can be written an elementary set in many different ways, for instance as

$$[0, 1] = \{0\} \bigcup (0, 2/3] \bigcup [1/2, 1] = [0, 1/4] \bigcup [1/4, 3/4) \bigcup \{3/4\} \bigcup (3/4, 1].$$

(iii) Note that it is **not assumed** that an elementary set is built up by a disjoint collection of boxes. However, the following lemma asserts that indeed we can always represent any elementary set in this way.

Lemma 6.3. *Every elementary set in $\mathbb{R}^d$ can be written as a finite **disjoint** union of boxes.*

Proof. The case $d = 1$ is a very simple exercise which is left to the reader. The general case for $d \geq 2$ can be treated easily, the only difficulty being the complicated notation. Let $E = \bigcup_{i=1}^{m} B_i \subseteq \mathbb{R}^d$ be an elementary set such that $B_i = I_i^{(1)} \times \cdots \times I_i^{(d)}$, where each $I_i^{(j)}$ is an interval. Apply the result for $d = 1$ to the collection of intervals $I_1^{(j)}, \ldots, I_m^{(j)}$ for a fixed "direction" j. Then we obtain disjoint intervals $J_1^{(j)}, \ldots, J_{p_j}^{(j)}$ and a set of indices $K_i^{(j)} \subseteq \{1, \ldots, M_j\}$ such that $I_i^{(j)} = \bigcup_{k \in K_i^{(j)}} J_k^{(j)}$ (see Figure 6.7).

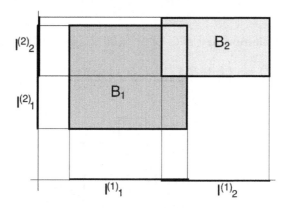

Figure 6.7: The idea of the proof of Lemma 6.3 for $d = 2$ and two boxes.

Then, by using the identity

$$\left(\bigcup_a A_a \right) \times \left(\bigcup_b B_b \right) = \bigcup_a \bigcup_b (A_a \times B_b)$$

relating Cartesian products and finite unions, we have

$$E = \bigcup_{i=1}^{m} \left(\left(\bigcup_{k \in K_i^{(1)}} J_k^{(1)} \right) \times \cdots \times \left(\bigcup_{k \in K_i^{(d)}} J_k^{(d)} \right) \right)$$

$$= \bigcup_{i=1}^{m} \bigcup_{k_1 \in K_i^{(1)}} \cdots \bigcup_{k_d \in K_i^{(d)}} \left(J_{k_1}^{(1)} \times \cdots \times J_{k_d}^{(d)} \right)$$

and all the boxes $J_{k_1}^{(1)} \times \cdots \times J_{k_d}^{(d)}$ are by construction disjoint. $\square$

The next simple result establishes that if a box can be written as a disjoint union of boxes, the volume of the box equals the sum of the volumes of the boxes comprising it.

Lemma 6.4. *Let $B \subseteq \mathbb{R}^d$ be a box. If there exist disjoint boxes $B^1, \ldots, B^N$ in $\mathbb{R}^d$ such that $B = B^1 \bigcup \cdots \bigcup B^N$ and $B^i \bigcap B^j = \emptyset$ for $i \neq j$, then*

$$\mathcal{V}^d(B) = \mathcal{V}^d(B^i) + \cdots + \mathcal{V}^d(B^N). \tag{⊞}$$

Proof. The desired conclusion is a consequence of the next steps:

Step 1. We may assume that the box B is non-degenerate. If it is degenerate, then its volume vanishes and so does the volume of every sub-box contained in it. Hence, both sides of (⊞) vanish and there is nothing to prove.

Step 2. Let $n \in \{1, \ldots, N\}$ be the number of non-degenerate boxes in the collection $\{B^1, \ldots, B^N\}$. By a relabelling of the boxes, we may assume that the boxes $\{B^1, \ldots, B^n\}$ are non-degenerate and hence have positive volume.

Step 3. By replacing all the boxes $\{B, B^1, \ldots, B^N\}$ by the respective closed boxes $\{\overline{B}, \overline{B}^1, \ldots, \overline{B}^N\}$, we have that the volume of all the boxes remains unaltered because the boundary of each box consists of a union of degenerate boxes, its sides, which have zero volume in $\mathbb{R}^d$:

$$\mathcal{V}^d(\overline{B}) = \mathcal{V}^d(B), \quad \mathcal{V}^d(\overline{B}^i) = \mathcal{V}^d(B^i), \quad i = 1, \ldots, N.$$

In addition, the degenerate boxes $\{B^{n+1}, \ldots, B^N\}$ can be discarded since they are contained in the union of boundaries of the non-degenerate boxes $\{\overline{B}^1, \ldots, \overline{B}^n\}$. Moreover, the non-degenerate closed boxes have disjoint interiors with overlaps only along their sides.

Step 4. By Steps 1–3, it suffices to prove: *Given a non-degenerate closed box B in $\mathbb{R}^d$ and non-degenerate closed boxes $\{B^1, \ldots, B^n\}$ with disjoint interiors (namely $B^{i\circ} \bigcap B^{j\circ} = \emptyset$ for $i \neq j$) such that $B = B^1 \bigcup \cdots \bigcup B^n$, we have*

$$\mathcal{V}^d(B) = \mathcal{V}^d(B^i) + \cdots + \mathcal{V}^d(B^n). \tag{⊞⊞}$$

Step 5. Given **any** non-degenerate closed box $\mathbf{B} = [a^1, b^1] \times \cdots \times [a^d, b^d]$ and, for each $k \in \{1, \ldots, d\}$, any partitions of points

$$a^k = a_0^k < a_1^k < a_2^k < \cdots < a_{n_k}^k = b^k, \quad n_k \in \mathbb{N}, \tag{⊠}$$

consider the family of closed boxes with disjoint interiors (see Figure 6.8)

$$\left\{ \mathbf{B}_{j_1 \ldots j_d} : j_k \in \{1, \ldots, n_k\}, \ k \in \{1, \ldots, d\} \right\} \tag{⊡}$$

(namely $\mathbf{B}_{j_1 \ldots j_d}^{\circ} \bigcap \mathbf{B}_{j_1' \ldots j_d'}^{\circ} = \emptyset$ when $j_i \neq j_i'$ for some i), where

$$\mathbf{B}_{j_1 \ldots j_d} := [a_{j_1-1}^1, a_{j_1}^1] \times \cdots \times [a_{j_d-1}^d, a_{j_d}^d].$$

Then, we have that

$$\mathcal{V}^d(\mathbf{B}) = \sum_{j_d=1}^{n_d} \cdots \sum_{j_1=1}^{n_1} \mathcal{V}^d(\mathbf{B}_{j_1 \ldots j_d}). \tag{⊟}$$

This is a trivial consequence of the definition of the volume of boxes.

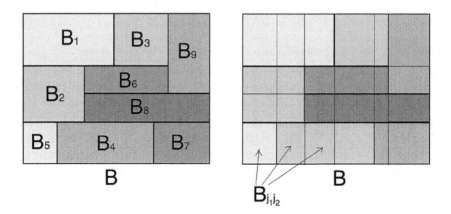

Figure 6.8: The idea of the proof of the splitting to sub-boxes with disjoint interiors in the special case of $d = 2$.

Step 6. For the closed boxes $\{B^1, \ldots, B^n\}$ with disjoint interiors whose union is B (*Step 4*), suppose that B has the form $B = [a^1, b^1] \times \cdots \times [a^d, b^d]$. For each k-th coordinate consider the partition ($\boxtimes$) of $[a^k, b^k]$ (*Step 5*) arising by ordering in a increasing way the k-th coordinates of the vertices of the boxes $\{B^1, \ldots, B^n\}$. By using the partitions ($\boxtimes$), we may split all the boxes $B, B^1, \ldots, B^n$ into sub-boxes $B_{j_1 \ldots j_d} \subseteq B$ of the form ($\boxdot$). Then, for each k there exist $M_i^k \leq N_i^k$ in $\mathbb{N}$ such that

$$\bigcup_{j_d=1}^{n_d} \cdots \bigcup_{j_1=1}^{n_1} B_{j_1 \ldots j_d} = B = \bigcup_{i=1}^{n} B^i = \bigcup_{i=1}^{n} \bigcup_{j_d=M_i^d}^{N_i^d} \cdots \bigcup_{j_1=M_i^1}^{N_i^1} B_{j_1 \ldots j_d}. \qquad (\boxminus\boxminus)$$

By construction, all the boxes $B_{j_1 \ldots j_d}$ are closed with disjoint interiors. Hence, by ($\boxminus$)-($\boxminus\boxminus$) we infer that

$$\mathcal{V}^d(B) = \sum_{j_d=1}^{n_d} \cdots \sum_{j_1=1}^{n_1} \mathcal{V}^d(B_{j_1 \ldots j_d}) = \sum_{i=1}^{n} \sum_{j_d=M_i^d}^{N_i^d} \cdots \sum_{j_1=M_i^1}^{N_i^1} \mathcal{V}^d(B_{j_1 \ldots j_d}) = \sum_{i=1}^{n} \mathcal{V}^d(B^i).$$

Thus, ($\boxplus\boxplus$) ensues and the result is now a consequence of *Steps 1–6.* □

We shall now establish some simple properties of elementary sets. To this aim, we need the next two simple concepts.

Definition 6.5. Given two sets $E, F \subseteq \mathbb{R}^d$ and $x \in \mathbb{R}^d$, the **symmetric difference** $E\Delta F$ of the sets E, F is defined as

$$E\Delta F := (E \setminus F) \bigcup (F \setminus E).$$

Moreover, for the **translate** $x + E$ of the set E (by x) is given by

$$x + E := \{x + y : y \in E\}.$$

Lemma 6.6. *Let $E, F \in \mathcal{E}(\mathbb{R}^d)$. Then, the following hold true as well:*

$$E \cup F, \ E \cap F, \ E \setminus F, \ E \triangle F \ \in \mathcal{E}(\mathbb{R}^d).$$

Proof. We shall prove for the sake of completeness only that $E \cap F \in \mathcal{E}(\mathbb{R}^d)$. The remaining claims are left as an exercise for the reader. Since $E, F \in \mathcal{E}(\mathbb{R}^d)$, we have $E = \bigcup_{i=1}^{m} B_i$ and $F = \bigcup_{l=1}^{n} C_l$, where $\{B_1, \dots, B_m\}$ and $\{C_1, \dots, C_n\}$ are finite families of boxes. Then

$$E \cap F = \bigcup_{i=1}^{m} \bigcup_{l=1}^{n} (B_i \cap C_l)$$

and as the intersection of any two boxes is a box as well, we deduce that indeed $E \cap F \in \mathcal{E}(\mathbb{R}^d)$. $\qquad \square$

If an elementary set has the representation $E = \bigcup_{k=1}^{m} B_k$ where the B_k are disjoint (such a representation is always possible by Lemma 6.3) then it is natural to define its volumes as $\mathcal{V}^d(E) = \sum_{k=1}^{m} \mathcal{V}^d(B_k)$ (sum of volumes), where the volume of each single box is defined as before.

Definition 6.7. We define the **d-dimensional volume function (or Jordan content)** on the family $\mathcal{E}(\mathbb{R}^d)$ of elementary sets

$$\mathcal{V}^d \ : \ \mathcal{E}(\mathbb{R}^d) \longrightarrow [0, \infty),$$

by taking

$$\mathcal{V}^d(E) := \begin{cases} \sum_{i=1}^{m} \mathcal{V}^d(B_i), & \text{if } E = \bigcup_{i=1}^{m} B_i, \ \{B_1, \dots, B_m\} \text{ \textbf{disjoint} boxes}, \\ 0, & \text{if } E = \emptyset. \end{cases}$$

Remark 6.8 ($\mathcal{V}^d$ well-defined on $\mathcal{E}(\mathbb{R}^d)$). Since the definition above is given in terms of the representation via boxes and this representation is not unique, we need to prove that $\mathcal{V}^d$ **is well-defined** on $\mathcal{E}(\mathbb{R}^d)$, in the sense of showing that different representations of E as disjoint unions of boxes do not give different results. This is indeed the case. Assume that

$$E = \bigcup_{i=1}^{m} B_i = \bigcup_{l=1}^{n} C_l$$

for two different families of **disjoint** boxes $\{B_1, \dots, B_m\}$, $\{C_1, \dots, C_n\}$. Note that each B_i can be written as $\bigcup_{l=1}^{n} (B_i \cap C_l)$ and each C_l as $\bigcup_{i=1}^{m} (C_l \cap B_i)$, whilst the family of boxes

$$\Big\{ B_i \cap C_l \ : \ i = 1, \dots, m, \ l = 1, \dots, n \Big\}$$

is disjoint. Then, by Definition 6.7 above and Lemma 6.4, we have

$$\sum_{i=1}^{m} \mathcal{V}^d(B_i) = \sum_{i=1}^{m} \mathcal{V}^d\left(\bigcup_{l=1}^{n}(B_i \cap C_l)\right) = \sum_{i=1}^{m}\sum_{l=1}^{n} \mathcal{V}^d(B_i \cap C_l) =$$

$$= \sum_{l=1}^{n}\sum_{i=1}^{m} \mathcal{V}^d(C_l \cap B_i) = \sum_{l=1}^{n} \mathcal{V}^d\left(\bigcup_{i=1}^{m}(C_l \cap B_i)\right) =$$

$$= \sum_{l=1}^{n} \mathcal{V}^d(C_l).$$

Hence, $\mathcal{V}^d(E)$ is well-defined regardless the different representations of any elementary set $E \in \mathcal{E}(\mathbb{R}^d)$.

Now we establish some important properties of the volume on the class of elementary sets.

Proposition 6.9 (Properties of the volume). *Let $E_1, E_2, ..., E_p$ be elementary sets in $\mathcal{E}(\mathbb{R}^d)$. Then the following hold:*

(i) *The volume is* **finitely-additive**. *Namely, if the sets $\{E_1, ...E_p\}$ are disjoint, i.e. if $E_k \cap E_l = \emptyset$ when $k \neq l$, then*

$$\mathcal{V}^d\left(\bigcup_{k=1}^{p} E_k\right) = \sum_{k=1}^{p} \mathcal{V}^d(E_k).$$

(ii) *The volume is* **monotone**. *Namely, if $E_1 \subseteq E_2$, then $\mathcal{V}^d(E_1) \leq \mathcal{V}^d(E_2)$.*

(iii) *The volume is* **finitely-sub-additive**. *Namely,*

$$\mathcal{V}^d\left(\bigcup_{k=1}^{p} E_k\right) \leq \sum_{k=1}^{p} \mathcal{V}^d(E_k).$$

(iv) *The volume is* **invariant under translations**. *Namely,*

$$\mathcal{V}^d(x + E) = \mathcal{V}^d(E) \quad \text{for any } x \in \mathbb{R}^d.$$

Proof. (i) By Lemma 6.3, each elementary set E_k can be written as the disjoint union of boxes $\bigcup_{i=1}^{m_k} B_i^{(k)}$. Since $E_k \cap E_l = \emptyset$ for $k \neq l$, we have $B_i^{(k)} \cap B_j^{(l)} = \emptyset$ for any $i \in \{1, ..., m_k\}, j \in \{1, ..., m_l\}$. Thus, we obtain

$$\bigcup_{k=1}^{p} E_k = \bigcup_{k=1}^{p}\bigcup_{i=1}^{m_k} B_i^{(k)}$$

and all the boxes are disjoint. As a consequence,

$$\mathcal{V}^d\left(\bigcup_{k=1}^{p} E_k\right) = \sum_{k=1}^{p}\sum_{i=1}^{m_k} \mathcal{V}^d(B_i^{(k)}) = \sum_{k=1}^{p} \mathcal{V}^d(E_k).$$

(ii) Since $E_2 = (E_2 \setminus E_1) \cup E_1$ and $(E_2 \setminus E_1) \cap E_1 = \emptyset$ we get from item (i) above that $\mathcal{V}^d(E_2) = \mathcal{V}^d(E_2 \setminus E_1) + \mathcal{V}^d(E_1) \geq \mathcal{V}^d(E_1)$.
The proofs of (iii) and (iv) are left as exercises for the reader. $\qquad\square$

Remark 6.10 (Uniqueness of the volume). The reader at this point might be wondering if there can exist any other function $\mathcal{V} : \mathcal{E}(\mathbb{R}^d) \longrightarrow [0, \infty)$ which can also be considered as volume, or if the volume as defined in Definition 6.7 is "essentially unique". In fact, the latter alternative is indeed the case. It can be shown that the properties (i) and (iv) of Proposition 6.9 (finite-additivity and invariance under translations) of a function $\mathcal{V} : \mathcal{E}(\mathbb{R}^d) \longrightarrow [0, \infty)$, together with the requirement that $\mathcal{V}(\emptyset) = 0$, uniquely determine the volume up to a multiplicative constant: there exists $C(d) > 0$ depending only on the dimension such that $\mathcal{V}(E) = C(d)\mathcal{V}^d(E)$ for any $E \in \mathcal{E}(\mathbb{R}^d)$. We refrain from delving into the details since we will not utilise this fact in the sequel.

6.4 The Lebesgue outer measure

We now extend the function $\mathcal{V}^d : \mathcal{E}(\mathbb{R}^d) \longrightarrow [0, \infty)$ constructed in the previous section from the class of elementary sets to **all subsets** of $\mathbb{R}^d$. As we have already explained, this function is not yet the desired Lebesgue measure because for some very irregular sets it is *not a well-defined "volume"*. We shall see such examples a bit later. The idea on which the extension of $\mathcal{V}^d$ from $\mathcal{E}(\mathbb{R}^d)$ to the powerset $\mathcal{P}(\mathbb{R}^d)$ goes as follows: given $A \subseteq \mathbb{R}^d$, we cover it with countably many boxes and we take the series of volumes of these boxes. Note that we cannot use finitely many boxes in general (see Figure 6.9).

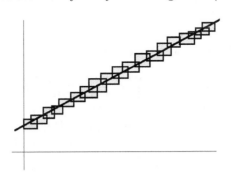

Figure 6.9: No line in $\mathbb{R}^2$ can be covered by any finite number of boxes.

For example, no unbounded subset of $\mathbb{R}^d$ can fit inside the union of finitely many boxes which is a bounded set. Then, we optimise this value over all such boxes by taking a finer-and-finer approximation from "the outside" by boxes (see Figure 6.10).

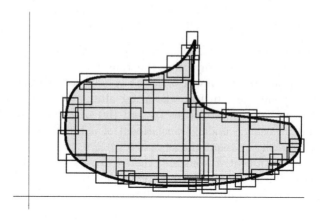

Figure 6.10: Illustration of the definition of the Lebesgue outer measure on $\mathbb{R}^2$. By taking "inf", we force the unions of boxes to become smaller and smaller, eventually approximating the set from the outside.

More precisely, we have the next definition:

Definition 6.11. The **Lebesgue Outer Measure** is the function

$$\mathcal{L}^{d*} \ : \ \mathcal{P}(\mathbb{R}^d) \longrightarrow [0, \infty],$$

defined on any set $A \subseteq \mathbb{R}^d$ as:

$$\mathcal{L}^{d*}(A) := \inf \left\{ \sum_{k=1}^{\infty} \mathcal{V}^d(B_k) \ \middle| \ (B_k)_1^{\infty} \text{ non-degenerate boxes, } A \subseteq \bigcup_{k=1}^{\infty} B_k \right\}.$$

We note that we may restrict our attention to non-degenerate boxes in the definition above, essentially because degenerate boxes have vanishing volume and do not contribute to the value of the infimum.

Remark 6.12. The modifier "outer" and the asterisk "$\mathcal{L}^{d*}$" are there to remind us that **the value $\mathcal{L}^{d*}(A)$ is not a meaningful volume for some bizarre subsets of $\mathbb{R}^d$** which we shall later have to exclude from our consideration by **restricting $\mathcal{L}^{d*}$ to the class of the so-called "measurable sets"**. This will give the final Lebesgue measure in the next section.

Now we shall study the properties of the Lebesgue outer measure. Before continuing, we record a simple technical lemma needed right next.

Lemma 6.13. *Let $(a_{n,m})_{n,m\in\mathbb{N}}$ be a double sequence of elements in $[0, \infty]$.*
(i) We have the commutativity property

$$\sum_{n=1}^{\infty} \sum_{m=1}^{\infty} a_{n,m} \ = \ \sum_{m=1}^{\infty} \sum_{n=1}^{\infty} a_{n,m}.$$

(ii) *We have the inequality*

$$\sup_{n\in\mathbb{N}} \sum_{m=1}^{\infty} a_{n,m} \leq \sum_{m=1}^{\infty} \sup_{n\in\mathbb{N}} a_{n,m}.$$

The meaning of the equality in (i) is that if one side is convergent, then the other side is also convergent to the same limit; on the other hand, if one side is divergent, then it is divergent to infinity, and in this case the other side is also divergent to infinity. An analogous interpretation is assigned to the inequality in (ii).

Proof. (i) By the associative law for (finite!) sums, we can interchange the order of summation:

$$\sum_{n=1}^{N} \sum_{m=1}^{M} a_{n,m} = \sum_{m=1}^{M} \sum_{n=1}^{N} a_{n,m}.$$

Since $a_{n,m} \geq 0$, we have the bound

$$\sum_{m=1}^{M} \sum_{n=1}^{N} a_{n,m} \leq \sum_{m=1}^{M} \sum_{n=1}^{\infty} a_{n,m}.$$

Both sums $\sum_{n=1}^{N} \sum_{m=1}^{M} a_{n,m}$ and $\sum_{m=1}^{M} \sum_{n=1}^{\infty} a_{n,m}$, considered as sequences with respect to M, are monotone increasing. Therefore, both limits as $M \to \infty$ exist in $[0, \infty]$, and we infer that

$$\sum_{n=1}^{N} \sum_{m=1}^{\infty} a_{n,m} \leq \sum_{m=1}^{\infty} \sum_{n=1}^{\infty} a_{n,m}.$$

Once again, both limits as $N \to \infty$ exist in $[0, \infty]$, and we obtain that

$$\sum_{n=1}^{\infty} \sum_{m=1}^{\infty} a_{n,m} \leq \sum_{m=1}^{\infty} \sum_{n=1}^{\infty} a_{n,m}.$$

By arguing symmetrically but starting from taking first the limit in N, and then that in M, the reader can easily verify as an exercise that the opposite inequality holds true as well. Item (ii) is also left as an easy exercise. □

The following lemma lists some very important properties of the Lebesgue outer measure on $\mathbb{R}^d$.

Lemma 6.14. *Let $\mathcal{L}^{d*} : \mathcal{P}(\mathbb{R}^d) \longrightarrow [0, \infty]$ be the Lebesgue outer measure on $\mathbb{R}^d$. Then the following hold:*

(i) $\mathcal{L}^{d*}(\emptyset) = 0$.

(ii) $\mathcal{L}^{d*}$ *is* **finitely sub-additive**. *Namely, for any $E, F \subseteq \mathbb{R}^d$ we have*

$$\mathcal{L}^{d*}(E \cup F) \leq \mathcal{L}^{d*}(E) + \mathcal{L}^{d*}(F).$$

(iii) $\mathcal{L}^{d*}$ is **monotone**. *Namely, if $E \subseteq F \subseteq \mathbb{R}^d$, then $\mathcal{L}^{d*}(E) \leq \mathcal{L}^{d*}(F)$.*

(iv) $\mathcal{L}^{d*}$ is **σ-sub-additive (or countably sub-additive)**. *Namely, for any sequence of sets $(E_n)_1^\infty \subseteq \mathbb{R}^d$, we have*

$$\mathcal{L}^{d*}\left(\bigcup_{n=1}^{\infty} E_n\right) \leq \sum_{n=1}^{\infty} \mathcal{L}^{d*}(E_n).$$

(v) $\mathcal{L}^{d*}$ **vanishes on countable sets**. *Namely, if $C \subseteq \mathbb{R}^d$ is countable (i.e. a sequence $\{c_n : n \in \mathbb{N}\}$), then $\mathcal{L}^{d*}(C) = 0$.*

(vi) *If B is a box, then $\mathcal{L}^{d*}(B) = \mathcal{V}^d(B)$.*

Remark 6.15. In the next section we shall generalise (vi) by proving that

$$\mathcal{L}^{d*}(E) = \mathcal{V}^d(E), \quad \text{for any elementary set } E \in \mathcal{E}(\mathbb{R}^d).$$

Hence, $\mathcal{L}^{d*}$ is just an extension of the volume from the class of elementary sets to all subsets of $\mathbb{R}^d$. Note further that property (ii) is actually contained in property (iv) (you can see this by taking $E_1 = E$, $E_2 = F$, $E_k = \emptyset$ for $k \geq 3$), but we demonstrate it separately for the sake of clarity[2].

Proof. (i) Let $\varepsilon > 0$ and consider the boxes $B_1 := [0, \varepsilon]^d$, $B_k := \emptyset$ for $k \geq 2$. Then, $\emptyset \subseteq \bigcup_{k=1}^\infty B_k$ and hence $\mathcal{L}^{d*}(\emptyset) \leq \varepsilon^d$. As this holds for all $\varepsilon > 0$, the assertion follows.

(ii) By the definition of $\mathcal{L}^{d*}$, for any $\varepsilon > 0$ there exists a covering of E by a countable family of boxes $\{B_k\}_{k=1}^\infty$ such that

$$\sum_{k=1}^{\infty} \mathcal{V}^d(B_k) < \mathcal{L}^{d*}(E) + \varepsilon.$$

Analogously, there exists a covering of F by a countable family of boxes $\{C_k\}_{k=1}^\infty$ with

$$\sum_{k=1}^{\infty} \mathcal{V}^d(C_k) < \mathcal{L}^{d*}(F) + \varepsilon.$$

We now define the sequence of boxes (see Figure 6.11)

$$D_k := \begin{cases} B_{(k+1)/2}, & \text{if } k \text{ is odd}, \quad k \in \mathbb{N}, \\ C_{k/2}, & \text{if } k \text{ is even}, \quad k \in \mathbb{N}. \end{cases}$$

This means that we have $\{D_1, D_2, D_3, D_4...\} = \{B_1, C_1, B_2, C_2, ...\}$. Note now that $\{D_k\}_1^\infty$ covers $E \cup F$, because

$$\bigcup_{k=1}^{\infty} D_k = \left(\bigcup_{k=1}^{\infty} B_k\right) \bigcup \left(\bigcup_{k=1}^{\infty} C_k\right) \supseteq E \cup F.$$

---

[2] At this point we recommend to recall the basic properties of suprema and infima.

By the definition of the outer measure we have

$$\mathcal{L}^{d*}(E \cup F) \leq \sum_{k=1}^{\infty} \mathcal{V}^d(D_k)$$
$$= \sum_{k=1}^{\infty} \mathcal{V}^d(B_k) + \sum_{k=1}^{\infty} \mathcal{V}^d(C_k)$$

and hence

$$\mathcal{L}^{d*}(E \cup F) < \mathcal{L}^{d*}(E) + \mathcal{L}^{d*}(F) + 2\varepsilon.$$

Given that this holds for all $\varepsilon > 0$, the claimed result follows.

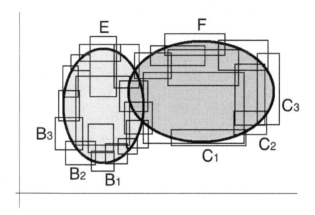

Figure 6.11: Illustration of the idea of the proof of the finite additivity.

(iii) is an exercise.

(iv) is similar to (ii), the only difference being that we need to adjust the "ε" appropriately because of the presence of infinitely-many sets. By the definition of $\mathcal{L}^{d*}(E_n)$, for any $n \in \mathbb{N}$ and $\varepsilon > 0$ there exists a covering of E_n by boxes $\{B_k^{(n)}\}_1^{\infty}$ such that

$$\sum_{k=1}^{\infty} \mathcal{V}^d(B_k^{(n)}) < \mathcal{L}^{d*}(E_n) + \varepsilon 2^{-n}.$$

Then $\{B_k^{(n)}\}_{n,k \in \mathbb{N}}$ is a covering of the set $\bigcup_{n=1}^{\infty} E_n$. According to Lemma 6.13 the value of the series $\sum_{n=1}^{\infty} \sum_{k=1}^{\infty} \mathcal{L}^{d*}(B_k^{(n)})$ is independent of the order of summation. Hence we obtain

$$\mathcal{L}^{d*}\left(\bigcup_{n=1}^{\infty} E_n\right) \leq \sum_{n=1}^{\infty} \sum_{k=1}^{\infty} \mathcal{V}^d(B_k^{(n)})$$

and as a result

$$\mathcal{L}^{d*}\left(\bigcup_{n=1}^{\infty} E_n\right) < \sum_{n=1}^{\infty} \left(\mathcal{L}^{d*}(E_n) + \varepsilon 2^{-n}\right)$$

$$= \sum_{n=1}^{\infty} \mathcal{L}^{d*}(E_n) + \varepsilon \left\{\sum_{n=1}^{\infty} 2^{-n}\right\}.$$

Since this holds for all $\varepsilon > 0$, the claimed result follows.

(v) Let $C \subseteq \mathbb{R}^d$ be a countable set, that is a sequence $\{c_1, c_2, \ldots\}$ in $\mathbb{R}^d$. For each point c_k, we consider the respective box (see Figure 6.12):

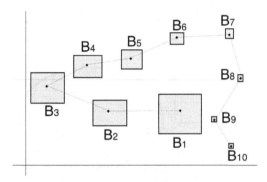

Figure 6.12: Illustration of the idea of the proof of vanishing of the outer measure for countable sets on $\mathbb{R}^2$ of part (v).

$$B_k := \left[c_k - 2^{-k-1}\varepsilon, \, c_k + 2^{-k-1}\varepsilon\right]^d \supseteq \{c_k\}.$$

Then, $\{B_k\}_{k \in \mathbb{N}}$ covers C, that is $C \subseteq \bigcup_{k=1}^{\infty} B_k$. Since this cover by boxes is one of the possible countable coverings, we obtain

$$\mathcal{L}^{d*}(C) \le \sum_{k=1}^{\infty} \mathcal{V}^d(B_k) = \sum_{k=1}^{\infty} \left(2 \, 2^{-k-1}\varepsilon\right)^d = \varepsilon^d \left\{\sum_{k=1}^{\infty} 2^{-dk}\right\}.$$

By letting $\varepsilon \to 0^+$, we deduce $\mathcal{L}^{d*}(C) = 0$.

(vi) Let B be any box in $\mathbb{R}^d$. We first show that $\mathcal{L}^{d*}(B) \le \mathcal{V}^d(B)$. Indeed, by considering the covering $\bigcup_{k=1}^{\infty} B_k$ of B with boxes given by $B_1 := B$, $B_k := \emptyset$ for $k \ge 2$, we obtain

$$\mathcal{L}^{d*}(B) \le \sum_{k=1}^{\infty} \mathcal{V}^d(B_k) = \mathcal{V}^d(B).$$

We now show that $\mathcal{V}^d(B) \le \mathcal{L}^{d*}(B)$. Let $B = I_1 \times \cdots \times I_d$ be a box, where each I_j is an interval with endpoints $a_j < b_j$ (either of (a_j, b_j), $(a_j, b_j]$, $[a_j, b_j)$,

$[a_j, b_j])$. It suffice to consider the case when B is non-degenerate. Fix $\delta > 0$, small, and consider the following concentric **closed** box $B^{(\delta)} \subseteq \mathbb{R}^d$ contained in B:

$$B^{(\delta)} := \prod_{j=1}^{d} \left[a_j + \frac{\delta}{2}(b_j - a_j),\, b_j - \frac{\delta}{2}(b_j - a_j) \right] \subseteq B.$$

Then, the volume of $B^{(\delta)}$ is

$$\begin{aligned}
\mathcal{V}^d(B^{(\delta)}) &= \prod_{j=1}^{d} \left((b_j - a_j) - \delta(b_j - a_j) \right) \\
&= (1 - \delta)^d \left[(b_1 - a_1) \cdots (b_d - a_d) \right]
\end{aligned}$$

and hence

$$\mathcal{V}^d(B^{(\delta)}) = (1 - \delta)^d\, \mathcal{V}^d(B). \tag{$*$}$$

Let $\bigcup_1^{\infty} B_k$ be an arbitrary countable covering of B by boxes $\{B_k\}$, i.e. $B \subseteq \bigcup_1^{\infty} B_k$. If each box B_k can be represented as $B_k = I_1^{(k)} \times \cdots \times I_d^{(k)}$ and each $I_j^{(k)}$ is an interval with endpoints $a_j^{(k)} < b_j^{(k)}$, similarly as before for each $k \in \mathbb{N}$ we consider the **open** box $B_{(\delta)k}$ which is concentric with B_k and contains it (see Figure 6.13):

$$B_{(\delta)k} := \prod_{i=1}^{d} \left(a_j^{(k)} - \frac{\delta}{2}(b_j^{(k)} - a_j^{(k)}),\, b_j + \frac{\delta}{2}(b_j^{(k)} - a_j^{(k)}) \right) \supseteq B_k, \quad k \in \mathbb{N}.$$

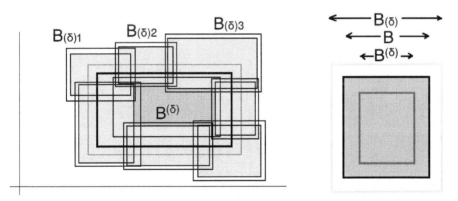

Figure 6.13: Illustration of the open and closed boxed "squeezing" the given box from the inside and the outside.

Similarly, the volume of each $B_{(\delta)k}$ is given by

$$\begin{aligned}
\mathcal{V}^d(B_{(\delta)k}) &= \prod_{i=1}^{d} \left((b_j^{(k)} - a_j^{(k)}) + \delta\,(b_j^{(k)} - a_j^{(k)}) \right) \\
&= (1 + \delta)^d \left[(b_1^{(k)} - a_1^{(k)}) \cdots (b_d^{(k)} - a_d^{(k)}) \right]
\end{aligned}$$

and hence

$$\mathcal{V}^d(B_{(\delta)k}) = (1+\delta)^d \mathcal{V}^d(B_k). \tag{$**$}$$

Moreover, $(B_{(\delta)k})_1^\infty$ is a covering by open boxes of the closed (and bounded) box $B^{(\delta)}$, that is

$$B^{(\delta)} \subseteq \bigcup_{k=1}^{\infty} B_k \subseteq \bigcup_{k=1}^{\infty} B_{(\delta)k}.$$

Since $B^{(\delta)}$ is a closed and bounded set, in view of Theorems 5.4 and 5.21 of Chapter 5, it is compact as well. Hence, it follows that our open covering $(B_{(\delta)k})_1^\infty$ has a finite subcovering and there exists $N = N(\delta) \in \mathbb{N}$ such that the finitely many $B_{(\delta)1}, \ldots, B_{(\delta)N}$ boxes cover $B^{(\delta)}$:

$$B^{(\delta)} \subseteq \bigcup_{k=1}^{N} B_{(\delta)k} =: E.$$

Note now that $E \in \mathcal{E}(\mathbb{R}^d)$, i.e. E is an elementary set. Hence, by Proposition 6.9 we have that

$$\mathcal{V}^d(B^{(\delta)}) \le \mathcal{V}^d(E) \le \sum_{k=1}^{N} \mathcal{V}^d(B_{(\delta)k}).$$

By using $(*)$ and $(**)$, the above estimate yields

$$(1-\delta)^d \mathcal{V}^d(B) = \mathcal{V}^d(B^{(\delta)}) \le \sum_{k=1}^{N} \mathcal{V}^d(B_{(\delta)k}).$$

Since we also have

$$\sum_{k=1}^{N} \mathcal{V}^d(B_{(\delta)k}) \le \sum_{k=1}^{\infty} \mathcal{V}^d(B_{(\delta)k}) = (1+\delta)^d \sum_{k=1}^{\infty} \mathcal{V}^d(B_k),$$

it follows that

$$\mathcal{V}^d(B) \le \left(\frac{1+\delta}{1-\delta}\right)^d \sum_{k=1}^{\infty} \mathcal{V}^d(B_k).$$

By letting $\delta \to 0^+$ and taking the infimum over all coverings $(B_k)_1^\infty$ of B by boxes, we get

$$\mathcal{V}^d(B) \le \inf \left\{ \sum_{k=1}^{\infty} \mathcal{V}^d(B_k) : (B_k)_1^\infty \text{ non-degenerate boxes, } B \subseteq \bigcup_{k=1}^{\infty} B_k \right\}$$
$$= \mathcal{L}^{d*}(B).$$

The lemma has been established.   □

6.5   Measurable and non-measurable sets

Recall from the previous section that if we wish to interpret the Lebesgue outer measure map

$$\mathcal{L}^{d*} \; : \; \mathcal{P}(\mathbb{R}^d) \longrightarrow [0, \infty]$$

as "volume", then the following property should be true: If $E_1, E_2, \ldots, E_m \subseteq \mathbb{R}^d$ are disjoint sets (i.e. $E_j \cap E_j = \emptyset$ for $i \neq j$), then $\mathcal{L}^{d*}$ should be additive, that is

$$\mathcal{L}^{d*}(E_1 \cup \cdots \cup E_m) = \mathcal{L}^{d*}(E_1) + \cdots + \mathcal{L}^{d*}(E_m). \tag{\#}$$

The equality (#) is a very natural feature which is the idea of "conservation of volume" and expresses the fact that

"if we split a set into finitely many parts, then the sum of the volumes of the parts equals the volume of the original set".

Unfortunately, there exists very bizarre sets for which this formula is false!!!

Example 6.16 (Banach-Tarski Paradox). Consider a ball of some fixed radius in $\mathbb{R}^d$, $d \geq 3$. Then, there exists a partition of the ball into finitely many pieces such that, by re-assembling the pieces appropriately (without changing their size or shape!)... we can construct two balls identical to the original (see Figure 6.14 for an artistic representation)!!!

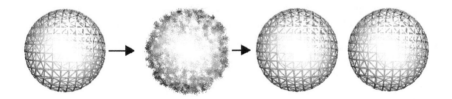

Figure 6.14: Graphic illustration of the Banach-Tarski "Paradox" theorem.

Obviously, the sets of the partition above **cannot have a well-defined notion of volume**, for otherwise (#) would be violated. Non-measurable sets which do not have volume (of a little less striking nature) exist in $\mathbb{R}$ and $\mathbb{R}^2$ as well. We refrain from presenting any precise construction of such sets because they rely on certain ideas from Algebra, and this would take us too far afield.[3] We merely mention that such constructions (and in fact any known construction of non-measurable sets) rely on the Axiom of Choice, since, roughly speaking, in order to create a non-measurable set, one has to make an "uncountable selection of points".

---

[3]The interested reader is invited to consult the references [6, 38, 42] for further information regarding the construction of non-measurable sets.

The next example, of a less exotic nature, shows that non-measurable sets exist even on the real line $\mathbb{R}$:

Example 6.17 (A Lebesgue non-measurable set on $\mathbb{R}$). Consider the equivalence relation on $[0, 1) \subseteq \mathbb{R}$ defined by

$$x \sim y \quad \Longleftrightarrow \quad x - y \in \mathbb{Q}.$$

Consider the quotient set

$$[0, 1)/_\sim := \big\{ [x]_\sim \ : \ x \in [0, 1) \big\}$$

of all equivalence classes. Let S be a subset of $[0, 1)$ which contains exactly one representative from each equivalence class. (For this to be done, we need to invoke the Axiom of Choice, since we have to make uncountably many selections!) For each $q \in [0, 1) \cap \mathbb{Q}$, we define the following set

$$S_q := \Big(q + \big(S \cap [0, 1 - q) \big) \Big) \bigcup \Big(q - 1 + \big(S \cap (1 - q, 1) \big) \Big).$$

That is, to obtain the set S_q we move S to the right by q and then move the part of S that sticks out of $[0, 1)$ by 1 to the left. It can be easily verified that every $x \in [0, 1)$ belongs exactly to one set $S_q \subseteq [0, 1)$. Remember also that $\mathbb{Q}$ is countable (and hence a set of zero outer measure). Thus, the sequence of sets

$$\big\{ S_q \ : \ q \in [0, 1) \cap \mathbb{Q} \big\}$$

is a countable partition of $[0, 1)$. We will now show that the countable family of sets $\{ S_q : q \in [0, 1) \cap \mathbb{Q} \}$ contains at least one non-measurable set. Indeed, if for the sake of contradiction all the sets were measurable, since the Lebesgue outer measure is invariant under translations, we would have

$$\mathcal{L}^{1*}(S) = \mathcal{L}^{1*}\Big(q + \big(S \cap [0, 1 - q) \big) \Big) + \mathcal{L}^{1*}\Big(q - 1 + \big(S \cap [1 - q, 1) \big) \Big)$$
$$= \mathcal{L}^{1*}\big(S \cap [0, 1 - q) \big) + \mathcal{L}^{1*}\big(S \cap [1 - q, 1) \big) = \mathcal{L}^{1*}(S_q)$$

and since $\{ S_q : q \in [0, 1) \cap \mathbb{Q} \}$ is a countable partition of $[0, 1)$ into disjoint sets, we have

$$1 = \mathcal{L}^{1*}\big([0, 1) \big) = \sum_{q \in [0, 1) \cap \mathbb{Q}} \mathcal{L}^{1*}(S_q) = \sum_{q \in [0, 1) \cap \mathbb{Q}} \mathcal{L}^{1*}(S).$$

The above equality is an obvious contradiction, since the value of the right hand side is either zero (if $\mathcal{L}^{1*}(S) = 0$) or infinity (if $\mathcal{L}^{1*}(S) > 0$) and hence, in either case it is different from 1.

In order to get rid of pathological sets as above, we restrict our attention to sets for which relation (#) holds. We shall call these sets "measurable", and regard them as those for which we can associate a well-defined volume value.

The idea is to call a set $A \subseteq \mathbb{R}^d$ measurable when the following happens: if we intersect it with any set $E \subseteq \mathbb{R}^d$, then for the part $E \cap A$ of E inside A and for the part $E \setminus A$ of E outside A the additivity property (#) holds, that is

$$\mathcal{L}^{d*}(E) = \mathcal{L}^{d*}(E \cap A) + \mathcal{L}^{d*}(E \setminus A),$$

since $E = (E \cap A) \cup (E \setminus A)$ (conservation of the volume when splitting E into $E \cap A$ and $E \setminus A$). Actually, we may require less: by Proposition 6.9 the Lebesgue outer measure is sub-additive, which means we *always* have

$$\mathcal{L}^{d*}(E) = \mathcal{L}^{d*}\left((E \cap A) \bigcup (E \setminus A)\right) \leq \mathcal{L}^{d*}(E \cap A) + \mathcal{L}^{d*}(E \setminus A).$$

Hence, it suffices to require only the opposite inequality.

Definition 6.18. A set $A \subseteq \mathbb{R}^d$ is called **(Lebesgue) measurable** if for every set $E \subseteq \mathbb{R}^d$ it holds that (see Figure 6.15)

$$\mathcal{L}^{d*}(E) \geq \mathcal{L}^{d*}(E \cap A) + \mathcal{L}^{d*}(E \setminus A).$$

We denote the class of all Lebesgue measurable sets by $\mathcal{L}(\mathbb{R}^d)$, that is

$$\mathcal{L}(\mathbb{R}^d) := \left\{ A \subseteq \mathbb{R}^d \;\middle|\; \mathcal{L}^{d*}(E) \geq \mathcal{L}^{d*}(E \cap A) + \mathcal{L}^{d*}(E \setminus A), \text{ for all } E \subseteq \mathbb{R}^d \right\}.$$

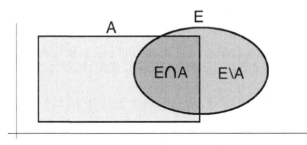

Figure 6.15: An illustration of the idea behind the definition of a Lebesgue measurable set on $\mathbb{R}^2$. The set A is measurable if whenever we intersect it by any other arbitrary set E, the Lebesgue outer measure of E behaves additively with respect to the partition $E \cap A$, $E \setminus A$.

Now we may define the central notion of volume for general sets in $\mathcal{L}(\mathbb{R}^d)$.

Definition 6.19. The **Lebesgue measure on $\mathbb{R}^d$** is the mapping

$$\mathcal{L}^d \;:\; \mathcal{L}(\mathbb{R}^d) \longrightarrow [0, \infty], \qquad \mathcal{L}^d := \mathcal{L}^{d*}\big|_{\mathcal{L}(\mathbb{R}^d)}.$$

That is, the Lebesgue measure $\mathcal{L}^d$ on $\mathbb{R}^d$ (Definition 6.11) is the **restriction** of the Lebesgue outer measure $\mathcal{L}^{d*}$ on the class of Lebesgue measurable sets $\mathcal{L}(\mathbb{R}^d)$ (Definition 6.18).

The main properties of the Lebesgue measure which illustrate its significance will be explored in the next section.

6.6   The Carathéodory Theorem

As we show right next, the Lebesgue outer measure defined in the previous section coincides with the usual volume on the class of elementary sets $\mathcal{E}(\mathbb{R}^d)$ and these sets are measurable, namely, $\mathcal{V}^d = \mathcal{L}^{d*}$ on $\mathcal{E}(\mathbb{R}^d)$ and $\mathcal{E}(\mathbb{R}^d) \subseteq \mathcal{L}(\mathbb{R}^d)$. We begin by identifying a **very important** class of Lebesgue measurable sets:

Definition 6.20. The family of **Borel sets of $\mathbb{R}^d$**, denoted by $\mathcal{B}(\mathbb{R}^d)$, is the family of sets that can be formed from open sets through (countably-many applications of) the operations of countable union, countable intersection and complement, in any order and/or number of applications.

The definition above is deliberately implicit for the sake of convenience. In the next chapter we shall characterise the class of Borel sets in a more explicit fashion and for general topological spaces, as the smallest "appropriate" class of sets which contains the family of open sets.

Example 6.21. Suppose that $(A_n)_1^\infty \subseteq \mathbb{R}^d$ is a sequence of open sets. Then,

$$\bigcap_{n=1}^{\infty} A_n \,,\ \bigcup_{n=1}^{\infty} A_n \,,\ \bigcup_{n=1}^{\infty} (\mathbb{R}^d \setminus A_n) \,,\ \bigcup_{n\in 2\mathbb{N}} \left(\mathbb{R}^d \setminus \left(\bigcap_{n\in 2\mathbb{N}+1} A_n \right) \right)$$

are Borel sets. Note that the sets $\bigcap_{n=1}^{\infty} A_n$ and $\bigcup_{n=1}^{\infty}(\mathbb{R}^d \setminus A_n)$ in general are neither open nor closed. To convince yourself, note that

$$(-1,1] = \bigcap_{n=1}^{\infty} \left(-1, 1+\frac{1}{n} \right) \,,\quad (-1,1] = \bigcup_{n=1}^{\infty} \left[-1+\frac{1}{n}, 1 \right].$$

The next **very important** result lists the main properties of the Lebesgue measure.[4]

Theorem 6.22 (The "Carathéodory theorem"). *The Lebesgue measure on $\mathbb{R}^d$ satisfies the following properties:*

(i) $\emptyset, \mathbb{R}^d \in \mathcal{L}(\mathbb{R}^d)$.

(ii) *If $A \in \mathcal{L}(\mathbb{R}^d)$, then $\mathbb{R}^d \setminus A \in \mathcal{L}(\mathbb{R}^d)$.*

(iii) *If $(A_k)_1^\infty \subseteq \mathcal{L}(\mathbb{R}^d)$, then $\bigcup_{k=1}^{\infty} A_k$ and $\bigcap_{k=1}^{\infty} A_k$ are in $\mathcal{L}(\mathbb{R}^d)$. If in addition the sequence is disjoint, that is $A_k \cap A_j = \emptyset$ when $j \neq k$, then*

$$\mathcal{L}^d \left(\bigcup_{k=1}^{\infty} A_k \right) = \sum_{k=1}^{\infty} \mathcal{L}^d(A_k).$$

---

[4]The proof of the result is fairly technical and could perhaps be omitted on the first reading of this book.

(iv) *Every countable set is measurable and has zero measure.*

(v) *Open subsets, closed subsets and boxes of* $\mathbb{R}^d$ *belong to* $\mathcal{L}(\mathbb{R}^d)$, *that is, they are measurable.*

(vi) *We have that*

$$\mathcal{E}(\mathbb{R}^d) \subseteq \mathcal{B}(\mathbb{R}^d) \subseteq \mathcal{L}(\mathbb{R}^d).$$

that is every elementary set is a Borel set and every Borel set is measurable.

(vii) *We have that*

$$\mathcal{L}^d\big|_{\mathcal{E}(\mathbb{R}^d)} = \mathcal{V}^d,$$

that is the measure of any elementary set coincides with its volume.

(viii) *[Completeness] Let* $A \subseteq \mathbb{R}^d$ *satisfy* $\mathcal{L}^{d*}(A) = 0$. *Then,* A *is measurable, i.e.* $A \in \mathcal{L}(\mathbb{R}^d)$.

The point of (viii) above is that if the Lebesgue outer measure of A vanishes, then A is measurable even though it was not assumed to be in $\mathcal{L}(\mathbb{R}^d)$.

The following very simple result will be used in the proof of part (iii) of Theorem 6.22, and will be invoked again in the next chapter. Its essence is that, given a sequence $(A_k)_1^\infty$, we can construct a new sequence $(B_k)_1^\infty \subseteq X$ which has the same union as the original sequence, but in addition it is **disjoint**.

Lemma 6.23. *Let* $(A_k)_1^\infty$ *be a sequence of subsets of a set* X. *Then, the sequence* $(B_k)_1^\infty$ *defined recursively by*

$$B_1 := A_1, \qquad B_k := A_k \setminus \bigcup_{j=1}^{k-1} A_j, \qquad k \geq 2,$$

is disjoint (i.e. $B_i \cap B_j = \emptyset$ *for* $i \neq j$), *satisfies* $B_k \subseteq A_k$ *and*

$$\bigcup_{k=1}^{n} A_k = \bigcup_{k=1}^{n} B_k \quad \text{for all } n \in \mathbb{N}, \qquad \text{and} \qquad \bigcup_{k=1}^{\infty} A_k = \bigcup_{k=1}^{\infty} B_k.$$

The proof of Lemma 6.23 is left as an exercise (but see Figure 6.16 for an illustration of the idea of this simple construction).

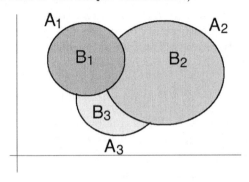

Figure 6.16: The idea of the definition the disjoint sets in Lemma 6.23.

Proof. (i) and (ii) are simple exercises. For (iii), we consider first the case of any two sets $A, B \in \mathcal{L}(\mathbb{R}^d)$, and prove that $A \cup B \in \mathcal{L}(\mathbb{R}^d)$ and, if $A \cap B = \emptyset$, then

$$\mathcal{L}^{d*}(A \cup B) = \mathcal{L}^{d*}(A) + \mathcal{L}^{d*}(B).$$

Let $E \subseteq \mathbb{R}^d$ be arbitrary. Then

$$\begin{aligned}
\mathcal{L}^{d*}(E) &= \mathcal{L}^{d*}(E \cap A) + \mathcal{L}^{d*}(E \cap A^c) \\
&\geq \mathcal{L}^{d*}((E \cap A) \cap B) + \mathcal{L}^{d*}((E \cap A) \cap B^c) \\
&\quad + \mathcal{L}^{d*}((E \cap A^c) \cap B) + \mathcal{L}^{d*}((E \cap A^c) \cap B^c).
\end{aligned}$$

Since

$$A \cup B = (A \cap B) \bigcup (A \cap B^c) \bigcup (A^c \cap B),$$

by sub-additivity of the Lebesgue outer measure, we have

$$\mathcal{L}^{d*}(E \cap (A \cup B)) \leq \mathcal{L}^{d*}(E \cap A \cap B) + \mathcal{L}^{d*}(E \cap A \cap B^c) + \mathcal{L}^{d*}(E \cap A^c \cap B)$$

and hence

$$\mathcal{L}^{d*}(E) \geq \mathcal{L}^{d*}(E \cap (A \cup B)) + \mathcal{L}^{d*}(E \cap (A \cup B)^c)$$

which gives that $A \cup B \in \mathcal{L}(\mathbb{R}^d)$. Moreover, if $A \cap B = \emptyset$, then

$$\begin{aligned}
\mathcal{L}^{d*}(A \cup B) &\geq \mathcal{L}^{d*}((A \cup B) \cap A) + \mathcal{L}^{d*}((A \cup B) \cap A^c) \\
&= \mathcal{L}^{d*}(A) + \mathcal{L}^{d*}(B).
\end{aligned}$$

Since the opposite of the above inequality is always true, we have proved that $\mathcal{L}^d$ is additive on $\mathcal{L}(\mathbb{R}^d)$.

Let now $(A_k)_1^\infty$ be an arbitrary sequence in $\mathcal{L}(\mathbb{R}^d)$. For brevity, let us set

$$C_n := \bigcup_{k=1}^n A_k, \quad C := \bigcup_{k=1}^\infty A_k.$$

To prove that $C \in \mathcal{L}(\mathbb{R}^d)$, it suffices to consider the case when the sets $(A_k)_1^\infty$ are pairwise disjoint. (Indeed, the general case would follow by applying the same argument to the sequence $(B_k)_1^\infty$ of pairwise disjoint associated to $(A_k)_1^\infty$ as in Lemma 6.23, the measurability of the sets $(B_k)_1^\infty$ being a consequence of what we have already proved.) Thus, let us consider any $E \subseteq \mathbb{R}^d$. Note that

$$\begin{aligned}
\mathcal{L}^{d*}(E \cap C_n) &\geq \mathcal{L}^{d*}(E \cap C_n \cap A_n) + \mathcal{L}^{d*}(E \cap C_n \cap A_n^c) \\
&= \mathcal{L}^{d*}(E \cap A_n) + \mathcal{L}^{d*}(E \cap C_{n-1})
\end{aligned}$$

and the opposite inequality is always true by sub-additivity. A simple induction argument applied to the above (this is an exercise!) shows that, for any $n \in \mathbb{N}$, we have

$$\mathcal{L}^{d*}(E \cap C_n) = \sum_{k=1}^n \mathcal{L}^{d*}(E \cap A_k).$$

Therefore, since $C_n \in \mathcal{L}(\mathbb{R}^d)$, we have

$$\mathcal{L}^{d*}(E) \geq \mathcal{L}^{d*}(E \cap C_n) + \mathcal{L}^{d*}(E \cap C_n^c)$$

$$\geq \sum_{k=1}^{n} \mathcal{L}^{d*}(E \cap A_k) + \mathcal{L}^{d*}(E \cap C^c)$$

and by letting $n \to \infty$ we obtain

$$\mathcal{L}^{d*}(E) \geq \sum_{k=1}^{\infty} \mathcal{L}^{d*}(E \cap A_k) + \mathcal{L}^{d*}(E \cap C^c)$$

or

$$\mathcal{L}^{d*}(E) \geq \mathcal{L}^{d*}\left(\bigcup_{k=1}^{\infty} (E \cap A_k) \right) + \mathcal{L}^{d*}(E \cap C^c)$$

$$\geq \mathcal{L}^{d*}(E \cap C) + \mathcal{L}^{d*}(E \cap C^c)$$

$$\geq \mathcal{L}^{d*}(E),$$

for any $E \subseteq \mathbb{R}^d$. It follows that $C \in \mathcal{L}(\mathbb{R}^d)$ and that all the inequalities above actually are equalities. By taking $E := C$ we obtain that $\mathcal{L}^d$ is σ-additive on $\mathcal{L}(\mathbb{R}^d)$. On the other hand, the fact that $\bigcap_{k=1}^{\infty} A_k \in \mathcal{L}(\mathbb{R}^d)$ is an immediate consequence of the fact that

$$\mathbb{R}^d \setminus \left(\bigcap_{k=1}^{\infty} A_k \right) = \bigcup_{k=1}^{\infty} (\mathbb{R}^d \setminus A_k),$$

combined with (ii) and what we proved already in (iii) for countable unions.

(iv) This is a consequence of the fact that singleton sets are in $\mathcal{L}(\mathbb{R}^d)$, (iii) and Lemma 6.14. In order to see the first claim, note that for any $E \subseteq \mathbb{R}^d$, $x \in \mathbb{R}^d$ we have

$$\mathcal{L}^{d*}(E \cap \{x\}) + \mathcal{L}^{d*}(E \setminus \{x\}) = \mathcal{L}^{d*}(E \setminus \{x\}) \leq \mathcal{L}^{d*}(E).$$

(v) We start by proving that every box $B \subseteq \mathbb{R}^d$ is measurable. To this end, fix any $E \subseteq \mathbb{R}^d$ and $\varepsilon > 0$. By the definition of the outer measure, there exist a countable family of boxes $\{B_k\}_{k=1}^{\infty}$ such that $E \subseteq \bigcup_1^{\infty} B_k$ and

$$\sum_{k=1}^{\infty} V^d(B_k) < \mathcal{L}^{d*}(E) + \varepsilon.$$

Since

$$E \cap B \subseteq \left(\bigcup_{k=1}^{\infty} B_k \right) \cap B = \bigcup_{k=1}^{\infty} (B_k \cap B),$$

and $B_k \cap B$ is a box for each $k \in \mathbb{N}$, we have

$$\mathcal{L}^{d*}(E \cap B) \le \sum_{k=1}^{\infty} \mathcal{V}^d(B_k \cap B).$$

Similarly,

$$E \cap B^c \subseteq \left(\bigcup_{k=1}^{\infty} B_k \right) \cap B^c = \bigcup_{k=1}^{\infty} (B_k \cap B^c).$$

Note that $B_k \cap B^c$ is, for each $k \in \mathbb{N}$, an elementary set, and therefore can be written as a union of disjoint boxes $C_1^{(k)}, \ldots, C_{n_k}^{(k)}$ for some $n_k \in \mathbb{N}$, i.e.

$$B_k \cap B^c = \bigcup_{j=1}^{n_k} C_j^{(k)}.$$

This implies that

$$E \cap B^c \subseteq \bigcup_{k=1}^{\infty} \bigcup_{j=1}^{n_k} C_j^{(k)}$$

and hence, using again the definition of the outer measure, we have

$$\mathcal{L}^{d*}(E \cap B^c) \le \sum_{k=1}^{\infty} \sum_{j=1}^{n_k} \mathcal{V}^d(C_j^{(k)}).$$

By Lemma 6.13 we may change the order of summation, and hence the last two estimates give

$$\mathcal{L}^{d*}(E \cap B) + \mathcal{L}^{d*}(E \cap B^c) \le \sum_{k=1}^{\infty} \mathcal{V}^d(B_k \cap B) + \sum_{k=1}^{\infty} \sum_{j=1}^{n_k} \mathcal{V}^d(C_j^{(k)})$$

$$= \sum_{k=1}^{\infty} \left(\mathcal{V}^d(B_k \cap B) + \sum_{j=1}^{n_k} \mathcal{V}^d(C_j^{(k)}) \right).$$

Note now that, for each $k \in \mathbb{N}$, we have

$$(B_k \cap B) \bigcup \left(\bigcup_{j=1}^{n_k} C_j^{(k)} \right) = (B \cap B_k) \bigcup (B^c \cap B_k) = B_k,$$

and thus, by Proposition 6.9 (i), we have that

$$\mathcal{V}^d(B_k \cap B) + \sum_{j=1}^{n_k} \mathcal{V}^d(C_j^{(k)}) = \mathcal{V}^d(B_k)$$

so that

$$\mathcal{L}^{d*}(E \cap B) + \mathcal{L}^{d*}(E \cap B^c) \le \sum_{k=1}^{\infty} \mathcal{V}^d(B_k) < \mathcal{L}^{d*}(E) + \varepsilon.$$

Since the above holds for all $\varepsilon > 0$, we deduce that the box B is measurable, i.e. $B \in \mathcal{L}(\mathbb{R}^d)$.

We now prove the measurability of all open sets. Let $U \subseteq \mathbb{R}^d$ be any open set. Consider, for each $j \in \mathbb{N}$, the grid $2^{-j}\mathbb{Z}^d \subseteq \mathbb{R}^d$, and let $\{B_i^{(j)} | i \in \mathbb{Z}\}$ be the family of boxes determined by the grid, that is, the family of disjoint (half-open) cubes of size 2^{-j} whose vertices are on the grid. We first take $j = 1$, consider the (at most countably many) boxes $B_i^{(1)}$ which lie inside U for some indices i, and denote by $U^1 \subseteq U$ the union of these boxes. Then we take $j = 2$, consider the (at most countably many) boxes $B_i^{(2)}$ which lie inside $U \setminus U^1$ for some indices i, and denote by U^2 the union of these boxes. Then, we take $j = 3$ and apply the same argument to the set $U \setminus (U^1 \cup U^2)$. We continue this process recursively, to arrive at the conclusion that the open set U can be written as a disjoint union of countably many boxes (the fact that U is open is essential for this argument!). Since all boxes are measurable, the fact that U is measurable follows from (iii). Finally, the measurability of all closed subsets of $\mathbb{R}^d$ follows from (ii).

(vi) Note that every elementary set can be written as a disjoint union of boxes. Moreover, each box is a Borel set. Indeed, this is immediate if the box is either open or closed. If a box is neither open nor closed, suppose first it is non-degenerate. Then, by using the idea of the proof of Lemma 6.14 it can be written as a countable union (or intersection) of closed (or open) boxes. If the box is degenerate, then it can be written in the Cartesian product form of a non-degenerate box in a vector subspace times a constant vector[5] and apply the same idea to the box on the subspace. In either case, we see that every elementary set is a Borel set. Similarly, since by (v) open sets are measurable and $\mathcal{L}(\mathbb{R}^d)$ is closed under complements and countable unions/intersections, it follows that each Borel set is measurable.

(vii) By (iii), $\mathcal{L}^{d*}$ is additive on $\mathcal{L}(\mathbb{R}^d)$. By (vi), every elementary set is measurable and by Lemma 6.3 it can be written as a disjoint union of boxes. By (v), every box is measurable and by Lemma 6.14 the (outer) measure of any box is equal to its volume. The conclusion follows.

(viii) Let $A \subseteq \mathbb{R}^d$ satisfy $\mathcal{L}^{d*}(A) = 0$. Then, for any $E \subseteq \mathbb{R}^d$ we have

$$\mathcal{L}^{d*}(E) \leq \mathcal{L}^{d*}(E \cap A) + \mathcal{L}^{d*}(E \cap A^c) = \mathcal{L}^{d*}(E \cap A^c) \leq \mathcal{L}^{d*}(E).$$

Hence, A is measurable. □

The above result shows that every countable set is measurable and has zero measure. However, the converse is **not true**, i.e. a set might be "measure-theoretically small" in the sense of having zero measure, whilst being uncountable! On the other hand, there exist "topologically large" sets in the sense of

---

[5]For example, one may write the degenerate box $[1,2] \times \{3\} \times [4,5] \times \{6\}$ in the space $\mathbb{R}^4$ as the Cartesian product (up to a relabelling of the coordinates) of the non-degenerate box $[1,2] \times [4,5]$ in $\mathbb{R}^2$ times the vector $\{(3,6)\}$ in $\mathbb{R}^2$.

being open and dense, whilst having as small measure as desired! The next examples demonstrate such subtleties.

Remark 6.24 (The Cantor set). Consider the interval $C_0 := [0,1] \subseteq \mathbb{R}$. We split C_0 into three subintervals

$$[0,1/3], \ [1/3,2/3], \ [2/3,1]$$

of length $1/3$ and choose $C_1 = [0,1/3] \cup [2/3,1]$. We divide $[0,1/3]$ and $[2/3,1]$ into three intervals of length $1/3^2$ and choose the first and the third of each, symbolising the union of these sets by C_2. We continue in this way recursively, constructing a decreasing sequence of compact sets $\{C_n : n \in \mathbb{N}\}$ (see Figure 6.17). Let $C \subseteq [0,1]$ denote the intersection of these compact sets:

$$C := \bigcap_{n=1}^{\infty} C_n.$$

It can be easily proved that C is a non-empty compact set of Lebesgue measure zero (this is an easy exercise). Moreover, it can be shown that it has the same cardinality as $\mathbb{R}$ and hence it is uncountable! This set is known as the **Cantor (middle third) set**.

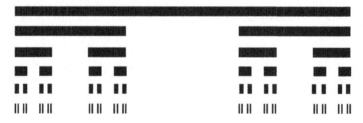

Figure 6.17: Depiction of the Cantor set (usually called Cantor middle third set).

On the other hand, a set might be topologically large (open and dense) but have very small measure! Accordingly, consider the next pathology:

Remark 6.25 ("Thin" open dense sets). Consider the interval $[0,1] \subseteq \mathbb{R}$ and fix $0 < \varepsilon < 1$. Let $(r_j)_1^{\infty}$ be an enumeration of the countable set $\mathbb{Q} \cap [0,1]$. For each $j \in \mathbb{N}$, consider the open interval $(r_j - \varepsilon 2^{-j}, r_j + \varepsilon 2^{-j})$ and set

$$U := \bigcup_{j=1}^{\infty} \left(r_j - \varepsilon 2^{-j}, r_j + \varepsilon 2^{-j} \right).$$

Then, U is obviously an open set which is dense in $[0,1]$ (since it contains a dense set), and

$$\mathcal{L}^1(U) \leq \sum_{j=1}^{\infty} \mathcal{L}^1\left(\left(r_j - \varepsilon 2^{-j}, r_j + \varepsilon 2^{-j} \right) \right) \leq 2\varepsilon \sum_{j=1}^{\infty} 2^{-j} \leq 2\varepsilon.$$

This shows that there exists open sets which are dense in $[0,1]$ and whose measure is as small as one desires!

We conclude this chapter with the following so-called *"outer regularity"* result for the Lebesgue outer measure. In a nutshell, it says that *every set of* $\mathbb{R}^d$ *can be approximated from the outside by a sequence of unions of open boxes*, and we can even select the boxes to have rational coordinates. An important consequence of this kind of regularity is that every Lebesgue measurable set can be written as the union of a Borel set and a nullset.

Proposition 6.26 (Outer regularity of $\mathcal{L}^{d*}$). *For any subset $A \subseteq \mathbb{R}^d$, the Lebesgue outer measure satisfies the following property:*

$$\mathcal{L}^{d*}(A) = \inf \left\{ \mathcal{L}^{d*}\left(\bigcup_{k=1}^{\infty} B_k \right) \,\middle|\, (B_k)_1^{\infty} \text{ open boxes, } A \subseteq \bigcup_{k=1}^{\infty} B_k \right\}.$$

In addition, the vertices of the sequence of open boxes above can be selected so that they have rational coordinates:

$$\mathcal{L}^{d*}(A) = \inf \left\{ \mathcal{L}^{d*}\left(\bigcup_{k=1}^{\infty} B_k \right) \,\middle|\, B_k = \prod_{j=1}^{d} \left(a_j^{(k)}, b_j^{(k)} \right), A \subseteq \bigcup_{k=1}^{\infty} B_k, \, a_j^{(k)}, b_j^{(k)} \in \mathbb{Q} \right\}.$$

The above result demonstrates a very significant property of the Lebesgue outer measure, particularly in view of the existence of very pathological "volumeless subsets" of $\mathbb{R}^d$, as we exhibit in the next section.

Proof. Let $A \subseteq \mathbb{R}^d$ be any set. We assume with no loss of generality that $\mathcal{L}^{d*}(A) < \infty$, the required result being trivial if this were not the case. Fix $\delta > 0$, arbitrary. Then, by the definition of $\mathcal{L}^{d*}(A)$, there exists a sequence $(B_k)_1^{\infty}$ of boxes (not necessarily open) such that

$$\mathcal{L}^{d*}(A) \leq \sum_{k=1}^{\infty} \mathcal{V}^d(B_k) \leq \mathcal{L}^{d*}(A) + \delta$$

and

$$A \subseteq \bigcup_{k=1}^{\infty} B_k.$$

Consider for each box B_k the concentric open box $B_{(\delta)k} \supseteq B_k$ constructed in the proof of Lemma 6.14(vi) above. Then, we have

$$A \subseteq \bigcup_{k=1}^{\infty} B_k \subseteq \bigcup_{k=1}^{\infty} B_{(\delta)k}$$

and by items (iii), (iv) and (vi) of Lemma 6.14, we have

$$\mathcal{L}^{d*}(A) \leq \mathcal{L}^{d*}\left(\bigcup_{k=1}^{\infty} B_{(\delta)k} \right) \leq \sum_{k=1}^{\infty} \mathcal{L}^{d*}\left(B_{(\delta)k} \right) = \sum_{k=1}^{\infty} \mathcal{V}^d\left(B_{(\delta)k} \right).$$

On the other hand, we also have, by construction,

$$\sum_{k=1}^{\infty} \mathcal{V}^d\big(B_{(\delta)k}\big) = (1+\delta)^d \sum_{k=1}^{\infty} \mathcal{V}^d\big(B_k\big).$$

Hence, by putting the above together, we infer that for any $\delta > 0$, there exists a sequence of open boxes $\big(B_{(\delta)k}\big)_1^{\infty} \subseteq \mathbb{R}^d$ covering A such that

$$\mathcal{L}^{d*}(A) \leq \mathcal{L}^{d*}\left(\bigcup_{k=1}^{\infty} B_{(\delta)k}\right)$$

$$\leq (1+\delta)^d \left(\sum_{k=1}^{\infty} \mathcal{V}^d\big(B_k\big)\right)$$

$$\leq (1+\delta)^d\big(\mathcal{L}^{d*}(A) + \delta\big).$$

Hence, the first desired equality of the statement ensues. For the second one, it suffices to note that since $\mathbb{Q}^d$ is dense in $\mathbb{R}^d$ (see Lemma 1.11), given for a fixed $k \in \mathbb{N}$ the boxes B_k and $B_{(\delta)k}$, due to the strict inclusion of B_k in $B_{(\delta)k}$, we can find a box $Q_{(\delta)k}$ whose *vertices have rational coordinates* such that

$$B_k \subseteq Q_{(\delta)k} \subseteq B_{(\delta)k}.$$

By repeating the argument above with $Q_{(\delta)k}$ is the place of $B_{(\delta)k}$, we see that the second desired equality has been established as well. □

An important consequence of the above regularity result is the following characterisation of the Lebesgue measurable sets.

Proposition 6.27 (Lebesgue set = Borel set − nullset). *Consider a set $A \subseteq \mathbb{R}^d$. Then, the following statements are equivalent:*

(i) $A \in \mathcal{L}(\mathbb{R}^d)$.

(ii) $A = G \backslash N$, *where $G \in \mathcal{B}(\mathbb{R}^d)$ is a Borel set and $N \in \mathcal{L}(\mathbb{R}^d)$ is a Lebesgue nullset with $\mathcal{L}^d(N) = 0$.*

Hence, every Lebesgue measurable set in $\mathbb{R}^d$ can be written as the set-theoretic difference between a Borel[6] set and a Lebesgue nullset.

Proof. (i) $\Longrightarrow$ (ii): By Proposition 6.26 and the properties of the infimum, there exists a sequence of open sets $(U_n)_1^{\infty} \subseteq \mathbb{R}^d$ such that $A \subseteq U_n$ for all $n \in \mathbb{N}$ and $\mathcal{L}^d(U_n) \longrightarrow \mathcal{L}^d(A)$ as $n \to \infty$. By setting

$$G := \bigcap_{n=1}^{\infty} U_n$$

---

[6] In addition, the Borel set claimed above can be chosen to be a countable intersection of open sets (a so-called G_δ-set). Our proof shows this only for sets of finite measure, but with a little more effort the result can be proved for any set A.

we have that $A \subseteq G$ and hence $\mathcal{L}^d(A) \leq \mathcal{L}^d(G) \leq \mathcal{L}^d(U_n)$ for all $n \in \mathbb{N}$, giving that $\mathcal{L}(A) = \mathcal{L}(G)$. If $\mathcal{L}^d(A) < \infty$, then we set $N := G \setminus A$ and it follows that $\mathcal{L}(N) = 0$ and $A = G \setminus N$.

If $\mathcal{L}^d(A) = \infty$, we fix $R \in \mathbb{N}$, define

$$A^R := A \cap \left(\mathbb{B}_{R+1}(0) \setminus \mathbb{B}_R(0) \right)$$

and apply the previous argument to A^R which has finite measure. By the previous case, there exists a Borel set $G^R \in \mathcal{B}(\mathbb{R}^d)$ with $G^R \supseteq A^R$ and such that $\mathcal{L}^d(G^R \setminus A^R) = 0$. Without loss of generality, we may assume that

$$G^R \subseteq \mathbb{B}_{R+1}(0) \setminus \mathbb{B}_R(0).$$

Indeed, if this is not the case, it suffice to replace G^R by

$$\tilde{G}^R := G^R \cap \left(\mathbb{B}_{R+1}(0) \setminus \mathbb{B}_R(0) \right)$$

which has the desired properties. We then set

$$N := \bigcup_{R=1}^{\infty} (G^R \setminus A^R), \quad G := \bigcup_{R=1}^{\infty} G^R.$$

Note that the sequence $(G^R \setminus A^R)_1^\infty$ is disjoint by construction. Also, $G \supseteq A$ and $A = G \setminus N$. Finally, by the σ-additivity of the Lebesgue measure,

$$\mathcal{L}^d(N) = \mathcal{L}^d \left(\bigcup_{R=1}^{\infty} (G^R \setminus A^R) \right) = \sum_{R=1}^{\infty} \mathcal{L}^d(G^R \setminus A^R) = 0.$$

(ii) $\implies$ (i): This is an immediate consequence of the completeness of the Lebesgue measure. $\qquad\square$

6.7 Exercises

Exercise 6.1. Let $E, F \subseteq \mathbb{R}^d$ be elementary sets. Show that $E \setminus F$ is an elementary set.

Exercise 6.2. Let $E, F \subseteq \mathbb{R}^d$ be a elementary sets and $x \in \mathbb{R}^d$. Show that

1. $E \cup F$ is an elementary set.

2. $E \Delta F = (E \setminus F) \cup (F \setminus E)$ is an elementary set.

3. $x + E = \{x + y \,|\, y \in E\}$ is an elementary set.

Exercise 6.3. Show that the triplet $\left(\mathcal{E}(\mathbb{R}^d), \Delta, \cap\right)$ is a ring in the sense of Algebra, that is with respect to the operations "Δ" and "$\cap$".

Exercise 6.4. Suppose that $E \subseteq F \subseteq \mathbb{R}^d$. By using the definition of the Lebesgue outer measure, show that $0 \leq \mathcal{L}^{d*}(E) \leq \mathcal{L}^{d*}(F)$.

Exercise 6.5. Let $E \in \mathcal{L}(\mathbb{R}^d)$. Show that $\mathbb{R}^d \setminus E \in \mathcal{L}(\mathbb{R}^d)$.

Exercise 6.6. Show that $\mathbb{R}^d \in \mathcal{L}(\mathbb{R}^d)$.

Exercise 6.7. Let $A \in \mathcal{L}(\mathbb{R}^d)$, $t > 0$ and $x \in \mathbb{R}^d$. Show that

$$\mathcal{L}^d(tA) = t^d \, \mathcal{L}^d(A), \qquad \mathcal{L}^d(x + A) = \mathcal{L}^d(A)$$

where tA is the dilation of the set A by t:

$$tA := \{ta \mid a \in A\},$$

and $x + A = \{x + a \mid x \in A\}$ is the translation of A by x. Hence, the **Lebesgue measure is invariant under translations and dilations**.

Exercise 6.8. Show that the definition of the Lebesgue outer measure remains the same if we replace the cover by boxes over which we take infimum by a cover by cubes (i.e. the sides of the box all have equal length). Namely,

$$\mathcal{L}^{d*}(A) = \inf \left\{ \sum_{i=1}^{\infty} \mathbb{V}^d(Q_i) \,\middle|\, (Q_i)_1^{\infty} \text{ non-degenerate cubes, } A \subseteq \bigcup_{i=1}^{\infty} Q_i \right\}.$$

Exercise 6.9. Show that every hyperplane of $\mathbb{R}^d$ has zero $\mathcal{L}^d$ measure in $\mathbb{R}^d$. It follows in particular that every line has zero measure in $\mathbb{R}^2$ and every plane has zero measure in $\mathbb{R}^3$.

[Hint: By a change of coordinates and the invariance of the Lebesgue measure under translations and dilation (Exercise 6.7), it suffices to consider the standard hyperplane $\{0\} \times \mathbb{R}^{d-1}$. Then, use that every $(d-1)$-cube in $\{0\} \times \mathbb{R}^{d-1}$ has zero $\mathcal{L}^d$-measure and the proof of Proposition 6.27.]

Exercise 6.10. Let $A, B \in \mathcal{L}^d(\mathbb{R}^d)$ satisfy $\mathcal{L}^d((A \cup B) \setminus (A \cap B)) = 0$. Then, prove that $\mathcal{L}^d(A) = \mathcal{L}^d(B)$. Does this result also holds for the outer measure $\mathcal{L}^{d*}$?

Exercise 6.11. Show that all countable subsets of $\mathbb{R}^d$ are Lebesgue measurable.

Exercise 6.12. Let $(a_{n,m})_{n,m \in \mathbb{N}} \subseteq [0, \infty]$ be a double sequence of non-negative extended real numbers. Show that

$$\sup_{n \in \mathbb{N}} \sum_{m=1}^{\infty} a_{n,m} \leq \sum_{m=1}^{\infty} \sup_{n \in \mathbb{N}} a_{n,m}$$

Exercise 6.13. Let $(a_{n,m})_{n,m \in \mathbb{N}} \subseteq [-\infty, \infty]$ be a double sequence of extended real numbers. Find an example showing that the non-negativity hypothesis is necessary for the series to commute:

$$\sum_{m=1}^{\infty} \sum_{n=1}^{\infty} a_{n,m} = \sum_{n=1}^{\infty} \sum_{m=1}^{\infty} a_{n,m}.$$

[Hint: Consider the sequence $a_{n,m} := \frac{m}{2^n} - \frac{2m}{3^n}$ and use the geometric series.]

Chapter 7

Measure theory on general spaces

7.1 Measuring beyond the Euclidean setting

In this chapter we commence our investigations of "measures" on abstract sets which in principle may have nothing to do with the Euclidean space. You are probably now wondering why we need this extra layer of abstraction. The answer is that even simple applications require to "measure" on some set different from the Euclidean space[1]. For instance, consider the problem of cartography of the Earth: we need to split the globe which is (almost) a sphere into pieces which obviously are curved. If the pieces are large enough, e.g. of size comparable to countries or continents, then they cannot be approximated by flat areas! (By the way, no map of the Earth can be fully accurate, because of the inherent distortion arising when trying to depict a sphere on a part of the plane with which it is not "isometric", i.e. distances are not preserved.) Another exemplary problem is that of measuring the area of a curved surface, like that below (Figure 7.1).

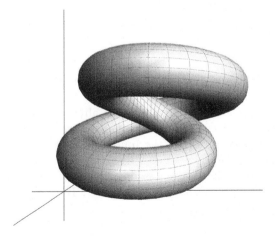

Figure 7.1: A strange shape of a so-called "toroidal" surface in $\mathbb{R}^3$ which looks like a stretched bagel.

---

[1]At this point it may be useful to review also the discussion in Section 6.1 of Chapter 6.

In this chapter we shall study functions called **measures** which **associate to *certain* subsets $A \subseteq X$ a non-negative (extended) number $\mu(A)$:**

$$\mu \; : \; X \supseteq A \; \mapsto \; \mu(A) \in [0, \infty].$$

This abstract notion of measure is a generalisation which axiomatises not just the volume, but many more concepts (like charges in Physics, etc.). Our only requirements will be

(i) $\mu(\emptyset) = 0$,

(ii) σ-additivity: for any disjoint sequence of sets $(A_k)_1^\infty$, namely any such that $A_i \cap A_j = \emptyset$ when $i \neq j$, we have

$$\mu \left(\bigcup_{k=1}^{\infty} A_k \right) = \sum_{k=1}^{\infty} \mu(A_k).$$

Such an object μ as above will be called a *measure*. Quite surprisingly, it turns out that these two axiomatic properties suffice to build a rich theory of abstract measures.

Note that **a measure μ is defined on subsets $A \in \mathcal{P}(X)$, namely on elements of the powerset $\mathcal{P}(X)$, which is the set of all subsets of X, not at points $x \in X$!** Hence, if $x \in X$, the hypothetical quantity "$\mu(x)$" **does not make sense.** However, if the singleton set $\{x\} \in \mathcal{P}(X)$ is considered (instead of the point $x \in X$), then talking about the measure $\mu(\{x\})$ of the set $\{x\}$ does indeed make sense.

Recapitulating, measures are functions with values in the extended non-negative numbers $[0, \infty]$, however their domain of definition is not X, but it is contained in the powerset $\mathcal{P}(X)$. Note though we deliberately *avoided* writing

$$\text{``} \mu \; : \; \mathcal{P}(X) \longrightarrow [0, \infty] \text{''}$$

because in general measures **cannot be defined on the entire powerset,** but only on certain classes of subsets of X, due to the possible existence of non-measurable sets. This phenomenon has already arisen in the case of the Lebesgue measure $\mathcal{L}^d$ of Chapter 6: not all subsets of $\mathbb{R}^d$ are Lebesgue measurable, and we had to restrict the (outer) measure to the class of sets $\mathcal{L}(\mathbb{R}^d)$, on which it is σ-additive.

In order to introduce our abstract axiomatic theory, we shall draw on the experience gained through our investigations in $\mathbb{R}^d$. The starting point will be to replace the triplet

$$\left(\mathbb{R}^d, \mathcal{L}(\mathbb{R}^d), \mathcal{L}^d \right)$$

of "Euclidean space – Lebesgue measurable sets – Lebesgue measure" which was studied in Chapter 6, by the more general triplet

$$\left(X, \mathfrak{M}, \mu \right)$$

of "abstract set – collection of (measurable) sets – abstract measure" which has the **essential properties** of the original triplet. (It turns out that neither the algebraic, nor topological structure, of the Euclidean space, are important.) This generality will allow us to develop a powerful **axiomatic theory** of very wide applicability.

After investigating the general properties of abstract measures, we shall also introduce and study an appropriate generalisation of continuous functions, called *measurable functions*. These are, roughly speaking, the functions "whose area under the graph can be measured" and for which, in the next chapter we shall be able to give a rigorous meaning to *integrals* of the form

$$ `` \int_X f(x)\, \mathrm{d}\mu(x) \text{''} $$

with respect to general measures μ. The mystery of the above object and of the associated notation will be revealed in Chapter 8.

7.2 Measurable spaces and measure spaces

We begin by defining an appropriate general setting for the axiomatic study of measures. The first concept below allows us to identify particular collections of sets which behave very nicely in terms of taking unions/intersections/complements of its elements. These properties are similar to some enjoyed by the class of Lebesgue measurable sets of $\mathbb{R}^d$.

Definition 7.1. Let X be a set and $\mathfrak{M}$ a class of subsets of X. Then, $\mathfrak{M}$ is called a σ-**algebra** on X if it satisfies the following properties:

(i) $\emptyset, X \in \mathfrak{M}$.

(ii) If $A \in \mathfrak{M}$, then $X \setminus A \in \mathfrak{M}$.

(iii) If $A_1, A_2, \ldots \in \mathfrak{M}$, then $\bigcup_{k=1}^{\infty} A_k \in \mathfrak{M}$.

The pair $(X, \mathfrak{M})$ is called a **measurable space**. The elements $A \in \mathfrak{M}$ are called **measurable sets**, or $\mathfrak{M}$-**measurable sets** if there is any danger of confusion.

The properties (ii)-(iii) above are usually referred to as that $\mathfrak{M}$ *is closed under these operations, in the sense that complements and (countable) unions of its elements also belong to the class.* The seemingly strange terminology "σ-algebra" (also occasionally called "σ-ring" by other authors) is inspired by the relevant terminology of Algebra regarding the properties of $\mathfrak{M}$ with respect to the operations of intersection and symmetric difference on X (compare with

Exercise 6.3). As we already said, the class $\mathfrak{M}$ will play the role of measurable subsets of X.

It follows directly from the definition that any σ-algebra is also closed under differences and countable intersections of sets:

Lemma 7.2. *Let $(X, \mathfrak{M})$ be a measurable space. Then the following hold:*

(i) *If $A, B \in \mathfrak{M}$, then $A \setminus B \in \mathfrak{M}$.*

(ii) *If $A_1, A_2, \ldots \in \mathfrak{M}$, then $\bigcap_{k=1}^{\infty} A_k \in \mathfrak{M}$.*

The proof of Lemma 7.2 is a simple exercise on the definition which we refrain from giving. Below we discuss some first examples of σ-algebras.

Example 7.3. (i) For any set X, the collections of sets[2] $\mathcal{P}(X)$ and $\{\emptyset, X\}$ are σ-algebras on X. Actually, $\mathcal{P}(X)$ and $\{\emptyset, X\}$ are the "maximal" and "minimal" σ-algebras on X respectively (with respect to the partial ordering generated by the inclusion relation $\subseteq$). Accordingly, any other σ-algebra $\mathfrak{M}$ on X satisfies

$$\{\emptyset, X\} \subseteq \mathfrak{M} \subseteq \mathcal{P}(X).$$

Therefore, any set can always be equipped with at least one σ-algebra, and in fact with at least two if it is non-empty.

(ii) In Chapter 6 we showed that $\mathcal{L}(\mathbb{R}^d)$ is a σ-algebra on $\mathbb{R}^d$.

(iii) In Chapter 6 we defined the *Borel sets* $\mathcal{B}(\mathbb{R}^d)$ as those sets that can be obtained from the open sets under (countably many) applications of the operations of countable unions, countable intersections and complement. It is immediate from the definition that $\mathcal{B}(\mathbb{R}^d)$ **is a σ-algebra on** $\mathbb{R}^d$ and in addition

$$\mathcal{B}(\mathbb{R}^d) \subseteq \mathcal{L}(\mathbb{R}^d).$$

We shall revisit the class of Borel sets a little later, after discussing the concept of a σ-algebra generated by a family of sets.

The archetypal examples of Borel sets and Lebesgue sets on the Euclidean space, motivate the next natural notion:

Definition 7.4. Let $(X, \mathfrak{M})$ be a measurable space. A class of sets $\mathfrak{N} \subseteq \mathcal{P}(X)$ is called a σ-**sub-algebra of** $\mathfrak{M}$ if it holds that $\mathfrak{N} \subseteq \mathfrak{M}$ and the pair $(X, \mathfrak{N})$ is a measurable space itself.

Namely, according to the above definition, $\mathfrak{N}$ is a σ-sub-algebra of $\mathfrak{M}$ when it is a σ-algebra itself on X, which is contained in $\mathfrak{M}$. Using this new terminology, we may reformulate Example 7.3(i) as follows: the σ-algebra $\{\emptyset, X\}$ is a σ-sub-algebra of any σ-algebra $\mathfrak{M}$ on a set X, whilst any σ-algebra $\mathfrak{M}$ is a σ-sub-algebra of the powerset $\mathcal{P}(X)$.

Now we give the following axiomatic definition of an abstract measure, which will play a fundamental role in the remainder of this book.

---

[2] We recall that $\mathcal{P}(X)$ symbolises the collection of all subsets of X.

Definition 7.5. Let $(X, \mathfrak{M})$ is a measurable space. A function $\mu : \mathfrak{M} \longrightarrow [0, \infty]$ is called a **measure on** X when

(i) $\mu(\emptyset) = 0$.

(ii) [σ-additivity] If $(A_k)_1^\infty \subseteq \mathfrak{M}$ is any disjoint sequence (that is, $A_i \cap A_j = \emptyset$ when $i \neq j$), then

$$\mu\left(\bigcup_{k=1}^\infty A_k\right) = \sum_{k=1}^\infty \mu(A_k).$$

We call the triplet $(X, \mathfrak{M}, \mu)$ a **measure space.** Further, sets in $\mathfrak{M}$ of zero μ-measure will be called μ-**nullsets, or merely nullsets** if the measure is understood.

Note that **a measure on** X **is a function whose domain is the class** $\mathfrak{M} \subseteq \mathcal{P}(X)$, **not** X itself. Let us now examine some examples of measures.

Example 7.6 (The Lebesgue measure). In Chapter 6 we established that $(\mathbb{R}^d, \mathcal{L}(\mathbb{R}^d), \mathcal{L}^d)$ is a measure space. Note that the triplet $(\mathbb{R}^d, \mathcal{P}(\mathbb{R}^d), \mathcal{L}^{d*})$ is **not** a measure space because the additivity property of $\mathcal{L}^{d*}$ does not hold on all of the powerset $\mathcal{P}(\mathbb{R}^d)$ (recall that we singled out non-measurable sets).

Although the Lebesgue measure is only defined on the σ-algebra of the Lebesgue measurable sets in $\mathbb{R}^d$ (and does not admit any natural extension to the powerset of $\mathbb{R}^d$), there exist other measures which are defined on the maximal σ-algebra of the powerset, and this is true not only for $\mathbb{R}^d$ but for any set X.

Example 7.7 (Dirac measures). Let X be a non-empty set. We define a measure on X

$$\delta_x : \quad \mathcal{P}(X) \longrightarrow [0, 1]$$

as follows: for any set $A \in \mathcal{P}(X)$, we set

$$\delta_x(A) := \begin{cases} 1, & \text{if } x \in A, \\ 0, & \text{if } x \in X \setminus A. \end{cases}$$

It is an exercise to confirm that δ_x is indeed a measure, which furthermore satisfies $\delta_x(X) = 1$, see Figure 7.2 (later we shall see that such measures are called "probabilities"). The measure δ_x is known as the **Dirac "delta" measure (or point mass)** at $x \in X$ (or with basepoint $x \in X$).

More generally, if $(x_i)_1^\infty$ is a sequence in the set X and $(a_j)_1^\infty$ is a sequence of non-negative numbers (perhaps even extended in $[0, \infty]$), the function

$$\mu : \quad \mathcal{P}(X) \longrightarrow [0, \infty]$$

given by

$$\mu(A) := \sum_{i=1}^\infty a_i \, \delta_{x_i}(A), \quad A \in \mathcal{P}(X),$$

is also a measure on the set X. We leave the verification of the details as an exercise for the reader.

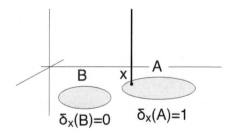

Figure 7.2: Illustration of the Dirac delta measure for $X = \mathbb{R}^2$: The set A has δ_x-measure equal to 1 since $x \in A$, whilst set B has δ_x-measure equal to 0 since $x \notin B$.

Example 7.8 (The counting measure). Let again X be a non-empty set and consider the measurable space $(X, \mathcal{P}(X))$. We define a measure

$$\sharp \ : \ \mathcal{P}(X) \longrightarrow [0, \infty]$$

as follows: for any set $A \in \mathcal{P}(X)$, we set

$$\sharp(A) \ := \ \begin{cases} \operatorname{card}(A), & \text{if } A \text{ is a finite set,} \\ \infty, & \text{otherwise,} \end{cases}$$

where we recall from Chapter 1 that "card" is the cardinality of the set, that is, the integer number of the elements (of a finite set). It can be checked that this is a measure on X, known as the **counting measure**. An alternative notation for the counting measure is "$\mathcal{H}^0$". The notation $\mathcal{H}^0$ relates to the so-called Hausdorff measure which we discuss later in Example 7.20.

Many more examples of measures will be examined in the sequel. Given now a measurable space $(X, \mathfrak{M})$ and a $\mathfrak{M}$-measurable subset $A \subseteq X$, there is an easy way to endow A itself with the structure of a measurable space, in a natural manner which is induced from $(X, \mathfrak{M})$. This can be done as follows:

Definition 7.9. Let $(X, \mathfrak{M})$ be a measurable space and consider a $\mathfrak{M}$-measurable subset $A \subseteq X$.

(i) The **restriction of the σ-algebra $\mathfrak{M}$ on A** is symbolised by $\mathfrak{M}|_A$ and is defined as the class of subsets of A given by

$$\mathfrak{M}|_A \ := \ \big\{ A \cap E \ : \ E \in \mathfrak{M} \big\}.$$

Then, $\mathfrak{M}|_A$ **is a σ-algebra of subsets on** A and the pair

$$(A, \mathfrak{M}|_A)$$

becomes a measurable space itself, which is said to be a **measurable subspace** of $(X, \mathfrak{M})$. Note that $\mathfrak{M}|_A \subseteq \mathfrak{M}$.

(ii) Suppose we additionally have a measure $\mu : \mathfrak{M} \longrightarrow [0, \infty]$ on the measurable space $(X, \mathfrak{M})$. Since $\mathfrak{M}|_A \subseteq \mathfrak{M}$, we may restrict μ to $\mathfrak{M}|_A$ and the restricted function[3]

$$\mu : \mathfrak{M}|_A \longrightarrow [0, \infty]$$

is a **measure on the measurable space** $(A, \mathfrak{M}|_A)$, making the triplet

$$(A, \mathfrak{M}|_A, \mu)$$

a measure space, which is said to be a **measure subspace** of $(X, \mathfrak{M}, \mu)$.

Remark 7.10. As the nomenclature suggests, the class of sets $\mathfrak{M}|_A$ defined in (i) above is a σ-algebra on A, making indeed the pair $(A, \mathfrak{M}|_A)$ a measurable space. To this end, note that we immediately have $\emptyset, A \in \mathfrak{M}|_A$, and let us show for the sake of illustration the closedness under complements. If $B \in \mathfrak{M}|_A$, then by the definition there exists $E \in \mathfrak{M}$ such that $B = A \cap E$. Hence, $A \setminus B = A \cap (X \setminus (A \cap E))$ which is in $\mathfrak{M}$ because $X \setminus (A \cap E) \in \mathfrak{M}$. The closedness under countable unions is similar. Further, it is immediate to check that $\mathfrak{M}|_A$ can also be characterised as

$$\mathfrak{M}|_A = \{B \in \mathfrak{M} : B \subseteq A\}.$$

Note however that $X \notin \mathfrak{M}|_A$ **and, although $\mathfrak{M}|_A$ is a σ-algebra on A, it is not a σ-sub-algebra of $\mathfrak{M}$!** Finally, one easily checks that $\mu : \mathfrak{M}|_A \longrightarrow [0, \infty]$ as defined in (ii) above is indeed a measure on A.

The next example sheds more light on the differences between restrictions of σ-algebras and σ-sub-algebras.

Example 7.11. Consider the space $\mathbb{R}^d$ and a Borel set $\Omega \subseteq \mathbb{R}^d$. Then the following hold:

(i) The triplet $(\mathbb{R}^d, \{\emptyset, \mathbb{R}^d\}, \mathcal{L}^d)$ is a measure subspace of $(\mathbb{R}^d, \mathcal{B}(\mathbb{R}^d), \mathcal{L}^d)$, and $(\mathbb{R}^d, \mathcal{B}(\mathbb{R}^d), \mathcal{L}^d)$ is a measure subspace of $(\mathbb{R}^d, \mathcal{L}(\mathbb{R}^d), \mathcal{L}^d)$.

(ii) The triplet $(\Omega, \mathcal{L}(\mathbb{R}^d)|_\Omega, \mathcal{L}^d)$ is a measure space and $(\Omega, \mathcal{B}(\mathbb{R}^d)|_\Omega, \mathcal{L}^d)$ is a measure subspace of $(\Omega, \mathcal{L}(\mathbb{R}^d)|_\Omega, \mathcal{L}^d)$, but not of $(\mathbb{R}^d, \mathcal{L}(\mathbb{R}^d), \mathcal{L}^d)$.

7.3 Outer measures and Carathéodory's Theorem

One of the ways to construct measures on spaces is by **restricting outer measures to their σ-algebra of measurable sets**. This idea generalises

---

[3]It would perhaps have been more appropriate to symbolise the restriction of the measure μ to the subset $\mathfrak{M}|_A$ of $\mathfrak{M}$ by "$\mu|_A$" or even by "$\mu|_{\mathfrak{M}|_A}$" (instead of merely μ). However, since the domain of definition is clear from the notation for the σ-algebra, we will insist on the less heavy notation we introduce here. There is also a further reason for this simplified notation; there could be a confusion regarding a slightly different (but compatible) concept of restriction of measures and outer measures, see Definition 7.14 that follows.

the construction of the Lebesgue measure from the Lebesgue outer measure achieved by means of the Carathéodory theorem. To this end, we begin by defining axiomatically the following concept of abstract outer measure on a set, following closely the situation in $\mathbb{R}^d$.

Definition 7.12. Let X be a set. A function $\mu^* : \mathcal{P}(X) \longrightarrow [0, \infty]$ is called an **outer measure** if it satisfies the following properties:

(i) $\mu^*(\emptyset) = 0$.

(ii) [Monotonicity] If $A \subseteq B \subseteq X$, then $\mu^*(A) \le \mu^*(B)$.

(iii) [σ-sub-additivity] If $A_1, A_2, \ldots \subseteq X$, then

$$\mu^*\left(\bigcup_{k=1}^{\infty} A_k \right) \le \sum_{k=1}^{\infty} \mu^*(A_k).$$

Note that, unlike the situation for measures, **outer measures are defined on the entire powerset $\mathcal{P}(X)$, no (other) σ-algebras being required.** Roughly speaking, this is possible since only σ-sub-additivity is required, and not σ-additivity.

Remark 7.13. The significance of outer measures and of Theorem 7.18 that follows is that, in general, it might be much simpler to specify an outer measure on a set, since it does not need to be σ-additive. Theorem 7.18 then will guarantee that any outer measure can be restricted to a class of sets, on which it becomes a proper (σ-additive) measure.

Immediate examples of this phenomenon of restriction of outer measures to classes of sets on which they become σ-additive are given by the Lebesgue outer measure and the Hausdorff outer measure, the latter being defined in Example 7.20 that follows.

We continue with the next concept which gives a tool allowing to define an (outer) measure on a subset, which arises from the localisation of another (outer) measure to the subset. This is done in such a way that it vanishes outside the given set.

Definition 7.14. Let X be a non-empty set.
(i) Let $\mu^* : \mathcal{P}(X) \longrightarrow [0, \infty]$ be an outer measure and consider a subset $A \subseteq X$. The **restriction of the outer measure μ^* on A** is the function

$$\mu^* \llcorner A : \mathcal{P}(X) \longrightarrow [0, \infty]$$

defined by
$$(\mu^* \llcorner A)(E) := \mu^*(A \cap E) \quad \text{for all } E \in \mathcal{P}(X).$$

(ii) Given a measure space $(X, \mathfrak{M}, \mu)$, the **restriction of the measure μ on a $\mathfrak{M}$-measurable subset $A \subseteq X$** is the function

$$\mu \llcorner A : \mathfrak{M} \longrightarrow [0, \infty]$$

defined by

$$(\mu \llcorner A)(E) := \mu(A \cap E) \quad \text{for all } E \in \mathfrak{M}.$$

It is immediate that $\mu^* \llcorner A : \mathcal{P}(X) \longrightarrow [0, \infty]$ is an outer measure and $\mu \llcorner A : \mathfrak{M} \longrightarrow [0, \infty]$ is a measure on X, which in addition satisfy

$$\begin{cases} (\mu^* \llcorner A)(E) = 0 & \text{for all } E \subseteq X \setminus A, \\ (\mu \llcorner A)(E) = 0 & \text{for all } E \in \mathfrak{M}, \ E \subseteq X \setminus A. \end{cases}$$

Next is a simple example of restrictions.

Example 7.15 (Restriction of the Lebesgue outer measure). Consider a set $A \subseteq \mathbb{R}^d$ and the Lebesgue outer measure $\mathcal{L}^{d*} : \mathcal{P}(\mathbb{R}^d) \longrightarrow [0, \infty]$. Then, by recalling the definition of $\mathcal{L}^{d*}$, the restriction $\mathcal{L}^{d*} \llcorner A$ is given by

$$(\mathcal{L}^{d*} \llcorner A)(E) = \inf \left\{ \sum_{k=1}^{\infty} \mathcal{V}^d(B_k) \ \middle| \ B_1, B_2, \dots \text{ boxes and } A \cap E \subseteq \bigcup_{k=1}^{\infty} B_k \right\}.$$

Note that $(\mathcal{L}^{d*} \llcorner A)(E) = 0$ for all $E \subseteq \mathbb{R}^d \setminus A$.

We now give a set of definitions identifying properties of measures which will be utilised and exploited in later chapters and in the exercises. The important properties of measures identified below are very frequent in theoretical and practical applications of Measure Theory.

Definition 7.16 (Types of measures). Let $(X, \mathfrak{M})$ be a measurable space and $\mu, \nu : \mathfrak{M} \longrightarrow [0, \infty]$ two measures on X (over the same σ-algebra $\mathfrak{M}$).

Then, the measure μ (or the measure space $(X, \mathfrak{M}, \mu)$) is called:

- **complete**, when every subset of a μ-nullset is a μ-nullset itself, that is when

$$A \in \mathfrak{M}, \ \mu(A) = 0 \ \text{ and } \ E \subseteq A \implies E \in \mathfrak{M} \ \text{ and } \ \mu(E) = 0.$$

(the point is that E may **not** a priori be in the σ-algebra).

- **finite**, when $\mu(X) < \infty$.

- **semi-finite**, when X contains a set of positive finite measure. Namely, when there exists $A \in \mathfrak{M}$ such that $0 < \mu(A) < \infty$.

- **σ-finite**, when X can be written as a countable union of sets of finite measure; namely, when there exists a sequence $(A_k)_1^{\infty} \subseteq \mathfrak{M}$ such that

$$X = \bigcup_{k=1}^{\infty} A_k, \quad 0 < \mu(A_k) < \infty.$$

Further, the measure μ is called:

- **probability** measure, when $\mu(X) = 1$.

- **absolutely continuous** with respect to ν, and this is symbolised as

$$\mu << \nu,$$

when any ν-nullset is a μ-nullset as well. Namely, if for any $A \in \mathfrak{M}$ for which $\nu(A) = 0$, we have $\mu(A) = 0$.

- **singular (or orthogonal)** with respect to ν, and this is symbolised as

$$\mu \perp \nu,$$

when there exists a set A such that A is a μ-nullset and its complement $X \setminus A$ is a ν-nullset; namely, when there exists $A \in \mathfrak{M}$ such that $\nu(A) = 0 = \mu(X \setminus A)$.

We now give some examples of different types of measures. The verification of the simple claims (i)-(vii) below is left as a straightforward exercise for the reader.

Example 7.17. (i) The measure space $(\mathbb{R}^d, \mathcal{B}(\mathbb{R}^d), \mathcal{L}^d)$ is neither complete nor finite, whilst the measure space $(\mathbb{R}^d, \mathcal{L}(\mathbb{R}^d), \mathcal{L}^d)$ is complete but it is not finite.

(ii) For any Lebesgue set $A \in \mathcal{L}(\mathbb{R}^d)$ with $\mathcal{L}^d(A) < \infty$, the measure space $(A, \{\emptyset, A\}, \mathcal{L}^d)$ is finite, but not complete.

(iii) The measure space $(\mathbb{R}^d, \mathcal{L}(\mathbb{R}^d), \mathcal{L}^d)$ is σ-finite (and semi-finite) since $\mathbb{R}^d$ can be written as an increasing union of balls:

$$\mathbb{R}^d = \bigcup_{r=1}^{\infty} \mathbb{B}_r(0) , \quad \mathcal{L}^d(\mathbb{B}_r(0)) < \infty.$$

(iv) For any measurable set $A \in \mathcal{L}(\mathbb{R}^d)$ with $\mathcal{L}^d(A) < \infty$, the measure

$$\mu := \frac{1}{\mathcal{L}^d(A)} \mathcal{L}^d \llcorner A$$

is a probability and $(\mathbb{R}^d, \mathcal{L}(\mathbb{R}^d), \mu)$ is a probability space.

(v) For any set $X \neq \emptyset$ and any $x, y \in X$, the measures $\mu := \delta_x$ and $\nu := \frac{1}{2}\delta_x + \frac{1}{2}\delta_y$ are probabilities and $(X, \mathcal{P}(X), \mu), (X, \mathcal{P}(X), \nu)$ are probability spaces.

(vi) Consider the measurable space $(\mathbb{R}^d, \mathcal{B}(\mathbb{R}^d))$ and the measures $\mu := a\mathcal{L}^d \llcorner \Omega$ and $\nu := b\delta_x + c\mathcal{L}^d$, where $a, b, c > 0$, $x \in \mathbb{R}^d$ and $\Omega \in \mathcal{L}(\mathbb{R}^d)$. Then, $\mu << \nu$.

(vii) Let $\Omega \subseteq \mathbb{R}^d$ be a Borel set. On the measurable space $(\Omega, \mathcal{B}(\mathbb{R}^d)|_\Omega)$, consider $\mu := a\mathcal{L}^d$ and $\nu := b\delta_x$, where $a, b > 0$ and $x \in \Omega$. Then, $\mu \perp \nu$.

Next is the main result of this section.

Theorem 7.18 (The general Carathéodory theorem). *Let X be a non-empty set, and let $\mu^* : \mathcal{P}(X) \longrightarrow [0, \infty]$ be an outer measure on X. Consider the next class of subsets of X:*

$$\mathfrak{M}_\mu := \left\{ A \in \mathcal{P}(X) \mid \mu^*(E) \geq \mu^*(E \cap A) + \mu^*(E \setminus A), \ \ \forall E \in \mathcal{P}(X) \right\}.$$

Then, $\mathfrak{M}_\mu$ is a σ-algebra of subsets on X and the restriction of μ^* to it

$$\mu \ : \ \mathfrak{M}_\mu \longrightarrow [0, \infty] \ , \qquad \mu := \mu^*\big|_{\mathfrak{M}_\mu},$$

is a measure on X. Namely, $(X, \mathfrak{M}_\mu, \mu^*)$ is a measure space which in addition is complete, that is, if $\mu^*(A) = 0$ then A is $\mathfrak{M}_\mu$-measurable.

The elements of the class $\mathfrak{M}_\mu$ will be called μ-**measurable sets**. Any set $A \in \mathfrak{M}_\mu$ for which $\mu(A) = 0$ will be called a μ-**nullset**.

The proof of the general theorem above can be done mutatis-mutandis to the proof of the special Carathéodory Theorem 6.22 on $\mathbb{R}^d$, but we nonetheless give the details for the sake of completeness.

Proof. Let $A, B \in \mathfrak{M}_\mu$ and $E \subseteq X$. Then, by the definition we have

$$\begin{aligned}
\mu^*(E) &= \mu^*(E \cap A) + \mu^*(E \cap A^c) \\
&\geq \mu^*\big((E \cap A) \cap B\big) + \mu^*\big((E \cap A) \cap B^c\big) \\
&\quad + \mu^*\big((E \cap A^c) \cap B\big) + \mu^*\big((E \cap A^c) \cap B^c\big).
\end{aligned}$$

Furthermore, it can be easily verified directly from the definition that $\emptyset, X \in \mathfrak{M}_\mu$, and also $\mathfrak{M}_\mu$ is closed under complements. Since now

$$A \cup B = (A \cap B) \bigcup (A \cap B^c) \bigcup (A^c \cap B),$$

by the sub-additivity property of the outer measure, we have

$$\mu^*\big(E \cap (A \cup B)\big) \leq \mu^*(E \cap A \cap B) + \mu^*(E \cap A \cap B^c) + \mu^*(E \cap A^c \cap B)$$

and hence

$$\mu^*(E) \geq \mu^*\big(E \cap (A \cup B)\big) + \mu^*\big(E \cap (A \cup B)^c\big)$$

which gives that $A \cup B \in \mathfrak{M}_\mu$. Moreover, if $A \cap B = \emptyset$, then

$$\begin{aligned}
\mu^*(A \cup B) &\geq \mu^*\big((A \cup B) \cap A\big) + \mu^*\big((A \cup B) \cap A^c\big) \\
&= \mu^*(A) + \mu^*(B).
\end{aligned}$$

Since the opposite of the above is always true, it follows that μ is additive on $\mathfrak{M}_\mu$.

Let now $(A_k)_1^\infty$ be an arbitrary sequence in $\mathfrak{M}_\mu$, and let us use the convenient abbreviation

$$C_n := \bigcup_{k=1}^n A_k, \quad C := \bigcup_{k=1}^\infty A_k.$$

To prove that $C \in \mathfrak{M}_\mu$, it suffices to consider the case when the sets $(A_k)_1^\infty$ are pairwise disjoint. (Indeed, the general case would follow by applying the same argument to the sequence $(B_k)_1^\infty$ of pairwise disjoint associated to $(A_k)_1^\infty$ as in Lemma 6.23, the measurability of the sets $(B_k)_1^\infty$ being a consequence of what we have already proved.) Thus, let us consider any $E \subseteq X$. Note that, for any $n \in \mathbb{N}$, we have

$$\mu^*(A \cap C_n) \geq \mu^*(E \cap C_n \cap A_n) + \mu^*(E \cap C_n \cap A_n^c)$$
$$= \mu^*(E \cap A_n) + \mu^*(E \cap C_{n-1})$$

and the opposite inequality is always true by the sub-additivity of the outer measure. A straightforward induction argument easily establishes that, for any $n \in \mathbb{N}$,

$$\mu^*(E \cap C_n) = \sum_{k=1}^n \mu^*(E \cap A_k).$$

Therefore, since $C_n \in \mathfrak{M}_\mu$, we have

$$\mu^*(E) \geq \mu^*(E \cap C_n) + \mu^*(E \cap C_n^c) \geq \sum_{k=1}^n \mu^*(E \cap A_k) + \mu^*(E \cap C^c).$$

By letting $n \to \infty$, we infer that

$$\mu^*(E) \geq \sum_{k=1}^\infty \mu^*(E \cap A_k) + \mu^*(E \cap C^c)$$
$$\geq \mu^*\left(\bigcup_{k=1}^\infty (E \cap A_k) \right) + \mu^*(E \cap C^c)$$
$$\geq \mu^*(E \cap C) + \mu^*(E \cap C^c)$$
$$\geq \mu^*(E),$$

for any $E \subseteq X$. Hence, it follows that $C \in \mathfrak{M}_\mu$ and that all our inequalities above are in fact equalities. In particular, $\mathfrak{M}_\mu$ is a σ-algebra. Further, by choosing $E := C$ we deduce that the map μ is σ-additive on $\mathfrak{M}_\mu$, as desired.

It remains to establish the completeness. To this end, suppose $A \subseteq X$ satisfies $\mu^*(A) = 0$. Then, for any $E \subseteq X$ we have

$$\mu^*(E) \leq \mu^*(E \cap A) + \mu^*(E \cap A^c) = \mu^*(E \cap A^c) \leq \mu^*(E).$$

In conclusion, A is a measurable set in the σ-algebra $\mathfrak{M}_\mu$. □

Remark 7.19. Let X be a set and $\mu^* : \mathcal{P}(X) \longrightarrow [0, \infty]$ be an outer measure on X. In general, there exist more than one σ-algebra $\mathfrak{M}$ such that $(X, \mathfrak{M}, \mu)$ is a measure space, where μ denotes the restriction of μ^* to $\mathfrak{M}$. However, **there exists exactly one σ-algebra which is maximal** (i.e. the largest possible) and **complete** (i.e. contains all nullsets): this is the σ-algebra $\mathfrak{M}_\mu$ given by the Carathéodory Theorem 7.18. **Any other σ-algebra one can define on the set X is in fact merely a σ-sub-algebra of the maximal one $\mathfrak{M}_\mu$.** If further it is a strict (proper) σ-sub-algebra, then it has to be incomplete.

We close this section with a very important (outer) measure on $\mathbb{R}^d$, the so-called s-**dimensional Hausdorff measure** $\mathcal{H}^s$, defined for any $0 \le s \le d$. Its main use is to measure "very small" sets in $\mathbb{R}^d$ of "dimension" less that d (perhaps even fractional, see Figures 7.3-7.5).

Example 7.20 (Hausdorff measure). Let $0 \le s \le d$. Then, for each $\delta > 0$ we define a function

$$\mathcal{H}^{s*} \ : \ \mathcal{P}(\mathbb{R}^d) \longrightarrow [0, \infty]$$

given by

$$\mathcal{H}^{s*}(A) := \lim_{\delta \to 0} \left(\inf \left\{ \sum_{k=1}^\infty \alpha(s) \left(\frac{\operatorname{diam}(C_k)}{2} \right)^s : \ \operatorname{diam}(C_k) \le \delta, \ A \subseteq \bigcup_{k=1}^\infty C_k \right\} \right).$$

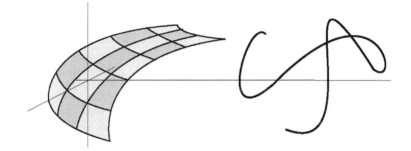

Figure 7.3: The Hausdorff measure $\mathcal{H}^2$ allows us, among other things, to measure the area of smooth surfaces in $\mathbb{R}^3$. On the other hand, the Hausdorff measure $\mathcal{H}^1$ allows us to measure the length of smooth curves in $\mathbb{R}^3$.

The δ-dependent infimum inside the limit is commonly denoted by "$\mathcal{H}^{s*}_\delta(A)$". Here "diam" denotes the diameter of the set, namely

$$\operatorname{diam}(C) := \sup \left\{ \|x - y\|_2 : \ x, y \in C \right\}$$

and $\alpha(s)$ is a constant depending only on s, given by

$$\alpha(s) := \frac{\pi^{s/2}}{\Gamma(1 + s/2)},$$

where

$$\Gamma(t) = \int_0^\infty x^{t-1} e^{-x} dx$$

is the so-called Gamma function. Note that we do not cover the set by boxes, but by **arbitrary** sets $\{C_1, C_2, ..\}$ of diameter bounded by δ. Further, note that the δ-dependent quantity given by the infimum over all series above is monotone in δ, so the limit always exists in $[0, \infty]$.

It can be shown that $\mathcal{H}^{s*}$ **is an outer measure** on $\mathbb{R}^d$. Hence, by the generalised Carathéodory Theorem 7.18, one may obtain a σ-algebra $\mathfrak{M}_{\mathcal{H}^s}(\mathbb{R}^d)$ and a measure $\mathcal{H}^s : \mathfrak{M}_{\mathcal{H}^s}(\mathbb{R}^d) \longrightarrow [0, \infty]$, which is just the restriction of $\mathcal{H}^{s*}$ to $\mathfrak{M}_{\mathcal{H}^s}(\mathbb{R}^d)$. This measure is known as the s-**dimensional Hausdorff measure**. It can be proved that, if $s = 0$, then $\mathcal{H}^0$ coincides with the counting measure $\sharp$ on $\mathbb{R}^d$. If $s = d$, then $\mathcal{H}^d$ coincides with the Lebesgue measure $\mathcal{L}^d$ on $\mathbb{R}^d$ (this is non trivial to prove and requires more advanced tools than those developed in the previous chapters).

Figure 7.4: The "Cantor dust" , the 2-dimensional generalisation of the Cantor set of Chapter 6, is a singular "fragmented" set of fractional dimension (such sets are called fractals), of which we can measure the dimension and the "fractional area" by using the Hausdorff measure(s) $\mathcal{H}^s$ with $s < 2$.

One of the applications of the Hausdorff measure in that one can define the **(fractional Hausdorff) dimension** of **any** set E as the infimum over all $s \in [0, \infty]$ for which $\mathcal{H}^s(E) = 0$. An equivalent way to define it is as the supremum over all $s \in [0, \infty]$ for which $\mathcal{H}^s(E) = \infty$. This concept of dimensionality for non-smooth "fragmented" sets of $\mathbb{R}^d$ coincides with the usual concept of dimension for smooth surfaces and curves.

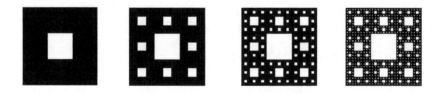

Figure 7.5: The "Sierpinski carpet" is another famous fractal set, of which we can measure the dimension and the "fractional area" by using the Hausdorff measure(s) $\mathcal{H}^s$ with $s < 2$.

7.4   Continuity and other properties of measures

In this section we record a set of important monotonicity, sub-additivity and continuity properties for measures. Although, strictly speaking, measures themselves are functions (defined not on X but on a subset of the powerset $\mathcal{P}(X)$, the σ-algebra $\mathfrak{M}$), these continuity properties are of different nature and should not be confused with the standard continuity properties of functions. We also define and study the concept of product σ-algebras, generalising along the way the notion of Borel sets from the Euclidean to an arbitrary topological space.

Theorem 7.21. *Let $(X, \mathfrak{M}, \mu)$ be a measure space.*

(i) **[Monotonicity]** *If $E \subseteq F$ and $E, F \in \mathfrak{M}$, then $\mu(E) \leq \mu(F)$.*

(ii) **[σ-subadditivity]** *If $(A_k)_1^\infty \subseteq \mathfrak{M}$, then*

$$\mu\left(\bigcup_{k=1}^{\infty} A_k\right) \leq \sum_{k=1}^{\infty} \mu(A_k).$$

(iii) **[Upper continuity]** *If $(A_k)_1^\infty \subseteq \mathfrak{M}$ is an increasing sequence of sets, that is $A_k \subseteq A_{k+1}$, $k \in \mathbb{N}$, then*

$$\mu\left(\bigcup_{k=1}^{\infty} A_k\right) = \lim_{k \to \infty} \mu(A_k).$$

(iv) **[Lower continuity]** *If $(A_k)_1^\infty \subseteq \mathfrak{M}$ is a decreasing sequence of sets, that is $A_k \supseteq A_{k+1}$, $k \in \mathbb{N}$, and in addition it holds that $\mu(A_1) < \infty$, then*

$$\mu\left(\bigcap_{k=1}^{\infty} A_k\right) = \lim_{k \to \infty} \mu(A_k).$$

Remark 7.22 (Existence of the limits). Note that since the sequence $(A_k)_1^\infty$ of (iii) is increasing, we have the identity

$$\lim_{k \to \infty} \mu(A_k) = \sup_{k \in \mathbb{N}} \mu(A_k).$$

Similarly, since the sequence $(A_k)_1^\infty$ of (iv) is decreasing, we have the identity

$$\lim_{k \to \infty} \mu(A_k) = \inf_{n \in \mathbb{N}} \mu(A_k).$$

Evidently, both the above limits exist due to the monotonicity of the corresponding sequences.

The following example shows that the finiteness hypothesis $\mu(A_1) < \infty$ in part (iv) of the theorem above is necessary:

Example 7.23 (Necessity of the finiteness assumption). Consider the measure space consisting of the natural numbers with the counting measure on the whole powerset:

$$(\mathbb{N}, \mathcal{P}(\mathbb{N}), \sharp).$$

Then, the sets

$$A_k := \{n \in \mathbb{N} : n \geq k\}, \quad k \in \mathbb{N},$$

satisfy $\bigcap_{k=1}^{\infty} A_k = \emptyset$, from which we infer that $\sharp(\bigcap_{k=1}^{\infty} A_k) = 0$. On the other hand, we have $\sharp(A_k) = \infty$ for all $k \in \mathbb{N}$ because each set A_k is infinite.

Proof of Theorem 7.21. (i) Since $E, F, F \setminus E \in \mathfrak{M}$, we have $\mu(F) = \mu(E) + \mu(F \setminus E) \geq \mu(E)$.

(ii) Given $(A_k)_1^{\infty} \subseteq \mathfrak{M}$, let $(B_k)_1^{\infty} \subseteq \mathfrak{M}$ be the disjoint sequence of Lemma 6.23. Then, we have

$$\mu\left(\bigcup_{k=1}^{\infty} A_k\right) = \mu\left(\bigcup_{k=1}^{\infty} B_k\right) = \sum_{k=1}^{\infty} \mu(B_k) \leq \sum_{k=1}^{\infty} \mu(A_k).$$

(iii) By setting $A_0 := \emptyset$ and by using the identities (following from the fact that $(A_k \setminus A_{k-1})_1^{\infty}$ is an increasing union)

$$\bigcup_{k=1}^{n} A_k = \bigcup_{k=1}^{n} (A_k \setminus A_{k-1}), \quad \sum_{k=1}^{n} \mu(A_k \setminus A_{k-1}) = \mu(A_n),$$

we have

$$\mu\left(\bigcup_{k=1}^{\infty} A_k\right) = \sum_{k=1}^{\infty} \mu(A_k \setminus A_{k-1}) = \lim_{n\to\infty} \sum_{k=1}^{n} \mu(A_k \setminus A_{k-1}) = \lim_{n\to\infty} \mu(A_n).$$

(iv) We set $C_n := A_1 \setminus A_n$, $n \in \mathbb{N}$. Then, since $(A_n)_1^{\infty}$ is a decreasing sequence of sets in $\mathfrak{M}$, it follows that the sequence $(C_n)_1^{\infty}$ is also in $\mathfrak{M}$ and it is increasing, that is $C_1 \subseteq C_2 \subseteq \cdots$. By item (iii) above, we have

$$\mu\left(\bigcup_{n=1}^{\infty} C_n\right) = \lim_{n\to\infty} \mu(C_n). \tag{$\triangle$}$$

Moreover, since $A_n \subseteq A_1$, we have the identities

$$\mu(C_n) = \mu(A_1) - \mu(A_n), \quad \bigcup_{n=1}^{\infty} C_n = A_1 \setminus \left(\bigcap_{n=1}^{\infty} A_n\right).$$

Therefore, by substituting these in ($\triangle$), we have

$$\mu\left(A_1 \setminus \left(\bigcap_{n=1}^{\infty} A_n\right)\right) = \lim_{n\to\infty} \left(\mu(A_1) - \mu(A_n)\right)$$

which is equivalent to

$$\mu\left(\bigcap_{n=1}^{\infty} A_n\right) - \mu(A_1) = \lim_{n\to\infty}\Big(\mu(A_n) - \mu(A_1)\Big).$$

Since $0 \le \mu(A_1) < \infty$, we may add $\mu(A_1)$ to the above equality in order to conclude the proof of the theorem. $\qquad\square$

We now discuss briefly the issue of possibly extending an incomplete measure to a complete one.

Remark 7.24 (Completion of measures). By using Theorem 7.18, it can be shown that every perhaps incomplete measure μ on a measure space $(X, \mathfrak{M}, \mu)$ can be **extended** to a complete measure by enlarging the σ-algebra to a maximal one. This can be achieved as follows (we refrain from giving the details which are left as an exercise): given $\mu : \mathfrak{M} \longrightarrow [0, \infty]$, we define

$$\mu^* : \quad \mathcal{P}(X) \longrightarrow [0, \infty]$$

by

$$\mu^*(A) := \inf\left\{\sum_{k=1}^{\infty}\mu(E_k) \ : \ (E_k)_{k=1}^{\infty} \subseteq \mathfrak{M}, \ A \subseteq \bigcup_{k=1}^{\infty} E_k\right\}.$$

Then, it can be shown that μ^* is an outer measure on X which satisfies

$$\mu(E) = \mu^*(E), \quad \text{for all } E \in \mathfrak{M}.$$

By Theorem 7.18, there exists a σ-algebra $\overline{\mathfrak{M}} \supseteq \mathfrak{M}$ and a complete measure $\overline{\mu}$ such that $(X, \overline{\mathfrak{M}}, \overline{\mu})$ extends the original measure μ, that is

$$\overline{\mu} = \mu^* = \mu, \quad \text{on } \mathfrak{M},$$

and in addition the measure $\overline{\mu}$ is complete on $\overline{\mathfrak{M}}$. The measure $\overline{\mu}$ is called the completion of μ.

We conclude with the very important concepts of product σ-algebra on a Cartesian product of measurable spaces and of Borel sets of a topological spaces. These will allow us later to define and study "multiple integrals" on product spaces. Before delving into that, we need the auxiliary notion of a *σ-algebra generated by a family of sets.*

Definition 7.25. Let X be a set and $\mathfrak{S} \subseteq \mathcal{P}(X)$ a family of subsets. We define the σ**-algebra generated by the family of sets** $\mathfrak{S}$ as the smallest σ-algebra (with respect to the inclusion) which contains $\mathfrak{S}$:

$$\mathcal{G}(\mathfrak{S}) := \bigcap\left\{\mathfrak{M} \ \middle| \ \mathfrak{M} \ \sigma\text{-algebra on } Y, \ \mathfrak{S} \subseteq \mathfrak{M}\right\}.$$

Namely, $\mathcal{G}(\mathfrak{S})$ is the intersection of all σ-algebras containing the (arbitrary) class of sets $\mathfrak{S}$.

Remark 7.26 ($\mathcal{G}(\mathfrak{S})$ well-defined). The above intersection $\mathcal{G}(\mathfrak{S})$ is well-defined because there exists at least one σ-algebra containing $\mathfrak{S}$, namely the powerset $\mathcal{P}(X)$ itself. Additionally, the intersection of (even **uncountably many!**) σ-algebras is a σ-algebra itself (you can verify this as an exercise).

The concept introduced in Definition 7.25 is very important for a variety of reasons. One of these is that it allows us to exhibit the existence of σ-algebras on arbitrary sets, by specifying a class of subsets which can be augmented to a "minimal" σ-algebra. In particular, it allows us to introduce the concept of **Borel sets on an arbitrary topological space**, thus allowing to extend Definition 6.20 from the case of $\mathbb{R}^d$ with its metric topology to a far greater generality.[4]

Definition 7.27. Let $(X, \mathcal{T})$ be a topological space. Then, the **Borel σ-algebra on the topological space** X is the smallest σ-algebra generated by the topology $\mathcal{T}$ on X (Definition 7.25):

$$\mathcal{B}(X) := \mathcal{G}(\mathcal{T}).$$

Remark 7.28 (Topology vs. σ-algebra). Note that in principle a σ-algebra **and a topology on a set are two totally independent structures** satisfying different properties! However, having specified a topology, one might introduce a "compatible" σ-algebra, as the smallest one containing all the open sets. Hence, **all open sets of a topological space can be made measurable if we endow the space with a σ-algebra containing the Borel sets** $\mathcal{B}(X)$. Note also that different topologies on a set might lead to different Borel σ-algebras! Hence, $\mathcal{B}(X)$ depends on the topology with respect to which it is considered and it should be denoted as $\mathcal{B}_{\mathcal{T}}(X)$ or $\mathcal{B}(X, \mathcal{T})$ when any ambiguity arises.

Now we may continue with the concept of product σ-algebra.

Definition 7.29. Let $\{(X_i, \mathfrak{M}_i)\}_{i=1}^{\infty}$ be countably many measurable spaces. The **product σ-algebra** arising from the σ-algebras $\{\mathfrak{M}_i\}_{i=1}^{\infty}$ on the Cartesian product set $\prod_{i=1}^{\infty} X_i$ is denoted by

$$\bigotimes_{i=1}^{\infty} \mathfrak{M}_i$$

and is the smallest σ-algebra on $\prod_{i=1}^{\infty} X_i$ (that is the intersection of all possible σ-algebras) which contains the "rectangular" sets of the form

$$\prod_{i=1}^{\infty} E_i = E_1 \times E_2 \times \cdots \; ; \quad E_i \in \mathfrak{M}_i \, , \quad i \in \mathbb{N}.$$

---

[4]At this point it might be useful to review the material of Chapter 4 on general topological spaces and continuity of mappings between them.

Namely, by Definition 7.25,

$$\bigotimes_{i=1}^{\infty} \mathfrak{M}_i := \mathcal{G}\left(\left\{\prod_{i=1}^{\infty} E_i \ : \ E_i \in \mathfrak{M}_i, \ i \in \mathbb{N}\right\}\right).$$

Remark 7.30. Note that the product σ-algebra is squeezed between the maximal and the minimal σ-algebras on the product space $\prod_{i=1}^{\infty} X_i$:

$$\left\{\emptyset, \ \prod_{i=1}^{\infty} X_i\right\} \subseteq \bigotimes_{i=1}^{\infty} \mathfrak{M}_i \subseteq \mathcal{P}\left(\prod_{i=1}^{\infty} X_i\right).$$

We finally record the following important relation which connects the Borel σ-algebra on the Cartesian product Euclidean space $\mathbb{R}^d$ to the d-th power of the Borel σ-algebra on $\mathbb{R}$, by using the notion of product σ-algebras. For any $d \in \mathbb{N}$, the following equality holds true

$$\mathcal{B}(\mathbb{R}^d) = \underbrace{\mathcal{B}(\mathbb{R}) \bigotimes \cdots \bigotimes \mathcal{B}(\mathbb{R})}_{d-\text{times}} \equiv \bigotimes^{d} \mathcal{B}(\mathbb{R}).$$

We refrain from presenting the details of the proof which are left as in an interesting exercise for the reader on the application of the definitions. We merely note that one needs to use that every open set in $\mathbb{R}^d$ can be written as the countable union of open cubes whose vertices have rational coordinates.

7.5 Measurable functions and mappings

In this section we define and study the appropriate notion of *mappings between measure spaces* which are "morphisms", that is they *respect the measurability properties of the spaces*[5]. For these maps, which generalise the usual continuous maps studied in the previous chapters, we shall define and study in the next chapter the notion of integral with respect to a general measure. Measurable functions are, roughly speaking, the function for which "the volume under their graph can be measured".

We begin by recalling from Chapter 1 the concept of the inverse image map. Let $f : X \longrightarrow Y$ be a mapping between two non-empty sets. Then, f induces the inverse image map $f^{-1} : \mathcal{P}(Y) \longrightarrow \mathcal{P}(X)$ defined as

$$f^{-1}(B) := \left\{x \in X \ : \ f(x) \in B\right\}, \quad B \subseteq Y.$$

---

[5]At the point it might be useful to recall the comments in Remark 4.31 regarding the case of appropriate "morphisms" in topology, namely the continuous mappings.

Note that f^{-1} is defined on sets regardless the invertibility or not of the mapping f and recall the formulas $(\star)$-$(\star\star)$ of Chapter 1 which list some of the nice properties of the inverse image map. If f is bijective, then the inverse image map $f^{-1} : \mathcal{P}(Y) \longrightarrow \mathcal{P}(X)$ and the usual inverse map $f^{-1} : Y \longrightarrow X$ coincide, in the sense that

$$f(x) = y \iff f^{-1}(\{y\}) = \{x\}.$$

Let us also recall that the Borel σ-algebra $\mathcal{B}(X)$ on a topological space $(X, \mathcal{T})$ is the smallest σ-algebra (intersection of all σ-algebras) containing the topology $\mathcal{T}$ (Definition 7.27, Remark 7.28).

Definition 7.31. Let $(X, \mathfrak{M})$ and $(Y, \mathfrak{N})$ be two measurable spaces and let also $f : X \longrightarrow Y$ be a map between sets.

(i) The mapping f will be called $(\mathfrak{M}, \mathfrak{N})$-**measurable** when

$$f^{-1}(\mathfrak{N}) \subseteq \mathfrak{M},$$

that is, when the inverse image map of f sends $\mathfrak{N}$-measurable subsets of Y to $\mathfrak{M}$-measurable subsets of X:

$$f^{-1}(E) \in \mathfrak{M}, \quad \text{for all } E \in \mathfrak{N}.$$

(ii) If Y is endowed with a topology and $\mathfrak{N} = \mathcal{B}(Y)$ (the Borel sets generated by the topology of Y), the map f will be called $\mathfrak{M}$-**measurable** if it is measurable in the sense that f is $(\mathfrak{M}, \mathcal{B}(Y))$-measurable. If in addition X is also endowed with a topology and $\mathfrak{M} = \mathcal{B}(X)$, then the mapping f will be called **Borel-measurable**.

(iii) If either of the measure spaces $(X, \mathfrak{M})$, $(Y, \mathfrak{N})$ is additionally endowed with a respective measure μ or ν, then we may alternatively **use the modifiers "μ-" or "ν-" in the place of "$\mathfrak{M}$-" or "$\mathfrak{N}$-"**, when referring to the measurability of the map f.

(iv) Particularly, if $(X, \mathfrak{M}) = (\mathbb{R}^n, \mathcal{L}(\mathbb{R}^n))$, then the mapping f will be called **Lebesgue-measurable**.

Remark 7.32 (Convention for the target σ-algebras). If $Y = \mathbb{R}^d$ ($d \in \mathbb{N}$), or more generally if Y is a topological space $(Y, \mathcal{S})$, then, unless stated otherwise, we shall **always** endow Y with the σ-algebra of Borel sets generated by the topology $\mathcal{S}$ (recall Definition 7.27 and Remark 7.28):

$$(Y, \mathcal{B}(Y)).$$

Note that since $\mathcal{B}(\mathbb{R}^n) \subseteq \mathcal{L}(\mathbb{R}^n)$, it follows directly from the definition of measurability that any *Borel-measurable map* $f : \mathbb{R}^n \longrightarrow \mathbb{R}^d$ *is automatically Lebesgue-measurable*. As we will see in the next chapter, measurable maps and functions are those which "can be integrated" against measures.

In the lemma below we show that measurable maps are indeed "morphisms", in the sense of preserving the structure:

Proposition 7.33 (Compositions of measurable mappings are measurable). *Let $(X, \mathfrak{M}), (Y, \mathfrak{N})$ and $(Z, \mathfrak{K})$ be measurable spaces. Consider also the mappings $f : X \longrightarrow Y$ and $g : Y \longrightarrow Z$. If f is $(\mathfrak{M}, \mathfrak{N})$-measurable and g is $(\mathfrak{N}, \mathfrak{K})$-measurable, then the composition $g \circ f : X \longrightarrow Z$ is $(\mathfrak{M}, \mathfrak{K})$-measurable.*

Proof. It suffices to note that the induced inverse image mappings satisfy the identity

$$(g \circ f)^{-1} = f^{-1} \circ g^{-1}$$

(why?). The remaining of the proof is left as an exercise on the use of the definition of measurability. $\qquad\square$

In the sequel we shall mostly consider the case $Y = \mathbb{R}^d$, $d \in \mathbb{N}$, which includes[6] the case of $Y = \mathbb{F}^d$, where $\mathbb{F}$ equal either $\mathbb{R}$ or $\mathbb{C}$. By using the properties of the inverse image map and of the Borel sets, it can be shown if the target space is a topological space endowed with the Borel σ-algebra (Definition 7.27), then the measurability of a map can be characterised by requiring only that the open sets of Y are inverted to measurable sets. Slightly more generally, we have the next result:

Proposition 7.34. *Let $(X, \mathfrak{M})$ and $(Y, \mathfrak{N})$ be measurable spaces. Let also $\mathfrak{S} \subseteq \mathcal{P}(Y)$ be a family of subsets of Y generating the σ-algebra $\mathfrak{N}$, namely*

$$\mathfrak{N} = \mathcal{G}(\mathfrak{S}),$$

Let also $f : X \longrightarrow Y$ be a mapping. Then the following hold:

$$f \text{ is } (\mathfrak{M}, \mathfrak{N})\text{-measurable} \quad \Longleftrightarrow \quad f^{-1}(S) \in \mathfrak{M}, \quad \text{for all } S \in \mathfrak{S}.$$

The importance of the above result lies in that it suffices f to check only that f^{-1} sends the elements of $\mathfrak{S}$ to $\mathfrak{M}$-measurable sets.

Proof. Since $\mathfrak{N} = \mathcal{G}(\mathfrak{S})$, it follows that $\mathfrak{N}$ contains all sets $S \in \mathfrak{S}$. If the map $f : X \longrightarrow Y$ is $(\mathfrak{M}, \mathfrak{N})$-measurable, then the inverse image map f^{-1} sends $\mathfrak{S}$ into $\mathfrak{M}$. Conversely, if $f^{-1}(S) \in \mathfrak{M}$ for any set $S \in \mathfrak{S}$, then by the formulas $(\star)$-$(\star\star)$ of Chapter 1, the collection of sets

$$\left\{ S \subseteq Y : f^{-1}(S) \in \mathfrak{M} \right\}$$

is a σ-algebra on Y containing $\mathfrak{S}$. Hence, it contains $\mathfrak{N} = \mathcal{G}(\mathfrak{S})$ as well because it is the smallest σ-algebra containing $\mathfrak{S}$. $\qquad\square$

---

[6] In this chapter we will consider only the field $\mathbb{F} = \mathbb{R}$ of real numbers. This is not a restriction whatsoever since $\mathbb{C}^d = \mathbb{R}^{2d}$. The same practice will be followed in any subsequent chapters where, like herein, the complex structure of the space plays no essential role.

In particular, by invoking Definition 7.27 we have the next consequence:

Corollary 7.35. *Let $(X, \mathfrak{M})$ be a measurable space and let Y be a topological space, endowed with the Borel σ-algebra $\mathcal{B}(Y)$ generated by its topology. Then, a mapping $f : X \longrightarrow Y$ is $\mathfrak{M}$-measurable if and only if $f^{-1}(V) \in \mathfrak{M}$, for all open sets $V \subseteq Y$.*

As the next result asserts, measurable mappings are a generalisation of continuous mappings.

Proposition 7.36 (Continuity implies measurability). *Let $(X, \mathcal{T})$, $(Y, \mathcal{S})$ be two topological spaces. Suppose further that we have two σ-algebras $\mathfrak{M}, \mathfrak{N}$ on X, Y respectively, such that $\mathfrak{M} \supseteq \mathcal{B}(X)$ and $\mathfrak{N} = \mathcal{B}(Y)$, where $\mathcal{B}(X), \mathcal{B}(Y)$ are the Borel sets on the spaces, generated by the respective topologies.*

Then, if $f : X \longrightarrow Y$ is a continuous map (with respect to $\mathcal{T}, \mathcal{S}$), it follows that f is $\mathfrak{M}$-measurable.

Proof. Since $f : X \longrightarrow Y$ is continuous, for any open set $V \in \mathcal{S} \subseteq \mathcal{B}(Y) = \mathfrak{N}$, its inverse image is open and hence $f^{-1}(V) \in \mathcal{T}$. In view of the inclusions $\mathcal{T} \subseteq \mathcal{B}(X) \subseteq \mathfrak{M}$, Proposition 7.34 yields that f is $\mathfrak{M}$-measurable. $\square$

Now we characterise **real** measurable functions in a more concrete way.

Proposition 7.37. *Let $f : X \longrightarrow \mathbb{R}$ be a function and $\mathfrak{M}$ a σ-algebra of sets on X. Then, the following statements are equivalent (see Figure 7.6):*

(i) $f : X \longrightarrow \mathbb{R}$ is $\mathfrak{M}$-measurable.

(ii) *For any $a \in \mathbb{R}$, we have $\{f > a\} := f^{-1}((a, \infty)) \in \mathfrak{M}$.*

(iii) *For any $a \in \mathbb{R}$, we have $\{f \geq a\} := f^{-1}([a, \infty)) \in \mathfrak{M}$.*

(iv) *For any $a \in \mathbb{R}$, we have $\{f \leq a\} := f^{-1}((-\infty, a]) \in \mathfrak{M}$.*

(v) *For any $a \in \mathbb{R}$, we have $\{f < a\} := f^{-1}((-\infty, a)) \in \mathfrak{M}$.*

(vi) *For any $a < b$ in $\mathbb{R}$, we have $\{a < f < b\} := f^{-1}((a, b)) \in \mathfrak{M}$.*

(vii) *For any $a < b$ in $\mathbb{R}$, we have $\{a \leq f < b\} := f^{-1}([a, b)) \in \mathfrak{M}$.*

(viii) *For any $a < b$ in $\mathbb{R}$, we have $\{a < f \leq b\} := f^{-1}((a, b]) \in \mathfrak{M}$.*

(ix) *For any $a < b$ in $\mathbb{R}$, we have $\{a \leq f \leq b\} := f^{-1}([a, b]) \in \mathfrak{M}$.*

Proof. By Proposition 7.34 above, it suffice to show that the Borel σ-algebra $\mathcal{B}(\mathbb{R})$ is the smallest σ-algebra containing the families of subsets of $\mathbb{R}$ appearing in any of the items (ii)-(ix) above, that is those classes which are rays or bounded intervals, half-open, open or closed. This however is an immediate consequence of the definition of $\mathcal{B}(\mathbb{R})$ and of the fact that *any open set in $\mathbb{R}$ can be written as a countable union of disjoint open intervals (including rays)*, as Lemma 7.38 that follows confirms. $\square$

The essence of Proposition 7.37 is that, in order for a function to be measurable, it suffices that it inverts intervals or rays (i.e. half-lines of the form $(-\infty, a)$ or (a, ∞) for some $a \in \mathbb{R}$) into measurable sets.

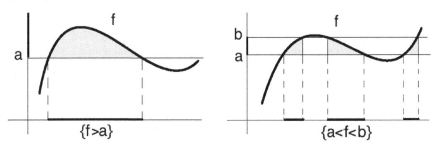

Figure 7.6: An illustrative example of the sup-level $\{f > a\}$ and sub-sup-level set $\{a < f < b\}$ for a real function f in one dimension.

The next result completes the proof of Proposition 7.37.

Lemma 7.38. *Every open set in $\mathbb{R}$ can be expressed as a countable disjoint union of intervals and rays.*[7]

Proof. Let $U \subseteq \mathbb{R}$ be a non-empty open set and consider the class $\mathcal{I}$ of all open intervals and rays contained in U. For any $I, J \in \mathcal{I}$, we define an equivalence relation $I \sim J$ if and only if there exist $K \in \mathcal{I}$ such that $I \cap K \neq \emptyset$ and $K \cap J \neq \emptyset$. Consider now the quotient set $\mathcal{I}/_{\sim} = \{[I]_{\sim} : I \in \mathcal{I}\}$ and, for each equivalence class in $\mathcal{I}/_{\sim}$, consider the union of all the intervals in that class. This union is also an element in that class, in fact a maximal one. The family of maximal intervals so obtained is a decomposition of U into pairwise disjoint open intervals and rays. The countability of this family of intervals and rays follows by the fact that each non-empty open interval contains a rational number. □

We now show that measurable functions behave very well under the usual algebraic operations.

Proposition 7.39 (Algebraic operations with measurable functions). *Suppose $(X, \mathfrak{M})$ is a measurable space and let $f, g : X \longrightarrow \mathbb{R}$ be two $\mathfrak{M}$-measurable functions. Then, for any $a, b \in \mathbb{R}$ the functions*

$$af + bg , \quad fg : \quad X \longrightarrow \mathbb{R}$$

are $\mathfrak{M}$-measurable as well. If $g(x) \neq 0$ for all $x \in X$ (so that the quotient f/g is well-defined), then f/g is a $\mathfrak{M}$-measurable function too.

---

[7]It is worth mentioning that the corresponding result in $\mathbb{R}^d$ for $d \geq 2$ **is not true**. Accordingly, the best that one can do is fill up an open sets with countably-many **disjoint** balls or cubes *except for a set of zero measure*. This is a particular consequence of the so-called Vitali and Besicovitch *covering theorems*. We refrain from getting into more details of this more advanced topic for which the interested reader might consult e.g. [14, 18].

Proof. Fix $a, b \in \mathbb{R}$ and consider the mappings

$$(f,g) \; : \; X \longrightarrow \mathbb{R} \times \mathbb{R}, \qquad x \mapsto \big(f(x), g(x)\big),$$
$$\Phi \; : \; \mathbb{R} \times \mathbb{R} \longrightarrow \mathbb{R}, \qquad (t, s) \mapsto as + bt.$$
$$\Psi \; : \; \mathbb{R} \times \mathbb{R} \longrightarrow \mathbb{R}, \qquad (t, s) \mapsto ts.$$

Then, Φ is continuous and by Proposition 7.36 it is Borel measurable as well. Further, we will now show that (f, g) is $\mathfrak{M}$-measurable. To this end, let $(a, b) \times (c, d) \subseteq \mathbb{R} \times \mathbb{R} = \mathbb{R}^2$ be an open rectangle. The inverse image of this set

$$(f,g)^{-1}\big((a, b) \times (c, d)\big) \; = \; \Big\{x \in X \, : \, \big(f(x), g(x)\big) \in (a, b) \times (c, d)\Big\}$$

equals the intersection of inverse images via f^{-1}, g^{-1}:

$$(f,g)^{-1}\big((a, b) \times (c, d)\big) \; = \; \Big\{x \in X \, : \, f(x) \in (a, b) \ \text{ and } \ g(x) \in (c, d)\Big\}$$
$$= \; f^{-1}\big((a, b)\big) \bigcap g^{-1}\big((c, d)\big). \qquad (\circ)$$

Let now $U \subseteq \mathbb{R}^2$ be any open set. Then, U can be written as the countable union of open rectangles contained in it with rational vertices:

$$U \; = \; \bigcup_{j=1}^{\infty} (a_j, b_j) \times (c_j, d_j), \qquad a_j, b_j, c_j, d_j \in \mathbb{Q}.$$

Then, by using the equality $(\circ)$ and the formulas $(\star)$-$(\star\star)$ of Chapter 1, we have that

$$(f,g)^{-1}(U) \; = \; \bigcup_{j=1}^{\infty} \Big(f^{-1}\big((a_j, b_j)\big) \bigcap g^{-1}\big((c_j, d_j)\big) \Big).$$

Since f, g are $\mathfrak{M}$-measurable, the above implies that $(f, g)^{-1}(U) \in \mathfrak{M}$ and hence Corollary 7.35 yields that (f, g) is $\mathfrak{M}$-measurable, as desired. Then, by Proposition 7.33 (and Definition 7.27), it follows that the function

$$af + bg \; = \; \Phi \circ (f, g)$$

is $\mathfrak{M}$-measurable as a composition of measurable maps. Similarly, the function

$$fg \; = \; \Psi \circ (f, g)$$

is $\mathfrak{M}$-measurable as a composition of measurable maps. Finally, if g is $\mathfrak{M}$-measurable, then for any $a \in \mathbb{R}$ we have that

$$\{1/g > a\} = \begin{cases} \{0 < g < 1/a\}, & \text{if } a > 0, \\ \{g < 1/a\} \cup \{g > 0\}, & \text{if } a < 0, \\ \{g > 0\}, & \text{if } a = 0, \end{cases}$$

and thus, by Proposition 7.37, the function $1/g$ is $\mathfrak{M}$-measurable as well. Therefore, the same is true for f/g also. $\qquad\square$

We now show that measurable functions behave very well under limit operations.

Proposition 7.40 (Limit operations with measurable functions). *Let $(X, \mathfrak{M})$ be a measurable space and $(f_n)_1^\infty$ a sequence of functions $X \longrightarrow \mathbb{R}$ which are $\mathfrak{M}$-measurable for all $n \in \mathbb{N}$. Then, the functions*

$$M, L, m, l \; : \; X \longrightarrow \mathbb{R}$$

given for any $x \in X$ by

$$M(x) := \sup_{n \in \mathbb{N}} f_n(x), \qquad L(x) := \limsup_{n \to \infty} f_n(x),$$

$$m(x) := \inf_{n \in \mathbb{N}} f_n(x), \qquad l(x) := \liminf_{n \to \infty} f_n(x),$$

are all $\mathfrak{M}$-measurable. If moreover the limit $\lim_{n \to \infty} f_n(x)$ exists for all $x \in X$, then the limit function

$$f(x) := \lim_{n \to \infty} f_n(x), \quad x \in X,$$

is also $\mathfrak{M}$-measurable.

We recall that for a sequence $(a_n)_1^\infty \subseteq \mathbb{R}$, by definition we have

$$\left\{ \begin{array}{l} \limsup_{n \to \infty} a_n := \inf_{k \in \mathbb{N}} \sup_{n \geq k} a_n, \\[2mm] \liminf_{n \to \infty} a_n := \sup_{k \in \mathbb{N}} \inf_{n \geq k} a_n. \end{array} \right. \qquad (\star)$$

The example in Figure 7.7 below shows that not all measurable functions are continuous.

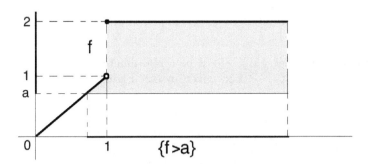

Figure 7.7: The function

$$f(x) := \left\{ \begin{array}{ll} x, & 0 \leq x < 1, \\ 2, & 2 \leq x < \infty, \end{array} \right.$$

is a discontinuous Borel measurable function on $[0, \infty)$. Indeed, for any $a \in \mathbb{R}$, it can be easily verified that the sup-level set $\{f > a\}$ is a Borel set. Actually, it can be proved that all piecewise continuous functions on $\mathbb{R}$ are in fact Borel measurable (this is left as an easy exercise for the reader on the use of the definition).

Proof. For any $a \in \mathbb{R}$ and $x \in X$, we have $M(x) \leq a$ if and only if $f_n(x) \leq a$ for all $n \in \mathbb{N}$. Hence,

$$\{M \leq a\} = \bigcap_{n=1}^{\infty} \{f_n \leq a\}.$$

Thus, for all $n \in \mathbb{N}$ we have

$$f_n^{-1}((-\infty, a]) = \{f_n \leq a\} \in \mathfrak{M}$$

and the same holds for the intersection of these sub-level sets. Hence, $M^{-1}((-\infty, a]) \in \mathfrak{M}$ and by Proposition 7.37 it follows that M is $\mathfrak{M}$-measurable.

In order to establish the measurability of the function $m : X \longrightarrow \mathbb{R}$, we argue similarly. In this case though we have to use that $m(x) \geq a$ if and only if $f_n(x) \geq a$ for all $n \in \mathbb{N}$.

For the function $L : X \longrightarrow \mathbb{R}$, we argue as follows: By the formulas $(\star)$ and our earlier reasoning, we have that the function

$$h_k := \sup_{n \geq k} f_n \; : \; X \longrightarrow \mathbb{R}$$

is $\mathfrak{M}$-measurable and hence the same holds for $\inf_{k \in \mathbb{N}} h_k$, whence L is $\mathfrak{M}$-measurable as well. For l we argue similarly and the proposition follows. Finally, for the limit function we have that, the pointwise limit exists if and only if the limsup and the liminf coincide. However, we have already shown that these are measurable. $\square$

Now we define a useful decomposition of any function $f : X \longrightarrow \mathbb{R}$ as the difference of two non-negative functions where $f^{\pm} : X \longrightarrow [0, \infty)$ which allows us to reduce arguments for f to simpler arguments for the non-negative functions $f^{\pm}$.

Definition 7.41. Let $f : X \longrightarrow \mathbb{R}$ be a real function defined on the set X. The functions $f^{\pm} : X \longrightarrow [0, \infty)$ given by (see Figure 7.8)

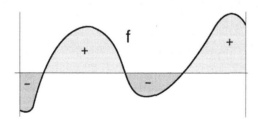

Figure 7.8: An illustration of the positive and negative part decomposition.

$$f^+(x) := \max\{f(x), 0\} = \begin{cases} f(x), & x \in \{f > 0\}, \\ 0, & x \in \{f \leq 0\}, \end{cases}$$

and

$$f^-(x) := \max\{-f(x), 0\} = \begin{cases} -f(x), & x \in \{f < 0\}, \\ 0, & x \in \{f \geq 0\}, \end{cases}$$

are called the **positive and the negative part of** f respectively.

In view of Definition 7.41, any function $f : X \longrightarrow \mathbb{R}$ and its absolute value $|f| : X \longrightarrow [0, \infty)$ can be decomposed as

$$f = f^+ - f^- , \quad |f| = f^+ + f^-.$$

The next result connects the measurability of f to that of $f^{\pm}$.

Lemma 7.42. *Let* $f : X \longrightarrow \mathbb{R}$ *be a function on the measurable space* $(X, \mathfrak{M})$. *Then,* f *is* $\mathfrak{M}$-*measurable if and only if* $f^{\pm}$ *are both* $\mathfrak{M}$-*measurable.*

Proof. Suppose that f is $\mathfrak{M}$-measurable. Since $f^+ = \max\{f, 0\}$ and $f^- = \max\{-f, 0\}$, by Proposition 7.40 it follows that $f^{\pm}$ are $\mathfrak{M}$-measurable as well. Conversely, if $f^{\pm}$ are $\mathfrak{M}$-measurable, then by Proposition 7.39 we have that $f = f^+ - f^-$ is $\mathfrak{M}$-measurable too. □

7.6 Simple functions and mappings

In this section we study a **very important** particular class of measurable functions and mappings. These will be the cornerstones of the definition of our generalised integral in the next chapter.

Definition 7.43. Let $(X, \mathfrak{M})$ be a measure space.

(i) Let $A \in \mathfrak{M}$. The function $\chi_A : X \longrightarrow \{0, 1\}$ given by

$$\chi_A(x) := \begin{cases} 1, & x \in A, \\ 0, & x \in X \setminus A, \end{cases}$$

is called the **characteristic function of the set** A (see Figure 7.9).

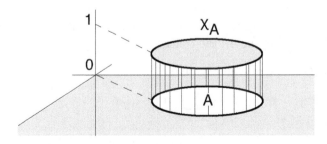

Figure 7.9: An illustration of the characteristic function of a set $A \subseteq \mathbb{R}^2$.

It follows directly from the definitions that the characteristic function χ_A of a set is $\mathfrak{M}$-measurable if and only if the set is measurable, that is $A \in \mathfrak{M}$.

(ii) A mapping $f : X \longrightarrow \mathbb{R}^d$ $(d \in \mathbb{N})$ is called **simple**[8] if it can be represented as a finite linear combination

$$f(x) = \sum_{j=1}^{p} a_j \, \chi_{A_j}(x), \quad x \in X,$$

for some $p \in \mathbb{N}$, where $\{a_1, \ldots, a_p\}$ are points in $\mathbb{R}^d$ and $\{A_1, \ldots, A_p\}$ are disjoint $\mathfrak{M}$-measurable subsets of X, that is, $A_k \cap A_l = \emptyset$ when $k \neq l$ and $\{A_1, \ldots, A_p\} \subseteq \mathfrak{M}$ (see Figure 7.10).

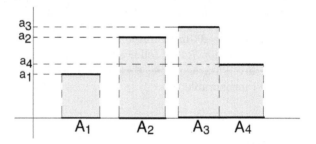

Figure 7.10: An illustration of a simple function on $\mathbb{R}$ with four constituent sets.

An equivalent way to characterise simple maps is by requiring that their image set must be finite. This can be stated very concisely by using the counting measure:

Corollary 7.44. *Let* $(X, \mathfrak{M})$ *be a measure space and* $f : X \longrightarrow \mathbb{R}^d$ *an* $\mathfrak{M}$-*measurable map,* $d \in \mathbb{N}$. *Then, we have the equivalence*

$$f \text{ is a simple map} \quad \Longleftrightarrow \quad \sharp\big(f(X)\big) < \infty.$$

Proof. The direct implication is obvious. Conversely, if $\sharp\big(f(X)\big) < \infty$, then $f(X) = \{a_1, \ldots, a_p\}$ for some $p \in \mathbb{N}$. Then, set

$$A_i := f^{-1}(\{a_i\}), \quad i = 1, \ldots, p,$$

and note that $\{A_1, \ldots, A_p\} \subseteq \mathfrak{M}$ by measurability, whilst they are disjoint by definition. Since $f \equiv a_i$ on A_i, the conclusion follows. $\qquad\square$

The representation of a simple map as

$$f = \sum_{a \in f(X)} a \, \chi_{f^{-1}(\{a\})}$$

---

[8]Note that simple mappings can be defined only when then the target is a vector space, since the linear structure is essential to write the linear combination $a_1 \, \chi_{A_1} + \cdots + a_p \, \chi_{A_p}$!

alluded to above, is in a sense the "canonical" one.

The next result is a fundamental approximation tool which claims that measurable maps can be characterised as those which can be approximated by simple maps in the pointwise sense.

Theorem 7.45 (Characterisation of measurability through pointwise approximation by simple maps). *Let $(X, \mathfrak{M})$ be a measurable space and $f : X \longrightarrow \mathbb{R}^d$ a map, $d \in \mathbb{N}$. Then, the following statements are equivalent:*

(i) *The map f is $\mathfrak{M}$-measurable.*

(ii) *The map f can be approximated by a sequence of simple maps in the pointwise sense. Namely, there exists a sequence $(f_n)_1^\infty$ of $\mathfrak{M}$-measurable maps where each $f_n : X \longrightarrow \mathbb{R}^d$ is simple and*

$$f_n(x) \longrightarrow f(x) \quad \text{as } n \to \infty, \quad \text{for every } x \in X.$$

(iii) *The map f can be approximated by a sequence of simple maps in the pointwise sense, with the additional property that their moduli are increasing and bounded by the modulus[9] of $|f|$. Further, the convergence to f is uniform on any measurable set on which f is bounded. Namely, there exists a sequence $(f_n)_1^\infty$ of $\mathfrak{M}$-measurable maps where each $f_n : X \longrightarrow \mathbb{R}^d$ is simple and*

$$\begin{cases} f_n(x) \longrightarrow f(x), & \text{as } n \to \infty, \\ |f_n(x)| \leq |f_{n+1}(x)| \leq |f(x)|, & \text{for } n \in \mathbb{N}, \\ \sup_A |f_n - f| \longrightarrow 0, & \text{as } n \to \infty, \end{cases}$$

for every $x \in X$ and any $A \in \mathfrak{M}$ such that $\sup_A |f| < \infty$.

Proof. (iii) $\Longrightarrow$ (ii) $\Longrightarrow$ (i): This is immediate from Proposition 7.40 applied to each of the d-many scalar-valued components $f_i : X \longrightarrow \mathbb{R}$ of the mapping $f = (f_1, \ldots, f_d) : X \longrightarrow \mathbb{R}^d$.

(i) $\Longrightarrow$ (iii): By decomposing f into its components $(f_1, \ldots, f_d)$, and then each component f_i into its positive and negative parts

$$f_i = f_i^+ - f_i^-, \quad i = 1, \ldots, d,$$

(see Definition 7.41) it suffices to assume that $f : X \longrightarrow [0, \infty)$ is real-valued and non-negative (it is an exercise to complete the details). Hence, we shall construct an approximation of a $\mathfrak{M}$-measurable function $f : X \longrightarrow [0, \infty)$ by simple functions.

---

[9]It is standard practise in the study of spaces of $\mathbb{F}^m$-valued maps to reserve the symbol "$\| \cdot \|$" for the norms on the (infinite-dimensional) functional vector spaces and to simplify the notation for the norm on the Euclidean space to merely "$| \cdot |$" and be referred to as "modulus". Unless indicated otherwise, from now on "$| \cdot |$" will always symbolise the Euclidean norm "$\| \cdot \|_2$" of $\mathbb{F}^d$. Recall though that, in view of Theorem 5.22, all norms are equivalent on $\mathbb{F}^m$. Hence, it plays no real role which norm we use on $\mathbb{F}^m$!

To this end, for each $n \in \mathbb{N}$ and $k = 0, ..., 2^{2n} - 1$, we define

$$A_n^k := \left\{ k2^{-n} < f \leq (k+1)2^{-n} \right\}$$

and

$$A_n := \{ f > 2^n \}$$

and define

$$f_n(x) := 2^n \chi_{A_n}(x) + \sum_{k=0}^{2^{2n}-1} k2^{-n} \chi_{A_n^k}(x), \qquad x \in X.$$

The formula above looks complicated but geometrically the idea is quite simple and intuitive, see Figure 7.11 below.

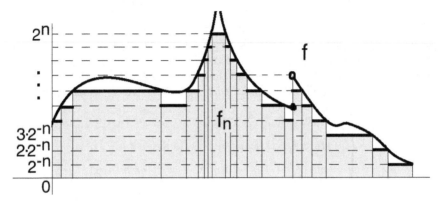

Figure 7.11: An illustration of the approximation result of a positive measurable function by simple functions in the case $X \subseteq \mathbb{R}$.

The idea is to slice the target $[0, \infty)$ into intervals of length 2^{-n} and define f_n on each such sup-sub-level set of f with the constant value f given by the lower end-point of the respective interval. The fact that we have to stop at 2^n (instead of going up to infinity) is a small technical inconvenience, since our function f may not be bounded.

By construction, the function f_n is simple, whilst we also have $f_n \leq f_{n+1}$ on X. In addition,

$$0 \leq f(x) - f_n(x) \leq 2^{-n},$$

for all x in the set $\{f \leq 2^n\}$ and any $n \in \mathbb{N}$. It is immediate to confirm that for any fixed $x \in X$ we have $f_n(x) \longrightarrow f(x)$ as $n \to \infty$. In addition, for any set $A \in \mathfrak{M}$ on which f is bounded, it follows that there exist $M \in (0, \infty)$ such that $0 \leq f \leq M$ on A. Since for each $n \in \mathbb{N}$ we have

$$X = A_n \bigcup \left(\bigcup_{k=0}^{2^{2n}-1} A_n^k \right)$$

and $A_n = \{f > 2^n\}$, it follows that if n is large enough, then

$$A \subseteq \bigcup_{k=0}^{2^{2n}-1} A_n^k$$

and hence

$$\sup_A |f_n - f| \le \max_{k=0,\ldots,2^{2n}-1} \left(\sup_{A_n^k} |f_n - f| \right) \le 2^{-n}.$$

The conclusion ensues. $\qquad\qquad\qquad\qquad\qquad\qquad\qquad\qquad\qquad$ $\square$

We conclude this chapter with the next observations:

Remark 7.46 (Almost everywhere convergence and completeness). It can be shown that if the measurable space $(X, \mathfrak{M})$ can be endowed with a *complete measure* μ, then the pointwise convergence above can be replaced by convergence on X except for a μ-nullset (a set of zero μ-measure). We refrain from getting into more details now, but this type of convergence (which later we shall call *almost everywhere convergence*) will play a very important role in the next chapters.

Remark 7.47 (Functions with values $\pm\infty$). In the exact same way that we studied real-valued measurable functions

$$f : X \longrightarrow \mathbb{R} = (-\infty, \infty),$$

one can equally efficiently study *extended real-valued* measurable functions

$$f : X \longrightarrow \overline{\mathbb{R}} = [-\infty, \infty].$$

Indeed, it is easy to see that the "compact line" $[-\infty, \infty]$ behaves the same as $[-1, 1]$ (strictly speaking, they are homeomorphic via the "shrinking" function

$$(-\infty, \infty) \ni t \mapsto \frac{t}{1 + |t|} \in (-1, 1)$$

extended in the obvious way at $\pm\infty$) and the σ-algebra of Borel sets $\mathcal{B}(\overline{\mathbb{R}})$ can be defined exactly as in the case of $\mathbb{R}$. Everything then carries over without significant changes except for the obvious ones. For instance, in Proposition 7.37, one has to replace the rays $[a, \infty), (a, \infty)$ by $[a, \infty], (a, \infty]$, etc. Some complications arises though when one tries to consider vector-valued functions with infinite values. Since we are primarily interested in vector-valued functions, as a rule we shall not consider functions attaining the values $\pm\infty$.

7.7 Exercises

Exercise 7.1 (Cantor set). Let $C \subseteq \mathbb{R}$ be the Cantor set, as defined in Remark 6.24 of Chapter 6.

1. Justify why C is compact in $\mathbb{R}$.

2. Show that C is uncountable.

   [Hint: Show by induction that all numbers in C have a triadic expansion $\sum_1^\infty \frac{a_k}{3^k}$ with no a_k taking the value one.]

3. Compute $\mathcal{L}^1(C_n)$.

4. Show that $\mathcal{L}^1(C) = 0$ by using the lower continuity of the Lebesgue measure with respect to intersections.

Exercise 7.2. Let $\mathfrak{M}$ be a σ-algebra on a set X. Show that:

- If $A_1, A_2, A_3, \ldots \in \mathfrak{M}$ then $\bigcap_{n=1}^\infty A_n \in \mathfrak{M}$.

- If $A, B \in \mathfrak{M}$ then $B \setminus A \in \mathfrak{M}$.

Exercise 7.3. Let $(X, \mathfrak{M}, \mu)$ be a measure space and $A, B \in \mathfrak{M}$. Show that:

- If $A \subseteq B$ then $\mu(A) \leq \mu(B)$.

- If $A \subseteq B$ and $\mu(A) < \infty$ then $\mu(B \setminus A) = \mu(B) - \mu(A)$.

Exercise 7.4. Show that the Hausdorff outer measure $\mathcal{H}^{s*}$ is indeed an outer measure on $\mathbb{R}^d$, $d \in \mathbb{N}$.

Exercise 7.5. Let $(X, \mathfrak{M})$ be a measurable space and $f, g : X \longrightarrow \mathbb{R}$ be $\mathfrak{M}$-measurable functions. Then the set $\{f \leq g\}$ is $\mathfrak{M}$-measurable.

Exercise 7.6. Let $X \neq \emptyset$ be a set and $x \in X$. Show that the Dirac mass δ_x is a measure on $(X, \mathcal{P}(X))$.

Exercise 7.7. Let $(X, \mathfrak{M})$ be a measurable space and f, g non-negative measurable functions and $\alpha \geq 0$. By using the characterisation of measurable functions as those that can be pointwise approximated by simple functions, prove that also the functions $f \cdot g$ and αf are non-negative and measurable.

Exercise 7.8. Let $(X, \mathfrak{M})$ be a measurable space and f, g non-negative $\mathfrak{M}$-measurable functions. By using the characterisation of measurable functions as those that can be pointwise approximated by simple functions, show that the functions $f + g$ and $\max\{f, g\}$ are also non-negative and $\mathfrak{M}$-measurable.

Exercise 7.9. Let $(X, \mathfrak{M}, \mu)$ be a measure space and $f, g \in L^+(X, \mathfrak{M})$. Show the following:

1. $\int_X (f + g)\, d\mu = \int_X f\, d\mu + \int_X g\, d\mu$

2. $\int_X (\alpha f)\, d\mu = \alpha \int_X f\, d\mu$, $a \geq 0$.

Exercise 7.10. Let φ, ψ be non-negative simple functions on the measure space $(X, \mathfrak{M}, \mu)$ and $\alpha \in [0, \infty)$. Show that then $a\varphi$ is a simple functions and also

1. $\int_X \chi_A\, d\mu = \mu(A)$.

2. $\int_X \alpha\varphi\, d\mu = \alpha \int_X \varphi\, d\mu$

3. $0 \leq \int_X \varphi\, d\mu \leq \int_X \psi\, d\mu$ whenever $0 \leq \varphi \leq \psi$.

Exercise 7.11. On the measure space $(X, \mathfrak{M}, \mu)$, let $(\varphi_n)_1^\infty$ and $(\psi_n)_1^\infty$ be monotone increasing sequences of non-negative simple functions. Show that, if $\lim\limits_{n \to \infty} \varphi_n = \lim\limits_{n \to \infty} \psi_n$, then

$$\lim_{n \to \infty} \int_X \varphi_n\, d\mu = \lim_{n \to \infty} \int_X \psi_n\, d\mu.$$

Exercise 7.12. Let $(X, \mathfrak{M})$ be a measurable space and $(f_n)_1^\infty$ a sequence of $\mathfrak{M}$-measurable functions $f_n : X \longrightarrow \mathbb{R}$, $n \in \mathbb{N}$. By using the characterisation of measurable functions as those that can be pointwise approximated by simple functions, prove that the functions $\inf\limits_{n \in \mathbb{N}} f_n$ and $\liminf\limits_{n \to \infty} f_n$ are also measurable.

Exercise 7.13. Let $(X, \mathfrak{M})$ be a measurable space and $(f_n)_1^\infty$ a sequence of $\mathfrak{M}$-measurable functions $f_n : X \longrightarrow \mathbb{R}$, $n \in \mathbb{N}$.

1. Show that the set

$$A := \left\{ x \in X \ : \ \lim_{n \to \infty} f_n(x) \text{ exists} \right\}$$

is in $\mathfrak{M}$.

2. Let us define

$$f(x) := \left(\chi_A \lim_{n \to \infty} f_n\right)(x) = \begin{cases} \lim\limits_{n \to \infty} f_n, & \text{if } x \in A, \\ 0, & \text{if } x \notin A. \end{cases}$$

Then, show that f is $\mathfrak{M}$-measurable.

Exercise 7.14. Let $(X, \mathfrak{M})$ be a measurable space, $f : X \longrightarrow \mathbb{R}$ a $\mathfrak{M}$-measurable function and fix $a \in \mathbb{R}$. Show that

$$f^a(x) := \min\{f(x), a\} = \begin{cases} a, & \text{if } f(x) > a, \\ f(x), & \text{if } f(x) \leq a, \end{cases}$$

is also $\mathfrak{M}$-measurable.

Exercise 7.15. Show that any $f : \mathbb{R} \longrightarrow \mathbb{R}$ is $\mathcal{L}^1$-measurable if it is monotone (non-decreasing or non-increasing).

Exercise 7.16. Show that any $f : [a, b] \longrightarrow [0, \infty]$ is in $L^+([a, b], \mathcal{L}(\mathbb{R}))$ if it is monotone (non-decreasing or non-increasing).

Exercise 7.17. Let $f : X \longrightarrow \mathbb{R}$ be $\mathfrak{M}$-measurable and $g : \mathbb{R} \longrightarrow \mathbb{R}$ be Borel-measurable. Show that $g \circ f$ is $\mathfrak{M}$-measurable.

Exercise 7.18. Show that $\mathbb{Q} \in \mathcal{B}(\mathbb{R})$.

Exercise 7.19. Show that:

1. The Borel σ-algebra $\mathcal{B}(\mathbb{R})$ is the smallest σ-algebra which contains all closed sets.

2. $\mathcal{B}(\mathbb{R})$ is the smallest σ-algebra which contains all open boxes.

3. $\mathcal{B}(\mathbb{R})$ is the smallest σ-algebra which contains all boxes.

Exercise 7.20. Let $(X, \mathfrak{M})$ be a measurable space and $f : X \longrightarrow \mathbb{R}$. Show that the following statements are equivalent by giving all the details:

1. f is $\mathfrak{M}$-measurable.

2. For all $a \in \mathbb{R}$ it holds that $\{f \leq a\} \in \mathfrak{M}$.

3. For all $a < b \in \mathbb{R}$ it holds $\{a < f < b\} \in \mathfrak{M}$.

Exercise 7.21. Let $(X, \mathfrak{M})$ be a measurable space. Show that χ_A is measurable if and only if $A \in \mathfrak{M}$.

Exercise 7.22. Let $(X, \mathfrak{M}, \mu)$ be a measure space and suppose $\mu(A \cap B) < \infty$. Then, show that $\mu(A \cup B) = \mu(A) + \mu(B) - \mu(A \cap B)$.

Exercise 7.23. Let $(X, \mathfrak{M})$ be a measurable space and let also $f : X \longrightarrow \mathbb{R}$ be $\mathfrak{M}$-measurable. Consider the collection of subsets

$$\mathcal{G} := \{A \subseteq \mathbb{R} \mid f^{-1}(A) \in \mathfrak{M}\}.$$

1. Show that for all $a \in \mathbb{R}$, $(a, \infty) \in \mathcal{G}$.

2. Show that $\mathcal{G}$ is a σ-algebra.

3. Conclude that for all $A \in \mathcal{B}(\mathbb{R})$ holds $f^{-1}(A) \in \mathfrak{M}$.

Chapter 8

The Lebesgue integration theory

8.1 From Riemann's to Lebesgue's integral

In this chapter we shall construct, for any measure μ on a measurable space $(X, \mathfrak{M})$, an integral with respect to that measure, the so-called *Lebesgue integral*. The latter applies to the so-called *integrable functions*, i.e. those which, roughly speaking, can be "integrated". Subsequently, we shall derive some powerful formulas and relations for these integrals.

We would like to emphasise that *this new integral is not unrelated the well-known Riemann integral*. Rather, it is a smart *generalisation* of it which applies to much more general spaces and to much less regular functions than its Riemann counterpart. In particular, we shall show later that all Riemann integrable functions (and a fortiori all continuous functions) are actually Lebesgue integrable as well. Further, for continuous functions the Lebesgue and the Riemann integral coincide, and thus in this case the Lebesgue integral recovers the familiar object known from Real Analysis.

On the other hand, though, the generalised integral opens up new horizons, since the class of functions that can be integrated is substantially expanded. What is even more remarkable (and to a large extent the reason for its success within modern Analysis) is that *it enjoys outstandingly useful approximation, compactness and convergence properties*.

The main idea which distinguishes the Lebesgue integral from the Riemann integral is the following:

- In the Riemann case, we construct the integral by approximating from below (and from above) by the upper and lower Riemann integrals. In our measure-theoretic language, these will amount to integrals of simple functions whose constituent sets are *boxes* (the partition).

- In the Lebesgue case, the idea will be to approximate from below (it suffices only from below) by integrals of simple functions whose constituent sets are *arbitrary measurable sets*, and not necessarily boxes.

The latter idea, apart from the fact that it can be generalised on any measurable space $(X, \mathfrak{M})$ (the Euclidean structure not being needed anymore for the definition of the partitions), has the advantage of leading to a definition of the integral for *all* non-negative measurable functions.

8.2   Integration of simple functions

Let $(X, \mathfrak{M}, \mu)$ be a measure space and $f : X \longrightarrow \mathbb{R}^d$ an $\mathfrak{M}$-measurable map, $d \in \mathbb{N}$. In this section we define and study the notion of *Lebesgue integral of f with respect to μ on X*, which will be symbolised as

$$\int_X f \, \mathrm{d}\mu \equiv \int_X f(x) \, \mathrm{d}\mu(x).$$

The *abbreviated* notation on the left hand side will be preferred if no arguments are needed to be present, but beware that **both symbols mean exactly the same thing**. As we have explained, this will be an extension of the Riemann integral to just measurable maps on an arbitrary measure space.

Following the method used in the construction of the Lebesgue measure starting from the "volume" of elementary sets, we shall define the integral in a stepwise fashion:

- First, we shall define the integral for simple non-negative functions.

- By approximation, we shall extend the integral to non-negative measurable functions.

- Finally, we shall extend the integral to scalar and to vector-valued measurable functions.

Throughout this section we consider a fixed underlying **measure space** $(X, \mathfrak{M}, \mu)$ which may not be explicitly mentioned in the statements. We begin with the first step of the definition for simple functions. First, we introduce some notation:

Definition 8.1. For any measurable space $(X, \mathfrak{M})$, the set of all **non-negative (extended-valued) measurable functions** is:

$$L^+(X, \mathfrak{M}) := \left\{ f : X \longrightarrow [0, \infty] \;\middle|\; f \text{ is } \mathfrak{M}\text{-measurable} \right\}$$

(recall Remark 7.47). If the σ-algebra is understood from the context, we shall occasionally abbreviate $L^+(X, \mathfrak{M})$ to merely $L^+(X)$.

Definition 8.2. Let $f \in L^+(X, \mathfrak{M})$ be a simple function with representation

$$f = \sum_{j=1}^p a_j \chi_{A_j}$$

where $a_1, \ldots, a_p \geq 0$, $A_1, \ldots, A_p \in \mathfrak{M}$, $A_k \cap A_l = \emptyset$ for $k \neq l$. We define the **integral of the simple function** $f \in L^+(X, \mathfrak{M})$ as

$$\int_X f \, \mathrm{d}\mu := \sum_{j=1}^p a_j \, \mu(A_j).$$

If $A \in \mathfrak{M}$, we define the **integral of the simple function** f **over** A as

$$\int_A f \, d\mu := \int_X (f\chi_A) \, d\mu = \sum_{j=1}^{p} a_j \, \mu(A \cap A_j).$$

Namely, we put $f \equiv 0$ outside A and integrate the "cut off" function

$$f\chi_A = \begin{cases} f, & \text{on } A, \\ 0, & \text{on } X \setminus A. \end{cases}$$

Remark 8.3 (A convention for indeterminate forms). In the above definition and subsequently, we have used and shall be using **a convention which is** *special* **to measure theory** regarding the indeterminate forms "$0 \cdot \infty$" and "$\infty \cdot 0$":

$$0 \cdot \infty = \infty \cdot 0 = 0.$$

This convention is very important and we impose it for the following reasons:

(i) Consider the function $f \equiv 0$ on $\mathbb{R}^2$. Then, intuitively *"there is no volume under the graph of f"* although the area of $\mathbb{R}^2$ is infinite (Figure 8.1). Hence, we should have

$$\int_{\mathbb{R}^2} f(x) \, dx = \infty \cdot 0 = 0.$$

(ii) Consider the function given by

$$g(x) = \begin{cases} 0, & x \in \mathbb{R}^2 \setminus \{0\}, \\ \infty, & x = 0. \end{cases}$$

Then, again *"there is no area under the graph of g"* although g attains the value infinity at the origin (Figure 8.1). Hence, we should have

$$\int_{[-1,1] \times [-1,1]} g(x) \, dx = 0 \cdot \infty = 0.$$

At this point, we would like to clarify a source of potential confusion. **The convention "$0 \cdot \infty = \infty \cdot 0 = 0$" will NOT be applied indistinguishably whenever convenient to any indeterminate form arising. It applies only to products of the form**

$$a \cdot \mu(A)$$

where $a \in [0, \infty]$ **is an extended real number and** $\mu(A) \in [0, \infty]$ **is the measure of a set** $A \in \mathfrak{M}$.

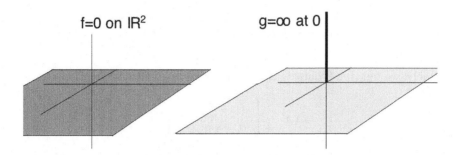

f=0 on IR²   g=∞ at 0

Figure 8.1: An illustration of the necessity of the convention $\infty \cdot 0 = 0$ for products of the form $a \cdot \mu(A)$ with $a \in [0, \infty]$ and $A \in \mathfrak{M}$.

Remark 8.4 (The integral for simple functions is well defined). Since *the representation of a simple function may not be unique*,[1] we need to check that the integral above is well defined and gives the same value irrespective of the representation considered. The argument follows similar lines to that used the definition of volume of elementary sets (Definition 6.7). Indeed, fix $f \in L^+(X, \mathfrak{M})$ and suppose

$$f = \sum_{j=1}^{p} a_j \chi_{A_j} = \sum_{k=1}^{q} b_k \chi_{B_k}$$

where $a_1, \ldots, a_p, b_1, \ldots b_q \geq 0$ and $A_1, \ldots, A_p, B_1, \ldots, B_q \in \mathfrak{M}$ are both disjoint families of $\mathfrak{M}$-measurable sets. We begin by adding to these families the sets

$$A_{p+1} := X \setminus \left(\bigcup_{j=1}^{p} A_j \right) , \quad B_{q+1} := X \setminus \left(\bigcup_{k=1}^{q} B_k \right)$$

and then then augmented families of sets $\{A_1, \ldots, A_{p+1}\}$ and $\{B_1, \ldots, B_{q+1}\}$ are $\mathfrak{M}$-measurable partitions of the space:

$$X = \bigcup_{j=1}^{p+1} A_j = \bigcup_{k=1}^{q+1} B_k.$$

It follows that the family of $\mathfrak{M}$-measurable sets

$$\left\{ A_j \cap B_k \mid j = 1, \ldots, p+1 \, ; \, k = 1, \ldots, q+1 \right\}$$

---

[1] A simple function may be represented in more than one form. For example, we may rewrite the characteristic $\chi_{(0,2)}$ on $\mathbb{R}$ in infinitely many different ways, including

$$\chi_{(0,2)} = \chi_{(0,1]} + \chi_{(1,2)} = \chi_{(0,1/2]} + \chi_{(1/2,1]} + \chi_{(1,2)} = 1 \cdot \chi_{(0,2)} + 0 \cdot \chi_{\mathbb{R} \setminus (0,2)}.$$

is also a partition of X, such that $a_j = b_k$ whenever $A_j \cap B_k \neq \emptyset$. By the additivity property of the measure μ, we have

$$\sum_{j=1}^{p} a_j \mu(A_j) = \sum_{j=1}^{p} a_j \left(\sum_{k=1}^{q} \mu(A_j \cap B_k) \right)$$

$$= \sum_{k=1}^{q} b_k \left(\sum_{j=1}^{p} \mu(A_j \cap B_k) \right)$$

$$= \sum_{k=1}^{q} b_k \, \mu(B_k).$$

Hence the integral is well defined, regardless of the possibly different representations of the simple function.

The integral defined above satisfies the following reasonable properties:

Lemma 8.5. *Let $(X, \mathfrak{M}, \mu)$ be a measure space and let also $f, g \in L^+(X, \mathfrak{M})$ be simple functions. Then, we have*

$$C \geq 0 \implies \int_X (Cf) \, d\mu = C \int_X f \, d\mu, \tag{i}$$

$$f \leq g \implies \int_X f \, d\mu \leq \int_X g \, d\mu, \tag{ii}$$

$$\int_X (f + g) \, d\mu = \int_X f \, d\mu + \int_X g \, d\mu. \tag{iii}$$

Proof. (i) and (ii) are straightforward exercises on the application of Definition 8.2 (by bearing in mind the decomposition method explained in Remark 8.4). For (iii), suppose f, g have representations given by

$$f = \sum_{j=1}^{p} a_j \chi_{A_j}, \quad g = \sum_{k=1}^{q} b_k \chi_{B_k}.$$

Then,

$$f + g = \sum_{j=1}^{p} a_j \chi_{A_j} + \sum_{k=1}^{q} b_k \chi_{B_k} = \sum_{j=1}^{p} \sum_{k=1}^{q} (a_j + b_k) \chi_{A_j \cap B_k}.$$

Since for $(j, k) \neq (m, n)$ the sets $A_j \cap B_k$ and $A_m \cap B_n$ are disjoint, we infer that

$$\int_X (f + g) \, d\mu = \sum_{j=1}^{p} \sum_{k=1}^{q} (a_j + b_k) \mu(A_j \cap B_k)$$

$$= \sum_{j=1}^{p} \sum_{k=1}^{q} a_j \, \mu(A_j \cap B_k) + \sum_{j=1}^{p} \sum_{k=1}^{q} b_k \, \mu(A_j \cap B_j)$$

$$= \int_X f \, d\mu + \int_X g \, d\mu,$$

as desired.   □

In Chapter 7 we defined the notion of absolutely continuous measure with respect to another measure (Definition 7.16). Now we present a construction which, given a measure, generates other non-trivial measures that are absolutely continuous with respect to it.

Lemma 8.6 (Absolutely continuous measures). *Let $f \in L^+(X, \mathfrak{M})$ be a simple function. Then, the function $\nu : \mathfrak{M} \longrightarrow [0, \infty]$ given by*

$$\nu(A) := \int_A f \, d\mu, \quad A \in \mathfrak{M},$$

*is a **measure** on the measurable space $(X, \mathfrak{M})$ and (in view of Definition 7.16) it is absolutely continuous with respect to μ:*

$$\nu \ll \mu.$$

Proof. Suppose that f has the representation

$$f = \sum_{j=1}^{p} a_j \chi_{A_j}.$$

Then, by definition we have

$$\int_A f d\mu = \sum_{j=1}^{p} a_j \mu(A \cap A_j)$$

and hence $\nu(\emptyset) = 0$. Further, if $\mu(A) = 0$, then $\nu(A) = 0$ as well. Finally, if $(E_m)_1^\infty \subseteq \mathfrak{M}$ is a disjoint sequence of $\mathfrak{M}$-measurable sets and, for notational convenience, we set

$$E := \bigcup_{m=1}^{\infty} E_m,$$

then we have

$$\int_E f \, d\mu = \sum_{j=1}^{p} a_j \, \mu(E \cap A_j) = \sum_{j=1}^{p} a_j \left(\sum_{m=1}^{\infty} \mu(E_m \cap A_j) \right).$$

By interchanging the order of summation of the series (all terms are non-negative, recall Lemma 6.13), we have

$$\int_E f \, d\mu = \sum_{m=1}^{\infty} \left(\sum_{j=1}^{p} a_j \, \mu(E_m \cap A_j) \right) = \sum_{m=1}^{\infty} \int_{E_m} f \, d\mu.$$

Hence, we obtain that

$$\nu \left(\bigcup_{m=1}^{\infty} E_m \right) = \sum_{m=1}^{\infty} \nu(E_m),$$

establishing that indeed ν is a measure on $(X, \mathfrak{M})$.   □

Remark 8.7 (Lebesgue-Radon-Nikodym decomposition theorem). The above result partly justifies the definition of absolute continuity. Actually, it can be shown that the opposite is true as well, namely if two (σ-finite) measures satisfy $\nu << \mu$, then actually ν can be recovered by integration of a fixed function f against the measure μ (in general f is, however, merely integrable and not necessarily simple, see the next section for the definition). In addition, much more is true: given two (σ-finite) measures μ, ν, we can decompose ν into an absolutely continuous measure $\nu_{ac} << \mu$ given by integration against a function and a singular measure $\nu_s \perp \mu$. This result is much more advanced than the scope of this introductory book and therefore we refrain from delving into this important topic any further.[2]

8.3 Integration of non-negative measurable functions

Now we move on to the next step, extending the integral from simple functions in $L^+(X, \mathfrak{M})$ to all the functions in $L^+(X, \mathfrak{M})$. We caution the reader to note that, unlike in the Riemann case, here the definition is "one-sided", namely we do not need to approximate from both above and below (although this superfluous alternative may also be carried out).

Definition 8.8. Let $(X, \mathfrak{M}, \mu)$ be a measure space and $f \in L^+(X, \mathfrak{M})$. We define the **integral of f with respect to μ over X** as (see Figure 8.2)

$$\int_X f \, d\mu := \sup \left\{ \int_X \phi \, d\mu \ \middle| \ \phi \text{ simple function on } X, 0 \le \phi \le f \text{ on } X \right\}.$$

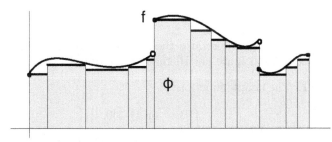

Figure 8.2: An illustration of the definition of the Lebesgue integral for non-negative simple functions on an interval in $\mathbb{R}$. Note that in this figure the measurable sets appear to be intervals, but in general this need not be the case!

If $A \in \mathfrak{M}$, we define the **integral of f with respect to μ over A** as

$$\int_A f \, d\mu := \int_X (f \chi_A) \, d\mu.$$

[2]For further reading in this direction, the reader may consult e.g. the books [14, 17, 18].

As noted in the introduction of this chapter, **we approximate from below by simple functions supported on arbitrary measurable sets, not just rectangular partitions, as in the Riemann case.** In any case, in the generality of an abstract measure space, there is no concept of partition into rectangles!

The following result lists some first properties of the integral for non-negative measurable functions.

Lemma 8.9. *Let $(X, \mathfrak{M}, \mu)$ be a measure space and $f, g \in L^+(X, \mathfrak{M})$. Then:*

$$C \geq 0 \quad \Longrightarrow \quad \int_X (Cf) \, \mathrm{d}\mu = C \int_X f \, \mathrm{d}\mu. \qquad \text{(i)}$$

$$f \leq g \quad \Longrightarrow \quad \int_X f \, \mathrm{d}\mu \leq \int_X g \, \mathrm{d}\mu. \qquad \text{(ii)}$$

Proof. This is a direct consequence of the definition and the properties of the supremum. The details are left as an exercise.   □

Based on our experience with the Riemann integral, a further anticipated property of the Lebsegue integral is its additivity, namely the fact that the integral of a sum of functions should equal the sum of the integrals. It turns out that this property, while true, is less trivial, and as such its proof is deferred until the next section, after we shall have established our first principal convergence theorem.

Remark 8.10 (Existence of Lebesgue integral). Note that (unlike the Riemann case) **every non-negative measurable $f \in L^+(X, \mathfrak{M})$ has integral** (although it might be $+\infty$)! Hence,

$$f \in L^+(X, \mathfrak{M}) \quad \Longrightarrow \quad \text{exists } \int_X f \, \mathrm{d}\mu \in [0, \infty].$$

The following example illustrates the wide applicability of the Lebesgue integral to functions which are not Riemann integrable:[3]

Example 8.11. The simple function

$$f := \chi_{\mathbb{Q} \cap [0,1]} = \begin{cases} 1, & \text{on } \mathbb{Q} \cap [0, 1], \\ 0, & \text{on } \mathbb{R} \setminus (\mathbb{Q} \cap [0, 1]), \end{cases}$$

is Lebesgue integrable and its integral is zero: since $\mathbb{Q} \cap [0, 1]$ is countable, we have

$$\int_{\mathbb{R}} f \, \mathrm{d}\mathcal{L}^1 = \int_{\mathbb{Q} \cap [0,1]} 1 \, \mathrm{d}\mathcal{L}^1 = \mathcal{L}^1(\mathbb{Q} \cap [0, 1]) = 0.$$

However, the function f is not Riemann integrable. Indeed, its upper and lower

---

[3]The definition of the Riemann integral is recalled in Section 8.9, so the reader may consult that section if they don't remember the particulars!

Riemann integrals are different: since $f \equiv 0$ on $\mathbb{R} \setminus [0,1]$, it suffices to consider Riemann integration on $[0,1]$ only. Since $f = 0$ on a dense subset of $[0,1]$, we have that any rectangle below the graph of f must be trivial. Therefore, by the definition of the Riemann integral (you may see also Section 8.9), we obtain

$$\underline{\int_0^1} f(x)\,dx = 0.$$

Similarly, since $f = 1$ on a dense subset of $[0,1]$, we have that any rectangle above the graph of f must have height at least equal to one, a fact which implies that

$$\overline{\int_0^1} f(x)\,dx = 1.$$

Since the upper and lower integrals of f over $[0,1]$ are different, f is not Riemann integrable on $[0,1]$.

8.4   The Monotone Convergence Theorem

The Lebesgue integral satisfies three main convergence theorems, which in a sense justify its handiness in Analysis. In this section we prove our first principal convergence result.

Theorem 8.12 (The Monotone Convergence Theorem). *Let $(X, \mathfrak{M}, \mu)$ be a measure space. Consider a non-decreasing sequence $(f_n)_1^\infty \subseteq L^+(X, \mathfrak{M})$ (see Figure 8.3), namely one for which*

$$f_n \leq f_{n+1} \text{ on } X, \quad \text{for all } n \in \mathbb{N}.$$

Then, the limit "commutes" with the integral:

$$\int_X \left(\lim_{n \to \infty} f_n \right) d\mu = \lim_{n \to \infty} \int_X f_n \, d\mu.$$

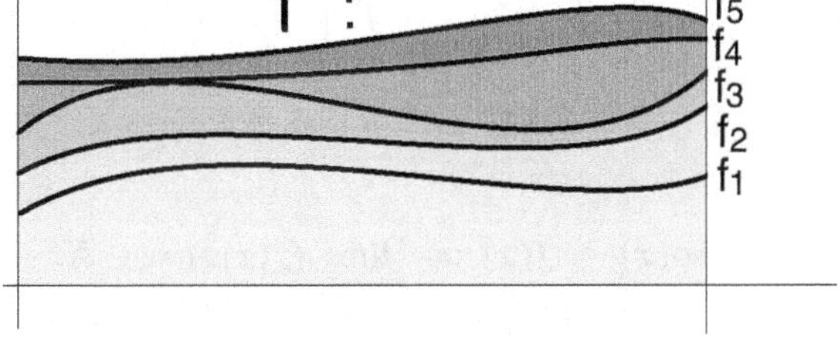

Figure 8.3: An illustration of the setting of the statement of the Monotone Convergence Theorem. The sequence $(f_n)_1^\infty$ is pointwise increasing.

Remark 8.13. Note that in the above the limit function

$$f := \lim_{n \to \infty} f_n \ : \ X \longrightarrow [0, \infty]$$

always exists because the sequence is non-decreasing. Further,

$$\lim_{n \to \infty} f_n(x) = \sup_{n \in \mathbb{N}} f_n(x), \quad x \in X. \tag{•}$$

Moreover, Theorem 8.12 can be compactly rephrased as

If $(f_n)_1^\infty \subseteq L^+(X, \mathfrak{M})$ and $f_n \nearrow f$, then $\displaystyle\int_X f_n \, d\mu \longrightarrow \int_X f \, d\mu$ as $n \to \infty$.

Remark 8.14 (Significance). Theorem 8.12 is **a very important result** and no analogous statement holds true for the Riemann integral (on $\mathbb{R}^d$). Apart from being an important computational tool, its theoretical significance lies in that, when combined with Theorem 7.45, it allows us to recover the integral through sequential convergence of (simple) functions, even though it has originally been defined as the supremum over a potentially *uncountable* family of simple functions.

Proof. We begin by observing that since $(f_n)_1^\infty$ is nondecreasing, integrating the inequality $f_n \leq f_{n+1}$ on X implies that the sequence of numbers

$$\left(\int_X f_n \, d\mu \right)_{n=1}^\infty \subseteq [0, \infty]$$

is non-decreasing and its limit (which coincides with its supremum) always exists, but could be $+\infty$. By using Remark 8.13 and integrating the inequality

$$f_n \leq \sup_{k \in \mathbb{N}} f_k$$

over X, in view of (•) we obtain

$$\int_X f_n \, d\mu \leq \int_X \left(\lim_{k \to \infty} f_k \right) d\mu, \quad n \in \mathbb{N}.$$

Hence, since the left hand side is non-decreasing, we obtain

$$\lim_{n \to \infty} \int_X f_n \, d\mu \leq \int_X \left(\lim_{k \to \infty} f_k \right) d\mu.$$

In order to show the converse, fix $\alpha \in (0, 1)$ and let $\phi \in L^+(X, \mathfrak{M})$ be a simple function such that

$$\phi(x) \leq f(x) := \lim_{n \to \infty} f_n(x), \quad x \in X.$$

We define the measurable sets

$$E_n := \left\{ x \in X \ : \ f_n(x) \geq \alpha\phi \right\}, \quad n \in \mathbb{N}.$$

Then, $\{E_n\}_1^\infty$ is an increasing sequence whose union is X. Also, since $f_n \geq 0$, we have

$$\int_X f_n \, d\mu \geq \int_{E_n} f_n \, d\mu \geq \alpha \int_{E_n} \phi \, d\mu.$$

By letting $n \to \infty$ and utilising the upper continuity of the (absolutely continuous) measure

$$\nu(A) := \int_A \phi \, d\mu, \quad \nu : \mathfrak{M} \longrightarrow [0, \infty]$$

(recall Theorem 7.21 and Lemma 8.6), we obtain

$$\lim_{n \to \infty} \int_X f_n \, d\mu \geq \alpha \int_X \phi \, d\mu.$$

We may now let $\alpha \nearrow 1$ and take the supremum with respect to all such simple functions ϕ with $0 \leq \phi \leq f$, to find

$$\lim_{n \to \infty} \int_X f_n \, d\mu \geq \int_X f \, d\mu.$$

The theorem ensues. $\qquad\qquad\qquad\qquad\qquad\qquad\qquad\qquad\qquad\qquad\qquad\quad$ $\square$

The next example shows that the monotonicity hypothesis in the Monotone Convergence Theorem is necessary:

Example 8.15. In the measure space $(\mathbb{R}, \mathcal{L}(\mathbb{R}), \mathcal{L}^1)$, consider the sequence

$$f_m := \chi_{[m, m+1]}, \quad m \in \mathbb{N}.$$

Then, we have that $f_m \longrightarrow 0$ (pointwise) as $m \to \infty$ (see Figure 8.4).

Figure 8.4: An illustration of the counter-example to the Monotone Convergence Theorem.

Indeed, for any $x \in \mathbb{R}$, choose $m_x \in \mathbb{N}$ with $m_x > |x| + 1$. Then, $f_m(x) = 0$ for all $m \geq m_x$. However, the convergence is not monotonic, that is

$$f_m \not\leq f_{m+1}, \quad m \in \mathbb{N}.$$

On the other hand, we have

$$\int_{\mathbb{R}} f_m(x) \, d\mathcal{L}^1(x) = 1, \quad m \in \mathbb{N}.$$

Hence, the conclusion of the Monotone Convergence Theorem fails. However, its monotonicity assumption is not satisfied.

We now utilise the Monotone Convergence Theorem to prove that the integral of non-negative functions is additive.

Theorem 8.16 (additivity of the integral on $L^+(X, \mathfrak{M})$). *Let $(X, \mathfrak{M}, \mu)$ be a measure space and $f, g \in L^+(X, \mathfrak{M})$. Then*

$$\int_X (f + g) \, \mathrm{d}\mu = \int_X f \, \mathrm{d}\mu + \int_X g \, \mathrm{d}\mu.$$

Proof. By Theorem 7.45 we can find non-decreasing sequences of simple functions $(\phi_j)_1^\infty, (\psi_j)_1^\infty$ such that $\phi_j \longrightarrow f$ and $\psi_j \longrightarrow g$ as $j \to \infty$. Hence, the sequence

$$\left(\phi_j + \psi_j \right)_{j=1}^\infty$$

consists of simple functions which pointwise increase to the function $f + g$. By the Monotone Convergence Theorem, we have

$$
\begin{aligned}
\int_X (f + g) \, \mathrm{d}\mu &= \lim_{j \to \infty} \int_X (\phi_j + \psi_j) \, \mathrm{d}\mu \\
&= \lim_{j \to \infty} \left(\int_X \phi_j \, \mathrm{d}\mu + \int_X \psi_j \, \mathrm{d}\mu \right) \\
&= \lim_{j \to \infty} \int_X \phi_j \, \mathrm{d}\mu + \lim_{j \to \infty} \int_X \psi_j \, \mathrm{d}\mu \\
&= \int_X f \, \mathrm{d}\mu + \int_X g \, \mathrm{d}\mu,
\end{aligned}
$$

as required. $\qquad\square$

Moreover, the integral constructed is σ-additive (i.e. countably additive), in the following sense.

Theorem 8.17 (σ-additivity of the integral on $L^+(X, \mathfrak{M})$). *Let $(X, \mathfrak{M}, \mu)$ be a measure space and $(f_n)_1^\infty \subseteq L^+(X, \mathfrak{M})$. Then,*

$$\sum_{n=1}^\infty \int_X f_n \, \mathrm{d}\mu = \int_X \left(\sum_{n=1}^\infty f_n \right) \mathrm{d}\mu.$$

Proof. It is an immediate consequence of the preceding result that, for each fixed $N \in \mathbb{N}$, we have

$$\int_X \left(\sum_{n=1}^N f_n \right) \mathrm{d}\mu = \sum_{n=1}^N \int_X f_n \, \mathrm{d}\mu.$$

The conclusion follows by letting $N \to \infty$ and applying the Monotone Convergence Theorem. $\qquad\square$

8.5 Almost everywhere properties and Fatou's Lemma

In order to continue we need a new notion which is prevalent in modern Analysis and will be systematically utilised hereafter.

Definition 8.18. Let $(X, \mathfrak{M}, \mu)$ be a measure space and $A \subseteq X$. We say that a property (typically one involving measurable functions) holds μ-**almost everywhere on a measurable set** $A \in \mathfrak{M}$, abbreviated as

$$\text{``}\mu\text{-a.e. on } A\text{''},$$

when there exists a μ-nullset $E \in \mathfrak{M}$, i.e. a $\mathfrak{M}$-measurable set with $\mu(E) = 0$, such that the property holds for all $x \in A \setminus E$.

If $(X, \mathfrak{M}, \mu) = \left(\mathbb{R}^d, \mathcal{L}(\mathbb{R}^d), \mathcal{L}^d\right)$, then the modifier "$\mu$" will be dropped.

The a.e. property is a generalisation of a property holding pointwise everywhere on a measurable set. Let us give some of the most typical examples where the a.e. properties arise:

Example 8.19 (Frequent "a.e. occurrences"). Suppose that $(f_n)_1^\infty, f, g :$ $X \longrightarrow \mathbb{R}$ are functions on the measure space $(X, \mathfrak{M}, \mu)$.

- We say $f \le g$ μ-a.e. on X if there exists $E \in \mathfrak{M}$ with $\mu(E) = 0$ such that $f(x) \le g(x)$, for all $x \in X \setminus E$.

- We say $f_n \longrightarrow f$ μ-a.e. on X as $n \to \infty$ if there exists $E \in \mathfrak{M}$ with $\mu(E) = 0$ such that $f_n(x) \longrightarrow f(x)$ as $n \to \infty$, for all $x \in X \setminus E$.

One of the main reasons we introduce the a.e. properties is the following: if a function is nonnegative, measurable and its integral vanishes, then **it is not true that the function vanishes identically.** For example, the function $\chi_{\mathbb{Q} \cap [0,1]}$ as seen in Example 8.11 has zero Lebesgue integral on $\mathbb{R}$, but it is not identically zero! In general, the function vanishes merely a.e. on the space. *In order for the function to vanish everywhere, we need additional hypotheses.* If we have a σ-algebra on a topological or metric space containing the Borel sets, then the additional assumption needed is **continuity**. In order to prove these facts, we need the next result which, even though it is merely an observation with an almost trivial proof, has numerous important consequences.

Lemma 8.20 (Chebyshev inequality). *For any measure space* $(X, \mathfrak{M}, \mu)$, *any* $f \in L^+(X, \mathfrak{M})$ *and any* $t \ge 0$, *we have the estimate*

$$\int_X f \, d\mu \ge t \, \mu(\{f > t\}).$$

Proof. For any $t \ge 0$, the set $\{f > t\}$ is in $\mathfrak{M}$ and also

$$f(x) \ge t \chi_{\{f > t\}}(x), \quad x \in X.$$

The conclusion follows by integrating this inequality. $\qquad\square$

We now come to the claimed equivalence result.

Proposition 8.21. *Let $f \in L^+(X, \mathfrak{M})$. Then*

$$\int_X f \, d\mu = 0 \iff f = 0 \ \mu\text{-a.e. on } X.$$

Proof. The claim above is obvious if f is a simple function by the definition of the integral. In general, if $f = 0$ μ-a.e. on X, let ϕ be a simple function with $0 \leq \phi \leq f$. Then, $\phi = 0$ μ-a.e. on X and hence

$$\int_X f \, d\mu = \sup\left\{ \int_X \phi \, d\mu \ \middle| \ 0 \leq \phi \leq f, \ \phi \text{ simple} \right\} = 0.$$

Conversely, suppose that $\int_X f \, d\mu = 0$ and consider the set $A_n := \{f \geq 1/n\}$. Then, $A_n \in \mathfrak{M}$ by the measurability of f. Note that by the Chebyshev inequality (Lemma 8.20) we have

$$\int_X f \, d\mu \geq \frac{1}{n} \mu(A_n), \quad n \in \mathbb{N}.$$

Hence, $\mu(A_n) = 0$ for all $n \in \mathbb{N}$. Since $A_n \subseteq A_{n+1}$ and also

$$\bigcup_{n=1}^{\infty} A_n = \{f > 0\} = \{f \neq 0\},$$

by using Theorem 7.21, we infer that

$$\mu(\{f \neq 0\}) = \mu\left(\bigcup_{n=1}^{\infty} A_n \right) = \lim_{n \to \infty} \mu(A_n) = 0.$$

In conclusion, we have shown that $f = 0$ μ-a.e. on X. $\qquad\square$

We now show how to strengthen the conclusion to "$f \equiv 0$ on X" in Proposition 8.21, by imposing additional hypotheses on the measure space and on the function. To this end, we need to recall Definition 7.27 and Remark 7.28.

Proposition 8.22. *Let $(X, \mathcal{T})$ be a topological space and let $\mathfrak{M}$ be a σ-algebra on X which contains the Borel sets $\mathcal{B}(X)$. Suppose also we are given a measure $\mu : \mathfrak{M} \longrightarrow [0, \infty]$ on X which gives positive values to any non-empty open set, that is $\mu(U) > 0$ for all $U \in \mathcal{T} \setminus \{\emptyset\}$. Let finally $f \in C(X, \mathbb{R}^d)$, $d \in \mathbb{N}$. Then*

$$\int_X |f| \, d\mu = 0 \iff f \equiv 0 \text{ on } X.$$

Proof. It suffices to show that if the integral vanishes, then $f \equiv 0$ on X. Hence, assume that

$$\int_X |f(x)| \, d\mu(x) = 0$$

and for the sake of contradiction suppose that there exists $x_0 \in X$ such that $f(x_0) \neq 0$. Then, by the continuity of f at x_0, there exists $U_0 \in \mathcal{T}$ with $U \ni x_0$ such that

$$\frac{|f(x_0)|}{2} \leq |f(x)| \quad \text{for all } x \in U_0.$$

Since open sets are μ-measurable, by integrating this inequality over U_0 we obtain that

$$\frac{|f(x_0)|}{2} \mu(U_0) \leq \int_{U_0} |f| \, \mathrm{d}\mu \leq \int_X |f| \, \mathrm{d}\mu = 0$$

which is a contradiction because $\mu(U_0) > 0$ and $f(x_0) \neq 0$. $\qquad \Box$

Corollary 8.23. *The conclusion of Proposition 8.22 is true in the Euclidean case of the measure space* $(X, \mathfrak{M}, \mu) = \left(\mathbb{R}^d, \mathcal{L}(\mathbb{R}^d), \mathcal{L}^d \right)$.

Remark 8.24 (Negation of an a.e. property). We now examine the **negation of a property holding μ-a.e. on A**. Let $(X, \mathfrak{M}, \mu)$ be a measure space with $\mu \not\equiv 0$ and let also P symbolise a property (involving measurable functions) on X. Then, for a point $x \in X$, "$P(x)$" refers to the property at x. Note first that, in view of Definition 8.18, the statement "P holds a.e. on X" can be written as

$$\exists \, E \in \mathfrak{M} \text{ with } \mu(E) = 0 : \quad \{x \in X : P(x) \text{ fails}\} \subseteq E. \qquad (S)$$

The negation then is "for no μ-nullset E is the set $\{x \in X : P(x) \text{ fails}\}$ contained in E":

$$\forall \, E \in \mathfrak{M} \text{ with } \mu(E) = 0 : \quad \{x \in X : P(x) \text{ fails}\} \setminus E \neq \emptyset.$$

However, the above apparently complex negation statement is hardly ever needed in practice, for the following reasons:

(i) It is typically the case that **the set $\{x \in X : P(x)\}$ is $\mathfrak{M}$-measurable at the outset**. This happens for instance when we have

$$\{x \in X : f(x) = g(x)\}, \quad \text{or} \quad \left\{x \in X : \lim_{n \to \infty} f_n(x) = g(x)\right\},$$

etc, where all the involved functions are measurable. Then, (S) becomes

$$\mu\left(\{x \in X : P(x) \text{ fails}\} \right) = 0, \qquad (S')$$

and thus the negation of (S') reads $\mu(\{x \in X : P(x) \text{ fails}\}) > 0$, which says that **the set whereon the property P fails has positive μ-measure**.

(ii) Even if the set $\{x \in X : P(x)\}$ is not $\mathfrak{M}$-measurable, **a statement of the form (S') is always possible if we replace μ by an outer measure μ^* extending μ from $\mathfrak{M}$ to the entire powerset $\mathcal{P}(X)$**. This process is explained in Remark 7.24. For the case of the Lebesgue measure $\mathcal{L}^d$ on $\mathbb{R}^d$, we may utilise the Lebesgue outer measure $\mathcal{L}^{d*}$ directly.

We close this section with our second main convergence theorem. Its significance lies in that, unlike the Monotone Convergence Theorem, here there is **no pointwise convergence hypothesis** assumed for the sequence of measurable functions.

Theorem 8.25 (The Fatou Lemma). *Let $(X, \mathfrak{M}, \mu)$ be a measure space and consider a sequence $(f_n)_1^\infty \subseteq L^+(X, \mathfrak{M})$. Then, we have*

$$\int_X \left(\liminf_{n \to \infty} f_n \right) \mathrm{d}\mu \leq \liminf_{n \to \infty} \int_X f_n \, \mathrm{d}\mu.$$

Proof. For any $n \in \mathbb{N}$ we have that

$$\inf_{k \geq n} f_k \leq f_j, \quad \text{for all } j \geq n.$$

Hence, by integration we obtain

$$\int_X \left(\inf_{k \geq n} f_k \right) \mathrm{d}\mu \leq \int_X f_j \, \mathrm{d}\mu$$

for all $j \geq n$. Thus,

$$\int_X \left(\inf_{k \geq n} f_k \right) \mathrm{d}\mu \leq \inf_{j \geq n} \left(\int_X f_j \, \mathrm{d}\mu \right).$$

Now we may let $n \to \infty$ and apply the Monotone Convergence Theorem:

$$\int_X \left(\liminf_{n \to \infty} f_n \right) \mathrm{d}\mu = \lim_{n \to \infty} \int_X \left(\inf_{k \geq n} f_k \right) \mathrm{d}\mu \leq \liminf_{n \to \infty} \int_X f_n \, \mathrm{d}\mu.$$

The result ensues. □

The Fatou Lemma[4] can be interpreted as a result of one-sided continuity (known as "lower semi-continuity") for the Lebesgue integral with respect to the pointwise convergence. We refrain from discussing this any further now, but we shall return to this topic of "one-sided" upper and lower continuity later in Chapters 12-13. For the moment, we merely confine ourselves to illustrating the next relevant corollary:

Corollary 8.26 (Lower semi-continuity of the integral). *Let $(X, \mathfrak{M}, \mu)$ be a measure space and consider a sequence $(f_n)_1^\infty \subseteq L^+(X, \mathfrak{M})$ such that*

$$f_n \longrightarrow f \quad \text{pointwise on } X \text{ as } n \to \infty.$$

Then, we have

$$\int_X f \, \mathrm{d}\mu \leq \liminf_{n \to \infty} \int_X f_n \, \mathrm{d}\mu.$$

---

[4]Exactly like in the case of Zorn's Lemma, the nomenclature "Lemma" is kept for historical reasons of tradition and has nothing to do with its significance.

The next example shows that the non-negativity hypothesis is necessary for the validity of the Fatou Lemma. However, this requirement is not optimal and it can be relaxed to weaker conditions. Nonetheless, *some* integrability assumption is needed and cannot be removed completely. For example, it is an interesting exercise for the reader to verify that it can be relaxed to a lower bound of the form $f_n \geq -g$, where $g \in L^+(X, \mathfrak{M})$ and $\int_X g \, d\mu < \infty$.

Example 8.27. On the measure space $(\mathbb{R}, \mathcal{L}(\mathbb{R}), \mathcal{L}^1)$, consider the sequence

$$f_m := -\chi_{[m,m+1]}, \quad m \in \mathbb{N}.$$

Then, we have that $f_m \longrightarrow 0$ (pointwise) as $m \to \infty$, but

$$\int_{\mathbb{R}} f_m(x) \, d\mathcal{L}^1(x) = -1, \quad m \in \mathbb{N}.$$

Hence, the conclusion of the Fatou Lemma fails. However, its non-negativity assumption is not satisfied since $f_m \ngeq 0$ on $\mathbb{R}$ (see Figure 8.5).

Figure 8.5: An illustration of the counterexample to the Fatou Lemma.

In addition, note that the smallest non-negative function $g : \mathbb{R} \longrightarrow [0, \infty)$ for which $f_m \geq -g$ for all $m \in \mathbb{N}$ equals $g = \chi_{[1,\infty)}$, which happens to have infinite integral as well.

8.6 Integration of measurable maps and the space L^1

Now we extend the Lebesgue integral to real-valued and vector-valued measurable functions. Unlike the case of non-negative functions, the class of functions to which one can assign a well-defined integral involves an additional suitable restriction.[5]

Definition 8.28. Let $(X, \mathfrak{M}, \mu)$ be a measure space.

---

[5]The problem is the possible emergence of indeterminate forms of the type $+\infty - \infty$ when the real functions attain both positive and negative values. This cannot be handled by our convention $0 \cdot \infty = \infty \cdot 0 = 0$ of Remark 8.3, nor by any other similar convention!

(i) Let $f : X \longrightarrow \mathbb{R}$ be a $\mathfrak{M}$-measurable function. We define the **integral of f with respect to μ over X** as

$$\int_X f \, d\mu := \int_X f^+ \, d\mu - \int_X f^- \, d\mu,$$

provided that **at least one** of the integrals

$$\int_X f^{\pm} \, d\mu$$

is finite. Here[6] $f^{\pm} : X \longrightarrow [0, \infty)$ are the positive and negative parts of $f = f^+ - f^-$ (Definition 7.41).

(ii) Let $f : X \longrightarrow \mathbb{R}^d$ be a $\mathfrak{M}$-measurable map, $d \in \mathbb{N}$. We define the **integral of f with respect to μ over X** as

$$\int_X f \, d\mu := \left(\int_X f_1 \, d\mu, \, \dots, \int_X f_d \, d\mu \right)$$

where $f = (f_1, \dots, f_d)$ and each $f_i : X \longrightarrow \mathbb{R}$ is the $\mathfrak{M}$-measurable i-th component of f. The **integral of f over a measurable subset $A \in \mathfrak{M}$** is defined as

$$\int_A f \, d\mu := \int_X (f \chi_A) \, d\mu.$$

(iii) A $\mathfrak{M}$-measurable function $f : X \longrightarrow \mathbb{R}$ is called μ**-integrable on X** if both the integrals

$$\int_X f^{\pm} \, d\mu$$

are finite. Equivalently, since $|f| = f^+ + f^-$, f is μ-integrable if and only if

$$\int_X |f| \, d\mu < \infty.$$

(iv) A $\mathfrak{M}$-measurable mapping $f : X \longrightarrow \mathbb{R}^d$ $(d \in \mathbb{N})$ is called μ**-integrable on X** if

$$\int_X |f| \, d\mu < \infty.$$

In the above, $|f|$ symbolises the Euclidean norm of the mapping f. Namely, for any $a \in \mathbb{R}^d$, we use $|a|$ as a **convenient abbreviation** for $\|a\|_2$:

$$|a| \equiv \|a\|_2 = \left(\sum_{i=1}^d |a_i|^2 \right)^{1/2} \qquad a = (a_1, \dots, a_d) \in \mathbb{R}^d.$$

---

[6]The functions $f^{\pm}$ may also be allowed to attain the values $\pm \infty$, but the integrability hypothesis forces that $\pm \infty$ be attained at most on a nullset, for at least one of them.

(v) The set of all integrable functions $f : X \longrightarrow \mathbb{R}$ is called the L^1 **space over** $(X, \mathfrak{M}, \mu)$ (or the space of μ-integrable functions) and is denoted by

$$L^1(X, \mathfrak{M}, \mu) := \left\{ f : X \longrightarrow \mathbb{R}, \ \mathfrak{M}\text{-measurable} \ \middle| \ \int_X |f| \, d\mu < \infty \right\}.$$

(vi) The set of all integrable mappings $f : X \longrightarrow \mathbb{R}^d$ is called the L^1 **space over** $(X, \mathfrak{M}, \mu)$ of **vector-valued functions** (or the space of vector-valued μ-integrable functions) and is defined as

$$L^1(X, \mathfrak{M}, \mu, \mathbb{R}^d) := \left\{ f : X \longrightarrow \mathbb{R}^d, \ \mathfrak{M}\text{-measurable} \ \middle| \ \int_X |f| \, d\mu < \infty \right\}.$$

Evidently, $L^1(X, \mathfrak{M}, \mu, \mathbb{R}) \equiv L^1(X, \mathfrak{M}, \mu)$. If the σ-algebra is understood from the context, we shall abbreviate the symbolisations of (v)-(vi) for the L^1 spaces respectively to

$$L^1(X, \mu), \quad L^1(X, \mu, \mathbb{R}^d).$$

Remark 8.29 (L^1 is a vector space). It is very easy to check that $L^1(X, \mu)$ and $L^1(X, \mu, \mathbb{R}^d)$ are **vector spaces**, with respect to the usual pointwise addition and scalar multiplication of functions with real numbers. Indeed, if $f, g \in L^1(X, \mu, \mathbb{R}^d)$, then af, bg are in $L^1(X, \mu, \mathbb{R}^d)$ for any $a, b \in \mathbb{R}$ and, by integrating the triangle inequality

$$\left| af(x) + bg(x) \right| \leq |a| |f(x)| + |b| |g(x)|, \quad x \in X,$$

we obtain that $af + bg \in L^1(X, \mu, \mathbb{R}^d)$, as a consequence of Lemma 8.9 and Theorem 8.17.

Let us now record for the sake of completeness some basic properties of L^1 functions. For simplicity we treat only the scalar case of $d = 1$. The vectorial case of $d \geq 2$ is completely analogous (except for part (i)(a) below which has no counterpart).

Theorem 8.30 (Properties of L^1 functions). *Let $(X, \mathfrak{M}, \mu)$ be a measure space.*
(i) *Let $f, g \in L^1(X, \mu)$. Then, we have:*

$$f \leq g \implies \int_X f \, d\mu \leq \int_X g \, d\mu; \tag{a}$$

$$C \in \mathbb{R} \implies \int_X (Cf) \, d\mu = C \int_X f \, d\mu; \tag{b}$$

$$\left| \int_X f \, d\mu \right| \leq \int_X |f| \, d\mu. \tag{c}$$

(ii) *For any $\{f_1, \ldots f_N\} \subseteq L^1(X, \mu)$, we have*

$$\sum_{n=1}^{N} \int_X f_n \, d\mu = \int_X \left(\sum_{n=1}^{N} f_n \right) d\mu.$$

The proof is left as a simple exercise on the definition of the integral. The next result is an immediate consequence of Proposition 8.21.

Corollary 8.31. *Let* $(X, \mathfrak{M}, \mu)$ *be a measure space and* $f, g \in L^1(X, \mu)$. *Then,*

$$\int_X |f - g| \, d\mu = 0 \iff f = g, \ \mu\text{-a.e. on } X.$$

Remark 8.32 (Arbitrary modification on nullsets). Note the important consequence of Corollary 8.31 that **if a measurable function is arbitrarily modified on a nullset in such a way that the new function is also measurable, then the integral of the modified function is the same as that of the original function.** Certainly, the pointwise values may not coincide everywhere, but the functions are "essentially the same" when seen through the integral. This fact, which is a standard trick of the trade of modern Analysis, will be used systematically hereafter whenever integrals are involved.

The terminology introduced right next makes the discussion of mappings which differ from each other at most on a set of zero measure a little less cumbersome.

Definition 8.33. Let $(X, \mathfrak{M}, \mu)$ be a measure space, Y a set and $f : X \longrightarrow Y$ a mapping. A mapping $\tilde{f} : X \longrightarrow Y$ is called **a version of** f if f and $\tilde{f}$ differ only at most on a μ-nullset, in the sense that there exists a set $E \in \mathfrak{M}$ with $\mu(E) = 0$ such that $\tilde{f}(x) = f(x)$ for all $x \in X \setminus E$. Equivalently,

$$\exists \, E \in \mathfrak{M} \ \text{with} \ \mu(E) = 0 : \quad \left\{ x \in X : f(x) \neq \tilde{f}(x) \right\} \subseteq E. \qquad (\odot)$$

Remark 8.34 (About versions). Note that it is not assumed that $f, \tilde{f}$ are μ-measurable. Further, the concept of versions is symmetrical, namely $\tilde{f}$ is a version of f if any only if f is a version of $\tilde{f}$. Hence, we may say that the two maps are **versions of each other**, or that they represent **different versions of the same map**. Note also that in general $(\odot)$ is **not equivalent** to

$$\mu\left(\left\{ x \in X : f(x) \neq \tilde{f}(x) \right\} \right) = 0,$$

because the set $\{f \neq \tilde{f}\}$ **may not be measurable!** However, the statements are equivalent if the measure μ is complete, because then $\{f \neq \tilde{f}\}$ is a subset of a μ-nullset (and hence it is measurable itself).

We continue with a result which shows that the convergence of a sequence of measurable functions which is valid a.e. (as opposed to pointwise, or "everywhere") suffices to transfer measurability, if we allow the functions to be altered on nullsets. If moreover the measure is complete (recall Definition 7.16), then there is no need for modifications. We state and prove the result in the scalar case of real functions, noting that it remains virtually identical in the vector-valued case. Accordingly, we have:

Proposition 8.35 (Inheritance of measurability through a.e. relations). *Let $(X, \mathfrak{M}, \mu)$ be a measure space, $\{f_n\}_1^\infty$ a sequence of $\mathfrak{M}$-measurable real functions on X, and $f : X \longrightarrow \mathbb{R}$ such that*

$$f_n \longrightarrow f \quad \mu\text{-a.e. on } X \text{ as } n \to \infty.$$

Then:

(i) *The function f admits an $\mathfrak{M}$-measurable version $\tilde{f} : X \longrightarrow \mathbb{R}$.*

(ii) *Moreover, if the measure μ is complete, i.e. if all subsets of nullsets are measurable themselves, then the function f is itself μ-measurable. In particular, this is true for the measure space $(\mathbb{R}^d, \mathcal{L}(\mathbb{R}^d), \mathcal{L}^d)$.*

In part (i) of Proposition 8.35, the conclusion may be stated informally that f becomes $\mathfrak{M}$-measurable after perhaps a modification on a μ-nullset. We stress once again that, in the setting of Proposition 8.35, it plays no essential role that perhaps $\tilde{f} \neq f$ at most on a nullset, since Definition 8.36 that follows will enable us to give a meaning to integrals involving f, while Corollary 8.31 shows that any two functions that are versions of each other are "the same" when seen under the integral. We shall return to this important topic several times in this book.

Proof. Both parts of the proof are based on the general observation that, for any function $g : X \longrightarrow \mathbb{R}$ and any set $B \subseteq \mathbb{R}$, the following identity involving inverse images holds true (recall formulas $(\star)$-$(\star\star)$ of Chapter 1):

$$g^{-1}(B) = \left(g^{-1}(B) \cap E\right) \bigcup \left(g^{-1}(B) \setminus E\right)$$
$$= \left(g|_E\right)^{-1}(B) \bigcup \left(g|_{X \setminus E}\right)^{-1}(B).$$

An immediate consequence of this is that, if $E \in \mathfrak{M}$, then g is μ-measurable on X whenever $g|_E$ is μ-measurable on the set $E \in \mathfrak{M}$ and $g|_{X \setminus E}$ is μ-measurable on the set $X \setminus E \in \mathfrak{M}$.

(i) Let $E \in \mathfrak{M}$ with $\mu(E) = 0$ be such that $f_n(x) \longrightarrow f(x)$ for all $x \in X \setminus E$. Since, for each $n \in \mathbb{N}$, we have that $f_n|_{X \setminus E}$ is μ-measurable, it follows from Proposition 7.40 that $f|_{X \setminus E}$ is μ-measurable on the set $X \setminus E \in \mathfrak{M}$. We define

$$\tilde{f}(x) := \begin{cases} f(x), & x \in X \setminus E, \\ 0, & x \in E. \end{cases}$$

Note that $\tilde{f} = f$ on $X \setminus E$, a fact which implies that $\tilde{f}$ is a version of f, whilst also ensuring that $\tilde{f}|_{X \setminus E}$ is μ-measurable on $X \setminus E \in \mathfrak{M}$. Since we also obviously have that $\tilde{f}|_E$ is μ-measurable on the set $E \in \mathfrak{M}$, the remark at the start of the proof yields that $\tilde{f}$ is μ-measurable on X.

(ii) Suppose now that μ is complete. As in part (i), the function $f|_{X \setminus E}$ is μ-measurable on the set $X \setminus E \in \mathfrak{M}$. On the other hand, the completeness of μ

ensures that we also have that $f|_E$ is μ-measurable on the set $E \in \mathfrak{M}$. Indeed, for any Borel set $B \subseteq \mathbb{R}$, we have that $(f|_E)^{-1}(B) \subseteq E$, and as $\mu(E) = 0$ and μ is complete, it follows that $(f|_E)^{-1}(B) \in \mathfrak{M}$. Now the remark at the start of the proof, this time applied directly to f, yields the μ-measurability of f on X. □

It is worth emphasising here that **a version of a μ-measurable mapping is not automatically μ-measurable itself**, unless of course the measure μ happens to be **complete**. Thus, a priori it does not make sense to speak of the integral of this version. Nonetheless, the preceding results enable us to generalise the Lebesgue integral to the class of mappings which admit an $\mathfrak{M}$-measurable version that is μ-integrable.

Definition 8.36. Let $(X, \mathfrak{M}, \mu)$ be a measure space and $f : X \longrightarrow \mathbb{R}^d$ be a function which admits an $\mathfrak{M}$-measurable version $\tilde{f} : X \longrightarrow \mathbb{R}^d$ that is μ-integrable. Then integral of f with respect to μ over X is defined as

$$\int_X f \, \mathrm{d}\mu := \int_X \tilde{f} \, \mathrm{d}\mu.$$

Remark 8.37. The correctness of the above definition, that is, the fact that the integral of f so defined is independent of the $\mathfrak{M}$-measurable version $\tilde{f}$ of f involved, is an immediate consequence of Corollary 8.31.

8.7 The Dominated Convergence Theorem in L^1

In this section we record our final central convergence theorem for Lebesgue integrals of integrable functions. The essential point of this result is that a.e. convergence of a sequence of integrable (real or vector) functions implies convergence of the integrals (and in fact the stronger property of convergence of the integral of the differences to zero), if we have the additional assumption of a pointwise bound by an L^1 function, which "prevents mass from escaping to infinity".

Theorem 8.38 (The Dominated Convergence Theorem). *Let $(X, \mathfrak{M}, \mu)$ be a measure space and consider a sequence $(f_n)_1^\infty \subseteq L^1(X, \mu)$ such that*

(i) $f_n \longrightarrow f$ μ-a.e. on X, as $n \to \infty$,

(ii) *there exists $g \in L^1(X, \mu)$ such that $|f_n| \leq g$, $n \in \mathbb{N}$ (see Figure 8.6).*

Then, we have

$$\int_X |f_n - f| \, \mathrm{d}\mu \longrightarrow 0, \quad \text{as } n \to \infty.$$

In particular,

$$\int_X f_n \, \mathrm{d}\mu \longrightarrow \int_X f \, \mathrm{d}\mu, \quad \text{as } n \to \infty.$$

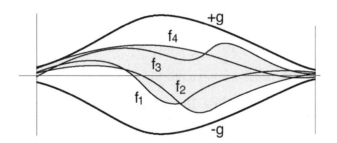

Figure 8.6: An illustration of the Dominated Convergence Theorem on $X \subseteq \mathbb{R}$. The graphs of $\{f_1, f_2, \ldots\}$ must be bounded above and below by the graphs of $\pm g$, in order to "prevent mass from escaping to infinity", see Example 8.40 that follows.

Proof. By Proposition 8.35, the limit function f is measurable after modification on a μ-nullset. Hence, there exists a version $\tilde{f} \in L^1(X, \mu)$ such that $f_n \longrightarrow \tilde{f}$ μ-a.e. on X as $n \to \infty$, whilst $\tilde{f} = f$ μ-a.e. on X.

Further, by modifying for each $n \in \mathbb{N}$ the function f_n to a version $\tilde{f}_n$ on a μ-nullset, we may arrange that $\tilde{f}_n \longrightarrow \tilde{f}$ pointwise **everywhere** on X as $n \to \infty$. Finally, since the countable union of nullsets is a nullset, we may further modify g to a version $\tilde{g}$ such that $|\tilde{f}_n|, |\tilde{f}| \leq \tilde{g}$ **everywhere** on X. By applying the triangle inequality, we infer that

$$2\tilde{g} - |\tilde{f}_n - \tilde{f}| \geq 2\tilde{g} - 2\tilde{g} = 0, \quad \text{on } X.$$

By invoking the Fatou Lemma, we estimate

$$\int_X 2\tilde{g}\, d\mu = \int_X \liminf_{n\to\infty} \left(2\tilde{g} - |\tilde{f}_n - \tilde{f}| \right) d\mu$$

$$\leq \liminf_{n\to\infty} \int_X \left(2\tilde{g} - |\tilde{f}_n - \tilde{f}| \right) d\mu$$

$$= \int_X 2\tilde{g}\, d\mu + \liminf_{n\to\infty} \int_X \left(-|\tilde{f}_n - \tilde{f}| \right) d\mu$$

$$= \int_X 2\tilde{g}\, d\mu - \limsup_{n\to\infty} \int_X |\tilde{f}_n - \tilde{f}|\, d\mu.$$

Therefore, we have obtained that

$$\limsup_{n\to\infty} \int_X |\tilde{f}_n - \tilde{f}|\, d\mu \leq 0.$$

Since $|\tilde{f}_n - \tilde{f}| = |f_n - f|$ μ-a.e. on X, by Corollary 8.31 we infer

$$\int_X |\tilde{f}_n - \tilde{f}|\, d\mu = \int_X |f_n - f|\, d\mu$$

and finally

$$\int_X |f_n - f|\, d\mu \longrightarrow 0, \quad \text{as } n \to \infty.$$

The final claim of the theorem is a consequence of the inequality

$$\left| \int_X f_n \, d\mu - \int_X f \, d\mu \right| \leq \int_X |f_n - f| \, d\mu, \quad n \in \mathbb{N}.$$

The proof of the theorem is now complete. □

As a direct consequence of the Dominated Convergence Theorem, we infer that the Lebesgue integral is σ-additive.

Theorem 8.39 (σ-additivity of the integral). *Suppose that $(X, \mathfrak{M}, \mu)$ is a measure space and let $(f_n)_1^\infty \subseteq L^1(X, \mu)$ be such that $\sum_1^\infty |f_n| \in L^1(X, \mu)$. Then,*

$$\sum_{n=1}^{\infty} \int_X f_n \, d\mu = \int_X \left(\sum_{n=1}^{\infty} f_n \right) d\mu.$$

Proof. By noting the following bound on the partial sums

$$\left| \sum_{n=1}^{N} f_n \right| \leq \sum_{n=1}^{\infty} |f_n|, \quad \text{for } N \in \mathbb{N},$$

the conclusion follows by invoking the finite additivity of the integral and the Dominated Convergence Theorem. □

The next example shows that the uniform boundedness hypothesis in the Dominated Convergence Theorem is necessary:

Example 8.40. Essentially the same example that showed the optimality of the Monotone Convergence Theorem, works here too. On the measure space $(\mathbb{R}, \mathcal{L}(\mathbb{R}), \mathcal{L}^1)$, consider the sequence of functions

$$f_m := \chi_{[m, m+1]}, \quad m \in \mathbb{N}.$$

Then, we have that $f_m \longrightarrow 0$ (pointwise) as $m \to \infty$ (see Figure 8.7).

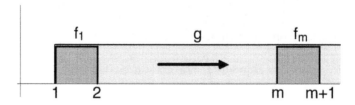

Figure 8.7: An illustration of the sequence of functions showing the necessity of the uniform bound to the Dominated Convergence Theorem.

However,

$$\int_{\mathbb{R}} |f_m(x)| \, d\mathcal{L}^1(x) = 1, \quad m \in \mathbb{N}.$$

Hence, the conclusion of the Dominated Convergence Theorem fails. However, its uniform boundedness assumption is not satisfied, since the smallest possible g for which $|f_m| \leq g$ holds true is $g = \chi_{[1,\infty)}$. Note though that $g \notin L^1(\mathbb{R}, \mathcal{L}^1)$ because the integral is infinite!

Similarly to the case of non-negativity hypothesis for the sequence in the Fatou Lemma, the uniform L^1 bound is not optimal here either. Nonetheless, *some* uniform boundedness assumption is necessary. The resulting optimal result is known in the literature as the **Vitali Convergence Theorem**. Since the latter result is relatively more advanced and requires some additional concepts to be introduced, we refrain from delving into more details.[7]

8.8 Almost uniform convergence and Egoroff's Theorem

In this section[8] we present a result which compares convergence a.e. with uniform convergence on "large" sets whose complement has small measure. The latter type of convergence is very important and has a special name:

Definition 8.41. Let $(f_m)_1^\infty$ be a sequence of $\mathbb{R}^d$-valued measurable maps defined on the measure space $(X, \mathfrak{M}, \mu)$, $d \in \mathbb{N}$. We say that $(f_m)_1^\infty$ **converges μ-almost uniformly** to the μ-measurable map f on X if, for any $\varepsilon > 0$, there exists a measurable subset $E \subseteq X$ with $\mu(E) < \varepsilon$ such that $f_m \longrightarrow f$ uniformly on $X \setminus E$, as $m \to \infty$. Namely,

$$\forall \varepsilon > 0, \ \exists E \subseteq X, \ \mu(E) < \varepsilon \ : \ \lim_{m\to\infty} \left(\sup_{X\setminus E} |f_m - f| \right) = 0.$$

In general, neither uniform nor almost uniform convergence are equivalent to a.e. convergence. However, if the space has finite measure, we have the next characterisation:

Theorem 8.42 (The Egoroff Theorem). *Suppose that $(X, \mathfrak{M}, \mu)$ is a finite measure space (i.e. $\mu(X) < \infty$) and $(f_m)_1^\infty$, f are measurable maps $X \longrightarrow \mathbb{R}^d$. Then, the following statements are equivalent:*

---

[7]The interested reader may look up this more general result in more advanced expositions, e.g. in the book [18]. The main idea of the Vitali theorem is to tame the possible loss of mass at infinity but still relax the uniform L^1 bound by imposing *equi-integrability and tightness*. Both these concepts refer to the sequence of measures on $(X, \mathfrak{M})$ given by

$$\mu_m := \int_{(\cdot)} |f_m| \, \mathrm{d}\mu, \quad m \in \mathbb{N}.$$

Equi-integrability requires that all $\{\mu_m : m \in \mathbb{N}\}$ are absolutely continuous with respect to μ uniformly in $m \in \mathbb{N}$. Tightness requires that for any $\varepsilon > 0$, we can find $A \in \mathfrak{M}$ with $\mu(A) < \infty$ such that the mass of $X \setminus A$ is uniformly small, that is $\mu_m(X \setminus A) < \varepsilon$, $m \in \mathbb{N}$.

[8]This section may perhaps be omitted on the first reading of this chapter.

(i) $f_m \longrightarrow f$ μ-almost uniformly on X, as $m \to \infty$.

(ii) $f_m \longrightarrow f$ μ-a.e. on X, as $m \to \infty$.

The finiteness assumption for the measure space X is necessary.[9]

Proof. (i) $\Longrightarrow$ (ii): Suppose first that $f_m \longrightarrow f$ μ-almost uniformly on X as $m \to \infty$. Fix $\varepsilon_0 := \mu(X)/2$. Then, there exists $E_0 \subseteq X$ with $\mu(E_0) < \varepsilon_0$ such that $f_m \longrightarrow f$ pointwise on $X \setminus E_0$ as $m \to \infty$.

Note now that by the μ-almost uniform convergence assumption, it follows that the sequence converges μ-almost uniformly on any measurable subset of X as well (this is a trivial exercise) and in particular on E_0. Hence, for $\varepsilon_1 := \mu(E_0)/2$, the exists $E_1 \subseteq E_0$ such that $f_m \longrightarrow f$ pointwise on $E_0 \setminus E_1$, as $m \to \infty$. Hence, we have

$$f_m \longrightarrow f \text{ pointwise on } X \setminus E_1 \text{ as } m \to \infty, \text{ and } \mu(E_1) < \frac{\mu(X)}{4}.$$

By repeating the same argument, we can recursively construct a decreasing sequence of measurable sets $E_0 \supseteq E_1 \supseteq E_2 \supseteq \cdots$, such that for any $k \in \mathbb{N}$ we have

$$f_m \longrightarrow f \text{ pointwise on } X \setminus E_k, \text{ as } m \to \infty, \text{ and } \mu(E_k) < \frac{\mu(X)}{2^{k+1}}.$$

By setting

$$E := \bigcap_{k=1}^{\infty} E_k,$$

the lower continuity of the measure and the finiteness of $\mu(X)$ yield that

$$\mu(E) = \mu\left(\bigcap_{k=1}^{\infty} E_k\right) = \lim_{k \to \infty} \mu(E_k) = 0.$$

Moreover, we have that $f_m \longrightarrow f$ pointwise on $X \setminus E$ as $m \to \infty$. Hence, the convergence is μ-a.e. on X and (ii) follows.

(ii) $\Longrightarrow$ (i): Suppose that $f_m \longrightarrow f$ μ-a.e. on X, as $m \to \infty$. We fix $k, m \in \mathbb{N}$ and define

$$E_m(k) := \bigcup_{s=m}^{\infty} \left\{ x \in X \;\middle|\; |f_s(x) - f(x)| \geq \frac{1}{k} \right\}.$$

For fixed $k \in \mathbb{N}$, the sequence $(E_m(k))_{m=1}^{\infty}$ decreases (that is $E_1(k) \supseteq E_2(k) \supseteq E_3(k) \supseteq \cdots$). Consider the intersection of the sequence:

$$\bigcap_{m=1}^{\infty} E_m(k) = \left\{ x \in X \;\middle|\; \forall m \in \mathbb{N}, \exists s \geq m : |f_s(x) - f(x)| \geq \frac{1}{k} \right\}.$$

---

[9]This time we leave it to the reader to construct a counterexample!

Recalling that by assumption $f_m \longrightarrow f$ μ-a.e. on X as $m \to \infty$, the intersection above is a nullset because it is exactly the set on which the a.e. convergence fails:

$$\mu\left(\bigcap_{m=1}^{\infty} E_m(k)\right) = 0.$$

By the lower continuity of the measure (and the finiteness of $\mu(X)$), it follows that $\mu(E_m(k)) \longrightarrow 0$ as $m \to \infty$, for any fixed $k \in \mathbb{N}$.

Given now any $\varepsilon > 0$ and $k \in \mathbb{N}$, we choose $m(k) \in \mathbb{N}$ large enough such that $\mu\big(E_{m(k)}(k)\big) < \varepsilon 2^{-k}$. We set

$$E := \bigcup_{k=1}^{\infty} E_{m(k)}(k).$$

Then, we have

$$\mu(E) \leq \mu\left(\bigcup_{k=1}^{\infty} E_{m(k)}(k)\right) \leq \sum_{k=1}^{\infty} \mu\big(E_{m(k)}(k)\big) \leq \sum_{k=1}^{\infty} \varepsilon 2^{-k} \leq \varepsilon$$

and by the definition of $E_m(k)$, for any $k \in \mathbb{N}$ we deduce that

$$\left|f_m - f\right| < \frac{1}{k} \quad \text{for } m > m(k), \quad \text{on } X \setminus E.$$

The conclusion ensues. $\qquad\square$

8.9 The Riemann versus the Lebesgue integral

In this section we compare the Lebesgue integral with the Riemann integral, in the case of the Euclidean space whereon both can be defined. The main conclusion will be that they coincide for all usual "good" functions, in particular all the Riemann integrable ones. The rationale of working with the Lebesgue integration theory is that *the Lebesgue integral works much more generally, it is easier to use and enjoys better properties overall.* Accordingly, we show that the Lebesgue integral is compatible with the Riemann integral, in the sense that

$$\int_{\mathbb{R}^d} f(x)\,dx = \int_{\mathbb{R}^d} f(x)\,d\mathcal{L}^d(x)$$

whenever $f : \mathbb{R}^d \longrightarrow \mathbb{R}$ is Riemann integrable. For simplicity we shall prove this only for $d = 1$ and on a compact interval. The general case can be handled with the same method, the only difference being the more complicated notation required for higher dimensional partitions.

Remark 8.43 (Reminder of the Riemann integral). Let us recall the definition of the Riemann integral. Let $f : [a, b] \longrightarrow \mathbb{R}$ be a bounded function, i.e. there is $M > 0$ such that $|f(x)| \leq M$ for all $x \in [a, b]$. For a partition[10] of $[a, b]$, say

$$D := \{x_0, ..., x_n\}$$

where $a = x_0 < x_1 < ... < x_n = b$, for some $n \in \mathbb{N}$, we define the *lower and upper Riemann integrals* of f as

$$\underline{\int_a^b} f(x)\,\mathrm{d}x := \sup_{D \text{ partition}} \left\{ \sum_{k=1}^n \left(\inf_{[x_{k-1}, x_k]} f \right) (x_k - x_{k-1}) \right\},$$

$$\overline{\int_a^b} f(x)\,\mathrm{d}x := \inf_{D \text{ partition}} \left\{ \sum_{k=1}^n \left(\sup_{[x_{k-1}, x_k]} f \right) (x_k - x_{k-1}) \right\}.$$

In general we have

$$-M(b-a) \leq \underline{\int_a^b} f(x)\,\mathrm{d}x \leq \overline{\int_a^b} f(x)\,\mathrm{d}x \leq M(b-a).$$

We say that the function f is Riemann-integrable if the lower and upper integrals above coincide. Their common value is called the *Riemann integral* of f:

$$\int_a^b f(x)\,\mathrm{d}x := \underline{\int_a^b} f(x)\,\mathrm{d}x = \overline{\int_a^b} f(x)\,\mathrm{d}x.$$

Theorem 8.44 (Lebesgue vs. Riemann integral on $\mathbb{R}$). *Consider the measure space $(\mathbb{R}, \mathcal{L}(\mathbb{R}), \mathcal{L})$ and let $f : [a, b] \longrightarrow \mathbb{R}$ be a bounded function. If f is Riemann-integrable, then f is Lebesgue-measurable and integrable. Moreover, the Riemann integral and the Lebesgue integral coincide:*

$$\int_a^b f(x)\,\mathrm{d}x = \int_{[a,b]} f(x)\,\mathrm{d}\mathcal{L}^1(x). \tag{$\natural$}$$

Remark 8.45. Further, it can be shown that the function f is Riemann-integrable if and only if f is $\mathcal{L}^1$-a.e. continuous, but we shall not prove that.

Proof. We begin by noting that there is no loss of generality to assume $f \geq 0$. Indeed, if this is not the case, we may replace f by the shifted function $\tilde{f} := f + M$ which is also bounded and in addition non-negative, since

$$0 \leq \tilde{f} \leq 2M, \quad \text{on } [a, b],$$

---

[10] In the context of Riemann integration, the term "partition" is used in a sense somewhat different from the meaning it has in other areas of Mathematics (e.g. in topology).

whilst $\tilde{f}$ differs from f only by an additive constant. Given a partition D, consider the simple functions

$$
\begin{cases}
\phi_D := f(a)\chi_{\{a\}} + \displaystyle\sum_{k=1}^{n} \left(\inf_{[x_{k-1},x_k]} f \right) \chi_{(x_{k-1},x_k]}, \\
\psi_D := f(a)\chi_{\{a\}} + \displaystyle\sum_{k=1}^{n} \left(\sup_{[x_{k-1},x_k]} f \right) \chi_{(x_{k-1},x_k]}.
\end{cases}
$$

Then we have

$$
0 \leq \phi_D \leq f \leq \psi_D \leq 2M.
$$

Let $(D^j)_{j=1}^{\infty}$ be a sequence of partitions determined by the points

$$
a = x_0^j < x_1^j < \ldots < x_{n_j-1}^j < x_{n_j}^j = b,
$$

that realises the supremum in the definition of $\underline{\int_a^b} f(x)\,dx$, that is

$$
\sum_{k=1}^{n_j} \left(\inf_{[x_{k-1}^j, x_k^j]} f \right) (x_k^j - x_{k-1}^j) \longrightarrow \underline{\int_a^b} f(x)\,dx,
$$

as $j \to \infty$. Further, this sequence of partitions can be chosen to be nondecreasing with respect to set inclusion, that is

$$
D^1 \subseteq D^2 \subseteq \cdots \subseteq D^j \subseteq D^{j+1} \subseteq \cdots,
$$

a condition which implies that the sequence $(\phi_{D^j})_{j=1}^{\infty}$ is nondecreasing. By the definition of the Lebesgue integral for the simple function ϕ_{D^j}, we have

$$
\int_{[a,b]} \phi_{D^j}(x)\,d\mathcal{L}^1(x) = \sum_{k=1}^{n_j} \left(\inf_{[x_{k-1}^j, x_k^j]} f \right) (x_k^j - x_{k-1}^j).
$$

By defining (see Figure 8.8)

$$
\phi := \lim_{j \to \infty} \phi_{D^j},
$$

the Monotone Convergence Theorem implies

$$
\int_{[a,b]} \phi_{D^j}(x)\,d\mathcal{L}^1(x) \longrightarrow \int_{[a,b]} \phi(x)\,d\mathcal{L}^1(x) = \underline{\int_a^b} f(x)\,dx,
$$

as $j \to \infty$. Similarly, let $(\psi_{D^j})_{j=1}^{\infty}$ be a nonincreasing sequence of simple functions realising the infimum in the definition $\overline{\int_a^b} f(x)\,dx$ and set (see again Figure 8.8)

$$
\psi := \lim_{j \to \infty} \psi_{D^j}.
$$

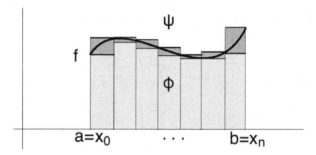

Figure 8.8: An illustration of the idea of the proof of Theorem 8.44.

The Monotone Convergence Theorem again implies

$$\int_{[a,b]} \psi_{D^j}(x)\, \mathrm{d}\mathcal{L}^1(x) \longrightarrow \int_{[a,b]} \psi(x)\, \mathrm{d}\mathcal{L}^1(x) = \overline{\int_a^b} f(x)\, \mathrm{d}x,$$

as $j \to \infty$. Since the lower and upper Riemann integrals of f coincide, the above imply that

$$\int_{[a,b]} \psi\, \mathrm{d}\mathcal{L}^1 = \int_{[a,b]} \phi\, \mathrm{d}\mathcal{L}^1,$$

from where we infer that

$$\int_{[a,b]} (\psi - \phi)\, \mathrm{d}\mathcal{L}^1 = 0.$$

Since $\phi \le f \le \psi$, by Proposition 8.21 it follows that f is Riemann-integrable if and only $\phi = \psi$, $\mathcal{L}^1$-a.e. on $[a,b]$. In this case, f is Lebesgue measurable as a consequence of Proposition 8.35, and also the desired identity ($\natural$) holds. $\square$

8.10   Product measures and the Fubini-Tonelli Theorem

We continue by considering the problem of how to define a "natural" product of measures given two (or finitely many) measure spaces. Namely, in the simplest case of two spaces, the question is, given $(X, \mathfrak{M}, \mu)$ and $(Y, \mathfrak{N}, \nu)$, is there a natural way to define a "product measure $\mu \times \nu$" on the Cartesian product $X \times Y$? Actually yes, as the next definitions assert. However, one needs to pass through outer measures and Carathéodory's Theorem to achieve this efficiently.

Definition 8.46. Let $(X, \mathfrak{M}, \mu)$ and $(Y, \mathfrak{N}, \nu)$ be measure spaces. The **product outer measure** $(\mu \times \nu)^*$ is defined on the Cartesian product $X \times Y$ in the following way (see Figure 8.9):

$$(\mu \times \nu)^* : \quad \mathcal{P}(X \times Y) \longrightarrow [0, \infty],$$

$$(\mu \times \nu)^*(A) := \inf \left\{ \sum_{j=1}^{\infty} \mu(E_j)\,\nu(F_j) \ \middle| \ A \subseteq \bigcup_{j=1}^{\infty} (E_j \times F_j),\ E_j \in \mathfrak{M},\ F_j \in \mathfrak{N} \right\}.$$

Remark 8.47. It is quite evident from its definition that $(\mu \times \nu)^*$ satisfies the properties of outer measures (Definition 7.12). By Carathéodory's Theorem 7.18, there exists a unique σ-algebra on $X \times Y$, symbolised by

$$\mathfrak{M} \times \mathfrak{N}$$

on which the restriction $(\mu \times \nu)^*\big|_{\mathfrak{M} \times \mathfrak{N}}$ is a complete measure. Further, the σ-algebra $\mathfrak{M} \times \mathfrak{N}$ consists of those $A \subseteq X \times Y$ for which $(\mu \times \nu)^*$ satisfies

$$(\mu \times \nu)^*(E) \geq (\mu \times \nu)^*(E \cap A) + (\mu \times \nu)^*(E \setminus A),$$

for all $E \subseteq X \times Y$.

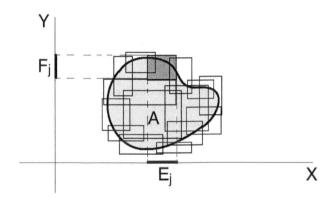

Figure 8.9: An illustration of the definition of the product outer measure on $X \times Y$ for $X = Y = \mathbb{R}$. The idea is to approximate "from the outside" the set $A \subseteq X \times Y$ by "generalised rectangles" of the form $E_j \times F_j$.

In view of the above observations, the product measure on $X \times Y$ can be defined as follows:

Definition 8.48. Let $(X, \mathfrak{M}, \mu)$ and $(Y, \mathfrak{N}, \nu)$ be measure spaces. The **product measure** $\mu \times \nu$ is defined as the restriction of the product outer measure $(\mu \times \nu)^*$ on the σ-algebra $\mathfrak{M} \times \mathfrak{N}$:

$$\mu \times \nu := (\mu \times \nu)^*\big|_{\mathfrak{M} \times \mathfrak{N}} : \quad \mathfrak{M} \times \mathfrak{N} \longrightarrow [0, \infty].$$

For reasons that will become apparent soon, we shall refer to the σ-algebra $\mathfrak{M} \times \mathfrak{N}$ as the **completion of the product σ-algebra**.

The terminology we have just used implicitly makes clear that $\mathfrak{M} \times \mathfrak{N}$ is not exactly what is usually called *"the product σ-algebra"*. Note in particular that the σ-algebra $\mathfrak{M} \times \mathfrak{N}$ tells nothing about the measurability or not of the types of sets one expects to be measurable on the product space, namely the Cartesian products $E \times F$ of measurable sets $E \in \mathfrak{M}$, $F \in \mathfrak{N}$. The product σ-algebra, symbolised as $\mathfrak{M} \otimes \mathfrak{N}$, is a slightly smaller σ-algebra on $X \times Y$ which by definition contains all these "rectangles" and is defined as follows (at this point it is recommended to review Definition 7.29):

Definition 8.49. Let $(X, \mathfrak{M}, \mu)$ and $(Y, \mathfrak{N}, \nu)$ be measure spaces. The **product σ-algebra on $X \times Y$** is defined as:

$$\mathfrak{M} \otimes \mathfrak{N} := \mathcal{G}\Big(\big\{E \times F \mid E \in \mathfrak{M}, \ F \in \mathfrak{N}\big\}\Big).$$

Namely, $\mathfrak{M} \otimes \mathfrak{N}$ is the smallest σ-algebra (intersection of all σ-algebras) containing all the Cartesian products of measurable sets.

Below we present a fundamental result which in particular shows that $\mathfrak{M} \otimes \mathfrak{N} \subseteq \mathfrak{M} \times \mathfrak{N}$ and **this inclusion of σ-algebras is always strict**, i.e.

$$\mathfrak{M} \otimes \mathfrak{N} \subsetneq \mathfrak{M} \times \mathfrak{N}.$$

This result, known as the **Fubini-Tonelli Theorem**, connects the "successive integrals" over X, Y separately with the "double integral" on the product space $X \times Y$, showing also that the **product outer measure has a nice regularity property, even if the particular measures composing it do not have it**. This property says that *every* subset of $X \times Y$ (possibly non-measurable) is contained in a measurable set of the same outer measure.[11]

Theorem 8.50 (Fubini-Tonelli Theorem). *Let $(X, \mathfrak{M}, \mu)$ and $(Y, \mathfrak{N}, \nu)$ be measure spaces and consider the product measure space*

$$\Big(X \times Y, \ \mathfrak{M} \times \mathfrak{N}, \ \mu \times \nu\Big).$$

(i) *If $A \in \mathfrak{M}$ and $B \in \mathfrak{N}$, then $A \times B \in \mathfrak{M} \times \mathfrak{N}$ and also*

$$(\mu \times \nu)(A \times B) = \mu(A)\,\nu(B).$$

Moreover, we have the inclusion $\mathfrak{M} \otimes \mathfrak{N} \subseteq \mathfrak{M} \times \mathfrak{N}$.

(ii) *The product outer measure $(\mu \times \nu)^*$ is **regular**, namely for any $S \subseteq X \times Y$ there exists a measurable set $A \in \mathfrak{M} \times \mathfrak{N}$ with $A \supseteq S$ such that*

$$(\mu \times \nu)^*(S) = (\mu \times \nu)^*(A) = (\mu \times \nu)(A).$$

(iii) *Suppose further that $(X, \mathfrak{M}, \mu)$ and $(Y, \mathfrak{N}, \nu)$ are σ-finite measure spaces. If $S \subseteq X \times Y$, then, for μ-a.e. $x \in X$, the slice (or cross-section)*

$$S_x := \big\{y \in Y \ : \ (x, y) \in S\big\}$$

---

[11]One might be tempted to describe the regularity property as that "we add a nullset to the given (possibly non-measurable) set and make it measurable". However, this is not the correct way to put it, since, the outer measure is not additive on non-measurable sets!

is ν-measurable (see Figure 8.10). Similarly, for ν-a.e. $y \in Y$, the slice

$$S_y := \{x \in X : (x, y) \in S\}$$

is μ-measurable. Further, the function $y \mapsto \mu(S_y)$ is in $L^+(Y, \mathfrak{N})$, the function $x \mapsto \nu(S_x)$ is in $L^+(X, \mathfrak{M})$ and also

$$(\mu \times \nu)(S) = \int_Y \mu(S_y) \, d\nu(y) = \int_X \mu(S_x) \, d\mu(x).$$

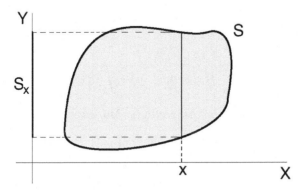

Figure 8.10: An illustration of the "slice" or "cross section" S_x.

(iv) *Under the same assumptions as in* (iii)*, if $f \in L^1(X \times Y, \mu \times \nu)$, then the partial functions*

$$x \mapsto \int_Y f(x, y) \, d\nu(y), \quad y \mapsto \int_X f(x, y) \, d\mu(x),$$

are in $L^1(X, \mu)$ and in $L^1(Y, \nu)$ respectively. Moreover,

$$\int_X \left(\int_Y f(x, y) \, d\nu(y) \right) d\mu(x) = \int_Y \left(\int_X f(x, y) \, d\mu(x) \right) d\nu(y) =$$

$$= \int_{X \times Y} f(x, y) \, d(\mu \times \nu)(x, y).$$

Before delving into the proof, it is important to understand the difference between the product σ-algebra and its completion, at least in the more familiar Euclidean case. To this end, let us isolate the relation of the Lebesgue measure on $\mathbb{R}^{n+m}$ with the Lebesgue measures on $\mathbb{R}^n, \mathbb{R}^m$, for $n, m \in \mathbb{N}$. In a nutshell,

$$\left(\mathbb{R}^{n+m}, \, \mathcal{L}(\mathbb{R}^{n+m}), \, \mathcal{L}^{n+m} \right) = \left(\mathbb{R}^n \times \mathbb{R}^m, \, \mathcal{L}(\mathbb{R}^n) \times \mathcal{L}(\mathbb{R}^m), \, \mathcal{L}^n \times \mathcal{L}^m \right).$$

However, the above equality requires some **attention. The crucial point is that the Lebesgue σ-algebra $\mathcal{L}(\mathbb{R}^{n+m})$ is not equal to the product of Lebesgue σ-algebras $\mathcal{L}(\mathbb{R}^n) \otimes \mathcal{L}(\mathbb{R}^m)$, but instead to its completion. In** fact,

$$\mathcal{L}(\mathbb{R}^{n+m}) \supsetneq \mathcal{L}(\mathbb{R}^n) \otimes \mathcal{L}(\mathbb{R}^m).$$

This can be seen by considering any non-measurable set $A \subseteq \mathbb{R}^n$ and any point $y \in \mathbb{R}^m$. Then, $A \times \{y\}$ is a measurable nullset on $\mathbb{R}^{n+m}$ (because $\mathcal{L}^m(\{y\}) = 0$), although it is not included in $\mathcal{L}(\mathbb{R}^n) \otimes \mathcal{L}(\mathbb{R}^m)$. On the other hand, as we have already seen, the respective Borel σ-algebras coincide:

$$\mathcal{B}(\mathbb{R}^{n+m}) = \mathcal{B}(\mathbb{R}^n) \otimes \mathcal{B}(\mathbb{R}^m).$$

We now isolate a particular corollary of the Fubini-Tonelli theorem for the case of the Lebesgue measure.

Corollary 8.51. *If f is a non-negative measurable function on $\mathbb{R}^n \times \mathbb{R}^m$, then the partial functions*

$$x \mapsto \int_{\mathbb{R}^m} f(x,y) \, dy, \quad y \mapsto \int_{\mathbb{R}^n} f(x,y) \, dx,$$

are measurable on $\mathbb{R}^n$ and $\mathbb{R}^m$ respectively. Moreover, we have the identities

$$\int_{\mathbb{R}^n \times \mathbb{R}^m} f(x,y) \, dx \, dy = \int_{\mathbb{R}^n} \left(\int_{\mathbb{R}^m} f(x,y) \, dy \right) dx =$$
$$= \int_{\mathbb{R}^m} \left(\int_{\mathbb{R}^n} f(x,y) \, dx \right) dy.$$

Proof of Theorem 8.50. [12] *Step 1.* Let $\mathcal{F}$ denote the family of all sets $S \subseteq X \times Y$ for which, for any $y \in Y$, the partial function

$$X \ni x \mapsto \chi_S(x,y) \in \mathbb{R}$$

is in $L^+(X,\mu)$ and the function

$$Y \ni y \mapsto \int_X \chi_S(x,y) \, d\mu(x) \in \mathbb{R}$$

is in $L^+(Y,\nu)$. For any $S \in \mathcal{F}$, we set

$$\rho(S) := \int_Y \left(\int_X \chi_S(x,y) \, d\mu(x) \right) d\nu(y).$$

Step 2. We define

$$\mathcal{A}_0 := \Big\{ E \times F \ : \ E \in \mathfrak{M}, \ F \in \mathfrak{N} \Big\},$$
$$\mathcal{A}_1 := \Big\{ \textstyle\bigcup_{j=1}^{\infty} S_j \ : \ S_j \in \mathcal{A}_0, \ j \in \mathbb{N} \Big\},$$
$$\mathcal{A}_2 := \Big\{ \textstyle\bigcap_{j=1}^{\infty} S_j \ : \ S_j \in \mathcal{A}_1, \ j \in \mathbb{N} \Big\}.$$

---

[12]The proof of the Fubini-Tonelli Theorem is relatively technical and may perhaps be omitted on the first reading of this book.

Note that $\mathcal{A}_0 \subseteq \mathcal{F}$ and also

$$\rho(E \times F) = \mu(E)\nu(F), \quad E \times F \in \mathcal{A}_0.$$

Further, if $E \times F, E' \times F' \in \mathcal{A}_0$, we have

$$(E \times F) \bigcap (E' \times F') = (E \cap E') \times (F \cap F') \in \mathcal{A}_0$$

and the set difference

$$(E \times F) \setminus (E' \times F') = \Big((E \setminus E') \times F \Big) \bigcup \Big((E \cap E') \times (F \setminus F') \Big)$$

is a disjoint union of members of $\mathcal{A}_0$. It follows that each set in $\mathcal{A}_1$ is a countable disjoint union of sets from $\mathcal{A}_0$. Therefore, we obtain that $\mathcal{A}_1 \subseteq \mathcal{F}$.

Step 3. Now we show that for any $S \subseteq X \times Y$ we have

$$(\mu \times \nu)^*(S) = \inf \Big\{ \rho(R) : R \supseteq S, \ R \in \mathcal{A}_1 \Big\}.$$

Indeed, to see this first note that if $R \supseteq S$, $R = \cup_{i=1}^{\infty} E_i \times F_i$, then

$$\rho(R) \leq \sum_{i=1}^{\infty} \rho(E_i \times F_i) = \sum_{i=1}^{\infty} \mu(E_i)\nu(F_i)$$

and hence

$$\inf \Big\{ \rho(R) : R \supseteq S, \ R \in \mathcal{A}_1 \Big\} \leq (\mu \times \nu)^*(S).$$

On the other hand, there exists a disjoint sequence of sets $(E_i' \times F_i')_{i=1}^{\infty} \subseteq \mathcal{A}_0$ such that

$$R = \bigcup_{i=1}^{\infty} E_i' \times F_i'.$$

Therefore,

$$\rho(R) = \sum_{i=1}^{\infty} \mu(E_i')\nu(F_i') \geq (\mu \times \nu)^*(S),$$

as desired.

Step 4. By Step 3 we have that

$$(\mu \times \nu)^*(E \times F) = \mu(E)\nu(F), \quad E \times F \in \mathcal{A}_0.$$

Indeed, it suffices to note that for all $R \in \mathcal{A}_1$ for which $R \supseteq E \times F$, we have

$$(\mu \times \nu)^*(E \times F) \leq \mu(E)\nu(F) = \rho(E \times F) \leq \rho(R).$$

Step 5. Now we show that $E \times F \in \mathfrak{M} \times \mathfrak{N}$ whenever $E \times F \in \mathcal{A}_0$. To this end, fix $T \subseteq X \times Y$ and $R \supseteq T$ with $R \in \mathcal{A}_1$. Then, $R \setminus (E \times F)$ and $R \cap (E \times F)$ are disjoint elements of $\mathcal{A}_1$. Hence,

$$(\mu \times \nu)^*(T \setminus (E \times F)) + (\mu \times \nu)^*(T \cap (E \times F))$$
$$\leq \rho(R \setminus (E \times F)) + \rho(R \cap (E \times F))$$
$$= \rho(R)$$

and by utilising Step 3, this leads to

$$(\mu \times \nu)^*(T \setminus (E \times F)) + (\mu \times \nu)^*(T \cap (E \times F)) \leq (\mu \times \nu)^*(T),$$

for any $T \subseteq X \times Y$. Hence, $E \times F \in \mathfrak{M} \times \mathfrak{N}$. This establishes (i).

Step 6. We now claim that for any $S \subseteq X \times Y$ there exists $A \in \mathcal{A}_2$ with $A \supseteq S$ such that

$$\rho(A) = (\mu \times \nu)^*(S).$$

Indeed, to see this first consider the case that $(\mu \times \nu)^*(S) = \infty$. Then, we may set $A := X \times Y$. If $(\mu \times \nu)^*(S) < \infty$, for each $j \in \mathbb{N}$ we invoke Step 3 to find a set $A_j \in \mathcal{A}_1$ such that $S \subseteq A_j$ and

$$\rho(R_j) < (\mu \times \nu)^*(S) + \frac{1}{j}.$$

We define

$$A := \bigcap_{j=1}^{\infty} A_j \in \mathcal{A}_2.$$

Then, $A \in \mathcal{F}$ and by the Dominated Convergence Theorem, we have

$$(\mu \times \nu)^*(S) \leq \rho(A) = \lim_{k \to \infty} \rho \left(\bigcap_{j=1}^{\infty} A_j \right) \leq (\mu \times \nu)^*(S).$$

Step 7. From item (i) we have the inclusion $\mathcal{A}_2 \subseteq \mathfrak{M} \times \mathfrak{N}$. Hence, item (ii) follows from Step 6 above.

Step 8. Note that for any $S \subseteq X \times Y$ for which $(\mu \times \nu)^*(S) = 0$, there exists an $A \in \mathcal{A}_2$ with $A \supseteq S$ and $\rho(A) = 0$. Hence, $S \in \mathcal{F}$ and $\rho(S) = 0$.

Suppose now that $S \in \mathfrak{M} \times \mathfrak{N}$ and $(\mu \times \nu)(S) < \infty$. Then, there exists $A \in \mathcal{A}_2$ such that $A \supseteq S$ and

$$(\mu \times \nu)(A \setminus S) = 0.$$

Hence, $\rho(A \setminus S) = 0$. This implies that for ν-a.e. $y \in Y$ we have

$$\mu\Big(\{x \in X : (x, y) \in S\}\Big) = \mu\Big(\{x \in X : (x, y) \in A\}\Big)$$

and also

$$(\mu \times \nu)(S) = \rho(A) = \int_Y \mu\Big(\{x \in X : (x, y) \in S\}\Big) \, d\nu(y).$$

Item (iii) now follows from the above when $(\mu \times \nu)(S) < \infty$. If $(\mu \times \nu)(S) = \infty$, by the σ-finiteness assumption we may decompose S to a countable union of sets of finite measure.

Step 9. Item (iv) reduces to (iii) in the particular case of $f = \chi_S$. In the general case, the conclusion follows from the approximation result of Theorem 7.45 and the Dominated Convergence Theorem. $\qquad\qquad\qquad\qquad\qquad\square$

8.11 The linear change of variables formula

In this brief section we prove a special case of a well-known result by utilising the machinery we developed earlier. This is the familiar **change of variables formula**, generalised to the realm of the Lebesgue integration theory. We state and prove it only in the case of a linear transformation of the variables, which is the only case that will be utilised[13] in Chapter 9.

Theorem 8.52 (Linear change of variables). *Consider the measure space*

$$\left(\mathbb{R}^d,\ \mathcal{L}(\mathbb{R}^d),\ \mathcal{L}^d \right)$$

and suppose that $f \in L^1(\mathbb{R}^d, \mathcal{L}^d)$. If $T : \mathbb{R}^d \longrightarrow \mathbb{R}^d$ is an invertible linear mapping[14], then we have

$$\int_{\mathbb{R}^d} f(y)\, d\mathcal{L}^d(y) = \left| \det(T) \right| \int_{\mathbb{R}^d} (f \circ T)(x)\, d\mathcal{L}^d(x).$$

Proof. Let $T : \mathbb{R}^d \longrightarrow \mathbb{R}^d$ be the invertible linear mapping of the statement. We begin by noting that the Lebesgue measure satisfies for any measurable set $E \in \mathcal{L}(\mathbb{R}^d)$ the following identity (see Figure 8.11)

$$\mathcal{L}^d(E) = \left| \det(T) \right| \mathcal{L}^d\left(T^{-1} E \right). \qquad (\clubsuit)$$

We first use ($\clubsuit$), and then justify its validity at the end of the proof.

Given $f \in L^1(\mathbb{R}^d, \mathcal{L}^d)$, by Theorem 7.45, there exists a sequence of simple functions $(f_m)_1^\infty$ given by

$$f_m = \sum_{i=1}^{N(m)} a_i^m \chi_{A_i^m}$$

approximating f pointwise on $\mathbb{R}^d$ as $m \to \infty$, and also $|f_m| \le |f|$ on $\mathbb{R}^d$. By the

---

[13] For the statement and the proof in the general case of a C^1 or even bi-Lipschitz transformation, the interested reader may consult for instance the more advanced books [14, 16].

[14] We recall that the linear map T is defined by a homonymous matrix $T \in \mathbb{R}^{d \times d}$ via $x \mapsto Tx$, whilst its invertibility means that the determinant is non-vanishing: $\det(T) \ne 0$.

Dominated Convergence Theorem, we also have that the integrals converge as well:

$$\lim_{m \to \infty} \int_{\mathbb{R}^d} f_m(y) \, d\mathcal{L}^d(y) = \int_{\mathbb{R}^d} f(y) \, d\mathcal{L}^d(y).$$

Then, we have

$$\int_{\mathbb{R}^d} f(y) \, d\mathcal{L}^d(y) = \lim_{m \to \infty} \int_{\mathbb{R}^d} \left(\sum_{i=1}^{N(m)} a_i^m \chi_{A_i^m}(y) \right) d\mathcal{L}^d(y)$$

$$= \lim_{m \to \infty} \sum_{i=1}^{N(m)} a_i^m \int_{\mathbb{R}^d} \chi_{A_i^m}(y) \, d\mathcal{L}^d(y),$$

which yields

$$\int_{\mathbb{R}^d} f(y) \, d\mathcal{L}^d(y) = \lim_{m \to \infty} \sum_{i=1}^{N(m)} a_i^m \mathcal{L}^d(A_i^m).$$

By ($\clubsuit$) and the fact that $\chi_E(Tx) = \chi_{T^{-1}E}(x)$ for any $E \in \mathcal{L}(\mathbb{R}^d)$, the above equality gives

$$\int_{\mathbb{R}^d} f(x) \, d\mathcal{L}^d(x) = \lim_{m \to \infty} \sum_{i=1}^{N(m)} a_i^m \mathcal{L}^d(A_i^m)$$

$$= |\det(T)| \lim_{m \to \infty} \sum_{i=1}^{N(m)} a_i^m \mathcal{L}^d(T^{-1} A_i^m)$$

which results in

$$\int_{\mathbb{R}^d} f(y) \, d\mathcal{L}^d(y) = |\det(T)| \lim_{m \to \infty} \int_{\mathbb{R}^d} \left(\sum_{i=1}^{N(m)} a_i^m \chi_{T^{-1} A_i^m}(x) \right) d\mathcal{L}^d(x)$$

$$= |\det(T)| \lim_{m \to \infty} \int_{\mathbb{R}^d} \left(\sum_{i=1}^{N(m)} a_i^m \chi_{A_i^m}(Tx) \right) d\mathcal{L}^d(x)$$

$$= |\det(T)| \lim_{m \to \infty} \int_{\mathbb{R}^d} f_m(Tx) \, d\mathcal{L}^d(x).$$

By quoting one more time the Dominated Convergence Theorem, the above equality allows us to infer that

$$\int_{\mathbb{R}^d} f(y) \, d\mathcal{L}^d(y) = |\det(T)| \int_{\mathbb{R}^d} \left(\lim_{m \to \infty} f_m(Tx) \right) d\mathcal{L}^d(x)$$

$$= |\det(T)| \int_{\mathbb{R}^d} (f \circ T)(x) \, d\mathcal{L}^d(x).$$

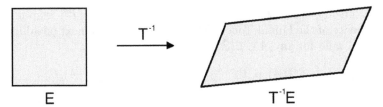

Figure 8.11: An illustration of the deformation of a box E of $\mathbb{R}^2$ under (the inverse mapping of) a linear invertible mapping $T : \mathbb{R}^2 \longrightarrow \mathbb{R}^2$.

Finally, it remains to justify the validity of (♣) for Lebesgue measurable sets. The essential point towards it is a fact from Linear Algebra regarding invertible linear maps $T : \mathbb{R}^d \longrightarrow \mathbb{R}^d$ for which $\det(T) \neq 0$ (identified in the standard way with their matrix with respect to the standard orthonormal basis of $\mathbb{R}^d$).

Step 1. It can be shown that every such T can be factorised as a composition of linear invertible maps

$$T = T^N \circ \cdots \circ T^1 \quad \text{for some } N \in \mathbb{N},$$

where each $T^i : \mathbb{R}^d \longrightarrow \mathbb{R}^d$ satisfies $\det(T^i) \neq 0$ and has either of the three forms

(i) $T'(x_1, \ldots, x_j, \ldots, x_d) := (x_1, \ldots, tx_j, \ldots, x_d), \quad t \neq 0,$

(ii) $T''(x_1, \ldots, x_j, \ldots, x_d) := (x_1, \ldots, x_j + tx_k, \ldots, x_d), \quad k \neq j, \, t \neq 0,$

(iii) $T'''(x_1, \ldots, x_j, \ldots, x_k, \ldots, x_d) := (x_1, \ldots, x_k, \ldots, x_j, \ldots, x_d), \, k \neq j.$

The validity of this factorisation is merely a restatement of the fact that every invertible matrix can be row-reduced to the identity matrix of $\mathbb{R}^d$. Further, by recalling that the matrix determinant of the product of two matrices equals the product of the determinants, then if $T = T^N \circ \cdots \circ T^1$, we have

$$\det(T) = \det(T^N) \cdots \det(T^1).$$

Step 2. It suffices to establish that the desired identity (♣) holds true merely for linear maps of the either of the form T', T'', T'''. To show this, we shall now show that if (♣) holds for two invertible linear maps $K, L : \mathbb{R}^d \longrightarrow \mathbb{R}^d$ (and any set $E \in \mathcal{L}(\mathbb{R}^d)$), then it holds for their composition. The claimed conclusion then follows by the factorisation of *Step 1.* Hence, if we suppose

$$\mathcal{L}^d(KE) = \left|\det(K)\right| \mathcal{L}^d(E), \quad \mathcal{L}^d(LE) = \left|\det(L)\right| \mathcal{L}^d(E),$$

then, we have

$$\begin{aligned}
\mathcal{L}^d(LKE) &= \mathcal{L}^d(L(KE)) \\
&= \left|\det(L)\right| \mathcal{L}^d(KE) \\
&= \left|\det(L)\right|\left|\det(K)\right| \mathcal{L}^d(E) \\
&= \left|\det(L)\det(K)\right| \mathcal{L}^d(E) \\
&= \left|\det(LK)\right| \mathcal{L}^d(E).
\end{aligned}$$

Step 3. The fact that ($\clubsuit$) holds true for the maps T', T'', T''' is an immediate consequence of the Fubini-Tonelli Theorem 8.50 and the next one-dimensional formulas, valid for any $A \in \mathcal{L}(\mathbb{R})$, $t \in \mathbb{R} \setminus \{0\}$ and $a \in \mathbb{R}$:

$$\mathcal{L}^1(tA) = |t| \mathcal{L}^1(A), \qquad \mathcal{L}^1(A + a) = \mathcal{L}^1(A).$$

The latter formulas have been shown in Exercise 6.7 of Chapter 6. Let us give the details in the case of operator of the form T', and in particular

$$T'(x_1, \ldots, x_j, \ldots, x_d) = (x_1, \ldots, x_{d-1}, t x_d),$$

with $t > 0$, the other cases being similar. Given $x = (x_1, \ldots, x_{d-1}, x_d) \in \mathbb{R}^d$, the slice subsets of $\mathbb{R}$

$$(T'E)_{(x_1, \ldots, x_{d-1})} \quad \text{and} \quad E_{(x_1, \ldots, x_{d-1})},$$

are related by

$$
\begin{aligned}
(T'E)_{(x_1, \ldots, x_{d-1})} &= \left\{ x_d \in \mathbb{R} \ : \ (x_1, \ldots, x_{d-1}, x_d) \in T'E \right\} \\
&= \left\{ x_d \in \mathbb{R} \ : \ \left(x_1, \ldots, x_{d-1}, \frac{x_d}{t} \right) \in E \right\} \\
&= t \left\{ x_d \in \mathbb{R} \ : \ (x_1, \ldots, x_{d-1}, x_d) \in E \right\} \\
&= t\, E_{(x_1, \ldots, x_{d-1})}.
\end{aligned}
$$

Hence, by the Fubini-Tonelli Theorem and the above relation,

$$
\begin{aligned}
\mathcal{L}^d(T'E) &= \int_{\mathbb{R}^{d-1}} \mathcal{L}^1 \left((T'E)_{(x_1, \ldots, x_{d-1})} \right) dx_1 \cdots dx_{d-1} \\
&= \int_{\mathbb{R}^{d-1}} t\, \mathcal{L}^1 \left(E_{(x_1, \ldots, x_{d-1})} \right) dx_1 \cdots dx_{d-1} \\
&= |\det(T')| \int_{\mathbb{R}^{d-1}} \mathcal{L}^1 \left(E_{(x_1, \ldots, x_{d-1})} \right) dx_1 \cdots dx_{d-1} \\
&= |\det(T')| \, \mathcal{L}^d(E).
\end{aligned}
$$

In conclusion, the theorem has now been established.   □

8.12   Exercises

Exercise 8.1. For any real number $x \in \mathbb{R}$, let us symbolise the largest integer smaller than x as $\lfloor x \rfloor$. For example, $\lfloor 2.7 \rfloor = 2$. Find the integral $\int_{\mathbb{R}} \varphi \, d\mathcal{L}^1$ of

$$
\varphi(x) := \begin{cases} \lfloor x^2 \rfloor, & \text{if } x \in [0, 2], \\ 0, & \text{otherwise.} \end{cases}
$$

Exercise 8.2. Let $f : [0,1] \longrightarrow \mathbb{R}$ be defined as follows: let C be the Cantor set (Remark 6.24) and let f be zero outside C. Let $f(x) := k$ on each of the intervals of length 3^{-k} which are excluded in the k-th. step. Show that f is $\mathcal{L}^1$-measurable and compute

$$\int_{[0,1]} f \, d\mathcal{L}.$$

Exercise 8.3. Let $(X, \mathfrak{M}, \mu)$ be a measure space and $f, g : X \longrightarrow \mathbb{R}$ be measurable functions. If both $\int_X f^+ \, d\mu$ and $\int_X g^+ \, d\mu$ are finite or both $\int_X f^- \, d\mu$ and $\int_X g^- \, d\mu$ are finite, then

$$\int_X (f + g) \, d\mu = \int_X f \, d\mu + \int_X g \, d\mu.$$

[Hint: Prove and use the estimate $(f + g)^+ \le f^+ + g^+$ and the identity

$$(f + g)^+ + f^- + g^- = (f + g)^- + f^+ + g^+. \ ]$$

Exercise 8.4. Compute the integral (here the exponent "-1" denotes the reciprocal)

$$\lim_{n \to \infty} \int_{-1}^1 \left(1 + \frac{x^{2017}}{n}\right)^{-1} dx.$$

Exercise 8.5. Consider the measure space $(\mathbb{R}, \mathcal{L}(\mathbb{R}), \mathcal{L}^1)$. Examine for which $\alpha \in \mathbb{R}$ the function

$$f_\alpha(x) := \chi_{(-\infty,0)}(x) \, x^\alpha$$

is integrable on the domain A (i.e. $f \in L^1(A, \mathcal{L}^1)$), in the cases that:

1. $A = [0, R], R > 0.$

2. $A = [0, \infty).$

3. $A = [-R, R]$ with $R > 0$ and $\alpha \in \mathbb{Z}.$

Exercise 8.6. Consider the measure space $(\mathbb{R}, \mathcal{B}(\mathbb{R}), \mathcal{L}^1)$. Is the function $f = \chi_\mathbb{Q}$ Borel-measurable? If yes, what is the integral of f?

Exercise 8.7. Let $(X, \mathfrak{M}, \mu)$ be a measure space and $f : X \longrightarrow \mathbb{R}^d$ a $\mathfrak{M}$-measurable function.

1. Let f be μ-integrable and $c \in \mathbb{R}$. Show that

$$\int_X (cf) \, d\mu = c \int_X f \, d\mu.$$

2. Show that f is integrable if and only if $\int_X |f| \, d\mu < \infty.$

3. Show that an integrable function f is μ-a.e. finite.

Exercise 8.8. Investigate the convergence of the sequence $(A_n)_1^\infty$, where

$$A_n := \int_a^\infty \frac{n^2 x e^{-n^2 x^2}}{1 + x^2} \, dx,$$

in the cases that either $a > 0$ or $a = 0$ separately.

Exercise 8.9. Let $(X, \mathfrak{M}, \mu)$ be a measure space and $f, g \in L^+(X, \mathfrak{M})$ such that $f \leq g$ μ-a.e. on X. Prove directly from the definition of the integral that

$$\int_X f \, d\mu \leq \int_X g \, d\mu.$$

Exercise 8.10. By using the Dominated Convergence Theorem, compute the integral

$$\lim_{n \to \infty} \int_1^\infty \frac{\sqrt{x}}{1 + nx^6} \, dx.$$

Exercise 8.11. Does the limit

$$\lim_{n \to \infty} \int_{-1}^1 x \sqrt[n]{1 + x^2} e^x \, dx$$

exist? In the case it exists, what is its value?

Exercise 8.12. Let $(f_n)_1^\infty, f$ be Riemann-integrable functions $[a, b] \longrightarrow \mathbb{R}$. Assume that

$$\lim_{n \to \infty} \left(\sup_{x \in [a,b]} |f_n(x) - f(x)| \right) = 0.$$

Show that f is Riemann-integrable as well and

$$\lim_{n \to \infty} \int_a^b f_n(x) \, dx = \int_a^b f(x) \, dx.$$

Chapter 9

The class of Lebesgue functional spaces

9.1 From integration to functional spaces

This chapter is devoted to an introductory study of the so-called Lebesgue spaces, or, (as they are more commonly referred to) the "L^p spaces". The mathematical content of this part of the book can be considered as a natural continuation of our measure-theoretic investigations of the Chapters 6, 7 and 8, but now we are also interested in the study of the structure of these spaces as a whole.

The L^p spaces form natural classes of $\mathfrak{M}$-measurable functions on a given measure space $(X, \mathfrak{M}, \mu)$, where the exponent p ranges[1] in $[1, \infty]$.

- If $p = 1$, the space L^1 is just the space of μ-integrable functions defined in Chapter 8.

- If $1 < p < \infty$, the space L^p consist of functions whose p-th power $|\cdot|^p$ is μ-integrable.

- If $p = \infty$, the space L^∞ consists of "essentially" bounded $\mathfrak{M}$-measurable functions.

This family of L^p spaces has a tremendously rich structure and is of fundamental importance in both theory and applications of Analysis. As we shall see, they have the structure of Banach spaces and are fundamental throughout Analysis, particularly for the theory of Ordinary and of Partial Differential Equations, for Calculus of Variations, for Operator Theory, for Numerical Analysis and for Harmonic Analysis. One could perhaps regard this chapter as the study of concrete exemplary Banach spaces, defined via the Lebesgue integration theory.

In the first four sections, we shall consider the class of L^p spaces of vector

---

[1] In fact, one can define the L^p space for $0 \leq p < 1$ as well, but in this case the resulting functional space in very pathological to be of interest in this introductory book. In particular, for $p < 1$ it does not have the structure of a Banach space but only of a topological vector space, that is, of a vector space endowed with a topology with respect to which the algebraic operations of addition and of multiplication by scalars are continuous.

valued mappings from X to $\mathbb{F}^d$, where the set X carries a general measure space structure $(X, \mathfrak{M}, \mu)$, d is a given positive integer and (as always) $\mathbb{F}$ equals either $\mathbb{R}$ or $\mathbb{C}$. The archetypal example of measure space is given by

$$\left(\Omega, \ \mathcal{L}(\mathbb{R}^n)|_\Omega, \ \mathcal{L}^n \right) \tag{$\star$}$$

where $\Omega \subseteq \mathbb{R}^n$ is a measurable subset of the n-dimensional Euclidean space equipped with the Lebesgue σ-algebra restricted to Ω and the Lebesgue measure $\mathcal{L}^n$ restricted to Ω. It is quite remarkable that restricting ourselves to the prototypical case of $(\star)$ for scalar valued functions does not offer any further insights or simplifications in the proofs. Accordingly, most of the analysis can be done in the generality of a measure space $(X, \mathfrak{M}, \mu)$ and for $\mathbb{F}^d$-valued mappings, at no extra cost and without any additional difficulties.

For the last two sections we consider a more delicate topic of "Euclidean nature" regarding the issue of density of spaces of continuous and smooth functions, namely whether we can approximate L^p functions by continuous or even infinitely differentiable ones. To this end, we shall restrict our attention to the case of $(\star)$ and additionally assume that Ω is an open set in $\mathbb{R}^n$, in order to be able to talk about derivatives of functions defined on Ω. For convenience, when consider the measure space $(\star)$, *we shall abbreviate it to merely Ω, bearing in mind the underlying measure space structure.*

9.2   The L^p spaces: Definition and first properties

We now begin our study of the class of Lebesgue functional spaces, over a given fixed measure space

$$(X, \mathfrak{M}, \mu)$$

where X is a non-empty set, $\mathfrak{M} \subseteq \mathcal{P}(X)$ a σ-algebra of subsets of X and $\mu : \mathfrak{M} \longrightarrow [0, \infty]$ a measure on X. Let $d \in \mathbb{N}$ be a given integer. Throughout this chapter, the triplet $(X, \mathfrak{M}, \mu)$ and the integer d will always be fixed.

Definition 9.1.

(i) Let $p \in [1, \infty)$. The **space of p-integrable mappings**[2] $X \longrightarrow \mathbb{F}^d$ **with respect to** μ (or the L^p **space of mappings** $X \longrightarrow \mathbb{F}^d$) is defined as

$$L^p(X, \mu, \mathbb{F}^d) := \left\{ f : X \longrightarrow \mathbb{F}^d \ \middle| \ f \text{ is } \mathfrak{M}\text{-measurable and } \int_X |f|^p \, d\mu < \infty \right\}.$$

---

[2] In this chapter we shall consider $\mathbb{F}^d$-valued maps (as opposed to $\mathbb{R}^d$-valued maps in the previous chapter) since occasionally the $\mathbb{F}$-linearity for the resulting functional vector space will become pertinent. We also recall from Chapter 8 (Definition 8.28) that the symbol "$|\cdot|$" is a simplified notation of the Euclidean norm "$\|\cdot\|_2$" on $\mathbb{F}^d$.

We also define the function

$$\|\cdot\|_{L^p(X,\mu,\mathbb{F}^d)} \ : \ L^p(X,\mu,\mathbb{F}^d) \longrightarrow [0,\infty)$$

given by

$$\|f\|_{L^p(X,\mu,\mathbb{F}^d)} := \left(\int_X |f(x)|^p \, \mathrm{d}\mu(x) \right)^{1/p}.$$

(ii) Let $p = \infty$. The **space of μ-essentially bounded mappings** $X \longrightarrow \mathbb{F}^d$ (or the **L^∞ space of mappings** $X \longrightarrow \mathbb{F}^d$) is defined as

$$L^\infty(X,\mu,\mathbb{F}^d) := \left\{ f : X \longrightarrow \mathbb{F}^d \ \middle| \ \begin{array}{l} f \text{ is } \mathfrak{M}\text{-measurable and } \exists\, c > 0 : \\ |f(x)| \leq c \text{ for } \mu\text{-a.e. } x \in X \end{array} \right\}.$$

We also define the function

$$\|\cdot\|_{L^\infty(X,\mu,\mathbb{F}^d)} \ : \ L^\infty(X,\mu,\mathbb{F}^d) \longrightarrow [0,\infty)$$

given by

$$\|f\|_{L^\infty(X,\mu,\mathbb{F}^d)} := \inf \left\{ a > 0 : \mu(\{x \in X : |f(x)| > a\}) = 0 \right\}.$$

We recall that we are using the standard Calculus conventions

$$\sup \emptyset = -\infty, \quad \inf \emptyset = +\infty.$$

The σ-algebra $\mathfrak{M}$ of the measure space $(X, \mathfrak{M}, \mu)$ is not denoted explicitly in the symbolisation for $L^p(X, \mu, \mathbb{F}^d)$ because it is subsumed into the symbolisation of the measure which is defined on $\mathfrak{M}$.

Remark 9.2 (Normed space structure). It will turn out in a while that for any $p \in [1, \infty]$, the L^p space is vector space and, the notation "$\|\cdot\|$" suggests, the functional $\|\cdot\|_{L^p(X,\mu,\mathbb{F}^d)}$ is a norm on $L^p(X, \mu, \mathbb{F}^d)$. Therefore, we take the liberty to call it a "norm" already, *even though we have not proved this yet*. We postpone the verification of these facts for the next section, after we shall have established some necessary important inequalities.

In addition, the case $p = 2$ is of particular importance since, except for a norm, the space $L^2(X, \mu, \mathbb{F}^d)$ has in addition a kind "generalised dot product structure" which gives rise to the L^2-norm:

Definition 9.3. We define a bilinear function (i.e., a function of two arguments which is linear with respect to each of them)

$$\langle \cdot, \cdot \rangle_{L^2(X,\mu,\mathbb{F}^d)} \ : \ L^2(X,\mu,\mathbb{F}^d) \times L^2(X,\mu,\mathbb{F}^d) \longrightarrow \mathbb{F},$$

by setting

$$\langle f, g \rangle_{L^2(X,\mu,\mathbb{F}^d)} := \int_X f \cdot \overline{g} \, \mathrm{d}\mu,$$

for any two functions $f, g \in L^2(X, \mu, \mathbb{F}^d)$, where the overline symbolises complex conjugation (recall the definition of the dot product on $\mathbb{F}^d$, Definition 2.73). Then, we have that the norm of L^2 is connected to the above bilinear mapping in the following fashion:

$$\|f\|_{L^2(X,\mu,\mathbb{F}^d)} = \left(\langle f, f \rangle_{L^2(X,\mu,\mathbb{F}^d)}\right)^{1/2}.$$

Remark 9.4 ("Dot product" structure). Again, we temporarily postpone to the next section the verification of the fact that the bilinear form $\langle \cdot, \cdot \rangle_{L^2(X,\mu,\mathbb{F}^d)}$ is well defined in the sense that fg is L^1 when f, g are L^2 (we have not proved this yet). This important additional structure[3] which emerges only for $p = 2$ plays the role of a "generalised dot product" along the lines of the Euclidean space.

For the sake of clarity, let us now specialise the Definition 9.1 to the Euclidean case and the Lebesgue measure, that is for the measure space $(\star)$.

Remark 9.5 (The case of the Lebesgue measure on $\mathbb{R}^n$).
(i) Let $p \in [1, \infty)$. The L^p space of real Lebesgue measurable functions is

$$L^p(\Omega) \equiv L^p(\Omega, \mathbb{R}) = \left\{ f : \Omega \longrightarrow \mathbb{R} \ \middle| \ f \text{ is measurable and } \int_\Omega |f|^p \, d\mathcal{L}^n < \infty \right\}.$$

and[4]

$$\|f\|_{L^p(\Omega)} = \left(\int_\Omega |f|^p \, d\mathcal{L}^n \right)^{1/p}.$$

(ii) Let $p = \infty$. The L^∞ space of real Lebesgue measurable functions is

$$L^\infty(\Omega) \equiv L^\infty(\Omega, \mathbb{R}) = \left\{ f : \Omega \longrightarrow \mathbb{R} \ \middle| \ \begin{array}{l} f \text{ is measurable and } \exists\, c > 0 : \\ |f(x)| \leq c \text{ for a.e. } x \in \Omega \end{array} \right\}$$

and

$$\|f\|_{L^\infty(\Omega)} = \inf \left\{ a > 0 : \mathcal{L}^n(\{|f| > a\}) = 0 \right\}.$$

We now consider the extreme case of $p = \infty$ and give two alternative formulations of the expression defining the L^∞ norm. To this aim, we introduce some terminology.

---

[3]The class of normed vector spaces which admit an additional geometric "dot product" structure compatible with their norm, called *"inner product spaces"*, will be studied quite systematically in the next chapter. The space $L^2(X, \mu, \mathbb{F}^d)$ is the prototypical example of such a space. Their importance of this particular lies in that we can define angles of vectors and study geometric questions, like projections, orthogonality, etc.

[4]In this and subsequent chapters we shall use the standard symbolisation "$\|\cdot\|_{L^p(\Omega)}$" for the (soon to be) norm on the space $L^p(\Omega)$. Recall that when we first introduced in Chapter 2 a toy version of this norm in Definition 2.79 on the subspace of continuous functions, we used the simplistic notation "$\|\cdot\|_p$". That was because at the time we could not explain the symbol "L" which tacitly implies that the integrals should be interpreted in the extended sense of Lebesgue, not in the restricted Riemann sense. Also, explicit dependence on the domain "Ω" was not needed, unlike it would be the case from now on in our arguments.

Definition 9.6. For any $\mathfrak{M}$-measurable real function $f : X \longrightarrow \mathbb{R}$, we define its μ-**essential supremum over** X as

$$\mu\text{-ess}\sup_{X} f := \inf\left\{a > 0 : \mu(\{f > a\}) = 0\right\}.$$

The essential supremum is an (extended) number in $[-\infty, \infty]$.

The notion of the essential supremum is a measure-theoretic extension of the usual supremum. One may define the *essential infimum* as well in a symmetric fashion. In view of Definition 9.6, we may rewrite the L^∞-norm as

$$\|f\|_{L^\infty(X,\mu,\mathbb{F}^d)} = \mu\text{-ess}\sup_{X} |f|.$$

Lemma 9.7 (Expressions of the L^∞-norm). *For any $f \in L^\infty(X, \mu, \mathbb{F}^d)$, we have the equivalent formulations*

$$\|f\|_{L^\infty(X,\mu,\mathbb{F}^d)} = \inf\left\{a > 0 : |f(x)| \le a \text{ for } \mu\text{-a.e. } x \in X\right\},$$

$$\|f\|_{L^\infty(X,\mu,\mathbb{F}^d)} = \sup\left\{a > 0 : \mu(\{|f| > a\}) > 0\right\}.$$

In particular, we have that

$$\left|f(x)\right| \le \|f\|_{L^\infty(X,\mu,\mathbb{F}^d)}, \quad \text{for } \mu\text{-a.e. } x \in X.$$

Therefore, the above lemma justifies the fact that the space $L^\infty(X, \mu, \mathbb{F}^d)$ consists of essentially bounded functions, namely function which are bounded if we ignore a nullset, that is a set of zero μ-measure (see Figure 9.1).

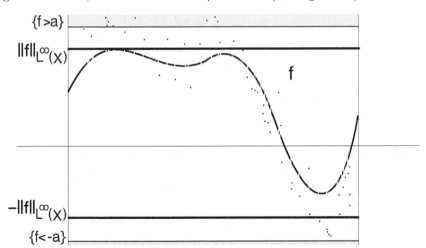

Figure 9.1: A function $f \in L^\infty(X, \mathcal{L}^1, \mathbb{R})$ on a Lebesgue measurable set $X \subseteq \mathbb{R}$ is bounded if we ignore a set $E \subseteq X$ of zero Lebesgue measure. Then, the L^∞ norm of f is just the supremum of f on $X \setminus E$.

Proof. Let us fix a function $f \in L^\infty(X, \mu, \mathbb{F}^d)$. In order to show that both formulas of the statement (with the "inf" and the "sup") coincide with the expression in the definition of the L^∞-norm, let us denote the right hand sides with "$i(f)$" and "$s(f)$" respectively. Hence, we need to show

$$\|f\|_{L^\infty(X, \mu, \mathbb{F}^d)} = i(f) , \quad \|f\|_{L^\infty(X, \mu, \mathbb{F}^d)} = s(f).$$

The first equality is immediate: indeed, if for some $a > 0$ we have $|f| \le a$ μ-a.e. on X, this is equivalent to saying that $|f|$ exceeds the value a at most on a set of zero μ-measure, that is $\mu(\{|f| > a\}) = 0$.

Now we show the second equality. By the definition of the essential supremum, if for some $a > 0$ we have that $\mu(\{|f| > a\}) > 0$, then necessarily $a < \|f\|_{L^\infty(X, \mu, \mathbb{F}^d)}$ since $\|f\|_{L^\infty(X, \mu, \mathbb{F}^d)}$ is the smallest possible value for which the sup-level set of f is a nullset. Hence, by taking supremum with respect to such a's, we have that $s(f) \le \|f\|_{L^\infty(X, \mu, \mathbb{F}^d)}$. If for the sake of contradiction we suppose that this inequality is strict, then there exists $b > 0$ such that

$$s(f) < b < \|f\|_{L^\infty(X, \mu, \mathbb{F}^d)}.$$

Since $b > s(f)$, then $\mu(\{|f| > b\}) = 0$ since $s(f)$ is the largest possible value for which the sup-level set of f has positive measure. Similarly, since $b < \|f\|_{L^\infty(X, \mu, \mathbb{F}^d)}$, we must have $\mu(\{|f| > b\}) > 0$. This contradiction establishes the conclusion.

In order to prove the last statement, let us take a decreasing sequence $(c_m)_1^\infty$ converging to the infimum $c_m \searrow \|f\|_{L^\infty(X, \mu, \mathbb{F}^d)}$ as $m \to \infty$. By using that $|f| \le c_m$ pointwise on $X \setminus E_m$ and $\mu(E_m) = 0$, we have that the set

$$E := \bigcup_{m=1}^\infty E_m$$

satisfies $\mu(E) = 0$ by the countable sub-additivity of the measure, and hence $|f| \le c_m$ on $X \setminus E$ for all $m \in \mathbb{N}$. □

We close this section by demonstrating that the "little" ℓ^p-spaces of sequences are nothing but a special case of the "capital" L^p spaces for a discrete measure.

Remark 9.8 (Little ℓ^p spaces). The space ℓ^p defined in Chapter 2 is just a special case of the space $L^p(X, \mu, \mathbb{F})$ when we take as μ the **counting measure** "$\sharp$" on $X = \mathbb{N}$ (with the entire powerset $\mathcal{P}(\mathbb{N})$ of $\mathbb{N}$ as our σ-algebra):

$$\ell^p = L^p(\mathbb{N}, \sharp, \mathbb{F}).$$

We recall that the counting measure (equivalently, the 0-dimensional Hausdorff measure $\mathcal{H}^0$) is just the cardinal:

$$\sharp : \mathcal{P}(\mathbb{N}) \longrightarrow [0, \infty], \quad \sharp(A) = \begin{cases} \text{card}(A), & A \text{ finite,} \\ +\infty, & A \text{ infinite.} \end{cases}$$

Accordingly, it can be shown that *series of numbers are nothing but integrals with respect to the counting measure*. We leave the details for the exercises (see Exercise 9.10).

In view of the above observations, **we may (and will) use from now on the augmented concise symbolisation "$\|\cdot\|_{\ell^p}$" for the norm of ℓ^p which is reminiscent of the fact it is essentially a "capital" L^p space for the counting measure, instead of the more simplistic "$\|\cdot\|_p$".**

9.3   The inequalities of Hölder and Minkowski

In this section we settle the pending matters raised in Remarks 9.2-9.4, namely that:

- for any $p \in [1, \infty]$, the set of functions $L^p(X, \mu, \mathbb{F}^d)$ is a vector space;

- for any $p \in [1, \infty]$, the quantity $\|\cdot\|_{L^p(X,\mu,\mathbb{F}^d)}$ is a norm on the space;

- the bilinear function $\langle \cdot, \cdot \rangle_{L^2(X,\mu,\mathbb{F}^d)}$ is well defined on $L^2(X, \mu, \mathbb{F}^d)$.

To achieve this goal, we shall establish some relevant inequalities, special cases of which have already been studied in Chapter 2, Section 2.8. Let us begin by recalling that for any $p \in [1, \infty]$, the number

$$p' := \frac{p}{p-1},$$

is called the **conjugate exponent** of p and we have the identity

$$\frac{1}{p} + \frac{1}{p'} = 1,$$

with the understanding that $1' = \infty$ and $\infty' = 1$. Note in particular that $2' = 2$. Next, we demonstrate a very important inequality which in particular implies that the bilinear function

$$\langle \cdot, \cdot \rangle_{L^2(X,\mu,\mathbb{F}^d)} \; : \; L^2(X, \mu, \mathbb{F}^d) \times L^2(X, \mu, \mathbb{F}^d) \longrightarrow \mathbb{F},$$

$$\langle f, g \rangle_{L^2(X,\mu,\mathbb{F}^d)} := \int_X f \cdot \overline{g} \, d\mu,$$

is well defined.

Lemma 9.9 (Hölder's Inequality). *Let $p \in [1, \infty]$, $f \in L^p(X, \mu, \mathbb{F}^d)$ and $g \in L^{p'}(X, \mu, \mathbb{F}^d)$. Then, we have that $f \cdot \overline{g} \in L^1(X, \mu, \mathbb{F})$ and in addition the following estimate holds true*

$$\left\| f \cdot \overline{g} \right\|_{L^1(X,\mu,\mathbb{F})} \leq \|f\|_{L^p(X,\mu,\mathbb{F}^d)} \|g\|_{L^{p'}(X,\mu,\mathbb{F}^d)}.$$

Proof. By Young's inequality for vectors (Corollary 2.74), for any $\varepsilon > 0$ and any fixed $x \in X$ we have

$$\left| f(x) \cdot \overline{g(x)} \right| = \left| \varepsilon^{1/p'} f(x) \cdot \overline{g(x)(1/\varepsilon^{1/p'})} \right| \leq \frac{|f(x)|^p}{p} \varepsilon^{p-1} + \frac{|g(x)|^{p'}}{\varepsilon p'}.$$

We then integrate this inequality over all $x \in X$ and choose the particular value:[5]

$$\varepsilon := \left(\|f\|_{L^p(X,\mu,\mathbb{F}^d)} \right)^{-1} \left(\|g\|_{L^{p'}(X,\mu,\mathbb{F}^d)} \right)^{p'/p}.$$

The conclusion ensues. $\qquad\square$

We are now ready to show that $L^p(X, \mu, \mathbb{F}^d)$ is a vector space. It is quite obvious from the definition that if a function f is in $L^p(X, \mu, \mathbb{F}^d)$, then, for any $a \in \mathbb{F}$, any scalar multiple af of f is also in $L^p(X, \mu, \mathbb{F}^d)$ and we have

$$\|af\|_{L^p(X,\mu,\mathbb{F}^d)} = |a| \, \|f\|_{L^p(X,\mu,\mathbb{F}^d)}.$$

What is far less obvious is whether sums of L^p-functions are also L^p functions themselves. This is indeed true and it is a consequence of the next inequality:

Lemma 9.10 (Minkowski's Inequality). *Let $p \in [1, \infty]$. If $f, g \in L^p(X, \mu, \mathbb{F}^d)$, then $f + g \in L^p(X, \mu, \mathbb{F}^d)$ and we have the estimate*

$$\|f + g\|_{L^p(X,\mu,\mathbb{F}^d)} \leq \|f\|_{L^p(X,\mu,\mathbb{F}^d)} + \|g\|_{L^p(X,\mu,\mathbb{F}^d)}. \qquad (\sharp)$$

We readily have the next consequence.

Corollary 9.11 (Linear structure). *For any $p \in [1, \infty]$, the set $L^p(X, \mu, \mathbb{F}^d)$ is a vector space over the field $\mathbb{F}$.*

Proof of Lemma 9.10. The inequality $(\sharp)$ is immediate if either $p = 1$ or $p = \infty$: for $p = 1$, it is just a restatement of the triangle inequality for the Lebesgue integral, whilst for $p = \infty$ it is an easy exercise on the definition of the essential supremum.

Hence, it suffices to prove $(\sharp)$ when $1 < p < \infty$. Let $f, g \in L^p(X, \mu, \mathbb{F}^d)$. By integrating on X with respect to μ the following elementary inequality

$$|f + g|^p \leq \left(|f| + |g| \right)^p \leq 2^p \left(|f|^p + |g|^p \right)$$

we have that $f + g \in L^p(X, \mu, \mathbb{F}^d)$. Moreover,

$$\|f + g\|_{L^p(X,\mu,\mathbb{F}^d)}^p \leq \int_X \left(|f + g|^{p-1} |f| \right) \mathrm{d}\mu + \int_X \left(|f + g|^{p-1} |g| \right) \mathrm{d}\mu$$

and since $|f + g|^{p-1} \in L^{p'}(X, \mu, \mathbb{F}^d)$, Hölder's inequality gives

$$\left(\|f + g\|_{L^p(X,\mu,\mathbb{F}^d)} \right)^p \leq \left(\|f + g\|_{L^p(X,\mu,\mathbb{F}^d)} \right)^{p-1} \|f\|_{L^p(X,\mu,\mathbb{F}^d)}$$
$$+ \left(\|f + g\|_{L^p(X,\mu,\mathbb{F}^d)} \right)^{p-1} \|g\|_{L^p(X,\mu,\mathbb{F}^d)}.$$

The desired inequality follows. $\qquad\square$

---

[5]In case you are wondering about this "magic" choice of ε, it is nothing but the value which minimises the expression on the right hand side with respect to ε.

Our result below consists of two useful consequences of Hölder's inequality, known as **interpolation inequalities**. They give relations regarding possible membership of an L^p map to some other L^q space. In general, it may well happen that neither $L^q(X, \mu, \mathbb{F}^d) \not\subseteq L^p(X, \mu, \mathbb{F}^d)$ nor $L^q(X, \mu, \mathbb{F}^d) \not\supseteq L^p(X, \mu, \mathbb{F}^d)$:

Remark 9.12 (How p-integrability may fail). For $X := (0, \infty) \subseteq \mathbb{R}$ and $\mu = \mathcal{L}^1$, consider the function

$$f(x) := |x|^{-a}, \quad a > 0.$$

Then (see Figure 9.2):

- We have $g := f\chi_{(0,1)} \in L^p(0, \infty)$ if and only if $p < \frac{1}{a}$.

- We have $h := f\chi_{(1,\infty)} \in L^p(0, \infty)$ if and only if $p > \frac{1}{a}$.

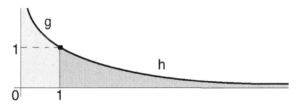

Figure 9.2: Illustration of the counter-examples to p-integrability.

Therefore, it appears that there are two main reasons that a function f may fail to be p-integrable: either

- the function $|f|^p$ grows too fast towards infinity in a neighbourhood of a point (and hence the area below the graph of $|f|^p$ is too large near the point), or

- the function $|f|^p$ fails to decay fast enough as we approach infinity (and again the area below the graph of $|f|^p$ is too large).

Hence, according to these examples, roughly speaking, the larger the p, the more "spread" over the domain the L^p functions are. The next results give some positive conditions regarding the above situation.

Lemma 9.13 (L^p Interpolation inequalities). *Let* $1 \leq p \leq q \leq \infty$.

(i) *For all* $r \in [p, q]$, *we have the inclusion*

$$L^p(X, \mu, \mathbb{F}^d) \bigcap L^q(X, \mu, \mathbb{F}^d) \subseteq L^r(X, \mu, \mathbb{F}^d).$$

Moreover, we have the estimate

$$\|f\|_{L^r(X,\mu,\mathbb{F}^d)} \leq \left(\|f\|_{L^p(X,\mu,\mathbb{F}^d)}\right)^a \left(\|f\|_{L^q(X,\mu,\mathbb{F}^d)}\right)^{1-a},$$

for all $f \in L^p(X, \mu, \mathbb{F}^d) \bigcap L^q(X, \mu, \mathbb{F}^d)$, *where a is given by*

$$\frac{1}{r} = \frac{a}{p} + \frac{1-a}{q}.$$

(ii) *If $\mu(X) < \infty$ (that is, X is a finite μ-measure space), then we have the inclusion*

$$L^q(X, \mu, \mathbb{F}^d) \subseteq L^p(X, \mu, \mathbb{F}^d)$$

Moreover, we have the estimate

$$\left(\frac{1}{\mu(X)} \int_X |f|^p \, d\mu \right)^{\frac{1}{p}} \leq \left(\frac{1}{\mu(X)} \int_X |f|^q \, d\mu \right)^{\frac{1}{p}},$$

for any $f \in L^q(X, \mu, \mathbb{F}^d)$. Equivalently,

$$\|f\|_{L^p(X,\mu,\mathbb{F}^d)} \leq \mu(X)^{\frac{1}{p} - \frac{1}{q}} \|f\|_{L^q(X,\mu,\mathbb{F}^d)}.$$

(iii) *If $(X, \mathfrak{M}, \mu)$ equals the "discrete" measure space $(X, \mathcal{P}(X), \sharp)$ where $\sharp$ is the counting measure on X, then $L^p(X, \sharp, \mathbb{F}^d) \subseteq L^q(X, \sharp, \mathbb{F}^d)$ and we have the estimate*

$$\|f\|_{L^q(X,\sharp,\mathbb{F}^d)} \leq \|f\|_{L^p(X,\sharp,\mathbb{F}^d)}.$$

Note the surprising fact that the **inclusion $L^q \subseteq L^p$ of (ii) which holds for finite measures is reversed for (iii) to $L^p \subseteq L^q$ when we have the counting measure.**

Proof. (i) We apply Hölder's inequality to the factorisation

$$|f|^r = \left(|f|^{ar} \right) \left(|f|^{(1-a)r} \right),$$

with conjugate exponents $\frac{p}{ar}, \frac{q}{(1-a)r}$:

$$\int_X |f|^r \, d\mu \leq \left(\int_X \left(|f|^{ar} \right)^{\frac{p}{ar}} d\mu \right)^{\frac{ar}{p}} \left(\int_X \left(|f|^{(1-a)r} \right)^{\frac{q}{(1-a)r}} d\mu \right)^{\frac{(1-a)r}{q}}$$

$$\leq \left(\int_X |f|^p \, d\mu \right)^{\frac{ar}{p}} \left(\int_X |f|^q \, d\mu \right)^{\frac{(1-a)r}{q}}.$$

(ii) We apply Hölder's inequality to the factorisation $|f|^p \cdot 1$, with conjugate exponents $\frac{q}{p}, \frac{q}{q-p}$:

$$\int_X |f|^p \, d\mu \leq \left(\int_X \left(|f|^p \right)^{\frac{q}{p}} d\mu \right)^{\frac{p}{q}} \left(\int_X 1^{\frac{q}{q-p}} d\mu \right)^{\frac{q-p}{q}}$$

$$\leq \left(\int_X |f|^q \, d\mu \right)^{\frac{p}{q}} \mu(X)^{1-\frac{p}{q}}.$$

(iii) If $q = \infty$, then by the definition of the essential supremum for non-negative functions[6] on $(X, \mathcal{P}(X), \sharp)$ and since the functions $t \mapsto t^{1/p}$ and $t \mapsto t^p$ are strictly increasing on $(0, \infty)$, we have

$$\|f\|_{L^\infty(X,\sharp,\mathbb{F}^d)} = \sup_{x \in X} |f(x)|$$

$$= \left(\sup_{x \in X} |f(x)|^p \right)^{1/p}$$

$$\leq \left(\sup_{E \subseteq X, \sharp(E) < \infty} \sum_{x \in E} |f(x)|^p \right)^{1/p}.$$

Hence, by the definition of the Lebesgue integral for non-negative functions on $(X, \mathcal{P}(X), \sharp)$ and the above estimate, we deduce

$$\|f\|_{L^\infty(X,\sharp,\mathbb{F}^d)} \leq \left(\int_X |f(x)|^p \, d\sharp(x) \right)^{1/p} = \|f\|_{L^p(X,\sharp,\mathbb{F}^d)}.$$

If $q < \infty$, then by part (i) and the previous estimate we have

$$\|f\|_{L^q(X,\sharp,\mathbb{F}^d)} \leq \left(\|f\|_{L^p(X,\sharp,\mathbb{F}^d)} \right)^\alpha \left(\|f\|_{L^\infty(X,\sharp,\mathbb{F}^d)} \right)^{1-\alpha} \leq \|f\|_{L^p(X,\sharp,\mathbb{F}^d)}.$$

Hence, the result ensues. $\qquad\square$

Remark 9.14 (Semi-norm structure of $L^p(X, \mu, \mathbb{F}^d)$). By gathering what we have established so far, we see that for any exponent $p \in [1, \infty]$:

(i) The set $L^p(X, \mu, \mathbb{F}^d)$ is an $\mathbb{F}$-vector space.

(ii) We have

    (a) $\|f + g\|_{L^p(X,\mu,\mathbb{F}^d)} \leq \|f\|_{L^p(X,\mu,\mathbb{F}^d)} + \|g\|_{L^p(X,\mu,\mathbb{F}^d)}$,

    (b) $\|af\|_{L^p(X,\mu,\mathbb{F}^d)} = |a| \, \|f\|_{L^p(X,\mu,\mathbb{F}^d)}$,

    (c) $\|0\|_{L^p(X,\mu,\mathbb{F}^d)} = 0$,

    when $f, g \in L^p(X, \mu, \mathbb{F}^d)$ and $a \in \mathbb{F}$.

The properties of the items (i)-(ii) above are very close to being the axioms defining a **normed vector space**. However, this is **not** the case! The problem is that we also need the validity of the positive definiteness of $\| \cdot \|_{L^p(X,\mu,\mathbb{F}^d)}$, namely that

$$\text{``}\|f\|_{L^p(X,\mu,\mathbb{F}^d)} = 0 \implies f = 0\text{''}. \tag{\dagger}$$

---

[6]Note that, interestingly, the only nullsets that exists with respect to the counting measure is just the empty set $\emptyset$. Therefore, $\sharp$-ess $\sup_X = \sup_X$. Further, the only sets of finite counting measure are the finite subsets of X and the non-negative simple functions have the form $\sum_{x \in E} a_x \chi_{A_x}$ with $a_x \geq 0$ and $x \in A_x \subseteq X$, for finite sets with $\sharp(E) < \infty$.

Nonetheless, (†) **need not always be satisfied in the case of discontinuous L^p functions!** Indeed, we have already seen examples of functions which are not identically zero but zero a.e. on the space. This may well happen even for the Lebesgue measure on the line $\mathbb{R}$ (e.g. Example 8.11). Actually, we have the following equivalence

$$\|f - g\|_{L^p(X,\mu,\mathbb{F}^d)} = 0 \iff f = g \ \mu\text{-a.e. on } X. \qquad (\ddagger)$$

If $p \in [1, \infty)$, ($\ddagger$) a consequence of Proposition 8.21, Proposition 8.22 and Corollary 8.23. For $p = \infty$, ($\ddagger$) is an immediate consequence of the definition of the L^∞ norm and of the essential supremum.

The above remark confronts us with the discouraging fact that the only way we can regard $L^p(X, \mu, \mathbb{F}^d)$ as a normed space is to abandon (†) and instead consider two mappings satisfying ($\ddagger$) as **different versions of the same element of** L^p (recall Definition 8.33 and Remark 8.34). Different versions certainly may not coincide pointwise, but can only differ on a set of zero measure, which plays no role from the point of view of integration. Hence it is possible, and in fact helpful, for different versions to be regarded as *representing the same element of the space*. The next definition makes these heuristics precise:

Definition 9.15. Let $p \in [1, \infty]$. We define the **equivalence relation for different versions of maps** on the space $L^p(X, \mu, \mathbb{F}^d)$ (Definition 8.33) by

$$f \sim g \iff f = g \ \mu\text{-a.e. on } X.$$

Hence $f \sim g$ if and only if f and g are different versions of the same map.

In view of Definition 9.15 and Remark 9.14, we arrive at the next result:

Lemma 9.16 (Normed space structure of the quotient L^p space). *For any $p \in [1, \infty]$, the quotient space of all the equivalence classes*

$$[f]_\sim := \left\{ \tilde{f} : \tilde{f} \sim f \in L^p(X, \mu, \mathbb{F}^d) \right\}$$

of different versions of maps

$$L^p(X, \mu, \mathbb{F}^d)/_\sim := \left\{ [f]_\sim : f \in L^p(X, \mu, \mathbb{F}^d) \right\}$$

forms a normed vector space over the field $\mathbb{F}$. Also, the function $\| \cdot \|_{L^p(X,\mu,\mathbb{F}^d)}$ given in Definition 9.1 is a norm on it.

The point of view of Lemma 9.16, despite being rigorous and precise, is quite inconvenient. Instead of talking about representatives of the above equivalence classes, we shall understand that the **elements of $L^p(X, \mu, \mathbb{F}^d)$ are actual measurable maps** which however are **defined μ-a.e. on X and can be modified at will on nullsets** without changing the element of $L^p(X, \mu, \mathbb{F}^d)$ they represent:

Theorem 9.17 (Normed space structure of $L^p(X, \mu, \mathbb{F}^d)$ modulo versions of maps). *For any $p \in [1, \infty]$, the pair*

$$\left(L^p(X, \mu, \mathbb{F}^d), \| \cdot \|_{L^p(X,\mu,\mathbb{F}^d)} \right)$$

*is a normed vector space over $\mathbb{F}$, if we **identify maps which coincide μ-a.e. on** X, namely if we identify different versions of maps.*

That is, when $f, g \in L^p(X, \mu, \mathbb{F}^d)$ and $f = g$ μ-a.e. on X, then f is identified with g as being different versions (in the sense of Definition 8.33) of the same map. This point of view, which is the standard way of seeing the L^p spaces in Analysis, will be adopted in this book as well. Hence, we shall *not* make any explicit reference to the quotient space $L^p(X, \mu, \mathbb{F}^d)/_\sim$ from now on.

9.4 Completeness and density of simple maps

The first main question we discuss in this section concerns the completeness of the L^p-spaces. Namely, whether sequences of maps which are Cauchy (with respect to the L^p norm), actually converge to some limit map. Further, we demonstrate a very useful approximation result, establishing that the vector subspace of simple mappings is actually dense in L^p.

We begin with the answer to the completeness question which turns out to be "yes", as the next result attests:

Theorem 9.18. *For any $p \in [1, \infty]$, $L^p(X, \mu, \mathbb{F}^d)$ is a Banach space.*

We recall that a Banach space is defined as a normed vector space which is complete with respect to the metric induced by the norm. We note that Theorem 9.18 generalises the particular case of Theorem 3.19 from Chapter 3, wherein we showed that the class of "little ℓ^p" spaces (corresponding to the counting measure on $\mathbb{N}$) are complete.

Proof. **Case 1:** $p = \infty$. Let $(f_m)_1^\infty \subseteq L^\infty(X, \mu, \mathbb{F}^d)$ be a Cauchy sequence. Then, by the definition of Cauchy sequences, for any fixed $k \in \mathbb{N}$, there exists $m(k) \in \mathbb{N}$, such that

$$\left\| f_m - f_{m'} \right\|_{L^\infty(X,\mu,\mathbb{F}^d)} \leq \frac{1}{k},$$

for all $m, m' \geq m(k)$. Hence, by the definition of the essential supremum, for each $k \in \mathbb{N}$, there exists a nullset $E_k \in \mathfrak{M}$ with $\mu(E_k) = 0$ such that

$$\left| f_m(x) - f_{m'}(x) \right| \leq \frac{1}{k}, \quad \text{for } x \in X \setminus E_k,$$

for all $m, m' \geq m(k)$. We set

$$E := \bigcup_{k=1}^{\infty} E_k.$$

Then, $E \in \mathfrak{M}$ and by the σ-sub-additivity of the measure, we have $\mu(E) = 0$ and

$$|f_m(x) - f_{m'}(x)| \leq \frac{1}{k}, \quad \text{for } x \in X \setminus E, \qquad (\spadesuit)$$

for all $m, m' \geq m(k)$. Hence, for any $x \in X \setminus E$, the sequence $(f_m(x))_1^{\infty}$ is Cauchy in $\mathbb{F}^d$. Since $\mathbb{F}^d$ is complete, for any fixed $x \in X \setminus E$, there exists a vector $f_x \in \mathbb{F}^d$ such that

$$f_m(x) \longrightarrow f_x \quad \text{as } m \to \infty.$$

Note that since the set on which the above convergence fails is a nullset, by Proposition 8.35 (recall also Remark 7.46) it follows that the version (in the sense of Definition 8.33) of the mapping defined by

$$f(x) := \begin{cases} f_x, & x \in X \setminus E, \\ 0, & x \in E, \end{cases}$$

is μ-measurable. By letting $m' \to \infty$ in $(\spadesuit)$, we have

$$|f_m(x) - f(x)| \leq \frac{1}{k}, \quad x \in X \setminus E.$$

Consequently, by the definition of the essential supremum, $f \in L^{\infty}(X, \mu, \mathbb{F}^d)$ and for the given arbitrary $k \in \mathbb{N}$,

$$\|f_m - f\|_{L^{\infty}(X,\mu,\mathbb{F}^d)} \leq \frac{1}{k}, \quad \text{for all } m \geq m(k).$$

Hence $f_m \longrightarrow f$ as $m \to \infty$ in (the norm of) $L^{\infty}(X, \mu, \mathbb{F}^d)$.

Case 2: $1 \leq p \leq \infty$. Let $(f_m)_1^{\infty} \subseteq L^p(X, \mu, \mathbb{F}^d)$ be a Cauchy sequence. The trick is that, as a consequence of the Cauchy condition, one can extract a "fast subsequence" $(f_{m_k})_1^{\infty} \subseteq (f_m)_1^{\infty}$ such that

$$\|f_{m_{k+1}} - f_{m_k}\|_{L^p(X,\mu,\mathbb{F}^d)} \leq \frac{1}{2^k},$$

for all $k \in \mathbb{N}$. We set

$$g_i := \sum_{k=1}^{i} |f_{m_{k+1}} - f_{m_k}|, \quad i \in \mathbb{N}.$$

Then $(g_i)_{i=1}^{\infty}$ satisfies that $g_i \leq g_{i+1}$ for all $i \in \mathbb{N}$. By the Monotone Convergence Theorem, by setting

$$g := \sup_{i \in \mathbb{N}} g_i = \lim_{i \to \infty} g_i$$

we have

$$\int_X (g_i)^p \, d\mu \longrightarrow \int_X (g)^p \, d\mu, \quad \text{as } i \to \infty.$$

We also have that g satisfies $\|g\|_{L^p(X,\mu,\mathbb{R})} \leq 1$, because

$$\|g_i\|_{L^p(X,\mu,\mathbb{R})} = \left\| \sum_{k=1}^{i} |f_{m_{k+1}} - f_{m_k}| \right\|_{L^p(X,\mu,\mathbb{F}^d)}$$

$$\leq \sum_{k=1}^{i} \|f_{m_{k+1}} - f_{m_k}\|_{L^p(X,\mu,\mathbb{F}^d)}$$

$$\leq 1.$$

Moreover, for $\ell \geq k \geq 2$, by the triangle inequality we have

$$|f_{m_\ell} - f_{m_k}| \leq \sum_{a=k}^{l-1} |f_{m_{a+1}} - f_{m_a}|$$

$$= \sum_{a=1}^{l-1} |f_{m_{a+1}} - f_{m_a}| - \sum_{a=1}^{k-1} |f_{m_{a+1}} - f_{m_a}|$$

$$\leq g - g_{k-1},$$

μ-a.e. on X. Therefore, for μ-a.e. $x \in X$, the sequence of vectors $(f_{m_k}(x))_1^\infty$ is Cauchy in $\mathbb{F}^d$. Using the completeness of $\mathbb{F}^d$ defining f in a similar way as in Case 1, we may let $\ell \to \infty$ to infer that

$$|f(x) - f_{m_k}(x)| \leq g(x) - g_{k-1}(x),$$

for μ-a.e. $x \in X$. Since

$$|f - f_{m_k}|^p \leq g^p, \ g^p \in L^1(X,\mu,\mathbb{R}) \ \text{ and } \ f_{m_k} \longrightarrow f \ \ \mu\text{-a.e. on } X \text{ as } k \to \infty,$$

the Dominated Convergence Theorem yields that $f_{m_k} \longrightarrow f$ as $k \to \infty$ in $L^p(X,\mu,\mathbb{F}^d)$. The desired conclusion that $f_m \longrightarrow f$ as $m \to \infty$ in $L^p(X,\mu,\mathbb{F}^d)$ now follows from Proposition 3.4 (ii). $\qquad\square$

We now present the following (partial) converse to the Dominated Convergence Theorem which essentially is a consequence of the proof of completeness we gave above.

Theorem 9.19 (Partial converse to the Dominated Convergence Theorem). *Let $p \in [1,\infty]$. Suppose that $(f_m)_1^\infty \subseteq L^p(X,\mu,\mathbb{F}^d)$ is a convergent sequence such that*

$$f_m \longrightarrow f \ \ as \ m \to \infty \ \ in \ \ L^p(X,\mu,\mathbb{F}^d).$$

Then, there exists a subsequence $(f_{m_k})_{k=1}^{\infty}$ of $(f_m)_{m=1}^{\infty}$ and a mapping $g \in L^p(X, \mu, \mathbb{F}^d)$ such that

$$|f_{m_k}| \le g \quad \text{and} \quad f_{m_k} \longrightarrow f \quad \text{as } k \to \infty,$$

both being true μ-a.e. on X.

Proof. The proof is almost identical to that of Case 2 above and we leave the completion of the details as an exercise for the reader. □

Corollary 9.20 (L^p **convergence implies subsequential a.e. convergence**). *For any $p \in [1, \infty]$, convergence of a sequence in L^p implies convergence a.e., up to perhaps the passage to a subsequence.*

The above results is general cannot be improved if $p < \infty$. If a sequence converges in L^p, then the full sequence may not converge a.e. on the domain:

Example 9.21. On the domain $X = (0,1) \subseteq \mathbb{R}$ and for $\mu = \mathcal{L}^1$, consider the sequence $(f_m)_1^{\infty}$ given by (see Figure 9.3)

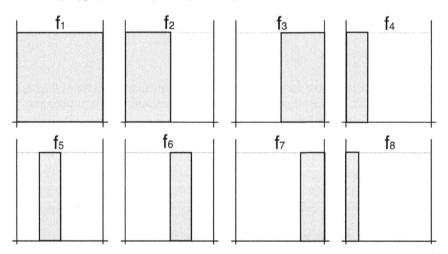

Figure 9.3: Illustration of the first eight terms of the sequence of the example.

$$f_1 := \chi_{[0,1]}, \quad f_2 := \chi_{[0,1/2]}, \quad f_3 := \chi_{[1/2,1]}, \quad f_4 := \chi_{[0,1/4]},$$
$$f_5 := \chi_{[1/4,1/2]}, \quad f_6 := \chi_{[1/2,3/4]}, \quad f_7 := \chi_{[3/4,1]}$$

and in general by

$$f_m := \chi_{\left[j2^{-k}, (j+1)2^{-k}\right]}, \quad m = 2^k + j, \ j = 0, 1, \ldots, 2^k - 1.$$

Then, $f_m \longrightarrow 0$ in $L^1((0,1))$ as $m \to \infty$, because

$$\|f_m\|_{L^1((0,1))} = 2^{-k}, \quad \text{for } 2^k \le m \le 2^{k+1}, \quad k \in \mathbb{N}.$$

However, $f_m(x) \not\to 0$ for **any** point $x \in (0,1)$! This happens because there are infinitely many values of $m \in \mathbb{N}$ for which $f_m(x) = 1$ and infinitely many values of $m \in \mathbb{N}$ for which $f_m(x) = 0$.

We shall now prove an important approximation result, showing that simple maps are dense in the space L^p for all $p \in [1, \infty]$. To this end, we shall use the next symbolisation:

Definition 9.22. The vector space of $\mathbb{F}^d$-**valued simple maps** on the measure space $(X, \mathfrak{M}, \mu)$ will be symbolised as:

$$\Sigma(X, \mu, \mathbb{F}^d) := \left\{ \phi = \sum_{j=1}^{m} a_j \chi_{A_j} \,\middle|\, (a_j)_1^p \subseteq \mathbb{F}^d, \ (A_j)_1^p \text{ disjoint } \mu\text{-measurable} \right\}.$$

We recall that, in view of Corollary 7.44, an equivalent characterisation of simple maps is as those whose image set is finite:

$$\Sigma(X, \mu, \mathbb{F}^d) = \left\{ \phi : X \longrightarrow \mathbb{F}^d \,\middle|\, \phi \text{ is } \mu\text{-measurable and } \sharp(\phi(X)) < \infty \right\}.$$

We may now state and prove the claimed result.

Theorem 9.23 (Density of simple maps in $L^p(X, \mu, \mathbb{F}^d)$).

(i) *Let $p \in [1, \infty)$. Then, the space of **integrable** simple maps is dense in $L^p(X, \mu, \mathbb{F}^d)$:*

$$\overline{(\Sigma \cap L^1)(X, \mu, \mathbb{F}^d)}^{L^p} = L^p(X, \mu, \mathbb{F}^d).$$

(ii) *Let $p = \infty$. Then, the space of simple maps is dense in $L^\infty(X, \mu, \mathbb{F}^d)$:*

$$\overline{\Sigma(X, \mu, \mathbb{F}^d)}^{L^\infty} = L^\infty(X, \mu, \mathbb{F}^d).$$

*If in addition X is a finite μ-measure space (that is $\mu(X) < \infty$), then **integrable** simple maps are also dense in $L^\infty(X, \mu, \mathbb{F}^d)$.*

Note that, unlike the case of $p \in [1, \infty)$, when $p = \infty$ in general we *cannot require the simple functions to be integrable*, as the next example shows:

Example 9.24 (Integrable simple maps not dense in L^∞). Consider the space $L^\infty(\mathbb{R}, \mathbb{R})$ (with $\mu = \mathcal{L}^1$) and the function $f \equiv 1$. Then, for **any** integrable simple function $\phi = \sum_1^m a_j \chi_{A_j}$, we have that $\phi \equiv 0$ on the set of infinite measure $\mathbb{R} \setminus \bigcup_{i=1}^m A_i$. Therefore,

$$\|f - \phi\|_{L^\infty(\mathbb{R}, \mathbb{R})} \geq \|f - \phi\|_{L^\infty(\mathbb{R} \setminus \bigcup_{i=1}^m A_i, \mathbb{R})} = |1 - 0| = 1$$

and hence f cannot be approximated in L^∞ by integrable simple functions (see Figure 9.4).

Remark 9.25. Let $p \in (1, \infty)$. Note that for any simple map with representation $\phi = \sum_{j=1}^{m} a_j \chi_{A_j}$, we have

$$\|\phi\|_{L^1(X,\mu,\mathbb{F}^d)} = \sum_{j=1}^{m} |a_j| \mu(A_j) < \infty \iff \|\phi\|_{L^p(X,\mu,\mathbb{F}^d)} = \sum_{j=1}^{m} |a_j| \mu(A_j)^{\frac{1}{p}} < \infty.$$

Hence, an equivalent requirement to the integrability of a simple map is its p-integrability. That is, if $\phi \in \Sigma(X, \mu, \mathbb{F}^d)$, then $\phi \in L^1(X, \mu, \mathbb{F}^d)$ if and only if $\phi \in L^p(X, \mu, \mathbb{F}^d)$ for some (or all) $p < \infty$.

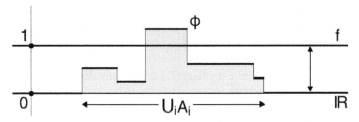

Figure 9.4: Illustration of the non-approximation Example 9.24. No matter how large the union of the finitely many sets $\bigcup_i A_i$ is, it will always have finite measure and hence on its complement (which has infinite measure), we have $\phi = 0$ and $f = 1$. Hence, the distance of f from ϕ in the norm of L^∞ is at least equal to one!

Proof. (i) Assume that $p \in [1, \infty)$. Fix $f \in L^p(X, \mu, \mathbb{F}^d)$ and choose a sequence $(f_i)_1^\infty$ of simple maps such that $f_i \longrightarrow f$ μ-a.e. on X as $i \to \infty$ and satisfying $f_i| \leq |f|$ on X for all $i \in \mathbb{N}$. Such a sequence is provided by Theorem 7.45. Since $f \in L^p(X, \mu, \mathbb{F}^d)$, it follows by the remark above that the sequence of simple maps $(f_i)_1^\infty$ is contained in $L^1(X, \mu, \mathbb{F}^d)$ as well. By applying the Dominated Convergence Theorem, we have $f_i \longrightarrow f$ in $L^p(X, \mu, \mathbb{F}^d)$ as $i \to \infty$. Hence, the space $\Sigma(X, \mu, \mathbb{F}^d) \cap L^1(X, \mu, \mathbb{F}^d)$ is indeed dense in $L^p(X, \mu, \mathbb{F}^d)$, as claimed.

(ii) It suffices to note that the case $p = \infty$ follows directly from our approximation Theorem 7.45. If in addition $\mu(X) < \infty$, then for any simple map with representation $\phi = \sum_{j=1}^{m} a_j \chi_{A_j}$, we have

$$\|\phi\|_{L^1(X,\mu,\mathbb{F}^d)} = \sum_{j=1}^{m} |a_j| \mu(A_j) \leq \mu(X) \sum_{j=1}^{m} |a_j| < \infty$$

and hence $\Sigma(X, \mu, \mathbb{F}^d) \subseteq L^1(X, \mu, \mathbb{F}^d)$ anyway. The theorem ensues. $\square$

9.5 Density of C^0 and separability in the Euclidean case

In this and the next section we shall restrict our attention to the case of the measure space $(\star)$. For the sake of brevity, we shall use the next standard abbreviations:

- $\Omega \equiv \left(\Omega,\, \mathcal{L}(\mathbb{R}^n)|_\Omega,\, \mathcal{L}^n \right)$

- $L^p(\Omega, \mathcal{L}^n, \mathbb{R}) \equiv L^p(\Omega),$

- $\displaystyle \int_\Omega f(x)\, d\mathcal{L}^n(x) \equiv \int_\Omega f,$ when $f \in L^+(\Omega)$ or $f \in L^1(\Omega),$

- $\mathcal{L}^n(A) \equiv |A|,$ when $A \in \mathcal{L}(\mathbb{R}^n)|_\Omega.$

Further, for the sake of simplicity **we will consider only $\mathbb{R}$-valued functions on Ω**. The vectorial case of $\mathbb{F}^d$-valued maps can be handled by arguing in a component-wise fashion for the real and imaginary components of maps

$$ f \;:\; \mathbb{R}^n \supseteq \Omega \longrightarrow \mathbb{F}^d\,, \quad f \;=\; \left(\mathrm{Re} f_1,\, \mathrm{Im} f_1;\, \ldots;\, \mathrm{Re} f_d,\, \mathrm{Im} f_d \right). $$

The analysis of the vectorial case remains essentially the same with only minor complications.

We begin by answering the question of separability of the space L^p. Outside the Euclidean realm of the measure space Ω, *the space L^p is not in general separable, not even in the more favourable case of $p = 2$*. Nonetheless, we recall that in Propositions 2.83 and 2.84 we have proved that in the "discrete case" of the counting measure on $\mathbb{N}$, the space $\ell^p = L^p(\mathbb{N}, \sharp, \mathbb{F})$ is separable for $p < \infty$ and non-separable for $p = \infty$. Similarly, in the Euclidean case of the measure space Ω, the L^p space is separable and contains a countable dense set of (simple) functions, when $p < \infty$, but it is non-separable for $p = \infty$.

Theorem 9.26 (Separability of L^p for $1 \le p < \infty$). *For any $p \in [1, \infty)$, the space $L^p(\Omega)$ is separable.*

Proof. We begin by noting that the set of simple functions with rational coefficients which are supported on boxes with rational vertices is countable (in fact this is a vector space over $\mathbb{Q}$):

$$ \mathbb{V} := \left\{ \phi = \sum_{i=1}^m q_i \chi_{R_i} \;\middle|\; q_i \in \mathbb{Q},\; R_i = \prod_{j=1}^n (a_j^{(i)}, b_j^{(i)}),\; a_j^{(i)}, b_j^{(i)} \in \mathbb{Q} \right\}. $$

The countability of $\mathbb{V}$ is a straightforward consequence of Lemmas 1.10 and 1.11. By using the density of integrable simple functions guaranteed by Theorem 9.23, it suffices to show that any $\phi \in L^1(\Omega) \cap \Sigma(\Omega)$ can be approximated by a sequence $(\phi_n)_1^\infty \subseteq \mathbb{V}$ in the L^p norm. Indeed, fix $\phi \in L^1(\Omega) \cap \Sigma(\Omega)$ such that

$$ \phi = \sum_{i=1}^m a_i \chi_{A_i}, \quad (a_i)_1^m \subseteq \mathbb{R},\; (A_i)_1^m \text{ disjoint measurable},\; \max_{i=1\ldots m} |A_i| < \infty. $$

By the density of $\mathbb{Q}$ in $\mathbb{R}$, each of the coefficients $a_1, \ldots, a_m$ can be approximated by a sequence $(a_i^n)_{n=1}^\infty \subseteq \mathbb{Q}$ such that

$$ a_i^n \longrightarrow a_i \quad \text{as } n \to \infty,\; i = 1, \ldots, m. $$

Further, by the outer regularity of the Lebesgue (outer) measure (Proposition 6.26) each of the measurable sets $\{A_1, \ldots, A_m\}$ with finite measure can be approximated from the outside by a sequence of open sets $(A_i^n)_{n=1}^\infty$ with $A_i^n \supseteq A_i$, each of these sets being a countable union of open boxes whose vertices have rational coordinates, and such that:

$$\left| A_i^n \setminus A_i \right| \longrightarrow 0 \ \text{ as } \ n \to \infty, \ \ i = 1, \ldots, m.$$

We define

$$\phi_n := \sum_{i=1}^m a_i^n \chi_{A_i^n}, \quad n \in \mathbb{N}.$$

Then, by using the triangle inequality and the trivial inequality

$$(a+b)^p \leq 2^p (a^p + b^p), \quad a, b \geq 0,$$

we have

$$\left\| \phi_n - \phi \right\|_{L^p(\Omega)} = \left(\int_\Omega \left| \sum_{i=1}^m a_i^n \chi_{A_i^n} - \sum_{i=1}^m a_i \chi_{A_i} \right|^p \right)^{1/p}$$

and hence we may estimate

$$\left\| \phi_n - \phi \right\|_{L^p(\Omega)}^p \leq 2^p \left(\sum_{i=1}^m |a_i^n - a_i|^p \int_\Omega \chi_{A_i} + \sum_{i=1}^m |a_i^n|^p \int_\Omega |\chi_{A_i^n} - \chi_{A_i}| \right)$$

$$\leq 2^p \left(\sum_{i=1}^m |a_i^n - a_i|^p \max_{i=1\ldots m} |A_i| + \sum_{i=1}^m |a_i^n|^p \max_{i=1\ldots m} |A_i^n \setminus A_i| \right).$$

The above estimate implies that $\phi_n \longrightarrow \phi$ in $L^p(\Omega)$ as $n \to \infty$. The conclusion ensues. □

We now show that, unlike L^p for $1 \leq p < \infty$, the space L^∞ is (a "large") **non-separable** Banach space even in the Euclidean case:

Example 9.27 (Non separability of L^∞). In the space $L^\infty(0,1)$, for each $t \in (0,1)$ consider the open ball centred at the characteristic function $\chi_{(0,t)}$ with radius $1/2$:

$$\mathbb{B}_{1/2}^{L^\infty}(\chi_{(0,t)}) = \left\{ f \in L^\infty(0,1) \ \middle| \ \|f - \chi_{(0,t)}\|_{L^\infty(0,1)} < \frac{1}{2} \right\}.$$

Then it can be easily verified that the family of open balls

$$\left\{ \mathbb{B}_{1/2}^{L^\infty}(\chi_{(0,t)}) : t \in (0,1) \right\} \subseteq L^\infty(0,1)$$

is mutually disjoint. Further, the balls are uncountably many since they are parameterised by the points of $(0,1)$, and this interval has the cardinality of $\mathbb{R}$. It follows that $L^\infty(0,1)$ cannot be separable, for otherwise we would have a contradiction to the fact that $(0,1)$ is uncountable (see Chapter 1).

As we have already discussed in the introduction to this chapter, we shall now establish that continuous functions are dense in the space L^p. Namely, every L^p function can be approximated by a sequence of continuous functions in the L^p norm. In addition, we will see that one may even require that the approximating functions are non-zero only within a compact set inside the domain Ω. Such functions are commonly called "compactly supported":

Definition 9.28. The space of **real continuous functions on Ω with compact support** is given by

$$C_c^0(\Omega) := \Big\{ f \in C^0(\Omega) \,\Big|\, \operatorname{supp}(f) \text{ is a compact set in } \Omega \Big\}.$$

Here "supp(f)" is the closure of the set on which the function is non zero, known as the **support** of the function (see Figure 9.5):

$$\operatorname{supp}(f) := \overline{\{ x \in \Omega \,:\, f(x) \neq 0 \}}.$$

Further, the symbol $C^0(\Omega)$ denotes the space of continuous (possibly unbounded) functions $f : \Omega \longrightarrow \mathbb{R}$.

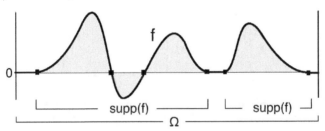

Figure 9.5: Illustration of the support of a function on a subset of $\mathbb{R}$.

In order to achieve our desired approximation result, we require a very strong tool of topology known as **"the Urysohn Lemma"**. This result is typically stated and proved in great generality, but, for the sake of clarity and for the simplicity of the exposition, we isolate and prove a simple special case which is actually all we need for our second main result in this section.

Lemma 9.29 (Urysohn). *Let $A, B \subseteq \mathbb{R}^n$ be two disjoint closed sets of $\mathbb{R}^n$. Then, there exists an $f \in C^0(\mathbb{R}^n)$ such that $f \equiv 1$ on A and $f \equiv 0$ on B. In particular, if K is compact and U is open with $U \supseteq K$, there exists a function $f \in C_c^0(\mathbb{R}^n)$ with $f \equiv 1$ on K and $f \equiv 0$ on $\mathbb{R}^n \setminus U$ (see Figure 9.6).*

Proof. (In the special case of K, U) Let $K \subseteq \mathbb{R}^n$ be compact and let also U be an open neighbourhood of K. Fix also $t > 0$ and set (see Figure 9.7)

$$f(x) := \max \big\{ 1 - t \operatorname{dist}(x, K), 0 \big\}, \quad x \in \mathbb{R}^n.$$

In the definition of f, "dist($\cdot, K$)" symbolises the distance from the set K:

$$\operatorname{dist}(x, K) := \inf_{k \in K} |x - k|.$$

It follows that $f \in C^0(\mathbb{R}^n)$, $0 \leq f \leq 1$ on $\mathbb{R}^n$ and $f \equiv 1$ on K. By choosing t large enough, we can achieve $f \equiv 0$ on $\mathbb{R}^n \setminus U$. $\qquad\square$

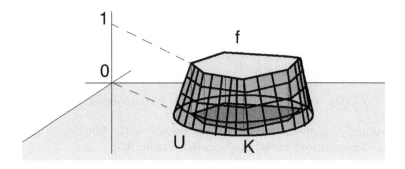

Figure 9.6: An illustration of the (special case of the) Urysohn Lemma on $\mathbb{R}^2$.

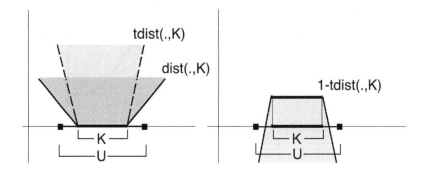

Figure 9.7: The idea of the construction the proof of the Urysohn Lemma.

Remark 9.30. An equivalent and compact way to state that "$f \equiv 0$ on $\mathbb{R}^n \setminus U$ and $f \equiv 1$ on K" is by means of the next inequality which involves only characteristic functions:

$$\chi_K \leq f \leq \chi_U \quad \text{on } \Omega.$$

We may now establish the second principal result of this section.

Theorem 9.31 (Continuous functions are dense in L^p for $1 \leq p < \infty$). *The space of continuous functions over Ω with compact support is dense in the space $L^p(\Omega)$, that is*

$$\overline{C_c^0(\Omega)}^{L^p(\Omega)} = L^p(\Omega),$$

in the case that $p \in [1, \infty)$.

Proof. Since by Theorem 9.23 the space of simple integrable functions is dense in $L^p(\Omega)$, it suffices to show that each simple function ϕ can be approximated as close as one wishes by a function $\psi \in C_c^0(\Omega)$. Fix such a ϕ with representation $\phi = \sum_{i=1}^m a_i \chi_{A_i}$ for some $(a_i)_1^m \subseteq \mathbb{R}$ and disjoint measurable sets $(A_i)_1^m$ of finite measure. Fix also an $\varepsilon > 0$. Note now that by the outer regularity of the Lebesgue (outer) measure (Proposition 6.26), for each of the measurable sets A_i, there exists an open set $U_i \supseteq A_i$ such that

$$|U_i \setminus A_i| = |U_i| - |A_i| < \frac{\varepsilon}{\sum_{j=1}^m |a_j|}.$$

Moreover, there is no loss of generality to assume that $U_i \subseteq \Omega$ for each $i \in \{1, ..., m\}$ (since otherwise we may just replace U_i by $U_i \cap \Omega$). By the Urysohn Lemma 9.29, for each $i = 1, \ldots, m$, we can choose $\psi_i \in C_c^0(\Omega)$ such that $\psi_i \equiv 1$ on A_i and $\psi_i \equiv 0$ on $\Omega \setminus U_i$, see Figure 9.8. Hence, by defining the function

$$\psi := \sum_{i=1}^m a_i \psi_i,$$

we see that $\psi \in C_c^0(\Omega)$ and also

$$\|\psi - \phi\|_{L^p(\Omega)} \leq \sum_{i=1}^m \left\| a_i \chi_{A_i} - a_i \psi_i \right\|_{L^p(U_i)}$$

$$\leq \sum_{i=1}^m |a_i| \, |U_i \setminus A_i|$$

$$< \varepsilon.$$

The theorem follows.

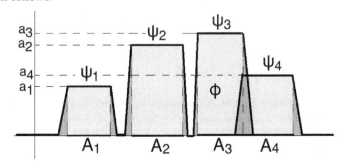

Figure 9.8: The idea of the construction of the continuous approximation to the simple function. We approximate each of the simple functions from above by a continuous one (in such a way that the measure of the difference of their supports is small) and then take their sum. □

We now show that unlike $L^p(\Omega)$ for $1 \leq p < \infty$, neither the space $C_b^0(\Omega)$ of bounded (real) continuous functions nor its subspace $C_0^0(\Omega)$ of functions vanishing on the boundary are dense in $L^\infty(\Omega)$. In fact, they are closed (recall also the relevant situation emerging for non-integrable simple functions in L^∞ we saw in Example 9.24).

Remark 9.32 (C_b^0 and C_0^0 are closed and not dense in L^∞).
• Consider the function $f \equiv 1 \in L^\infty(0,1)$ and choose any $\phi \in C_0^0(0,1)$ (a continuous function vanishing at the endpoints $\{0,1\}$). Then,

$$\|f - \phi\|_{L^\infty(0,1)} = \sup_{0<x<1} |1 - \phi(x)| \geq |1 - \phi(0)| = 1.$$

Hence, f cannot be approximated by functions from $C_0^0(0,1)$ since the distance (in L^∞) of it from the subspace $C_0^0(0,1)$ is greater than or equal to one (see Figure 9.9).

• Let Ω be a *bounded* open set. The space $C_b^0(\Omega)$ is actually **closed** in $L^\infty(\Omega)$:

$$\overline{C_b^0(\Omega)}^{\|\cdot\|_{L^\infty(\Omega)}} = C_b^0(\Omega).$$

This follows easily by using that for any $f \in C_b^0(\Omega)$ we have

$$\|f\|_{L^\infty(\Omega)} = \sup_\Omega |f|$$

and that Cauchy sequences of continuous functions in $L^\infty(\Omega)$ must converge uniformly to a function which is *continuous*.

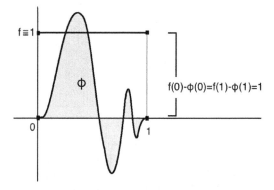

Figure 9.9: An illustration of the example. No matter how close the graph of ϕ is to that of f (this will make the area under the graph of $|f - \phi|$ small and hence the L^q norm small as well for any $q < \infty$), the maximum of $|f - \phi|$ will always be greater or equal to one.

9.6 Mollification and subspaces of smooth functions

For this last section,[7] we shall assume some familiarity with the elementary concepts of Vector Calculus on the Euclidean space $\mathbb{R}^n$, in particular with

---

[7]This material in this section is slightly more advanced that the previous developments in this chapter and could perhaps be omitted on the first reading of the book.

differentiation. The end goal will be to show that, when restricting our attention to the measure space $(\star)$, where **the domain $\Omega \subseteq \mathbb{R}^n$ is open**, smooth (infinitely differentiable) maps from Ω to $\mathbb{F}^d$ are dense in the space L^p when $p \in [1, \infty)$. In addition, one may even require that the maps are non-zero only within a compact set inside Ω, therefore vanishing on the boundary $\partial\Omega$ (if it is non-empty). This will strengthen the density result obtained in Section 9.5 quite substantially.

The usefulness of having a density result for smooth maps should be apparent: L^p maps (which generally are discontinuous, not even pointwise defined and difficult to handle) can be approximated in the L^p-norm by infinitely differentiable maps. **This approximation result is crucial when trying to prove results involving L^p maps because, roughly speaking, one can prove the result for smooth maps and pass to the limit.**

The idea of the smooth approximation scheme lies in showing that, given any map in L^p, one can "mollify" (i.e., regularise) the map to a new smooth one, in such a way that it remains as close as desired to the original map when measured in the L^p norm. This is possible through a by-now standard method in Analysis, known as **mollification by convolution**. The "convolution" is, as we shall see in the next subsection, a product operation between two maps f, g which produces a new map, symbolised as $f * g$.

A central feature of the convolution product is that *$f*g$ becomes as regular as the most regular of the maps f, g*. For instance, if f is C^1 and g is C^k for $k \in \mathbb{N} \cup \{\infty\}$, then $f * g$ becomes C^k too. Showing convergence is more tricky; one has to choose the map with which we multiply by convolution appropriately and we shall delve into this topic a little later.

9.6.1 Convolution and Young inequality

In this subsection we introduce and study the auxiliary notion of convolution product of maps on which our systematic regularisation technique is based on. **Here and until the end of this chapter, Ω will denote an open subset of $\mathbb{R}^n$ and we understand that Ω is endowed with the Lebesgue measure.** As it is common practice, until the end of the chapter we shall continue using the standard abbreviations we introduced in the previous section for the measure space $(\star)$. Further, for the sake of simplicity we will again consider only real-valued scalar functions.

We begin by reminding some notation from Calculus in several variables. Let Ω be an open set in $\mathbb{R}^n$. The space of continuous (not necessarily bounded) real functions $f : \Omega \longrightarrow \mathbb{R}$ is denoted by $C^0(\Omega)$. If f is k-times differentiable, its first, second and higher order partial derivatives (which by assumption exist) will be denoted respectively by

$$\mathrm{D}_i f, \quad \mathrm{D}_{ij}^2 f, \quad \dots, \mathrm{D}_{i_1 \dots i_k}^k f, \quad i, j, i_1, \dots, i_k \in \{1, \dots, n\},$$

where we are using the abbreviation $\mathrm{D}_i \equiv \partial/\partial x_i$. For any $k \in \mathbb{N}$, the space of **k-times continuously differentiable** functions (or just C^k space) is the space of those functions which are k-times differentiable with continuous derivatives:

$$C^k(\Omega) := \left\{ f \in C^0(\Omega) \ \middle| \ \exists \, D^k_{i_1 \ldots i_k} f \in C^0(\Omega), \ i_s = 1, \ldots, n, \ s = 1, \ldots, k \right\}.$$

The space of **infinitely differentiable** (or smooth or just C^∞) functions is

$$C^\infty(\Omega) := \bigcap_{k \in \mathbb{N}} C^k(\Omega).$$

Definition 9.33. The **subspace of real smooth functions on** Ω **with compact support** is given by

$$C^\infty_c(\Omega) := C^\infty(\Omega) \bigcap C^0_c(\Omega).$$

We recall that $C^0_c(\Omega)$ symbolises the space of real continuous functions $f : \Omega \longrightarrow \mathbb{R}$ with compact support in Ω, where the support of the function is the closure of the set on which the function is non zero:

$$\mathrm{supp}(f) = \overline{\left\{ x \in \Omega \ : \ f(x) \neq 0 \right\}}.$$

Recapping and utilising our newly introduced notation, the end goal of this section is to establish the density result

$$\overline{C^\infty_c(\Omega)}^{L^p(\Omega)} = L^p(\Omega),$$

whenever $p \in [1, \infty)$. We continue by introducing the convolution (algebraic) operation.

Definition 9.34. Let $f, g : \mathbb{R}^n \longrightarrow \mathbb{R}$ be measurable functions, such that, for a.e. $x \in \mathbb{R}^n$ the function $f(x - \cdot)g(\cdot)$ is integrable. We define the (measurable) function

$$f * g \ : \ \mathbb{R}^n \longrightarrow \mathbb{R}$$

by setting

$$(f * g)(x) := \int_{\mathbb{R}^n} f(x - y)g(y) \, \mathrm{d}y, \quad x \in \mathbb{R}^n.$$

The function $f * g$ is called the **convolution** of the functions f, g.

Remark 9.35. Note that the **convolution is defined only for functions living on the entire space** $\mathbb{R}^n$ **and not on bounded domains.** However, if the functions are defined on certain measurable subsets $A, B \subseteq \mathbb{R}^n$, one can extend them outside A, B by zero (namely by replacing f by $f\chi_A$ and g by $g\chi_B$) and consider their convolution as $(f\chi_A) * (g\chi_B)$.

The following important inequality provides sufficient conditions about the constraints we need to impose on f, g in order to guarantee that the function $f(x - \cdot)g(\cdot)$ is integrable for a.e. $x \in \mathbb{R}^n$.

Theorem 9.36 (Young inequality for convolutions). *Let $p \in [1, \infty]$, $f \in L^1(\mathbb{R}^n)$ and $g \in L^p(\mathbb{R})$. Then, the function $f * g$ is well defined a.e. on $\mathbb{R}^n$ and $f * g \in L^p(\mathbb{R}^n)$. Moreover,*

$$\|f * g\|_{L^p(\mathbb{R}^n)} \leq \|f\|_{L^1(\mathbb{R}^n)} \|g\|_{L^p(\mathbb{R}^n)}. \tag{$\bullet$}$$

As we shall see right after this theorem, the product $f * g$ is actually commutative and it hence the result remains true if instead $f \in L^p(\mathbb{R}^n)$ and $g \in L^1(\mathbb{R})$.

Proof. The desired inequality $(\bullet)$ is obvious if $g \in L^\infty(\mathbb{R}^n)$, because

$$|(f * g)(x)| \leq \|g\|_{L^\infty(\mathbb{R}^n)} \int_{\mathbb{R}^n} |f(x - y)| \, dy$$

for a.e. $x \in \mathbb{R}^n$, and the change of variables $z = x - y$ leads to $(\bullet)$. Hence it suffices to consider only the case $p < \infty$. Suppose first that $p = 1$ and for brevity set

$$h(x, y) := f(x - y)g(y).$$

Then, for a.e. $y \in \mathbb{R}^n$, by a translation of $f(\cdot - y)$ we get

$$\int_{\mathbb{R}^n} |h(x, y)| \, dx = \|f\|_{L^1(\mathbb{R}^n)} |g(y)|$$

and hence

$$\int_{\mathbb{R}^n} \int_{\mathbb{R}^n} |h(x, y)| \, dx \, dy = \|f\|_{L^1(\mathbb{R}^n)} \|g\|_{L^1(\mathbb{R}^n)}.$$

By the Fubini-Tonelli Theorem (Theorem 8.50), for a.e. $x \in \mathbb{R}^n$ we have that

$$\int_{\mathbb{R}^n} |h(x, y)| \, dy < \infty$$

and also

$$\|f * g\|_{L^1(\mathbb{R}^n)} = \int_{\mathbb{R}^n} \int_{\mathbb{R}^n} |h(x, y)| \, dy \, dx = \|f\|_{L^1(\mathbb{R}^n)} \|g\|_{L^1(\mathbb{R}^n)}.$$

Suppose now $p \in (1, \infty)$. For a.e. $x \in \mathbb{R}^n$ fixed, by the above we have that the next function is in $L^p(\mathbb{R}^n)$:

$$y \mapsto |f(x - y)|^{\frac{1}{p}} g(y).$$

By the Hölder inequality, the function $y \mapsto |f(x - y)| \|g(y)|$ is in $L^1(\mathbb{R}^n)$ and hence

$$
\begin{aligned}
|(f * g)(x)| &\leq \int_{\mathbb{R}^n} |f(x - y)| \|g(y)| \, dy \\
&\leq \left(\int_{\mathbb{R}^n} |f(x - y)| \|g(y)|^p \, dy \right)^{\frac{1}{p}} \left(\int_{\mathbb{R}^n} |f| \right)^{\frac{1}{p'}} \\
&\leq \left[\left(|f| * |g|^p \right)(x) \right]^{\frac{1}{p}} \left(\|f\|_{L^1(\mathbb{R}^n)} \right)^{\frac{1}{p'}}.
\end{aligned}
$$

By the case $p = 1$ proved previously, we have

$$\|f * g\|_{L^p(\mathbb{R}^n)}^p = \left\| |f * g|^p \right\|_{L^1(\mathbb{R}^n)}$$

$$\leq \|f\|_{L^1(\mathbb{R}^n)} \|g\|_{L^p(\mathbb{R}^n)}^p \left(\|f\|_{L^1(\mathbb{R}^n)} \right)^{\frac{1}{p'}}.$$

Since $1 - (1/p') = 1/p$, the desired inequality (•) follows and so does the theorem. □

By a change of variables formula (Theorem 8.52), the next result can also be easily proved:

Proposition 9.37. *Let $f \in L^1(\mathbb{R}^n)$, $g \in L^p(\mathbb{R}^n)$ and $h \in L^{p'}(\mathbb{R}^n)$.*

(i) *We have that*

$$f * g = g * f.$$

(ii) *We set*

$$\tilde{f}(x) := f(-x), \quad x \in \mathbb{R}^n.$$

(this is the composition of f with the reflection with respect to the origin). Then, we have

$$\int_{\mathbb{R}^n} (f * g)h = \int_{\mathbb{R}^n} (\tilde{f} * h)g.$$

We have already defined the support of a **continuous** function. This definition can be extended to measurable functions which are defined a.e. on $\mathbb{R}^n$ as follows:

Definition 9.38. Let $f : \mathbb{R}^n \longrightarrow \mathbb{R}$ be measurable. We define

$$\operatorname{supp}(f) := \mathbb{R}^n \setminus \bigcup \left\{ U \subseteq \mathbb{R}^n \text{ open} \,\middle|\, f = 0 \text{ a.e. on } U \right\}.$$

It can be shown[8] that this definition coincides with the one for continuous functions when $f \in C^0(\mathbb{R}^n)$. Moreover, it is independent of the pointwise values, being the same for all different versions of the function, in the sense of Definition 8.33. Namely, if $f = g$ a.e. on $\mathbb{R}^n$, then $\operatorname{supp}(f) = \operatorname{supp}(g)$. Further, the support is a closed set.

Remark 9.39. If $f \in L^1(\mathbb{R}^n)$, $g \in L^p(\mathbb{R}^n)$, then

$$\operatorname{supp}(f * g) \subseteq \operatorname{supp}(f) + \operatorname{supp}(g).$$

In particular, $f * g$ has compact support if **both** f, g have compact support. If one of the functions does not have compact support, the same in general happens for their convolution as well.

---

[8]The key is that $\mathbb{R}^n$ satisfies the first axiom of countability (recall Definition 4.26), which yields that any arbitrary union of open sets (as in the definition of the support) can be rewritten as a **countable** union of open balls with centres on $\mathbb{Q}^n$ and rational radii.

9.6.2 Mollification by convolution

In this section we show that by convoluting an L^1 function g with a (compactly supported) C^k function, then $f * g$ gains the regularity of f. Actually, the results hold more generally if g is not in $L^1(\mathbb{R}^n)$, but instead integrable only when restricted to *compact* sets. There exist numerous such functions: for instance, the function $f \equiv 1$ is not in $L^1(\mathbb{R}^n)$, but $f \in L^1(K)$ for any compact set K. This so-called **space of locally L^1 functions** is so abundant that there exists a notation for it:

$$L^1_{\text{loc}}(\mathbb{R}^n) := \left\{ g : \mathbb{R}^n \longrightarrow \mathbb{R} \text{ measurable} \,\Big|\, \forall R > 0, \ g \in L^1(\mathbb{B}_R(0)) \right\}.$$

The next result attests the claimed regularising effect on locally L^1 functions when convoluting them with smooth functions.

Theorem 9.40 (Regularisation by convolution). *Let $g \in L^1_{\text{loc}}(\mathbb{R}^n)$. Then the following hold:*

(i) *If $f \in C^0_c(\mathbb{R}^n)$, we have $f * g \in C^0(\mathbb{R}^n)$.*

(ii) *If $f \in C^1_c(\mathbb{R}^n)$, we have $f * g \in C^1(\mathbb{R}^n)$ and*

$$D_i(f * g) = (D_i f) * g. \tag{$\spadesuit$}$$

(iii) *If $f \in C^\infty_c(\mathbb{R}^n)$, then $f * g \in C^\infty(\mathbb{R}^n)$ and*

$$D^k_{i_1 \ldots i_k}(f * g) = \left(D^k_{i_1 \ldots i_k} f\right) * g,$$

for any partial k-th order derivative.

Proof. (i) For any point $x \in \mathbb{R}^n$ (not just a.e. $x \in \mathbb{R}^n$), the measurable function $y \mapsto f(x-y)g(y)$ is in $L^1(\mathbb{R}^n)$, as a result of Hölder's inequality. Indeed, $f(x-\cdot)$ vanishes outside the compact set $K_x := \text{supp}(f(x-\cdot))$ and $g \in L^1(K_x)$. Hence,

$$\int_{\mathbb{R}^n} |f(x - \cdot)g| \leq \|f\|_{L^\infty(K_x)} \int_{K_x} |g|.$$

Thus, $f * g$ is well defined everywhere on $\mathbb{R}^n$. Now we show continuity. Suppose that $x_m \longrightarrow x$ as $m \to \infty$ and set

$$\begin{cases} F_m(y) := f(x_m - y)g(y), \\ F(y) := f(x - y)g(y). \end{cases}$$

Since $f \in C^0(\mathbb{R}^n)$, we have $F_m \longrightarrow F$ a.e. on $\mathbb{R}^n$ as $m \to \infty$. Since $\text{supp}(f)$ is compact and $x_m \longrightarrow x$, there exists a compact set $K \subseteq \mathbb{R}^n$ such that

$$\text{supp}(f(x_m - \cdot)) \subseteq K.$$

Hence,

$$|F_m(y)| \leq \left(\max_K |f|\right) g(y)\, \chi_K(y) =: h(y)$$

and $h \in L^1(\mathbb{R}^n)$. By the Dominated Convergence Theorem, we have

$$(f * g)(x_m) \longrightarrow (f * g)(x) \text{ as } m \to \infty.$$

Hence, $f * g \in C^0(\mathbb{R}^n)$.

(ii) Let $D = (D_1, \ldots, D_n)$ denote the gradient, namely the vector of partial derivatives. Since Df is continuous and compactly supported in $C_c^0(\mathbb{R}^n, \mathbb{R}^n)$, it can be shown that it is uniformly continuous. Hence, there exists[9] a function (modulus of continuity) $\omega \in C^0[0, \infty)$, with $\omega(0) = 0$ and $\omega \geq 0$ such that

$$\left| Df(w) - Df(z) \right| \leq \omega\big(|w - z|\big), \quad w, z \in \mathbb{R}^n.$$

By Taylor's theorem, we have

$$\left| f(w + h) - f(w) - Df(w) \cdot h \right| \leq \int_0^1 \left| Df(w + th) - Df(w) \right| |h| \, dt$$
$$\leq |h| \, \omega(|h|).$$

Since both f, Df have compact supports, there exists a compact set $K \subseteq \mathbb{R}^n$ such that, for $w = x - y$ and x fixed,

$$\left| f(x + h - y) - f(x - y) - h \cdot Df(x - y) \right| \leq |h| \, \omega(|h|) \, \chi_K(y).$$

We multiply by $|g(y)|$ and integrate to find

$$\left| (f * g)(x + h) - (f * g)(x) - h \cdot \big((Df) * g\big)(x) \right| \leq |h| \, \omega(|h|) \int_K g.$$

Consequently, $f * g$ is differentiable at $x \in \mathbb{R}^n$ and in addition the identity (♠) holds.

(iii) Follows by induction. $\qquad \square$

By utilising convolution with smooth compactly supported functions, we may define the next systematic regularisation of a function, which we shall show that actually approximates the function. The idea is to average the values of the function near each point by convoluting with a sequence of smooth functions which have "small" supports, asymptotically concentrating near that point.

---

[9]This is an equivalent **extremely convenient** way to express uniform continuity. Indeed, for any function $F : \mathbb{R}^n \longrightarrow \mathbb{R}^d$, the proof boils down to defining

$$\omega(t) := \sup \left\{ |F(w) - F(z)| : |w - z| \leq t \right\}, \quad t \geq 0,$$

and checking that F is uniformly continuous on $\mathbb{R}^n$ if and only if the non-negative increasing function ω is continuous at 0. We leave the verification of the details for the reader.

Definition 9.41. Let $f \in L^1_{\text{loc}}(\mathbb{R}^n)$. We call **mollifier of** f every sequence of the form

$$\left(f * \eta_m\right)_{m=1}^\infty \subseteq C^\infty(\mathbb{R}^n)$$

where for each $m \in \mathbb{N}$,

$$\eta_m \in C_c^\infty(\mathbb{R}^n), \quad \text{supp}(\eta_m) \subseteq \mathbb{B}_{1/m}(0), \quad \int_{\mathbb{R}^n} \eta_m = 1, \quad \eta_m \geq 0.$$

The terms of the sequence $(\eta_m)_{m=1}^\infty \subseteq C_c^\infty(\mathbb{R}^n)$ itself will also be referred to as **mollifiers**.

There are many possible choices of $(\eta_m)_1^\infty$. One of those (the most popular) is shown in Figure 9.10.

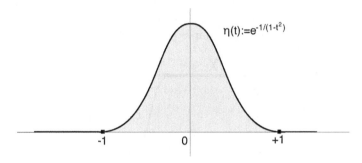

Figure 9.10: Illustration of the function defining the standard mollifier.

More precisely, $(\eta_m)_1^\infty$ is constructed by choosing

$$\eta(t) := \begin{cases} e^{-1/(1-t^2)}, & |t| < 1, \\ 0, & |t| \geq 1, \end{cases}$$

and then, we take

$$\eta_m(x) := \frac{m^n}{\int_{-1}^1 \eta} \eta(m|x|).$$

Herein we shall also make this "canonical" choice of sequence of mollifiers[10]. Recall that by Theorem 9.40, we know that indeed $f * \eta_m \in C^\infty(\mathbb{R}^n)$, when $f \in L^1_{\text{loc}}(\mathbb{R}^n)$.

We now finally show that the mollifier $(f * \eta_m)_1^\infty$ indeed approximates f as $m \to \infty$, in the following way:

---

[10]We note that it is very common in the literature to consider families $\{\eta^\varepsilon : \varepsilon > 0\} \subseteq C_c^\infty(\mathbb{R}^n)$ of mollifiers instead of just sequences, defined by replacing $1/m$ by ε, that is as

$$\eta^\varepsilon(x) := \frac{1}{\varepsilon^n \int_{-1}^1 \eta} \eta\left(\frac{|x|}{\varepsilon}\right), \quad x \in \mathbb{R}^n, \quad \varepsilon > 0.$$

Theorem 9.42 (Approximation by mollifiers).

(i) *For any $f \in C^0(\mathbb{R}^n)$, $f * \eta_m$ converges to f uniformly on compact subsets[11] of $\mathbb{R}^n$ as $m \to \infty$:*

$$\text{for any compact } K \subseteq \mathbb{R}^n, \quad \max_K \left| f * \eta_m - f \right| \longrightarrow 0 \text{ as } m \to \infty.$$

(ii) *Suppose that $p \in [1, \infty)$. For any $f \in L^p(\mathbb{R}^n)$, we have*

$$f * \eta_m \longrightarrow f \text{ in } L^p(\mathbb{R}^n), \text{ as } m \to \infty.$$

Remark 9.43. Part (ii) of Theorem 9.42 does **not** extend to the case of $p = \infty$. If it were true, the sequence of smooth functions $(f * \eta_m)_1^\infty$ would be Cauchy uniformly on each compact subset of $\mathbb{R}^n$ and hence we would have $f \in C^0(\mathbb{R}^n)$. This is an obvious contradiction if we attempt to approximate any $f \in L^\infty(\mathbb{R}^n)$ which is **discontinuous**. The characteristic function of an interval $[a, b] \subseteq \mathbb{R}$ offers such an example (see Figure 9.11).

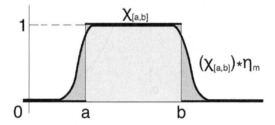

Figure 9.11: Illustration of the mollification of the characteristic function of an interval $[a, b] \subseteq \mathbb{R}$. For any $m \in \mathbb{N}$, the mollified function $(\chi_{[a,b]}) * \eta_m$ is smooth and as $m \to \infty$ it approximates a.e. and in $L^p(\mathbb{R})$ the discontinuous function $\chi_{[a,b]}$. However, the convergence cannot be uniform!

Proof. (i) Recall that every continuous function is uniformly continuous on compact sets. Hence, if $K \subseteq \mathbb{R}^n$ is compact, for any $\varepsilon > 0$, there exists $\delta > 0$ such that, for $x \in K$ and $y \in \mathbb{B}_\delta(0)$, we have

$$\left| f(x - y) - f(x) \right| < \varepsilon.$$

Since $\eta_m(y) = 0$ when $|y| > 1/m$, for any $m > 1/\delta$ we have

$$\left| (f * \eta_m - f)(x) \right| = \left| \int_{\mathbb{R}^n} \left(f(x - y) - f(x) \right) \eta_m(y) \, dy \right|$$

$$\leq \varepsilon \int_{\mathbb{R}^n} \eta_m(y) \, dy$$

$$= \varepsilon,$$

as desired.

---

[11] This is the convergence over compact subsets, or "locally uniform convergence".

(ii) Fix a function $f \in L^p(\mathbb{R}^n)$ and $\varepsilon > 0$. Then, by the density of compactly supported continuous functions in $L^p(\mathbb{R}^n)$, there exists $g \in C_c^0(\mathbb{R}^n)$ such that

$$\|f - g\|_{L^p(\mathbb{R}^n)} < \varepsilon.$$

Moreover, $g * \eta_m \in C_c^\infty(\mathbb{R}^n)$ and by Remark 9.39, there exists a compact set $K \subseteq \mathbb{R}^n$ such that

$$\operatorname{supp}(g * \eta_m), \ \operatorname{supp}(g) \subseteq K, \ \text{for } m \in \mathbb{N}.$$

Hence, for m large enough, by part (i) we have

$$\|g * \eta_m - g\|_{L^p(\mathbb{R}^n)} \leq |K|^{\frac{1}{p}} \|g * \eta_m - g\|_{L^\infty(K)} \leq \varepsilon.$$

Finally, by the Young inequality for convolutions, for m large we obtain

$$\|f * \eta_m - f\|_{L^p(\mathbb{R}^n)} \leq \|\eta_m * (f - g)\|_{L^p(\mathbb{R}^n)} + \|g * \eta_m - g\|_{L^p(\mathbb{R}^n)}$$
$$+ \|g - f\|_{L^p(\mathbb{R}^n)}$$
$$\leq 3\varepsilon.$$

The conclusion ensues and hence the theorem has been established. $\square$

We finally utilise the above convergence result to establish the desired density consequence.

Theorem 9.44 (Density of smooth functions with compact support).
For any $p \in [1, \infty)$ and any open domain $\Omega \subseteq \mathbb{R}^n$, the space of smooth functions on Ω with compact support is dense in the space of p-integrable functions on Ω in the L^p-norm:

$$\overline{C_c^\infty(\Omega)}^{L^p(\Omega)} = L^p(\Omega).$$

Proof. Fix a function $f \in L^p(\Omega)$. The idea is, roughly speaking, to extend f by zero on $\mathbb{R}^n \setminus \Omega$ and mollify it. However, without some care, this method is too coarse to produce the desired approximations because the supports of the mollified function may not be contained in Ω (unless f happens to have compact support itself). The remedy is to approximate first f by compactly supported cut-offs, namely functions that coincide with f except in an inner zone narrow zone near the boundary (whose width asymptotically decreases to zero) where they vanish.

To this end, we fix $\varepsilon > 0$ small and define (see Figure 9.12):

$$\Omega^\varepsilon := \mathbb{B}_{1/\varepsilon}(0) \bigcap \left\{ x \in \Omega : \operatorname{dist}(x, \partial\Omega) > \varepsilon \right\}.$$

The reason we intersect with the ball of radius $1/\varepsilon$ is that the domain may not be bounded. Next, consider the extension $f\chi_{\Omega^\varepsilon}$ of f from Ω^ε to $\mathbb{R}^n$ by defining it as zero on $\mathbb{R}^n \setminus \Omega^\varepsilon$, that is

$$f\chi_{\Omega^\varepsilon} = \begin{cases} f, & \text{on } \Omega^\varepsilon, \\ 0, & \text{on } \mathbb{R}^n \setminus \Omega^\varepsilon. \end{cases}$$

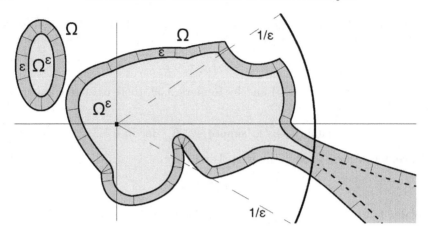

Figure 9.12: Illustration of the inner approximation Ω^ε.

Then, $f\chi_{\Omega^\varepsilon} \in L^p(\mathbb{R}^n)$. Moreover, for any $x \in \Omega$,

$$f\chi_{\Omega^\varepsilon}(x) \longrightarrow f(x), \quad \text{as } \varepsilon \to 0,$$

because for the given $x \in \Omega$, if we choose $\varepsilon < \text{dist}(x, \partial\Omega)$ we have $f\chi_{\Omega^\varepsilon}(x) = f(x)$. Furthermore, by construction we have $|f\chi_{\Omega^\varepsilon}| \leq |f|$ on Ω. By choosing $\varepsilon := 1/k$ for some large $k \in \mathbb{N}$, the Dominated Convergence Theorem implies that

$$f\chi_{\Omega^{1/k}} \longrightarrow f \quad \text{in } L^p(\Omega), \text{ as } k \to \infty. \tag{$\circledast$}$$

Fix now $k \in \mathbb{N}$ and consider the mollification of the inner approximation:

$$f_{m,k} := \left(f\chi_{\Omega^{1/k}}\right) * \eta_m, \quad m \in \mathbb{N}.$$

By Theorem 9.42, we have $(f_{m,k})_{m=1}^\infty \subseteq C_c^\infty(\mathbb{R}^n)$ and for any **fixed** $k \in \mathbb{N}$, we have

$$f_{m,k} \longrightarrow f\chi_{\Omega^{1/k}} \quad \text{in } L^p(\mathbb{R}^n), \text{ as } m \to \infty. \tag{$\circledast\circledast$}$$

Further, by Remark 9.39 we have that, there exists $m(k) \in \mathbb{N}$ is large enough, such that

$$\text{supp}(f_{m,k}) \subseteq \Omega, \quad m \geq m(k).$$

Hence, $(f_{m,k})_{m=m(k)}^\infty \subseteq C_c^\infty(\Omega)$ as well. Finally, we have

$$\left\|f - f_{m,k}\right\|_{L^p(\Omega)} \leq \left\|f - f\chi_{\Omega^{1/k}}\right\|_{L^p(\Omega)} + \left\|f\chi_{\Omega^{1/k}} - f_{m,k}\right\|_{L^p(\mathbb{R}^n)}.$$

For any $\delta > 0$, we may choose $k(\varepsilon) \in \mathbb{N}$ so that that (by ($\circledast$)) the first term on the right hand side is less that $\delta/2$, for any $k \geq k(\varepsilon)$. Further, for $k = k(\varepsilon)$, we may choose $m(\varepsilon) \geq m(k(\varepsilon)) \in \mathbb{N}$ so that (by ($\circledast\circledast$)) the second term on the right hand side is less that $\delta/2$, for any $m \geq m(\varepsilon)$. The theorem ensues. $\qquad\square$

9.7 Exercises

Exercise 9.1. Let p, r be such that $1 \leq p < r \leq \infty$. It is known that $\ell^p \subseteq \ell^r$. Show that on the vector space ℓ^p (where both norms $\|\cdot\|_{\ell^p}, \|\cdot\|_{\ell^r}$ can be defined) the norms $\|\cdot\|_{\ell^p}$ and $\|\cdot\|_{\ell^r}$ are not equivalent.

Exercise 9.2.

(i) Let $x^{(0)} = (1, 1/2, 1/3, \ldots, 1/k, \ldots)$. Show that $x^{(0)} \in \ell^2$ and $x^{(0)} \notin \ell^1$.

(ii) By suitably modifying the example in part (i), or otherwise, prove that for any p, r with $1 \leq p < r \leq \infty$, we have that $\ell^p \neq \ell^r$.

Exercise 9.3. Consider the normed space $(\ell^2, \|\cdot\|_2)$, and let

$$Y := \left\{ x = (x_1, x_2, \ldots) \in \ell^2 \,\middle|\, \exists\, K \in \mathbb{N} : \ x_k = 0, \ \forall\, k > K \right\}.$$

Show that Y is a vector subspace of ℓ^2, and find $\overline{Y}$, the closure of Y in $(\ell^2, \|\cdot\|_2)$.

Exercise 9.4. Let $f \in L^\infty(\Omega)$, $\Omega \subseteq \mathbb{R}^n$ measurable with $|\Omega| < \infty$. Show that

$$\lim_{p \to \infty} \|f\|_{L^p(\Omega)} = \|f\|_{L^\infty(\Omega)}.$$

Exercise 9.5. Let $f \in L^\infty(\Omega)$, $\Omega \subseteq \mathbb{R}^n$ measurable with $|\Omega| < \infty$. Show that

$$\lim_{p \to 1+} \|f\|_{L^p(\Omega)} = \|f\|_{L^1(\Omega)}.$$

Exercise 9.6. Let f, $(f_m)_1^\infty$ be in $L^p(\Omega)$ with $\Omega \subseteq \mathbb{R}^n$ measurable and $1 \leq p < \infty$. Show that if $f_m \longrightarrow f$ as $m \to \infty$ in $L^p(\Omega)$, then for each $\varepsilon > 0$, we have

$$\left| \{ |f_m - f| > \varepsilon \} \right| \longrightarrow 0, \quad \text{as } m \to \infty.$$

The above kind of convergence is called **convergence in measure**.

Exercise 9.7. Let $\Omega \subseteq \mathbb{R}^n$ be open, bounded and $1 \leq p < \infty$. Let also $w \in C^0(\overline{\Omega})$ with $w > 0$ on $\overline{\Omega}$. We consider the normed space

$$L^p(\Omega, w) := \left\{ f \in L^1_{\mathrm{loc}}(\Omega) \,\middle|\, \|f\|_{L^p(\Omega, w)} := \left(\int_\Omega |f(x)|^p w(x)\, \mathrm{d}x \right)^{1/p} < \infty \right\}$$

Show that $L^p(\Omega, w)$ is a Banach space. This space is called the **weighted L^p space** with weight function w.

Exercise 9.8. (a) Let $f \in (C^0 \cap L^1)(0, \infty)$ with $f \geq 0$. Show that there exists $t_m \to \infty$ as $m \to \infty$ such that $f(t_m) \longrightarrow 0$.

(b) Show that there is $g \in (C^0 \cap L^1)(0, \infty)$ with $g \geq 0$ such that

$$\limsup_{t \to \infty} g(t) = 1.$$

Exercise 9.9. Modify the function g of (b) above appropriately to some $h \in (C^0 \cap L^1)(0, \infty)$ with $h \geq 0$ in order to achieve

$$\limsup_{t \to \infty} g(t) = \infty.$$

Exercise 9.10. On the measurable space $(\mathbb{N}, \mathcal{P}(\mathbb{N}))$ consider the counting measure $\sharp : \mathcal{P}(X) \longrightarrow [0, \infty]$.

(1) Show that $\mathcal{P}(X)$ is indeed a σ-algebra.

(2) Show that $\sharp$ is a measure on $(\mathbb{N}, \mathcal{P}(\mathbb{N}))$.

(3) Show that all functions $f : \mathbb{N} \longrightarrow \mathbb{R}$ are measurable.

(4) Show that for all $f : \mathbb{N} \longrightarrow [0, \infty)$ it holds that

$$\int_{\mathbb{N}} f(n) \, d\sharp(n) = \sum_{n=1}^{\infty} f(n).$$

(Namely "series do not exist", in the sense that they actually are integrals over $\mathbb{N}$ with respect to the counting measure!)

(5) Show that f is integrable if and only if $\sum_{n=1}^{\infty} |f(n)| < \infty$ (absolutely convergent).

(6) Let $(a_{k,n})_{k,n}$ be a double sequence. Prove the following assertion: Assume there exists a non-negative sequence $(b_n)_1^{\infty}$ with $\sum_{n=1}^{\infty} b_n < \infty$ and for all k and n holds $|a_{k,n}| \leq b_n$. If for each n the limit $\lim_{k \to \infty} a_{k,n}$ exists then

$$\lim_{k \to \infty} \sum_{n=1}^{\infty} a_{k,n} = \sum_{n=1}^{\infty} \lim_{k \to \infty} a_{k,n}$$

[Hint: Use the Dominated Convergence Theorem].

(7) By using (6), deduce that

$$\sum_{n=1}^{\infty} \sum_{k=1}^{\infty} a_{k,n} = \sum_{k=1}^{\infty} \sum_{n=1}^{\infty} a_{k,n}.$$

Exercise 9.11 (Mean value theorem I). Let $(X, \mathfrak{M}, \mu)$ be a measure space and $f \in L^1(X, \mu)$. Let also $A \in \mathfrak{M}$ and a, b be real numbers. If $a \leq f(x) \leq b$, for all $x \in A$ then

$$a \, \mu(A) \leq \int_X f \, d\mu \leq b \, \mu(A).$$

Exercise 9.12 (Mean value theorem II). Consider the measure space $(X, \mathfrak{M}, \mu) = (\mathbb{R}, \mathcal{L}(\mathbb{R}), \mathcal{L}^1)$. Show by an example that if f is a discontinuous measurable function in $L^1((a, b)) \setminus C^0([a, b])$ on then compact interval $[a, b] \subseteq \mathbb{R}$, then it may happen that there exists no point $\xi \in [a, b]$ such that

$$\int_a^b f(x)\, \mathrm{d}x = f(\xi)(b - a).$$

Hence, the second form of the mean value theorem is **not** in general true for discontinuous functions.

Exercise 9.13. Show that the mollification by convolution respect functional inequalities in the following sense: if $f \leq g$ a.e. on $\mathbb{R}^n$ and $f, g \in L^1_{\mathrm{loc}}(\mathbb{R}^n)$, then $f^\varepsilon \leq g^\varepsilon$ on $\mathbb{R}^n$ for any $\varepsilon > 0$. Deduce that if you mollify a function $f \in L^\infty(\mathbb{R}^n)$, then

$$\|f^\varepsilon\|_{L^\infty(\mathbb{R}^n)} \leq \|f\|_{L^\infty(\mathbb{R}^n)}.$$

Chapter 10

Inner product spaces and Hilbert spaces

10.1 Euclidean structures past the Euclidean realm

In the previous chapters we have studied many instances of infinite-dimensional normed vector spaces quite systematically. A normed vector space $(X, \|\cdot\|)$ is an axiomatic generalisation of the Euclidean space, in the sense that the existence of a norm which allows us to define *distances* between vectors is assumed. However, the existence of norms is not the unique property of the Euclidean spaces. In particular, there is no immediate way one can talk about *angles* between vectors on a general normed space, a central geometric feature of the Euclidean structure.

This chapter is devoted to the study of infinite-dimensional normed spaces which admit more structure than a mere norm and on them we may study geometric questions, like angles between vectors. A central theme in this regard is an extension to arbitrary dimensions of the concept of **orthogonality**, known to the reader from classical Euclidean three-dimensional geometry. The path to such extensions is provided by the observation that orthogonality of two three-dimensional vectors can be characterised analytically by the fact that their **inner product** (also known as **dot product**) vanishes. On the vector space $\mathbb{F}^m$, $m \in \mathbb{N}$, the standard inner product of any two vectors $x = (x_1, \ldots, x_m), y = (y_1, \ldots, y_m)$ is given by

$$\langle x, y \rangle := \sum_{k=1}^{m} x_k \overline{y_k},$$

where the overline denotes complex conjugation[1] in the case $\mathbb{F} = \mathbb{C}$. Note that the Euclidean norm $\|\cdot\|_2$ itself may be defined by means of the inner product:

$$\|x\|_2 = \langle x, x \rangle^{1/2} \quad \text{for any } x \in \mathbb{F}^m.$$

In this chapter we shall concern ourselves with the study of **inner product**

---

[1] Let us recall that the reason the complex conjugation is required in the definition of the inner product on $\mathbb{C}^m$ is because the inner product of any vector with itself has to be a real nonnegative number, hence allowing the Euclidean norm to be recovered from it.

spaces. These are vector spaces endowed with an **inner product**, which is a function assigning a certain scalar to any pair of vectors. The concept is motivated by the case of $\mathbb{F}^m$, where the inner product can be seen as a function in two variables

$$\langle \cdot, \cdot \rangle \; : \; \mathbb{F}^m \times \mathbb{F}^m \longrightarrow \mathbb{F}, \quad (x, y) \mapsto \langle x, y \rangle$$

which is linear in the first variable and conjugate linear in the second variable. The abstract definition is carefully crafted in such a way that the essential properties of the Euclidean dot product are now postulated as axioms. Any general inner product space is automatically a normed space and, in the prominent case that the inner product space is also complete, it is called a **Hilbert space**.

Among the Banach spaces, the Hilbert spaces are, due to their additional geometric structure, the most accessible to our intuition. Many of the results proved in this chapter are extensions of results familiar from the Euclidean setting. For example, we shall prove that, given any closed vector subspace of a Hilbert space, any vector in the space admits an **orthogonal projection** onto it, i.e., it can be decomposed uniquely into a part which belongs to the subspace and a part that is orthogonal to it. We have already encountered examples of Hilbert spaces in previous chapters. The space ℓ^2 defined in Chapter 2 and the space $L^2(X, \mu, \mathbb{F}^d)$ over a measure space $(X, \mathfrak{M}, \mu)$ defined in Chapter 9 are such prototypical examples of paramount importance for applications.[2]

A central feature of the Euclidean space which makes life easier in Linear Algebra is the existence of **orthonormal bases**. We shall show that a generalisation of this idea exists for inner product spaces as well, although it is essentially different from its algebraic counterpart since we necessarily need *infinitely many basis vectors* and therefore the analytic concept of convergence comes into play. Orthonormal bases provide an infinite family of directions along which any vector in the space can be decomposed, in the sense of being equal to the sum of the series of its orthogonal projections along these directions.

When encountering a class of new mathematical objects, a natural question is their **classification**. Namely, we wish to know how many different "kinds" of such objects exist. Hilbert spaces are the "next best thing" after finite-dimensional vector spaces in regards to the question of their classification. To this end, we establish the so-called **Riesz–Fischer Theorem**, an important result which states that any separable Hilbert space is isometrically isomorphic to the space ℓ^2 (whose elements are square summable sequence of numbers). Therefore, any separable Hilbert space is indistinguishable from ℓ^2 in regards to their linear structure, although they may consist of totally different building stones! We will also close by discussing briefly the non-separable case.

---

[2]For instance, the axiomatic study of Hilbert spaces and linear mapping between them was largely motivated in the early 20th century by developments in Theoretical Physics, in particular the discovery of the so-called wavefunctions in Quantum Mechanics.

10.2 Inner products on vector spaces

We start with the definition of an **inner product** on a vector space, which, as pointed out in the introduction, represents an axiomatisation of some essential properties of the familiar **dot product** in Euclidean spaces. We note that, throughout what follows, $\mathbb{F}$ denotes either of the fields of real or complex numbers ($\mathbb{F}$ equals either $\mathbb{R}$ or $\mathbb{C}$).

Definition 10.1. Let X be a vector space over the field $\mathbb{F}$. An **inner product** on X is a function

$$\langle \cdot, \cdot \rangle \;:\; X \times X \longrightarrow \mathbb{F}$$

with the following properties:

(IP1) $\langle x, x \rangle \geq 0$ for all $x \in X$, and $\langle x, x \rangle = 0$ if and only if $x = 0$;

(IP2) $\langle \lambda x + \mu y, z \rangle = \lambda \langle x, z \rangle + \mu \langle y, z \rangle$ for all $x, y, z \in X, \lambda, \mu \in \mathbb{F}$;

(IP3) $\langle y, x \rangle = \overline{\langle x, y \rangle}$ for all $x, y \in X$.

An **inner product space** is a pair $(X, \langle \cdot, \cdot \rangle)$, where X is a vector space and $\langle \cdot, \cdot \rangle$ is an inner product on X.

The first, and at the same time, the simplest example is that of the **Euclidean space** itself, endowed with the usual dot product.

Example 10.2. On the vector space $\mathbb{F}^m$, $m \in \mathbb{N}$, let[3]

$$\langle x, y \rangle := \sum_{k=1}^{m} x_k \overline{y_k} \quad \text{for all } x = (x_1, \ldots, x_m), y = (y_1, \ldots, y_m) \in \mathbb{F}^m.$$

Then we have that $\langle \cdot, \cdot \rangle$ is an inner product on $\mathbb{F}^m$.

A similarly simple construction of an inner product can be carried out in any finite-dimensional vector space, once an algebraic basis is chosen. Note also that the inner product in the example below depends in an essential way on the basis chosen.

Example 10.3. Let X be a finite-dimensional vector space, and let $\{e_1, \ldots, e_m\}$ be a basis for X. Then, for any

$$x = \sum_{k=1}^{m} \lambda_k e_k, \quad y = \sum_{k=1}^{m} \mu_k e_k, \quad \text{where } \lambda_1, \ldots, \lambda_m, \mu_1, \ldots, \mu_m \in \mathbb{F},$$

---

[3]Note that in Chapter 2 (recall Remark 2.73) we used the alternative notation "$x \cdot \overline{y}$" for the exact same inner product $\langle x, y \rangle$ we give here. The notation "$x \cdot \overline{y}$" will still be used in this Chapter too, but will be employed in the case we have functional spaces in order to avoid confusion due to the clash of the different inner product notations on $\mathbb{F}^m$ and on the infinite-dimensional functional space, see Example 10.6 that follows.

we define $\langle \cdot, \cdot \rangle : X \times X \longrightarrow \mathbb{F}$ by taking

$$\langle x, y \rangle := \sum_{k=1}^{m} \lambda_k \overline{\mu_k}.$$

Then $\langle \cdot, \cdot \rangle$ is an inner product on X.

The extension to an infinite-dimensional setting of the above constructions leads us to an inner product on the sequence space ℓ^2.

Example 10.4. On the space ℓ^2 of square-summable sequences $\mathbb{N} \longrightarrow \mathbb{F}$ given in Definition 2.76, consider the function $\langle \cdot, \cdot \rangle_{\ell^2} : \ell^2 \times \ell^2 \longrightarrow \mathbb{F}$ defined as

$$\langle x, y \rangle_{\ell^2} := \sum_{k=1}^{\infty} x_k \overline{y_k} \quad \text{for all } x = (x_1, x_2, \ldots), \, y = (y_1, y_2, \ldots) \in \ell^2.$$

Note that the above series converges, as a consequence of Hölder's Inequality (Lemma 2.72) and of the fact that absolute convergence implies convergence. It can be easily shown that $\langle \cdot, \cdot \rangle_{\ell^2}$ is an inner product on the space ℓ^2.

The next two examples of inner product spaces involve integrals. Example 10.5 deals with continuous functions on an interval, and the inner product involves the Riemann integral. Example 10.6 provides a natural inner product on the Lebesgue space L^2 of square-integrable mappings $X \longrightarrow \mathbb{F}^d$ defined on a measure space $(X, \mathfrak{M}, \mu)$. The L^p spaces have already been studied quite extensively in the previous chapter and will be revisited again in Chapter 14. The particular case of $L^2(X, \mu, \mathbb{F}^d)$ is the most important example of an inner product space, since in fact it encompasses as particular cases all the preceding examples, by choosing appropriately the measure μ (recall Definition 9.1, Remark 9.8 and Exercise 9.10). In particular,

$$\ell^2 = L^2(\mathbb{N}, \sharp, \mathbb{F})$$

where "$\sharp$" is the counting measure on $\mathbb{N}$ (equivalently, the zero-dimensional Hausdorff measure $\mathcal{H}^0$).

Example 10.5. On the space $C([a, b], \mathbb{F})$ of continuous functions $[a, b] \longrightarrow \mathbb{F}$ given in Definition 2.79, consider the function

$$\langle \cdot, \cdot \rangle \; : \; C([a, b], \mathbb{F}) \times C([a, b], \mathbb{F}) \longrightarrow \mathbb{F}$$

given by

$$\langle f, g \rangle := \int_a^b f(x) \overline{g(x)} \, dx \quad \text{for all } f, g \in C([a, b], \mathbb{F}).$$

Then it can easily checked that $\langle \cdot, \cdot \rangle$ is an inner product on $C([a, b], \mathbb{F})$.

Example 10.6. Let $(X, \mathfrak{M}, \mu)$ be a measure space and $d \in \mathbb{N}$. Consider the vector space $L^2(X, \mu, \mathbb{F}^d)$ of square μ-integrable mappings $X \longrightarrow \mathbb{F}^d$, given by Definition 9.1. Then, the function

$$\langle \cdot, \cdot \rangle_{L^2} \; : \; L^2(X, \mu, \mathbb{F}^d) \times L^2(X, \mu, \mathbb{F}^d) \longrightarrow \mathbb{R}$$

which is defined by taking

$$\langle f, g \rangle_{L^2(X, \mu, \mathbb{F}^d)} := \int_X f \cdot \overline{g} \, d\mu$$

is an inner product on $L^2(X, \mu, \mathbb{F}^d)$. Here $f \cdot \overline{g}$ symbolises the dot product on $\mathbb{F}^d$ (recall Remark 2.73, Corollary 2.74 and the comments in Example 10.2). Note that $f \cdot \overline{g}$ is a scalar integrable function on X, as a consequence of the vector-valued version of Hölder's Inequality, Corollary 2.81 and Lemma 9.9. Further, the norm $\| \cdot \|_{L^2(X, \mu, \mathbb{F}^d)}$ we gave in Definition 9.1 arises from the inner product, in the sense that:

$$\|f\|_{L^2(X, \mu, \mathbb{F}^d)} = \left(\langle f, f \rangle_{L^2(X, \mu, \mathbb{F}^d)} \right)^{1/2} \quad f \in L^2(X, \mu, \mathbb{F}^d).$$

The next result shows that **any inner product on a vector space naturally induces a norm** on the same space. A preliminary step in the proof of this important fact is the validity of the **Cauchy–Schwarz Inequality**, a result of great interest also in itself. **Throughout this chapter, any inner product space will automatically be considered as a normed space with the induced norm.**

Proposition 10.7. *Let $(X, \langle \cdot, \cdot \rangle)$ be an inner product space. Then the following hold:*

(i) $|\langle x, y \rangle|^2 \leq \langle x, x \rangle \langle y, y \rangle$ *for all $x, y \in X$.*

(ii) *The function $\| \cdot \| : X \longrightarrow \mathbb{R}$ given by*

$$\|x\| := \langle x, x \rangle^{1/2} \quad \text{for all } x \in X,$$

is a norm on X.

The objects introduced above have the following particular names:

Definition 10.8. In the setting of Proposition 10.7:

• The inequality in (i) is called the **Cauchy–Schwarz Inequality**.

• The norm in (ii) is called the **norm induced by the inner product**.

Proof. (i) The required inequality is obviously true if $y = 0$. We assume in what follows that $y \neq 0$. For any $\lambda \in \mathbb{F}$, by the axioms defining an inner product, we have that

$$0 \leq \langle x - \lambda y, x - \lambda y \rangle = \langle x, x \rangle - \overline{\lambda} \langle x, y \rangle - \lambda \langle y, x \rangle + |\lambda|^2 \langle y, y \rangle$$
$$= \langle x, x \rangle - 2 \operatorname{Re}(\lambda \langle y, x \rangle) + |\lambda|^2 \langle y, y \rangle.$$

Note that, in the case when X is real vector space, the above expression is a quadratic binomial function of λ, which is minimised at $\lambda_0 := \langle x, y \rangle / \langle y, y \rangle$. Evaluating the expression at λ_0, irrespectively of whether the scalar field is $\mathbb{R}$ or $\mathbb{C}$, we obtain

$$0 \leq \langle x, x \rangle - \frac{|\langle x, y \rangle|^2}{\langle y, y \rangle},$$

which is precisely the claimed inequality.

(ii) For any $x, y \in X$, by using (i), we have

$$\|x + y\|^2 = \|x\|^2 + 2\mathrm{Re}(\langle x, y \rangle) + \|y\|^2 \leq \left(\|x\| + \|y\|\right)^2.$$

Thus, the triangle inequality is satisfied. The other axioms of a norm can be easily verified by the reader and hence we leave the details as an exercise. □

Remark 10.9. In an inner product space $(X, \langle \cdot, \cdot \rangle)$, the **Cauchy–Schwarz Inequality** is usually written in the equivalent form

$$|\langle x, y \rangle| \leq \|x\| \|y\| \quad \text{for all } x, y \in X.$$

In any inner product space, the inner product and the induced norm satisfy some nice algebraic identities which we record in the next result. These identities turn out to play an important role in certain subsequent calculations involving inner products.

Theorem 10.10. *Let* $(X, \langle \cdot, \cdot \rangle)$ *be an inner product space with induced norm* $\| \cdot \|$. *Then the following hold:*

(i) $\|x + y\|^2 + \|x - y\|^2 = 2(\|x\|^2 + \|y\|^2)$ *for all* $x, y \in X$.

(ii) *If X is a real inner product space, then*

$$4\langle x, y \rangle = \|x + y\|^2 - \|x - y\|^2 \quad \text{for all } x, y \in X.$$

(iii) *If X is a complex inner product space, then*

$$4\langle x, y \rangle = \|x+y\|^2 - \|x-y\|^2 + i\|x+iy\|^2 - i\|x-iy\|^2 \quad \text{for all } x, y \in X.$$

Proof. The proof is a direct calculation and hence the verification of the details is left as an easy exercise for the reader. □

The identities given in Theorem 10.10 have the next particular names:

Definition 10.11. In the setting of Theorem 10.10:

- The identity (i) is called the **parallelogram identity** (see Figure 10.1).

- The identity (ii) is called the **real polarisation identity**.

- The identity (iii) is called the **complex polarisation identity**.

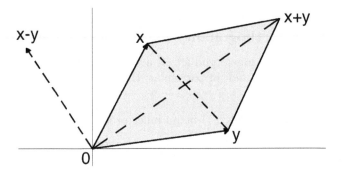

Figure 10.1: An illustration of the familiar parallelogram law on $\mathbb{R}^2$.

Remark 10.12. It can be shown that a certain converse of part (i) in Theorem 10.10 holds: *If $(X, \| \cdot \|)$ is a normed vector space such that the parallelogram identity holds, then there exists an inner product $\langle \cdot, \cdot \rangle$ on X for which the induced norm is exactly $\| \cdot \|$.* In the proof, $\langle \cdot, \cdot \rangle$ is defined by means of the polarisation identity, and then the axioms of an inner product are verified. Since, despite its elegant form, this result does not play a very important role in the theory of inner product spaces, we refrain from giving any details of the proof.

The following result deals with the passage to the limit in inner products along sequences of vectors.

Proposition 10.13. *Let $(X, \langle \cdot, \cdot \rangle)$ be an inner product space. Then, for any $(x_n)_1^\infty, (y_n)_1^\infty \subseteq X$ and any $x_0, y_0 \in X$, the following holds: if $x_n \longrightarrow x_0$ and $y_n \longrightarrow y_0$ in X as $n \to \infty$, then $\langle x_n, y_n \rangle \longrightarrow \langle x_0, y_0 \rangle$ in $\mathbb{F}$ as $n \to \infty$.*

Proof. By utilising the Cauchy–Schwarz Inequality, we obtain

$$\left| \langle x_n, y_n \rangle - \langle x_0, y_0 \rangle \right| = \left| \langle x_n - x_0, y_n \rangle + \langle x_0, y_0 - y_n \rangle \right|$$
$$\leq \|x_n - x_0\| \|y_n\| + \|x_0\| \|y_n - y_0\|,$$

from which the claimed result is immediate. $\qquad \square$

As in the case of normed spaces, the inner product spaces which are **complete** (with respect to the induced norm) play such an important role in Analysis that are given a special name.

Definition 10.14. A **Hilbert space** is an inner product space $(X, \langle \cdot, \cdot \rangle)$ which is complete with respect to the norm induced by the inner product.

We close this section by noting that the spaces in Examples 10.4 and 10.6 are Hilbert spaces since they are complete with respect to the convergence coming from the induced norm (see Chapters 3 and 9). However, the space in Example 10.5 is *not* a complete inner product space (see Proposition 3.22).

10.3   Angles, orthogonality and orthogonal splittings

The presence of an inner product on a vector space enables us to define **angles** between vectors and, in particular, the concept of **orthogonality** of two vectors.

Definition 10.15. Let $(X, \langle \cdot, \cdot \rangle)$ be an inner product space.

• Two elements $x, y \in X$ are said to be **orthogonal** to each other and we symbolise this as

$$x \perp y,$$

when their inner product vanishes: $\langle x, y \rangle = 0$.

• The **angle** $\angle(x, y)$ between two non-zero vectors $x, y \in X$ is defined as the number in the interval $[0, \pi] \subseteq \mathbb{R}$ given by

$$\angle(x, y) := \arccos \left(\frac{\mathrm{Re}\big(\langle x, y \rangle\big)}{\|x\| \|y\|} \right),$$

where $\arccos : [-1, 1] \longrightarrow [0, \pi]$ is the inverse of the cosine function.[4]

Remark 10.16. (i) Evidently, x is orthogonal to y if and only if y is orthogonal to x. Hence, orthogonality is a symmetric relation.[5]

(ii) The definition of the concept of angle might look complicated, but is very natural and is based on the following observation: by the Cauchy-Schwarz inequality, we have $|\langle x, y \rangle| \leq \|x\| \|y\|$. Hence, if $x, y \in X \setminus \{0\}$, the real part of the inner product satisfies

$$-1 \leq \frac{\mathrm{Re}\big(\langle x, y \rangle\big)}{\|x\| \|y\|} \leq 1.$$

By following our Euclidean intuition, the fraction above which takes values in $[-1, 1]$ qualifies to be taken as the cosine of the angle $\cos(\angle(x, y))$ between the vectors x, y (this is indeed the case on $\mathbb{F}^m$ for $m \in \mathbb{N}$).

(iii) Utilising the concept of angle between two vectors, we might restate the orthogonality property as that the angle $\angle(x, y)$ formed by the vectors x and y equals $\pi/2$:

$$x \perp y \iff \angle(x, y) = \frac{\pi}{2}.$$

---

[4]We recall that the cosine function is monotonically strictly decreasing on $[0, \pi]$ with image equal to $[-1, 1]$, therefore a (partial) inverse $\arccos : [-1, 1] \longrightarrow [0, \pi]$ is defined by taking $\cos(x) = y \Leftrightarrow y = \arccos(x)$. Note that the cosine is periodic on $\mathbb{R}$ and therefore it cannot be inverted globally!

[5]However, orthogonality is not an equivalence relation because $x \perp y$ and $y \perp z$ does not implies $x \perp z$ (think of $\{x, y, z\} = \{e_1, e_2, -e_1\}$ on $\mathbb{R}^2$).

(iv) Following the familiar Euclidean situation, we may call two vectors $x, y \in X$ in an inner product space **parallel** (or **collinear**) when $\angle(x, y) \in \{0, \pi\}$. In particular, when $\angle(x, y) = 0$ then we may say that they have the same direction and when $\angle(x, y) = \pi$ that they have opposite directions.

Utilising the concept of orthogonality, it is possible to consider the set of vectors that are orthogonal to all the elements of a given set.

Definition 10.17. Let $(X, \langle \cdot, \cdot \rangle)$ be an inner product space, and A be a subset of X. The **orthogonal complement** of A, denoted by $A^\perp$, is the set:

$$A^\perp := \{x \in X : x \perp a, \text{ for all } a \in A\}.$$

Although the set A may have no linear structure whatsoever, $A^\perp$ is always a vector subspace (see Figure 10.2).

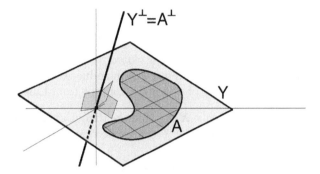

Figure 10.2: An illustration of the orthogonal complements in $\mathbb{R}^3$. The set A is contained in a 2-dimensional vector subspace of $Y \subseteq \mathbb{R}^3$ and the orthogonal complements of Y and A coincide.

The following result summarises some basic properties of orthogonal complements. Its proof is left as an easy exercise. We merely note that one needs to be careful when considering complements of subspaces (see Figure 10.3).

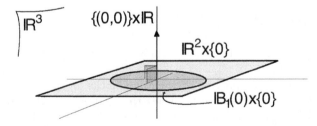

Figure 10.3: The orthogonal complement of a set depends on the space we consider it in. For instance, consider $\mathbb{R}^2$ as being identified with the subspace $\mathbb{R}^2 \times \{0\} \subseteq \mathbb{R}^3$. Then, the unit disc $\mathbb{B}_1(0) \times \{0\}$ has trivial complement in $\mathbb{R}^2 \times \{0\}$ (because there exists no non-zero vector of $\mathbb{R}^2 \times \{0\}$ orthogonal to it), but its orthogonal complement when considered as a subset of $\mathbb{R}^3$ is the on-dimensional subspace $\{(0, 0)\} \times \mathbb{R}$!

Proposition 10.18. *Let $(X, \langle \cdot, \cdot \rangle)$ be an inner product space, and A be a subset of X. Then the following properties hold true:*

(i) $A^\perp$ *is a closed vector subspace of X.*

(ii) $A \subseteq (A^\perp)^\perp$.

(iii) $A \cap A^\perp = \begin{cases} \{0\}, & \text{if } 0 \in A, \\ \emptyset, & \text{if } 0 \notin A. \end{cases}$

We now extend one of the basic constructions of Euclidean geometry to the setting of inner product spaces. As we know, given any (two-dimensional) plane in the (three-dimensional) Euclidean space, and any point in space, there exists a unique line passing through that point which is "perpendicular" to the plane. If the plane passes through the origin, then the vector determined by the given point can thus be decomposed as a sum of two parts, one which is contained in the given plane, and the other which is orthogonal to it.

Let $(X, \langle \cdot, \cdot \rangle)$ be an inner product space and Y be a vector subspace of X. Then, by Proposition 10.18, $Y^\perp$ is a closed vector subspace of X and $Y \cap Y^\perp = \{0\}$. In order to continue we need to introduce the following notion, which is an immediate extension of a familiar concept from Linear Algebra:

Definition 10.19. A vector space X is called the **direct sum of two vector subspaces** Y and Z of it, if any $x \in X$ can be decomposed in a unique way in the form

$$x = y + z \quad \text{where } y \in Y, \ z \in Z.$$

Then, we may write $X = Y \oplus Z$. Equivalently, $X = Y \oplus Z$ when for $Y, Z \subseteq X$ we have $X = Y + Z$ and the subspaces intersect only at the origin: $Y \cap Z = \{0\}$.

Note that a direct sum decomposition is symmetric, that is $Y \oplus Z = Z \oplus Y$. This is a direct consequence of the commutativity of the addition operation in the vector space.

In what follows, we address the question of whether X is the direct sum between Y and $Y^\perp$. It turns out that the answer is negative in general: if Y is a dense subspace of X and $Y \neq X$, then $Y^\perp = \{0\}$, and thus X is not the direct sum of Y and $Y^\perp$. Remarkably, as Theorem 10.23 shows, the answer is positive if X is a Hilbert space (namely, complete) and also Y is a *closed* vector subspace.

The next two results provide the building blocks for the proof of Theorem 10.23. The first result below establishes that, among all the line segments joining a given point to any of the points of a given vector subspace, the ones orthogonal to the subspace, if they exist, are exactly those of shortest length (with respect to the distance induced by the norm, see Figure 10.4).

Proposition 10.20. *Let Y be a vector subspace of an inner product space $(X, \langle \cdot, \cdot \rangle)$, $x \in X$ and $y \in Y$. Then the following statements are equivalent:*

(i) $x - y \in Y^{\perp}$.

(ii) $\|x - y\| \leq \|x - u\|$ *for all* $u \in Y$.

Note that (ii) above is equivalent to $\|x - y\| = \inf \{\|x - u\| : u \in Y\}$.

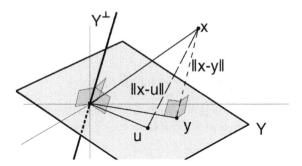

Figure 10.4: A graphic illustration of the idea in the statement of Proposition 10.20 when Y is a 2-dimensional subspace of $\mathbb{R}^3$.

Proof. Let us set $z := x - y$. Note that, for any $u \in Y$,

$$x - u = (x - y) - (u - y) = z - v,$$

where $v := u - y$, and, for fixed $y \in Y$, the mapping

$$Y \ni u \mapsto v := u - y \in Y$$

is a bijection. Therefore, in the newly introduced notation, (i) can be written as $z \in Y^{\perp}$, while (ii) can be written as

$$\|z\| \leq \|z - v\| \quad \text{for all } v \in Y.$$

The above inequality may thus be equivalently rewritten as

$$\|z\|^2 \leq \|z\|^2 - 2\operatorname{Re}\langle v, z \rangle + \|v\|^2 \quad \text{for all } v \in Y$$

or, furthermore, as

$$2\operatorname{Re}\langle v, z \rangle \leq \|v\|^2 \quad \text{for all } v \in Y. \tag{$*$}$$

Suppose first that (i) holds, so that $z \in Y^{\perp}$. Then obviously $(*)$ holds, and therefore (ii) also holds.

Suppose now that (ii) holds, so that, in view of the preceding considerations, $(*)$ also holds. Then, since Y is a vector subspace, we may apply $(*)$ to λw in the role of v, for any $\lambda \in \mathbb{F}$ and any $w \in Y$, we obtain

$$2\operatorname{Re}(\lambda \langle w, z \rangle) \leq |\lambda|^2 \|w\|^2 \quad \text{for all } \lambda \in \mathbb{F} \text{ and } w \in Y.$$

In particular, taking $\lambda = t\overline{\langle w, z \rangle}$, for any $t > 0$ and any $w \in Y$, to obtain

$$2t |\langle w, z \rangle|^2 \leq t^2 |\langle w, z \rangle|^2 \|w\|^2 \quad \text{for all } t > 0 \text{ and } w \in Y.$$

Suppose now, for a contradiction, that $z \notin Y^\perp$, so that there exists $w_0 \in Y$ with $\langle w_0, z \rangle \neq 0$. It then follows that

$$2t \leq t^2 \|w_0\|^2 \quad \text{for all } t > 0,$$

so that

$$\frac{2}{t} \leq \|w_0\|^2 \quad \text{for all } t > 0,$$

which is obviously a contradiction, since $w_0 \neq 0$ (as a consequence of the fact that $\langle w_0, z \rangle \neq 0$). It therefore follows that $z \in Y^\perp$, and hence (i) holds, as required. □

The next result, Theorem 10.22, addresses the questions of existence and uniqueness for the elements that minimise the distance between a given point and a given subset of a Hilbert space. Although of most interest here is the case of a closed vector subspace, we give the result in a more general form, which is useful in other contexts as well. To this aim, let us first recall the notion of a **convex set** in a vector space.

Definition 10.21. Let X be a vector space. A subset K of X is said to be **convex** if

$$(1 - \lambda)z + \lambda w \in K \quad \text{for any } z, w \in K \text{ and for any } \lambda \in [0, 1].$$

We may now state and prove the following important result.

Theorem 10.22. *Let $(X, \langle \cdot, \cdot \rangle)$ be a Hilbert space and K a closed, convex subset of X. Then, for any $x \in X$ there exists a unique $y \in K$ such that (see Figure 10.5)*

$$\|x - y\| = \inf \{ \|x - u\| : u \in K \}.$$

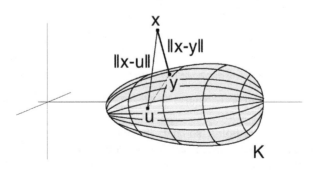

Figure 10.5: An illustration of the statement of Theorem 10.22 in the case of a closed convex subset K in $\mathbb{R}^3$. The point y realises the shortest distance of the point x from the set K.

Proof. The required inequality may also be written as

$$\|y - x\| = \inf \{\|u - x\| : u \in K\}.$$

Let $C = K - x = \{u - x : u \in K\}$. Then C is merely a translate of K, and is therefore also a closed convex set. It is therefore sufficient (and necessary) to show that there exists a unique $z_0 \in C$ such that

$$\|z_0\| = \inf \{\|z\| : z \in C\}.$$

Let $\gamma := \inf\{\|z\| : z \in C\}$. Then $\|z\| \geq \gamma$ for all $z \in C$, and for every $n \in \mathbb{N}$ there exists $z_n \in C$ such that

$$\gamma^2 \leq \|z_n\|^2 \leq \gamma^2 + \frac{1}{n}.$$

We now show that $(z_n)_1^\infty$ is a Cauchy sequence, and is therefore convergent.[6] A direct proof that $(z_n)_1^\infty$ is a Cauchy sequence is not straightforward, since

$$\|z_n - z_m\| \geq \big| \|z_n\| - \|z_m\| \big| \quad \text{for } n, m \in \mathbb{N},$$

and thus our estimates for the latter quantity are not sufficient to ensure a control of the former. However, what saves the day here is the parallelogram identity! Indeed, it is a consequence of that identity that, for all $n, m \in \mathbb{N}$, we have

$$\begin{aligned}
\|z_n - z_m\|^2 &= 2\big(\|z_n\|^2 + \|z_m\|^2\big) - 4\left\|\frac{1}{2}z_n + \frac{1}{2}z_m\right\|^2 \\
&\leq 2\left(\gamma^2 + \frac{1}{n} + \gamma^2 + \frac{1}{m}\right) - 4\gamma^2 \\
&= 2\left(\frac{1}{n} + \frac{1}{m}\right),
\end{aligned}$$

where we have also used the fact that

$$\frac{1}{2}z_n + \frac{1}{2}z_m \in C,$$

which follows by the convexity of C and the fact that $z_n, z_m \in C$. It ensues that $(z_n)_1^\infty$ is a Cauchy sequence, and therefore convergent, since X is a Hilbert space. Let us symbolise the limit as

$$z_0 := \lim_{n \to \infty} z_n.$$

---

[6]Note that the inequality $\|z_n\|^2 \leq \gamma^2 + 1/n$ implies that $(z_n)_1^\infty$ is bounded in X, but this is *not* sufficient to ensure the existence of a convergent subsequence, since X may be infinite-dimensional! We recall that, as we have seen in Theorem 5.25, closed and bounded sets need not be sequentially compact in infinite-dimensional normed spaces!

Since C is closed and $(z_n)_1^\infty \subseteq C$, it follows that $z_0 \in C$. Note also that

$$\|z_0\| = \lim_{n \to \infty} \|z_n\| = \gamma,$$

which proves the existence part of the claimed result. Suppose now that there also exists $\tilde{z}_0 \in C$ such that $\|\tilde{z}_0\| = \gamma$. By the parallelogram identity, the convexity of C and that $z_0, \tilde{z}_0 \in C$, we have

$$\frac{1}{2}z_0 + \frac{1}{2}\tilde{z}_0 \in C.$$

By utilising this inclusion, we infer that

$$\|z_0 - \tilde{z}_0\|^2 = 2\Big(\|z_0\|^2 + \|\tilde{z}_0\|^2\Big) - 4\left\|\frac{1}{2}z_0 + \frac{1}{2}\tilde{z}_0\right\|^2 \leq 0.$$

This yields $z_0 = \tilde{z}_0$, thus proving the uniqueness part of the claimed result. $\square$

We now come to the main result of this section, the existence of the orthogonal decomposition of a given Hilbert space determined by a closed vector space (and its orthogonal complement).

Theorem 10.23. *Let $(X, \langle \cdot, \cdot \rangle)$ be a Hilbert space and Y be a closed vector subspace of X. Then the following direct sum decomposition holds true:*

$$X = Y \oplus Y^\perp.$$

Namely, for every $x \in X$ there exist unique $y \in Y$, $z \in Y^\perp$ such that

$$x = y + z.$$

In addition, for every $v \in Y$ we have the identity (see Figure 10.6)

$$\|x - v\|^2 = \|x - y\|^2 + \|y - v\|^2.$$

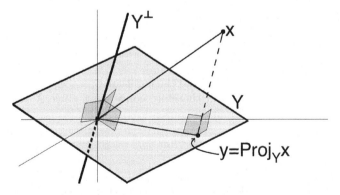

Figure 10.6: An illustration of the orthogonal projection $y = \mathrm{Proj}_Y x$ of a vector $x \in \mathbb{R}^3$ on a 2-dimensional subspace Y.

Definition 10.24. In the setting of Theorem 10.23, the direct sum decomposition $X = Y \oplus Y^\perp$ will be referred to as an **orthogonal decomposition**, since $y \perp z$. For each $x \in X$ the corresponding element $y \in Y$ is called the **orthogonal projection of x on the subspace Y** and is typically denoted by

$$y =: \operatorname{Proj}_Y x.$$

Similar notation and terminology obviously applies to the **projection** $\operatorname{Proj}_{Y^\perp} x =: z$ **of x on the subspace $Y^\perp$**.

Proof. Let $x \in X$ be an arbitrary point. Since Y is a closed convex set, by Theorem 10.22 there exists $y \in Y$ such that

$$\|x - y\| \leq \|x - u\| \quad \text{for all } u \in Y.$$

By Proposition 10.20, this is equivalent to $(x - y) \in Y^\perp$. Setting $z := x - y$, we can thus write

$$x = y + z,$$

where $y \in Y$ and $z \in Y^\perp$. For the uniqueness of such a decomposition, suppose that there exist also $\tilde{y} \in Y$ and $\tilde{z} \in Y^\perp$ such that

$$x = \tilde{y} + \tilde{z}.$$

It follows that

$$y - \tilde{y} = \tilde{z} - z.$$

Since $(y - \tilde{y}) \in Y$ and $(\tilde{z} - z) \in Y^\perp$, whilst $Y \cap Y^\perp = \{0\}$ by Proposition 10.18, it follows that $y = \tilde{y}$ and $z = \tilde{z}$, thus proving the uniqueness claim. Finally, the fact that, for any $v \in Y$,

$$\|x - v\|^2 = \|x - y\|^2 + \|y - v\|^2,$$

follows immediately from the fact that

$$x - v = (x - y) + (y - v),$$

taking into account that $(y - v) \in Y$ and $(x - y) \in Y^\perp$. $\qquad\square$

The next example shows that our Euclidean-originating intuition of orthogonality in Hilbert spaces is correct when we consider *closed* subspaces, but not necessarily arbitrary subspaces.

Example 10.25. Consider the Hilbert space $L^2(\mathbb{R}^n)$ (of square-integrable Lebesgue measurable functions $\mathbb{R}^n \longrightarrow \mathbb{R}$), endowed with the inner product

$$\langle f, g \rangle_{L^2(\mathbb{R}^n)} = \int_{\mathbb{R}^n} f\, g \, \mathrm{d}\mathcal{L}^n \quad f, g \in L^2(\mathbb{R}^n).$$

Let $\Omega \subseteq \mathbb{R}^n$ be a measurable set of positive Lebesgue measure with complement having also positive Lebesgue measure. Then, the vector spaces $L^2(\Omega)$

and $L^2(\mathbb{R}^n \setminus \Omega)$ are Hilbert spaces as well, when endowed with the respective inner products

$$\langle f, g \rangle_{L^2(\Omega)} = \int_\Omega f \, g \, \mathrm{d}\mathcal{L}^n , \qquad \langle f, g \rangle_{L^2(\mathbb{R}^n \setminus \Omega)} = \int_{\mathbb{R}^n \setminus \Omega} f \, g \, \mathrm{d}\mathcal{L}^n.$$

Further, we have

$$\|f\|^2_{L^2(\mathbb{R}^n)} = \|f\|^2_{L^2(\Omega)} + \|f\|^2_{L^2(\mathbb{R}^n \setminus \Omega)}.$$

which implies[7] that the map $f \mapsto f \chi_\Omega + f \chi_{\mathbb{R}^n \setminus \Omega}$ defines the orthogonal decomposition

$$L^2(\mathbb{R}^n) = L^2(\Omega) \bigoplus L^2(\mathbb{R}^n \setminus \Omega).$$

Hence, $f \chi_\Omega$ is projection of f on $L^2(\Omega)$ and similar comments apply to $f \chi_{\mathbb{R}^n \setminus \Omega}$. It follows that

$$\left(L^2(\Omega) \right)^\perp = L^2(\mathbb{R}^n \setminus \Omega) , \qquad \left(L^2(\mathbb{R}^n \setminus \Omega) \right)^\perp = L^2(\Omega).$$

However, for the dense subspace $C_c^\infty(\Omega)$ of $L^2(\Omega)$ (recall Theorem 9.44), we also have $\left(C_c^\infty(\Omega) \right)^\perp = L^2(\mathbb{R}^n \setminus \Omega)$ but $\left(L^2(\mathbb{R}^n \setminus \Omega) \right)^\perp \neq C_c^\infty(\Omega)$.

An immediate consequence of Theorem 10.23 is the following.

Proposition 10.26. *If Y is a closed vector subspace of a Hilbert space X, then $(Y^\perp)^\perp = Y$.*

Proof. By Proposition 10.18, we have that $Y \subseteq (Y^\perp)^\perp$. It remains therefore to prove the reverse inclusion. Let $x \in (Y^\perp)^\perp$, arbitrary. Then, by Theorem 10.23, one can write

$$x = y + z, \quad \text{where } y \in Y, z \in Y^\perp.$$

Taking the inner product with z in the above, we obtain that $z = 0$. This happens because $\langle x, z \rangle = 0$ due to the fact that $x \in (Y^\perp)^\perp, z \in Y^\perp$, while it holds that $\langle y, z \rangle = 0$ due to the fact that $y \in Y, z \in Y^\perp$. In conclusion, $x = y \in Y$, as required. $\square$

10.4 Orthonormal sets and orthonormal bases

The cornerstone of three-dimensional Analytic Geometry is the existence in $\mathbb{R}^3$ of the **canonical basis** $\{e_1, e_2, e_3\}$, where

$$e_1 = (1, 0, 0) \qquad e_2 = (0, 1, 0) \qquad e_3 = (0, 0, 1).$$

---

[7] Note that for any Lebesgue measurable subset $A \subseteq \mathbb{R}^n$, the space $L^2(A)$ can be identified with the subspace of function of $L^2(\mathbb{R}^n)$ which vanish a.e. outside A (by extending any $f \in L^2(A)$ by zero on $\mathbb{R}^n \setminus A$).

These three vectors not only form an algebraic basis for $\mathbb{R}^3$, but they are also mutually orthogonal and each of them has length one. Such a collection of vectors is called an **orthonormal basis** for $\mathbb{R}^3$. Note that every vector $x = (x_1, x_2, x_3)$ can be expressed in a very simple way in terms of this basis, namely

$$x = x_1 e_1 + x_2 e_2 + x_3 e_3,$$

while the coefficients of x with respect to these basis can be very easily obtained by means of the inner product (see Figure 10.7):

$$x_k = \langle x, e_k \rangle \quad \text{for all } k \in \{1, 2, 3\}.$$

Analogous results are obviously valid in the Euclidean space $\mathbb{R}^m$ for any dimension $m \in \mathbb{N}$, although it is impossible to visualise orthogonality as accurately!

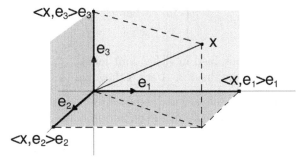

Figure 10.7: An illustration of the orthogonal expansion of a vector $x \in \mathbb{R}^3$ with respect to the usual Euclidean structure.

In this section we aim at considering the question of existence of so-called **orthonormal bases** which have properties similar to those above but are defined on arbitrary inner product spaces. It is instructive to start with the simplest example of an infinite-dimensional Hilbert space, the space ℓ^2. In that setting, the most obvious analogue of the canonical basis in $\mathbb{R}^3$ (or the analogous one in $\mathbb{R}^m$) is the set $A = \{e_k : k \in \mathbb{N}\}$ whose elements are given, for each $k \in \mathbb{N}$, by

$$e_k = (0, 0, \ldots, 0, 1, 0, \ldots),$$

namely, where the kth entry is 1 and the rest are 0. We observe that the elements of A are mutually orthogonal, and each of them has norm 1. One can thus imagine them as determining a infinite (in fact, countable) set of orthogonal directions in ℓ^2. Moreover, any element $x \in \ell^2$, with $x = (x_1, x_2, \ldots)$, may be decomposed along this family of directions in the form of a convergent series

$$x = \sum_{k=1}^{\infty} x_k e_k,$$

where each coefficient can be expressed as an inner product:

$$x_k = \langle x, e_k \rangle \quad \text{for each } k \in \mathbb{N}.$$

Then, in addition, as a consequence of the orthonormality of A, the length of any vector can be very easily expressed in terms of its respective coefficients as

$$\|x\|_{\ell^2}^2 = \sum_{k=1}^{\infty} |\langle x, e_k \rangle|^2.$$

Presumably, the possibility of obtaining similar orthogonal representations in a general Hilbert space should be useful in other contexts as well. A preliminary step in the definition and study of **orthonormal bases** is the study of the more general concept of **orthonormal sets**, which we now define.

Definition 10.27. Let $(X, \langle \cdot, \cdot \rangle)$ be an inner product space. A subset A of X is called **orthonormal** if for all $e, f \in A$, we have

$$\langle e, f \rangle = \begin{cases} 0, & \text{when } e \neq f, \\ 1, & \text{when } e = f. \end{cases}$$

Hence, every vector e contained in A has unit length ($\|e\| = 1$) and any two different vectors $e \neq f$ contained in A are orthogonal ($e \perp f$).

Note that orthonormal sets are not assumed to be neither finite nor countable.[8] An immediate consequence of the definition is the linear independence of any orthonormal set.

Proposition 10.28. *Any orthonormal set A in an inner product space X is linearly independent.*

Proof. Suppose that for some $n \in \mathbb{N}$, $\{e_1, \ldots, e_n\} \subseteq A$ and $\{\alpha_1, \ldots, \alpha_n\} \subseteq \mathbb{F}$, we have

$$\sum_{k=1}^{n} \alpha_k e_k = 0,$$

For any fixed $j \in \{1, \ldots, n\}$, taking the inner product with e_j in the above, we obtain that $\alpha_j = 0$. This shows that A is linearly independent. $\square$

A further immediate consequence of orthonormality is the following simple formula for the norm of any linear combination of the elements of a finite orthonormal set. This result, whose proof is left as an exercise, may be regarded as an extension of **Pythagoras's Theorem**.

Proposition 10.29 (The Pythagorean theorem in inner product spaces). *Let $\{e_1, \ldots, e_n\}$ be a finite orthonormal set in an inner product space X. Then, for every $\{\alpha_1, \ldots, \alpha_n\} \subseteq \mathbb{F}$, we have*

$$\left\| \sum_{k=1}^{n} \alpha_k e_k \right\|^2 = \sum_{k=1}^{n} |\alpha_k|^2.$$

---

[8]In fact, as we will see in the sequel, in non-separable Hilbert spaces the always exist uncountable orthonormal sets!

Given a finite-dimensional vector subspace generated by a (necessarily finite) orthonormal set of an inner product space X, the orthogonal decomposition $X = Y \oplus Y^\perp$ takes a particularly simple form, as follows.

Proposition 10.30. *Let* $(X, \langle \cdot, \cdot \rangle)$ *be an inner product space,* $A = \{e_1, \dots, e_n\}$ *a finite orthonormal set and consider the set* $Y := \operatorname{span}[A]$. *Then* X *has the orthogonal decomposition*

$$X = Y \oplus Y^\perp.$$

In addition, any $x \in X$ *can be written as*

$$x = \sum_{k=1}^{n} \langle x, e_k \rangle e_k + \left(x - \sum_{k=1}^{n} \langle x, e_k \rangle e_k \right) =: y + z,$$

where $y \in Y$ *and* $z \in Y^\perp$. *Further, for every* $v \in Y$ *we have the identity*

$$\|x - v\|^2 = \|x - y\|^2 + \|y - v\|^2.$$

Proof. Fix any $x \in X$, arbitrary. We define

$$y := \sum_{k=1}^{n} \langle x, e_k \rangle e_k, \quad z := x - \sum_{k=1}^{n} \langle x, e_k \rangle e_k.$$

It is obvious that $y \in Y$. Also, it can be easily checked that $\langle z, e_j \rangle = 0$ for any $j \in \{1, \dots, n\}$. It follows that $z \in Y^\perp$, as desired. The other claims of the statement are verified as in the proof of Theorem 10.23. $\square$

Remark 10.31. Note that the points $y \in Y$ and $z \in Y^\perp$ are in fact the orthogonal projections $\operatorname{Proj}_Y x$ and $\operatorname{Proj}_{Y^\perp} x$ of x on Y and $Y^\perp$ respectively. We also call the reader's attention to that **Proposition 10.30 is not a particular case of Theorem 10.23**. Indeed, here we do not assume completeness, namely that X is a Hilbert space. Also, the proof of Proposition 10.30 is simpler and more direct.

We now turn our attention to **countable orthonormal sets**. Motivated by the properties of the "standard" orthonormal set $A = \{e_k : k \in \mathbb{N}\}$ in ℓ^2 described earlier (with $e_k = (0, \dots, 0, 1, 0, \dots)$ and having the unity in the kth position), we shall study the following natural question:

Let $(X, \langle \cdot, \cdot \rangle)$ be an inner product space or a Hilbert space and $A = \{e_k : k \in \mathbb{N}\}$ a countable orthonormal set in X.

(Q1) Is it true that for any $x \in X$, the series of vectors $\sum_{k=1}^{\infty} \langle x, e_k \rangle e_k$ converges in the space X ?

(Q2) If the answer to (Q1) is affirmative, is it also true that $x = \sum_{k=1}^{\infty} \langle x, e_k \rangle e_k$ and $\|x\|^2 = \sum_{k=1}^{\infty} |\langle x, e_k \rangle|^2$?

In order to answer these questions, we begin with an auxiliary result, Proposition 10.32 below. This result provides necessary and sufficient conditions for the convergence of a series whose terms are scalar multiples of the elements of a countable orthonormal set.

Proposition 10.32. *Let* $(X, \langle \cdot, \cdot \rangle)$ *be a Hilbert space,* $A = \{e_k : k \in \mathbb{N}\}$ *a countable orthonormal set and* $(\alpha_n)_1^\infty$ *a sequence in* $\mathbb{F}$. *Then the following holds:*

$$\sum_{k=1}^{\infty} \alpha_k e_k \quad \text{converges in } X \quad \text{if and only if} \quad \sum_{k=1}^{\infty} |\alpha_k|^2 < \infty.$$

Moreover, if either of these series converges, the following identity is satisfied:

$$\left\| \sum_{k=1}^{\infty} \alpha_k e_k \right\|^2 = \sum_{k=1}^{\infty} |\alpha_k|^2.$$

Proof. We begin by defining the partial sums

$$s_n := \sum_{k=1}^{n} \alpha_k e_k, \quad \text{for all } n \in \mathbb{N}.$$

Then, as a consequence of the orthonormality of A, we have

$$\|s_n\|^2 = \sum_{k=1}^{n} |\alpha_k|^2.$$

Suppose first that $(s_n)_1^\infty$ converges in X. Then $(s_n)_1^\infty$ is bounded, and hence $\sum_{k=1}^{\infty} |\alpha_k|^2 < +\infty$. Conversely, suppose that $\sum_{k=1}^{\infty} |\alpha_k|^2 < +\infty$. We define the partial sums

$$t_n := \sum_{k=1}^{n} |\alpha_k|^2, \quad \text{for all } n \in \mathbb{N}.$$

Then $(t_n)_1^\infty$ is a convergent sequence in $\mathbb{R}$, and hence a Cauchy sequence. On the other hand, note that, for every $n, m \in \mathbb{N}$ with $n > m$, we have

$$\|s_n - s_m\|^2 = \sum_{k=m+1}^{n} |\alpha_k|^2 = t_n - t_m.$$

It follows easily that $(s_n)_1^\infty$ is a Cauchy sequence in X and therefore a convergent one, since X is a Hilbert space. Finally, in the event that either of the series converges, the desired identity ensues as follows:

$$\left\| \sum_{k=1}^{\infty} \alpha_k e_k \right\|^2 = \lim_{n \to \infty} \|s_n\|^2 = \sum_{k=1}^{\infty} |\alpha_k|^2.$$

The proposition has been established.   □

Let $(X, \langle \cdot, \cdot \rangle)$ be an inner product space and $\{e_k : k \in \mathbb{N}\}$ an orthonormal sequence in X. Proposition 10.32 makes clear that, at least in the case when X is Hilbert space, the convergence of a series of vectors of the form

$$\sum_{k=1}^{\infty} \langle x, e_k \rangle e_k$$

arising in the orthonormal expansion of a point $x \in X$, is interconnected to the convergence of the series of numbers

$$\sum_{k=1}^{\infty} |\langle x, e_k \rangle|^2.$$

The following remarkable result shows that the latter series is always convergent, even when X is not complete, and provides an explicit upper bound for the sum of the series.

Proposition 10.33 (Bessel's Inequality). *Let $(X, \langle \cdot, \cdot \rangle)$ be an inner product space and $\{e_k : k \in \mathbb{N}\}$ an orthonormal sequence. Then, for any $x \in X$, the series $\sum_{k=1}^{\infty} |\langle x, e_k \rangle|^2$ converges and we have the estimate*

$$\sum_{k=1}^{\infty} |\langle x, e_k \rangle|^2 \leq \|x\|^2.$$

Proof. Fix an arbitrary point $x \in X$. It can be easily seen that the desired conclusion is equivalent to the next estimate being true for any $n \in \mathbb{N}$:

$$\sum_{k=1}^{n} |\langle x, e_k \rangle|^2 \leq \|x\|^2.$$

Let us fix $n \in \mathbb{N}$, consider the subspace $Y_n := \text{span}[\{e_1, \ldots, e_n\}]$ and set

$$s_n(x) := \sum_{k=1}^{n} \langle x, e_k \rangle e_k.$$

Then, by Proposition 10.30 we have the splitting $x = s_n(x) + (x - s_n(x))$, where $s_n(x) \in Y_n$ and $x - s_n(x) \in Y_n^{\perp}$. It therefore follows that

$$\|x\|^2 = \|s_n(x)\|^2 + \|x - s_n(x)\|^2.$$

This implies the bound $\|s_n(x)\|^2 \leq \|x\|^2$, which combined with the identity

$$\|s_n(x)\|^2 = \sum_{k=1}^{n} |\langle x, e_k \rangle|^2,$$

yields the required conclusion. $\qquad\square$

By simply combining Proposition 10.32 and Proposition 10.33, we obtain the following important result which provides an explicit affirmative answer to the question (Q1) posed earlier in this section.

Corollary 10.34. *Let $(X, \langle \cdot, \cdot \rangle)$ be a Hilbert space and $A = \{e_k : k \in \mathbb{N}\}$ an orthonormal sequence. Then, for any $x \in X$, the series of vectors*

$$\sum_{k=1}^{\infty} \langle x, e_k \rangle e_k$$

converges in the space X.

The following important result provides necessary and sufficient conditions on A for the representation "$x = \sum_{k=1}^{\infty} \langle x, e_k \rangle e_k$" to hold for all $x \in X$. As it might have been suspected, **the series $\sum_{k=1}^{\infty} \langle x, e_k \rangle e_k$ may not always converge to** x. The key fact is that, roughly speaking, the set of vectors A needs to "span" the space X, in an analytic sense, to be made precise right below, that is slightly different from the algebraic spanning (remember that A may not be a finite set and limiting processes are required). In particular, we provide criteria guaranteeing an affirmative answer to question (Q2) posed earlier in this section.

Theorem 10.35. *Let $(X, \langle \cdot, \cdot \rangle)$ be an inner product space and $A = \{e_k : k \in \mathbb{N}\}$ a countable orthonormal set. Then the following statements are equivalent:*

(i) $\displaystyle x = \sum_{k=1}^{\infty} \langle x, e_k \rangle e_k \quad$ *for all $x \in X$.*

(ii) $\displaystyle \|x\|^2 = \sum_{k=1}^{\infty} |\langle x, e_k \rangle|^2 \quad$ *for all $x \in X$.*

(iii) $\overline{\mathrm{span}[A]} = X$.

If in addition X is a Hilbert space, each of (i), (ii), (iii) is also equivalent to:

(iv) $A^{\perp} = \{0\}$.

Theorem 10.35 implicitly introduces some very important new concepts:

Definition 10.36. Let $(X, \langle \cdot, \cdot \rangle)$ be a Hilbert space and $A = \{e_k : k \in \mathbb{N}\}$ a countable orthonormal set.

- The equation in Theorem 10.35(ii) is called **Parseval's identity**.

- If any/all statements (i)–(iv) in Theorem 10.35 hold true, then A is called an **orthonormal basis** of X.[9]

---

[9]The use of the word "basis" here should not be confused with the case of a Hamel

- If A is an orthonormal basis of X, then for any $x \in X$ the scalars $\{\langle x, e_k \rangle : k \in \mathbb{N}\}$ are called the **components** of x with respect to the orthonormal basis A (or **Fourier coefficients** of x with respect to A).

Proof. We begin with some general considerations. For each $n \in \mathbb{N}$, we set

$$Y_n := \operatorname{span}[\{e_1, \ldots, e_n\}].$$

For each $x \in X$, let us consider again the sequence of partial sums

$$s_n(x) := \sum_{k=1}^{n} \langle x, e_k \rangle e_k, \quad n \in \mathbb{N}.$$

Then, by Proposition 10.30, we may decompose x as $x = s_n(x) + \big(x - s_n(x)\big)$, where $s_n(x) \in Y_n$ and $x - s_n(x) \in Y_n^{\perp}$. It therefore follows that

$$\|x\|^2 = \|s_n(x)\|^2 + \|x - s_n(x)\|^2, \tag{$\diamond$}$$

and, moreover, for every $v \in Y_n$,

$$\|x - v\|^2 = \big\|x - s_n(x)\big\|^2 + \big\|s_n(x) - v\big\|^2. \tag{$\dagger$}$$

Note also that

$$\|s_n(x)\|^2 = \sum_{k=1}^{n} |\langle x, e_k \rangle|^2. \tag{$\ddagger$}$$

We may now proceed to establish the desired equivalences.

(i) $\Longrightarrow$ (ii): Fix $x \in X$ and suppose that (i) holds. Then, $x = \lim_{n \to \infty} s_n(x)$. This readily implies (ii), since we have

$$\|x\|^2 = \lim_{n \to \infty} \|s_n(x)\|^2, \quad \text{for any } x \in X.$$

(ii) $\Longrightarrow$ (i): Suppose that (ii) holds. This means

$$\|x\|^2 = \lim_{n \to \infty} \|s_n(x)\|^2, \quad \text{for any } x \in X.$$

But then ($\diamond$) yields that $\lim_{n \to \infty} \|x - s_n(x)\|^2 = 0$ which is precisely (i):

$$x = \lim_{n \to \infty} s_n(x), \quad \text{for any } x \in X.$$

---

algebraic basis of a vector space. Let us stress once again that an orthonormal expansion is not an algebraic finite sum, but instead the limit value of a convergent series. In fact, countable orthonormal bases are a particular case of a more general concept in the study of Banach spaces, that of **Schauder bases**. *A linearly independent sequence $\{e_i : i \in \mathbb{N}\}$ in a Banach space $(X, \| \cdot \|)$ is called a Schauder basis when for any $x \in X$ there exists a unique sequence of scalars $\{a_i(x) : i \in \mathbb{N}\}$ such that $x = \sum_{i=1}^{\infty} a_i(x) e_i$*, with the limit taken with respect to the norm of X. However, unlike Hamel bases of vector spaces and orthonormal bases of Hilbert spaces, a Schauder basis of a Banach space may or may not exist. If it does, though, it imposes the restriction that the space must be separable! Hence, not all Banach spaces have Schauder bases and in fact even separable spaces may not possess one.

(i) $\implies$ (iii): This is obvious, since for any $x \in X$ and $n \in \mathbb{N}$, we have that $s_n(x) \in Y_n \subseteq \operatorname{span}[A]$.

(iii) $\implies$ (i): Suppose that (iii) holds. Fix $x \in X$, arbitrary. Since $x \in \overline{\operatorname{span}[A]}$, it follows that there exists a sequence $(y_n)_1^\infty \subseteq \operatorname{span}[A]$ such that $y_n \longrightarrow x$ in X as $n \to \infty$. Since $\operatorname{span}[A] = \bigcup_1^\infty Y_n$, it follows that for every $n \in \mathbb{N}$ there exists $p_n \in \mathbb{N}$ such that $y_n \in Y_{p_n}$. It is now a consequence of (†) that, for every $n \in \mathbb{N}$,

$$\left\| x - s_{p_n}(x) \right\| \leq \left\| x - y_n \right\|.$$

Since $y_n \longrightarrow x$ as $n \to \infty$, this yields

$$\left\| x - s_{p_n}(x) \right\| \longrightarrow 0 \quad \text{as } n \to \infty. \tag{♮}$$

It is immediate from (‡) that $n \mapsto \|s_n(x)\|^2$ is nondecreasing, and hence, in view of ($\Diamond$), that $n \mapsto \|x - s_n(x)\|^2$ is non-increasing. Directly from (♮) we infer that (i) holds true, because

$$\left\| x - s_n(x) \right\| \longrightarrow 0 \quad \text{as } n \to \infty.$$

Suppose now that X is additionally complete, namely a Hilbert space. We shall prove that (i) is equivalent to (iv), therefore completing the proof.

(i) $\implies$ (iv): Let $z \in A^\perp$, arbitrary. Then $\langle z, e_k \rangle = 0$ for all $k \in \mathbb{N}$. It follows from (i) that $z = 0$. Hence $A^\perp = \{0\}$, so that (iv) holds.

(iv) $\implies$ (i): Let $x \in X$, arbitrary. Since X is a Hilbert space, Corollary 10.34 yields that the series $\sum_{k=1}^\infty \langle x, e_k \rangle e_k$ converges in X and let y symbolise its limit. Note now that $x - y \in A^\perp$, because

$$\langle y, e_j \rangle = \langle x, e_j \rangle \quad \text{for all } j \in \mathbb{N}.$$

However, since $A^\perp = \{0\}$, it follows that $x = y$. This proves (i). $\qquad\square$

Example 10.37. Consider the Hilbert space $L^2((-\pi, \pi), \mathbb{C})$ equipped with the standard inner product (with respect to the Lebesgue measure $\mathcal{L}^1$), see Example 10.6. Then, a direct calculation yields that the sequence of functions $A = \{f_n : n \in \mathbb{N}\}$, where

$$f_n(t) := \frac{1}{\sqrt{\pi}} \sin(nt) \quad \text{for all } t \in (-\pi, \pi),$$

is orthonormal. However, A is not an orthonormal basis (and the Bessel inequality is strict) since no non-trivial *even* function can be represented through an expansion with respect to A.

Remark 10.38. Let $(X, \langle \cdot, \cdot \rangle)$ be a Hilbert space and $A = \{e_k : k \in \mathbb{N}\}$ a countable orthonormal set. It may be easily confirmed that, for every $x \in X$,

$$x = \sum_{k=1}^\infty \langle x, e_k \rangle e_k + \left(x - \sum_{k=1}^\infty \langle x, e_k \rangle e_k \right) =: y + z,$$

where $y \in \overline{\text{span}[A]}$ and $z \in (\overline{\text{span}[A]})^{\perp} = A^{\perp}$. This decomposition actually provides more information than Theorem 10.35, but only in the case when X is a Hilbert space. However, in Theorem 10.35 we proved the equivalence of (i)–(iii) without the completeness assumption, and the proof given provides some useful insights.

We close this section by noting that the results herein admit suitable generalisations for orthonormal sets A which are **not necessarily countable**. However, due to the fact that the case of **countable** orthonormal sets and bases is of far greater importance in applications, and also to avoid some potentially distracting technicalities, we refrain from giving any further details concerning the uncountable case. However, at the end of the chapter (Remark 10.44) we do discuss briefly the general case for the sake of comparison with the countable case.

10.5 Isometries and classification of Hilbert spaces

In this section we address, even though only partially, the important question of classification of Hilbert spaces.[10] As we explained in the introduction of this chapter, mathematicians are interested in "classifying" structures and objects, in the sense of determining how many distinct such structures exist which behave differently from each other.

Given the enormous abundance of possibilities, in order to succeed in this attempt one needs a means to identify two different objects in terms of their properties, even if the objects are not truly identical. The means by which this is achieved is the universal mathematical concept of **isomorphism**. Roughly speaking, two sets endowed with certain mathematical structures are said to be **isomorphic** if there exists a bijection between them which "preserves" the structure in question. This means, intuitively, that the two sets behave in exactly the same way with respect to that structure and any properties satisfied by the elements in one of them are satisfied also by their images in the other set.

In this section we are particularly interested in classifying inner product spaces and especially the prominent class of Hilbert spaces.[11] Quite surprisingly, Hilbert spaces can be classified completely, by analogy to the Euclidean

---

[10]At this point it might be instructive to recall the comments regarding "morphisms" we made in Remark 4.31 about topological equivalence and the analogies we drew to the algebraic counterparts of morphisms and isomorphisms, as well as the remarks about morphisms in Measure Theory given in Section 7.5.

[11]Unfortunately it is generally impossible to classify neither topological, nor metric, not even Banach spaces in general, this is why we did not discuss the classification question earlier in this book. A major obstruction in this regard is that the problem of deciding whether two (topological) spaces are homeomorphic is not solvable algorithmically.

situation: every Hilbert space is essentially "the same" as some space ℓ^2 of sequences.

In order to make the foregoing discussion precise, we need some terminology. Before that, let us recall that if X and Y are vector spaces over the same scalar field $\mathbb{F}$, a mapping (also commonly referred to as "operator"[12]) $T : X \longrightarrow Y$ is called **linear** if it satisfies

$$T(\lambda x + \mu y) = \lambda T(x) + \mu T(y) \quad \text{for all } x, y \in X \text{ and } \lambda, \mu \in \mathbb{F}.$$

A systematic study of linear operators between normed spaces will be undertaken in the next chapter. For the moment, let us just briefly mention a standard notational convention: if $T : X \longrightarrow Y$ is a *linear* operator (and only in that case) and $x \in X$, we shall use the shorthand notation Tx instead of $T(x)$.

Definition 10.39.
• A linear mapping $T : X \longrightarrow Y$ between normed spaces is called an **isometry (or isometric embedding)** if it satisfies[13]

$$\|Tx\| = \|x\| \quad \text{for all } x \in X.$$

• Two normed spaces X and Y are said to be **isometric** (or **isomorphic as normed spaces**) if there exist a **surjective** isometry $T : X \longrightarrow Y$. The isometry T is then called a **isometric isomorphism between X and Y.**

• A linear mapping $T : X \longrightarrow Y$ between inner product spaces is called a **unitary map** if

$$\langle Tx, Ty \rangle = \langle x, y \rangle \quad \text{for all } x, y \in X.$$

• Two inner product spaces X and Y are said to be **unitarily equivalent** (or **isomorphic as inner product space**) if there exist a **surjective** unitary map $T : X \longrightarrow Y$. The unitary map T is then called a **unitary isomorphism between X and Y.**

The meaning of isometries of normed spaces is that "lengths are preserved" and of unitary maps of normed spaces is that "lengths *and* angles are preserved". Note that the equalities are more precisely written respectively as

$$\|Tx\|_Y = \|x\|_X \quad \text{and} \quad \langle Tx, Ty \rangle_Y = \langle x, y \rangle_X,$$

namely, the norms and inner products are taken on different spaces on the left and right hand side.

---

[12]It is customary for historical reasons to refer to linear mappings between normed or inner product spaces as **linear operators** and will also occasionally follow the same practise without any further warning, since this will not cause any confusion.

[13]We caution the reader that sometimes in the literature isometries may be required to be surjective and one has to be careful which variant of the definition is in effect. It is however immediate from the definition and the linearity of T that isometries are always injective mappings, because $\|Tx - Ty\| = \|T(x - y)\| = \|x - y\|$.

It is immediate from the definition that every unitary map is an isometry (just take $y = x$). Remarkably, two inner product spaces are isomorphic (i.e. unitarily equivalent) if and only if they isomorphic as normed spaces (i.e. isometric), as the next result implies.

Proposition 10.40. *Let $(X, \langle \cdot, \cdot \rangle)$ and $(Y, \langle \cdot, \cdot \rangle)$ be inner product spaces, and $T : X \longrightarrow Y$ a linear operator. Then the following statements are equivalent:*

(i) $\|Tx\| = \|x\|$ *for all $x \in X$;*

(ii) $\langle Tx, Ty \rangle = \langle x, y \rangle$ *for all $x, y \in X$.*

Proof. Establishing the equivalence is left as an easy exercise for the reader on the use of the polarisation identity. □

The following result is the main classification theorem, establishing that any Hilbert space which has a countable orthonormal basis is isomorphic to the Hilbert space ℓ^2, which is the simplest infinite-dimensional Hilbert space. As will see right next, requiring a Hilbert space to have a countable orthonormal basis is not a huge restriction and most spaces of interest do indeed satisfy this condition.

Theorem 10.41 (Riesz–Fischer Theorem). *Let $(X, \langle \cdot, \cdot \rangle)$ be a Hilbert space and suppose that $A = \{e_k : k \in \mathbb{N}\}$ is a countable orthonormal basis of X. Consider the linear mapping $T : X \longrightarrow \ell^2$ given by*

$$Tx := (\langle x, e_1 \rangle, \langle x, e_2 \rangle, \dots) \quad \text{for all } x \in X.$$

Then T is a unitary isomorphism between X and ℓ^2, identifying X with ℓ^2.

Proof. We begin by noting that, as a consequence of Bessel's Inequality (Proposition 10.33), the mapping $T : X \longrightarrow \ell^2$ is well defined, in the sense that $Tx \in \ell^2$ for any $x \in X$. Also, T is obviously a linear operator.

To check that T is injective, suppose that $Tx = Ty$ for some $x, y \in X$. Then $\langle x, e_k \rangle = \langle y, e_k \rangle$ for all $k \in \mathbb{N}$, which means that $x - y \in A^\perp$, and therefore, by Theorem 10.35(iv), this yields that $x = y$. Hence, T is indeed injective.

To check that T is surjective, consider an arbitrary $\underline{\lambda} := (\lambda_1, \dots, \lambda_k, \dots) \in \ell^2$. Let $x \in X$ be given by $x := \sum_{k=1}^\infty \lambda_k e_k$. This is well defined because the series converges in X as a result of Proposition 10.32 since $\underline{\lambda} \in \ell^2$. It is then immediate that $\langle x, e_j \rangle = \lambda_j$ for all $j \in \mathbb{N}$, and therefore $Tx = \underline{\lambda}$. Hence, T is indeed surjective. Finally, for any $x \in X$, by Parseval's Identity we have that

$$\|Tx\|_{\ell^2}^2 = \sum_{k=1}^\infty |\langle x, e_k \rangle|^2 = \|x\|^2.$$

In conclusion, T is an isometry between X and ℓ^2 and by Proposition 10.40 it is a unitary isomorphism as well, therefore identifying X with ℓ^2. □

In the light of the Riesz–Fischer Theorem, it becomes a fundamental issue to determine which Hilbert spaces admit a countable orthonormal basis. The next result provides a necessary and sufficient condition for their existence.

Theorem 10.42 (Existence of orthonormal bases). *Let $(X, \langle \cdot, \cdot \rangle)$ be an infinite-dimensional Hilbert space. Then X has a countable orthonormal basis if and only if X is separable.*

We recall that a metric space is called *separable* if it contains a countable dense subset. In the proof of Theorem 10.42, the more subtle part is the implication that the separability of X implies the existence of a countable orthonormal basis. The method we give involves an explicit construction of a basis, by means of an infinite-dimensional generalisation of the familiar algebraic Gram–Schmidt orthogonalisation process.

Proposition 10.43 (Gram–Schmidt process). *Let $(X, \langle \cdot, \cdot \rangle)$ be an inner product space, and $\{y_k : k \in \mathbb{N}\}$ be a countable linearly independent subset of X. Then there exists a countable orthonormal subset $\{e_k : k \in \mathbb{N}\}$ of X such that*

$$\operatorname{span}[\{y_1, \ldots, y_n\}] = \operatorname{span}[\{e_1, \ldots, e_n\}] \quad \text{for all } n \in \mathbb{N}.$$

Proof of Proposition 10.43. In order to construct the set $\{e_k : k \in \mathbb{N}\}$ with the required properties, we begin by setting $e_1 := y_1/\|y_1\|$. Suppose now that, for some $n \in \mathbb{N}$, the elements $\{e_1, \ldots, e_n\}$ with the required properties have been constructed. We define

$$v_{n+1} := y_{n+1} - \sum_{k=1}^{n} \langle y_{n+1}, e_k \rangle e_k.$$

Then, the vector v_{n+1} is orthogonal to each of $e_1, \ldots, e_n$. Moreover, since $y_{n+1} \notin \operatorname{span}[\{y_1, \ldots, y_n\}] = \operatorname{span}[\{e_1, \ldots, e_n\}]$, it follows that $v_{n+1} \neq 0$. We define $e_{n+1} := v_{n+1}/\|v_{n+1}\|$. It is immediate that $\{e_1, \ldots, e_{n+1}\}$ is an orthonormal set. Also, it is not difficult to check that

$$\operatorname{span}[\{y_1, \ldots, y_{n+1}\}] = \operatorname{span}[\{e_1, \ldots, e_{n+1}\}].$$

Therefore, this recursive construction provides a set of vectors $\{e_k : k \in \mathbb{N}\}$ with the required properties. $\qquad\square$

We now turn to the proof of Theorem 10.42.

Proof. Suppose first that X has a countable orthonormal basis which we denote by $B = \{e_n : n \in \mathbb{N}\}$. We define the set of linear combinations of elements of B with coefficients taken from the field $\mathbb{Q}_\mathbb{F}$, where we set $\mathbb{Q}_\mathbb{R} := \mathbb{Q}$ and also $\mathbb{Q}_\mathbb{C} := \mathbb{Q} + i\mathbb{Q}$:

$$A := \left\{ \sum_{k=1}^{n} \lambda_k e_k : n \in \mathbb{N}, \lambda_1, \ldots, \lambda_n \in \mathbb{Q}_\mathbb{F} \right\} = \operatorname{span}_{\mathbb{Q}_\mathbb{F}}[B],$$

Then, A is a countable union of countable sets and therefore is a countable set itself. We now show that A is dense in X. Fix $x \in X$ and $\varepsilon > 0$, both arbitrary. For any $n \in \mathbb{N}$ and $\lambda_1, \ldots, \lambda_n \in \mathbb{Q}_{\mathbb{F}}$, we have

$$\left\| x - \sum_{k=1}^{n} \lambda_k e_k \right\|^2 = \sum_{k=1}^{n} \left| \langle x, e_k \rangle - \lambda_k \right|^2 + \sum_{k=n+1}^{\infty} \left| \langle x, e_k \rangle \right|^2,$$

because B is an orthonormal basis. We choose $n \in \mathbb{N}$ large enough such that

$$\sum_{k=n+1}^{\infty} \left| \langle x, e_k \rangle \right|^2 < \frac{\varepsilon^2}{2},$$

which is possible because by Parseval's identity we have

$$\sum_{k=n+1}^{\infty} \left| \langle x, e_k \rangle \right|^2 = \sum_{k=1}^{\infty} \left| \langle x, e_k \rangle \right|^2 - \sum_{k=1}^{n} \left| \langle x, e_k \rangle \right|^2 \longrightarrow \|x\|^2 - \|x\|^2 = 0,$$

as $n \to \infty$. Then, for each $k \in \{1, \ldots, n\}$, we choose $\lambda_1, \ldots, \lambda_n \in \mathbb{Q}_{\mathbb{F}}$ such that

$$\left| \langle x, e_k \rangle - \lambda_k \right|^2 < \frac{\varepsilon^2}{2n},$$

which is possible because $\mathbb{Q}_{\mathbb{F}}$ is dense in $\mathbb{F}$. With this choice of n and of $\lambda_1, \ldots, \lambda_n \in \mathbb{Q}_{\mathbb{F}}$, we may define $y := \sum_{k=1}^{n} \lambda_k e_k$ to obtain that $y \in A$ and $\|x - y\| < \varepsilon$. Since $x \in X$ and $\varepsilon > 0$ were arbitrary, this shows that A is dense in X and since A is countable, it follows that X is separable.

Conversely, suppose that X is separable and let A be a countable dense subset of X which we symbolise as $A = \{x_n : n \in \mathbb{N}\}$. We define the next subset (subsequence) C of A, by invoking the induction principle:

$$C := \left\{ x_{n_k} \in A \,\middle|\, \begin{array}{l} \text{For each } k \in \mathbb{N}, \, x_{n_k} \text{ cannot be written} \\ \text{as a linear combination of } x_1, \ldots, x_{n_{k-1}} \end{array} \right\}.$$

It is immediate that C is linearly independent, and we shall now show that

$$\text{span}[A] = \text{span}[C].$$

The key idea for this argument is that, if $x_n \notin C$, then x_n can be written as a linear combination of some elements $x_k \in C$ with $k \in \{1, \ldots, n-1\}$. Since A is dense, it follows that $\overline{\text{span}[A]} = X$, and therefore $\overline{\text{span}[C]} = X$. Hence, C is an infinite set since, otherwise, $\text{span}[C]$ would be a finite-dimensional vector subspace of X and, being closed, X itself would be finite-dimensional. To conclude, let us set $y_k := x_{n_k}$ and therefore write $C = \{y_k : k \in \mathbb{N}\}$. Let B be the countable orthonormal set obtained from C by the Gram–Schmidt process. Then $\text{span}[C] = \text{span}[B]$, and thus

$$\overline{\text{span}[B]} = X,$$

which yields that B is a countable orthonormal basis, as required. $\qquad \square$

We close this chapter with some comments regarding the more complicated case of classifying general (i.e. possibly non-separable) Hilbert spaces.

Remark 10.44 (The non-separable case). All Hilbert spaces, regardless of their separability, can be classified up to isomorphism as being "some space ℓ^2", in the sense explained below.

By using the Axiom of Choice (through Zorn's Lemma 1.22) instead the Gram-Schmidt process, it can be proved that every Hilbert space X has an orthonormal basis $\{e_s : s \in S\}$ for *some* set S. This is not too difficult to do and follows similar lines to those of Theorem 1.24. In view of Theorem 10.42, S is necessarily uncountable if X is non-separable. On the other hand, for any non-empty set S, one may define the space $\ell^2(S)$ (and more generally $\ell^p(S)$) as the space L^2 over the measure space $(S, \mathcal{P}(S), \sharp)$ where "$\sharp$" is the counting measure on the set S and the σ-algebra is the entire powerset $\mathcal{P}(S)$ (Example 10.6, Definition 9.1, Remark 9.8 and Exercise 9.10):

$$\ell^2(S) := \left\{ f : S \longrightarrow \mathbb{F} \ \middle| \ \int_S |f(s)|^2 \, \mathrm{d}\sharp(s) < \infty \right\}.$$

Then, $\ell^2(S)$ is a Hilbert space when endowed with the inner product

$$\langle f, g \rangle_{\ell^2(S)} := \int_S f(s) \, \overline{g(s)} \, \mathrm{d}\sharp(s).$$

When S is a finite set (namely $\sharp(S) < \infty$), we have $\ell^2(S) = \mathbb{F}^m$. If $S = \mathbb{N}$, then $\ell^2(\mathbb{N})$ reduces to the standard space or square summable sequences ℓ^2 given in Definition 2.76. As you might have already suspected, it can be shown that X is unitarily equivalent to $\ell^2(S)$.

10.6 Exercises

Exercise 10.1. Let $(X, \langle \cdot, \cdot \rangle)$ be an inner product space. Let $(x_n)_1^\infty$ be a sequence of elements of X and let $x_0 \in X$. Show that if $\langle x_n, x_0 \rangle \longrightarrow \langle x_0, x_0 \rangle$ and $\|x_n\| \longrightarrow \|x_0\|$ as $n \to \infty$, then $x_n \longrightarrow x_0$ in X as $n \to \infty$.

Exercise 10.2. Consider the subset of ℓ^2 given by

$$Y := \left\{ x = (x_1, x_2, \ldots) \in \ell^2 \ : \ x_{3k} = 0 \text{ for all } k \in \mathbb{N} \right\}.$$

Find $Y^\perp$, the orthogonal complement of Y in the Hilbert space ℓ^2 with the usual inner product.

Exercise 10.3. Let $A = \{e_k : k \in \mathbb{N}\}$ be a countable orthonormal set in an inner product space $(X, \langle \cdot, \cdot \rangle)$. Show that

$$\sum_{k=1}^{\infty} |\langle x, e_k \rangle \langle y, e_k \rangle| \leq \|x\| \|y\| \qquad \text{for all } x, y \in X.$$

Exercise 10.4. (This exercise explains why we did not consider measures on Hilbert/Banach spaces in the previous chapters.) Let H be an infinite-dimensional Hilbert space. Show that there exists no Borel measure, namely no measure $\mu : \mathcal{B}(H) \longrightarrow [0, \infty]$ defined on the measurable space

$$\big(H, \mathcal{B}(H)\big)$$

such that:

(1) $\mu(U) > 0$ for all non-empty (norm) open sets $U \subseteq H$,

(2) $\mu(U) < \infty$ for all at least one (norm) open set $U \subseteq H$,

(3) $\mu(E) = \mu(x + E)$, for all $E \in \mathcal{B}(H)$ and $x \in H$.

Exercise 10.5. Let H be an infinite-dimensional separable Hilbert space. Show that H admits a norm which is not equivalent to the original norm.

Exercise 10.6. Let H be a Hilbert space. Prove the generalised parallelogram equality:

$$\sum_{\varepsilon_i \in \{-1, +1\}} \left\| \sum_{i=1}^{n} \varepsilon_i \, x_i \right\|^2 = 2^n \sum_{i=1}^{n} \|x_i\|^2,$$

for all $x_1, \ldots, x_n \in H$.

Exercise 10.7. let $(X, \| \cdot \|)$ be a real Banach space whose norm satisfies the parallelogram equality. We define the function

$$\langle \cdot, \cdot \rangle \ : \ X \times X \longrightarrow \mathbb{R}$$

by taking as definition the real polarisation identity:

$$\langle x, y \rangle \ := \ \frac{1}{4} \left(\|x + y\|^2 - \|x - y\|^2 \right), \quad x, y \in X.$$

Show that $\langle \cdot, \cdot \rangle$ is an inner product on X. State and prove also the corresponding result for complex Hilbert spaces.

Exercise 10.8. Demonstrate an example of a Hilbert space H and of a vector subspace $Y \subseteq H$ of it such that the following is true:

$$H \neq Y \bigoplus Y^{\perp}.$$

This shows that closedness assumption is necessary for the direct sum decomposition to be true.

Exercise 10.9. Let $\{a_i : i \in \mathbb{N}\}$ be an orthonormal set in a Hilbert space $(H, \langle \cdot, \cdot \rangle)$. If each a_i can be written as $(a_{i1}, a_{i2}, \ldots)$ with respect to an orthonormal basis, show that we have

$$\lim_{k \to \infty} e_{ik} = 0.$$

Exercise 10.10. Consider the Hilbert space ℓ^2 and the set

$$S := \left\{ x \in \ell^2 \ : \ \sum_{i=1}^{\infty} \left(1 + \frac{1}{i}\right) |\langle x, e_i \rangle|^2 \leq 1 \right\}$$

where $\{e_i : i \in \mathbb{N}\}$ is the standard orthonormal basis. Show that the exists no point $x \in S$ realising the supremum of the norm over S:

$$\nexists\, x \in S \ : \ \|x\| = \sup_{s \in S} \|s\|.$$

Exercise 10.11. Let H be an infinite-dimensional Hilbert space. Show that H does not admit any countable Hamel (algebraic) basis, namely any set $A = \{a_i : i \in I\}$ with $I = \mathbb{N}$ such that for any $x \in H$ there exist $\alpha_1, \ldots, \alpha_N \in \mathbb{F}$ and $a_{i_1}, \ldots, a_{i_N} \in A$ such that

$$x = \sum_{k=1}^{N} \alpha_k\, a_{i_k}.$$

Note that this does not contradict that H admits a countable orthonormal basis, if it is separable!

Chapter 11

Linear operators on normed spaces

11.1 What is Functional Analysis all about

Among the mappings between any two vector spaces, the simplest are those which preserve the linear structure, i.e., the **linear mappings**. Recall that, when X and Y are vector spaces over the same scalar field $\mathbb{F}$, a mapping $T : X \longrightarrow Y$ is said to be **linear** if

$$T(\lambda x + \mu y) = \lambda T(x) + \mu T(y) \quad \text{for all } x, y \in X \text{ and } \lambda, \mu \in \mathbb{F}.$$

The general properties of such mappings should be familiar from Linear Algebra. This chapter is devoted to the study of these mappings, that will most often be referred to as **linear operators**, in the case when X and Y are **normed spaces**. In this setting, the existence of the metric (and hence topological) structures on X and Y renders important the question of **continuity** of these operators.

Unlike the case of finite-dimensional normed spaces in which any linear operator is trivially continuous, linear operators between infinite-dimensional normed spaces may **not** be continuous. Our main interest here will be to study the properties of those linear operators which are **continuous**. In fact, for linear operators, their (global) continuity turns out to be equivalent to a property that is easier to work with analytically, that of "**linear boundedness**".[1] Hence, as it should have become already apparent from our investigations in Chapter 10, the study of infinite-dimensional normed vector spaces and of the continuous mappings between them is not mere Algebra.

Bounded linear operators between normed spaces, and especially those between Banach spaces, have some interesting properties of paramount importance throughout Analysis and its applications. Among the results to be proved in this chapter there are some famous ones in Mathematics, usually referred to as the **cornerstones of Functional Analysis**[2] (sometimes called "Linear Analysis"):

---

[1] The use of the term "linear boundedness" is this setting is potentially confusing and has to be understood in a special sense applying to linear operators **only**. In particular, "linear boundedness" does **not** coincide with the usual meaning of "boundedness"! However, this nomenclature is standard for historical reasons and we will not attempt to change it herein.

[2] "Functional Analysis" today describes the area of Analysis which studies vector spaces, particularly infinite-dimensional ones, and linear mappings, particularly continuous ones,

- the **Banach-Steinhaus Theorem** (known in the literature also as the "**Uniform Boundedness Principle**"),

- the **Open Mapping Theorem**,

- the **Riesz Representation Theorem**,

- the analytic **Hahn-Banach (extension) Theorem**,

- the geometric **Hahn-Banach (separation) Theorem**.

We shall refrain from delving into the specifics of what the above famous results exactly involve since it might be difficult to appreciate their significance at this initial stage. We confine ourselves with merely mentioning that these notorious theorems describe subtle properties involving continuous linear operators.

The study of normed spaces and of linear operators depends essentially on the concept of **dual space** of a normed space, which will have a pivotal role in our study. Given a normed space $(X, \|\cdot\|)$, its dual space X^* is the vector space of linear operators $X \longrightarrow \mathbb{F}$ (these operators are also known as **linear functionals**). The study of dual spaces and of their properties, to be initiated in this chapter, will be continued quite systematically in Chapters 12, 13, and 14, particularly in conjunction with compactness properties of normed spaces for weak topologies (as defined in Chapter 4).

11.2   Bounded linear operators on normed spaces

In this section we commence our journey in the subject of linear continuous operators on normed spaces. We begin with some standard notational conventions that will make our life a little easier.

Let $(X, \|\cdot\|)$, $(Y, \|\cdot\|)$ be normed spaces and $T : X \longrightarrow Y$ a linear operator. The linear structures of the spaces and of the operator are always considered over the field $\mathbb{F} \in \{\mathbb{R}, \mathbb{C}\}$. We will be using the same generic symbol $\|\cdot\|$ for both norms $\|\cdot\|_X$ on X and $\|\cdot\|_Y$ on Y, even though the two norms (or, even the two spaces) may be different, unless of course there is a danger of confusion. In general, however, the nature of the elements under consideration will make clear to which norm one is referring to. Another standard convention we shall adopt concerns the use of the notation "Tx" instead of "$T(x)$" for $x \in X$, when T is a *linear* operator **only**. For putative nonlinear maps $Q : X \longrightarrow Y$, we shall retain the universal functional notation "$Q(x)$". However, in many

---

between them. The terminology has its origins at the early stage of development when all the spaces of interest were spaces of functions, like those we presented in earlier chapters.

cases we shall retain the notation "$T(x)$" even for linear maps, particularly when it is notationally convenient in that particular context.

A first observation is that, for a linear operator between normed spaces, global continuity is equivalent to continuity at the origin.

Proposition 11.1. *Let* $(X, \|\cdot\|)$, $(Y, \|\cdot\|)$ *be normed spaces, and* $T : X \longrightarrow Y$ *a linear operator. The following statements are equivalent:*

(i) *T is continuous on X;*

(ii) *T is continuous at 0.*

Proof. (i) $\Longrightarrow$ (ii): obvious.

(ii) $\Longrightarrow$ (i): Suppose that T is continuous at 0. Let $x_0 \in X$ be arbitrary. We need to show that, for any $\varepsilon > 0$ there exists $\delta > 0$ such that

$$\|Tx - Tx_0\| < \varepsilon \quad \text{for all} \quad x \in X \text{ with } \|x - x_0\| < \delta.$$

Fix $\varepsilon > 0$, arbitrary. Since T is continuous at 0, it follows that there exists $\delta > 0$ such that $\|Tz\| < \varepsilon$ for every $z \in X$ with $\|z\| < \delta$. With this choice of δ, we have, using the linearity of T, that, for any $x \in X$ with $\|x - x_0\| < \delta$,

$$\|Tx - Tx_0\| = \|T(x - x_0)\| < \varepsilon.$$

This shows that T is continuous at the point x_0, for any $x_0 \in X$. Therefore, T is continuous on X. $\qquad\square$

We now give some simple examples of linear operators, the first two of which are almost trivial.

Example 11.2. Let $(X, \|\cdot\|)$ be a normed space. The mapping $I_X : X \longrightarrow X$ given by

$$I_X x := x \quad \text{for all } x \in X$$

is a linear operator which is continuous on X and is called the **identity operator** on X.

Example 11.3. Let $(X, \|\cdot\|)$ and $(Y, \|\cdot\|)$ be normed spaces. The mapping $O : X \longrightarrow Y$ given by

$$Ox := 0 \quad \text{for all } x \in X$$

is a linear operator which is continuous on X and is called the **zero operator**.

Example 11.4. Let $a < b$ in $\mathbb{R}$ and take $T : \big(C([a, b], \mathbb{R}), \|\cdot\|_\infty\big) \longrightarrow (\mathbb{R}, |\cdot|)$ given by

$$Tf := \int_a^b f(t) \, dt \quad \text{for all } f \in C([a, b], \mathbb{R}).$$

By using the mean value theorem, it follows that T is a continuous linear operator satisfying $|Tf| \leq (b - a)\|f\|_\infty$, for all $f \in C([a, b], \mathbb{R})$.

Example 11.5. Let $(X, \mathfrak{M}, \mu)$ be a measure space, d an integer and consider the operator $T : \left(L^1(X, \mu, \mathbb{F}^d), \| \cdot \|_{L^1(X, \mu, \mathbb{F}^d)}\right) \longrightarrow (\mathbb{F}^d, | \cdot |)$ given by

$$Tf := \int_X f(x) \, d\mu(x) \quad \text{for all } f \in L^1(X, \mu, \mathbb{F}^d).$$

Then, it follows that T is a continuous linear operator which satisfies $|Tf| \leq \|f\|_{L^1(X, \mu, \mathbb{F}^d)}$, for all $f \in L^1(X, \mu, \mathbb{F}^d)$.

Example 11.6. Consider the vector space

$$Y := \left\{ p : [0, 1] \longrightarrow \mathbb{R} \,\middle|\, p \text{ is a polynomial with coefficients in } \mathbb{R} \right\},$$

regarded as a normed subspace of $\left(C([a, b], \mathbb{R}), \| \cdot \|_\infty\right)$. We define an operator $T : Y \longrightarrow Y$ by taking

$$(Tp)(t) := p'(t) \quad \text{for all } t \in [0, 1].$$

Then T is a linear operator which is not continuous at 0. Indeed, let us consider the sequence $(p_n)_1^\infty \subseteq Y$ given by

$$p_n(t) := \frac{1}{n} t^n \quad \text{for all } t \in [0, 1].$$

Then $\|p_n\|_\infty = 1/n$ for all $n \in \mathbb{N}$ and hence $p_n \longrightarrow 0$ in $(Y, \| \cdot \|_\infty)$ as $n \to \infty$, while $\|Tp_n\|_\infty = 1$ for all $n \in \mathbb{N}$, so that $Tp_n \nrightarrow 0$ in $(Y, \| \cdot \|_\infty)$ as $n \to \infty$.

We now introduce, for linear operators between normed spaces, the notion of **(linear) boundedness**. This notion can be expressed in two equivalent ways, one which is analytically more convenient to work with and one whose intuitive meaning is clearer (and justifies the nomenclature "bounded"). Below is the statement regarding the equivalence of the two versions of the notion, followed by the relevant definition.

Proposition 11.7. *Let $T : X \longrightarrow Y$ be a linear operator between the normed spaces $(X, \| \cdot \|)$ and $(Y, \| \cdot \|)$. The following statements are equivalent:*

(i) *There exists $M > 0$ such that*

$$\|Tx\| \leq M\|x\| \quad \text{for all } x \in X.$$

(ii) *T maps any bounded subset of X into a bounded subset of Y.*

Definition 11.8. Let $(X, \|\cdot\|)$ and $(Y, \|\cdot\|)$ be normed spaces and $T : X \longrightarrow Y$ a linear operator.

• We say that T is a **bounded linear operator** if it satisfies either of the equivalent properties (i) or (ii) in Proposition 11.7.

• The **set of all bounded linear operators** from X to Y will be symbolised as

$$B(X, Y) := \left\{ T : X \longrightarrow Y \,\middle|\, T \text{ is bounded linear operator} \right\}.$$

Remark 11.9 (Linear boundedness vs. boundedness). We draw the reader's attention to the fact that **the definition of boundedness for linear operators is different from the definition of boundedness commonly used for general mappings** (which, in a metric-space setting, requires that the image of the domain is a bounded subset of the target). This is an unfortunate fact of life, since both meanings of boundedness are well established in the Mathematics literature. However, the risk of confusion is minimal, since, except for the zero operator in Example 11.3, no linear mapping can be bounded in the sense of mapping the entire domain into a bounded set of the range (why?). Thus, whenever we deal with linear mappings, their boundedness will be understood in the sense of Definition 11.8, while for other mappings boundedness will be understood in the sense of the other definition.

We recall that in the setting of normed spaces, the notations "$\mathbb{D}_R(0)$" and "$\overline{\mathbb{B}}_R(0)$" can be used interchangeably for the closed balls of radius R centred at the origin since by Proposition 2.67 they coincide with the closures of the open balls.

Proof of Proposition 11.7. (i) $\implies$ (ii): Suppose that (i) holds. Let A be an arbitrary bounded subset of X. Then there exists $R > 0$ such that $A \subseteq \mathbb{D}_R(0)$. Then, using (i), we obtain that

$$T(A) \subseteq T(\mathbb{D}_R(0)) \subseteq \mathbb{D}_{MR}(0),$$

so that $T(A)$ is bounded. Hence (ii) holds.

(ii) $\implies$ (i): Suppose now that (ii) holds. Then, in particular, there exists $M > 0$ such that $T(\mathbb{D}_1(0)) \subseteq \mathbb{D}_M(0)$. Then, for any $x \in X$ with $x \neq 0$, we have, using what we may call a **homogeneity argument**, that

$$\|Tx\| = \left\| T\left(\|x\| \frac{x}{\|x\|} \right) \right\| = \|x\| \left\| T\left(\frac{x}{\|x\|} \right) \right\| \leq M\|x\|,$$

where we have used the fact that $x/\|x\|$ is a unit vector and hence it lies in $\mathbb{D}_1(0)$. This proves that (i) holds. $\qquad\square$

Remarkably, for linear operators the notions of continuity and boundedness are equivalent.

Proposition 11.10. *Let X and Y be normed spaces and let also $T : X \longrightarrow Y$ be a linear operator. The following statement are equivalent:*

(i) *T is continuous on X.*

(ii) *T is a bounded linear operator.*

Proof. (i) $\implies$ (ii): Suppose that T is continuous on X. Then, in particular, for every $\varepsilon > 0$ there exists $\delta > 0$ such that

$$\|Tx\| \leq \varepsilon \quad \text{for all } x \in X \text{ with } \|x\| \leq \delta.$$

Fix some $\varepsilon_0 > 0$ (for example $\varepsilon_0 = 1$), and let $\delta_0 > 0$ be as above. Then, for any $x \in X$ with $x \neq 0$, we have:

$$Tx = T\left(\frac{\|x\|}{\delta_0}\frac{\delta_0 x}{\|x\|}\right) = \frac{\|x\|}{\delta_0}T\left(\frac{\delta_0 x}{\|x\|}\right).$$

Now observe that $\left\|\frac{\delta_0 x}{\|x\|}\right\| = \delta_0$, which implies that $\left\|T\left(\frac{\delta_0 x}{\|x\|}\right)\right\| \leq \varepsilon_0$. It therefore follows that

$$\|Tx\| = \frac{\|x\|}{\delta_0}\left\|T\left(\frac{\delta_0 x}{\|x\|}\right)\right\| \leq \frac{\varepsilon_0}{\delta_0}\|x\| \quad \text{for all } x \neq 0.$$

This implies that T is a bounded linear operator (with $M := \varepsilon_0/\delta_0$).

(ii) $\Longrightarrow$ (i): Suppose now that T is a bounded linear operator. Let $M > 0$ be such that
$$\|Tx\| \leq M\|x\| \quad \text{for all } x \in X.$$

For any fixed $\varepsilon > 0$, let $\delta := \varepsilon/M$. It is then clear that, for any $x \in X$ with $\|x\| < \delta$, we have that

$$\|Tx\| \leq M\|x\| < M\delta = \varepsilon.$$

This shows that T is continuous at 0, and therefore, by Proposition 11.1, T is continuous on X. $\qquad \square$

The next result shows that any linear operator on a **finite-dimensional** normed space is necessarily bounded, and therefore continuous.

Proposition 11.11. *Suppose that $(X, \|\cdot\|)$ is a finite-dimensional normed space, $(Y, \|\cdot\|)$ is a normed space and $T : X \longrightarrow Y$ is a linear operator. Then T is a bounded linear operator.*

Proof. Let $\{e_1, \ldots, e_n\}$ be a basis for X, for some $n \in \mathbb{N}$. Recall that any two norms on X are equivalent (Theorem 5.22), and note that the boundedness of T does not depend on the norm chosen on X. Then, the function $\|\cdot\|_1 : X \longrightarrow \mathbb{R}$ defined by means of representations with respect to this basis by

$$\|\alpha_1 e_1 + \cdots + \alpha_n e_n\|_1 := |\alpha_1| + \cdots + |\alpha_n| \quad \text{for all } \alpha_1, \ldots, \alpha_n \in \mathbb{F}$$

is a norm on X. We set

$$M := \max\{\|Te_1\|, \ldots, \|Te_n\|\}.$$

Then, for any $\alpha_1, \ldots, \alpha_n \in \mathbb{F}$, we have

$$\left\|T(\alpha_1 e_1 + \cdots + \alpha_n e_n)\right\| \leq |\alpha_1|\|Te_1\| + \cdots + |\alpha_n|\|Te_n\|$$
$$\leq M\|\alpha_1 e_1 + \cdots + \alpha_n e_n\|_1.$$

It follows that T is a bounded linear operator, as required. $\qquad \square$

We now give some further examples of linear operators and study their boundedness.

Example 11.12. Consider $a < b$ in $\mathbb{R}$ and $k : [a, b] \times [a, b] \longrightarrow \mathbb{C}$ a given continuous function, and set $M := \sup_{[a,b]^2} |k|$. Consider the space $C([a, b], \mathbb{C})$ of complex continuous (bounded) functions which, as proved in Chapter 3, is a Banach space when endowed with the norm $\|f\|_\infty = \sup_{[a,b]} |f|$. For any $f \in C([a, b], \mathbb{C})$, we define the function

$$(Tf)(x) := \int_a^b k(x, y) f(y) \, dy \quad \text{for all } x \in [a, b],$$

Note that, for any $x \in [a, b]$, $k(x, \cdot) f(\cdot)$ is Riemann integrable on $[a, b]$ and hence Tf is a well-defined function. Also, for any $f \in C([a, b], \mathbb{C})$ and any $x \in [a, b]$, we have

$$|(Tf)(x)| = \left| \int_a^b k(x, y) f(y) \, dy \right| \leq \int_a^b |k(x, y) f(y)| \, dy \leq M(b - a) \|f\|_\infty,$$

and hence Tf is a bounded function on $[a, b]$. Moreover, the fact that $x \mapsto (Tf)(x)$ is a continuous function on $[a, b]$ may be seen as a consequence of the Dominated Convergence Theorem. However, since in this example we are dealing we Riemann integrals, let us give a direct proof of this fact. Indeed, for any $x', x'' \in [a, b]$, we have

$$\left| (Tf)(x'') - (Tf)(x') \right| = \left| \int_a^b \left[k(x'', y) - k(x', y) \right] f(y) \, dy \right|$$

$$\leq \int_a^b \left| k(x'', y) - k(x', y) \right| |f(y)| \, dy$$

$$\leq \max_{y \in [a,b]} \left| k(x'', y) - k(x', y) \right| \int_a^b |f(y)| \, dy.$$

Since $[a, b]^2$ is compact, k is uniformly continuous on $[a, b]^2$. Hence for any $\varepsilon > 0$, there exists a $\delta = \delta(\varepsilon) > 0$ such that $|k(x'', y) - k(x', y)| < \varepsilon$, whenever $x', x'', y \in [a, b]$ with $|x'' - x'| < \delta$. The claimed result follows. Further, the linearity of T is obvious. Also, the fact that the map $f \mapsto Tf$ defines a bounded linear operator $T : C([a, b], \mathbb{C}) \longrightarrow C([a, b], \mathbb{C})$ is a consequence of the first estimate, which can be rewritten as $\|Tf\|_\infty \leq M(b - a) \|f\|_\infty$.

Example 11.13. More generally, let $(X, \mathfrak{M}, \mu)$ be a measure space. Let further $k : X \times X \longrightarrow \mathbb{F}$ be a given bounded function which is measurable with respect to the (completion of the) product σ-algebra. Set $M := \sup_{X^2} |k|$. Let $d \in \mathbb{N}$ and consider the space $L^p(X, \mu, \mathbb{F}^d)$ endowed with the L^p norm, where $p \in \{1, \infty\}$. We define a linear operator $T : L^1(X, \mu, \mathbb{F}^d) \longrightarrow L^\infty(X, \mu, \mathbb{F}^d)$ by setting

$$(Tf)(x) := \int_X k(x, y) f(y) \, d\mu(y) \quad \text{for } \mu\text{-a.e. } x \in X,$$

for any $f \in L^1(X, \mu, \mathbb{F}^d)$. For any $x \in X$, the map $k(x, \cdot)f(\cdot)$ is $\mathfrak{M}$-measurable and by Hölder's inequality, it is μ-integrable as well. Also, the map $Tf(\cdot)$ is $\mathfrak{M}$-measurable as a consequence of the Fubini–Tonelli theorem. Moreover, it follows easily as in Example 11.12 that

$$\|Tf\|_{L^\infty(X,\mu,\mathbb{F}^d)} \leq M\|f\|_{L^1(X,\mu,\mathbb{F}^d)}, \quad \text{for } f \in L^1(X, \mu, \mathbb{F}^d),$$

and hence T is a well-defined bounded linear operator.

Example 11.14. Given a fixed $a = (a_1, a_2, \ldots) \in \ell^\infty$, we consider the mapping $T_a : \ell^2 \to \ell^2$ given by

$$T_a(x_1, x_2, \ldots) := (a_1 x_1, a_2 x_2, \ldots) \quad \text{for all } x = (x_1, x_2, \ldots) \in \ell^2.$$

Then T is a well-defined bounded linear operator on ℓ^2. Indeed, for any $x \in \ell^2$ and all $k \in \mathbb{N}$, we have the bound $|a_k x_k| \leq \|a\|_\infty |x_k|$. It follows that the series $\sum_{k=1}^\infty |a_k x_k|^2$ converges and $Tx \in \ell^2$, since

$$\sum_{k=1}^\infty |a_k x_k|^2 \leq \|a\|_{\ell^\infty}^2 \|x\|_{\ell^2}^2.$$

Therefore, T is well defined and obviously it is linear too, whilst the preceding inequality can be rewritten as $\|Tx\|_{\ell^\infty} \leq \|a\|_{\ell^\infty} \|x\|_{\ell^2}$.

We now examine in more detail the structure of the set $B(X, Y)$ of all bounded linear operators between the normed spaces $(X, \|\cdot\|)$ and $(Y, \|\cdot\|)$. We begin by noting that $B(X, Y)$ has the structure of a vector space over $\mathbb{F}$ with the usual algebraic operations defined in Chapter 2, that is we define $\lambda T + \mu S \in B(X, Y)$ for any $T, S \in B(X, Y)$ and $\lambda, \mu \in \mathbb{F}$ by taking

$$(\lambda T + \mu S)x := \lambda(Tx) + \mu(Sx) \quad \text{for all } x \in X.$$

Let us verify this for the sake of completeness. Let $M_T, M_S > 0$ be such that

$$\|Tx\| \leq M_T\|x\|, \quad \|Sx\| \leq M_S\|x\| \quad \text{for all } x \in X.$$

Then, for any $x \in X$, we have

$$\big\|(\lambda T + \mu S)x\big\| \leq |\lambda|\|Tx\| + |\mu|\|Sx\| \leq \big(|\lambda|M_T + |\mu|M_S\big)\|x\|,$$

which shows that the linear operator $\lambda T + \mu S$ is bounded.

Also, it is worth mentioning that, in applications, the boundedness condition for a linear operator $T : X \longrightarrow Y$, where $(X, \|\cdot\|)$ and $(Y, \|\cdot\|)$ are normed spaces, usually comes into play by enabling one to estimate $\|Tx\|$ in terms of $\|x\|$, for any $x \in X$ in the very simple way provided by Proposition 11.7(i): there exists $M > 0$ such that

$$\|Tx\| \leq M\|x\| \quad \text{for all } x \in X. \tag{$\triangleleft$}$$

It is obviously of interest to find the smallest possible constant M for which (◁) holds true, if T is bounded. Since one easily sees that

$$\min\left\{M \geq 0 \,:\, \|Tx\| \leq M\|x\| \text{ for all } x \in X\right\} = \sup\left\{\frac{\|Tx\|}{\|x\|} \,:\, x \in X \setminus \{0\}\right\},$$

by the definition of the supremum, we are led to the following concept:

Definition 11.15. Let $(X, \|\cdot\|), (Y, \|\cdot\|)$ be normed spaces. For any bounded linear operator $T \in B(X, Y)$, the non-negative number

$$\|T\| := \sup\left\{\frac{\|Tx\|}{\|x\|} \,:\, x \in X \setminus \{0\}\right\},$$

is called **the (operator) norm** of T.

The nomenclature and the symbolisation used in Definition 11.15 are not at all accidental. As Proposition 11.16 below shows, the assignment $T \mapsto \|T\|$ is indeed a **norm** on the space $B(X, Y)$ of bounded linear operators $X \longrightarrow Y$. Note that we are following our previously mentioned convention of **using the symbol $\|\cdot\|$ for any norm under consideration** (here on both X and Y), as long as there is no great danger of confusion.

Proposition 11.16. *Let $(X, \|\cdot\|)$ and $(Y, \|\cdot\|)$ be normed spaces. The function*

$$\|\cdot\| \,:\, B(X, Y) \longrightarrow \mathbb{R}$$

given in Definition 11.15 is a norm on the vector space $B(X, Y)$. Therefore, $(B(X, Y), \|\cdot\|)$ is a normed space.

From now on we shall follow the convention that whenever $(X, \|\cdot\|)$ and $(Y, \|\cdot\|)$ are normed spaces, **the vector space $B(X, Y)$ will always be understood as being endowed with the operator norm.**

Remark 11.17 (Generalised "Cauchy-Schwarz" inequality). Directly from Definition 11.15 and the properties of the supremum, the following inequality is valid:

$$\|Tx\| \leq \|T\|\|x\| \quad \text{for all } x \in X.$$

Although seemingly trivial, this inequality can be seen as a generalisation of the Cauchy-Schwarz inequality. It will be used systematically throughout this chapter, as well as in the subsequent chapters.

Proof of Proposition 11.16. We content ourselves with showing only the triangle inequality, the other properties of a norm being trivial. Let $T, S \in B(X, Y)$. As observed earlier, the definition of $\|\cdot\|$ on $B(X, Y)$ implies that

$$\|Tx\| \leq \|T\|\|x\| \quad \text{for all } x \in X,$$
$$\|Sx\| \leq \|S\|\|x\| \quad \text{for all } x \in X.$$

Let us fix an $x \in X \setminus \{0\}$. Addition of these inequalities combined with the triangle inequality for $\| \cdot \|$ on Y, leads to

$$\|(T + S)x\| = \|Tx + Sx\| \leq \|Tx\| + \|Sx\| \leq \|T\|\|x\| + \|S\|\|x\|$$

and hence $\|(T + S)x\| \leq (\|T\| + \|S\|)\|x\|$. By the definition of $\|T + S\|$ and the last inequality, we infer that

$$\|T + S\| = \sup_{x \in X \setminus \{0\}} \frac{\|(T + S)x\|}{\|x\|} \leq \sup_{x \in X \setminus \{0\}} \frac{(\|T\| + \|S\|)\|x\|}{\|x\|} = \|T\| + \|S\|,$$

as required. $\qquad\square$

There exist alternative expressions of the operator norm which occasionally are very useful and will tacitly be used in the place of Definition 11.15 without perhaps any further notice. These are given in the next result.

Lemma 11.18 (Alternative expressions of the operator norm). *Let $(X, \| \cdot \|)$ and $(Y, \| \cdot \|)$ be normed spaces, and $T \in B(X,Y)$. Then, $\|T\|$ satisfies*

$$\|T\| = \sup \{\|Tx\| : \|x\| = 1\},$$
$$\|T\| = \sup \{\|Tx\| : \|x\| \leq 1\}.$$

Proof. Note that the first equality is an immediate consequence of the identity

$$\frac{\|Tx\|}{\|x\|} = \left\| T\left(\frac{x}{\|x\|}\right) \right\| \quad \text{for all } x \neq 0.$$

For the second, it is obvious that the supremum of $\{\|Tx\| : \|x\| \leq 1\}$ is greater than or equal to that of $\{\|Tx\| : \|x\| = 1\}$. These suprema actually coincide though, as one may verify by using again the identity above. The details are left as an exercise for the reader. $\qquad\square$

Given that $B(X,Y)$ is a normed space, it is a natural question how to identify conditions on the normed spaces $(X, \| \cdot \|)$ and $(Y, \| \cdot \|)$ that would ensure the completeness of $B(X,Y)$. Quite surprisingly and as the next result shows, the completeness of $(Y, \| \cdot \|)$ is sufficient, while the completeness of $(X, \| \cdot \|)$ plays no role in that respect.

Theorem 11.19. *Let $(X, \| \cdot \|)$ be a normed space and $(Y, \| \cdot \|)$ be a Banach space. Then $B(X,Y)$ is a Banach space.*

Proof. The argument is similar in spirit to those used in some of the completeness proofs given in Chapter 3. Let $(T_n)_1^\infty$ be an arbitrary Cauchy sequence in $B(X,Y)$. The Cauchy condition can be written analytically as:

$$\forall \, \varepsilon > 0, \; \exists \, N \in \mathbb{N} : \|T_n - T_m\| \leq \varepsilon, \quad \forall \, n, m \geq N.$$

By Definition 11.15, for any $x \in X$ we have

$$\|T_n x - T_m x\| \leq \|T_n - T_m\| \|x\|.$$

Using this inequality in the Cauchy condition, we obtain that

$$\forall \, \varepsilon > 0, \ \exists \, N \in \mathbb{N} : \ \|T_n x - T_m x\| \leq \varepsilon \|x\|, \quad \forall \, n, m \geq N, \forall \, x \in X. \qquad (\triangleright)$$

This clearly implies that, for any $x \in X$, $(T_n x)_1^\infty$ is a Cauchy sequence in $(Y, \| \cdot \|)$. Since $(Y, \| \cdot \|)$ is complete, it follows that for any fixed $x \in X$, the limit $\lim\limits_{n \to \infty} T_n x$ exists. Therefore, we may define $T : X \longrightarrow Y$ by

$$Tx := \lim_{n \to \infty} T_n x \quad \text{for any } x \in X.$$

It is easy to check that T is a linear operator. Since $(T_n)_1^\infty$ is a Cauchy sequence in $B(X, Y)$ and Cauchy sequences are bounded (this is valid in any metric space), there exists $M \geq 0$ such that $\|T_n\| \leq M$ for all $n \in \mathbb{N}$. By the Definition 11.15, this implies that

$$\|T_n x\| \leq M \|x\| \quad \text{for any } n \in \mathbb{N} \text{ and } x \in X.$$

For any fixed $x \in X$, passing to the limit as $n \to \infty$ in the above inequality, we obtain

$$\|Tx\| \leq M \|x\| \quad \text{for all } x \in X.$$

This shows that $T \in B(X, Y)$. We finally prove $T_n \longrightarrow T$ in $(B(X, Y), \| \cdot \|)$ as $n \to \infty$. Fix $\varepsilon > 0$, arbitrary, and let N be as in $(\triangleright)$. For any fixed $x \in X$ and $n \geq N$, passing to the limit as $m \to \infty$ in $(\triangleright)$ yields

$$\|T_n x - Tx\| \leq \varepsilon \|x\|.$$

For any fixed $n \geq N$, the validity of the above inequality for all $x \in X$ implies by the Definition 11.15 that

$$\|T_n - T\| \leq \varepsilon.$$

Since $\varepsilon > 0$ was arbitrary, this shows that indeed $T_n \longrightarrow T$ in $(B(X, Y), \| \cdot \|)$ as $n \to \infty$. In conclusion, $(B(X, Y), \| \cdot \|)$ is a Banach space. $\qquad \square$

We close this section with some comments relevant to applications involving the problem of calculating the norm of a given bounded linear operator. Unfortunately, given two normed spaces X, Y and $T \in B(X, Y)$, in most cases it might be very difficult, or even impossible, to calculate $\|T\|$ explicitly. However, finding $\|T\|$ may be possible if T has a sufficiently simple form. The next example treats such an instance.

Example 11.20. Given $a \leq c \leq b$ in $\mathbb{R}$, we consider the space $C([a, b], \mathbb{R})$ (endowed with the sup-norm) and let $T : C([a, b], \mathbb{R}) \longrightarrow \mathbb{R}$ be given by

$$Tf := f(c) \quad \text{for all } f \in C([a, b], \mathbb{R}).$$

Then T is a bounded linear operator and in fact $\|T\| = 1$. Indeed, linearity is obvious and for any $f \in C([a,b], \mathbb{R})$, we have

$$|Tf| = |f(c)| \leq \sup_{t \in [a,b]} |f(t)| = \|f\|_\infty,$$

which implies $\|T\| \leq 1$. On the other hand, for the constant function $f_0 \equiv 1$ we have $\|f_0\|_\infty = 1$ and by Lemma 11.18 we obtain $\|T\| \geq |Tf_0| = 1$.

In the method we used in Example 11.20, the general idea is first to find an upper bound on $\|Tx\|$ as sharp as possible and depending "linearly" on $\|x\|$ for arbitrary $x \in X$. Then, the goal is to either identify a point $x \in X$ realising the supremum, or, if this is not possible, find a maximising sequence $(x_n)_1^\infty$ for which $\|Tx_n\|/\|x_n\|$ approaches it. The latter is of course merely the method of proving that a certain constant is the value of supremum of a set.

11.3   The Banach-Steinhaus and Open Mapping theorems

In this section we expound on two deep result of Functional Analysis regarding bounded linear operators, the so-called **Banach–Steinhaus Theorem** and the **Open Mapping Theorem**. These results enjoy several common features. Firstly, their statements appear striking to a certain extent because the conclusions do not immediately seem to be plausible consequences of the assumptions. Secondly, their proofs, despite being elegant and perspicuous, they are purely analytical and, unlike Linear Algebra, they are not based on matrices and expansions on bases. In fact, the underlying ideas go beyond a mere applications of the relevant definitions and involve an important tool we discussed in Chapter 3, Baire's Theorem 3.15.

We begin be discussing the Banach–Steinhaus Theorem. The quintessence of it may be stated informally as: **any family of bounded linear mappings from a Banach space to a normed space which is pointwise bounded, actually is uniformly bounded.** More precisely, the exact statement reads as follows.

Theorem 11.21 (Banach–Steinhaus Theorem or Uniform Boundedness Principle). *Let* $(X, \|\cdot\|)$ *be a Banach space and* $(Y, \|\cdot\|)$ *a normed space. Suppose* $\{T_i : i \in I\} \subseteq B(X, Y)$ *is a family of bounded linear operators satisfying that for every* $x \in X$ *there exists an* $M_x \geq 0$ *such that*

$$\|T_i x\| \leq M_x \quad \text{for all } i \in I.$$

Then, there exists an $M \geq 0$ *such that*

$$\|T_i\| \leq M \quad \text{for all } i \in I.$$

Remark 11.22. Note that the conclusion of Theorem 11.21 may be equivalently rewritten as the existence of $M \geq 0$ such that

$$\|T_i x\| \leq M \|x\| \quad \text{for all } x \in X \text{ and } i \in I.$$

This yields that the original pointwise bounds $\{M_x : x \in X\}$ are bounded uniformly by some $M > 0$ which is independent of x. Further, it can be easily seen that the assumption of Theorem 11.21 is equivalent to the (seemingly weaker) condition that the family is pointwise bounded on the unit ball of $(X, \|\cdot\|)$, while the conclusion is that the family is uniformly bounded thereon.

Proof. As a motivation for the approach to be taken shortly, note first that the required conclusion may be restated as

$$\|T_i x\| \leq M \quad \text{for all } x \in \mathbb{D}_1(0) \text{ and for all } i \in I,$$

or, furthermore, as

$$\mathbb{D}_1(0) \subseteq \left\{ x \in X : \|T_i x\| \leq M \text{ for all } i \in I \right\},$$

for some $M \geq 0$ whose existence needs to be established. This motivates to define

$$F_n := \left\{ x \in X : \|T_i x\| \leq n \text{ for all } i \in I \right\}, \quad n \in \mathbb{N},$$

and to aim at showing that for some index $n \in \mathbb{N}$, the set F_n has non-empty interior. This will establish the desired conclusion.

We begin by noting that, for each $n \in \mathbb{N}$, the set F_n is closed in X as a consequence of the fact that each T_i is continuous. In addition, the pointwise boundedness of the family $\{T_i : i \in I\}$ ensures that any $x \in X$ belongs to some set F_n, therefore

$$X = \bigcup_{n=1}^{\infty} F_n.$$

Since the space X is complete, Baire's Theorem 3.15 implies that there exists $N \in \mathbb{N}$ such that F_N has non-empty interior. Hence, there exist $z_0 \in X$ and $s > 0$ such that

$$F_N \supseteq \mathbb{B}_s(z_0).$$

By taking closures in this relation (or by considering a closed ball of smaller radius inside the open ball), we obtain that

$$F_N \supseteq \mathbb{D}_s(z_0).$$

By the definition of F_N, we see that for all $z \in \mathbb{D}_s(z_0)$ we have

$$\|T_i z\| \leq N \quad \text{for all } i \in I.$$

Now observe that, for every $x \in \mathbb{D}_s(0)$, one may expand x as

$$x = \frac{1}{2}(z_0 + x) - \frac{1}{2}(z_0 - x),$$

where $z_0 \pm x \in \mathbb{D}_s(z_0)$. It follows that, for every $x \in \mathbb{D}_s(0)$ and $i \in I$,

$$\|T_i x\| \leq \frac{1}{2}\|T_i(z_0 + x)\| + \frac{1}{2}\|T_i(z_0 - x)\| \leq N.$$

This implies that for every $x \in X \setminus \{0\}$ and every $i \in I$, we have the bound

$$\|T_i x\| = \frac{\|x\|}{s}\left\|T_i\left(\frac{sx}{\|x\|}\right)\right\| \leq \frac{N}{s}\|x\|.$$

By the definition of the operator norm, we infer that

$$\|T_i\| \leq \frac{N}{s} \quad \text{for all } i \in I,$$

which is the required conclusion. $\square$

Remark 11.23. The result of Theorem 11.21 is particularly useful in (at least) two different situations:

• Given a family of bounded linear operators, the task of establishing pointwise bounds for the family may be much easier than that of estimating directly the norms of the operators, and thus the theorem implies the existence of an uniform bound that may be difficult to deduce otherwise.

• Given a family of linear operators which is *not uniformly bounded*, one can deduce from the theorem the existence of points at which the family is not bounded in situations in which such points cannot be exhibited explicitly.

We now turn to the other main result of this section, the so-called **Open Mapping Theorem**. We first justify the name of this result, by making precise the notion of an "open mapping". Recall from Chapter 4 that, if $(X, \mathcal{T})$ and $(Y, \mathcal{S})$ are topological spaces, a mapping $f : X \longrightarrow Y$ is continuous on X if and only if the *inverse image* of any open set in $(Y, \mathcal{S})$ is an open set in $(X, \mathcal{T})$. On the other hand, in the same setting it makes perfect sense to consider also those mappings with the opposite property that the *image* of any open set of $(X, \mathcal{T})$ is an open set in $(Y, \mathcal{S})$, a functional property called **openness**:

Definition 11.24. Let $(X, \mathcal{T})$ and $(Y, \mathcal{S})$ be topological spaces and suppose $f : X \longrightarrow Y$ is a given mapping. Then, f is called **open mapping** when it maps open sets of X to open sets of Y, that is when $f(U) \in \mathcal{S}$ for all $U \in \mathcal{T}$.

In general, there is no connection between openness and continuity of a map. An arbitrary map may well be continuous without being open and the other way around. For example, the function $f : \mathbb{R} \longrightarrow \mathbb{R}$ with $f \equiv 1$ is continuous but not open. The following result for linear operators between Banach spaces is therefore striking, since a strong topological conclusion is obtained from assumptions which are to a substantial extent algebraic in nature. The essential point is that surjective continuous linear mappings are always open.

Theorem 11.25 (Open Mapping Theorem). *Let $(X, \|\cdot\|)$ and $(Y, \|\cdot\|)$ be Banach spaces and $T \in B(X, Y)$ a bounded linear operator. If T is surjective, then T is an open mapping.*

Remark 11.26. The surjectivity assumption is necessary, even if X, Y are finite-dimensional. For instance, the linear operator $T : \mathbb{R} \longrightarrow \mathbb{R}^2$ given by $Tx := (x, 0)$ is not open but it is not surjective either since $T\mathbb{R} = \mathbb{R} \times \{0\}$.

It will be technically convenient for the upcoming proof to recall and use the following notation introduced in Chapter 6. Given any vector space Z, any subset $A \subseteq Z$, any $z \in Z$ and any number $\alpha > 0$, the *translation by z* and the *dilation by α* of the set A are the sets given by:

$$z + A := \{z + x : x \in A\}, \qquad \alpha A := \{\alpha x : x \in A\}.$$

Proof. We begin our reasoning by exhibiting a statement that, as we show, would imply the required conclusion. It then remains to deduce the validity of this statement from the surjectivity assumptions. We follow this approach because this initial step is rather easy and settling it at the beginning will enable us to concentrate on the essential difficulty of the proof from then on. Thus, we begin with the next claim:

Claim 11.27. *Let $(X, \|\cdot\|)$ and $(Y, \|\cdot\|)$ be Banach spaces and $T \in B(X, Y)$ a bounded linear operator. Assume there exists $\delta > 0$ such that*

$$T(\mathbb{B}_1(0)) \supseteq \mathbb{B}_\delta(0). \qquad (\clubsuit)$$

Then, $(\clubsuit)$ implies that T is an open mapping.

Proof of Claim 11.27. Suppose that $(\clubsuit)$ holds. Then, by using the linearity of T, it is clear that, for any $r > 0$,

$$T(\mathbb{B}_r(0)) = T(r\mathbb{B}_1(0)) = rT(\mathbb{B}_1(0)) \supseteq r\mathbb{B}_\delta(0) = \mathbb{B}_{r\delta}(0),$$

and also that, for any $x_0 \in X$ and $r > 0$,

$$\begin{aligned}
T(\mathbb{B}_r(x_0)) &= T\big(x_0 + \mathbb{B}_r(0)\big) \\
&= Tx_0 + T(\mathbb{B}_r(0)) \\
&\supseteq Tx_0 + rT(\mathbb{B}_1(0)) \\
&= Tx_0 + r\mathbb{B}_\delta(0) = \mathbb{B}_{r\delta}(Tx_0).
\end{aligned}$$

Let V be an arbitrary open set in $(X, \|\cdot\|)$, and let y_0 be any element of $T(V)$. Then there exists $x_0 \in V$ such that $y_0 = Tx_0$. Also, since V is open and $x_0 \in V$, it follows that there exists $r > 0$ such that $V \supseteq \mathbb{B}_r(x_0)$. We then obtain that

$$T(V) \supseteq T(\mathbb{B}_r(x_0)) \supseteq \mathbb{B}_{r\delta}(Tx_0) = \mathbb{B}_{r\delta}(y_0).$$

This show that $T(V)$ is open in $(Y, \|\cdot\|)$. $\qquad\square$

It therefore remains to prove the validity of the inclusion (♣) when T is surjective, which is exactly what we do in the remaining of the proof. By the surjectivity of T, we have

$$Y = T(X) = T\left(\bigcup_{n=1}^{\infty} \mathbb{B}_n(0)\right) = \bigcup_{n=1}^{\infty} T(\mathbb{B}_n(0)) = \bigcup_{n=1}^{\infty} \overline{T(\mathbb{B}_n(0))}.$$

Since Y is complete, Baire's Theorem 3.15 yields that there exist $N \in \mathbb{N}$, $z_0 \in Y$ and $s > 0$ such that

$$\overline{T(\mathbb{B}_N(0))} \supseteq \mathbb{B}_s(z_0),$$

and in fact, by taking closures in this relation,

$$\overline{T(\mathbb{B}_N(0))} \supseteq \mathbb{D}_s(z_0).$$

We now continue with the next:

Claim 11.28. *Under the assumptions of Theorem 11.25, we have*

$$\overline{T(\mathbb{B}_N(0))} \supseteq \mathbb{D}_s(0). \tag{♠}$$

Proof of Claim 11.28. Indeed, let $y \in \mathbb{D}_s(0)$, arbitrary. Then we have the following decomposition

$$y = \frac{1}{2}(z_0 + y) - \frac{1}{2}(z_0 - y).$$

Note that $z_0 \pm y \in \mathbb{D}_s(z_0)$. Since $\mathbb{D}_s(z_0) \subseteq \overline{T(\mathbb{B}_N(0))}$, it follows that there exist sequences $(x_n')_1^\infty, (x_n'')_1^\infty \subseteq \mathbb{B}_N(0)$ such that

$$Tx_n' \longrightarrow z_0 + y, \qquad Tx_n'' \longrightarrow z_0 - y,$$

as $n \to \infty$. For each $n \in \mathbb{N}$, we define

$$x_n = \frac{1}{2}x_n' - \frac{1}{2}x_n''.$$

It is clear that $(x_n)_1^\infty \subseteq \mathbb{B}_N(0)$ and

$$Tx_n \longrightarrow \frac{1}{2}(z_0 + y) - \frac{1}{2}(z_0 - y) = y,$$

as $n \to \infty$. Since $y \in \mathbb{D}_s(0)$ was arbitrary, this shows that the inclusion (♠) indeed holds true, as desired. □

Claim 11.29. *Under the assumptions of Theorem 11.25, if we set $\delta := s/N$ then for any $r > 0$ we have the inclusion*

$$\overline{T(\mathbb{D}_{\frac{r}{\delta}}(0))} \supseteq \mathbb{D}_r(0). \tag{♭}$$

Proof of Claim 11.29. Indeed, this is readily obtained from (♠) by an easy scaling argument:

$$\overline{T(\mathbb{D}_{\frac{r}{\delta}}(0))} \supseteq \overline{T(\mathbb{B}_{\frac{Nr}{s}}(0))} = \frac{r}{s}\overline{T(\mathbb{B}_N(0))} \supseteq \frac{r}{s}\mathbb{D}_s(0) = \mathbb{D}_r(0). \quad \square$$

Note that (♭) implies in particular that $\overline{T(\mathbb{D}_1(0))} \supseteq \mathbb{D}_\delta(0)$, which is fairly close to the required conclusion (♣). Let us now observe that an immediate consequence of (♭) is the following statement:

$$\forall\, w \in Y,\ \forall\, \varepsilon > 0,\ \exists\, z \in X\ :\ \|z\| \leq \frac{1}{\delta}\|w\|,\ \|w - Tz\| < \varepsilon. \tag{♮}$$

Indeed, for any $w \in Y$ with $w \neq 0$, the above statement follows by considering $r := \|w\|$ in (♭) and using the definition of the closure of a set.

We shall now conclude the proof by proving the relation (♣) by means of an argument based on repeated applications of (♮). Fix $y \in \mathbb{B}_\delta(0)$, arbitrary. Let α be a positive constant to be specified more precisely later on. We first apply (♮) with $w := y$ and $\varepsilon := \alpha$ to obtain $x_0 \in X$ with $\|x_0\| \leq \|y\|/\delta$ such that

$$\|y - Tx_0\| < \alpha.$$

We then apply (♮) with $w := y - Tx_0$ and $\varepsilon := 2^{-1}\alpha$ to obtain $x_1 \in X$ with $\|x_1\| \leq \alpha/\delta$ such that

$$\|y - Tx_0 - Tx_1\| < 2^{-1}\alpha.$$

We then apply (♮) with $w := y - Tx_0 - Tx_1$ and $\varepsilon := 2^{-2}\alpha$ to obtain $x_2 \in X$ with $\|x_2\| \leq 2^{-1}\alpha/\delta$ such that

$$\left\|y - Tx_0 - Tx_1 - Tx_2\right\| < 2^{-2}\alpha.$$

Assuming that for some $n \in \mathbb{N}$, $x_0,\dots,x_{n-1}$ have been constructed in the above manner, we apply (♮) with

$$w := y - Tx_0 - Tx_1 - \cdots - Tx_{n-1}$$

and $\varepsilon := 2^{-n}\alpha$ to obtain $x_n \in X$ with $\|x_n\| \leq 2^{-(n-1)}\alpha/\delta$ and such that

$$\left\|y - Tx_0 - Tx_1 - \cdots - Tx_{n-1} - Tx_n\right\| < 2^{-n}\alpha.$$

In this way, we construct a sequence $(x_n)_1^\infty$ in $(X, \|\cdot\|)$. Note that the series $\sum_{k=0}^\infty x_k$ converges absolutely, since

$$\sum_{k=0}^\infty \|x_k\| \leq \frac{\|y\|}{\delta} + \sum_{k=1}^\infty 2^{-(k-1)}\frac{\alpha}{\delta} = \frac{\|y\| + 2\alpha}{\delta}.$$

Therefore, since $(X, \|\cdot\|)$ is a Banach space, the series $\sum_{k=0}^\infty x_k$ converges. Let us denote, for any $n \in \mathbb{N}$,

$$s_n := x_0 + \cdots + x_n,$$

and let

$$x := \sum_{k=0}^{\infty} x_k = \lim_{n \to \infty} s_n.$$

Then, since

$$\|y - Ts_n\| < 2^{-n}\alpha \longrightarrow 0 \quad \text{as } n \to \infty,$$

and T is continuous, we obtain that $y = Tx$. Finally, since $\|y\| < \delta$, one may choose $\alpha > 0$ sufficiently small so that $\|y\| + 2\alpha < \delta$, which ensures that $\|x\| < 1$. Therefore $y = Tx \in T(\mathbb{B}_1(0))$. This proves the inclusion ($\clubsuit$), and thus, in view of our earlier considerations, completes the proof of the theorem. $\qquad \square$

Perhaps the most important application of Theorem 11.25 is the following result, which ensures the boundedness of the inverse of a bijective linear mapping between Banach spaces. Note that the result also provides, seemingly out of thin air, a quantitative estimate for the mapping itself.

Theorem 11.30 (Bounded Inverse Theorem). *Let* $(X, \|\cdot\|)$ *and* $(Y, \|\cdot\|)$ *be Banach spaces. If* $T \in B(X,Y)$ *is bijective, then* $T^{-1} \in B(Y,X)$. *Moreover, there exists* $m > 0$ *such that*

$$\|Tx\| \geq m\|x\| \quad \text{for all } x \in X.$$

Proof. It is straightforward to check that the inverse of a bijective linear operator between vector spaces is also a linear operator. To prove that $T^{-1} : (Y, \|\cdot\|) \longrightarrow (X, \|\cdot\|)$ is continuous on Y, we make use of the characterisation of global continuity by means of open sets given in Chapter 2. Indeed, for any open set V in $(X, \|\cdot\|)$, we have that $(T^{-1})^{-1}(V) = T(V)$, which is an open set in $(Y, \|\cdot\|)$ by Theorem 11.25. This shows that T^{-1} is a continuous mapping on Y. Since $T^{-1} : Y \longrightarrow X$ is also linear, Proposition 11.10 yields that $T^{-1} \in B(X,Y)$. Let $M > 0$ be such that

$$\|T^{-1}y\| \leq M\|y\| \quad \text{for all } y \in Y.$$

By selecting $y := Tx$ for some arbitrary $x \in X$ and denoting $m := 1/M$, we conclude that

$$\|Tx\| \geq m\|x\| \quad \text{for all } x \in X,$$

as required. $\qquad \square$

11.4   Dual spaces and the Riesz Representation Theorem

A special case of bounded linear operators is obtained in the case when the target space $(Y, \|\cdot\|)$ is the scalar field $(\mathbb{F}, |\cdot|)$ associated to the normed space $(X, \|\cdot\|)$. This particular case is of fundamental importance and therefore it is customary to designate a particular name and notation for it.

Definition 11.31. Let $(X, \|\cdot\|)$ be a normed space over the field $\mathbb{F}$. The space $B(X, \mathbb{F})$ is called the **(topological) dual space** of X and is denoted by X^*:

$$X^* := B(X, \mathbb{F}).$$

The elements of X^* are called **bounded linear functionals on** $(X, \|\cdot\|)$. For the **(operator) norm** on X^* we will use the symbolisation

$$\|f\|_* := \sup\left\{ \frac{|f(x)|}{\|x\|} \ : \ x \in X \setminus \{0\} \right\} \quad \text{for all } f \in X^*.$$

If the need arises, we will occasionally augment the symbolisations to "$\|\cdot\|_{X^*}$" in order to guarantee precision in case there appears any greater danger of confusion.

Remark 11.32 (Dual spaces are always complete). In view of Theorem 11.19, $(X^*, \|\cdot\|)$ **is always a Banach space**, even if the normed space $(X, \|\cdot\|)$ might not complete. Hence, the dual of a space always has better properties that the original space. The modifier "topological" is occasionally used to separate it from the (much larger and less useful) **algebraic dual space** X', which is the vector space of **all linear functionals** $X \longrightarrow \mathbb{F}$. In finite dimensions, the algebraic and the topological dual space coincide: $X' = X^*$. In infinite dimensions, however, the algebraic dual space X' is strictly larger since it contains numerous discontinuous functionals as well (see Example 11.33 that follows) and therefore it will be used very sparsely.

Example 11.33 (A discontinuous linear functional). Consider the space $C_b^1(\mathbb{R}, \mathbb{C})$ which consists of bounded continuously differentiable functions $f : \mathbb{R} \longrightarrow \mathbb{C}$ with bounded derivative, endowed with the supremum norm. Then, the linear functional

$$\phi \ : \ C_b^1(\mathbb{R}, \mathbb{C}) \longrightarrow \mathbb{C}, \quad \phi(f) := f'(0)$$

is discontinuous, as for the sequence $(f_n)_1^\infty$ given by $f_n := \frac{1}{n}\sin(n^2(\cdot))$, we have $\|f_n\|_\infty \longrightarrow 0$ but on the other hand $\phi(f_n) = n \longrightarrow \infty$, as $n \to \infty$.

The dual of a normed space $(X, \|\cdot\|)$ is a basic object associated to the normed space in question, whilst bounded linear functionals on $(X, \|\cdot\|)$ occur in applications very frequently. Moreover, a good understanding of the structure of $(X^*, \|\cdot\|)$ leads to a better understanding of the normed space $(X, \|\cdot\|)$ itself.

A very important and natural question which will concern us in this section regards the **explicit description of the dual space**. To this end, we shall exhibit some situations in which the duals of various normed spaces can be identified explicitly. We begin with the case of finite-dimensional spaces. In this case, the dual is also finite-dimensional, of the same dimension as the space itself, and also one can exhibit convenient bases of it.

Proposition 11.34 (Representation of the dual of $\mathbb{F}^n$). *Given* $n \in \mathbb{N}$, *let* $(X, \| \cdot \|)$ *be an* n-*dimensional normed space and let also* $\{e_1, \ldots, e_n\}$ *be a basis for* X. *Then,* X^* *is also* n-*dimensional and has a basis* $\{f_1, \ldots, f_n\}$, *such that*

$$f_i(e_j) = \delta_{ij} \quad \text{for all } i, j \in \{1, \ldots, n\},$$

where $\delta_{ij} = 0$ *if* $i \neq j$, *and* $\delta_{ij} = 1$ *if* $i = j$.

The above results shows that, essentially, **the dual space** $(\mathbb{F}^n)^*$ **of** $\mathbb{F}^n$ **is "the same space" as** $\mathbb{F}^m$ **itself,** up to the linear isomorphism identifying their basis elements. As we will see, this is not true in general in infinite dimensions, but in some cases, as those described right next, we can indeed glean comparably detailed information for the dual space.

Proof. For each $i \in \{1, \ldots, n\}$, we define a scalar function $f_i : X \longrightarrow \mathbb{F}$ by setting

$$f_i \left(\sum_{k=1}^n \alpha_k e_k \right) := \alpha_i, \quad \text{for any } (\alpha_1, \ldots, \alpha_n) \in \mathbb{F}^n.$$

It is easy to check that, for each i, f_i is a linear functional. Since $(X, \| \cdot \|)$ is finite-dimensional, Proposition 11.11 ensures that any linear functional on $(X, \| \cdot \|)$ is bounded. Hence, $f_i \in X^*$. Moreover, by construction we have the relations

$$f_i(e_j) = \delta_{ij} \quad \text{for all } i, j \in \{1, \ldots, n\}.$$

We now prove that $f_1, \ldots, f_n$ are linearly independent. Let $\{\beta_1, \ldots, \beta_n\} \subseteq \mathbb{F}$ be such that

$$\beta_1 f_1 + \cdots + \beta_n f_n = 0.$$

For each $k \in \{1, \ldots, n\}$, evaluation at e_k of the above linear combination of linear functionals yields that $\beta_k = 0$ because $f_i(e_k) = 0$ if $i \neq k$. This proves the linear independence. We now prove that $X^* = \mathrm{span}[\{f_1, \ldots, f_n\}]$. Let $f \in X^*$ be an arbitrary (bounded) linear functional. For any index $k \in \{1, \ldots, n\}$, we set $c_k := f(e_k)$. Then, for any vector $(\alpha_1, \ldots, \alpha_n) \in \mathbb{F}^n$, we have that f gives

$$f \left(\sum_{k=1}^n \alpha_k e_k \right) = \sum_{k=1}^n \alpha_k f(e_k) = \sum_{i=1}^n c_i \alpha_i = \sum_{i=1}^n c_i f_i \left(\sum_{k=1}^n \alpha_k e_k \right).$$

Since this is true for all $(\alpha_1, \ldots, \alpha_n) \in \mathbb{F}^n$, it follows that

$$f = \sum_{i=1}^n c_i f_i,$$

and hence $f \in \mathrm{span}[\{f_1, \ldots, f_n\}]$. This proves that $X^* = \mathrm{span}[\{f_1, \ldots, f_n\}]$, thus completing the proof. $\qquad \square$

Another instance of a normed space whose dual can be explicitly exhibited is the sequence space ℓ^p, where $p \in (1, \infty)$, whose dual is isometrically isomorphic with the sequence space $\ell^{p'}$, where $p' = p/(p-1)$ is the conjugate exponent of p (Definition 2.71). The idea is roughly as follows. Suppose that we wish to identify some functionals in $(\ell^p)^*$. To this end, it makes sense to consider functionals of the multiplicative form

$$\ell^p \ni x \mapsto \sum_{k=1}^{\infty} a_k x_k \in \mathbb{F}$$

where $a = (a_1, a_2, \ldots) \in \ell^q$ is a fixed element. Then, in view of Hölder inequality, if $q = p'$, the above functional is continuous on ℓ^p. The main point of Proposition 11.35 below is that **all** functionals arise indeed in this way and, in addition, there is an isometry which allows us to identify $\ell^{p'}$ with $(\ell^p)^*$.

Proposition 11.35 (Representation of the dual of ℓ^p). *Let $p \in (1, \infty)$ and let $p' = p/(p-1)$ be its conjugate exponent (Definition 2.71). For any $a = (a_1, a_2, \ldots) \in \ell^{p'}$, consider the functional $f_a : \ell^p \longrightarrow \mathbb{F}$ given by*

$$f_a(x) := \sum_{k=1}^{\infty} a_k x_k \quad \text{for all } x = (x_1, x_2, \ldots) \in \ell^p.$$

Then $f_a \in (\ell^p)^*$. Moreover, for any functional $f \in (\ell^p)^*$, there exist a unique $a \in \ell^{p'}$ such that $f = f_a$. In addition, the mapping $T : \ell^{p'} \longrightarrow (\ell^p)^*$ given by

$$Ta := f_a \quad \text{for all } a = (a_1, a_2, \ldots) \in \ell^{p'},$$

is an isometric isomorphism between $\ell^{p'}$ and $(\ell^p)^*$ and $\|f_a\|_{(\ell^p)^*} = \|a\|_{\ell^{p'}}$.

We note that Proposition 11.35 is a particular case of general representation results we will establish in Chapter 14 regarding the class of L^p spaces for $p \in [1, \infty)$ over an abstract measure space $(X, \mathfrak{M}, \mu)$.

Proof. Given any $a \in \ell^{p'}$ and any $x \in \ell^p$, Hölder's inequality (Definition 2.78) yields that $\sum_{k=1}^{\infty} |a_k||x_k|$ converges and

$$\sum_{k=1}^{\infty} |a_k||x_k| \leq \|a\|_{\ell^{p'}} \|x\|_{\ell^p}.$$

Since $\mathbb{F}$ is complete, absolute convergence of a series implies its convergence. Thus, $\sum_{k=1}^{\infty} a_k x_k$ converges and

$$\left| \sum_{k=1}^{\infty} a_k x_k \right| \leq \sum_{k=1}^{\infty} |a_k||x_k| \leq \|a\|_{\ell^{p'}} \|x\|_{\ell^p}.$$

This means that $f_a(x)$ is well defined for all $x \in \ell^p$, and, moreover,

$$|f_a(x)| \leq \|a\|_{\ell^{p'}} \|x\|_{\ell^p} \quad \text{for all } x \in \ell^p. \tag{♯}$$

Since f_a is obviously linear, it follows that $f_a \in (\ell^p)^*$ and $\|f_a\|_{(\ell^p)^*} \leq \|a\|_{\ell^{p'}}$. To prove that this inequality is actually an equality, let us consider, given any $a \in \ell^{p'}$, the element $\tilde{x} = (\tilde{x}_1, \tilde{x}_2, \dots)$ given by

$$\tilde{x}_k := \begin{cases} |a_k|^{p'}/a_k, & \text{if } a_k \neq 0, \\ 0, & \text{if } a_k = 0. \end{cases}$$

It then follows from the definition that $|\tilde{x}_k|^p = |a_k|^{p'}$ and $a_k \tilde{x}_k = |a_k|^{p'}$, for all $k \in \mathbb{N}$. Hence,

$$|f_a(\tilde{x})| = \sum_{k=1}^{\infty} |a_k|^{p'} = \|a\|_{\ell^{p'}} \|\tilde{x}\|_{\ell^p}.$$

In combination with ($\sharp$), this relation shows that

$$\|f_a\|_{(\ell^p)^*} = \|a\|_{\ell^{p'}},$$

as required. Now we show the surjectivity of T. Fix an arbitrary $f \in (\ell^p)^*$ and let us denote by $e_k \in \ell^p$ the vector $(0, \dots, 0, 1, 0, \dots)$ whose kth entry equals 1 and all other entries are 0, $k \in \mathbb{N}$. We define

$$a := (a_1, a_2, \dots), \quad \text{where } a_k := f(e_k), \ k \in \mathbb{N}.$$

We now show that $a \in \ell^{p'}$ and $f = f_a$. Observe first that

$$|f_a(x)| \leq \|f\|_{(\ell^p)^*} \|x\|_{\ell^p} \quad \text{for all } x \in \ell^p. \tag{†}$$

Note also that, for any $n \in \mathbb{N}$ and any $\{x_1, \dots, x_n\} \subseteq \mathbb{F}$, by the linearity of f we have

$$f\left(\sum_{k=1}^{n} x_k e_k\right) = \sum_{k=1}^{n} a_k x_k. \tag{‡}$$

We now consider, for each $n \in \mathbb{N}$, the element $\tilde{x}^{(n)} = \left(\tilde{x}_1^{(n)}, \tilde{x}_2^{(n)}, \dots\right)$ given by

$$\tilde{x}_k^{(n)} := \begin{cases} |a_k|^{p'}/a_k, & \text{if } k \leq n \text{ and } a_k \neq 0, \\ 0, & \text{if } k > n \text{ or } a_k = 0. \end{cases}$$

It follows, as a consequence of (†) and (‡) that, for each $n \in \mathbb{N}$,

$$f\left(\tilde{x}^{(n)}\right) = \sum_{k=1}^{n} |a_k|^{p'} \leq \|f\|_{(\ell^p)^*} \left(\sum_{k=1}^{n} |a_k|^{p'}\right)^{1/p}.$$

Equivalently, this can be written as

$$\left(\sum_{k=1}^{n} |a_k|^{p'}\right)^{1/p'} \leq \|f\|_{(\ell^p)^*}.$$

The validity of this inequality for all $n \in \mathbb{N}$ shows that $a \in \ell^{p'}$. Moreover, since any $x \in \ell^p$ can be represented as $x = \sum_{k=1}^{\infty} x_k e_k$, the continuity of f shows that, for any such x, we have

$$f(x) = \lim_{n \to \infty} f\left(\sum_{k=1}^{n} x_k e_k\right) = \lim_{n \to \infty} \left(\sum_{k=1}^{n} a_k x_k\right) = \sum_{k=1}^{\infty} a_k x_k = f_a(x).$$

The (absolute) convergence of the above series is guaranteed by Hölder's Inequality, since $x \in \ell^p$ and $a \in \ell^{p'}$. It follows that $f = f_a$. The uniqueness of such an $a \in \ell^{p'}$ is trivial. The fact that the mapping $T : \ell^q \longrightarrow (\ell^p)^*$ is an isometry is an immediate consequence of the preceding considerations. □

Remarkably, the dual of any Hilbert space can also be described explicitly. The result is known as the **Riesz–Fréchet Theorem**, or the **Riesz Representation Theorem**, and it states that any bounded linear functionals can be obtained by taking inner product with a certain element of the space. Similarly to Proposition 11.35, the main idea is that the map

$$X \ni y \mapsto \langle \cdot, y \rangle \in X^*$$

is an isometric isomorphism. In this case we have a situation similar to the Euclidean with **the dual X^* being isomorphic to X itself**. Before the statement, let us recall some familiar terminology from Linear Algebra.

Definition 11.36. Let X, Y be vector spaces over the field $\mathbb{F}$ and $T : X \longrightarrow Y$ a linear mapping. The subset of X given by

$$\mathrm{N}(T) := \left\{x \in X : Tx = 0\right\}$$

is called the **nullspace (or the kernel) of T** and is a vector subspace of X.

Theorem 11.37 (Riesz–Fréchet or Riesz Representation Theorem). *Let $(X, \langle \cdot, \cdot \rangle)$ be a Hilbert space over $\mathbb{F}$. For any $y \in X$, the function $f_y :$ $X \longrightarrow \mathbb{F}$ given by*

$$f_y(x) := \langle x, y \rangle \quad \text{for all } x \in X$$

satisfies $f_y \in X^*$ and $\|f_y\|_* = \|y\|$. Conversely, for any $f \in X^*$ there exists a unique $y \in X$ such that $f = f_y$.

Proof. For any fixed $y \in X$, the mapping $f_y : X \longrightarrow \mathbb{F}$ is obviously linear. Moreover, for any $x \in X$, we have, by the Cauchy–Schwarz Inequality, that

$$|f_y(x)| = |\langle x, y \rangle| \leq \|y\| \|x\|.$$

This shows that $f_y \in X^*$ and $\|f_y\|_* \leq \|y\|$. On the other hand, we also have

$$\|y\|^2 = |f_y(y)| \leq \|f_y\|_* \|y\|,$$

which implies that $\|f_y\|_* \geq \|y\|$. In conclusion, $\|f_y\|_* = \|y\|$. Consider now an arbitrary $f \in X^*$. If $f \equiv 0$, then $f = f_0$ (by which we mean that f corresponds to $y = 0$). Suppose in what follows that $f \not\equiv 0$. Since f is linear and continuous on X, the nullspace $N(f)$ is a closed vector subspace of X and $N(f) \neq X$. Therefore, $(N(f))^{\perp} \neq \{0\}$. Let $z \in (N(f))^{\perp} \setminus \{0\}$. Then necessarily $f(z) \neq 0$. By a direct calculation, we can easily check that for every $x \in X$, we have

$$f\left(x - \frac{f(x)}{f(z)}z\right) = 0,$$

which shows that $x - (f(x)/f(z))z \in N(f)$. Since $z \in (N(f))^{\perp}$, it follows that

$$\left\langle x - \frac{f(x)}{f(z)}z, z \right\rangle = 0 \quad \text{for all } x \in X.$$

By rearranging this equality, we see that

$$\langle x, z \rangle = \frac{f(x)}{f(z)}\|z\|^2 \quad \text{for all } x \in X,$$

which in turn implies that

$$f(x) = \frac{f(z)}{\|z\|^2}\langle x, z \rangle = \left\langle x, \frac{\overline{f(z)}}{\|z\|^2}z \right\rangle \quad \text{for all } x \in X.$$

This shows that $f = f_y$, with $y := (\overline{f(z)}/\|z\|^2)z$. We have thus proved that, for every $f \in X^*$, there exists $y \in X$ such that $f = f_y$. Finally, we show the uniqueness of y. Suppose to the contrary that we have $f_{y_1} = f_{y_2}$ for some $y_1, y_2 \in X$. This means that

$$\langle x, y_1 \rangle = \langle x, y_2 \rangle \quad \text{for all } x \in X.$$

In particular, the choice $x := y_1 - y_2$ in the above implies that $\|y_1 - y_2\|^2 = 0$, therefore completing the proof. □

11.5   The analytic Hahn-Banach extension theorems

In what follows we are interested in studying and identifying general properties of bounded linear functionals on abstract normed spaces. We begin the discussion with the observation that a priori **it is not at all clear that an abstract normed space** $(X, \|\cdot\|)$ **has a non-trivial (topological) dual space**. Namely, we do not know whether in general the dual necessarily contains continuous functionals other than zero (which maps every element of X into 0):

$$X^* \supsetneq \{0\}.$$

On the other hand, it is not particularly difficult to find **discontinuous** linear functionals $X \longrightarrow \mathbb{F}$ on a normed space $(X, \| \cdot \|)$, as we demonstrated in Example 11.33.

However, the discouraging possibility that perhaps $X^* = \{0\}$ we alluded to above, actually does not happen. Instead, it turns out that **the topological dual of any normed space is actually quite rich.** This is merely one of the many consequences of a fundamental result, known as the **Hahn–Banach Theorem**, which is the main theme of this section. The Hahn–Banach theorem belongs to a category of results called **extension theorems.** Such types of results attempt to answer question of the following kind:

(Q) Given a map with certain properties defined on a subset of given set, can it be extended to the entire set so as to maintain those properties?

The question (Q) is too abstract and too vague to be of actual interest. In the setting of bounded linear functionals defined on normed spaces, however, the extension problem takes the next more concrete form.

(Q*) Given a bounded linear functional defined on a vector subspace of a normed space, can this functional be extended linearly to the entire normed space (i.e., in such a way as to remain linear) and, moreover, so that its dual norm remains the same?

The answer to (Q*) is in fact affirmative, and reads as follows:

Theorem 11.38 (Hahn–Banach Theorem for normed spaces). *Let $(X, \| \cdot \|)$ be a normed space over $\mathbb{F}$, Y a vector subspace of X and $f \in Y^*$. Then, there exists $F \in X^*$ which is an extension of f (namely $F|_Y = f$), such that*

$$\|F\|_{X^*} = \|f\|_{Y^*}.$$

In the case $\mathbb{F} = \mathbb{R}$ of **real** scalars, Theorem 11.38 is a consequence of a purely algebraic result, Theorem 11.40 below. The case $\mathbb{F} = \mathbb{C}$ of **complex** scalars involves an additional trick and will be given afterwards[3]. In order to state the algebraic result, we need to recall the next familiar concept.

Definition 11.39. Let K be a convex subset of a vector space X. A (real) function $p : K \longrightarrow \mathbb{R}$ is said to be **convex** on K if

$$p\big((1 - \lambda)x + \lambda y\big) \leq (1 - \lambda)p(x) + \lambda p(y),$$

for all $x, y \in K$ and all $\lambda \in [0, 1]$.

Theorem 11.40 (Hahn–Banach Theorem: algebraic version). *Let X be a real vector space, and Y be a vector subspace of X. Let also $p : X \longrightarrow \mathbb{R}$ be a convex function and $f : Y \longrightarrow \mathbb{R}$ a linear function, such that*

$$f(y) \leq p(y) \quad \text{for all } y \in Y.$$

---

[3] By the way, the Hahn–Banach Theorem is one of the very few instances in which the nature of the scalar field ($\mathbb{R}$ or $\mathbb{C}$) makes a real difference in the proof of a result.

Then, there exists a linear function $F : X \longrightarrow \mathbb{R}$ which is an extension of f (that is, $F|_Y = f$) and satisfies

$$F(x) \leq p(x) \quad \text{for all } x \in X.$$

The following nontrivial lemma is the first step towards the proof of Theorem 11.40. It may be regarded as a weak form of it, in that we extend the given linear functional to a subspace of one more dimension than that of the original subspace it was defined on (this is a slight abuse of language, but it is rather suggestive).

Lemma 11.41. *Let X be a real vector space and Z a proper vector subspace of X, namely $Z \subsetneq X$. Let $p : X \longrightarrow \mathbb{R}$ be a convex function and $g : Z \longrightarrow \mathbb{R}$ a linear function, such that*

$$g(z) \leq p(z) \quad \text{for all } z \in Z.$$

Fix any $w_0 \in X \setminus Z$, and consider the direct sum $W := Z \oplus \text{span}[\{w_0\}]$. Then there exists a linear extension $G : W \longrightarrow \mathbb{R}$ of g with $G|_Z = g$, such that

$$G(w) \leq p(w) \quad \text{for all } w \in W.$$

Proof. Note that any linear extension G of g to W is uniquely determined by its value at w_0. Indeed, since any $w \in W$ can be written in an unique way in the form $w = z + \lambda w_0$, for some $z \in Z$ and $\lambda \in \mathbb{R}$, then, for any $\gamma \in \mathbb{R}$, the functional $G_\gamma : W \longrightarrow \mathbb{R}$ given, for any such $w \in W$, by

$$G_\gamma(w) = G_\gamma(z + \lambda w_0) := g(z) + \lambda \gamma,$$

is the unique linear extension of g to W with $G_\gamma(w_0) = \gamma$. Thus, the problem that needs to be solved is that of finding a suitable $\gamma \in \mathbb{R}$ such that

$$G_\gamma(w) \leq p(w) \quad \text{for all } w \in W.$$

Since G_γ is an extension of g, it is enough to require the above inequality to be satisfied for all $w \in W \setminus Z$. Now note that any $w \in W \setminus Z$ can be written as either $w = x + \alpha w_0$, for some $x \in Z$ and $\alpha > 0$, or $w = y - \beta w_0$, for some $y \in Z$ and $\beta > 0$. Thus, the inequality

$$G_\gamma(w) \leq p(w) \quad \text{for all } w \in W \setminus Z$$

can be equivalently rewritten as

$$\begin{cases} g(x) + \alpha\gamma \leq p(x + \alpha w_0) & \text{for all } x \in Z \text{ and all } \alpha > 0, \\ g(y) - \beta\gamma \leq p(y - \beta w_0) & \text{for all } y \in Z \text{ and all } \beta > 0, \end{cases}$$

or, furthermore, as

$$\begin{cases} \gamma \leq \dfrac{1}{\alpha}\big(p(x + \alpha w_0) - g(x)\big) & \text{for all } x \in Z \text{ and all } \alpha > 0, \\ \gamma \geq \dfrac{1}{\beta}\big(g(y) - p(y - \beta w_0)\big) & \text{for all } y \in Z \text{ and all } \beta > 0. \end{cases}$$

A necessary and sufficient condition for the existence of $\gamma \in \mathbb{R}$ with the above properties is that

$$\frac{1}{\beta}\left(g(y) - p(y - \beta w_0)\right) \leq \frac{1}{\alpha}\left(p(x + \alpha w_0) - g(x)\right) \qquad (\natural)$$

for all $x, y \in Z$ and all $\alpha, \beta > 0$. We now show that $(\natural)$ is indeed true. Note that $(\natural)$ may be rearranged, using the linearity of g, as

$$A := g\left(\frac{\beta}{\alpha + \beta}x + \frac{\alpha}{\alpha + \beta}y\right)$$

$$\leq \frac{\beta}{\alpha + \beta}\,p(x + \alpha w_0) + \frac{\alpha}{\alpha + \beta}\,p(y - \beta w_0)$$

$$=: B.$$

However, the convexity of p and the assumption that $g \leq p$ on Z ensure that

$$B \geq p\left(\frac{\beta}{\alpha + \beta}\left(x + \alpha w_0\right) + \frac{\alpha}{\alpha + \beta}\left(y - \beta w_0\right)\right)$$

$$= p\left(\frac{\beta}{\alpha + \beta}x + \frac{\alpha}{\alpha + \beta}y\right)$$

$$\geq A.$$

This shows that $(\natural)$ indeed holds for all $x, y \in Z$ and all $\alpha, \beta > 0$. Therefore, there exists $\gamma \in \mathbb{R}$ so that G_γ has all the required properties. This completes the proof. $\qquad\square$

The proof of Theorem 11.40 below involves an argument that could perhaps be regarded as sophisticated because it invokes the Axiom of Choice (recall Section 1.3), through an application of Zorn's Lemma. The underlying non-constructive nature of the reasoning is similar in spirit to the proof of existence of a Hamel basis for any vector space given in Chapter 1, Theorem 1.24.

Proof of Theorem 11.40. Let $\mathcal{E}$ be the family of all pairs (Z, g) such that

(i) $Y \subseteq Z \subseteq X$;

(ii) Z is a vector space;

(iii) $g : Z \longrightarrow \mathbb{R}$ is linear;

(iv) $g|_Y = f$;

(v) $g(z) \leq p(z)$ for all $z \in Z$.

Our goal is to show that $\mathcal{E}$ contains (at least) one pair of the form (X, F). Define a relation $\prec$ on $\mathcal{E}$ by declaring

$$(Z', g') \prec (Z'', g'') \qquad \text{if and only if} \qquad Z' \subseteq Z'' \quad \text{and} \quad g''|_{Z'} = g'.$$

It is easy to verify that $\prec$ is a partial ordering on $\mathcal{E}$. Let us now check the assumptions of Zorn's Lemma on $(\mathcal{E}, \prec)$. Let $\{(Z_\alpha, g_\alpha) : \alpha \in A\}$ be a totally ordered subset of $\mathcal{E}$. We shall show that it has an upper bound. Let

$$Z := \bigcup_{\alpha \in A} Z_\alpha.$$

Then, by using the total ordering of the family, we see that (i) and (ii) hold true. Let $g : Z \longrightarrow \mathbb{R}$ be given, for any $x \in Z$, by

$$g(z) := g_\alpha(z), \quad \text{where } \alpha \in A \text{ is such that } z \in Z_\alpha$$

(there might exist more than one such α). By using the total ordering, it follows that g is well defined and that (iii), (iv) and (v) hold true. Hence, $(Z, g) \in \mathcal{E}$. One can also easily check that (Z, g) is an upper bound for the totally ordered set $\{(Z_\alpha, g_\alpha) : \alpha \in A\}$. By Zorn's Lemma 1.22, there exists a maximal element $(Z_0, g_0) \in \mathcal{E}$. We now claim that necessarily $Z_0 = X$. Indeed, suppose for a contradiction that $Z_0 \subsetneq X$. In that case, let $w_0 \in X \setminus Z$. By applying Lemma 11.41, we can construct a pair $(W, G_0) \in \mathcal{E}$, with

$$W := Z_0 \oplus \text{span}[\{w_0\}],$$

such that $(Z_0, g_0) \prec (W, G_0)$ and $Z_0 \subsetneq W$, thus contradicting the maximality of (Z_0, g_0). Therefore, we have $Z_0 = X$. By defining $F := g_0$, the validity of (i)–(v) for (X, F) yields the required result. $\qquad\square$

Remark 11.42 (On the necessity of choice). The reader might be wondering whether a simpler constructive proof could not be given in the proof of Theorem 11.40, and if not, why not?

A seemingly natural alternative would perhaps be to attempt to apply Lemma 11.41 recursively, constructing a sequence of extensions of f with the required properties by "induction on the dimension of the subspace", and then possibly attempt an extension by continuity to the closure of the union of these subspaces.[4] The problem is that the space X may be "too large" for an extension of f to the whole of X to be obtainable in this way![5]

Thus, the non-constructive approach taken in the proof becomes a most natural one in this generality: among all possible apt extensions of f to subspaces of X larger that Y, we consider one whose domain is maximal, namely "as large as possible". This extension would then necessarily have to be defined on the entire space, since otherwise, by Lemma 11.41, one would then be

---

[4]It is quite possible that the reader may not understand what exactly we mean by this, as we describe an argument without really giving it. Fear not, though! This approach does not work anyway and this point is discussed mostly in order to justify the necessity of Zorn's Lemma.

[5]A useful analogy is the impossibility, no matter how hard we try, to enumerate in a sequence all the elements of an uncountable set! The uncountability problem here essentially the same, but refers to the "dimension" of the space to which the functional needs to be extended.

able to construct a further extension on an even larger subspace, which would contradict maximality. Zorn's Lemma is required to make the above informal approach rigorous.

We now give the proof of Theorem 11.38 in the case of $\mathbb{F} = \mathbb{R}$, which is a fairly straightforward application of Theorem 11.40.

Proof of Theorem 11.38 in the case $\mathbb{F} = \mathbb{R}$. Let $(X, \|\cdot\|)$ be a real normed space. We begin by noting that, for any functional $G \in X^*$ with $G|_Y = f$, one has the inequality $\|G\|_{X^*} \geq \|f\|_{Y^*}$, because

$$\|G\|_{X^*} = \sup\left\{ \frac{|G(x)|}{\|x\|} : x \in X \setminus \{0\} \right\}$$

$$\geq \sup\left\{ \frac{|G(x)|}{\|x\|} : x \in Y \setminus \{0\} \right\} = \|f\|_{Y^*}.$$

This shows in particular that the extension F of f that we are aiming at constructing will have a minimal (i.e. the smallest possible) norm in X^* among all possible bounded linear global extensions. Since $f \in Y^*$, we have

$$|f(y)| \leq \|f\|_{Y^*}\|y\| \quad \text{for all } y \in Y.$$

Let $p : X \longrightarrow \mathbb{R}$ be the convex function given by

$$p(x) := \|f\|_{Y^*}\|x\| \quad \text{for all } x \in X.$$

It is easy to check that it satisfies

$$f(y) \leq p(y) \quad \text{for all } y \in Y.$$

By applying Theorem 11.40, we obtain the existence of a linear extension $F : X \longrightarrow \mathbb{R}$, with $F|_Y = f$, such that

$$F(x) \leq \|f\|_{Y^*}\|x\| \quad \text{for all } x \in X.$$

However, application of this inequality to $-x$ instead of x, yields that

$$-F(x) = F(-x) \leq \|f\|_{Y^*}\|-x\| = \|f\|_{Y^*}\|x\| \quad \text{for all } x \in X.$$

Combining the two inequalities, we therefore obtain that

$$|F(x)| \leq \|f\|_{Y^*}\|x\| \quad \text{for all } x \in X.$$

This shows that $F \in X^*$ and $\|F\|_{X^*} \leq \|f\|_{Y^*}$. But, as argued at the start of the proof, the reverse inequality also holds, and thus, in conclusion,

$$\|F\|_{X^*} = \|f\|_{Y^*}.$$

This completes the proof of the claimed result in the case $\mathbb{F} = \mathbb{R}$. $\qquad\square$

To prove Theorem 11.38 in the case when $\mathbb{F} = \mathbb{C}$, we need some additional considerations which are of independent interest. Any complex vector space X is also a real vector space, if we restrict multiplication by scalars to $\mathbb{R} \times X$. For any complex vector space X, let us denote the **corresponding real vector space** by $X_{\mathbb{R}}$. For finite-dimensional complex spaces, this amounts to setting

$$\mathbb{C}_{\mathbb{R}} := \mathbb{R}^2 \quad \text{and generally} \quad \mathbb{C}_{\mathbb{R}}^m := \mathbb{R}^{2m} \quad \text{for } m \in \mathbb{N}.$$

Obviously X and $X_{\mathbb{R}}$ coincide as sets, but the distinction between X and $X_{\mathbb{R}}$ is important when considering the linearity of operators. For example, the mapping $z \mapsto \bar{z}$ is linear when seen as $(x, y) \mapsto (x, -y) : \mathbb{R}^2 \longrightarrow \mathbb{R}^2$, but since $\overline{\lambda z} \neq \lambda \bar{z}$ for $\lambda \in i\mathbb{R}$, it is not linear when seen as a function $\mathbb{C} \longrightarrow \mathbb{C}$.

When $(X, \|\cdot\|)$ is a complex normed space, the following lemma completely elucidates and clarifies the relation between X^* and $X_{\mathbb{R}}^*$.

Lemma 11.43. *Let X be a complex vector space and let $X_{\mathbb{R}}$ be the corresponding real vector space. Suppose also that $f : X \longrightarrow \mathbb{C}$ and $u : X_{\mathbb{R}} \longrightarrow \mathbb{R}$ are two given functions. Then,*

$$f \text{ is } \mathbb{C}\text{-linear and } u = \operatorname{Re} f \quad \Longleftrightarrow \quad u \text{ is } \mathbb{R}\text{-linear and } f = u(\cdot) - iu(i(\cdot)).$$

If, in addition, $(X, \|\cdot\|)$ is a complex normed space, then $f \in X^*$ if and only if $u \in X_{\mathbb{R}}^*$, and in that case

$$\|f\|_{X^*} = \|u\|_{X_{\mathbb{R}}^*}.$$

The proof of this lemma is left as an easy exercise. We are now in a position to give the Proof of Theorem 11.38 when $\mathbb{F} = \mathbb{C}$ too.

Proof of Theorem 11.38 when $\mathbb{F} = \mathbb{C}$. Let $(X, \|\cdot\|)$ be a complex normed space, and Y a vector subspace of X. Fix a functional $f \in Y^*$ and let $u : Y_{\mathbb{R}} \longrightarrow \mathbb{R}$ be given by

$$u(y) := \operatorname{Re} f(y) \quad \text{for all } y \in Y.$$

Then, by Lemma 11.43, we have $u \in Y_{\mathbb{R}}^*$ and also

$$\|u\|_{Y_{\mathbb{R}}^*} = \|f\|_{Y^*}.$$

We now apply Theorem 11.38 in the case of **real** normed spaces (which has already been proven earlier) to $u \in Y_{\mathbb{R}}^*$. Hence, there exists a bounded linear extension $U \in X_{\mathbb{R}}^*$ such that $U|_Y = u$ and also

$$\|U\|_{X_{\mathbb{R}}^*} = \|u\|_{Y_{\mathbb{R}}^*}.$$

We define $F : X \longrightarrow \mathbb{C}$ by setting

$$F(x) := U(x) - iU(ix) \quad \text{for all } x \in X.$$

By invoking again Lemma 11.43, we have that $F \in X^*$ and

$$\|F\|_{X^*} = \|U\|_{X_\mathbb{R}^*}.$$

Moreover, for every $y \in Y$,

$$F(y) = U(y) - iU(iy) = u(y) - iu(iy) = f(y).$$

This shows that $F|_Y = f$, and since we also have that $\|F\|_{X^*} = \|f\|_{Y^*}$, the proof of the theorem is completed (also) in the case when $\mathbb{F} = \mathbb{C}$. $\square$

We now present several consequences of the Hahn–Banach Theorem. The first one involves the construction of a bounded linear functional which vanishes on a given subspace and takes a certain specified value at a given point which does not belong to the subspace. Without loss of generality, the value may be normalised to one.

Proposition 11.44. *Let $(X, \|\cdot\|)$ be a normed space. Let also Y be a vector subspace of X, and suppose $x_0 \in X \setminus Y$ is such that*

$$d := \inf \left\{ \|x_0 - y\| : y \in Y \right\} > 0.$$

Then there exists $F_0 \in X^*$ such that

$$F_0 \equiv 0 \ \ on \ Y, \quad F_0(x_0) = 1 \quad and \quad \|F_0\|_{X^*} = \frac{1}{d}.$$

Proof. The first step of the proof involves the construction of a suitable linear functional on the direct sum vector subspace

$$Z := Y \oplus \operatorname{span}[\{x_0\}].$$

Recall that any element $z \in Z$ can be written in a unique way as $z = y + \lambda x_0$, where $y \in Y$ and $\lambda \in \mathbb{F}$. Let $f_0 : Z \longrightarrow \mathbb{F}$ be given by

$$f_0(y + \lambda x_0) := \lambda \quad \text{for all } y \in Y, \lambda \in \mathbb{F}.$$

It is easy to check that f_0 is linear on Z. To show that $f_0 \in Z^*$ and to calculate its norm, we express $\|f_0\|_{Z^*}$ as

$$\|f_0\|_{Z^*} = \sup \left\{ \frac{|f_0(z)|}{\|z\|} : z \in Z \setminus \{0\} \right\}$$

$$= \sup \left\{ \frac{|f_0(y + \lambda x_0)|}{\|y + \lambda x_0\|} : y \in Y, \lambda \in \mathbb{F}, y + \lambda x_0 \neq 0 \right\}$$

$$= \sup \left\{ \frac{|\lambda|}{\|y + \lambda x_0\|} : y \in Y, \lambda \in \mathbb{F} \setminus \{0\} \right\}$$

$$= \sup \left\{ \frac{1}{\|x_0 - (-1/\lambda)y\|} : y \in Y, \lambda \in \mathbb{F} \setminus \{0\} \right\}$$

and hence we have

$$\|f_0\|_{Z^*} \; = \; \frac{1}{\inf\{\|x_0 - w\| \; : \; w \in Y\}} \; = \; \frac{1}{d}.$$

This shows that $f_0 \in Z^*$ and $\|f_0\|_{Z^*} = 1/d$. By applying the Hahn–Banach Theorem for normed spaces, we see that there exists $F_0 \in X^*$ with $F_0|_Z = f_0$ and $\|F_0\|_{X^*} = \|f_0\|_{Z^*} = 1/d$. It is clear then that F_0 has all the required properties. $\qquad\square$

An immediate important consequence of Proposition 11.44 is obtained by taking $Y := \{0\}$ and by multiplying the resulting functional by a constant:

Corollary 11.45. *Let $(X, \|\cdot\|)$ be a normed space and $x_0 \in X \setminus \{0\}$. Then, there exists $f_0 \in X^*$ such that*

$$f_0(x_0) \; = \; \|x_0\| \quad and \quad \|f_0\|_* = 1.$$

Given a normed space $(X, \|\cdot\|)$ and any $f \in X^*$, recall that the Definition 11.31 of the dual norm says

$$\|f\|_* \; = \; \sup\{|f(x)| \; : \; \|x\| \le 1\}.$$

An interesting counterpart to the above formula is provided by the next result which is a consequence of Corollary 11.45 (and ultimately of the Hahn–Banach Theorem).

Corollary 11.46. *Let $(X, \|\cdot\|)$ be a normed space. Then, for any $x \in X$, the norm $\|x\|$ can be represented in terms of the norm on the dual space:*

$$\|x\| \; = \; \max\{|f(x)| \; : \; \|f\|_* \le 1\}.$$

*In particular, the supremum in the expression for $\|x\|$ above is always **realised** by some functional $f \in X^*$ and hence it is a **maximum**.*

Note the important difference between Definition 11.31 and Corollary 11.46, which is a result. In particular, the **supremum in the formula of Definition 11.31 in general is not realised**.

Proof. The claimed result is obviously true for $x = 0$. Suppose therefore that $x \ne 0$. Then, for any $f \in X^*$ with $\|f\|_* \le 1$, we have that

$$|f(x)| \; \le \; \|f\|_* \|x\| \; \le \; \|x\|.$$

On the other hand, let f_0 be given by Corollary 11.45, namely being such that $\|f_0\|_* = 1$ and $|f_0(x)| = \|x\|$. The required conclusion follows. $\qquad\square$

Let $(X, \|\cdot\|)$ be a normed space. Then its dual $(X^*, \|\cdot\|_*)$ is a normed space in itself (in fact, a Banach space), and therefore has a dual space of its own, as well as a corresponding operator norm:

Definition 11.47. Let $(X, \|\cdot\|)$ be a normed space. The dual space of the dual space of X, namely the space $(X^*)^*$ is called the **bidual (or second dual) space of** X:

$$X^{**} := (X^*)^* = B\big(B(X, \mathbb{F}), \mathbb{F}\big).$$

The **bidual norm on** X^{**} is given by

$$\|\phi\|_{**} := \max\big\{|\phi(f)| : \|f\|_* \leq 1\big\} \quad \text{for all } \phi \in X^{**}.$$

Remark 11.48.

• As in the case of the dual norm, if there are more than one spaces involved or if there is any danger of confusion, we will augment the symbolisation of the bidual norm from "$\|\cdot\|_{**}$" to "$\|\cdot\|_{X^{**}}$".

• By iterating Definition 11.31, one can define higher dual spaces like X^{***}, etc. As we will see in Chapter 12, though, it has little interest to go any further than the third dual space in order to glean information for the original space.

Interestingly, the preceding consequences of the Hahn–Banach Theorem yield a significant, and perhaps unexpected relation: **every normed space** $(X, \|\cdot\|)$ **is isometrically isomorphic to a subspace of its bidual space!** This result will in fact be one of our starting points regarding compactness issues in Chapter 12 that follows. The putative isomorphism is the mapping from X into X^{**} defined right below.

Definition 11.49. Let $(X, \|\cdot\|)$ be a normed space. Consider the mapping

$$\widehat{\cdot} \; : \; X \longrightarrow X^{**}, \quad X \ni x \mapsto \widehat{x} \in X^{**},$$

where each bounded linear functional $\widehat{x} : X^* \longrightarrow \mathbb{F}$ (acting on the first dual space) is defined as

$$\widehat{x}(f) := f(x) \quad \text{for all } f \in X^*.$$

Then, $\widehat{\cdot} : X \longrightarrow X^{**}$ is called **the evaluation mapping** from the space into its bidual space.

Proposition 11.50 (Isometric embedding into the bidual space). *Let* $(X, \|\cdot\|)$ *be a normed space. Then, the evaluation mapping is an isometric embedding of* X *into its bidual* X^{**}*, that is*

$$\|\widehat{x}\|_{**} = \|x\| \quad \text{for all } x \in X.$$

Hence, the evaluation map $\widehat{\cdot} : X \longrightarrow \widehat{X}$ *is an isometric isomorphism between* X *and the its image* $\widehat{X} \subseteq X^{**}$ *into the bidual space.*

Proof. For any $x \in X$, the mapping $\widehat{x} : X^* \longrightarrow \mathbb{F}$ is obviously linear. Moreover, by Corollary 11.46, we have that

$$\sup\big\{|\widehat{x}(f)| : \|f\|_* \leq 1\big\} = \sup\big\{|f(x)| : \|f\|_* \leq 1\big\} = \|x\|.$$

This shows that $\widehat{x} \in X^{**}$ and $\|\widehat{x}\|_{**} = \|x\|$. The remaining claims are immediate to be verified and hence the proof ensues. $\qquad\square$

It can be proved that the evaluation map is surjective for certain normed spaces, like for example the Euclidean spaces $\mathbb{F}^n$, the class of ℓ^p spaces for $p \in (1, \infty)$, and the class of the Hilbert spaces (these facts follow from Proposition 11.34, Proposition 11.35 and Theorem 11.37 respectively). The evaluation map is also surjective for the class of L^p spaces over a general measure space for $p \in (1, \infty)$ (we shall prove this in Chapter 14). However, the evaluation map is not surjective in general. It fails, for example, for the spaces L^1 and L^∞. Since spaces for which the evaluation map is surjective have nicer properties, we designate a name for this class of spaces.

Definition 11.51. A normed space $(X, \|\cdot\|)$ is said to be **reflexive** when the evaluation mapping of Definition 11.49 is surjective, that is when $\widehat{X} = X^{**}$.

Remark 11.52. Note that any reflexive normed space is necessarily a Banach space (because dual spaces are always complete). Further, **for reflexive spaces we may (and will) identify X with its isometric image $\widehat{X} = X^{**}$ via the evaluation map.**

Rather than being a mere mathematical curiosity, the property of a Banach space of being reflexive leads to better **compactness properties** of the space endowed with its weak topology, already defined in Chapter 4, and this will be investigated systematically in Chapters 12 and 13.

11.6   The geometric Hahn-Banach separation theorems

In the previous section we saw that the Hahn-Banach theorem is an important tool for infinite-dimensional spaces, allowing to establish *existence of functionals* on the entire space as extensions of functionals from a subspace, preserving also certain properties of the original functional on the subspace. In this final section we take up a seemingly unrelated topic of geometric flavour which, quite surprisingly, turns out to be related to the Hahn-Banach theorem.

We begin by motivating the subsequent developments in the special case of the space[6] $\mathbb{R}^n$. Suppose that we have two disjoint sets A, B in $\mathbb{R}^n$. Then, it is of great interest for many applications to know if one can **separate the sets by a hyperplane**, namely if one can find an *affine hyperplane*, namely a set of the form

$$H := \left\{ x \in \mathbb{R}^n \ \ x \cdot \xi = a \right\}$$

where $\xi \in \mathbb{R}^n$, $a \in \mathbb{R}$ and "$\cdot$" is the Euclidean inner product, so that

$$A \subseteq H^+ , \quad B \subseteq H^- .$$

---

[6]In this section we shall consider only **real** vector spaces over the field $\mathbb{R}$.

Here $H^\pm$ symbolise the respective half-spaces determined by the hyperplane:

$$H^\pm := \left\{ x \in \mathbb{R}^n \ : \ x \cdot \xi \gtreqless a \right\}.$$

Said differently, we would like A to lie on one side of the hyperplane and B on the other side of the hyperplane. Unfortunately, this is not in general possible without extra assumptions, not even for $n = 1$. Indeed, hyperplanes in $\mathbb{R}$ are just points and for instance the sets $A = (-\infty, -1) \bigcup (1, \infty)$ and $B = (-1, 1)$ cannot be separated since A lies "on both sides" of the complement of B. In higher dimensions, the situation is perhaps intuitively clearer, see Figures 11.1–11.2.

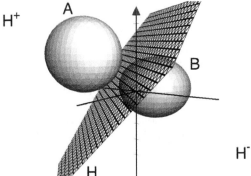

Figure 11.1: The separation of two convex sets in $\mathbb{R}^3$ by an affine hyperplane.

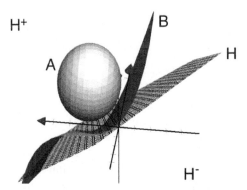

Figure 11.2: In general, we cannot separate non convex sets!

It turns out that the key assumption to assure that separation of two sets is possible is **convexity**. In addition, if one of the sets is closed and the other is compact, then we may even separate them strictly, namely we can show that

$$A \subseteq (H^+)^\circ, \qquad B \subseteq (H^-)^\circ$$

where $(H^\pm)^\circ$ symbolises the interior of the half-spaces, which is the same as the respective open half-spaces:

$$(H^\pm)^\circ = \left\{ x \in \mathbb{R}^n \ : \ x \cdot \xi \gtrless a \right\}.$$

However, even if both A, B are closed, convex and disjoint, they cannot in general be separated strictly, as the example in Figure 11.3 demonstrates.

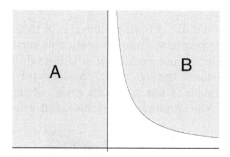

Figure 11.3: One cannot separate strictly the disjoint convex closed subsets of $\mathbb{R}^2$ given by $A := \{(x, y) \in \mathbb{R}^2 : x \leq 0, y \geq 0\}$ and $B := \{(x, y) \in \mathbb{R}^2 : x > 0, y \geq 1/x\}$.

The purpose of the present section is to perform a relevant investigation regarding the possible separation of convex sets in infinite-dimensional vector spaces. As we shall see, the situation is much more delicate and for this task a geometric form of the Hahn-Banach theorem is required. We begin with some necessary terminology.

Definition 11.53. Let X be a real vector space. An **affine hyperplane** H in X is a set of the form $f^{-1}(\{a\})$, that is

$$H = \big\{x \in X \ : \ f(x) = a\big\}$$

where $f : X \longrightarrow \mathbb{R}$ is a non-zero linear functional[7] and $a \in \mathbb{R}$.

For convenience, we shall occasionally symbolise the hyperplane in the abbreviated level-set form $H = \{f = a\}$. In the case of $\mathbb{R}^n$, the above definition reconciles with the definition of hyperplane used earlier because by the Riesz Representation Theorem, any linear functional on $\mathbb{R}^n$ is continuous and has the form $f(x) := x \cdot \xi$ for some $\xi \in \mathbb{R}^n$.

We begin by showing that, if $(X, \| \cdot \|)$ is a normed vector space, then the hyperplane is closed if and only if f is continuous, i.e. an element of the dual space X^*. We recall that in finite dimensions, all vector subspaces are automatically closed.

Proposition 11.54. *Let $(X, \| \cdot \|)$ be a real normed vector space, $f : X \longrightarrow \mathbb{R}$ a non-zero linear functional and $a \in \mathbb{R}$. Then, the hyperplane $H = \{f = a\}$ is closed in X if and only if $f \in X^*$.*

Proof. If f is continuous, then H is closed since it is the inverse image through f of the closed set $\{a\}$ in $\mathbb{R}$.

---

[7]Note that f is not assumed to be continuous in any sense, and in any case X is not assumed to be endowed with any topology!

Conversely, suppose that H is closed. Then, $X \setminus H$ is open and hence the union $\{f < a\} \bigcup \{f > a\}$ is open in X. Fix $x_0 \in \{f < a\}$ and $\delta > 0$ such that

$$\mathbb{B}_\delta(x_0) \subseteq \{f < a\} \bigcup \{f > a\}.$$

Then, since the sets $\{f < a\}$ and $\{f > a\}$ are convex and disjoint, we claim that $\mathbb{B}_\delta(x_0) \subseteq \{f < a\}$, i.e. the ball is contained entirely in the connected component that its centre lies. Indeed, if this were not the case and there existed $x \in \mathbb{B}_\delta(x_0)$ such that $f(x) > a$, then the straight line segment

$$\{tx + (1 - t)x_0 : t \in [0, 1]\}$$

would lie inside the ball $\mathbb{B}_\delta(x_0)$ and the scalar affine function

$$F : [0, 1] \longrightarrow \mathbb{R}, \quad F(t) := f\big(tx + (1 - t)x_0\big) = tf(x) + (1 - t)f(x_0),$$

would necessarily be continuous and satisfying $F(0) < a$ and $F(1) > a$. Then, by the Bolzano theorem there would exists $t_0 \in (0, 1)$ with $F(t_0) = 0$, or equivalently $f\big(t_0 x + (1 - t_0)x_0\big) = 0$ which is a contradiction because

$$t_0 x + (1 - t_0)x_0 \in \mathbb{B}_\delta(x_0) \subseteq \{f \neq a\}.$$

It therefore follows that $f(x_0 \pm w) < a$, for all $w \in X$ with $\|w\| < \delta$, or equivalently, $|f(w)| < a - f(x_0)$ for all $w \in \mathbb{B}_\delta(0)$. Since f is linear, its boundedness (and therefore, continuity) follows by the standard homogeneity argument. $\square$

We now introduce some further terminology, making the notion of separation of sets in infinite dimensions more precise.

Definition 11.55. Let $(X, \|\cdot\|)$ be a real normed vector space, A, B subsets of X, $f : X \longrightarrow \mathbb{R}$ a non-zero linear functional and $a \in \mathbb{R}$. We shall say that **the hyperplane $\{f = a\}$ separates the sets A, B** if

$$f(x) \leq a \text{ for all } x \in A \quad \text{and} \quad f(x) \geq a \text{ for all } x \in B.$$

We will say that **the hyperplane $\{f = a\}$ separates the sets A, B strictly** if there exists $\varepsilon > 0$ such that

$$f(x) \leq a - \varepsilon \text{ for all } x \in A \quad \text{and} \quad f(x) \geq a + \varepsilon \text{ for all } x \in B.$$

Note that the geometric intuition behind the above definition is the same as in the finite-dimensional case (see Figure 11.1) because "$f(x) \leq a$ for $x \in A$" is merely a restatement of $A \subseteq \{f \leq a\}$. As we said earlier, the principal aim of this section is to obtain separation theorems for convex sets. To this end, we need the next two lemmas which are also of independent interest.

Lemma 11.56 (Minkowski functional). *Let $(X, \|\cdot\|)$ be a real normed vector* *space and $C \subseteq X$ a convex open set containing the origin $0 \in X$. We define the function*

$$p : X \longrightarrow \mathbb{R}, \qquad p(x) := \inf\left\{t > 0 : \frac{x}{t} \in C\right\}, \ x \in X.$$

*The function p is typically called the **Minkowski functional** or the **gauge** of the convex set C and satisfies:*

(i) $p(\lambda x) = \lambda p(x)$, *for all $\lambda > 0$, $x \in X$;*

(ii) *there exists $M > 0$ such that $0 \le p(x) \le M\|x\|$, for all $x \in X$;*

(iii) $C = \{p < 1\}$;

(iv) $p(x + y) \le p(x) + p(y)$, *for all $x, y \in X$.*

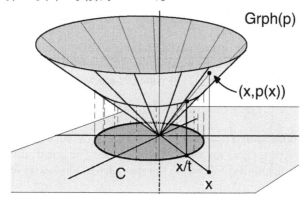

Figure 11.4: An illustration of a convex set $C \subseteq \mathbb{R}^2$ and its corresponding gauge p. The graph of the function p is essentially a cone and p is "almost" like a norm, except for that it might not be positive definite (i.e. $p(x) = 0$ may not imply $x = 0$) and it may also happen that $p(x) \ne p(-x)$. The latter together with (i) amounts to $p(\lambda x) = |\lambda| p(x)$ for all $\lambda \in \mathbb{R}$.

Proof. (i) By substituting $t = \lambda s$ for $s > 0$, we have

$$p(\lambda x) := \inf\left\{t > 0 : \frac{\lambda x}{t} \in C\right\} = \inf\left\{\lambda s > 0 : \frac{\lambda x}{\lambda s} \in C\right\} = \lambda p(x).$$

(ii) Since $0 \in C$ and C is open, there exists[8] an $r > 0$ such that $\overline{\mathbb{B}}_r(0) \subseteq C$. Hence, for any $x \in X$ we have

$$p(x) = \inf\left\{t > 0 : \frac{x}{t} \in C\right\}$$

$$\le \inf\left\{t > 0 : \frac{x}{t} \in \overline{\mathbb{B}}_r(0)\right\}$$

$$= \inf\left\{t > 0 : \frac{\|x\|}{t} \le r\right\}$$

---

[8]Because every open ball contains a closed ball with slightly smaller radius.

and hence $p(x) \le \|x\|/r$.

(iii) Fix $x \in C$. Since C is open, there exists $\varepsilon > 0$ small such that $x + \varepsilon x \in C$ which implies by the definition of p that $p(x) \le 1/(1+\varepsilon) < 1$. Conversely, if $p(x) < 1$, there exists $a \in (0,1)$ such that $x/a \in C$ and hence by convexity we have

$$x = a\left(\frac{x}{a}\right) + (1-a)\,0 \in C.$$

The above reasoning establishes (iii).

(iv) Fix $x, y \in X$ and $\varepsilon > 0$. Then, by (i) and (iii) we infer that

$$\frac{x}{p(x)+\varepsilon}, \ \frac{y}{p(y)+\varepsilon} \in C.$$

By the convexity of C, it follows that

$$t\left(\frac{x}{p(x)+\varepsilon}\right) + (1-t)\left(\frac{y}{p(y)+\varepsilon}\right) \in C$$

for all $t \in [0,1]$. The particular choice

$$t_\varepsilon := \frac{p(x)+\varepsilon}{p(x)+p(y)+2\varepsilon}$$

yields that

$$\frac{x+y+\varepsilon}{p(x)+p(y)+2\varepsilon} \in C, \qquad \text{for all } \varepsilon > 0.$$

By the definition of p, the above implies that $p(x+y) \le p(x) + p(y) + 2\varepsilon$, for any $\varepsilon > 0$, from where then conclusion ensues. $\qquad\square$

The next and final auxiliary result we need is a particular case of the desired separation theorem in the case when one of the sets is a singleton.

Lemma 11.57 (Separation of an open convex set from a singleton set). *Let $(X, \|\cdot\|)$ be a real normed space and $C \subseteq X$ a convex open non-empty subset. Consider also $x_0 \in X \setminus C$. Then, there exists a closed hyperplane of the form $\{f = f(x_0)\}$ for some $f \in X^*$, which separates C from $\{x_0\}$ (Definition 11.55). In addition, $f(x) < f(x_0)$ for all $x \in C$.*

Proof. By a translation of C and x_0, we may assume that $0 \in C$. Consider the one-dimensional space $Y := \text{span}[\{x_0\}] \subseteq X$ and the linear functional on Y given by

$$g \ : \ Y \longrightarrow \mathbb{R}, \quad g(tx_0) := t, \quad t \in \mathbb{R}.$$

Then, it follows from the definition of p that $g(x) \le p(x)$ for $x \in Y$. By the Hahn-Banach Theorems 11.40 and 11.38, it follows that there exists a continuous linear extension $f \in X^*$ of g from Y to X such that $f \le p$ on X and also $f(x_0) = 1$. Moreover, by Lemma 11.56(iii) we have $f < 1$ on C. The lemma has been established. $\qquad\square$

We are now ready to state and prove the first principal result of this section, which allows us to separate two disjoint convex sets with a closed hyperplane under the hypothesis that at least one of them is open.

Theorem 11.58 (Geometric Hahn-Banach Theorem–version 1). *Let $(X, \|\cdot\|)$ be a real normed space and $A, B \subseteq X$ convex, non-empty disjoint sets with A open. Then, there exists a closed hyperplane H separating A from B (Definition 11.55).*

Proof. Given $A, B \subseteq X$ as in the statement, we define the set

$$C := A - B = \{a - b : a \in A, b \in B\}$$

Then, it can be easily confirmed that C is convex. Further, C is open because

$$C = \bigcup_{b \in B} (A - b).$$

Moreover, $0 \notin C$ because A, B are disjoint: $A \cap B = \emptyset$. By Lemma 11.57 applied to C and $\{0\}$, there exists a closed hyperplane H separating C from $\{0\}$ and in particular we can find $f \in X^*$ such that $f(x) < 0$ for all $x \in C$ and $f(0) = 0$. By the definition of C, it follows that $f(a) < f(b)$ for all $a \in A$ and all $b \in B$. By choosing a number $c \in \mathbb{R}$ lying in the interval

$$\left[\sup_{a \in A} f(x), \sup_{b \in B} f(b) \right] \subseteq \mathbb{R},$$

we have that the hyperplane $\{f = c\}$ separates A from B, as desired. $\qquad\square$

Now we come to the second main result of this section which allows us to separate *strictly* two disjoint convex sets if one of them is compact (see also Figure 11.5 below).

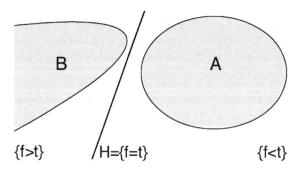

Figure 11.5: An illustration of the strict separation of two closed disjoint convex sets A, B in $\mathbb{R}^2$, with B being compact and A being unbounded. Note that the compactness of at least one of the sets excludes pathological examples of non-strict separation, like that demonstrated in Figure 11.3.

Theorem 11.59 (Geometric Hahn-Banach Theorem–version 2). *Let $(X, \|\cdot\|)$ be a real normed vector space and $A, B \subseteq X$ convex, non-empty disjoint sets with A closed and B compact. Then, there exists a closed hyperplane H separating A from B strictly (Definition 11.55).*

Proof. Given sets A, B as in the statement, let $\varepsilon > 0$ and define the open sets

$$A_\varepsilon := \bigcup_{a \in A} \mathbb{B}_\varepsilon(a), \quad B_\varepsilon := \bigcup_{b \in B} \mathbb{B}_\varepsilon(b).$$

Then, the sets $A_\varepsilon, B_\varepsilon$ are also convex, since

$$\bigcup_{a \in A} \mathbb{B}_\varepsilon(a) = \{a + z : a \in A, \|z\| < \varepsilon\} = A + \mathbb{B}_\varepsilon(0)$$

and similarly for B_ε. Further, if ε is small enough, then the sets $A_\varepsilon, B_\varepsilon$ are disjoint. Indeed, if this were not the case, there would exist sequences $(a_n)_1^\infty \subseteq A$ and $(b_n)_1^\infty \subseteq B$ such that $\|a_n - b_n\| \longrightarrow 0$ as $n \to \infty$. Since compact and sequentially compact sets coincide in normed spaces, we would have that there is $b \in B$ such that $b_{n_k} \longrightarrow b$ as $k \to \infty$ and since A is closed we would get $b \in \overline{A} \cap \overline{B} = A \cap B$, which is a contradiction since $A \cap B = \emptyset$. Fix such a value of ε, and denote it by ε_0. By Theorem 11.58, there exist $f \in X^*$ and $t \in \mathbb{R}$ such that the hyperplane $\{f = t\}$ separates A_{ε_0} from B_{ε_0}. Therefore,

$$f(a + \varepsilon_0 z) \le t \le f(b + \varepsilon_0 z), \quad \text{for all } a \in A, \, b \in B, \, \|z\| < 1,$$

or, by the linearity of f,

$$f(a) + \varepsilon_0 f(z) \le t \le f(b) + \varepsilon_0 f(z), \quad \text{for all } a \in A, \, b \in B, \, \|z\| < 1.$$

By taking supremum in the left hand side inequality with respect to all $z \in \mathbb{B}_1(0)$, we see that $f(a) + \varepsilon_0 \|f\|_* \le t$. By taking infimum in the right hand side inequality with respect to all $z \in \mathbb{B}_1(0)$, we see that $t \le f(b) - \varepsilon_0 \|f\|_*$. Hence, we conclude that

$$f(a) + \varepsilon_0 \|f\|_* \le t \le f(b) - \varepsilon_0 \|f\|_*,$$

for all $a \in A$ and $b \in B$. Thus, $\{f = t\}$ separates A, B strictly, as claimed. $\square$

Without the extra hypothesis of compactness, in general it is **not** always possible to separate two disjoint closed convex sets with a hyperplane, not even in the non-strict sense! Such an example is given as follows:

Example 11.60. Consider the Hilbert space $X = L^2(0,1)$ (with the Lebesgue measure) and the closed convex sets

$$A := \{h \in L^2(0,1) : h \ge 1 \text{ a.e. on } (0,1)\}, \quad B := \{t \, \mathrm{id} : t \in \mathbb{R}, \, \mathrm{id}(x) = x\}.$$

Then, there exists no hyperplane separating A from B! We leave the verification of the details as an exercise (Exercise 11.12).

If however the space X is finite dimensional, it is always possible to separate two closed disjoint convex sets.

11.7   Exercises

Exercise 11.1. Let $(X, \|\cdot\|)$ be a normed space and Y a vector subspace of X.

(1) Show that $\overline{Y}$ is also a vector subspace of X.

(2) Show that if $Y^\circ \neq \emptyset$, then $Y = X$.

Exercise 11.2. Let $(C([a, b], \mathbb{R}), \|\cdot\|_\infty)$ be the Banach space of all continuous functions $f : [a, b] \longrightarrow \mathbb{R}$, where $[a, b]$ is a closed bounded interval of $\mathbb{R}$ and

$$\|f\|_\infty = \sup_{x \in [a,b]} |f(x)| \quad \text{for all } f \in C([a, b], \mathbb{R}).$$

Show that:

(1) The function $T : C([a, b], \mathbb{R}) \longrightarrow \mathbb{R}$ given by

$$T(f) = f(a) + f(b) \quad \text{for all } f \in C([a, b], \mathbb{R})$$

is a bounded linear functional, that is $T \in (C([a, b], \mathbb{R}))^*$.

(2) The set

$$F = \Big\{ f \in C([a, b], \mathbb{R}) \ : \ f(a) + f(b) = 0 \Big\}$$

is a closed affine hyperplane in $C([a, b], \mathbb{R})$.

Exercise 11.3. Let $k \in \mathbb{N}$. We consider the space $C^k([0, 1])$, which is defined as the subspace of the space $C([0, 1])$ (i.e., of all real-valued continuous functions $f : [0, 1] \longrightarrow \mathbb{R}$) of those functions which have k continuous derivatives on $[0, 1]$, interpreted as one-sided derivatives as the endpoints $\{0, 1\}$, namely

$$f'(0^+) = \lim_{t \to 0^+} \frac{f(t) - f(0)}{t} \ , \quad f'(1^-) = \lim_{t \to 0^-} \frac{f(1 + t) - f(1)}{t}.$$

We endow the space $C^k([0, 1])$ with the function $\|\cdot\|_{C^k} : X \longrightarrow \mathbb{R}$, given by

$$\|f\|_{C^k} := \max_{p \in \{0,\ldots,k\}} \max_{t \in [0,1]} \big| f^{(p)}(t) \big|.$$

(1) Show that $\|\cdot\|_{C^k}$ is a norm on $C^k([0, 1])$ and that the space is complete (i.e. a Banach space) with respect to this norm.

(2) Consider for $k \in \mathbb{N}$ the linear operator $T : C^k([0, 1]) \longrightarrow C^{k-1}([0, 1])$ given by

$$(Tf)(x) := f'(x), \quad x \in [0, 1].$$

Show that T is a bounded linear operator. (Recall that, if $k = 0$, then $C^0([0, 1]) \equiv C([0, 1])$.)

Exercise 11.4. (Sobolev spaces and strong derivatives) For this exercise you will perhaps need to revisit the Chapters 6-9 on measure theory and L^p spaces. Let $\Omega \subseteq \mathbb{R}^n$ be an open set. Consider the functional space

$$W^{1,\infty}(\Omega) := \left\{ f \in L^\infty(\Omega) : Df(x) \text{ exists for a.e. } x \in \Omega \text{ \& } Df \in L^\infty(\Omega, \mathbb{R}^n) \right\}$$

where Df denotes the gradient of f and the space $W^{1,\infty}(\Omega)$ is understood with the usual identification of functions which coincide a.e. on Ω. We endow $W^{1,\infty}(\Omega)$ with the function $\|\cdot\|_{W^{1,\infty}(\Omega)} : W^{1,\infty}(\Omega) \longrightarrow \mathbb{R}$ given by

$$\|f\|_{W^{1,\infty}(\Omega)} := \|f\|_{L^\infty(\Omega)} + \|Df\|_{L^\infty(\Omega)}.$$

(1) Confirm that, as the notation suggests, $\|\cdot\|_{W^{1,\infty}(\Omega)}$ is a norm on $W^{1,\infty}(\Omega)$.

(2) Consider for $k \in \mathbb{N}$ the linear operator $T : W^{1,\infty}(\Omega) \longrightarrow L^\infty(\Omega, \mathbb{R}^n)$ given by

$$(Tf)(x) := Df(x), \quad \text{a.e. } x \in \Omega.$$

Show that T is a bounded linear operator.

Exercise 11.5. Let $p \in [1, \infty)$. Show that the space ℓ^p is linearly isometric to a subspace of $L^p(0,1)$ (endowed with the Lebesgue measure $\mathcal{L}^1$ on $(0,1)$).

[Hint: Consider the sequence $f_n := (n(n+1))^{1/p} \chi_{(\frac{1}{n+1}, \frac{1}{n})}, n \in \mathbb{N}$.]

Exercise 11.6. Assume that $T : X \longrightarrow Y$ is a linear operator from a normed space $(X, \|\cdot\|)$ into a normed space $(Y, \|\cdot\|)$ such that $(Tx_n)_1^\infty$ is a bounded sequence in Y for every sequence $(x_n)_1^\infty$ in X such that $\|x_n\| \longrightarrow 0$ as $n \to \infty$. Is T necessarily continuous?

Exercise 11.7. Let $T : X \longrightarrow Y$ be an injective operator between normed spaces. Show that the following statements are equivalent:

(1) T is an isometry onto Y.

(2) $T(\mathbb{S}^X) = \mathbb{S}^Y$.

(3) $T(\overline{\mathbb{B}}^X) = \overline{\mathbb{B}}^Y$.

Exercise 11.8. Let X, Y be Banach spaces and $T : X \longrightarrow Y$ a bounded linear operator. If T is bounded from below, namely if there exists $\delta > 0$ such that $\|Tx\| \geq \delta\|x\|$ for all $x \in X$, then the image of T is closed in Y and T is an isomorphism from X to Y.

Exercise 11.9. Let M be a dense subset (not necessarily countable) of a Banach space $(X, \|\cdot\|)$. Show that for every $x \in X \setminus \{0\}$ there exists a sequence $(x_n)_1^\infty$ such that $x = \sum_1^\infty x_k$ and $\|x_k\| \leq 3 \cdot 2^{-k}\|x\|$.

Exercise 11.10. Show that a Banach space $(X, \|\cdot\|)$ is separable if and only if $\mathbb{S}^X$ is separable.

Exercise 11.11. Find two (necessarily not closed) dense subspaces $Z, Y \subseteq X$ of a Banach space $(X, \| \cdot \|)$ which intersect only at the origin, namely such that $Z \cap Y = \{0\}$.

Exercise 11.12. Complete the details of the counterexample to the geometric Hahn-Banach theorem given in Example 11.60.

Exercise 11.13. Consider the Banach space $(B([-\pi, \pi], \mathbb{R}), \| \cdot \|_\infty)$, where

$$B([-\pi, \pi], \mathbb{R}) := \left\{ f : [-\pi, \pi] \longrightarrow \mathbb{R} \,\middle|\, f \text{ is bounded on } [-\pi, \pi] \right\},$$

and

$$\|f\|_\infty := \sup_{x \in [-\pi, \pi]} |f(x)| \quad \text{for all } f \in B([-\pi, \pi], \mathbb{R}).$$

(1) For the functions $f, g \in B([-\pi, \pi], \mathbb{R})$ given by $f(x) = \sin x$ and $g(x) = \cos x$, for all $x \in [-\pi, \pi]$, compute the distance $\|f - g\|_\infty$.

(2) Consider for some $r > 0$ the sets

$$\overline{\mathbb{B}}_r(f), \quad \{g\},$$

where $\overline{\mathbb{B}}_r(f)$ is the closed ball of radius r in $B([-\pi, \pi], \mathbb{R})$. Determine if it is possible or not to separate with a hyperplane of the form $\{\phi = a\}$ (for some bounded linear functional $\phi \in (B([-\pi, \pi], \mathbb{R}))^*$ and some $a \in \mathbb{R}$) strictly these sets and, if yes, for which admissible radii $r > 0$.

Exercise 11.14. This exercise gives a construction, known as **Banach limit**, which allows us to define in a consistent manner a **limit for any non-convergent bounded sequence**, by utilising the Hahn-Banach theorem. This **generalised sequential limit** is **not unique**, however. Consider the space ℓ^∞ and its subspace of all convergent sequences

$$c := \left\{ x = (x_1, x_2, \ldots) \in \ell^\infty \,\middle|\, \exists\, x_\infty \in \mathbb{F} : x_n \longrightarrow x_\infty \text{ as } n \to \infty. \right\}.$$

Show that there exists a linear functional $\mathrm{Lim} \in (\ell^\infty)^*$ such that:

(1) $\|\mathrm{Lim}\|_{(\ell^\infty)^*} = 1$.

(2) Let $S : \ell^\infty \longrightarrow \ell^\infty$ be the so-called **shift operator**:

$$Sx := (x_2, x_3, \ldots), \quad \text{when } x = (x_1, x_2, \ldots).$$

Then, we have

$$\mathrm{Lim}\,(Sx) = \mathrm{Lim}\,x.$$

(3) For any constant sequence $(c, c, \ldots) \in \ell^\infty$, we have

$$\mathrm{Lim}\,(c, c, \ldots) = c.$$

(4) For any $(x_1, x_2, \ldots) \in \ell^\infty$ with $x_n \geq 0$ for all $n \in \mathbb{N}$, we have

$$\mathrm{Lim}\,(x_1, x_2, \ldots) \geq 0.$$

(5) For any sequence $(x_1, x_2, \ldots) \in \ell^\infty$, we have

$$\liminf_{n \to \infty} x_n \leq \mathrm{Lim}\,(x_1, x_2, \ldots) \leq \limsup_{n \to \infty} x_n.$$

(6) For any convergent sequence $(x_1, x_2, \ldots) \in c$, we have

$$\mathrm{Lim}\,(x_1, x_2, \ldots) = \lim_{n \to \infty} x_n.$$

[Hint: Extend the functional

$$\phi \;:\; \left((S - I)\ell^\infty\right) \bigoplus \mathrm{span}[\{(1, 1, \ldots)\}] \longrightarrow \mathbb{R},$$

given by $\phi\Big((x_2, x_2, \ldots) + (c, c, \ldots)\Big) := c$, to the entire space ℓ^∞, where I is the identity operator.]

Chapter 12

Weak topologies on Banach spaces

12.1 The raison d'être for weakening topologies

In this chapter we continue with our study of Banach spaces. However, contrary to Chapter 11 in which we examined the main structures and relevant results associated with bounded linear operators between Banach spaces, herein we change our focus to a topic more closely related to the spaces themselves.

The main theme of this chapter combines ideas from the study of general topological spaces (Chapter 4), compactness (Chapter 5) and linear operators (Chapter 11) in order to solve the fundamental problem already highlighted in Theorem 5.25 and Lemma 5.26: **closed and bounded sets in infinite-dimensional normed spaces are neither compact, nor sequentially compact**.

Our quest for compactness stems from the fundamental fact of life that **compactness is usually the only means by which we can establish the existence of certain mathematical objects via a limiting process**, typically because no explicit construction of them is possible whatsoever. Such phenomena are now the trademark of the modern theory of **nonlinear Partial Differential Equations** and of the **Calculus of Variations**.

In most cases, sequential compactness is more useful than compactness itself, since it allows us to construct algorithmically an approximation in countably-many steps. Therefore, it allows us to prove that a limiting object exists, by merely constructing a bounded sequence of "confined objects", or, as we typically say, by obtaining an **a priori uniform estimate**. Then, the property of sequential compactness, if it is satisfied, guarantees that a subsequence of these objects cannot wander around the space indefinitely, but has to eventually settle down to a limit object.

Even in the dull Euclidean setting, the compactness notions underlie numerous Analysis theorems. For example, it is essentially the only known means (which we actually used in Theorem 5.30) to prove that a function $f \in C(X, \mathbb{R})$ on a compact space X attains its maximum and its minimum value:

$$\exists\, x_m, x_M \in X: \quad \inf\big\{f(x) : x \in X\big\} = f(x_m), \quad \sup\big\{f(x) : x \in X\big\} = f(x_M).$$

The sequential version of compactness, known as the "Bolzano-Weierstrass property", holds true in (the elite, very small, class of locally) sequentially compact topological spaces. In the setting of a normed space $(X, \|\cdot\|)$, it can be formulated as follows:

$$(BW) \begin{cases} \text{Every bounded sequence } (x_m)_{m=1}^{\infty} \subseteq (X, \|\cdot\|) \text{ which satisfies} \\ \sup\{\|x_m\| : m \in \mathbb{N}\} < \infty, \text{ has a convergent subsequence in } X: \\ \exists\, (x_{m_k})_{k=1}^{\infty} \subseteq (x_m)_{m=1}^{\infty} \text{ and } \exists\, x \in X: \ x_{m_k} \longrightarrow x \text{ as } k \to \infty. \end{cases}$$

The principal example of normed space in which (BW) holds true is the Euclidean space, as Theorem 5.3 confirmed. However, as Theorem 5.25 and Lemma 5.26 certify, this is all we can ascertain! (BW) **always fails for any normed space** $(X, \|\cdot\|)$ **for which** $\dim(X) = \infty$.

The end goal in this and the next chapter is to show that **we can obtain compactness in an infinite-dimensional normed space if we compromise on our requirements** of the desired conclusion in (BW), in the sense of perhaps weakening the conclusion of convergence "$x_{m_k} \longrightarrow x$". It turns out that the only viable option is to attempt to **modify the natural norm topology of the space in hopes of restoring (some kind of) sequential compactness for this new topology**. Indeed, certain new topologies which are homonymous to the title of this chapter have the desired compactness properties.

The topologies we alluded to above have **fewer open sets** than the natural norm topologies of the spaces and have already been defined in the abstract setting of topological spaces[1] in Chapter 4. A toll that has to be paid for the gain in compactness is that these topologies are **never metrisable**. Hence, the material of Chapter 2 cannot be utilised, and the full power of the theory of topological space is required.

Along the way we shall also explore new and fascinating structures which emerge from the weakening of the usual norm topologies. However, the reader will have to bear with us until the next chapter in which our compactness problem is eventually resolved, through a further and final necessary ultra-weakening of these topologies. Finally, we note that in the present and the next chapter we shall consider **only real vector spaces**.

12.2   The weak topology of a Banach space

In this section we define the notion of weak topology on a normed space, which, as we explained in Section 12.1, will be the central object of study in the present and the next chapter. The weak topology (arising from a family of

---

[1]At this point the reader is advised to review the material of Chapter 4 and, in particular, Section 4.6 regarding the basic definitions and properties of weak topologies associated to families of mappings on general topological spaces.

maps) has already been defined and discussed in Section 4.6, in the abstract setting of general topological spaces. Herein we shall consider a particular case of it (arising from the family of bounded linear functionals), with the aim of resolving the compactness problem on normed spaces.

We begin by recalling from Chapter 4 the concept of the (norm) topology on a normed space. Let $(X, \|\cdot\|)$ be a normed space. For our purposes, it will be convenient to make explicit reference to the space X in our notation for balls and spheres. Also, since in most cases the balls will be of unit radius and centred at the origin, the dependence on the centre and the radius may be omitted. Hence, we shall be using the following symbolisations:[2]

$$\mathbb{B}_r^X(x) \equiv \{y \in X \,:\, \|x - y\| < r\}\,, \quad \mathbb{B}_1^X(0) \equiv \mathbb{B}^X,$$
$$\overline{\mathbb{B}}_r^X(x) \equiv \{y \in X \,:\, \|x - y\| \leq r\}\,, \quad \overline{\mathbb{B}}_1^X(0) \equiv \overline{\mathbb{B}}^X,$$
$$\mathbb{S}_r^X(x) \equiv \{y \in X \,:\, \|x - y\| = r\}\,, \quad \mathbb{S}_1^X(0) \equiv \mathbb{S}^X.$$

The **strong (or norm) topology** on X is the topology generated by the open balls with respect to the norm, namely

$$U \in \mathcal{T}^{\|\cdot\|} \quad \Longleftrightarrow \quad \forall\, x \in U,\; \exists\, r > 0 \,:\, \mathbb{B}_r^X(x) \subseteq U.$$

For emphatic reasons, the elements of this standard topology will be called **norm open sets**, in order to distinguish them from open sets with respect to other topologies which might be involved in our study. The norm topology is exactly the topology that we need to "weaken" i.e. make it smaller by removing some of its elements. This will work because, roughly speaking,

the fewer the open sets of a topological space are,
the more numerous the compact sets of the space are.

Before delving into that, let us recall for the sake of convenience some concepts from Chapter 11. The (topological) dual space $X^* = B(X, \mathbb{R})$ is the space of all continuous linear functionals (that is, of the bounded linear ones):

$$X^* = \{\, f : X \longrightarrow \mathbb{R} \text{ linear} \mid \|f\|_* < \infty \,\}\,, \quad \|f\|_* = \sup\{|f(x)| \,:\, x \in \mathbb{B}^X\}.$$

Here $\|\cdot\|_*$ is the dual norm which makes $(X^*, \|\cdot\|_*)$ a Banach space. Directly from the definition, the following ("generalised Cauchy-Schwarz") inequality also holds true:

$$|f(x)| \leq \|f\|_*\|x\|, \quad f \in X^*,\; x \in X.$$

We may now define the notion of weak topology on a normed space, which is a special case of Definition 4.51 already given in Chapter 4.

---

[2]Recall that, by Proposition 2.67, closed balls and norm closures of open balls coincide and the notation $\overline{\mathbb{B}}_r^X(x)$ may be safely used to symbolise either.

Definition 12.1. Let $(X, \|\cdot\|)$ be a normed space. The **weak topology** $\mathcal{T}^{\mathrm{w}}$ of the space X (symbolised by $\mathcal{T}_X^{\mathrm{w}}$ if there is any danger of confusion) is the smallest topology which contains the family of sets

$$\mathcal{E} := \left\{ f^{-1}(V) \;\middle|\; f \in X^* \text{ and } V \subseteq \mathbb{R} \text{ open} \right\},$$

that is, the class of all inverse images of open sets in $\mathbb{R}$ through bounded linear functionals in X^*:

$$\mathcal{T}^{\mathrm{w}} := \bigcap \left\{ \mathcal{T} \;\middle|\; \mathcal{T} \text{ is a topology on } X : \mathcal{T} \supseteq \mathcal{E} \right\}.$$

Equivalently, the **weak topology of X is the weakest (smallest) topology which makes all the functionals $f \in X^*$, continuous.** The elements $\mathcal{U} \in \mathcal{T}^{\mathrm{w}}$ of the weak topology will be called **weakly open sets.**

Remark 12.2.
(i) [Uniqueness] In view of Lemma 4.44, the weak topology $\mathcal{T}^{\mathrm{w}}$ of any normed space $(X, \|\cdot\|)$ is unique.

(ii) [Meaning of continuity] All functionals $f \in X^*$ are by definition **already continuous with respect to the norm**, that is when considered as mappings[3] between topological spaces

$$f \;:\; (X, \mathcal{T}^{\|\cdot\|}) \longrightarrow (\mathbb{R}, \mathcal{T}^{|\cdot|}).$$

What we do here is to define a new topology with fewer open sets (by removing certain open sets from the norm topology $\mathcal{T}^{\|\cdot\|}$) such that each $f \in X^*$ remains continuous with respect to the weakened topology, namely when seen as a mapping

$$f \;:\; (X, \mathcal{T}^{\mathrm{w}}) \longrightarrow (\mathbb{R}, \mathcal{T}^{|\cdot|}).$$

Note also that, in view of Proposition 4.49, the second part of the definition is equivalent to the first because $f : (X, \mathcal{T}^{\mathrm{w}}) \longrightarrow \mathbb{R}$ is continuous if and only if $f^{-1}(V)$ is a weakly open set in $\mathcal{T}^{\mathrm{w}}$, for any open set $V \subseteq \mathbb{R}$.

(iii) [Terminology] The modifiers **"weak/weakly"** will more generally be used to specify that the topological notions following them (e.g. weakly open, weakly closed, weakly compact, weakly dense, weak convergence, weak continuity, ...) are considered with respect to the weak topology, namely, they refer to the topological space $(X, \mathcal{T}^{\mathrm{w}})$. In particular, recalling Definition 4.34 and Proposition 4.35, the symbolisations

$$\text{``int}_{\mathrm{w}}(\cdot)\text{''} \quad \text{and} \quad \overline{(\cdot)}^{\mathrm{w}}$$

will be used to denote the **weak interior** and the **weak closure** respectively, namely with respect to the weak topology $\mathcal{T}^{\mathrm{w}}$ on the normed space X.

---

[3]Here we write $(\mathbb{R}, \mathcal{T}^{|\cdot|})$ just for emphatic reasons, the one and only topology we are using on $\mathbb{R}$ is the standard one.

The next result characterises the weak topology in a more concrete manner, by utilising Definition 4.47.

Proposition 12.3 (Neighbourhood basis of the weak topology). *Let $(X, \|\cdot\|)$ be a normed space. The weak topology of X consists of $\emptyset$, X, and all unions of finite intersections of elements of $\mathcal{E}$.*

In addition, weakly open sets can be characterised in the following fashion:

$$\mathcal{U} \in \mathcal{T}^w \iff \begin{cases} \forall\, x \in \mathcal{U}, \ \exists\, \varepsilon > 0 \ \& \ \exists\, f_1, \ldots, f_N \in X^* : \\[2mm] \displaystyle\bigcap_{k=1}^{N} f_k^{-1}\Big(\big(f_k(x) - \varepsilon, \ f_k(x) + \varepsilon\big)\Big) \subseteq \mathcal{U}. \end{cases}$$

Further, any weakly open set can be expressed as the arbitrary union of such finite intersections of inverse images (see Figure 12.1).

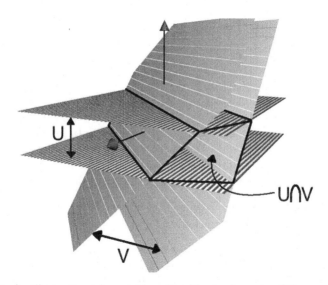

Figure 12.1: An illustration of a neighbourhood basis elements of the weak topology around the origin of $\mathbb{R}^3$ for $N = 2$. The four planes are the level-sets $\{f = \pm 1\}$ and $\{g = \pm 2\}$ of the functionals $f(x, y, z) := z$ and $g(x, y, z) := x + y + z$ (the sets $\{g = \pm 2\}$ may not seem parallel, but this is an optical illusion). These planes are the lateral boundaries of the weakly open neighbourhoods of $(0, 0, 0)$, given by

$$U := \big\{(x, y, z) \in \mathbb{R}^3 : |f(x, y, z)| < 1\big\}, \quad V := \big\{(x, y, z) \in \mathbb{R}^3 : |f(x, y, z)| < 2\big\}$$

and the set $U \cap V$ is the unbounded parallelogram enclosed by these four pairwise parallel planes. Note that, although $U \cap V$ is unbounded, it may well happen that this kind of intersections are bounded sets in this finite-dimensional reduction (if e.g. we had three such sets $U \cap V \cap W$ with the extra set W "transversal" to U, V). However, by Theorem 12.6 that follows, this can never happen in an infinite-dimensional normed space, no matter how many the intersecting inverse images might be, provided there are only finitely many of them!

Proof. The desired conclusion is an immediate consequence of Proposition 4.45 and Remark 4.52. $\qquad\square$

In topological terms, the meaning of the characterisation in Proposition 12.3 is that the sets which have the form of finite intersections of inverse images of intervals form a *neighbourhood basis for the weak topology.*

Remark 12.4. For the sake of clarity, let us note that the intersection of the inverse images in Proposition 12.3 can be written in level-set notation as follows

$$\bigcap_{k=1}^{N} f_k^{-1}\Big(\big(f_k(x) - \varepsilon,\ f_k(x) + \varepsilon \big) \Big) = \Big\{ y \in X\ :\ \big| f_k(x - y) \big| < \varepsilon,\ k = 1, \ldots, N \Big\}.$$

Also, it is a very simple exercise for the reader to check that one might rephrase the definition as

$$\mathcal{U} \in \mathcal{T}^{\mathrm{w}} \iff \left\{ \begin{array}{l} \forall\, x \subset \mathcal{U},\ \exists\, \varepsilon > 0,\ \exists\, (a_k)_1^N \subseteq \mathbb{R}\ \&\ \exists\, (f_k)_1^N \subseteq X^* : \\[2mm] \displaystyle\bigcap_{k=1}^{N} f_k^{-1}\Big(\big(a_k - \varepsilon,\ a_k + \varepsilon \big) \Big) \subseteq \mathcal{U}. \end{array} \right.$$

Note the obvious fact that, exactly like the norm topology, the **weak topology of a normed space is translation invariant**. In particular, $\mathcal{U} \in \mathcal{T}^{\mathrm{w}}$ if and only if $x + \mathcal{U} \in \mathcal{T}^{\mathrm{w}}$, $x \in X$. Hence, we may (and will) occasionally restrict our attention to weakly open sets containing the origin since all the arguments can be transferred to an arbitrary point in the space by a translation.

We close this section by showing that the weak topology of any normed space is Hausdorff (recall Definition 4.17).

Proposition 12.5 ($\mathcal{T}^{\mathrm{w}}$ is a Hausdorff topology). *For any normed space* $(X, \|\cdot\|)$, *its weak topology* $\mathcal{T}^{\mathrm{w}}$ *satisfies the Hausdorff axiom. Hence, for any two points* $x \neq y$ *in* X, *there exist two disjoint weakly open neighbourhoods* $\mathcal{U}_x, \mathcal{U}_y \in \mathcal{T}^{\mathrm{w}}$, *namely* $\mathcal{U}_x \ni x$, $\mathcal{U}_y \ni y$, *and* $\mathcal{U}_x \bigcap \mathcal{U}_x = \emptyset$.

Proof. Fix two points $x \neq y$ in X. Since $\{x\}, \{y\}$ are disjoint compact convex sets, by the geometric Hahn-Banach Theorem 11.59, there exists a functional $f \in X^*$, an $a \in \mathbb{R}$ and an associated hyperplane $\{f = a\}$ which separates $\{x\}, \{y\}$ strictly, namely $f(x) < a < f(y)$. Then, the sets

$$\mathcal{U}_x := \big\{ z \in X\ :\ f(z) < a \big\}, \quad \mathcal{U}_y := \big\{ z \in X\ :\ f(z) > a \big\},$$

are disjoint weakly open neighbourhoods of x, y, as desired. $\qquad\square$

In the next section we shall discuss some striking properties of this new concept of topology on a normed space, which are completely against our finite- dimensional Euclidean intuition.

12.3 On the nature of weakly open sets

Let $(X, \|\cdot\|)$ be a normed space. Quite surprisingly, if $\dim(X) = \infty$, **the weak topology of X is strictly weaker than its norm topology.** As we shall see right next, no open ball of X (which is trivially $\neq$ norm open) can be weakly open! Moreover, any nonempty weakly open set is necessarily unbounded and contains a finite-dimensional subspace! These counter-intuitive properties of the weak topology are confirmed by the following result.

Theorem 12.6 (Weakly open vs. open sets). *Let $(X, \|\cdot\|)$ be an infinite-dimensional normed space. Then the following hold:*

(i) **Every nonempty weakly open set is unbounded and contains an affine line:**

$$\mathcal{U} \in \mathcal{T}^{\mathrm{w}} \setminus \{\emptyset\} \implies \exists\, x, x_0 \in X,\ x_0 \neq 0 :\ x + \mathrm{span}[\{x_0\}] \subseteq \mathcal{U}.$$

(ii) **Every bounded norm open set has empty weak interior:**

$$\mathcal{U} \in \mathcal{T}^{\|\cdot\|},\ \ \mathrm{diam}(\mathcal{U}) < \infty \implies \mathrm{int}_{\mathrm{w}}(\mathcal{U}) = \emptyset.$$

Remark 12.7. We record the next comments regarding Theorem 12.6:

• In (i), if $\mathcal{U}$ contains the origin $0 \in X$, the proof below shows that one may choose the line to pass through it, namely

$$0 \in \mathcal{U} \in \mathcal{T}^{\mathrm{w}} \implies \exists\, x_0 \in X \setminus \{0\} :\ \mathrm{span}[\{x_0\}] \subseteq \mathcal{U}.$$

• In (ii), by the translation invariance of the weak topology, no bounded open set (and in particular no open ball) can be weakly open in an infinite-dimensional normed space. In particular, no ball $\mathbb{B}_r^X(x)$ is weakly open for any $x \in X$ and any $r > 0$, whilst its weak interior is empty.

Proof. (i) Let $\mathcal{U} \in \mathcal{T}^{\mathrm{w}} \setminus \{\emptyset\}$ be a nonempty weakly open set and choose $x \in \mathcal{U}$. By Proposition 12.3, there exists an $\varepsilon > 0$ and $f_1, \ldots, f_m \in X^*$ such that

$$\bigcap_{k=1}^{m} \{y \in X :\ |f_k(y - x)| < \varepsilon\} \subseteq \mathcal{U}.$$

We claim that there exists $x_0 \in X \setminus \{0\}$ such that

$$f_1(x_0) = \cdots = f_m(x_0) = 0.$$

This means that $x_0 \in \bigcap_{i=1}^{m} \{f_i = 0\}$, which is the intersection of the nullspaces of the functionals $f_1, \ldots, f_m$. If hypothetically no such $x_0 \neq 0$ existed, then the map from X to $\mathbb{R}^m$ given by

$$x \mapsto \big(f_1(x), \ldots, f_m(x)\big) \quad \text{for all } x \in X$$

would be a linear injection into a finite-dimensional space, because its nullspace satisfies

$$\Big\{ x \in X \; : \; \big(f_1(x), \ldots, f_m(x) \big) = 0 \Big\} = \bigcap_{i=1}^{m} \{ x \in X : f_i(x) = 0 \} = \{0\}.$$

Then, we would in turn have $\dim(X) \leq m$, which contradicts the assumption $\dim(X) = \infty$. Therefore,

$$f_1(x_0) = f_2(x_0) = \cdots = f_m(x_0) = 0.$$

By linearity, this yields $f_k(x + tx_0) = f_k(x)$ for any $t \in \mathbb{R}$ and all indices $k \in \{1, \ldots, m\}$. By the formulas $(\star)$-$(\star\star)$ of Chapter 1, we have

$$x + tx_0 \in \bigcap_{k=1}^{m} f_k^{-1}\Big(\{f_k(x)\} \Big) \subseteq \bigcap_{k=1}^{m} f_k^{-1}\Big(\big(f_k(x) - \varepsilon, \; f_k(x) + \varepsilon \big) \Big)$$

which in turn allows us to infer that

$$x + \operatorname{span}[\{x_0\}] \subseteq \mathcal{U}.$$

(ii) Let $\mathcal{U}$ be a bounded norm open set. By the definition of the interior and Proposition 4.35, we have[4]

$$\operatorname{int}_{\mathrm{w}}(\mathcal{U}) = \bigcup \Big\{ \mathcal{V} \in \mathcal{P}(\mathcal{U}) \; \Big| \; \mathcal{V} \in \mathcal{T}^{\mathrm{w}} \Big\}.$$

However, by part (i), the only weakly open set contained in $\mathcal{U}$ is the empty set. Hence, $\operatorname{int}_{\mathrm{w}}(\mathcal{U}) = \emptyset$. $\square$

The result above is not true in finite dimensions. Note the interesting fact that one cannot obtain the contradiction in the construction of the point x_0 as we did above. In fact, weak and strong topologies coincide in finite dimensions! We could prove this directly, but we prefer instead to obtain it as a consequence of the relevant notion of **weak convergence** we shall define a little later. By utilising Theorem 12.6, we may also show the following counter-intuitive fact about closed sets:

Proposition 12.8 (On weakly closed sets). *Let $(X, \| \cdot \|)$ be an infinite-dimensional normed space. Then,* **the unit sphere $\mathbb{S}^X$ of X is not weakly closed***. Moreover, the* **weak closure of the sphere equals the respective closed ball***:*

$$\overline{\mathbb{S}^X}^{\mathrm{w}} = \overline{\mathbb{B}}^X.$$

Proof. We begin by noting that by Theorem 12.6 we have $\operatorname{int}_{\mathrm{w}}\big(\mathbb{B}^X \big) = \emptyset$. Further, note that the complement of the sphere can be written as the disjoint

---

[4]We recall that, for any set X, the symbol $\mathcal{P}(X)$ stands for the powerset of the set X.

union $X \setminus \mathbb{S}^X = \mathbb{B}^X \bigcup (X \setminus \overline{\mathbb{B}}^X)$. By recalling the properties of the topological interior (Proposition 4.35), we have

$$
\begin{aligned}
X \setminus \left(\overline{\mathbb{S}^X}^{\mathrm{w}} \right) &= \mathrm{int}_{\mathrm{w}} \left(X \setminus \mathbb{S}^X \right) \\
&= \mathrm{int}_{\mathrm{w}} \left(\mathbb{B}^X \bigcup (X \setminus \overline{\mathbb{B}}^X) \right) \\
&= \mathrm{int}_{\mathrm{w}} \left(\mathbb{B}^X \right) \bigcup \mathrm{int}_{\mathrm{w}} \left(X \setminus \overline{\mathbb{B}}^X \right) \\
&= \emptyset \bigcup \mathrm{int}_{\mathrm{w}} \left(X \setminus \overline{\mathbb{B}}^X \right) \\
&= \mathrm{int}_{\mathrm{w}} \left(X \setminus \overline{\mathbb{B}}^X \right) \\
&= X \setminus \left(\overline{\overline{\mathbb{B}}^X}^{\mathrm{w}} \right),
\end{aligned}
$$

which in turn yields $\overline{\mathbb{S}^X}^{\mathrm{w}} = \overline{\overline{\mathbb{B}}^X}^{\mathrm{w}}$. In order to conclude, we invoke the Mazur Theorem 12.10 that follows, which implies that the weak closure and the norm closure of the closed ball coincide:

$$
\overline{\overline{\mathbb{B}}^X}^{\mathrm{w}} = \overline{\overline{\mathbb{B}}^X}^{\|\cdot\|} = \overline{\mathbb{B}}^X.
$$

The proposition therefore ensues. □

Remark 12.9 (Weakly closed balls vs. weak closures of open balls). By Proposition 2.67, on any normed space the closed balls coincides with the (norm) closures of the respective open balls. However, this is not true for the weak closures of the balls.

Several more properties and examples of weakly open and weakly closed sets will be examined in the sequel. In particular, in the next section, we show the important general fact, already used above, that a convex set is weakly closed if and only if it is norm closed.

12.4   The Mazur Theorem for convex sets

In this brief section we establish a very important result concerning the weak topology of a Banach space in relation to convex subsets of the space. Given a subset $F \subseteq X$ of a Banach space $(X, \|\cdot\|)$, it follows from the very definitions that in general we have the implication

$$F \textbf{ weakly-closed} \implies F \textbf{ norm-closed}$$

and as we saw earlier, in infinite-dimensional Banach spaces the opposite is **not** true in general. However, for **convex sets** the reverse implication is true as well, as the next result attests. The proof is based essentially on the geometric Hahn-Banach theorem we expounded in Chapter 11. Let us recall that a subset C of a vector space X is convex when for all $x, y \in C$, the straight line segment connecting x, y is contained in C, namely $(1 - t)x + ty \in C$ for all $t \in [0, 1]$.

Theorem 12.10 (Mazur). *Let $C \subseteq X$ be a convex subset of the Banach space $(X, \| \cdot \|)$. Then C is weakly closed if and only if it is norm closed. In addition, for any convex set C which is either closed or weakly closed, we have*

$$\overline{C}^{\| \cdot \|} = C = \overline{C}^{\mathrm{w}}.$$

Proof. We readily have that C is norm closed if it is weakly closed. Conversely, assume C is norm closed. The idea is to utilise the (geometric form of the) Hahn-Banach separation Theorem 11.58 to show that $X \setminus C$ is weakly open. To this end, fix $x \in X \setminus C$. Then, there exists a functional $f \in X^*$ and an associated hyperplane $\{f = a\}$ for some $a \in \mathbb{R}$ (Definition 11.53) which separates strictly the compact set $\{x\}$ from the closed set C (see also Figure 12.2):

$$f(x) < a < f(y), \quad \forall y \in C.$$

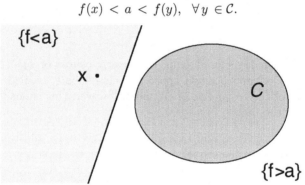

Figure 12.2: An illustration of the idea of the proof of the Mazur Theorem for $X = \mathbb{R}^2$. By the geometric Hahn-Banach Theorem, every point $x \in X \setminus C$ in the exterior of C is contained in a (weakly open) half-space of the form $\{f < a\}$, where f is a bounded linear functional.

Then, the set defined as

$$\mathcal{U}_x := f^{-1}\big((-\infty, a)\big) = \{f < a\}$$

contains x and by the definition of $\mathcal{T}^{\mathrm{w}}$ it is weakly open. Hence $\mathcal{U}_x$ is a weakly open neighbourhood of the point x inside $X \setminus C$. Recapitulating, we have shown that any point $x \notin C$ in the complement of the convex set C, is contained in a weakly open set $\mathcal{U}_x \subseteq X \setminus C$. By the general properties of the interior (Proposition 4.35), this establishes that it is a weakly open set because

$$X \setminus C = \mathrm{int}_{\mathrm{w}}(X \setminus C).$$

Therefore, the complement C of the set $X \setminus C$ is weakly closed. Finally, the desired equality of closures can be seen as follows. If C is weakly closed, then by Proposition 4.35 we have $C = \overline{C}^{\mathrm{w}}$. By the result we just proved, C is norm closed too and thus $C = \overline{C}^{\|\cdot\|}$. The other direction is similar. The theorem follows. □

The Mazur Theorem has many fundamental applications, some of which will be examined in subsequent sections and in the next chapters.

12.5 Weak convergence and its properties

The concept that we are actually mostly interested in is the notion of convergence associated with the weak topology, which we introduce and study in this section:

Definition 12.11. Let $(X, \|\cdot\|)$ be a normed space. We say that the sequence $(x_m)_1^\infty \subseteq X$ **converges weakly** to some $x \in X$ and we write

$$x_m \rightharpoonup x \quad \text{as} \quad m \to \infty,$$

if

$$\text{for any } f \in X^*, \text{ we have } f(x_m) \longrightarrow f(x) \text{ as } m \to \infty.$$

Remark 12.12. Note that Definition 12.11 requires that for any **fixed** functional $f \in X^*$, we have the *pointwise* convergence $f(x_m) \longrightarrow f(x)$ as $m \to \infty$ of the respective values (which are real numbers).

We now confirm what our terminology suggests, namely that the notion of weak convergence arises as the notion of convergence associated to the weak topology of X. Let us first recall from Chapter 4 the definition of convergence of a sequence[5] on a general topological space $(X, \mathcal{T})$ (Definition 4.15). We say

$$x_m \longrightarrow x \quad \text{as} \quad m \to \infty, \quad \text{with respect to } \mathcal{T},$$

when for any neighbourhood $\mathcal{U}_x \in \mathcal{T}$ of x (i.e. any element of the topology $\mathcal{T}$ containing x), there exists $m_0 = m_0(\mathcal{U}_x) \in \mathbb{N}$, such that, for all $m \geq m_0$, we have $x_m \in \mathcal{U}_x$.

---

[5] As already announced in Section 12.1, we will see later that the weak topology is not metrisable. Since the standard notion of sequence is not really adequate for the description of convergence in non-metrisable topological spaces, to this end more advanced tools have been developed which generalise it. Primarily, these concepts are those of *nets* and of *filters*. As we noted in Chapter 4, in the present quite elementary approach to Analysis, we have made every possible effort to avoid using these peculiar notions which are less pertinent to applications. However, little is lost by restricting our attention to sequences in this chapter because, as we shall see later, metrisability of the weak topology indeed holds for *bounded* sets when the dual space is separable, a representative occurrence in most cases.

Proposition 12.13. *Let $(X, \|\cdot\|)$ be a normed space. Then, the next statements are equivalent:*

(i) $x_m \longrightarrow x$ *as* $m \to \infty$, *weakly in* X *(Definition 12.11).*

(ii) $x_m \longrightarrow x$ *as* $m \to \infty$, *with respect to the weak topology* $\mathcal{T}^w$ *of* X.

Proof. (ii) $\Longrightarrow$ (i): By definition, the weak topology on X makes all functionals $f \in X^*$ continuous. Suppose that $x_m \longrightarrow x$ with respect to $\mathcal{T}^w$, as $m \to \infty$. By the sequential characterisation of continuity (Theorem 4.28), we infer that $f(x_m) \longrightarrow f(x)$, as $m \to \infty$.

(i) $\Longrightarrow$ (ii): Let $\mathcal{U}_x \in \mathcal{T}^w$ be a weakly open neighbourhood of $x \in X$. Then, $\mathcal{U}_0 := \mathcal{U}_x - x \in \mathcal{T}^w$ is a weakly open neighbourhood of the origin $0 \in X$. Hence, there exists an $\varepsilon > 0$ and $f_1, \ldots, f_N \in X^*$ such that

$$\mathcal{U}_0 \supseteq \bigcap_{k=1}^{N} f_k^{-1}\big((-\varepsilon, \varepsilon)\big).$$

Since $x_m \longrightarrow x$, by linearity we have $f(x_m - x) \longrightarrow 0$ as $m \to \infty$ for any $f \in X^*$. Since the family $\{f_1, \ldots, f_N\}$ is finite, we have

$$\max_{k=1,\ldots,N} \big|f_k(x_m - x)\big| \longrightarrow 0$$

as $m \to \infty$. By choosing m_0 large, we have, for $m \geq m_0$,

$$\big|f_k(x_m - x)\big| < \varepsilon, \quad k = 1, \ldots, N$$

and hence we obtain $x_m - x \in \mathcal{U}_0$, or equivalently $x_m \in \mathcal{U}_x$, as desired. The proposition ensues. $\qquad\square$

The next result asserts the plausible fact that weak limits, if they exist, must be unique:

Lemma 12.14 (Uniqueness of weak limits). *Let $(x_m)_1^\infty$ be a sequence in a normed space $(X, \|\cdot\|)$. If there exists $x \in X$ such that $x_m \longrightarrow x$ as $m \to \infty$, then the weak limit x is unique.*

The uniqueness of weak limits is a consequence of the Hausdorff property of the weak topology (Proposition 12.5) and the results in Chapter 4, but we may also prove it directly.

Proof. Suppose for the sake of contradiction that there exists another point $y \neq x$ in X such that $x_m \longrightarrow y$ as $m \to \infty$ too. Then, for any $f \in X^*$ we have $f(x_m) \longrightarrow f(x)$ and $f(x_m) \longrightarrow f(y)$, as $m \to \infty$. Therefore, $f(x - y) = f(x) - f(y) = 0$ by the uniqueness of limits on $\mathbb{R}$. By the next representation formula[6] for the norm of X

$$\|z\| = \sup\big\{|f(z)| : f \in \mathbb{B}^{X^*}\big\}, \quad z \in X, \tag{$\natural$}$$

---

[6]We recall from Corollary 11.46 (recall also Definition 11.47 and Proposition 11.50) that the formula ($\natural$) is a result of isometric embedding of X into its bidual X^{**} via the evaluation mapping $\widehat{\cdot} : X \longrightarrow X^{**}$ given at any $f \in X^*$ by $\widehat{x}(f) := f(x)$.

the choice $z := x - y$ yields $\|x - y\| = 0$, contradicting that $x \neq y$. Hence, the uniqueness of weak limits ensues. $\qquad\square$

Next we demonstrate a fact hinted at by the terminology used, namely that strong convergence implies weak convergence:

Lemma 12.15 (Strong convergence implies weak convergence). *Let $(X, \|\cdot\|)$ be a normed space, $(x_m)_1^\infty$ a sequence in it and $x \in X$. If $x_m \longrightarrow x$ strongly in X as $m \to \infty$ (i.e., with respect to the norm), then $x_m \longrightarrow x$ weakly in X, as $m \to \infty$.*

Proof. Fix a sequence $(x_m)_1^\infty \subseteq X$ and a point $x \in X$. For any $f \in X^*$, we have the estimate

$$\left| f(x_m) - f(x) \right| = |f(x_m - x)| \leq \|f\|_* \|x_m - x\|,$$

for any $m \in \mathbb{N}$. Hence, if $x_m \longrightarrow x$ as $m \to \infty$, then we have $x_m \longrightarrow x$, as $m \to \infty$. $\qquad\square$

Now we show that in finite-dimensional Banach spaces, weak and strong modes of convergence coincide. In particular, **there exists no "genuinely weak" convergence**[7] **in the Euclidean space $\mathbb{R}^N$.**

Proposition 12.16 (No weak notions in finite dimensions). *Let $(X, \|\cdot\|)$ be a finite-dimensional normed space, $(x_m)_1^\infty$ be a sequence in X, and $x \in X$. Then, $x_m \longrightarrow x$ weakly as $m \to \infty$ if and only if $x_m \longrightarrow x$ strongly as $m \to \infty$.*

Proof. We give the proof only in the particular case of the space $\mathbb{R}^N$, and leave as an exercise for the reader its adaptation to the general case, in which the description of the dual of X given by Proposition 11.34 is relevant.

Assume therefore that X is the Euclidean space $\mathbb{R}^N$. It suffices to show that if $x_m \longrightarrow x$, then $x_m \longrightarrow x$. Consider the linear functionals

$$f_1, \ldots, f_N \; : \; \mathbb{R}^N \longrightarrow \mathbb{R}, \quad f_i(x) := x^i \,, \quad i = 1, \ldots, N.$$

These are merely the projections on the coordinate components of vectors $x = (x^1, \ldots, x^N)$. Then, since $x_m \longrightarrow x$, we have

$$\left| f_i(x_m - x) \right| \longrightarrow 0, \quad \text{as } m \to \infty.$$

Hence, by the definition of the Euclidean norm

$$|x_m - x| = \left(\sum_{i=1}^N \left(x_m^i - x^i \right)^2 \right)^{1/2} \leq N \max_{i=1,\ldots,N} \left| f_i(x_m - x) \right|$$

and since $f_1, \ldots, f_N$ are finitely many, the right hand side of the above estimate tends to zero as $m \to \infty$. In conclusion, we have that $x_m \longrightarrow x$ as $m \to \infty$. $\qquad\square$

---

[7]This is the reason you never heard of weak convergence or weak topology in Real Analysis on $\mathbb{R}^N$! These are genuinely infinite-dimensional features of normed spaces.

Remark 12.17 (Weak convergence in Hilbert spaces). We now particularise the notion of weak convergence in the case of (real) separable Hilbert spaces

$$\big(H, \langle\cdot,\cdot\rangle\big).$$

By the Riesz Representation Theorem for Hilbert spaces (Theorem 11.37), we know that the dual space H^* is linearly isometric to H itself and actually H^* consists exactly of those functionals of the form $\langle y, \cdot\rangle$ for some $y \in H$. Hence, by the definition of weak convergence, we have as $m \to \infty$ that

$$x_m \xrightarrow{\quad} x \ \text{ weakly in } H \quad \Longleftrightarrow \quad \langle y, x_m\rangle \longrightarrow \langle y, x\rangle \ \text{ for all } y \in H.$$

Our previous results lead naturally to the following questions:

(Q1) What is the essential difference between weakly convergent and strongly convergent sequences?

(Q2) When does a weakly convergent sequence fail to converge strongly?

Remark 12.17 allows us to examine a non-trivial class of important examples where such interesting phenomena occur and partly answers (Q1), (Q2), by utilising concepts from Chapter 10. We will revisit these questions again in Chapter 14, to discuss them in the prominent class of L^p spaces.

Lemma 12.18 (Orthonormal sequences in inner product spaces are weakly null). *Let $(H, \langle\cdot,\cdot\rangle)$ be a (real) inner product space and let $(e_i)_1^\infty$ be a countable orthonormal subset of H. By Definition 10.27, this means*

$$\langle e_i, e_j\rangle = \delta_{ij}, \quad \text{for all } i, j \in \mathbb{N}$$

(where $\delta_{ii} = 1$ and $\delta_{ij} = 0$ for $i \neq j$). Then, the sequence $(e_i)_1^\infty$ converges weakly to zero, that is $e_i \longrightarrow 0$ weakly in H as $i \to \infty$. However, $(e_i)_1^\infty$ is not strongly Cauchy.

Proof. Since $(e_i)_1^\infty$ is an orthonormal sequence, by Bessel's inequality (Proposition 10.33), for any $x \in H$ we have the estimate

$$\|x\|^2 \geq \sum_{i=1}^\infty |\langle x, e_i\rangle|^2.$$

By the convergence of the series above, we have that

$$|\langle x, e_i\rangle|^2 \longrightarrow 0 \quad \text{as } i \to \infty,$$

for any given $x \in X$. The latter is exactly the notion of weak convergence on H. On the other hand, $(e_i)_1^\infty$ does not converge strongly to any limit, since it is not Cauchy in norm. Indeed, by the Pythagorean theorem for $i \neq j$ we have

$$\|e_i - e_j\|^2 = \|e_i\|^2 + \|e_j\|^2 - 2\langle e_i, e_j\rangle = 2.$$

Hence, the terms of the sequence $(e_i)_1^\infty$ have pairwise distances between them equal to $\sqrt{2}$. □

We close this section with the next observation regarding weak convergence on a Banach spaces which is the dual space of some other space (the former being usually called pre-dual space).

Remark 12.19 (Weak convergence in dual Banach spaces). The dual space X^* of a normed space, being a normed space itself, has a weak topology of its own which we shall symbolised as $\mathcal{T}_{X^*}^w$. Hence, a sequence of functionals $(f_m)_1^\infty \subseteq X^*$ converges weakly in X^*, namely

$$f_m \longrightarrow f \text{ in } X^* \text{ as } m \to \infty,$$

when for all elements $\phi \in X^{**}$ we have $\phi(f_m) \longrightarrow \phi(f)$, as $m \to \infty$. Here X^{**} is the bidual (or second dual) space of X (recall Definition 11.47):

$$X^{**} = (X^*)^* = B\big(B(X, \mathbb{R}), \mathbb{R}\big), \quad \|\phi\|_{**} = \sup\big\{|\phi(f)| : f \in \mathbb{B}^{X^*}\big\}.$$

We underline that weakened topologies on dual Banach spaces are particularly important and will be the main subject of the next chapter. As we shall see, they have the desired compactness properties we are seeking.

12.6 Weak lower semi-continuity and convexity

In this section we continue studying the notion of weak convergence. The results herein are motivated by the following questions. Let $(X, \|\cdot\|)$, $(Y, \|\cdot\|)$ be normed spaces and $F : X \longrightarrow Y$ a (possibly nonlinear) mapping. Suppose F is continuous with respect to the norm topologies of X, Y (this is sometimes referred to as "norm-norm" continuity). Fix a sequence $(x_m)_1^\infty \subseteq X$ and a point $x \in X$.

(Q1) Is it true that the continuous mapping F is weakly continuous, namely

$$x_m \longrightarrow x \text{ in } X \overset{?}{\Longrightarrow} F(x_m) \longrightarrow F(x) \text{ in } Y,$$

as $m \to \infty$? In particular:

(Q2) If $Y = \mathbb{R}$, is it true that F is weakly continuous on X, namely that

$$x_m \longrightarrow x \text{ in } X \overset{?}{\Longrightarrow} F(x_m) \longrightarrow F(x) \text{ in } \mathbb{R},$$

as $m \to \infty$? Is this true for $F = \|\cdot\|$?

Quite unexpectedly, for either of the questions the answer is affirmative if and only F **is affine**, that is, linear up to a shift by a constant! Hence, (genuinely)

nonlinear functionals are always weakly discontinuous[8] on infinite-dimensional normed spaces.

The above is a fundamental difficulty in Analysis. However, not everything is lost since in many cases the full strength of weak continuity is not needed and an one-sided version of it suffices instead. In fact, it suffices to discuss only the sequential version. To this end, recall the concept of continuity of a function $F : (X, \mathcal{T}) \longrightarrow \mathbb{R}$, which by Proposition 4.28 yields that for any $x \in X$ and any sequence $(x_m)_1^\infty \subseteq X$ that

$$x_m \xrightarrow{\mathcal{T}} x \text{ as } m \to \infty \implies F(x) = \lim_{m\to\infty} F(x_m). \qquad (\diamond)$$

Then, the key idea is that $(\diamond)$ is equivalent to the pair of one-sided requirements $(\triangle)$-$(\triangledown)$ in the next definition.

Definition 12.20. Let $(X, \mathcal{T})$ be a topological space and let also $F : X \longrightarrow \mathbb{R}$ be a function. Fix a point $x \in X$ and any sequence $(x_m)_1^\infty \subseteq X$. Consider the statements:

$$x_m \xrightarrow{\mathcal{T}} x \text{ as } m \to \infty \implies F(x) \leq \liminf_{m\to\infty} F(x_m), \qquad (\triangle)$$

$$x_m \xrightarrow{\mathcal{T}} x \text{ as } m \to \infty \implies F(x) \geq \limsup_{m\to\infty} F(x_m). \qquad (\triangledown)$$

If $(\triangle)$ holds for F for all $x \in X$, we say that F **is sequentially Lower Semi-Continuous on** X with respect to $\mathcal{T}$ (abbreviated as **LSC**).

If $(\triangledown)$ holds for F for all $x \in X$, we say that F **is sequentially Upper Semi-Continuous on** X with respect to $\mathcal{T}$ (abbreviated as **USC**).

The next sequential version of lower semi-continuity[9] isolates the particular case that is actually of interest to us in relation to the weak topology.

Definition 12.21. Let $f : X \longrightarrow \mathbb{R}$ be a function on a normed space $(X, \|\cdot\|)$. We say that f **is sequentially weakly lower semi-continuous**, if for any sequence $(x_m)_1^\infty \subseteq X$ and any $x \in X$ such that

$$x_m \longrightarrow x \text{ weakly as } m \to \infty,$$

we have

$$f(x) \leq \liminf_{m\to\infty} f(x_m).$$

---

[8]There exists precisely one notable exception of weakly continuous nonlinear functionals known in Calculus of Variations as "quasi-affine" or "null Lagrangians". These arise only in the so-called Sobolev spaces of vector-valued functions. These spaces are subspaces of the L^p spaces consisting of "weakly differentiable" functions. Any null Lagrangian essentially consist of the composition of a linear mapping with the vector of all subdeterminants of the gradient matrix. However, even null Lagrangians do not qualify as putative solutions to (Q1)-(Q2) because what is actually true is a weaker statement: weak continuity holds true for a different topology of the target space which is always weaker than that of the domain. In this book we refrain from discussing this more advanced topic any further.

[9]At this point it is instructive to revisit the corollary of the Fatou Lemma in Chapter 8 (Corollary 8.26) and convince yourselves that indeed, as we claimed there, the integral is LSC with respect to the a.e. convergence of functions on any measure space.

One may obviously define sequential weak USC in a symmetric fashion, but we will not need that in the sequel. We are now able to give the next very important result in relation to the question (Q2) we posed at the beginning of this section.

Theorem 12.22 (Weak LSC, weak boundedness and strong-weak continuity of the duality pairing). *Let* $(X, \|\cdot\|)$ *be a Banach space with dual space* X^*. *Consider also* $(x_m)_1^\infty \subseteq X$, $x \in X$, $(f_m)_1^\infty \subseteq X^*$ *and* $f \in X^*$. *Then, the following statements hold true:*

(i) **Weakly convergent sequences in** X **are norm bounded;** *that is, if* $x_m \rightharpoonup x$ *weakly as* $m \to \infty$, *then* $(\|x_m\|)_1^\infty$ *is bounded in* $\mathbb{R}$.

(ii) **The norm** $\|\cdot\|$ **of** X **is sequentially weakly LSC;** *that is, if* $x_m \rightharpoonup x$ *weakly as* $m \to \infty$, *then*

$$\|x\| \leq \liminf_{m \to \infty} \|x_m\|.$$

(iii) **The duality pairing between** X^* **and** X, **namely the function**

$$X^* \times X \longrightarrow \mathbb{R} \ : \quad (f, x) \mapsto f(x),$$

is strongly-weakly continuous; *that is, if we have* $f_m \longrightarrow f$ *strongly in* X^* *and* $x_m \rightharpoonup x$ *weakly in* X, *then*

$$f_m(x_m) \longrightarrow f(x), \quad \text{as } m \to \infty.$$

Item (iii) of Theorem 12.22 **is not true when both convergences are weak**, that is when $f_m \rightharpoonup f$ in X* and $x_m \rightharpoonup x$ in X. This happens because (like the inner product of Hilbert spaces) the duality pairing is a quadratic nonlinear function. Accordingly, we provide the next representative example:

Remark 12.23 (Quadratic functionals need not be weakly continuous). Consider the Hilbert space $L^2((-\pi, \pi))$ with the standard inner product (following the conventions and the notation of Chapter 9)

$$\langle f, g \rangle := \int_{-\pi}^{\pi} f(t)\, g(t) \, \mathrm{d}\mathcal{L}^1(t).$$

Then, the sequence of functions $(f_m)_1^\infty$ given by

$$f_m(t) := \frac{1}{\sqrt{\pi}} \sin(mt), \quad t \in (-\pi, \pi), \quad m \in \mathbb{N},$$

is orthonormal in $L^2((-\pi, \pi))$. By Remark 12.17 and Lemma 12.18, we know that $f_m \rightharpoonup 0$ weakly as $m \to \infty$. What this sequence actually does it that is oscillates faster-and-faster around zero, with zero average[10], see Figure 12.3. However, even the quadratic operation is weakly discontinuous because

$$(f_m)^2 \not\rightharpoonup 0 \quad \text{as } m \to \infty.$$

---

[10]In Chapter 14 we shall see that weak convergence in all L^p spaces can be characterised as convergence of the average over measurable subsets, coupled by a norm bound.

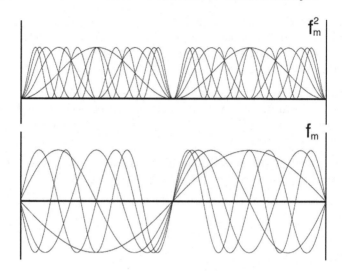

Figure 12.3: At the lower level are the first five terms of the sequence $(f_m)_1^\infty$ and at the higher level are the first five terms of the sequence $(f_m^2)_1^\infty$. The squared sequence oscillates around a positive value, which in effect is its weak limit.

Indeed, for the linear functional $\langle \cdot, \mathbf{1} \rangle \in \left(L^2((-\pi, \pi)) \right)^*$, where $\mathbf{1}$ is the function $\mathbf{1}(t) \equiv 1$, we have

$$\langle (f_m)^2, \mathbf{1} \rangle = \int_{-\pi}^{\pi} f_m^2(t)\, d\mathcal{L}^1(t) = \|f_m\|_{L^2((-\pi,\pi))}^2 = 1.$$

We will examine more examples of weak convergence in Chapter 14 in the setting of the L^p functional spaces.

Proof. (Theorem 12.22) (i) Suppose $x_m \rightharpoonup x$ in X as $m \to \infty$. By the Uniform Boundedness Principle (Banach-Steinhaus Theorem 11.21), it suffices to show that for each $f \in X^*$, the set

$$\{ f(x_m) : m \in \mathbb{N} \} \subseteq \mathbb{R}$$

is bounded. This however follows directly from the fact that $f(x_m) \longrightarrow f(x)$ as $m \to \infty$ because convergent sequences are bounded.

(ii) Suppose again that $x_m \rightharpoonup x$ in X, as $m \to \infty$. We have

$$|f(x)| \leq |f(x - x_m)| + |f(x_m)|$$
$$\leq |f(x - x_m)| + \|f\|_* \|x_m\|$$

and hence, in the limit

$$|f(x)| \leq \liminf_{m \to \infty} \left(|f(x - x_m)| + \|f\|_* \|x_m\| \right)$$
$$\leq \|f\|_* \liminf_{m \to \infty} \|x_m\|.$$

By utilising the representation formula (♮) for the norm of X (recall Definition 11.47) and the estimate just obtained, we have

$$\|x\| = \sup \left\{ |f(x)| : f \in \mathbb{B}^{X^*} \right\}$$
$$\leq \sup \left\{ \|f\|_* \liminf_{m \to \infty} \|x_m\| : f \in \mathbb{B}^{X^*} \right\}$$
$$\leq \liminf_{m \to \infty} \|x_m\|.$$

Hence, the norm of X is indeed weakly LSC.

(iii) Suppose that $f_m \longrightarrow f$ strongly in X^* and $x_m \rightharpoonup x$ weakly in X, both as $m \to \infty$. By the triangle inequality, we have

$$\left| f_m(x_m) - f(x) \right| \leq \|f_m - f\|_* \|x_m\| + \left| f(x_m - x) \right|.$$

Note that $|f(x_m - x)| \longrightarrow 0$ as $m \to \infty$ and by item (i) we have

$$\sup_{m \in \mathbb{N}} \|x_m\| < \infty.$$

Also, we have $\|f_m - f\|_* \longrightarrow 0$ as $m \to \infty$. In conclusion, we see that

$$\limsup_{m \to \infty} \left| f_m(x_m) - f(x) \right| \leq \left(\sup_{m \in \mathbb{N}} \|x_m\| \right) \limsup_{m \to \infty} \|f_m - f\|_*$$
$$+ \limsup_{m \to \infty} |f(x_m - x)|$$
$$= 0.$$

Thus, $f_m(x_m) \longrightarrow f(x)$, as $m \to \infty$. The theorem has been established. $\quad\square$

Remark 12.24. At this point the reader might be wondering what is so special about the norm of X that makes it one-sided continuous from below. This answer is: nothing but its **convexity** property. We recall from Definition 11.39 that a function $F : X \longrightarrow \mathbb{R}$ on a vector space X is called **convex** if

$$F\big((1 - \lambda)x + \lambda y\big) \leq (1 - \lambda)F(x) + \lambda F(y),$$

for all $x, y \in X$ and all $\lambda \in [0, 1]$, whilst F is called **concave** if $-F$ is convex. In fact, the splitting of weak continuity to the one-sided notions LSC and USC corresponds to the splitting of the affinity of a function to the one-sided notions of convexity and concavity, as Figure 12.4 illustrates.

The next result, which is a consequence of Mazur's theorem, claims that any convex strongly LSC function is indeed weakly LSC.

Theorem 12.25 (Convex functions are sequentially weakly LSC). *Let $F : X \longrightarrow \mathbb{R}$ be a convex function on the normed space $(X, \|\cdot\|)$, and which is sequentially strongly LSC, in the sense that for any $(x_m)_1^\infty \subseteq X$ and $x \in X$, we have*

$$x_m \longrightarrow x \quad as \ m \to \infty \quad \Longrightarrow \quad F(x) \leq \liminf_{m \to \infty} F(x_m). \qquad (\circledast)$$

Then, F is sequentially weakly LSC on X. That is, for any $(x_m)_1^\infty \subseteq X$ and $x \in X$, we have[11]

$$x_m \rightharpoonup x \quad as \ m \to \infty \quad \Longrightarrow \quad F(x) \leq \liminf_{m \to \infty} F(x_m). \qquad (\circledast\circledast)$$

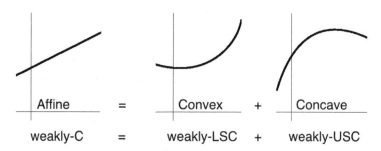

Affine	=	Convex	+	Concave
weakly-C	=	weakly-LSC	+	weakly-USC

Figure 12.4: A graphic and hopefully easy-to-memorise illustration of the connection between weak lower-upper continuity and convexity-concavity.

Proof. Since F is convex, the sub-level set $\{F \leq t\} = F^{-1}((-\infty, t])$ is a convex subset of X for any $t \in \mathbb{R}$ (Figure 12.5). By utilising ($\circledast$), the first step is to show that $\{F \leq t\}$ is (strongly) closed as well. Indeed, fix a number $t \in \mathbb{R}$ such that $\{F \leq t\} \neq \emptyset$ and consider a sequence $(x_m)_1^\infty \subseteq \{F \leq t\}$ such that $x_m \longrightarrow x$ as $m \to \infty$, for some $x \in X$. Then, $F(x_m) \leq t$ for all $m \in \mathbb{N}$ and by ($\circledast$),

$$F(x) \leq \liminf_{m \to \infty} F(x_m) \leq t,$$

yielding that $x \in \{F \leq t\}$, which in turn shows $\{F \leq t\}$ is norm closed. Since $\{F \leq t\}$ is also convex, Mazur's Theorem 12.10 implies it is weakly closed. Hence, $\{F > t\}$ is weakly open. Let $(x_m)_1^\infty$ be a sequence such that $x_m \rightharpoonup x$, as $m \to \infty$. If ($\circledast\circledast$) fails, there exists some $t_0 > 0$ such that

$$F(x) > t_0 > \liminf_{m \to \infty} F(x_m).$$

Note that $\{F > t_0\}$ belongs to $\mathcal{T}^w$ and $x \in \{F > t_0\}$. By Proposition 12.13, $x_m \rightharpoonup x$ as $m \to \infty$ if and only if $x_m \longrightarrow x$ with respect to the topology $\mathcal{T}^w$ as $m \to \infty$. Hence, there exists an index $m_0 \in \mathbb{N}$ such that, for all $m \geq m_0$, we have that $x_m \in \{F > t_0\}$. Therefore,

$$\liminf_{m \to \infty} F(x_m) = \sup_{k \in \mathbb{N}} \left(\inf_{m \geq k} F(x_m) \right)$$
$$\geq \inf_{m \geq m_0} F(x_m)$$
$$\geq t_0.$$

---

[11]Note the "paradoxical" fact that the "strong LSC" property is generally weaker than the "weak LSC" property! This happens because in the respective implications ($\circledast$) and ($\circledast\circledast$) defining these properties, the assumption "$x_m \longrightarrow x$" is stronger than "$x_m \rightharpoonup x$", which in turn makes the "strong LSC" statement weaker! On the other hand, every weakly LSC function is automatically a strongly LSC one, for the exact same reason.

This is a contradiction, and hence (⊛⊛) ensues. □

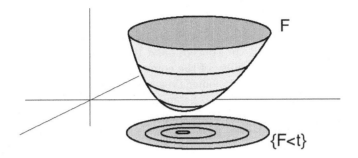

Figure 12.5: An illustration of the sub-level sets of a convex function on $\mathbb{R}^2$. The sub-level sets are convex as a direct consequence of the definition of convexity itself.

We note that Theorem 12.25 is particularly important for existence theorems, especially in the Calculus of Variations and the theory of nonlinear Partial Differential Equations. We close this chapter by proving that the weak topology on any infinite-dimensional Banach space is not metrisable, by showing that it does not satisfy the first axiom of countability.

Proposition 12.26 (The weak topology is not metrisable). *The weak topology of any infinite-dimensional Banach space does not satisfy the first axiom of countability (Definition 4.26). Hence, in view of Remark 4.27, the weak topology is not metrisable.*

Proof. Let $(X, \|\cdot\|)$ be a Banach space with $\dim(X) = \infty$ and suppose for the sake of contradiction that the weak topology $\mathcal{T}^w$ is first countable. Since $\mathcal{T}^w$ is translation invariant, it suffices to consider only weakly open neighbourhoods of the origin.

Step 1. Suppose that $(\mathcal{U}_n)_1^\infty$ is a neighbourhood basis of the origin such that $\mathcal{U}_1 \supseteq \mathcal{U}_2 \supseteq \cdots \supseteq \{0\}$, and for any $\mathcal{U} \in \mathcal{T}^w$ with $\mathcal{U} \ni 0$, there exists $n \in \mathbb{N}$ such that $\mathcal{U}_n \subseteq \mathcal{U}$. Since $\dim(X) = \infty$, by Theorem 12.6 and Remark 12.7, every $\mathcal{U}_n$ contains an one-dimensional subspace $\operatorname{span}[\{x_n\}] \subseteq \mathcal{U}_n$ for some $x_n \in X \setminus \{0\}$. We set $y_n := n x_n / \|x_n\|$ and consider the sequence $(y_n)_1^\infty$, noting that $y_n \in \mathcal{U}_n$.

Step 2. Fix now any $f \in X^*$ and $\varepsilon > 0$. Then, the set $\{|f| < \varepsilon\} = f^{-1}((-\varepsilon, \varepsilon))$ is a weakly open neighbourhood of the origin and hence there exists $n_\varepsilon \in \mathbb{N}$ such that $\mathcal{U}_{n_\varepsilon} \subseteq \{|f| < \varepsilon\}$. Since $y_n \in \mathcal{U}_n$, we have $|f(y_n)| < \varepsilon$. Note also that since $(\mathcal{U}_n)_1^\infty$ is decreasing, we actually have that $y_n \in \mathcal{U}_n \subseteq \{|f| < \varepsilon\}$ for all $n \geq n_\varepsilon$.

Step 3. By *Step 1*, we see that Theorem 12.22(ii) implies that the sequence $(y_n)_1^\infty$ cannot be weakly convergent because $\|y_n\| = n \longrightarrow \infty$ as $n \to \infty$. On the other hand, in *Step 2* we have shown that for any $\varepsilon > 0$ and any $f \in X^*$, there exists $n_\varepsilon \in \mathbb{N}$ such that $|f(y_n)| < \varepsilon$ for all $n \geq n_\varepsilon$. Thus, $y_n \longrightarrow 0$ as $n \to \infty$. This contradiction establishes the conclusion. $\qquad\square$

Our investigations in this chapter regarding the necessity to weaken the norm topology were motivated by the lack of strong compactness in infinite-dimensional spaces. However, the concepts and results we have presented so far are nowhere near the solvability of the compactness problem. Instead, we have encountered new peculiar phenomena regarding the weakened topological and analytical concepts. The problem of compactness still persists even for the weak topology simply because for most Banach spaces *it is not weak enough* and it has to be weakened even more. As we have promised, this will be achieved in the next chapter.

12.7 Exercises

Exercise 12.1. Let $(X, \|\cdot\|)$ be a Banach space and fix also $N \in \mathbb{N}$. We define a function $F : X \longrightarrow \mathbb{R}$ by setting

$$F(x) := a_0\|x\| + a_2\|x\|^2 + a_4\|x\|^4 + \cdots + a_{2N}\|x\|^{2N}$$

where $a_0, a_2, \ldots, a_{2N} \geq 0$. Is F sequentially weakly lower semi-continuous on X? That is, does it hold that

$$F(x) \leq \liminf_{n \to \infty} F(x_n)$$

whenever $x_n \longrightarrow x$, as $n \to \infty$?

Exercise 12.2. Given a sequence $\{x_m\}_1^\infty$ in a Banach space $(X, \|\cdot\|)$, construct a *separable* Banach space Y such that $\{x_m\}_1^\infty \subseteq Y \subseteq X$.

Exercise 12.3. Let $x, y \in X$ be two linearly independent unit vectors of a Banach space $(X, \|\cdot\|)$, i.e. $\|x\| = \|y\| = 1$. Consider the vector space

$$Y := \mathrm{span}[\{x, y\}] = \{z \in X : z = ax + by, \ a, b \in \mathbb{R}\}.$$

Show that the weak topology of the Banach space $(Y, \|\cdot\|)$ coincides with its norm topology.

Exercise 12.4. Let $(X, \mathcal{T})$ be a compact topological space. Consider the Banach space $C(X) = C(X, \mathbb{R})$, endowed with the sup-norm

$$\|f\|_\infty := \sup_{x \in X} |f(x)|, \quad f \in C(X).$$

Let $\Phi : [0, \infty) \longrightarrow [0, \infty)$ be a strictly increasing continuous function. Consider the function

$$E : C(X) \longrightarrow \mathbb{R}, \qquad E(f) := \sup_{x \in X} \Phi(|f(x)|).$$

Show that E is sequentially weakly lower-semicontinuous, namely if $f_m \rightharpoonup f$ as $m \to \infty$ in $C(K)$, then

$$E(f) \leq \liminf_{m \to \infty} E(f_m).$$

Exercise 12.5. Along the lines of Exercise 12.4, let $(X, \mathfrak{M}, \mu)$ be a measure space. Consider the Banach space $L^\infty(X, \mu, \mathbb{F})$, endowed with the essential supremum norm $\|\cdot\|_{L^\infty(X, \mu, \mathbb{F})}$. Let $\Phi : [0, \infty) \longrightarrow [0, \infty)$ be a strictly increasing continuous function. Consider the function

$$E_\infty : L^\infty(X, \mu, \mathbb{F}) \longrightarrow \mathbb{R}, \qquad E_\infty(f) := \operatorname*{ess\,sup}_{x \in X} \Phi(|f(x)|).$$

Show that E_∞ is sequentially weakly lower-semicontinuous, namely if $f_m \rightharpoonup f$ as $m \to \infty$ in $L^\infty(X, \mu, \mathbb{F})$, then

$$E_\infty(f) \leq \liminf_{m \to \infty} E_\infty(f_m).$$

Exercise 12.6. Similarly to Exercise 12.5, let again $(X, \mathfrak{M}, \mu)$ be a measure space. Consider the Banach space $L^p(X, \mu, \mathbb{F})$, endowed with the L^p norm, where $1 \leq p < \infty$. Let $\Phi : [0, \infty) \longrightarrow [0, \infty)$ be a strictly increasing continuous function. We set

$$E_p(f) := \int_X \Phi(|f(x)|)^p \, d\mu(x), \qquad f \in L^p(X, \mu, \mathbb{F}).$$

Can we infer the same conclusion as in Exercise 12.5?

[Hint: E_p maps $L^p(X, \mu, \mathbb{F}))$ into $(-\infty, \infty]$, not $\mathbb{R}$.]

Exercise 12.7. Let $(x_n)_1^\infty$ be a sequence in a Banach space $(X, \|\cdot\|)$ with dual space X^*. Let also $x \in X$. Show that $x_n \rightharpoonup x$ as $n \to \infty$ if and only if $(x_n)_1^\infty$ is bounded and the set

$$\left\{ f \in X^* : f(x_n) \longrightarrow f(x) \text{ as } n \to \infty \right\}$$

is dense in X^*.

Exercise 12.8. Let $(X, \|\cdot\|)$ be a Banach space. Show that if $(x_n)_1^\infty$ is a norm Cauchy sequence which is weakly null, that is $x_n \rightharpoonup 0$ as $n \to \infty$, then it is norm null, that is $x_n \longrightarrow 0$ as $n \to \infty$.

Exercise 12.9. Show directly that if X is an infinite-dimensional normed space, then $0 \in \overline{\mathbb{S}^X}^w$.

Exercise 12.10. Show that in the space ℓ^∞, the set $S := \{e_i : i \in \mathbb{N}\} \bigcup \{0\}$ (where $e_i = (0, \ldots, 0, 1, 0, \ldots)$ with 1 in the ith position) is weakly sequentially compact, but not norm compact.

Exercise 12.11. Consider the (real version of the) space ℓ^2 and the standard orthonormal basis $e_n = (0, \ldots, 0, 1, 0, \ldots)$ with 1 in the nth position. Show that

$$0 \in \overline{\{\sqrt{n}e_n : n \in \mathbb{N}\}}^w$$

but no subsequence of $(\sqrt{n}e_n)_1^\infty$ converges weakly to $0 \in \ell^2$. Deduce that the weak topology of ℓ^2 is not metrisable.

Exercise 12.12. Let $\{x\}, (x_n)_1^\infty \subseteq \ell^2$ be such that that $x_n \longrightarrow x$ as $n \to \infty$. Show that there exists a subsequence $(x_{n_k})_1^\infty$ such that the Cesáro means converges **strongly** to x:

$$\frac{x_{n_1} + \cdots + x_{n_k}}{k} \longrightarrow x \quad \text{in } \ell^2 \text{ as } k \to \infty.$$

Exercise 12.13. Let $(X, \|\cdot\|)$ be an infinite-dimensional Banach space. Show that the weak topology does not satisfy the first axiom of countability by utilising that the dual space X^* has no countable (Hamel) basis. Conclude that it is not metrisable.

[Hint: Since the dual space X^* has no countable (Hamel) basis, given any sequence of functionals on X, there exists at least one functional which is not the in the linear span of the sequence.]

Chapter 13

Weak* topologies and compactness

13.1 Weak topologies are not weak enough

In this chapter we continue our study of "weakened" topologies on Banach spaces, with our end goal being the quest for (sequential) compactness results for these topologies.

As we discussed in the previous chapter, the lack of compactness in infinite-dimensional vector spaces has already been highlighted in Theorem 5.25 and Lemma 5.26: *closed and bounded sets in infinite-dimensional Banach spaces are not compact.* In particular, they are not sequentially compact and the Bolzano-Weierstrass property fails, namely, bounded sequences may not have converging subsequences. Compactness, particularly its sequential counterpart, is essential for existence theorems of limiting object which can not be constructed explicitly.

The incentive to introduce weak topologies on Banach spaces stemmed from the perhaps unclear aspiration that by reducing the class of open sets, this would enlarge the class of the compact sets. This weakening of the norm topology moves indeed towards the right direction, but it is not in general sufficient. As we shall prove in this chapter, for *reflexive Banach spaces*, namely for those which coincide with their bidual space, weak compactness of closed bounded sets holds indeed true. For general Banach spaces, there is no remedy and it might happen that closed bounded sets may not have weak compactness properties.

On the other hand, for *dual Banach spaces*, there exists an even weaker topology, known as the **weak* topology** (sometimes referred to as the **ultra weak topology**) which indeed has the desired compactness properties. This ultra weak topology is obtained by removing additional open sets from the weak topology of the dual Banach space. The study of this topology and of its features is the main theme of this chapter. By exploring some extremely interesting infinite-dimensional phenomena, the end goal will be to establish the much-sought sequential compactness theorems for Banach spaces with respect to this topology.

13.2 The weak* topology of a dual Banach space

We now introduce a topology on the dual space X^* of a normed space $(X, \|\cdot\|)$ which is weaker than the weak topology $\mathcal{T}_{X^*}^{\mathrm{w}}$ of the dual space X^*. This topology has the desired compactness properties and is a stepping stone towards ensuring the compactness properties of the weak topology of X itself (under as extra condition). Let us begin by noting that **in this chapter we will be using the same notational and working conventions we used in Chapters 11 and 12**. In particular, we will consider normed spaces only over the field $\mathbb{R}$. Let us recall also from Definition 11.31 that the dual space X^* of bounded linear functionals on X comes equipped with the natural dual norm

$$\|f\|_* = \sup\{|f(x)| : x \in \mathbb{B}^X\} \quad \text{for all } f \in X^*.$$

The bidual space $X^{**} = (X^*)^* = B(B(X,\mathbb{R}),\mathbb{R})$ given in Definition 11.47 is also a Banach space when endowed with the bidual norm

$$\|\phi\|_{**} = \sup\{|\phi(f)| : f \in \mathbb{B}^{X^*}\} \quad \text{for all } \phi \in X^{**}.$$

Moreover, as we have already seen (recall Definition 11.47 and Proposition 11.50), there exists a canonical embedding of X into X^{**} via the evaluation mapping (Definition 11.49), which is an isometry:

$$\widehat{\cdot} : X \longrightarrow X^{**}, \quad x \mapsto \widehat{x}, \quad \widehat{x}(f) := f(x), \quad f \in X^*.$$

The image of the above embedding is denoted by $\widehat{X}$. Let us also recall that the weak topology on X^* is the topology with the fewest possible open sets which makes continuous all linear functionals $z : X^* \longrightarrow \mathbb{R}$ which lie in X^{**}. The weak* topology on X^* is the (even weaker) topology which *makes continuous only* the functionals contained in the subspace $\widehat{X} \subseteq X^{**}$:

Definition 13.1. Let $(X, \|\cdot\|)$ be a normed space, with dual space X^* and bidual space X^{**}. The **weak* topology** $\mathcal{T}^{\mathrm{w}*}$ **on** X^* (symbolised by $\mathcal{T}_{X^*}^{\mathrm{w}*}$ if there is any danger of confusion) is the smallest topology (intersection of all topologies) which contains the family of sets

$$\mathcal{E}^* := \left\{\widehat{x}^{-1}(V) \,\middle|\, \widehat{x} \in \widehat{X} \subseteq X^{**}, \ V \subseteq \mathbb{R} \text{ open}\right\},$$

that is, all inverse images of open sets in $\mathbb{R}$ via functionals in $\widehat{X} \subseteq X^{**}$:

$$\mathcal{T}^{\mathrm{w}*} := \bigcap\left\{\mathcal{T} \,\middle|\, \mathcal{T} \text{ is a topology on } X^* : \ \mathcal{T} \supseteq \mathcal{E}^*\right\}.$$

Equivalently, **the weak* topology of X^* is the weakest (smallest) topology which makes all the bounded linear functionals in $\widehat{X} \subseteq X^{**}$, continuous.** The elements of the weak* topology will be called **weakly* open sets.**

Remark 13.2. In the terminology of Definition 4.51, the weak* topology is a weak topology on the dual space generated by a **smaller** family of functionals. Its uniqueness follows from Lemma 4.44. As in the case of the weak topology, note that all functionals in $\widehat{X} \subseteq X^{**}$ are by definition continuous with respect to the norm. What we do here is to define a new, even weaker topology on X^*, with even fewer open sets, by the additional removal of some open sets from the weak topology $\mathcal{T}^{\mathrm{w}}_{X^*}$ of X^*. This is done in such a way that all $\widehat{x} \in \widehat{X}$ are still continuous with respect to the weak* topology (recall Proposition 4.49).

Similarly to the case of the weak topology, we will follow similar conventions regarding the usage of the modifiers "**weakly*/weak***" as in the case of the weak topology, meaning that e.g. weakly* closed, weak* continuity, etc. they will refer to the topological notion arising with respect to the weak* topology.

Similarly to the weak topology, the next analogue of Proposition 12.3 characterises the weak* topology in a more concrete manner.

Proposition 13.3 (Neighbourhood basis of the weak* topology). *The weak* topology of any dual Banach space X^* consists of $\emptyset$, X^*, and all unions of finite intersections of elements of $\mathcal{E}^*$.*

In addition, weakly* open sets can be characterised by:

$$\mathcal{U} \in \mathcal{T}^{\mathrm{w}*} \iff \left\{ \begin{array}{l} \forall\, f \in \mathcal{U}, \ \exists\, \varepsilon > 0 \ \&\ \exists\, \widehat{x}_1, \ldots, \widehat{x}_N \in \widehat{X} : \\ \displaystyle \bigcap_{k=1}^{N} \widehat{x}_k^{-1}\Big(\big(\widehat{x}_k(f) - \varepsilon,\ \widehat{x}_k(f) + \varepsilon \big) \Big) \subseteq \mathcal{U}. \end{array} \right.$$

$$\iff \left\{ \begin{array}{l} \forall\, f \in \mathcal{U}, \ \exists\, \varepsilon > 0 \ \&\ \exists\, x_1, \ldots, x_N \in X : \\ \Big\{ g \in X^* : \ \big|(g - f)(x_k)\big| < \varepsilon, \ k = 1, \ldots, N \Big\} \subseteq \mathcal{U}. \end{array} \right.$$

Further, any weakly* open set in $\mathcal{T}^{\mathrm{w}*}$ can be written as the arbitrary union of such finite intersections of inverse images.

Proof. The proof is a straightforward consequence of Proposition 4.45 and Remark 4.52. □

We note that, exactly like in the case of the weak topology of a normed space $(X, \|\cdot\|)$ we saw in Chapter 12, it can be shown that the weak* topology of a dual Banach space X^* is **translation invariant**. Moreover:

Remark 13.4 (The weak* topology is Hausdorff). The weak* topology $\mathcal{T}^{\mathrm{w}*}$ on a dual Banach space X^* of a normed space $(X, \|\cdot\|)$ satisfies the Hausdorff axiom. This is actually easier to prove than in the case of the weak topology and does not require the Hahn-Banach theorem. Let $f \neq g$ be two elements in X^*. Then, there exists at least one point $x_0 \in X$ such that $f(x_0) \neq g(x_0)$.

Without loss of generality, we may suppose that $f(x_0) < g(x_0)$. Then, there exists $a \in \mathbb{R}$ with $f(x_0) < a < g(x_0)$. It follows that the sets

$$\mathcal{U}_f := \left\{ h \in X^* : h(x) < a \right\}, \quad \mathcal{U}_g := \left\{ h \in X^* : h(x) > a \right\}$$

are disjoint weakly* open neighbourhoods of f and g, respectively.

13.3   On the nature of weakly* open sets

Similar observations to those made in the previous chapter regarding the nature of weakly* open sets can be made here as well. To begin with, using similar arguments to some used previously, it can be shown that the **weak* topology is generally strictly weaker than the weak topology of X^***:

Remark 13.5. Given a normed space $(X, \| \cdot \|)$, the three different topologies which can be defined on its dual space X^* (weak*, weak, norm) satisfy the inclusions

$$\mathcal{T}^{\mathrm{w}*} \subseteq \mathcal{T}^{\mathrm{w}} \subseteq \mathcal{T}^{\| \cdot \|*}$$

and the inclusions are in general strict. In particular, by the results of Chapter 12, it follows that $\mathcal{T}^{\mathrm{w}} = \mathcal{T}^{\| \cdot \|*}$ if and only if $\dim(X^*) < \infty$. The relation between $\mathcal{T}^{\mathrm{w}*}$ and $\mathcal{T}^{\mathrm{w}}$ is slightly trickier if $\dim(X^*) = \infty$. We will see later that the property of the pre-dual space X being (or not) reflexive is relevant.

Further, as we shall show in this section, in general **the Mazur Theorem 12.10 holds only for the weak topology and has no counterpart for the weak* topology** on dual spaces.

Remark 13.6 (Types of closed convex sets in dual Banach spaces). In fact, if a Banach space $(X, \| \cdot \|)$ is not reflexive (i.e. if the evaluation map $\widehat{} :$ $X \longrightarrow X^{**}$ is not surjective), then the weak* topology on X^* is strictly weaker than the weak topology even when restricted to convex sets and **there exist convex weakly closed sets in X^* which are not weakly* closed.** For example, if $\phi \in X^{**} \setminus \widehat{X}$, then the hyperplane in X^* given by the nullspace

$$\mathrm{N}(\phi) = \left\{ f \in X^* : \phi(f) = 0 \right\}$$

is **weakly* closed in X^*, but not weakly closed in X^*!** This is a consequence of the result below. Actually, closed convex sets in dual spaces can be classified into two categories:

- Convex sets in X^* which are norm closed, or equivalently weakly closed (by Mazur's theorem).

- Convex sets in X^* which are weakly* closed.

Theorem 13.7 and Proposition 13.9 that follow confirm the assertions of Remark 13.6. We begin by showing that the class of **linear** functionals $\phi : X^* \longrightarrow \mathbb{R}$ (i.e., the elements $\phi \in (X^*)'$ of the algebraic dual space of the topological dual space of X) which are weakly* continuous is precisely the class of the elements $\phi \in \widehat{X} \subseteq X^{**}$.

Theorem 13.7 (Representation of weakly* continuous linear functionals). *Let $(X, \| \cdot \|)$ be a Banach space with dual space X^*, bidual space X^{**} and evaluation mapping $\widehat{\cdot} : X \longrightarrow X^{**}$. Let further $\phi : X^* \longrightarrow \mathbb{R}$ be a linear functional on X^*. Then, the following statements are equivalent:*

(i) *The functional ϕ is weakly* continuous on X^*.*

(ii) *The functional ϕ belongs to $\widehat{X} \subseteq X^{**}$, i.e. $\phi = \widehat{x}$ for some $x \in X$.*

Namely, the **only weakly* continuous functionals on X^* are precisely the elements of $\widehat{X} \subseteq X^{**}$** *(which are norm continuous too).*

The proof of Theorem 13.7 is based on the next algebraic lemma.

Lemma 13.8. *Let W be a real vector space and $\phi_0, \phi_1, \ldots, \phi_N$ linear functionals, i.e. $\{\phi_0, \ldots, \phi_N\} \subseteq W'$ (the algebraic dual space). If*

$$\bigcap_{i=1}^{N} \{\phi_i = 0\} \subseteq \{\phi_0 = 0\},$$

then, there exist $\{c_1, \ldots, c_N\} \subseteq \mathbb{R}$ such that

$$\phi_0 = \sum_{i=1}^{N} c_i \, \phi_i.$$

Proof of Lemma 13.8. Consider the linear mapping given by

$$\Phi : W \longrightarrow \mathbb{R}^{N+1} , \quad \Phi(w) := \big(\phi_0(w), \ldots, \phi_N(w)\big).$$

By standard linear algebra, the assumption on the kernels (nullspaces) of the functionals implies that the vector $e_1 := (1, 0, \ldots, 0)$ of $\mathbb{R}^{N+1}$ is not contained in the image $\Phi(W) \subseteq \mathbb{R}^{N+1}$ of Φ. Since $\Phi(W)$ is a convex subset of $\mathbb{R}^{N+1}$ (in fact a vector subspace), there exists an affine hyperplane of $\mathbb{R}^{N+1}$ separating $\Phi(W)$ from $\{e_1\}$. This means that there exist $\{a_0, \ldots, a_N\} \subseteq \mathbb{R}$ and $b \in \mathbb{R}$ such that

$$a_0 < b < a_0 \, \phi_0(w) + \sum_{i=1}^{N} a_i \, \phi_i(w) \quad \text{for all } w \in W.$$

Hence, we have

$$a_0 \, \phi_0(w) + \sum_{i=1}^{N} a_i \, \phi_i(w) = 0 \quad \text{for all } w \in W,$$

and we conclude by setting $c_i := -a_i/a_0$, because $a_0 \neq 0$. $\qquad\square$

Now we may prove Theorem 13.7.

Proof. (i) $\Longrightarrow$ (ii): If $\phi : X^* \longrightarrow \mathbb{R}$ is weakly* continuous, then for any $\varepsilon > 0$, there exists a weakly* open set $\mathcal{U}_0$ containing the origin of X^*, such that

$$|\phi(f)| < \varepsilon, \quad f \in \mathcal{U}_0.$$

Moreover, by Proposition 13.3, we may assume $\mathcal{U}_0$ has the form

$$\mathcal{U}_0 = \Big\{ f \in X^* : \big|f(x_k)\big| < \delta, \ k = 1, \dots, N \Big\},$$

for some (small enough) $\delta \in (0, \varepsilon)$ and elements $\{x_1, \dots, x_N\} \subseteq X$. It follows that $\phi(f) = 0$ when $f(x_1) = \cdots f(x_N) = 0$. Hence,

$$\bigcap_{i=1}^{N} \{\widehat{x}_i = 0\} \subseteq \{\phi = 0\} \quad \text{in } X^*$$

and by applying Lemma 13.8, we obtain that there exist $\{c_1, \dots, c_N\} \subseteq \mathbb{R}$ such that

$$\phi = \sum_{i=1}^{N} c_i \, \widehat{x}_i, \quad \text{on } X^*.$$

Since the evaluation mapping $\widehat{} : X \longrightarrow X^{**}$ is linear, the above gives

$$\phi = \Big(\sum_{i=1}^{N} c_i \, x_i \Big)^{\widehat{}}, \quad \text{on } X^*,$$

as claimed.

(ii) $\Longrightarrow$ (i): This is immediate from the definitions. $\qquad\square$

The above result has the next consequence:

Proposition 13.9 (Weakly* closed hyperplanes in dual Banach spaces). *Let* $(X, \|\cdot\|)$ *be a Banach space with dual space* X^*, *bidual space* X^{**} *and evaluation mapping* $\widehat{} : X \longrightarrow X^{**}$. *Suppose that* $\phi \in (X^*)'$, *namely* $\phi : X^* \longrightarrow \mathbb{R}$ *is a linear functional (possibly discontinuous). If the hyperplane*

$$\{\phi = 0\} \subseteq X^*$$

is a weakly closed subspace of* X^*, *then* $\phi \in \widehat{X} \subseteq X^{**}$. *Namely, there exists* $x \in X$ *such that* $\phi = \widehat{x}$.

Proof. It suffices to show that ϕ is weakly* continuous on X^* and then the conclusion will follow directly from Theorem 13.7. By assumption, the set $\{\phi \neq 0\}$ is weakly* open in X^*, consisting of the disjoint union of weakly* open sets $\{\phi > 0\}$ and $\{\phi < 0\}$. Fix $\varepsilon > 0$ and some $f \in \{\phi \neq 0\}$ such that

$|\phi(f)| < \varepsilon$. Then, there exists a weakly* open neighbourhood $\mathcal{U}_f$ of f inside the set $\{\phi \neq 0\}$, that is

$$\mathcal{U}_f \ni f, \quad \mathcal{U}_f \subseteq \{\phi \neq 0\}.$$

By Proposition 13.3, the weak* neighbourhood $\mathcal{U}_f$ can be chosen to be of the form

$$\mathcal{U}_f = \left\{ g \in X^* : \; |(g - f)(x_k)| < \delta, \; k = 1, \ldots, N \right\},$$

for some $\delta > 0$ and some $\{x_1, \ldots, x_N\} \subseteq X$. Note that the set $\mathcal{U}_f$ is convex in X^* as the intersection of convex sets[1]. Consider the image set $\phi(\mathcal{U}_f)$ of the convex set $\mathcal{U}_f$ under the linear function $\phi : X^* \longrightarrow \mathbb{R}$. Since $\phi(\mathcal{U}_f)$ is a convex subset[2] of $\mathbb{R}$, namely an interval, we have the next dichotomy:

$$\text{either} \quad \phi(\mathcal{U}_f) \subseteq (0, \infty), \quad \text{or} \quad \phi(\mathcal{U}_f) \subseteq (-\infty, 0).$$

Suppose first $\phi(\mathcal{U}_f) \subseteq (-\infty, 0)$, which means that $\phi(g) < 0$ for any $g \in \mathcal{U}_f$. Then, the translated weakly* open set

$$\mathcal{U}_0 := -f + \mathcal{U}_f = \left\{ h \in X^* : \; |h(x_k)| < \delta, \; k = 1, \ldots, N \right\}$$

is a weak* neighbourhood of the origin. Since any $h \in \mathcal{U}_0$ can be written as $h = g - f$ for some $g \in \mathcal{U}_f$, we have

$$\phi(h) = \phi(g) + \phi(-f) < \phi(-f) \leq |\phi(f)| < \varepsilon.$$

On the other hand, by the form of $\mathcal{U}_0$ we have $\mathcal{U}_0 = -\mathcal{U}_0$. This means that $h \in \mathcal{U}_0$ if and only if $-h \in \mathcal{U}_0$. Thus, by replacing h by $-h$ in the last inequality, we have

$$-\phi(h) = \phi(-h) < \varepsilon.$$

In conclusion, for any $\varepsilon > 0$, there exists a weakly* open neighbourhood $\mathcal{U}_0$ of the origin such that $|\phi(h)| < \varepsilon$, for all $h \in \mathcal{U}_0$. The same conclusion can be reached if instead we start from $\phi(\mathcal{U}_f) \subseteq (0, \infty)$. In either case, we see that ϕ is weakly* continuous at the origin of X^*. By invoking the linearity of ϕ and the translation invariance of the weak* topology, it follows ϕ is continuous on X^*. The proposition ensues. $\qquad \square$

We refrain from delving any further into the study of the weak* topology and instead we turn our attention to the associated notion of weak* convergence in dual Banach spaces.

---

[1] In any vector space, the intersection of convex sets is convex. This is a general fact regarding convex sets whose proof is a simple exercise, recall also Figure 12.1 in Chapter 12.

[2] It is a general fact that the image of a convex subset $C \subseteq Z$ under a linear mapping $F : Z \longrightarrow W$ between vector spaces is a convex subset of W. Indeed, the linearity of F imples

$$(1 - t)F(x) + tF(y) = F\big((1 - t)x + ty\big),$$

which holds true for all $x, y \in C$ and $t \in [0, 1]$. It follows that for any $F(x), F(y) \in F(C) \subseteq W$, their convex combination $(1 - t)F(x) + tF(y)$ is again an element of $F(C)$, whence the convexity ensues.

13.4   Weak* convergence and its properties

In this section we define and study the relevant notion of (sequential) convergence emerging from the weak* topology. By analogy, Definition 12.11 and Proposition 12.13 become as follows:

Definition 13.10. Let $(f_m)_1^\infty \subseteq X^*$ be a sequence in a dual Banach space X^*. We say that $(f_m)_1^\infty$ **converges weakly*** to $f \in X^*$ and write

$$f_m \xrightarrow{\ *\ } f \quad \text{as} \ \ m \to \infty,$$

if for all $x \in X$, we have $\widehat{x}(f_m) \longrightarrow \widehat{x}(f)$ as $m \to \infty$. Namely, this means that

$$f_m(x) \longrightarrow f(x) \ \ \text{as} \ m \to \infty, \quad \text{for any} \ x \in X.$$

Proposition 13.11. *Let X^* be a dual Banach space. The next statements are equivalent:*

(i) $f_m \xrightarrow{\ *\ } f$ *weakly* as $m \to \infty$, in X^* (Definition 13.10).*

(ii) $f_m \longrightarrow f$ *as $m \to \infty$, with respect to the weak* topology, namely*

for any $\mathcal{U}_f \in \mathcal{T}^{w*}$, exists $m(\mathcal{U}_f) \in \mathbb{N}$: $f_m \in \mathcal{U}_f$, for all $m \geq m(\mathcal{U}_f)$.

Proof. We refrain from providing the details because they follow the exact same lines to those of Proposition 12.13 and therefore are left as an exercise for the reader. □

The next lemma confirms the plausible fact that weak* limits, if they exist, must be unique. As in the case of the weak topology, this can be deduced from the Hausdorff property we established earlier, but instead we prefer to prove it directly.

Lemma 13.12 (Uniqueness of weak* limits). *Let $(f_m)_1^\infty$ be a sequence in a dual Banach space X^*. If there exists $f \in X^*$ such that $f_m \xrightarrow{\ *\ } f$ as $m \to \infty$, then the weak* limit f is unique.*

Proof. If there exists $g \in X^*$ such that $f_m \xrightarrow{\ *\ } g$ as $m \to \infty$ as well, then, for any $x \in X$ we have $f_m(x) \longrightarrow f(x)$ and $f_m(x) \longrightarrow g(x)$, as $m \to \infty$. Therefore, $f(x) = g(x)$ for any $x \in X$, by the uniqueness of limits on $\mathbb{R}$. Hence, $f = g$. □

Next we record the obvious fact that weak convergence implies weak* convergence:

Lemma 13.13 (Weak implies weak* convergence). *Let X^* be a dual Banach space, $(f_m)_1^\infty$ a sequence in it and $f \in X^*$. If we have $f_m \longrightarrow f$ weakly in X^* as $m \to \infty$, then $f_m \xrightarrow{\ *\ } f$ weakly* in X^* as $m \to \infty$.*

By analogy with the case of sequential weak lower semi-continuity of Definition 12.21 on Banach spaces, one might define the one-sided notion of sequential weak* lower semi-continuity on dual Banach spaces as follows:

Definition 13.14. Suppose that $\phi : X^* \longrightarrow \mathbb{R}$ is a function on a dual Banach space X^*. We say that ϕ **is sequentially weakly* lower semi-continuous (abbreviated as LSC)**, if for any sequence $(f_m)_1^\infty \subseteq X^*$ and any $f \in X^*$ such that

$$f_m \xrightarrow{*} f \quad \text{weakly}^* \text{ as } m \to \infty,$$

we have

$$\phi(f) \leq \liminf_{m \to \infty} \phi(f_m).$$

We may now present the following weak* counterpart of Theorem 12.22:

Theorem 13.15 (Weak* LSC, weak* boundedness and weak*-strong continuity of the duality pairing). *Let $(X, \|\cdot\|)$ be a Banach space with dual space $(X^*, \|\cdot\|_*)$. Then, the following statements hold true:*

(i) **Weakly* convergent sequences in X^* are norm bounded**; *that is, if $f_m \xrightarrow{*} f$ weakly as $m \to \infty$, then $(\|f_m\|_*)_1^\infty$ is bounded in $\mathbb{R}$.*

(ii) **The norm $\|\cdot\|_*$ of the space X^* is sequentially weakly* LSC**; *that is, if $f_m \longrightarrow f$ weakly as $m \to \infty$, then*

$$\|f\|_* \leq \liminf_{m \to \infty} \|f_m\|_*.$$

(iii) **The duality pairing between X^* and X** *given by*

$$X^* \times X \longrightarrow \mathbb{R}, \quad (f, x) \mapsto f(x),$$

is **weakly*-strong continuous**; *that is, if we have $f_m \xrightarrow{*} f$ weakly* in X^* and $x_m \longrightarrow x$ strongly in X, then*

$$f_m(x_m) \longrightarrow f(x), \quad \text{as } m \to \infty.$$

Proof. The proof of Theorem 13.15 follows the exact same lines as those of Theorem 12.22 and therefore it is omitted (see the exercises). □

Remark 13.16 (Quadratic functionals need not be weakly* continuous). Similarly to the case of the weak topology where we saw that quadratic functionals need not be weakly continuous in infinite dimensions (Example 12.23), essentially the same example can be utilised here to show that weak* continuity need not hold either. We postpone a detailed discussion of this matter until Chapter 14, when we will have first identified the dual spaces in the general class of L^p spaces over a general measure space.

In the next section we finally establish the sequential compactness of closed and bounded sets in dual Banach spaces with respect to the weak* topology.

13.5   Weak* compactness in dual Banach spaces

We are now ready to prove our much sought-after first compactness result for weakened topologies on Banach spaces.

Theorem 13.17 (Sequential Alaoglu-Banach theorem). *Let $(X, \| \cdot \|)$ be a separable Banach space. Then, the **unit closed ball of the dual space** X^* is **sequentially weakly* compact**[3], that is with respect to $\mathcal{T}^{w*}$.*

Equivalently, for any bounded sequence $(f_m)_1^\infty$ in X^*, there exists a subsequence $(f_{m_k})_1^\infty \subseteq (f_m)_1^\infty$ and an $f \in X^*$ such that $f_{m_k} \overset{*}{\rightharpoonup} f$, as $k \to \infty$.

Let us recall that a normed space $(X, \| \cdot \|)$ is called separable when it contains a dense sequence (in the norm topology).

Proof. Let $D := \{x_k : k \in \mathbb{N}\}$ be a dense sequence of points in X. Suppose that $(f_m)_1^\infty \subseteq X^*$ is a bounded sequence of functionals and set

$$M := \sup_{m \in \mathbb{N}} \|f_m\|_* < \infty.$$

Let $Z \subseteq X$ be the subspace generated by D, that is $Z := \mathrm{span}[D]$. This means

$$Z = \left\{ z = \sum_{\alpha \in A} \lambda_a x_{i_\alpha} \ \middle| \ (\lambda_\alpha)_{\alpha \in A} \subseteq \mathbb{R}, \ (x_{i_\alpha})_{\alpha \in A} \subseteq D, \ \sharp(A) < \infty \right\}.$$

Here "$\sharp$" symbolises the counting measure and thus the statement "$\sharp(A) < \infty$" abbreviates that A is a finite set. For each fixed $x \in X$, we have

$$|f_m(x)| \leq \|f_m\|_* \|x\| \leq M \|x\|.$$

Thus, the sequence $(f_m(x_1))_1^\infty$ is bounded in $\mathbb{R}$. By the Bolzano-Weierstrass Theorem 5.3, there exists a subsequence $(f_{1_m}(x_1))_1^\infty$ and an $a_1 \in \mathbb{R}$ such that

$$f_{1_m}(x_1) \longrightarrow a_1, \quad \text{as } m \to \infty.$$

---

[3]We note that the closed unit ball of X^* is compact even if the pre-dual space $(X, \| \cdot \|)$ is *not* separable. However, *the closed unit ball may not be sequentially compact in this case.* This more general result is most commonly referred to as the "Alaoglu-Banach theorem". However, compactness (as opposed to sequential compactness, since the notions are not in general equivalent in topological spaces) is rarely useful in most applications because bounded sequences (that can be constructed recursively) may not have any converging subsequences! This strange nature of non-sequential compactness is reflected on the fact that the general theorem cannot be proved constructively, as we do below, but instead its proof relies on a separate axiom which we discussed in Chapter 1, the Axiom of Choice. The proof of the general result is based on another statement which is equivalent to the Axiom of Choice and the Zorn Lemma, known as the **Tychonoff theorem**. The latter claims that an arbitrary Cartesian product of compact topological spaces is compact when equipped with the so-called product topology. For the above reasons we refrain from stating and proving this more abstract and less useful result. For details see e.g. [4, 15, 16].

By passing to a further subsequence $(f_{2_m})_1^\infty \subseteq (f_{1_m})_1^\infty$, there exists an $a_2 \in \mathbb{R}$ such that

$$f_{2_m}(x_j) \longrightarrow a_j , \quad \text{as } m \to \infty, \quad j = 1, 2.$$

By induction, for each $k \in \mathbb{N}$ there exists subsequences

$$(f_{k_m})_1^\infty \subseteq (f_{(k-1)_m})_1^\infty \subseteq \cdots \subseteq (f_{1_m}(x_1))_1^\infty$$

and numbers $a_1, \ldots, a_k$ in $\mathbb{R}$ such that

$$f_{k_m}(x_j) \longrightarrow a_j , \quad \text{as } m \to \infty, \quad j = 1, 2, \ldots, k.$$

We consider the **diagonal sequence**[4] defined by:

$$f^m := f_{m_m}, \quad m \in \mathbb{N}.$$

This sequence has as its mth term the mth term of the mth subsequence. (It is worth trying to convince yourself that this construction produces a proper subsequence of the original sequence!) Then,

$$f^m(x_k) \longrightarrow a_k \quad \text{as } m \to \infty, \quad \text{for all } k \in \mathbb{N}.$$

We now define a candidate limit functional f on X. This will be done step-wise: first we define it on the dense set D, then on its linear span Z and finally on its closure which is the entire space X.

Step 1. We define $f : D \longrightarrow \mathbb{R}$ by setting

$$f(x_k) := \lim_{m \to \infty} f^m(x_k) = a_k,$$

Step 2. We extend f as a function $f : Z = \text{span}[D] \longrightarrow \mathbb{R}$ by setting

$$f(z) := \sum_{\alpha \in A} \lambda_\alpha \, f(x_{i_\alpha})$$

when

$$z = \sum_{a \in A} \lambda_\alpha \, x_{i_\alpha}, \quad (\lambda_\alpha)_{\alpha \in A} \subseteq \mathbb{R}, \ (\lambda_\alpha)_{\alpha \in A} \subseteq D, \ \sharp(A) < \infty.$$

We need to show that f is well defined on Z, regardless the possibly different representations that the point $x \in Z$ might have. Namely, we need to show that if

$$\sum_{\alpha \in A} \lambda_\alpha \, x_{i_\alpha} = \sum_{\beta \in B} \mu_\beta \, x_{i_\beta}$$

for some coefficients $(\lambda_\alpha)_{\alpha \in A}$, $(\mu_\beta)_{\beta \in B}$ and vectors $(x_{i_\alpha})_{\alpha \in A}$, $(x_{i_\beta})_{\beta \in B}$, then,

$$f\left(\sum_{\alpha \in A} \lambda_\alpha \, x_{i_\alpha} \right) = f\left(\sum_{\beta \in B} \mu_\beta \, x_{i_\beta} \right).$$

---

[4]This is commonly referred to as Cantor's diagonal method of extracting a subsequence.

Indeed, we have

$$f\left(\sum_{\alpha \in A} \lambda_\alpha x_{i_\alpha}\right) = \sum_{\alpha \in A} \lambda_\alpha \lim_{m \to \infty} f^m(x_{i_\alpha}) = \lim_{m \to \infty}\left(\sum_{a \in A} \lambda_a f^m(x_{i_a})\right).$$

Since each f^m is linear, the above calculation gives

$$f\left(\sum_{\alpha \in A} \lambda_\alpha x_{i_\alpha}\right) = \lim_{m \to \infty} f^m\left(\sum_{a \in A} \lambda_a x_{i_a}\right)$$

$$= \lim_{m \to \infty} f_m\left(\sum_{\beta \in B} \mu_\beta x_{i_\beta}\right)$$

$$= \lim_{m \to \infty}\left(\sum_{\beta \in B} \mu_\beta f^m(x_{i_\beta})\right).$$

Since

$$\lim_{m \to \infty}\left(\sum_{\beta \in B} \mu_\beta f^m(x_{i_\beta})\right) = f\left(\sum_{\beta \in B} \mu_\beta x_{i_\beta}\right)$$

it ensues that f is indeed well defined on Z. Further, a trivial modification of the above argument shows that f is also linear on Z.

Step 3. Now we show that f can be extended to a continuous linear functional on X. Since $Z = \mathrm{span}[D]$ is a vector subspace of X containing the dense set D, we may define $f : X \longrightarrow \mathbb{R}$ by setting

$$f(x) := \lim_{k \to \infty} f(x_{i_k}), \quad \text{when} \quad x = \lim_{k \to \infty} x_{i_k}, \ (x_{i_k})_1^\infty \subseteq D.$$

The definition is actually independent of the choice of approximating sequence, and f is continuous on X. In order to show these facts, note that by the uniform boundedness of the functionals $(f^m)_1^\infty$, we have

$$\left|f^m(x_k) - f^m(x_l)\right| \le M \|x_k - x_l\|, \quad x_k, x_l \in D.$$

Hence, by letting $m \to \infty$ for fixed $k, l \in \mathbb{N}$, we see that

$$\left|f(x_k) - f(x_l)\right| \le M \|x_k - x_l\|, \quad x_k, x_l \in D.$$

This shows that f is continuous on D and, by the linearity of f on $Z = \mathrm{span}[D]$, it follows that actually f is continuous on Z. Moreover, if $(x_{j_k})_1^\infty$ is any other subsequence in D which converges to the same point $x \in X$, by the linearity of f on Z we have

$$\left|f(x_{i_k}) - f(x_{j_k})\right| = \left|f(x_{i_k} - x_{j_k})\right| \le M \|x_{i_k} - x_{j_k}\|,$$

and hence the possibly different values for $f(x)$ actually coincide. The above argument also shows that the continuous extension to X is unique.

Finally, we show that $f^m \xrightarrow{*} f$ as $m \to \infty$. Indeed, we have

$$
\begin{aligned}
\left|f^m(x) - f(x)\right| &\leq \left|f^m(x - x_k)\right| + \left|f(x - x_k)\right| + \left|f^m(x_k) - f(x_k)\right| \\
&\leq \left(\|f^m\|_* + \|f\|_*\right)\|x - x_k\| + \left|f^m(x_k) - f(x_k)\right| \\
&\leq 2M\|x - x_k\| + \left|f^m(x_k) - f(x_k)\right|,
\end{aligned}
$$

for any $x \in X$, $x_k \in D$. Hence, for k fixed we let $m \to \infty$

$$
\limsup_{m \to \infty} \left|f^m(x) - f(x)\right| \leq 2M\|x - x_k\|
$$

and subsequently take $k \to \infty$. Since D is dense, we get $f^m(x) \longrightarrow f(x)$ for any $x \in X$. The theorem ensues. $\qquad\square$

The method of proof of our compactness theorem above establishes in particular that **the weak* topology on a dual space is metrisable on bounded sets, if the pre-dual space is separable**, as the next result confirms.

Proposition 13.18 (Metrisability of the weak* topology on bounded sets). *Let $(X, \|\cdot\|)$ be a **separable** Banach space with dual space X^*. Let $B \subseteq X^*$ be a bounded set. Then, the **restriction** $\mathcal{T}^{w*}|_B$ of the **weak* topology** to the bounded set B is metrisable. A particular metric which induces the weak* topology on B is given by*

$$
d : B \times B \longrightarrow \mathbb{R}, \quad d(f, g) := \sum_{i=1}^{\infty} 2^{-i} \min\left\{\left|f(x_i) - g(x_i)\right|, 1\right\},
$$

where $(x_i)_1^\infty \subseteq X$ is any dense sequence.

Proof. Suppose that $(x_i)_1^\infty$ is a norm dense sequence in X and that d is given by the formula above. We first confirm that d is a metric, making (B, d) a metric space. In view of Exercises 2.18 and 2.20, we have that it is non-negative, symmetric, $d(f, f) = 0$ and satisfies the triangle inequality. Moreover, if $d(f, g) = 0$, then $f = g$ on the dense subset $\{x_i : i \in \mathbb{N}\}$ of X, from where it follows by continuity that $f \equiv g$ on X.

Now we show that the metric d induces the weak* topology on the bounded set $B \subseteq X^*$. Since d is translation invariant, it suffices to assume that $0 \in B$ and show that the metric topology of d and the weak* topology have the same neighbourhood basis elements at the origin of X^*. The respective neighbourhood bases are

$$
\mathbb{B}(0, \rho) := \left\{f \in X^* : \sum_{i=1}^{\infty} 2^{-i} \min\left\{\left|f(x_i)\right|, 1\right\} < \rho\right\},
$$

$$
\mathcal{U}\left(0, \varepsilon, \{y_k\}_1^N\right) := \left\{f \in X^* : \left|f(y_k)\right| < \varepsilon, \ k = 1, \ldots, N\right\},
$$

where the parameters above satisfy $\rho, \varepsilon > 0$ and $\{y_k\}_1^N \subseteq X$.[5] Fix $\rho > 0$ and suppose that $f \in \mathcal{U}(0, \rho/2, \{x_k\}_1^N)$, where $\{x_1, \ldots, x_N\}$ are the first N terms of the given dense sequence. Then, we have that $|f(x_i)| < \rho/2$ for $i = 1, \ldots, N$. This allows us to compute

$$\sum_{i=1}^{\infty} 2^{-i} \min\{|f(x_i)|, 1\} = \sum_{i=1}^{N} 2^{-i} \min\{|f(x_i)|, 1\}$$

$$+ \sum_{i=N+1}^{\infty} 2^{-i} \min\{|f(x_i)|, 1\}$$

$$\leq \sum_{i=1}^{N} 2^{-i} |f(x_i)| + \sum_{i=N+1}^{\infty} 2^{-i}$$

$$\leq \frac{\rho}{2} + \sum_{i=N+1}^{\infty} 2^{-i}.$$

By choosing $N = N(\rho) \in \mathbb{N}$ large enough, we can arrange the geometric series in the right hand side of the above estimate to be less that $\rho/2$. Hence,

$$\mathcal{U}\left(0, \rho/2, \{x_k\}_1^{N(\rho)}\right) \subseteq \mathbb{B}(0, \rho).$$

Conversely, fix $\varepsilon > 0$ and $\{y_1, \ldots, y_N\} \subseteq X$. Since the sequence $(x_i)_1^{\infty}$ is dense, we can find N-many terms of the sequence such that each one is close to the given points $\{y_1, \ldots, y_N\}$. Moreover, by a relabelling of the sequence, we may assume that these are the first N terms $\{x_1, \ldots, x_N\}$. Thus, we may assume that

$$\|x_i - y_i\| < \frac{\varepsilon}{2 \operatorname{diam}(B)}, \quad \text{for } i = 1, \ldots, N.$$

Then, for any $f \in B$ we have the estimate

$$|f(y_i)| \leq |f(x_i)| + |f(y_i - x_i)|$$

$$\leq |f(x_i)| + \|f\|_* \|x_i - y_i\|$$

$$\leq |f(x_i)| + \operatorname{diam}(B) \frac{\varepsilon}{2 \operatorname{diam}(B)}$$

$$= |f(x_i)| + \frac{\varepsilon}{2},$$

for $i = 1, \ldots, N$. Suppose now that $f \in \mathbb{B}(0, 2^{-N-1} \min\{\varepsilon, 1\})$. Then, we have

$$2^{-i} \min\{|f(x_i)|, 1\} < 2^{-N-1} \min\{\varepsilon, 1\}, \quad \text{for } i = 1, \ldots, N,$$

which gives that

$$\min\{|f(x_i)|, 1\} < \frac{\min\{\varepsilon, 1\}}{2}, \quad \text{for } i = 1, \ldots, N.$$

---

[5] Note that $\mathbb{B}(0, \rho)$ is the ball around the origin of X^* with respect to the metric d.

Since $(\min\{\varepsilon, 1\})/2 < 1$, the above inequality yields that

$$|f(x_i)| < \frac{\min\{\varepsilon, 1\}}{2} \le \frac{\varepsilon}{2}, \quad \text{for } i = 1, \ldots, N.$$

The previous estimates now allow us to deduce that

$$|f(y_i)| \le |f(x_i)| + \frac{\varepsilon}{2} < \varepsilon, \quad \text{for } i = 1, \ldots, N.$$

Hence, we have established

$$\mathbb{B}\left(0, 2^{-N-1} \min\{\varepsilon, 1\}\right) \subseteq \mathcal{U}\left(0, \varepsilon, \{y_k\}_1^N\right)$$

and therefore the desired conclusion ensues. $\qquad\square$

Although bounded sets are metrisable with respect to the weak* topology, the entire space itself is not:

Remark 13.19 (The weak* topology is not metrisable). By following similar methods to those we used in Proposition 12.26, it can be shown that the entire dual space is not metrisable.

Separability is a desirable property for an infinite-dimensional normed vector space to have. One of the possible ways one can conclude that a Banach space is separable is by establishing that its dual space is separable, as the next result asserts.

Theorem 13.20 (X^* separable $\implies X$ separable). *Let $(X, \|\cdot\|)$ be a Banach space, such that $(X^*, \|\cdot\|_*)$ is separable. Then, X is separable itself.*

Proof. Let $(f_m)_1^\infty$ be a dense sequence in X^*. Since

$$\|f_m\|_* = \sup\left\{|f_m(x)| : x \in \mathbb{B}^X\right\},$$

there exists a maximising sequence $(x_m)_1^\infty \subseteq X$, $\|x_m\| \le 1$, such that

$$|f_m(x_m)| \ge \frac{1}{2}\|f_m\|_*.$$

We set

$$Y := \operatorname{span}\left[\{x_1, x_2, \ldots\}\right].$$

i.e. Y is the vector space whose elements are (finite) linear combinations of the form $\sum_{a \in A} \lambda_a x_{i_a}$, $\lambda_a \in \mathbb{R}$, $x_{i_a} \in (x_k)_1^\infty$ and $\sharp(A) < \infty$. We note that Y contains a countable dense set, i.e. the set of linear combinations with rational coefficients:

$$Y_0 := \operatorname{span}_{\mathbb{Q}}\left[\{x_1, x_2, \ldots\}\right] = \left\{\sum_{a \in A} \lambda_\alpha x_{i_\alpha} \,\middle|\, \lambda_\alpha \in \mathbb{Q}, \ x_{i_\alpha} \in (x_k)_1^\infty, \ \sharp(A) < \infty\right\}.$$

Then, note that Y_0 is countable because it is a countable union of countable sets:

$$Y_0 = \bigcup_{k \in \mathbb{N}} Y_0^k, \quad Y_0^k := \mathrm{span}_{\mathbb{Q}}[\{x_1, \ldots, x_k\}] \cong \mathbb{Q}^k.$$

We conclude by showing that Y is dense in X. Let $f \in X^*$ such that $f \equiv 0$ on Y. Since $(f_m)_1^\infty$ is dense in X, there exists an $m \in \mathbb{N}$ such that $\|f_m - f\|_* \leq \varepsilon$. Thus,

$$\begin{aligned}
\|f\|_* &\leq \|f - f_m\|_* + \|f_m\|_* \\
&\leq \varepsilon + 2|f_m(x_m)| \\
&\leq \varepsilon + 2|f_m(x_m) - f(x_m)| + 2|f(x_m)| \\
&\leq \varepsilon + 2\|f_m - f\|_* \\
&\leq 3\varepsilon.
\end{aligned}$$

Hence $f \equiv 0$ on X, whence Y is dense in X. The proof is complete. $\qquad\square$

13.6 Weak compactness in reflexive Banach spaces

We may now answer the question of compactness of the weak topology of a Banach space. In general, the unit ball $\overline{\mathbb{B}}^X$ is **not** weakly sequentially compact. This may happen for those spaces X for which the canonical embedding into their bidual space given by the evaluation mapping (recall Definition 11.49),

$$\widehat{\cdot} : X \longrightarrow X^{**}, \quad \widehat{x}(f) := f(x), \quad \text{for all } f \in X^*,$$

is not surjective. When the evaluation map is surjective, namely when $\widehat{X} = X^{**}$, then this prominent class of spaces satisfies this desired weak compactness property. We recall that when $\widehat{X} = X^{**}$ X is called reflexive and we tacitly identify X with its image $\widehat{X}$ in X^{**}.

We begin with some simple auxiliary results. The first one below is essentially a consequence of the definitions and the isometry property of the evaluation map.

Lemma 13.21. *Let $(X, \|\cdot\|)$ be a reflexive Banach space and let also $Y \subseteq X$ be a closed subspace. Then, Y is a reflexive Banach space.*

Proof. We need to show that the evaluation mapping

$$\widehat{\cdot} : Y \longrightarrow Y^{**}, \quad \widehat{y}(g) := g(y), \quad \text{for all } g \in Y^*,$$

is surjective.[6] This means that for any $\psi \in Y^{**}$, there exists $y \in Y$ such that

---

[6]Note that we are using the same symbol to denote two different evaluation mappings, that of the space X itself and that of its closed subspace Y.

$\widehat{y} = \psi$. Given a fixed $\psi \in Y^{**}$, we define $\phi \in X^{**}$ by setting

$$\phi(f) := \psi(f|_Y), \quad \text{for all } f \in X^*.$$

Since X is by assumption reflexive, there exists $x \in X$ such that $\widehat{x} = \phi$. It suffices now to show that $x \in Y$. Indeed, if this is the case, then

$$f(x) = \widehat{x}(f) = \psi(f|_Y) \quad \text{for all } f \in X^*.$$

The conclusion then follows by invoking the analytic Hahn-Banach Theorem 11.58 to show that any $g \in Y^*$ can be extended to some $f \in X^*$ with $f|_Y = g$. We finally show that $x \in Y$. If for the sake of contradiction this were not the case, by the geometric Hahn-Banach Theorem 11.59 there would exist a hyperplane separating the closed subspace Y from the disjoint compact set $\{x\}$. Hence, there would exist $f \in X^*$ such that $f|_Y = 0$ and $f(x) > 0$. For this choice of f, the definition of ϕ above gives

$$\phi(f) = \psi(f|_Y) = \psi(0) = 0,$$

whilst also $\phi(f) = \widehat{x}(f) = f(x) > 0$. This contradiction completes the proof. $\square$

Lemma 13.22. *The Banach space $(X, \|\cdot\|)$ is reflexive if and only if $(X^*, \|\cdot\|_*)$ is reflexive. Namely,*

$$X = X^{**} \iff X^* = X^{***}. \tag{$\diamond$}$$

Roughly speaking, Lemma 13.22 says that "we can cancel stars" from the right hand side of $(\diamond)$ above. More precisely, note that the statements above says $\widehat{X} = X^{**}$ if and only if $\widehat{X^*} = X^{***}$, where "$\widehat{}$" symbolises two different different evaluation maps, namely $\widehat{} : X \longrightarrow X^{**}$ and $\widehat{} : X^* \longrightarrow X^{***}$.

Proof. Suppose first that $\widehat{X} = X^{**}$, where $\widehat{} : X \longrightarrow X^{**}$ is the evaluation map of X. Fix $\mathbf{x} \in X^{***}$ and note that the function given by

$$f : X \longrightarrow \mathbb{R}, \quad f(x) := \mathbf{x}(\widehat{x}),$$

is is fact continuous and linear on X. Therefore, $f \in X^*$. Since $f(x) = \widehat{x}(f)$, the definition of f implies $\mathbf{x}(\widehat{x}) = \widehat{x}(f)$ for all $x \in X$. Since the evaluation map of X is surjective, we infer that $\mathbf{x}(\phi) = \phi(f)$ for all $\phi \in X^{**}$. The latter means that the evaluation map $\widehat{} : X^* \longrightarrow X^{***}$ is surjective.

Conversely, suppose that X^* is reflexive. From the previous reasoning, it follows that X^{**} is reflexive. Since $\widehat{X}$ is (norm) closed in X^{**}, by Lemma 13.21 it follows that the subspace $\widehat{X}$ is reflexive. $\square$

We finally have the next simple result.

Lemma 13.23. *Let $(X, \|\cdot\|)$ be a Banach space. Then,*

X is reflexive and separable $\iff$ X^* is reflexive and separable.

Proof. If X^* is reflexive and separable, then by Lemma 13.22 and Theorem 13.20 it follows that X is reflexive and separable. Conversely, if X is reflexive and separable, the same is true for $X^{**} = \widehat{X}$. Hence, X^* is reflexive and separable. □

The following theorem is our second main compactness assertion for the weak topology on reflexive Banach spaces.

Theorem 13.24 (Sequential compactness of the weak topology for reflexive spaces). *Let $(X, \|\cdot\|)$ be a reflexive Banach space. Then, the closed unit ball $\overline{\mathbb{B}}^X$ of X is weakly sequentially compact.[7]*

Equivalently, for any bounded sequence $(x_m)_1^\infty \subseteq X$, there exists $x \in X$ and a subsequence $(x_{m_k})_1^\infty$ such that $x_{m_k} \longrightarrow x$, as $k \to \infty$.

Proof. Let $(x_m)_1^\infty$ be a bounded sequence in X. We set

$$Y := \overline{\text{span}[\{x_1, x_2, ...\}]}^{\|\cdot\|}.$$

Then, by Lemmas 13.21-13.23, since Y is a closed, reflexive and separable subspace of X, it follows that Y^* is also reflexive and separable. By using that

$$\overline{\mathbb{B}}^Y = \overline{\mathbb{B}}^{Y^{**}}$$

and that the weak topology of Y coincides with the weak* topology on $Y^{**} = Y$, i.e. that

$$\mathcal{T}_{Y^{**}}^{\text{w}^*} = \mathcal{T}_Y^{\text{w}},$$

we deduce that $\overline{\mathbb{B}}^Y$ is sequentially weakly compact. Therefore, we see that the theorem ensues. □

13.7   Uniformly convex Banach spaces

For general Banach spaces, it is usually very difficult to check directly from the definition whether a space is reflexive or not. We conclude our study of weak topologies on Banach spaces by introducing a geometric criterion which is much easier to check and, as we shall show later, it is a sufficient condition for reflexivity. This geometric condition, roughly speaking, expresses the fact that *the balls of the space must be "round" enough*. Quite surprisingly, it depends on the properties of the norm on the space and is not invariant under the possible change from the given to an equivalent norm.

---

[7]It can be shown that the converse of Theorem 13.24 is also true, but this is considerably more difficult to prove, see e.g. [11, 21, 43].

Definition 13.25. Let $(X, \| \cdot \|)$ be a Banach space. We say that the norm $\| \cdot \|$ of the space (or the pair $(X, \| \cdot \|)$) is **uniformly convex** when for any $\varepsilon > 0$, there exists a $\delta = \delta(\varepsilon) > 0$ such that the next implication holds true:

$$\left. \begin{array}{l} \|x\| \leq 1, \ \|y\| \leq 1 \\ \text{and} \ \ \|x - y\| \geq \varepsilon \end{array} \right\} \quad \Longrightarrow \quad \left\| \frac{x+y}{2} \right\| \leq 1 - \delta.$$

Remark 13.26 (Geometric meaning). Uniform convexity is a geometric condition which says that the ball $\mathbb{B}^X$ is round and has "no corners", see Figure 13.1 below.

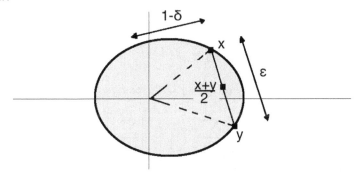

Figure 13.1: The idea of the definition of uniform convexity.

The condition excludes regions of flatness on the boundary of the ball (the unit sphere) and in particular it excludes the existence of straight line segments on it.

Remark 13.27 (Non-topological nature). Note that uniform convexity **it is not a topological condition on X, but instead depends on the choice of norm** $\| \cdot \|$ on X. For, instance, a space might be uniformly convex for one norm, but not so for some other norm equivalent to it. This might happen even in finite dimensions, see Figure 13.2. However, if a Banach space possesses **one** norm which is uniformly convex, then as we shall show below it is reflexive.

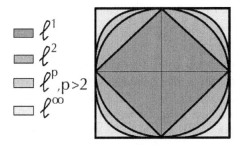

Figure 13.2: The space $(\mathbb{R}^2, \| \cdot \|_{\ell^2})$ is uniformly convex and the same holds for $(\mathbb{R}^2, \| \cdot \|_{\ell^p})$ for all $p \in (1, \infty)$. However, $(\mathbb{R}^2, \| \cdot \|_{\ell^1})$ and $(\mathbb{R}^2, \| \cdot \|_{\ell^\infty})$ are not uniformly convex. A Banach space might be uniformly convex for one norm but not for some other, even if the norms are equivalent and produce the same topology.

We may now establish the main result of this section.

Theorem 13.28 (Milman-Pettis). *Every uniformly convex Banach space* $(X, \|\cdot\|)$ *is reflexive.*

The converse of Theorem 13.28 is false in general. For the proof of this result we need the next two important lemmas.

Lemma 13.29 (Helly). *Let* $(X, \|\cdot\|)$ *be a Banach space with dual space* $(X^*, \|\cdot\|_*)$. *Let* $f_1, \ldots, f_n \in X^*$ *be given functionals, and let* $a_1, \ldots, a_n \in \mathbb{R}$ *be given constants. Then, the following statements are equivalent:*

(i) *For any* $\varepsilon > 0$, *there exists* $x_\varepsilon \in X$ *with* $\|x_\varepsilon\| \leq 1$ *and such that*

$$\max_{i=1,\ldots,n} \left| f_i(x_\varepsilon) - a_i \right| < \varepsilon.$$

(ii) *For any constants* $\{b_1, \ldots, b_n\} \subseteq \mathbb{R}$, *we have*

$$\left| \sum_{i=1}^n b_i a_i \right| \leq \left\| \sum_{i=1}^n b_i f_i \right\|_*.$$

Proof. (i) $\Longrightarrow$ (ii): Fix $\varepsilon > 0$, constants $\{b_1, \ldots, b_n\} \subseteq \mathbb{R}$ and set $S := \sum_{i=1}^n |b_i|$. Then, by assumption there exists x_ε as in the statement such that

$$\left| \sum_{i=1}^n b_i f_i(x_\varepsilon) - \sum_{i=1}^n b_i a_i \right| < \varepsilon S.$$

Therefore,

$$\left| \sum_{i=1}^n b_i a_i \right| \leq \left\| \sum_{i=1}^n b_i f_i \right\|_* \|x_\varepsilon\| + \varepsilon S \leq \left\| \sum_{i=1}^n b_i f_i \right\|_* + \varepsilon S$$

and we conclude by letting $\varepsilon \to 0$.

(ii) $\Longrightarrow$ (i): Let us set $a := (a_1, \ldots, a_n) \in \mathbb{R}^n$ and consider the bounded linear map given by

$$\Phi : X \longrightarrow \mathbb{R}^n, \quad \Phi(x) := \left(f_1(x), \ldots, f_n(x) \right).$$

Then, with the above notation, the desired statement (i) expresses the fact that $a \in \overline{\Phi(\mathbb{B}^X)}$. Further, the set $\overline{\Phi(\mathbb{B}^X)} \subseteq \mathbb{R}^n$ is convex because it is the closure of the image of a convex set, the ball of X, under a linear mapping. Suppose for the sake of contradiction that $a \notin \overline{\Phi(\mathbb{B}^X)}$. By the Hahn-Banach Theorem 11.59 we may separate the convex compact sets $\{a\}$ and $\overline{\Phi(\mathbb{B}^X)}$ strictly by a hyperplane, that is, there exists $b = (b_1, \ldots, b_n) \in \mathbb{R}^n$ and $c \in \mathbb{R}$ such that

$$\Phi(x) \cdot b < c < a \cdot b \quad \text{for all } x \in X \text{ with } \|x\| \leq 1,$$

where "·" is the Euclidean inner product on $\mathbb{R}^n$. The above implies that

$$\left| \sum_{i=1}^n b_i f_i(x) \right| \le c < \sum_{i=1}^n b_i a_i, \quad x \in X, \ \|x\| \le 1$$

which, by taking "supremum" over all such x, contradicts (ii). Hence, the conclusion follows. $\qquad\square$

Lemma 13.30 (Goldstine). *Let $(X, \|\cdot\|)$ be a Banach space. Then, the image of the unit ball $\mathbb{B}^X$ under the isometry given by the evaluation mapping (Definition 11.49) is dense in $\mathbb{B}^{X^{**}}$ with respect to the weak* topology of X^{**}. Namely,*

$$\overline{\widehat{\mathbb{B}^X}}^{\mathrm{w}*} \supseteq \mathbb{B}^{X^{**}}.$$

Equivalently, for any functional $\phi \in \mathbb{B}^{X^{**}}$ and any weakly* open neighbourhood $\mathcal{U} \in \mathcal{T}_{X^{**}}^{\mathrm{w}*}$ of ϕ, there exists $x \in \mathbb{B}^X$ such that $\widehat{x} \in \mathcal{U}$.

Remark 13.31. Since the evaluation mapping is an isometry, $\overline{\widehat{\mathbb{B}^X}}$ **is closed** in $\mathbb{B}^{X^{**}}$ (with respect to the norm topology of X^{**})!

Proof. Fix $\phi \in \mathbb{B}^{X^{**}}$ and a weakly* open set $\mathcal{U} \subseteq X^{**}$ with $\mathcal{U} \ni \phi$. It suffices to show that $\mathcal{U} \cap \widehat{\mathbb{B}^X} \ne \emptyset$. We may assume that $\mathcal{U}$ has the form

$$\mathcal{U} = \left\{ \psi \in X^{**} : \ \left| \psi(f_i) - \phi(f_i) \right| < \varepsilon, \ i = 1, \dots, n \right\},$$

for some $\varepsilon > 0$ and elements $\{f_1, \dots, f_n\} \subseteq X^*$. Hence, we need to find $x \in \mathbb{B}^X$ such that

$$\left| f_i(x) - \phi(f_i) \right| < \varepsilon, \quad i = 1, \dots, n.$$

We set $a_i := \phi(f_i)$ and note that for any $\{b_1, \dots, b_n\} \subseteq \mathbb{R}$ we have

$$\left| \sum_{i=1}^n b_i a_i \right| = \left| \phi\left(\sum_{i=1}^n b_i f_i \right) \right| \le \left\| \sum_{i=1}^n b_i f_i \right\|_*,$$

because $\|\phi\|_{**} \le 1$. By invoking Helly's Lemma 13.29, there exists $x_\varepsilon \in \mathbb{B}^X$ such that $\left| f_i(x_\varepsilon) - a_i \right| < \varepsilon$, $i = 1, \dots, n$. Hence, $\widehat{x_\varepsilon} \in \widehat{\mathbb{B}^X} \cap \mathcal{U}$. $\qquad\square$

Now we may prove the main theorem.

Proof of Theorem 13.28. Fix any $\phi \in X^{**}$ with $\|\phi\|_{**} = 1$. By Remark 13.31, it suffices to show that for all $\varepsilon > 0$, there exists $x \in \mathbb{B}^X$ such that $\|\widehat{x} - \phi\|_{**} < \varepsilon$. Fix $\varepsilon, \delta > 0$ as in Definition 13.25 and $f \in X^*$ such that $\|f\|_* = 1$ and

$$\|\phi\|_{**} \ge \phi(f) > 1 - \frac{\delta}{2}. \tag{$\odot$}$$

We set

$$\mathcal{V} := \left\{ \psi \in X^{**} : \ |\psi(f) - \phi(f)| < \frac{\delta}{2} \right\}.$$

Then, $\mathcal{V} \in \mathcal{T}^{w^*}$ and $\mathcal{V} \ni \phi$. By Lemma 13.30, $\mathcal{V} \cap \widehat{\mathbb{B}^X} \neq \emptyset$. Hence, we can choose $x \in \mathbb{B}^X$ such that $\widehat{x} \in \mathcal{V}$. We shall now show that $\phi \in \mathbb{B}_\varepsilon^{X^{**}}(\widehat{x})$, and that will finish the proof. Assume for the sake of contradiction that

$$\phi \in X^{**} \setminus \mathbb{B}_\varepsilon^{X^{**}}(\widehat{x}) =: \mathcal{W}.$$

By Lemma 13.30, we have $(\mathcal{V} \cap \mathcal{W}) \cap \widehat{\mathbb{B}_\varepsilon^X} \neq \emptyset$, hence there exists $y \in \mathbb{B}^X$ such that $\widehat{y} \in \mathcal{V} \cap \mathcal{W}$. Since $\widehat{x}, \widehat{y} \in \mathcal{V}$, we have

$$\begin{cases} |\phi(f) - \widehat{x}(f)| < \dfrac{\delta}{2}, \\[2mm] |\phi(f) - \widehat{y}(f)| < \dfrac{\delta}{2}, \end{cases}$$

and hence

$$2\,\phi(f) \leq f(x+y) + \delta \leq \|x+y\| + \delta.$$

Thus, by $(\odot)$ we infer that

$$\left\| \frac{x+y}{2} \right\| \geq 1 - \delta.$$

By the uniform convexity of the norm, we have $\|x - y\| \leq \varepsilon$. Since $\widehat{x} \in \mathcal{W}$, we also have $\|x - y\| > \varepsilon$, and we have a contradiction. Hence, if X is uniformly convex for some norm, then it is reflexive. $\qquad\square$

The following result is a very useful property of uniformly convex Banach spaces which provides a sufficient condition we can impose on weakly converging sequences in order to guarantee that "the oscillations are dampened" and actually it converges strongly. In a nutshell, if the norms of the terms of the sequence also converge to the norm of the limit, then this strengthens weak to strong convergence. In fact, one can show that the same is true if we replace the norm by any *uniformly convex* function on the space. (The latter notion is of course defined in the obvious manner.)

Proposition 13.32 (Strengthening weak to strong convergence). *Let* $(X, \|\cdot\|)$ *be a uniformly convex Banach space. Then, the next statements are equivalent:*

(i) $x_m \longrightarrow x$ *strongly as* $m \to \infty$;

(ii) $x_m \longrightarrow x$ *weakly as* $m \to \infty$ *and* $\limsup\limits_{m \to \infty} \|x_m\| \leq \|x\|$;

(iii) $x_m \longrightarrow x$ *weakly as* $m \to \infty$ *and* $\lim\limits_{m \to \infty} \|x_m\| = \|x\|$.

Proof. By Theorem 12.22, (i) implies (ii). Moreover, by Theorem 12.22, (ii) implies (iii) since

$$\|x\| \leq \liminf\limits_{m \to \infty} \|x_m\|$$

when $x_m \longrightarrow x$ as $m \to \infty$. As a result,

$$\|x\| \leq \liminf_{m \to \infty} \|x_m\| \leq \limsup_{m \to \infty} \|x_m\| \leq \|x\|.$$

We now use Theorem 13.28 to prove that (iii) implies (i). Assume $x \neq 0$ (otherwise $\|x_m\| \longrightarrow 0$ and we are done) and set

$$\lambda_m := \max\{\|x_m\|, \|x\|\}, \quad m \in \mathbb{N}.$$

Then, $\lambda_m \longrightarrow \|x\|$ as $m \to \infty$. We also set

$$y_m := \frac{x_m}{\lambda_m}, \quad y := \frac{x}{\|x\|}.$$

Since $y_m \longrightarrow y$ as $m \to \infty$, by Theorem 12.22, we have

$$\|y\| \leq \liminf_{m \to \infty} \left\| \frac{y_m + y}{2} \right\|$$

because $y_m + y \longrightarrow 2y$ as $m \to \infty$. Since $\|y\| = 1$, $\|y_m\| \leq 1$, we have

$$\left\| \frac{y_m + y}{2} \right\| \longrightarrow 1 \quad \text{as } m \to \infty.$$

By the uniform convexity of X, we have $\|y - y_m\| \longrightarrow 0$, as $m \to \infty$. As a result, $x_m \longrightarrow x$ as $m \to \infty$ and the result follows. $\qquad\square$

13.8   Exercises

Exercise 13.1. Show that the norm of every (real) Hilbert space $(X, \langle \cdot, \cdot \rangle)$ which is induced by the inner product, is uniformly convex. Deduce the following implication: if $x_n \longrightarrow x$ and $\langle x_n, x_n \rangle \longrightarrow \langle x, x \rangle$ as $n \to \infty$, then $x_n \longrightarrow x$ as $n \to \infty$.

Exercise 13.2. Prove Proposition 13.11.

Exercise 13.3. Prove Theorem 13.15.

Exercise 13.4. Let $(X, \|\cdot\|)$ be a Banach space, let X^* be its dual and let also $D := (f_k)_1^\infty \subseteq X^*$ a (strongly) dense sequence, that is $\overline{D} = X^*$. Suppose that $(x_m)_1^\infty \subseteq X$ is a bounded sequence, namely there exists $M > 0$ such that $\|x_m\| \leq M$ for all $m \in \mathbb{N}$. Fix also $x \in X$. Show that the next statements are equivalent:

(1) We have $x_m \longrightarrow x$ as $m \to \infty$ in X.

(2) For any $f \in D$, we have $f(x_m) \longrightarrow f(x)$ as $m \to \infty$.

Exercise 13.5. Let $(X, \| \cdot \|)$ be a Banach space, let X^* be its dual and let also $D := (x_k)_1^\infty \subseteq X^*$ a (strongly) dense sequence, that is $\overline{D} = X$. Suppose that $(f_m)_1^\infty \subseteq X^*$ is a bounded sequence, namely there exists $M > 0$ such that $\|f_m\|_* \leq M$ for all $m \in \mathbb{N}$. Fix also $f \in X^*$. Show that the next statements are equivalent:

(1) We have $f_m \xrightarrow{*} f$ as $m \to \infty$ in X^*.

(2) For any $x \in D$, we have $f_m(x) \longrightarrow f(x)$ as $m \to \infty$.

Exercise 13.6. Find a Banach space $(X, \| \cdot \|)$ and a sequence $(f_n)_1^\infty \subseteq X^*$ such that $f_n \xrightarrow{*} 0$ but for any $h \in X^*$ generated by a convex combination of terms of the sequence $(f_n)_1^\infty$, we have $\|h\|_* = 1$. This confirms that the Mazur Theorem is not true for the weak* topology.

Exercise 13.7. Let X^* be an infinite-dimensional dual Banach space. Show that the weak* topology does not satisfy the first axiom of countability. Therefore, it is not metrisable.

[Hint: Use as known that the pre-dual space $(X, \|\cdot\|)$ has no countable (Hamel) basis and hence, given any sequence of elements of X, there exists at least one element which is not the in the linear span of the sequence.]

Exercise 13.8. Consider the space

$$c_0 := \left\{ x = (x_1, x_2, \ldots) \in \ell^\infty \ : \ x_k \longrightarrow 0 \text{ as } k \to \infty \right\}.$$

Then, it can be shown that $(c_0)^* \cong \ell^1$, in the following sense which follows the same lines as the Riesz Representation Theorem: we have $f \in (c_0)^*$ it and only if f has a unique representation as

$$f(x) = \sum_{i=1}^{\infty} u_i \, x_i, \quad \text{for } x \in c_0,$$

for some $u = (u_1, u_2, \ldots) \in \ell^1$, whilst the next isometry condition is also satisfied:

$$\|f\|_{(c_0)^*} = \|u\|_{\ell^1}.$$

Using the above fact (namely that ℓ^1 is a dual space), show that the standard sequence $(e_i)_1^\infty$ with $e_i = (0, \ldots, 1, 0, \ldots)$ (having 1 in the ith position) satisfies

$$e_i \xrightarrow{*} 0 \text{ as } i \to \infty, \quad \text{but } e_i \not\to 0 \text{ as } i \to \infty.$$

[Hint: Use that, similarly to the above isomorphism, one has $(c_0)^{**} \cong \ell^\infty$.]

Exercise 13.9. Recall the duality isomorphism $(c_0)^* = \ell^1$ of Exercise 13.8 and define $f \in (\ell^1)^*$ by taking

$$f(x) := \sum_{i=1}^{\infty} x_i, \quad \text{for } x = (x_1, x_2, \ldots) \in \ell^1.$$

Show that the linear functional f is norm continuous on ℓ^1, but it is not weakly* continuous. Thus, $f \in \ell^\infty \setminus c_0$ (having in mind the usual identification via the evaluation mapping).

Exercise 13.10. Let $(X, \|\cdot\|)$ be an infinite-dimensional Banach space. Show that the unit sphere $\mathbb{S}^X$ is a weakly dense and weakly G_δ set in the closed unit ball $\overline{\mathbb{B}}^X$.

[Hint: We recall that, in a topological space, a set is called "G_δ" if it can be written as the countable intersection of open sets.]

Chapter 14

Functional properties of the Lebesgue spaces

14.1 The Lebesgue spaces are Banach spaces after all

In this chapter we revisit the class of the L^p spaces defined in Chapter 9 in order to continue their study, having now in our toolbox the machinery developed in Chapters 10-13. The main goal is to study now the functional-analytic properties of these spaces, rather than the measure-theoretic properties of their elements.

Accordingly, the plan is to examine whether the L^p spaces have nice properties, like being uniformly convex (namely, whether their balls are "round enough"), or reflexive (namely, whether they coincide with their second dual space). In particular, herein we are interested in identifying and characterising their dual spaces. We shall see that different phenomena emerge in the space L^p, depending on whether

- $p = 1$,
- $1 < p < \infty$,
- $p = \infty$.

For the sake of added clarity, in this chapter we shall restrict our attention to the case of **real-valued** functions $f : X \longrightarrow \mathbb{R}$ defined on a measure space $(X, \mathfrak{M}, \mu)$. The general vector-valued case of mappings $f : X \longrightarrow \mathbb{F}^d$ can be handled by arguing in a component-wise fashion and noting the isomorphism

$$L^p(X, \mu, \mathbb{F}^d) \cong \left(L^p(X, \mu, \mathbb{F}) \right)^d$$

which allows us to identify the L^p space of vector-valued functions with the d-fold Cartesian power of the space of scalar-valued functions. For the sake of brevity we shall simplify the notation accordingly to

$$L^p(X, \mu, \mathbb{R}) \equiv L^p(X, \mu)$$

where, as in Chapter 9, the dependence on the σ-algebra $\mathfrak{M}$ (which is always fixed) is subsumed in the notation for the measure.

We begin our investigations with Section 14.2 wherein we show that when the exponent p lies in the range $(1, \infty)$, then the (norm of the) space L^p is uniformly convex and therefore the space in this case is reflexive. This is an easy consequence of a pair of inequalities which generalise the parallelogram equality to $p \neq 2$.

Next, in Section 14.3 we establish two central results, the so-called **Riesz Representation Theorems** for the L^p spaces. These results combined establish that, when $p \in [1, \infty)$, the dual of L^p equals (up to an isometry) the space $L^{p'}$ where p' is the conjugate exponent of p. If $p > 1$ this holds true for any measure space, whilst for $p = 1$ the measure space needs to be σ-finite (Definition 7.16).

In Section 14.4 we discuss what the abstract concept of weak/weak* convergence actually means in the concrete setting of the L^p spaces. As we shall see, weak convergence essentially consists of **convergence of the averages over any measurable set of finite measure**. To this end, we identify through examples the two principal reasons for a weakly convergent sequence to fail to be strongly (norm) convergent: either the phenomenon of **concentrations**, or that of **oscillations**, or a combination of both.

Finally, in Section 14.5 we present the **very important** consequences of our earlier results, combined with those of Chapter 13, that, since $L^{p'}$ is separable when $p \in (1, \infty]$, then **bounded sets of its dual space L^p are sequentially weakly precompact (weakly* if $p = \infty$)**.

14.2   Uniform convexity and reflexivity for $1 < p < \infty$

Let $(X, \mathfrak{M}, \mu)$ be a fixed arbitrary measure space. In this brief section we show that the standard norm of the space $L^p(X, \mu)$ is uniformly convex in the range of exponents $1 < p < \infty$. Hence, by the results of Chapter 13, it follows that $L^p(X, \mu)$ is reflexive when $p \neq \{1, \infty\}$. The proof of this fact is based on the following inequalities which generalise the Parallelogram Law of Hilbert spaces from $p = 2$ to $1 < p < \infty$.

Lemma 14.1 (Clarkson Inequalities). *Let $p \in (1, \infty)$ and fix $f, g \in L^p(X, \mu)$.*

(i) *If $2 \leq p < \infty$, we have*

$$\left\| \frac{f+g}{2} \right\|^p_{L^p(X,\mu)} + \left\| \frac{f-g}{2} \right\|^p_{L^p(X,\mu)} \leq \frac{1}{2} \left(\|f\|^p_{L^p(X,\mu)} + \|g\|^p_{L^p(X,\mu)} \right). \quad (\Diamond)$$

(ii) *If $1 < p < 2$, we have*

$$\left\| \frac{f+g}{2} \right\|^{p'}_{L^p(X,\mu)} + \left\| \frac{f-g}{2} \right\|^{p'}_{L^p(X,\mu)} \leq \left[\frac{1}{2} \left(\|f\|^p_{L^p(X,\mu)} + \|g\|^p_{L^p(X,\mu)} \right) \right]^{\frac{1}{p-1}}.$$

$$(\Diamond\Diamond)$$

Proof. We only show (i), since the proof of (ii) is more complicated and would take us too far afield. Note that it suffices to show

$$\left|\frac{a+b}{2}\right|^p + \left|\frac{a-b}{2}\right|^p \leq \frac{1}{2}\left(|a|^p + |b|^p\right)$$

for all $a, b \in \mathbb{R}$. To show the above, we note that the real function

$$t \mapsto \left(t^2 + 1\right)^{p/2} - t^p - 1, \quad t \geq 0,$$

is increasing, which implies for $t := B/A$, $B > A > 0$, that

$$A^p + B^p \leq \left(A^2 + B^2\right)^{p/2}.$$

We conclude by selecting the values $A := |a+b|/2$, $B := |a-b|/2$ and noting the convexity of the real function $t \mapsto t^{p/2}$ for $t \geq 0$. $\qquad\square$

Now we come to the main result of this section.

Theorem 14.2 (Reflexivity of L^p). *If $1 < p < \infty$, then the space $L^p(X, \mu)$ is reflexive, that is:*

$$L^p(X, \mu) = \left(L^p(X, \mu)\right)^{**}.$$

Note that, although we follow standard conventions of Analysis and write the equality sign, the actual meaning of the result above is that the evaluation mapping $\hat{\cdot} : L^p(X, \mu) \longrightarrow \left(L^p(X, \mu)\right)^{**}$ is surjective (Definition 11.49).

Proof. It suffices to show that $L^p(X, \mu)$ is uniformly convex, since by Theorem 13.28 we know that uniform convexity of the norm of a Banach space implies its reflexivity. Indeed, if $p \geq 2$, suppose that

$$\|f\|_{L^p(X,\mu)} \leq 1, \quad \|g\|_{L^p(X,\mu)} \leq 1, \quad \|f - g\|_{L^p(X,\mu)} \geq \varepsilon,$$

for some $\varepsilon > 0$ small enough. We set

$$\delta := 1 - \left(1 - \left(\frac{\varepsilon}{2}\right)^p\right)^{1/p}.$$

Then, by the Clarkson inequality ($\Diamond$) we have

$$\left\|\frac{f+g}{2}\right\|_{L^p(X,\mu)} \leq \left(1 - \left(\frac{\varepsilon}{2}\right)^p\right)^{\frac{1}{p}}$$

$$= 1 - \delta.$$

Hence, indeed the space is uniformly convex. The case $1 < p < 2$ is similar and follows by using the inequality ($\Diamond\Diamond$) instead of the inequality ($\Diamond$). The details are left as an exercise. $\qquad\square$

14.3   Identifying the duals of the Lebesgue spaces

Let again $(X, \mathfrak{M}, \mu)$ be a fixed arbitrary measure space. In this section we characterise the dual space of the space L^p when $p \in [1, \infty)$. The case of the dual space of L^∞ is **significantly** more complicated and we refrain from giving any proofs (but see Remark 14.6). The two relevant results we present, Theorems 14.4 and 14.5, distinguish the case when $1 < p < \infty$ which is simpler and the case $p = 1$ which is more complicated (and needs an extra assumption).

Remark 14.3 (Riesz Theorems in L^p explained). The main idea behind the upcoming Riesz-type Theorems 14.4 and 14.5 begins with the question of how to find *some* explicit functionals acting on L^p. It is not too difficult to imagine that integration against a function might do the trick. Indeed, fix $p \in [1, \infty]$ and let $p' = p/(p-1)$ be the conjugate exponent of p, namely the one for which

$$\frac{1}{p} + \frac{1}{p'} = 1$$

with the standard convention that $1' = \infty$ and $\infty' = 1$. Fix $u \in L^{p'}(X, \mu)$ and consider the linear functional defined as

$$\phi_u(f) := \int_X u f \, d\mu \quad \text{for all } f \in L^p(X, \mu).$$

Then, $\phi_u : L^p(X, \mu) \longrightarrow \mathbb{R}$ is continuous and thus contained in $(L^p(X, \mu))^*$, as a result of Hölder's inequality:

$$\left| \phi_u(f) \right| \leq \|u\|_{L^{p'}(X,\mu)} \|f\|_{L^p(X,\mu)}.$$

Hence, for any $u \in L^{p'}(X, \mu)$, we have a functional $\phi_u \in (L^p(X, \mu))^*$ arising from a function in $L^{p'}$ through integration against it. Further, it is not too difficult to show, as we shall do, that the map $u \mapsto \phi_u$ is actually an isometry (recall Definition 10.39). The main point of Theorems 14.4 and 14.5 is that **all functionals arise in this way when $p < \infty$**. Quite surprisingly, **when $p = \infty$ the isometry is not surjective** and there exist functionals in $(L^\infty(X, \mu))^*$ which do not arise from an L^1 function! See Remark 14.6 and Example 14.15.

We begin by characterising the dual of L^p when $1 < p < \infty$.

Theorem 14.4 (Riesz Theorem in L^p for $1 < p < \infty$). *Let $p \in (1, \infty)$. Then, the dual space of $L^p(X, \mu)$ is isometrically isomorphic[1] to $L^{p'}(X, \mu)$,*

---

[1] Let us stress once again that, as opposed to Algebra, in Analysis it is customary and convenient to write $L^{p'}(X, \mu) = (L^p(X, \mu))^*$ (as we did above) instead of the more precise

$$L^{p'}(X, \mu) \cong (L^p(X, \mu))^*,$$

but we must keep in mind the map $u \mapsto \int_X u \, (\cdot) \, d\mu$ which identifies the spaces.

where $p' = p/(p-1)$ is the conjugate exponent:

$$\left(L^p(X,\mu)\right)^* = L^{p'}(X,\mu).$$

In addition, for any functional $\phi \in \left(L^p(X,\mu)\right)^*$, there exists a unique function $u \in L^{p'}(X,\mu)$ such that

$$\phi(f) = \int_X uf\,d\mu \quad \text{for all } f \in L^p(X,\mu),$$

and also

$$\|\phi\|_{(L^p(X,\mu))^*} = \|u\|_{L^{p'}(X,\mu)}.$$

Note that if $p = 2$, then $p' = 2$ and we have that the dual space of $L^2(X,\mu)$ coincides with itself. Since all Hilbert spaces are isometric to some $L^2(X,\mu)$ (recall Theorem 10.41 and Remark 10.44), we recover as a consequence the Riesz theorem for Hilbert spaces.

Proof. We define the linear map $T : L^{p'}(X,\mu) \longrightarrow (L^p(X,\mu))^*$ by setting

$$L^{p'}(X,\mu) \ni u \mapsto Tu \in (L^p(X,\mu))^*,$$

$$(Tu)(f) := \int_X uf\,d\mu \quad \text{for all } f \in L^p(X,\mu).$$

Then, T is a bounded linear map and (by Hölder's inequality)

$$\left|(Tu)(f)\right| \le \int_\Omega |u||f|\,d\mu \le \|u\|_{L^{p'}(X,\mu)}\|f\|_{L^p(X,\mu)}$$

which gives

$$\|Tu\|_{L^{p'}(X,\mu)^*} \le \|u\|_{L^{p'}(X,\mu)}.$$

Moreover, for the choice

$$f_0 := |u|^{p'-2}\,u\,\chi_{\{u \ne 0\}},$$

we have

$$(Tu)(f_0) = \int_{\{u \ne 0\} \cap X} u\,|u|^{p'-2}\,u\,d\mu = \left(\|u\|_{L^{p'}(X,\mu)}\right)^{p'}$$

and since $p' - 1 = 1/(p-1)$, we also have

$$\|f_0\|_{L^p(X,\mu)} = \left(\int_{\{u \ne 0\}} \left(|u|^{p'-1}\right)^p d\mu\right)^{\frac{1}{p}} = \left(\|u\|_{L^{p'}(X,\mu)}\right)^{p'-1}.$$

In conclusion, we have the estimate

$$\|Tu\|_{(L^p(X,\mu))^*} = \sup_{f \ne 0} \frac{|(Tu)f|}{\|f\|_{L^p(X,\mu)}} \ge \frac{|(Tu)f_0|}{\|f_0\|_{L^p(X,\mu)}} = \|u\|_{L^{p'}(X,\mu)}$$

which establishes that the inequality is actually an equality. As a result, T is a linear isometry (embedding). Recall now that by the results of Chapter 11 we already know that $T\big(L^p(X,\mu)\big)$ is closed in $(L^p(X,\mu))^*$, since T is an isometry. Hence, it suffices to show it is dense too. To this end, let h be an element of the orthogonal complement[2] $T(L^{p'}(X,\mu))^\perp$. Remember though that

$$T\big(L^{p'}(X,\mu)\big)^\perp \subseteq (L^p(X,\mu))^{**} = L^p(X,\mu)$$

where the inequality above is by the definition of orthogonal complements and the equality is a consequence of the reflexivity of $L^p(X,\mu)$ shown earlier. Therefore, by the definition of T we have

$$\int_X uh\,d\mu \,=\, (Tu)(h) \,=\, 0, \quad \text{for all } u \in L^{p'}(X,\mu).$$

The particular choice

$$u_0 := |h|^{p-2}h\,\chi_{\{h\neq0\}},$$

allows us to infer that $\|h\|_{L^p(X,\mu)} = 0$ and thus $h = 0 \in L^p(X,\mu)$. Consequently,[3]

$$T\big(L^{p'}(X,\mu)\big) \,=\, \Big(T\big(L^{p'}(X,\mu)\big)^\perp\Big)^\perp \,=\, \{0\}^\perp \,=\, (L^p(X,\mu))^*,$$

and as a result T is onto as well. The theorem ensues. $\qquad\square$

We now continue with characterising the dual of the space $L^1(X,\mu)$. Contrary to the case of $L^p(X,\mu)$ for $p > 1$, in this case it can be shown that the analogue of Theorem 14.4 cannot be true in general. A sufficient condition (which is nearly optimal) is that the measure space is σ-finite (recall Definition 7.16).

Theorem 14.5 (Riesz Theorem in L^1). *Suppose that the measure space $(X,\mathfrak{M},\mu)$ is σ-finite. Then, the dual space $\big(L^1(X,\mu)\big)^*$ is isometrically isomorphic to (and hence can be identified with) $L^\infty(X,\mu)$:*

$$\big(L^1(X,\mu)\big)^* = L^\infty(X,\mu).$$

In addition, for any $\phi \in \big(L^1(X,\mu)\big)^*$, there exists a unique $u \in L^\infty(X,\mu)$ such that

$$\phi(f) = \int_X fu\,d\mu \quad \text{for all } f \in L^1(X,\mu),$$

---

[2]If $\mathbb{X}$ is a Banach space and $A \subseteq \mathbb{X}$ is *any* subset, by analogy with the Hilbert spaces, the orthogonal complement $A^\perp$ is the set of bounded linear functionals vanishing on A:

$$A^\perp := \{f \in \mathbb{X}^* : f(a) = 0, \ \forall\, a \in A\}.$$

Note that $A^\perp \subseteq \mathbb{X}^*$ and generally $A^\perp \not\subseteq \mathbb{X}$.

[3]In case this argument is confusing, an equivalent formulation of our reasoning in the proof above without involving orthogonal complements is as follows: let $h \in (L^p(X,\mu))^{**}$ be a bounded linear functional on $(L^p(X,\mu))^*$ such that $h = 0$ on $T(L^{p'}(X,\mu))$. Since we showed that $h \equiv 0$, it follows that $T(L^{p'}(X,\mu))$ has to be dense (by the continuity of h).

and also

$$\|\phi\|_{(L^1(X,\mu))^*} = \|u\|_{L^\infty(X,\mu)}.$$

Proof. The idea is based on a reduction to the Riesz theorem in the case of $p = 2$ by constructing an artificial "weight" which gives us the appropriate improved integrability, by utilising the assumption of σ-finiteness of the measure.

Step 1. By Definition 7.16, X can be written as the countable union $\bigcup_{n=1}^\infty X_n$ of sets $\{X_n : n \in \mathbb{N}\} \subseteq \mathfrak{M}$ such that $0 < \mu(X_n) < \infty$ for all $n \in \mathbb{N}$. Without loss of generality, we may assume that the union is increasing, that is

$$X_1 \subseteq X_2 \subseteq \cdots \subseteq X.$$

Indeed, if this not the case, we define $Z_1 := X_1, \ldots, Z_n := \bigcup_{k=1}^n X_k$ and the new sequence $\{Z_n : n \in \mathbb{N}\} \subseteq \mathfrak{M}$ also consists of sets of finite measure and has the same union but is increasing.

Step 2. We shall now construct a function which satisfies

$$w \in L^2(X,\mu), \quad w > 0 \text{ on } X \quad \text{and} \quad \frac{1}{w} \in L^\infty(X_n, \mu), \ n \in \mathbb{N}.$$

To this aim, we define a μ-measurable function $w : X \longrightarrow \mathbb{R}$ by setting

$$w(x) := \lim_{N \to \infty} w_N(x) \quad \text{for } x \in X,$$

where $(w_N)_1^\infty$ is the sequence of μ-measurable functions on X given by

$$w_N := \sum_{n=1}^N \frac{2^{-n}}{\sqrt{\mu(X_n)}} \chi_{X_n} \quad \text{for } N \in \mathbb{N}.$$

We claim that w as defined above has the required properties. First note that by Minkowski's inequality, for any $n \in \mathbb{N}$ we have

$$\|w_N\|_{L^2(X,\mu)} \leq \sum_{n=1}^N \left(\int_X \left| \frac{2^{-n}}{\sqrt{\mu(X_n)}} \chi_{X_n} \right|^2 d\mu \right)^{1/2} = \sum_{n=1}^N 2^{-n} \leq 1.$$

Since $(w_N)^2$ increases monotonically to w^2 as $N \to \infty$, the Monotone Convergence Theorem yields that $\|w\|_{L^2(X,\mu)} \leq 1$. Further, for any fixed $k \in \mathbb{N}$, on X_k we have $\chi_{X_n}|_{X_k} = 1$ for all $n \geq k$ and hence

$$w|_{X_k} = \sum_{n=1}^\infty \left(\frac{2^{-n}}{\sqrt{\mu(X_n)}} \chi_{X_n} \right)\Big|_{X_k} \geq \frac{2^{-k}}{\sqrt{\mu(X_k)}} > 0.$$

Note in particular that $w > 0$ on X since $X = \bigcup_{k=1}^\infty X_k$. In addition,

$$0 < \frac{1}{w}\Big|_{X_k} \leq 2^k \sqrt{\mu(X_k)} < \infty$$

whence we infer that $1/w \in L^\infty(X_k, \mu)$ for any $k \in \mathbb{N}$, as claimed. Fix now a functional $\phi \in \left(L^1(X, \mu)\right)^*$.

Step 3. We claim that the formula

$$\tilde{\phi}(f) := \phi(wf) \quad \text{for all } f \in L^2(X, \mu),$$

defines a bounded linear functional $L^2(X, \mu) \longrightarrow \mathbb{R}$, that is $\phi \in \left(L^2(X, \mu)\right)^*$. Indeed, linearity is obvious, whilst by Hölder's inequality

$$
\begin{aligned}
\left|\tilde{\phi}(f)\right| &\leq \|\phi\|_{(L^1(\Omega))^*} \|wf\|_{L^1(X,\mu)} \\
&\leq \left(\|\phi\|_{(L^1(X,\mu))^*} \|w\|_{L^2(X,\mu)}\right) \|f\|_{L^2(X,\mu)}.
\end{aligned}
\tag{♯}
$$

By Riesz's Theorem 14.4, there exists a unique $v \in L^2(X, \mu)$ such that

$$\phi(wf) = \int_X fv \, d\mu \quad \text{for all } f \in L^2(X, \mu). \tag{♭}$$

We define the μ-measurable function

$$u := \frac{v}{w} \quad : \quad X \longrightarrow \mathbb{R}$$

to obtain from (♯)-(♭) that

$$
\left.
\begin{aligned}
\phi(fw) &= \int_X fuw \, d\mu, \\
|\phi(fw)| &\leq \|\phi\|_{(L^1(X,\mu))^*} \|wf\|_{L^1(X,\mu)},
\end{aligned}
\right\} \quad \text{for all } f \in L^2(X, \mu). \tag{⊖}
$$

Step 4. Fix any $c > \|\phi\|_{(L^1(X,\mu))^*}$ and consider the set

$$A := \{|u| > c\}.$$

We will show that A is a μ-nullset and from the equality $\mu(A) = 0$ and the definition of the essential supremum it will follow that $u \in L^\infty(X, \mu)$ and

$$\|u\|_{L^\infty(X,\mu)} \leq \|\phi\|_{(L^1(X,\mu))^*}.$$

If for the sake of contradiction we suppose that $\mu(A) > 0$, we may select as f the function

$$\tilde{f} := \chi_{\tilde{A}\bigcup\{u>0\}} - \chi_{\tilde{A}\bigcup\{u<0\}}$$

where $\tilde{A}$ is a measurable subset of A such that $0 < \mu(\tilde{A}) < \infty$. The existence of such a set is a consequence of the σ-finiteness of the measure (just take $\tilde{A} := A \bigcap X_n$ for a large enough $n \in \mathbb{N}$). Then, we calculate

$$
\|w\tilde{f}\|_{L^1(X,\mu)} = \int_{\tilde{A}\bigcup\{u\geq0\}} |w| \, d\mu + \int_{\tilde{A}\bigcup\{u<0\}} |w| \, d\mu = \int_{\tilde{A}} w \, d\mu,
$$

$$
\left|\int_X \tilde{f} uw \, d\mu\right| = \int_{\tilde{A}\bigcup\{u\geq0\}} uw \, d\mu + \int_{\tilde{A}\bigcup\{u<0\}} (-u)w \, d\mu = \int_{\tilde{A}} |u|w \, d\mu.
$$

Consequently, in view of the above estimates, $(\ominus)$ for the choice $f = \tilde{f}$ yields

$$\int_{\tilde{A}} |u| w \, d\mu \leq \|\phi\|_{(L^1(X,\mu))^*} \int_{\tilde{A}} w \, d\mu$$

and hence, by the definition of the sets A and $\tilde{A}$,

$$c \int_{\tilde{A}} w \, d\mu \leq \|\phi\|_{(L^1(X,\mu))^*} \int_{\tilde{A}} w \, d\mu.$$

The later is a contradiction, because $w > 0$. Therefore, $\mu(A) = 0$.

Step 5. Fix a function $g \in L^1(X,\mu)$ and set $g_n := g\chi_{X_n}$ for any $n \in \mathbb{N}$. Then, we have that $g_n/w \in L^1(X_n, \mu)$ as a result of Hölder's inequality, because $1/w \in L^\infty(X_n, \mu)$. By extending g_n by zero on $X \setminus X_n$, we have $g_n/w \in L^1(X,\mu)$. By $(\ominus)$ and since $u = v/w$, we calculate

$$\phi(g_n) = \phi\left(w\frac{g_n}{w}\right) = \int_X vs\frac{g_n}{w} \, d\mu = \int_X ug_n \, d\mu.$$

Hence, we may estimate

$$|\phi(g_n)| \leq \left| \int_X g_n u \, d\mu \right| \leq \|u\|_{L^\infty(X,\mu)} \|g_n\|_{L^1(X,\mu)}. \qquad (\oplus)$$

Note now that since $|g_n| \leq |g|$ and $g_n \longrightarrow g$ μ-a.e. on X as $n \to \infty$, by the Dominated Convergence Theorem we have $g_n \longrightarrow g$ in $L^1(X,\mu)$ as $n \to \infty$. Since $\phi : L^1(X,\mu) \longrightarrow \mathbb{R}$ is continuous, by passing to the limit as $n \to \infty$ in the estimate $(\oplus)$, we infer that

$$|\phi(g)| \leq \|u\|_{L^\infty(X,\mu)} \|g\|_{L^1(X,\mu)}.$$

Therefore, we have

$$\|\phi\|_{(L^1(X,\mu))^*} = \sup\left\{|\phi(g)| : g \in L^1(X,\mu), \|g\|_{L^1(X,\mu)} \leq 1\right\} \leq \|u\|_{L^\infty(X,\mu)}.$$

The theorem ensues. $\qquad\qquad\qquad\qquad\qquad\qquad\qquad\qquad\qquad\qquad\qquad\qquad\quad\square$

Remark 14.6 (The dual of L^∞). It can be proved that the dual space $\left(L^\infty(X,\mu)\right)^*$ is a very large space of so-called *finitely-additive signed (real-valued) bounded measures* $\mu : \mathfrak{M} \longrightarrow \mathbb{R}$ which are absolutely continuous with respect to μ, commonly denoted as $\mathrm{ba}_\mu(X)$. Real-valued measures can be defined as differences of positive measures, namely as $\nu = \nu^+ - \nu^-$, where $\nu^\pm$ are positive (finite) measures. These finitely-additive measures have been studied quite systematically in recent years but they are *very* pathological and they do not share the nice properties of the countably additive measures.[4] See also Example 14.15 and Exercise 14.10 that follow.

---

[4]For this topic and many more advanced questions, as for instance the optimal condition on $(X, \mathfrak{M}, \mu)$ in order to have $\left(L^1(X,\mu)\right)^* = L^\infty(X,\mu)$– the interested reader may consult the (much more advanced) book [18].

14.4   The true meaning of weak convergence in L^p

In this section we discuss the special meaning the notion of weak/weak* convergence takes in the class of the Lebesgue spaces. We shall again discuss the general case of an arbitrary measure space

$$(X, \mathfrak{M}, \mu)$$

but which we will now assume to be σ-**finite**. Recall that by our representation Theorems 14.4 and 14.5, every bounded continuous functional on L^p for $p \in [1, \infty)$ can be represented as an $L^{p'}$ function by means of the integral which (in the terminology of Chapter 12) plays the role of a duality pairing between L^p and $L^{p'}$.

Remark 14.7 (Weak convergence in L^p, $1 < p < \infty$). Let $(X, \mathfrak{M}, \mu)$ be a σ-finite measure space. By the definition of weak convergence and Riesz's Theorem 14.4, the notion of weak convergence in $L^p(X, \mu)$ takes the form

$$f_m \longrightarrow f \text{ in } L^p(X, \mu) \quad \Longleftrightarrow \quad \int_X f_m \phi \, d\mu \longrightarrow \int_X f \phi \, d\mu,$$

for all $\phi \in L^{p'}(X, \mu)$ as $m \to \infty$. The particular choice of

$$\phi = \chi_E \text{ with } E \in \mathfrak{M} \text{ such that } 0 < \mu(E) < \infty,$$

implies that if $f_m \longrightarrow f$ in $L^p(X, \mu)$ as $m \to \infty$, then we have **convergence of the averages** over any measurable set of finite measure:

$$\frac{1}{\mu(E)} \int_E f_m \, d\mu \longrightarrow \frac{1}{\mu(E)} \int_E f \, d\mu \quad \text{as } m \to \infty.$$

Actually, it can be shown that the opposite of the above statement is true as well if we impose the *uniform bound*

$$\sup_{m \in \mathbb{N}} \|f_m\|_{L^p(X,\mu)} < \infty.$$

This fact will be discussed further in the exercises and later below we shall also discuss how weak convergence may fail to be strong. We now consider the case of L^1.

Remark 14.8 (Weak convergence in L^1). Again from the definition of weak convergence and Riesz's Theorem 14.5, the notion weak convergence in L^1 takes the form

$$f_m \longrightarrow f \text{ in } L^1(X, \mu) \quad \Longleftrightarrow \quad \int_X f_m \phi \, d\mu \longrightarrow \int_X f \phi \, d\mu,$$

for all $\phi \in L^\infty(X, \mu)$ as $m \to \infty$. The particular choice of

$$\phi = \chi_E \quad \text{with } E \in \mathfrak{M} \text{ such that } 0 < \mu(E) < \infty,$$

implies that if $f_m \longrightarrow f$ in $L^1(X, \mu)$ as $m \to \infty$, then we have **convergence of the averages** over any measurable set of finite measure:

$$\frac{1}{\mu(E)} \int_E f_m \, d\mu \longrightarrow \frac{1}{\mu(E)} \int_E f \, d\mu \quad \text{as } m \to \infty.$$

However, in this case it can be show that the opposite is true as well, **even in the absence of a uniform norm bound**[5] (unlike in the case $1 < p < \infty$).

Remark 14.9 (Weak* convergence in L^∞). By the definition of the weak* convergence and Riesz's Theorem 14.5, the notion of weak* convergence in L^∞ takes the form

$$f_m \overset{*}{\longrightarrow} f \text{ in } L^\infty(X, \mu) \quad \Longleftrightarrow \quad \int_X f_m \phi \, d\mu \longrightarrow \int_X f \phi \, d\mu,$$

for all $\phi \in L^1(X, \mu)$ as $m \to \infty$. As in the case of $1 < p < \infty$, the choice of

$$\phi = \chi_E \quad \text{with } E \in \mathfrak{M} \text{ such that } 0 < \mu(E) < \infty,$$

implies that if $f_m \overset{*}{\longrightarrow} f$ in $L^\infty(X, \mu)$ as $m \to \infty$, then we have **convergence of the averages** over any measurable set of finite measure:

$$\frac{1}{\mu(E)} \int_E f_m \, d\mu \longrightarrow \frac{1}{\mu(E)} \int_E f \, d\mu \quad \text{as } m \to \infty.$$

Here again is can be shown that again the opposite is true as well, if we impose the *uniform bound*

$$\sup_{m \in \mathbb{N}} \|f_m\|_{L^\infty(X, \mu)} < \infty.$$

The main difference however between weak convergence in $L^p(X, \mu)$ for $1 \le p < \infty$ and weak* convergence in $L^\infty(X, \mu)$ is that the weak* convergence may fail to be strong only due to "oscillations of mass", since the essential boundedness excludes "mass escaping to ∞", see Example 14.10 below. **On the other hand, for $1 \le p < \infty$ the weak convergence might fail to be strong additionally due to "concentrations of mass",** see the upcoming Example 14.12.

Example 14.10 (Oscillations). Let $X = (0, 1)$, $\mu = \mathcal{L}^1$, $p \in (1, \infty]$ and consider the sequence (see Figure 14.1)

$$f_m(x) := \sin(2m\pi x), \quad m \in \mathbb{N}.$$

Then, it can be shown that $f_m \longrightarrow 0$ in $L^p((0, 1))$ and $f_m \overset{*}{\longrightarrow} 0$ in $L^\infty((0, 1))$ as $m \to \infty$, essentially because the averages converge to zero. However, strong convergence fails and $f_m \not\longrightarrow 0$ because the functions oscillate faster-and-faster and the pointwise values in the limit "diffuse".

---

[5]The interested reader might consult e.g. [17] for a proof of this fact.

Remark 14.11 (Young measures). The presence oscillations is a serious problem when one is dealing with nonlinear L^p functionals, which are typically **not** weakly continuous. In fact, there exists a deep tool known as Young measures, which is far beyond the scope of this book, that allows one to detect these oscillations quite precisely, by constructing for a.e. $x \in \Omega$ a probability measure μ_x on $\mathbb{R}$. For this advanced topic the interested reader may e.g. consult [17, 18, 24, 31]. The family of measures $(\mu_x)_{x \in \Omega}$ describes the asymptotic behaviour of the sequence $(f_m)_1^\infty$ which in general may not converge pointwise. For the particular sequence of Example 14.10, μ_x is a multiple of the Lebesgue measure on $[-1, 1]$ for a.e. $x \in (0, 1)$. It can be shown that the sequence $(f_m)_1^\infty$ converges a.e. on Ω to some f if and only if the Young measure is a.e. equal to the constant Young measure given by the Dirac measure $\mu_x = \delta_{f(x)}$ for a.e. $x \in \Omega$. This represents the lack of diffusion. On the other hand, if the sequence fails to converge pointwise on a set of positive measure, then thereon the probability measures necessarily diffuse and cannot be concentration measures anymore.

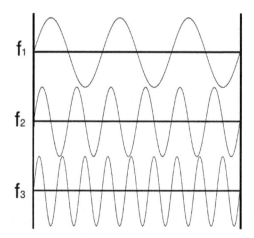

Figure 14.1: The first three terms of the sequence of functions of the Example 14.10. The oscillations around zero become faster-and-faster as $m \to \infty$ and the pointwise limit of the sequence does not exist.

Example 14.12 (concentrations). Let $X = \mathbb{R}$, $\mu = \mathcal{L}^1$, $p \in (1, \infty)$ and consider the sequence (see Figure 14.2)

$$f_m := \left(\frac{m}{2}\right)^p \chi_{[-1/m,\, 1/m]}, \quad m \in \mathbb{N}.$$

Then, it is easy to check that $\|f_m\|_{L^p(\mathbb{R})} = 1$ for all $m \in \mathbb{N}$. Since $L^p(\mathbb{R})$ is reflexive, there exists $f \in L^p(\mathbb{R})$ such that

$$f_m \longrightarrow f \text{ along a subsequence } (m_k)_1^\infty, \text{ as } m \to \infty.$$

Since
$$f_m \equiv 0, \quad \text{on} \quad \mathbb{R} \setminus \left[-\frac{1}{m}, \frac{1}{m} \right],$$

we have that $f_m \longrightarrow 0$ on $\mathbb{R} \setminus \{0\}$. In addition, for any $\varepsilon > 0$ fixed, we have

$$f_m \longrightarrow 0 \quad \text{in} \quad L^p \left(\mathbb{R} \setminus (-\varepsilon, \varepsilon) \right) \quad \text{as} \quad m \to \infty.$$

Hence,
$$f \equiv 0 \quad \text{on} \quad \mathbb{R} \setminus (-\varepsilon, \varepsilon) \quad \text{for all } \varepsilon > 0,$$

which implies that $f \equiv 0$ on $\mathbb{R} \setminus \{0\}$. It follows that $f = 0$ a.e. on $\mathbb{R}$. However, we also have that $f_m \nrightarrow 0$ because $\|f_m\|_{L^p(\mathbb{R})} = 1$ and it is not possible to converge strongly to zero.

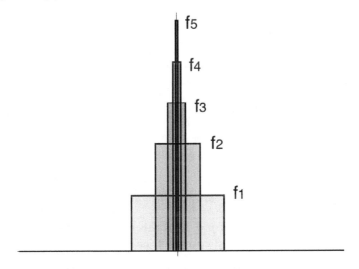

Figure 14.2: The first five terms of the sequence of functions of the example. The sequence becomes "taller and thinner" as $m \to \infty$, concentrating at the origin.

14.5   Sequential weak compactness and weak* compactness

In this final section we present some sequential compactness results for the (norm closed) bounded sets of the class of Lebesgue spaces with respect to the weak and the weak* topology. The compactness results below are direct consequences of the results of Chapter 13, combined with the duality representation Theorems 14.4 and 14.5.

We begin with the case of $p = \infty$. Let $(X, \mathfrak{M}, \mu)$ be a σ-finite measure space. We already saw that $L^\infty(X, \mu)$ is a dual space and that $\left(L^1(X, \mu)\right)^* = L^\infty(X, \mu)$. Moreover, $L^1(X, \mu)$ **is separable for a variety of measure spaces**[6], including, as we saw in Section 9.5, the archetypal example of $(\Omega, \mathcal{L}(\mathbb{R}^n)|_\Omega, \mathcal{L}^n)$ with $\Omega \subseteq \mathbb{R}^n$ open. Hence, by the sequential Alaoglu-Banach Theorem 13.17, we have the next very important sequential compactness result:

Theorem 14.13 (Weak* sequential compactness of $\overline{\mathbb{B}}^{L^\infty(X,\mu)}$). *Let* $(X, \mathfrak{M}, \mu)$ *be a σ-finite measure space, such that $L^1(X, \mu)$ is separable (e.g., $(\Omega, \mathcal{L}(\mathbb{R}^n)|_\Omega, \mathcal{L}^n)$). Then, the closed unit ball of the space $L^\infty(X, \mu)$ is sequentially weakly* compact.*

Equivalently, when $(f_m)_1^\infty \subseteq L^\infty(X, \mu)$ is a bounded sequence, there exist $f \in L^\infty(X, \mu)$ and a subsequence $(f_{m_k})_1^\infty \subseteq (f_m)_1^\infty$ such that

$$f_{m_k} \overset{*}{\longrightarrow} f \quad in\ L^\infty(X, \mu) \ as\ k \to \infty.$$

We now turn to the case of $1 < p < \infty$. By using the reflexivity of $L^p(X, \mu)$, the sequential compactness of closed bounded sets for the weak topology (Theorem 13.24) yields the next very important compactness result:

Theorem 14.14 (Weak sequential compactness of $\overline{\mathbb{B}}^{L^p(X,\mu)}$). *Let* $(X, \mathfrak{M}, \mu)$ *be a measure space. Then, the unit closed ball of the space $L^p(X, \mu)$ is sequentially weakly compact, when $p \in (1, \infty)$. Equivalently, when $(f_m)_1^\infty \subseteq L^p(X, \mu)$ is a bounded sequence, there exist $f \in L^p(X, \mu)$ and a subsequence $(f_{m_k})_1^\infty \subseteq (f_m)_1^\infty$ such that*

$$f_{m_k} \longrightarrow f \quad in\ L^p(X, \mu) \ as\ k \to \infty.$$

We now consider the case of $p = 1$. In this case **the closed unit ball of L^1 is not sequentially weakly compact**, as Example 14.15 below certifies:

Example 14.15 ($L^1(X, \mu)$ is not reflexive)**.** The dual space of $L^\infty(X, \mu)$

$$\left(L^\infty(X, \mu)\right)^* = \left(L^1(X, \mu)\right)^{**}$$

is strictly larger than $L^1(X, \mu)$ (Recall Remark 14.6) and the evaluation mapping is never a surjection. In particular, $L^1(X, \mu)$ is a proper (strict) subspace of its bidual space, even in the case of the Lebesgue measure on the Euclidean space. To this end, let $X = \mathbb{R}$, $\mu = \mathcal{L}^1$ and take the sequence (see Figure 14.7)

$$f_m := \frac{m}{2} \chi_{[-1/m, 1/m]}, \quad m \in \mathbb{N}.$$

---

[6]It can be shown that a sufficient condition for the general space $L^p(X, \mu)$ for $p \in [1, \infty)$ to be separable is that the **measure space** $(X, \mathfrak{M}, \mu)$ **is separable**, in a different sense from the topological separability. In the measure-theoretic sense, a measure space $(X, \mathfrak{M}, \mu)$ (without any topology on X whatsoever) is called separable if $\mathfrak{M}$ is countably generated, namely when there exists a sequence $\mathcal{S} = \{S_m : m \in \mathbb{N}\} \subseteq \mathfrak{M}$ of measurable sets which generates $\mathfrak{M} = \mathcal{G}(\mathcal{S})$ (Definition 7.25). We refrain from giving the details of this result, for which the reader is referred e.g. to the more advanced book [18].

Then, $\|f_m\|_{L^1(\mathbb{R})} = 1$ for all $m \in \mathbb{N}$. If hypothetically $L^1(\mathbb{R})$ were reflexive, then we would have $f_m \longrightarrow f$ as $m \to \infty$ along a subsequence $(m_k)_1^\infty$. Since

$$f_m \equiv 0, \quad \text{on } \mathbb{R} \setminus \left[-\frac{1}{m}, \frac{1}{m} \right],$$

we have that $f_m \longrightarrow 0$ on $\mathbb{R} \setminus \{0\}$. In addition, for any $\varepsilon > 0$ fixed, we have

$$f_m \longrightarrow 0, \quad \text{in } L^1\big(\mathbb{R} \setminus (-\varepsilon, \varepsilon)\big), \quad \text{as } m \to \infty.$$

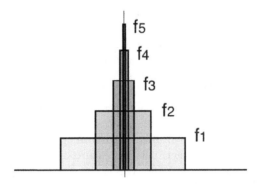

Figure 14.7: The first five terms of the sequence of functions, here for $p = 1$. The sequence again becomes "taller and thinner" as $m \to \infty$, concentrating around the origin, but surprisingly in this case the limit is not a function!

On the other hand, by the contradiction hypothesis we have that

$$\int_{\mathbb{R}} \phi f_m \, d\mathcal{L}^1 \longrightarrow \int_{\mathbb{R}} \phi f \, d\mathcal{L}^1, \quad \text{as } m \to \infty,$$

for any $\phi \in L^\infty(\mathbb{R})$. The choice of

$$\phi := \chi_{\{f>0\}\setminus(-\varepsilon,\varepsilon)} - \chi_{\{f<0\}\setminus(-\varepsilon,\varepsilon)}$$

yields that

$$\int_{\mathbb{R}\setminus(-\varepsilon,\varepsilon)} |f| \, d\mathcal{L}^1 = \lim_{m\to\infty} \int_{\mathbb{R}\setminus(-\varepsilon,\varepsilon)} f_m \, d\mathcal{L}^1 = 0.$$

Hence, $f \equiv 0$ on $\mathbb{R} \setminus (-\varepsilon, \varepsilon)$, for all $\varepsilon > 0$, which implies that

$$f \equiv 0 \ \text{ on } \mathbb{R} \setminus \{0\}. \tag{♮}$$

On the other hand, the choice $\phi := \chi_{\mathbb{R}}$ yields

$$\int_{\mathbb{R}} f \, d\mathcal{L}^1 = \lim_{m\to\infty} \int_{\mathbb{R}} f_m \, d\mathcal{L}^1 = 1. \tag{♮♮}$$

However, (♮)-(♮♮) contradict each other, since no non-negative function exists with integral equal to one and identically zero except for one point.

Remark 14.16 (Embedding of L^1 into the space of signed Radon measures).

(i) Continuing in the setting of Example 14.15, we note that it can be shown that this sequence of functions converges in a *"very weak"* sense to the Dirac measure δ_0. This very weak convergence is defined as: for any $\phi \in C_c^0(\mathbb{R})$, we have

$$\int_{\mathbb{R}} \phi f_m \, d\mathcal{L}^1 \longrightarrow \int_{\mathbb{R}} \phi \, d\delta_0 = \phi(0) \quad \text{as} \quad m \to \infty.$$

In fact, this convergence is precisely the weak* convergence of the dual Banach space $\left(C_0^0(\mathbb{R})\right)^*$, where $C_0^0(\mathbb{R})$ is the space of continuous functions vanishing at infinity. It can be shown through a "Riesz-type theorem" that this dual space actually consists of real-valued finite measures, which can be understood as differences $\mu = \mu^+ - \mu^-$ of positive measures $\mu^\pm \geq 0$ (by analogy to the decomposition $f = f^+ - f^-$ for functions). In this relatively elementary book we refrain from discussing the more advanced topic of spaces of measures and the interested reader may consult the more advanced books [4, 16, 18]. The limiting object is exactly what in **non-rigorous** applications of Mathematics call the "Dirac function": in fact, it is a measure, not a function! Completely analogous remarks can be made for the general case regarding the lack of compactness of the space $L^1(X, \mu)$.

(ii) Consider the measure space $(\Omega, \mathcal{L}(\mathbb{R}^n)|_\Omega, \mathcal{L}^n)$ with $\Omega \subseteq \mathbb{R}^n$ open. A standard "remedy" of the lack of weak compactness of closed bounded sets in $L^1(\Omega)$ is to imbed $L^1(\Omega)$ in the space $\mathcal{M}(\Omega) = \left(C_0^0(\Omega)\right)^*$ of real-valued regular (so-called *Radon*) measures via the mapping

$$L^1(\Omega) \ni u \mapsto \int_{(\cdot)} u \, d\mathcal{L}^n \in \mathcal{M}(\Omega).$$

This process "restores" weak* compactness in the sense that any bounded sequence of $L^1(\Omega)$ functions converges subsequentially to a measure! Then one needs to examine whether the limit is a genuine measure with singular part with respect to $\mathcal{L}^n$, or if it is essentially a function, namely an absolutely continuous measure with respect to $\mathcal{L}^n$. We will not discuss this more advanced topic any further and the reader might see e.g. [14, 18] for further reading.

According to Remark 14.7, the main point is that **weak convergence in L^p means convergence of the averages over arbitrary measurable sets of finite measure** of the functions to the corresponding averages of their weak limit. On the other hand, strong (norm) convergence in $L^p(X, \mu)$ means that the area between the graphs of the functions and their limit converges to zero and we know that along a subsequence we have pointwise convergence for a.e. point was well. As we have already seen in the abstract functional setting, in general **weak convergence does not imply strong convergence**. The examples we gave earlier show that the typical problem is the **occurrence of oscillations and/or concentrations** in the approximating sequences, whilst in general arbitrarily complicated entwined combinations of both phenomena can be observed.

14.6 Exercises

Exercise 14.1. Let $(X, \mathfrak{M}, \mu)$ be a finite measure space (i.e. $\mu(X) < \infty$), let also $\Phi : \mathbb{R}^d \longrightarrow \mathbb{R}$ be a convex continuously differentiable function and $d \in \mathbb{N}$. Consider the functional

$$E \; : \;\; L^\infty(X, \mu, \mathbb{R}^d) \longrightarrow \mathbb{R}, \qquad E(f) := \int_X \Phi(f) \, d\mu.$$

Show that E is sequentially weakly* LSC (lower semi-continuous) on the space $L^\infty(X, \mu, \mathbb{R}^d)$, that is

$$E(f) \leq \liminf_{m \to \infty} E(f_m),$$

whenever $f_m \overset{*}{\rightharpoonup} f$ in $L^\infty(X, \mu, \mathbb{R}^d)$ as $m \to \infty$.

[Hint: Use that the hyperplane which is tangent to the graph of Φ at any $p \in \mathbb{R}^d$ (determined by the range of the gradient $D\Phi(p)$, seen a linear functional $\mathbb{R}^d \ni q \mapsto D\Phi(p) \cdot q \in \mathbb{R}$) lies below the graph of Φ, as a consequence of its convexity.]

Exercise 14.2. Consider the Hilbert space $L^2(0, 2\pi)$ with the standard inner product. By using that $C_c^\infty(0, 2\pi)$ is dense in $L^2(0, 2\pi)$ and the characterisation of weak convergence in L^2 (via Riesz's theorem), show that, if $f_m(t) := \cos(mt)$, $t \in (0, 2\pi)$, $m \in \mathbb{N}$, then

$$f_m \rightharpoonup 0 \quad \text{but} \quad f_m^2 \rightharpoonup \frac{1}{2},$$

as $m \to \infty$ in $L^2(0, 2\pi)$. Deduce that in general the next implication is FALSE:

$$\left. \begin{array}{c} f_m \rightharpoonup f \\ g_m \rightharpoonup g \end{array} \right\} \;\; \Longrightarrow \;\; \int_0^{2\pi} f_m g_m \longrightarrow \int_0^{2\pi} fg.$$

Exercise 14.3. Let $\Omega \subseteq \mathbb{R}^n$ be open and $1 < p < \infty$. Show that

$$f_m \rightharpoonup f \;\; \text{in } L^p(\Omega) \text{ as } m \to \infty \;\; \Longleftrightarrow \;\; \begin{cases} \|f_m\|_{L^p(\Omega)} \text{ is bounded and} \\ \text{for all } E \subseteq \Omega \text{ measurable} \\ \text{with } |E| < \infty, \\ \displaystyle\int_E f_m \longrightarrow \int_E f \text{ as } m \to \infty. \end{cases}$$

Exercise 14.4. Let $\Omega \subseteq \mathbb{R}^n$ be open. Show that

$$f_m \overset{*}{\rightharpoonup} f \text{ in } L^\infty(\Omega), \text{ as } m \to \infty \iff \begin{cases} \|f_m\|_{L^\infty(\Omega)} \text{ is bounded and} \\ \text{for all } E \subseteq \Omega \text{ measurable} \\ \text{with } |E| < \infty, \\ \displaystyle\int_E f_m \longrightarrow \int_E f, \text{ as } m \to \infty. \end{cases}$$

Exercise 14.5. Show that the space of smooth bounded functions $C_b^\infty(\mathbb{R}^n)$ is weakly* dense in $L^\infty(\Omega)$:

$$\overline{C_b^\infty(\mathbb{R}^n)}^{w*} = L^\infty(\mathbb{R}^n).$$

Exercise 14.6. Let $\Omega \subseteq \mathbb{R}^n$ be open. Show that the space of smooth compactly supported functions $C_c^\infty(\Omega)$ is weakly* dense in $L^\infty(\Omega)$:

$$\overline{C_c^\infty(\Omega)}^{w*} = L^\infty(\Omega).$$

Exercise 14.7. Suppose that $1 < p < \infty$ and that $u_m \rightharpoonup u$ in $L^p(\Omega)$ as $m \to \infty$. If in addition

$$\|u_m\|_{L^p(\Omega)} \longrightarrow \|u\|_{L^p(\Omega)} \quad \text{as } m \to \infty,$$

deduce that the convergence is actually strong:

$$u_m \longrightarrow u \text{ in } L^p(\Omega) \text{ as } m \to \infty.$$

Exercise 14.8 (Poincaré inequality). Let $\Omega \subseteq \mathbb{R}^n$ be a convex bounded open set. Show that there exists a constant $C > 0$ depending only on Ω so that for any $u \in C_c^1(\Omega)$ we have the estimate

$$\|u\|_{L^\infty(\Omega)} \leq C\|Du\|_{L^\infty(\Omega)},$$

where D symbolises the gradient of u.

Exercise 14.9. Suppose that $1 < p < \infty$ and that $u_m \rightharpoonup u$ in $L^p(\mathbb{R}^n)$ as $m \to \infty$. Consider the mollified sequence $(u_m * \eta^\varepsilon)_1^\infty$ for some $\varepsilon > 0$. Show that

$$u_m * \eta^\varepsilon \longrightarrow u * \eta^\varepsilon \quad \text{a.e. on } \mathbb{R}^n, \text{ as } m \to \infty.$$

Does this hold if $p = 1$? Does it hold if $p = \infty$ but when $u_m \overset{*}{\rightharpoonup} u$ in $L^\infty(\mathbb{R}^n)$ as $m \to \infty$? Justify your answer in both cases.

Exercise 14.10. Here we construct an example of a **finitely additive** Borel measure on $\mathbb{R}$ which is absolutely continuous with respect to the Lebesgue measure $\mathcal{L}^1$, but which is not a countably additive. Such "measures" lie in the space $\text{ba}_{\mathcal{L}^1}(\mathbb{R}) \cong (L^\infty(\mathbb{R}))^*$. We define

$$\mu \; : \; \mathcal{L}(\mathbb{R}) \longrightarrow [0, \infty]$$

by setting

$$\mu(A) := \lim_{n\to\infty} \frac{\mathcal{L}^1(A \cap (0, n))}{n}$$

where the limit above is understood as a specified **Banach limit**, in the way defined in Exercise 11.14, of the bounded sequence $(\mathcal{L}^1(A \cap (0, n))/n)_{n=1}^{\infty} \in \ell^{\infty}$ (Banach limits exist even for non-convergent bounded sequences!). Show that:

(1) For any Lebesgue nullset on $\mathbb{R}$, namely for any subset $A \subseteq \mathbb{R}$ satisfying $\mathcal{L}^1(A) = 0$, we have $\mu(A) = 0$. In short,

$$\mu \ll \mathcal{L}^1.$$

(2) For any disjoint sets $A, B \in \mathcal{L}(\mathbb{R})$ with $A \cap B = \emptyset$, we have

$$\mu(A \cup B) = \mu(A) + \mu(B).$$

(3) There exists a disjoint sequence $(A_n)_1^{\infty} \subseteq \mathcal{L}(\mathbb{R})$ such that

$$\mu\left(\bigcup_{n=1}^{\infty} A_n\right) \neq \sum_{n=1}^{\infty} \mu(A_n)$$

[Hint: Consider $A_k := (k-1, k]$, $k \in \mathbb{N}$.]

Chapter 15

Solutions to the exercises

15.1 Solutions to the exercises in Chapter 1

Solution to Exercise 1.1. We argue in the same notation as in the main text:

$$f^{-1}\left(\bigcap_{i \in I} A_i\right) = \left\{x \in X \ \middle|\ f(x) \in \bigcap_{i \in I} A_i\right\}$$
$$= \left\{x \in X \ \middle|\ \text{for all } i \in I: \ f(x) \in A_i\right\}$$
$$= \left\{x \in X \ \middle|\ \text{for all } i \in I: \ x \in f^{-1}(A_i)\right\}$$
$$= \bigcap_{i \in I} f^{-1}(A_i),$$

$$f^{-1}\left(\bigcup_{i \in I} A_i\right) = \left\{x \in X \ \middle|\ f(x) \in \bigcup_{i \in I} A_i\right\}$$
$$= \left\{x \in X \ \middle|\ \text{exists } i \in I: \ f(x) \in A_i\right\}$$
$$= \left\{x \in X \ \middle|\ \text{exists } i \in I: \ x \in f^{-1}(A_i)\right\}$$
$$= \bigcup_{i \in I} f^{-1}(A_i).$$

Further,

$$f^{-1}(Y \setminus X) = \{x \in X \ : \ f(x) \in Y \setminus B\}$$
$$= \{x \in X \ : \ f(x) \notin B\}$$
$$= X \setminus \{x \in X \ : \ f(x) \in B\}$$
$$= X \setminus f^1(B).$$

If now $B \subseteq C$, then $f^{-1}(B) = \{x \in X : f(x) \in B\} \subseteq \{x \in X : f(x) \in C\} = f^{-1}(C)$. Next, we have

$$(f|_A)^{-1}(B) = \{x \in X \ : \ (f|_A)(x) \in B\}$$
$$= \{x \in A \ : \ f(x) \in B\}$$
$$= A \bigcap \{x \in X \ : \ f(x) \in B\}$$
$$= A \bigcap f^1(B).$$

Also,

$$f\left(\bigcup_{i \in I} A_i\right) = \left\{y \in Y \,\middle|\, \exists\, x \in \bigcup_{i \in I} A_i : \ f(x) = y\right\}$$

$$= \left\{y \in Y \,\middle|\, \exists\, i \in I \ \& \ \exists\, x \in A_i : \ f(x) = y\right\}$$

$$= \bigcup_{i \in I} \left\{y \in Y \,\middle|\, \exists\, x \in A_i : \ f(x) = y\right\}$$

$$= \bigcup_{i \in I} f(A_i),$$

$$f\left(\bigcap_{i \in I} A_i\right) = \left\{y \in Y \,\middle|\, \exists\, x \in \bigcap_{i \in I} A_i : \ f(x) = y\right\}$$

$$= \left\{y \in Y \,\middle|\, \exists\, x \in X : \ x \in A_i, \forall\, i \in I \ \& \ f(x) = y\right\}$$

$$\subseteq \left\{y \in Y \,\middle|\, \forall\, i \in I, \ \exists\, x \in A_i : \ f(x) = y\right\}$$

$$= \bigcup_{i \in I} \left\{y \in Y \,\middle|\, \exists\, x \in A_i : \ f(x) = y\right\}$$

$$= \bigcap_{i \in I} f(A_i).$$

In addition, if $A \subseteq D$ then

$$f(A) = \left\{y \in Y \,\middle|\, \exists\, x \in A : \ f(x) = y\right\}$$

$$\subseteq \left\{y \in Y \,\middle|\, \exists\, x \in D : \ f(x) = y\right\}$$

$$= f(D).$$

Now, we have

$$f^{-1}(f(A)) = \left\{x \in X \,\middle|\, \exists\, f(x) \in f(A)\right\}$$

$$\supseteq A,$$

$$f\big(f^{-1}(B)\big) = \left\{y \in Y \,\middle|\, \exists\, x \in f^{-1}(B) : \ f(x) = y\right\}$$

$$= \left\{y \in Y \,\middle|\, \exists\, x \in X : \ f(x) \in B \ \& \ f(x) = y\right\}$$

$$\subseteq B.$$

Finally, the equivalence $f^{-1}(f(A)) \supseteq A$ if and only if $f\big(f^{-1}(B)\big) \subseteq B$ follows from the previous. □

Solution to Exercise 1.2 1. Let X be a countable set. Then, there exists an injective map $f : X \longrightarrow \mathbb{N}$. If $A \subseteq X$ is a subset, then the restriction $f|_A : A \longrightarrow \mathbb{N}$ is also injective. Hence A is countable.
2. Suppose now $h : X \longrightarrow Y$ is an injective map between sets and $h(X)$ a is countable set. Then, the inverse function $h^{-1} : h(X) \longrightarrow X$ exists and is bijective. Since $h(X)$ is countable, there exists $f : \mathbb{N} \longrightarrow h(X)$ which is surjective. Hence, the

map $h^{-1} \circ f : \mathbb{N} \longrightarrow h(X) \longrightarrow X$ is surjective, showing that X is countable, as desired. □

Solution to Exercise 1.3. Suppose there exists a surjective map $g : \mathbb{N} \longrightarrow \{0,1\}^{\mathbb{N}}$. We set $x^{(n)} := g(n)$, $n \in \mathbb{N}$. Then each $x^{(n)}$ is a sequence $(x_1^{(n)}, x_2^{(n)}, x_3^{(n)}, \ldots)$. Then we have to construct a sequence which is not in the image. Define

$$ y_n := \begin{cases} 1, & \text{if } x_n^{(n)} = 0, \\ 0, & \text{if } x_n^{(n)} = 1. \end{cases} $$

Then y and $x^{(n)}$ are different, because they differ at the n-th entry. □

Solution to Exercise 1.4. $\mathcal{P}(X)$ is countable if and only if X is finite. Firstly, if X is finite, then also $\mathcal{P}(X)$ is finite and hence countable. In particular, $\mathcal{P}(X)$ has 2^N elements when X has N elements. Secondly, X cannot be countable and have $\mathcal{P}(X)$ countable too. Indeed, we know that a countable infinite set has the same cardinal as $\mathbb{N}$ and by a known lemma in the main text we know that $\mathcal{P}(\mathbb{N})$ has the same cardinality as $\mathbb{R}$ which is uncountable. □

Solution to Exercise 1.5. Suppose for the sake of contradiction that the family $\{f_k : k \in \mathbb{N} \cup \{0\}\}$ is linearly dependent. Then there exists $n \in \mathbb{N}$, and scalars $\lambda_0, \lambda_1, \ldots, \lambda_n \in \mathbb{R}$ with $\lambda_n \neq 0$, such that

$$ \lambda_0 f_0 + \lambda_1 f_1 + \ldots + \lambda_n f_n = 0, $$

meaning that

$$ \lambda_0 + \lambda_1 x + \ldots + \lambda_n x^n = 0 \quad \text{for all } x \in [0,1]. \qquad (*) $$

However, it is well known that any polynomial of degree n with real coefficients cannot have more than n real roots, and hence in particular cannot vanish identically on an interval with different end-points. Hence in $(*)$ a contradiction has been reached with the fact that $\lambda_n \neq 0$. In conclusion, the family $\{f_k : k \in \mathbb{N} \cup \{0\}\}$ is linearly independent.

If the vector space $C([0,1], \mathbb{R})$ were finite-dimensional, then we know from Linear Algebra that it would have a certain dimension, say $m \in \mathbb{N}$, and that then it could not contain more than m linearly independent elements. However, the fact that the infinite family $\{f_k : k \in \mathbb{N} \cup \{0\}\}$ is linearly independent shows that the vector space $C([0,1], \mathbb{R})$ is infinite-dimensional. □

15.2 Solutions to the exercises in Chapter 2

Solution to Exercise 2.1. We need to check, for each of the functions $\| \cdot \|_1$, $\| \cdot \|_2$ and $\| \cdot \|_\infty$, that the axioms $(N1)$–$(N3)$ of a norm are satisfied. In what follows, $x = (x_1, \ldots, x_m)$, $y = (y_1, \ldots, y_m)$, $z = (z_1, \ldots, z_m)$ are arbitrary elements of $\mathbb{F}^m$. Since $|z_k| \geq 0$ for all $k \in \{1, \ldots, m\}$, it follows that $\|z\| \geq 0$ for all $\| \cdot \| \in \{\|\cdot\|_1, \|\cdot\|_2, \|\cdot\|_\infty\}$, whilst for each such $\| \cdot \|$, $\|z\| = 0$ if and only if $|z_k| = 0$ for all k, which is equivalent to $z = 0$. Thus $(N1)$ holds. Since $|\lambda z_k| = |\lambda| |z_k|$ for all $\lambda \in \mathbb{R}$ and all $k \in \{1, \ldots, m\}$, $(N2)$ follows for each $\| \cdot \| \in \{\| \cdot \|_1, \| \cdot \|_2, \| \cdot \|_\infty\}$. Note now that

$$|x_k + y_k| \leq |x_k| + |y_k| \quad \text{for all } k \in \{1, \ldots, m\}.$$

Summing the above inequalities gives $(N3)$ for $\| \cdot \|_1$. It is also immediate from the above inequality that

$$|x_k + y_k| \leq \max_{i \in \{1, \ldots, m\}} |x_i| + \max_{i \in \{1, \ldots, m\}} |y_i| \quad \text{for all } k \in \{1, \ldots, m\}.$$

This clearly implies $(N3)$ for $\| \cdot \|_\infty$. To prove $(N3)$ for $\| \cdot \|_2$, we need to show that

$$\left(\sum_{k=1}^m |x_k + y_k|^2 \right)^{1/2} \leq \left(\sum_{k=1}^m |x_k|^2 \right)^{1/2} + \left(\sum_{k=1}^m |y_k|^2 \right)^{1/2}.$$

This is equivalent, upon taking squares, expanding, and using $|\alpha|^2 = \alpha \overline{\alpha}$ for any $\alpha \in \mathbb{F}$, to

$$\sum_{k=1}^m |x_k|^2 + \sum_{k=1}^m |y_k|^2 + 2 \sum_{k=1}^m \operatorname{Re}(x_k \overline{y_k})$$

$$\leq \sum_{k=1}^m |x_k|^2 + \sum_{k=1}^m |y_k|^2 + 2 \left(\sum_{k=1}^m |x_k|^2 \right)^{1/2} \left(\sum_{k=1}^m |y_k|^2 \right)^{1/2},$$

One therefore needs to prove that

$$\sum_{k=1}^m \operatorname{Re}(x_k \overline{y_k}) \leq \left(\sum_{k=1}^m |x_k|^2 \right)^{1/2} \left(\sum_{k=1}^m |y_k|^2 \right)^{1/2}.$$

This is the well known Cauchy–Schwarz Inequality, whose proof can be obtained starting from the inequality

$$\sum_{k=1}^m |x_k - \lambda y_k|^2 \geq 0 \quad \text{for all } \lambda \in \mathbb{F},$$

expanding, and making a suitable choice of λ in the resulting quadratic expression. It follows that $(N3)$ holds for $\| \cdot \|_2$. $\qquad \square$

Solution to Exercise 2.2. (i) Taking into account that for every real number α, $|\alpha| = \max\{\alpha, -\alpha\}$, what needs to be proved is exactly that

$$d(x, y) - d(a, b) \leq d(x, a) + d(y, b) \quad \text{and} \quad d(a, b) - d(x, y) \leq d(x, a) + d(y, b).$$

The first inequality may be rearranged as

$$d(x, y) \leq d(x, a) + d(a, b) + d(b, y),$$

and can be obtained by applying $(M3)$ to the triplet x, a, y, and then to the triplet a, b, y, namely

$$d(x, y) \leq d(x, a) + d(a, y) \leq d(x, a) + d(a, b) + d(b, y).$$

The second inequality may be rearranged as

$$d(a, b) \leq d(a, x) + d(x, y) + d(y, b).$$

and can be obtained by applying $(M3)$ to the triplet a, x, b, and then to the triplet x, b, y, namely

$$d(a, b) \leq d(a, x) + d(x, b) \leq d(a, x) + d(x, y) + d(y, b).$$

(ii) The condition $x_n \longrightarrow x_0$ in (X, d) is equivalent to $d(x_n, x_0) \longrightarrow 0$ in $\mathbb{R}$, while the condition $y_n \longrightarrow y_0$ in (X, d) is equivalent to $d(y_n, y_0) \longrightarrow 0$ in $\mathbb{R}$. Using Part (i), we obtain that, for all $n \in \mathbb{N}$,

$$0 \leq |d(x_n, y_n) - d(x_0, y_0)| \leq d(x_n, x_0) + d(y_n, y_0) \longrightarrow 0 + 0 = 0 \quad \text{as } n \to \infty.$$

By the Squeeze Theorem, it follows that $|d(x_n, y_n) - d(x_0, y_0)| \longrightarrow 0$ in $\mathbb{R}$, which is the same as $d(x_n, y_n) \longrightarrow d(x_0, y_0)$ in $\mathbb{R}$ as $n \to \infty$. $\qquad\square$

Solution to Exercise 2.3. The fact that $x_n \longrightarrow x_0 \in (X, \|\cdot\|)$ is equivalent to $\|x_n - x_0\| \longrightarrow 0$ in $\mathbb{R}$. Since any convergent sequence is bounded, it follows that there exists $M \geq 0$ such that $\|x_n - x_0\| \leq M$. Fix $\varepsilon > 0$, arbitrary. We need to show that there exists $N \in \mathbb{N}$ such that $\|y_n - y_0\| < \varepsilon$ for all $n \geq N$. Note that, for any $n \in \mathbb{N}$, we have

$$\|y_n - x_0\| = \frac{1}{n}\|(x_1 + x_2 + \ldots + x_n) - nx_0\|$$

$$= \frac{1}{n}\|(x_1 - x_0) + \ldots + (x_n - x_0)\|$$

$$\leq \frac{1}{n}(\|x_1 - x_0\| + \ldots + \|x_n - x_0\|)$$

Since $x_n \longrightarrow x_0$ in $(X, \|\cdot\|)$, we have that there exists $N_0 \in \mathbb{N}$ such that $\|x_n - x_0\| < \varepsilon/2$ for all $n \geq N_0$. It now follows from the preceding inequality that, for all $n \geq N_0$, we have

$$\|y_n - x_0\| \leq \frac{1}{n}(\|x_1 - x_0\| + \ldots + \|x_{N_0} - x_0\|)$$

$$+ \frac{1}{n}(\|x_{N_0+1} - x_0\| + \ldots + \|x_n - x_0\|)$$

$$\leq \frac{N_0 M}{n} + \frac{n - N_0}{n}\frac{\varepsilon}{2}$$

Choosing now $N \in \mathbb{N}$ such that $N \geq N_0$ and $N \geq (2N_0 M)/\varepsilon$ it follows from the preceding inequality that, for all $n \geq N$, we have

$$\|y_n - x_0\| < \varepsilon/2 + \varepsilon/2 = \varepsilon.$$

Since $\varepsilon > 0$ was arbitrary, this completes the proof that $y_n \longrightarrow x_0$ in $(X, \|\cdot\|)$. $\quad\square$

Solution to Exercise 2.4. Let $A_1, A_2, \ldots, A_n$ be bounded sets in (X, d), where $n \in \mathbb{N}$. Since the point x_0 in the definition of boundedness can be chosen to be any point of X (provided the constant M is chosen appropriately), we choose x_0 to be *the same* point for all the sets $A_1, \ldots, A_n$. Thus, for each $k \in \{1, \ldots, n\}$ there exists

$M_k \geq 0$ such that $d(x, x_0) \leq M_k$ for all $x \in A_k$. It is then obvious that, if we define $M := \max\{M_k : k \in \{1, \ldots, n\}\}$, then $d(x, x_0) \leq M$ for all $x \in \bigcup_{k=1}^{n} A_k$. This shows that $\bigcup_{k=1}^{n} A_k$ is a bounded set, as required. □

Solution to Exercise 2.5. (i) Since C is bounded, there exist $x_0 \in A$ and $M \geq 0$ such that
$$d(x, x_0) \leq M \quad \text{for all } x \in C.$$
This implies, by using $(M3)$, that, for any $x, y \in C$,
$$d(x, y) \leq d(x, x_0) + d(y, x_0) \leq M + M = 2M,$$
and hence the set $\{d(x, y) : x, y \in C\}$ is bounded in $\mathbb{R}$.
(ii) Let $z_0 \in A \bigcap B$. Then, for any $x, y \in A \bigcup B$, we have that
$$d(x, y) \leq \text{diam} A \quad \text{if } x \in A, y \in A,$$
$$d(x, y) \leq \text{diam} B \quad \text{if } x \in B, y \in B,$$
and also
$$d(x, y) \leq d(x, z_0) + d(y, z_0) \leq \text{diam} A + \text{diam} B \quad \text{if } x \in A, y \in B,$$
$$d(x, y) \leq d(x, z_0) + d(y, z_0) \leq \text{diam} B + \text{diam} A \quad \text{if } x \in B, y \in A.$$
This implies that, in all cases,
$$d(x, y) \leq \text{diam} A + \text{diam} B.$$
Since $x, y \in A \bigcup B$ were arbitrary, it follows that
$$\text{diam}(A \bigcup B) \leq \text{diam} A + \text{diam} B.$$

(iii) Let $A := [1, 2]$ and $B := [-2, -1]$. Then $\text{diam} A = 1$, $\text{diam} B = 1$, but $\text{diam}(A \bigcup B) = 4$, so that indeed
$$\text{diam}(A \bigcup B) > \text{diam} A + \text{diam} B,$$
as desired. □

Solution to Exercise 2.6. It is immediate from $(M3)$ that, for any x and x_0 in X,
$$|f(x) - f(x_0)| = |d(x, a) - d(x_0, a)| \leq d(x, x_0).$$
This implies that, for any $\varepsilon > 0$, choosing $\delta := \varepsilon$, we have
$$\forall x \in X : d(x, x_0) < \delta \Longrightarrow |f(x) - f(x_0)| < \varepsilon.$$
Hence f is continuous at x_0 for every $x_0 \in X$, and therefore f is continuous on X. □

Solution to Exercise 2.7. (i) The first inequality is a consequence of the definition of $d(x, A)$ as an infimum and the fact that $b \in A$. The second inequality is just $(M3)$.
(ii) We rearrange (i) as
$$d(x, A) - d(x, y) \leq d(y, b) \quad \text{for all } b \in A.$$

This means that, for fixed x and y, $d(x, A) - d(x, y)$ is a lower bound for $\{d(y, b) : b \in A\}$. Hence, the definition of $d(y, A)$ as an infimum shows that

$$d(x, A) - d(x, y) \leq d(y, A),$$

which is the required inequality.

(iii) Rearranging the preceding inequality, we obtain

$$d(x, A) - d(y, A) \leq d(x, y).$$

Interchanging the roles of x and y, we obtain in a similar way

$$d(y, A) - d(x, A) \leq d(x, y).$$

Combining the two inequalities, we thus obtain

$$|d(x, A) - d(y, A)| \leq d(x, y).$$

Note that this is true for every x and y in X.

(iv) The preceding inequality can be rewritten as

$$|g(x) - g(y)| \leq d(x, y) \quad \text{for all } x, y \in X.$$

This shows that, for each fixed $x_0 \in X$ and $\varepsilon > 0$, both arbitrary, by choosing $\delta := \varepsilon$, we have

$$\forall x \in X : d(x, x_0) < \delta \implies |g(x) - g(x_0)| < \varepsilon.$$

Thus g is continuous at every $x_0 \in X$, and therefore g is continuous on X. $\quad\square$

Solution to Exercise 2.8. Let $x_0 \in X$, arbitrary. We prove that g is continuous at x_0. Note that, for any $x \in X$, we have, using Exercise 2.2(i), that

$$|g(x) - g(x_0)| = |d(x, f(x)) - d(x_0, f(x_0))| \leq d(x, x_0) + d(f(x), f(x_0)).$$

Fix $\varepsilon > 0$, arbitrary. Since f is continuous at x_0, it follows that there exists $\delta > 0$ such that $d(f(x), f(x_0)) < \varepsilon/2$ for all $x \in X$ with $d(x, x_0) < \delta$. We deduce therefore that, for any $x \in X$ with $d(x, x_0) < \min\{\delta, \varepsilon/2\}$, that

$$|g(x) - g(x_0)| \leq d(x, x_0) + d(f(x), f(x_0)) < \varepsilon/2 + \varepsilon/2 = \varepsilon.$$

This shows that g is continuous at x_0. Since $x_0 \in X$ was arbitrary, it follows that g is continuous on X. $\quad\square$

Solution to Exercise 2.9. The fact that $y_0 \in \mathbb{B}_{r/2}(x_0)$ means that $d(x_0, y_0) < r/2$. Let z be an arbitrary point of $\mathbb{B}_{r/2}(y_0)$. Then $d(y_0, z) < r/2$. Using now the triangle inequality, it follows that

$$d(x_0, z) \leq d(x_0, y_0) + d(y_0, z_0) < \frac{r}{2} + \frac{r}{2} = r,$$

which means that $z \in \mathbb{B}_r(x_0)$. This shows that $\mathbb{B}_{r/2}(y_0) \subseteq \mathbb{B}_r(x_0)$, as required. $\quad\square$

Solution to Exercise 2.10. (i) Since $x_0 \neq y_0$, it follows that $d(x_0, y_0) > 0$. A picture suggests that it might be true that $\mathbb{B}_r(x_0) \bigcap \mathbb{B}_r(y_0) = \emptyset$ for any r with $0 < r \leq d(x_0, y_0)/2$. To prove this claim, take some arbitrary r with that property and suppose for a contradiction that $\mathbb{B}_r(x_0) \bigcap \mathbb{B}_r(y_0) \neq \emptyset$. Then one can pick some $z \in \mathbb{B}_r(x_0) \bigcap \mathbb{B}_r(y_0)$, for which $d(x_0, z) < r$ and $d(y_0, z) < r$. It follows by using $(M3)$ that

$$d(x_0, y_0) \leq d(x_0, z) + d(y_0, z) < r + r = 2r,$$

a contradiction to our choice for the value of r. This proves our claim, and hence the required result.

(ii) Let $x_0, y_0 \in X$ with $x_0 \neq y_0$. By part (i), there exists $r > 0$ such that $\mathbb{B}_r(x_0) \bigcap \mathbb{B}_r(y_0) = \emptyset$. This implies, in particular, that $y_0 \notin \mathbb{B}_r(x_0)$. Then $\mathbb{B}_r(x_0)$ is an example of an open set which is not empty, since it contains x_0, and also is not equal to the entire space, since it does not contain y_0. Thus, it is impossible that $\emptyset$ and X are the only open subsets of X. $\qquad\square$

Solution to Exercise 2.11. Let $x, y \in \mathbb{D}_r(x_0)$, and $\lambda \in [0, 1]$, arbitrary. Then $\|x - x_0\| \leq r$ and $\|y - x_0\| \leq r$. It follows that

$$\begin{aligned}
\|(1 - \lambda)x + \lambda y - x_0\| &= \|(1 - \lambda)(x - x_0) + \lambda(y - x_0)\| \\
&\leq (1 - \lambda)\|x - x_0\| + \lambda\|y - x_0\| \\
&\leq (1 - \lambda)r + \lambda r = r.
\end{aligned}$$

This shows that $((1 - \lambda)x + \lambda y) \in \mathbb{D}_r(x_0)$. In conclusion, $\mathbb{D}_r(x_0)$ is a convex set. $\quad\square$

Solution to Exercise 2.12. Let F be an arbitrary finite subset of X. Then one can write $F = \{x_1, \ldots, x_n\}$, for some $n \in \mathbb{N}$ and $x_1, \ldots, x_n \in X$. We need to show that $X \setminus F$ is open. Let x be an arbitrary point of $X \setminus F$. Since $x \notin F$, it follows that $d(x, x_k) > 0$ for all $k \in \{1, \ldots, n\}$. Let $r := \min\{d(x, x_k) : k \in \{1, \ldots, n\}\}$. Then $r > 0$. For any $k \in \{1, \ldots, n\}$, since $d(x, x_k) \geq r$, it follows that $x_k \notin \mathbb{B}_r(x)$. Hence $F \bigcap \mathbb{B}_r(x) = \emptyset$, and therefore $\mathbb{B}_r(x) \subseteq X \setminus F$. In conclusion, the set $X \setminus F$ is open, and hence F is closed. $\qquad\square$

Solution to Exercise 2.13. Let $x \in X$, arbitrary. Clearly, $x \in \mathbb{B}_r(x)$ for all $r > 0$. On the other hand, for any $y \in X$ with $y \neq x$ we have $d_0(x, y) = 1$, which implies that $y \in \mathbb{B}_r(x)$ if $r > 1$ and $y \notin \mathbb{B}_r(x)$ if $r \leq 1$. Thus $\mathbb{B}_r(x) = \{x\}$ if $r \leq 1$ and $\mathbb{B}_r(x) = X$ if $r > 1$, which proves the required result. $\qquad\square$

Solution to Exercise 2.14. (i) It is obvious that d_2 satisfies $(M1)$ and $(M2)$. To prove $(M3)$, let us consider arbitrary (x_1, y_1), (x_2, y_2) and (x_3, y_3) in $X \times Y$. Then

$$\begin{aligned}
d_2((x_1, y_1), (x_3, y_3)) &= \left((d_X(x_1, x_3))^2 + (d_Y(y_1, y_3))^2\right)^{1/2} \\
&\leq \big((d_X(x_1, x_2) + d_X(x_2, x_3))^2 \\
&\quad + (d_Y(y_1, y_2) + d_Y(y_2, y_3))^2\big)^{1/2} \\
&\leq \left((d_X(x_1, x_2))^2 + (d_Y(y_1, y_2))^2\right)^{1/2} \\
&\quad + \left((d_X(x_2, x_3))^2 + (d_Y(y_2, y_3))^2\right)^{1/2} \\
&= d_2((x_1, y_1), (x_2, y_2)) + d_2((x_2, y_2), (x_3, y_3)),
\end{aligned}$$

where we have used $(M3)$ for d_X and d_Y, and the fact that $\|\cdot\|_2$ is a norm on $\mathbb{R}^2$. This proves $(M3)$ for d_2.

(ii) Let U be an open set in (X, d_X) and V be an open set in (Y, d_Y). Let (x_0, y_0) be an arbitrary point of $U \times V$. Then $x_0 \in U$ and $y_0 \in V$, and since U is open in X and V is open in Y, it follows that there exist $r > 0$ and $s > 0$ such that $\mathbb{B}_r^X(x_0) \subseteq U$ and $\mathbb{B}_s^Y(y_0) \subseteq V$. We claim that the ball of centre (x_0, y_0) and radius $\min\{r, s\}$ in $(X \times Y, d_2)$ is contained in $U \times V$. Indeed, for any point (x, y) in that ball, we have that $d_2((x_0, y_0), (x, y)) < \min\{r, s\}$, which implies that

$$d_X(x_0, x) \leq d_2((x_0, y_0), (x, y)) < \min\{r, s\} \leq r,$$
$$d_Y(y_0, y) \leq d_2((x_0, y_0), (x, y)) < \min\{r, s\} \leq s,$$

so that $x \in \mathbb{B}_r^X(x_0) \subseteq U$ and $y \in \mathbb{B}_s^Y(x_0) \subseteq V$, which means that $(x, y) \in U \times V$. This proves that $U \times V$ is open in $(X \times Y, d_2)$.

(iii) In the case when $X = Y = \mathbb{R}$ and d_X, d_Y are the usual metric on $\mathbb{R}$, the metric d_2 defined in this exercise coincides with the usual Euclidean metric d_2 on $\mathbb{R}^2$. Let us consider the open ball $\mathbb{B}_4((0, 0))$ in $(\mathbb{R}^2, d_2)$, which is obviously an open set. Suppose for a contradiction that $\mathbb{B}_4((0, 0))$ can be written as $U \times V$, for some (open) sets U, V in $\mathbb{R}$. Since $(3, 0)$ and $(0, 3)$ belong to $\mathbb{B}_4((0, 0))$, it follows that $3 \in U$ and $3 \in V$. But then $(3, 3) \in U \times V$ and on the other hand $(3, 3) \notin \mathbb{B}_4((0, 0))$, a contradiction to the assumption $\mathbb{B}_4((0, 0)) = U \times V$. This shows that $\mathbb{B}_4((0, 0))$ is an open set in $(\mathbb{R}^2, d_2)$ which is not of the form $U \times V$ for any U and V open sets in $\mathbb{R}$. $\qquad\square$

Solution to Exercise 2.15. (i) The definition of $x_n \longrightarrow x_0$ in (X, d) may be recast using open balls as follows:

$$\forall \varepsilon > 0 \; \exists N \in \mathbb{N} \text{ such that } \forall n \in \mathbb{N} : n \geq N \Longrightarrow x_n \in \mathbb{B}_\varepsilon(x_0).$$

Let $(*)$ denote the property in the statement of the question.

Suppose first that $x_n \longrightarrow x_0$ in (X, d). Let U be an arbitrary open set containing x_0. Then, since U is open, it follows that there exists $\varepsilon > 0$ such that $\mathbb{B}_\varepsilon(x_0) \subseteq U$. The fact that $x_n \longrightarrow x_0$ in (X, d) as $n \to \infty$ implies that there exists $N \in \mathbb{N}$ such that $x_n \in \mathbb{B}_\varepsilon(x_0)$ for all $n \geq N$, and therefore, since $\mathbb{B}_\varepsilon(x_0) \subseteq U$, that $x_n \in U$ for all $n \geq N$. Since U was an arbitrary open set containing x_0, it follows that property $(*)$ holds.

Suppose now that property $(*)$ holds. Since, for any $\varepsilon > 0$, $\mathbb{B}_\varepsilon(x_0)$ is an open set containing x_0, we may apply $(*)$ with $\mathbb{B}_\varepsilon(x_0)$ in the role of U to obtain the existence of $N \in \mathbb{N}$ such that $x_n \in \mathbb{B}_\varepsilon(x_0)$ for all $n \geq N$. Since $\varepsilon > 0$ was arbitrary, it follows that $x_n \longrightarrow x_0$ in (X, d).

(ii) The definition of f being continuous at x_0 may be recast using open balls as follows:

$$\forall \varepsilon > 0 \; \exists \delta > 0 \text{ such that } f(\mathbb{B}_\delta^X(x_0)) \subseteq \mathbb{B}_\varepsilon^Y(f(x_0)).$$

Let $(**)$ denote the property in the statement of the question.

Suppose first that f is continuous at x_0. Let V be an arbitrary open set in Y containing $f(x_0)$. Then there exists $\varepsilon > 0$ such that

$$\mathbb{B}_\varepsilon^Y(f(x_0)) \subseteq V.$$

Since f is continuous at x_0, it follows that there exists $\delta > 0$ such that

$$f(\mathbb{B}_\delta^X(x_0)) \subseteq \mathbb{B}_\varepsilon^Y(f(x_0)).$$

This implies that $f(\mathbb{B}_\delta^X(x_0)) \subseteq V$. Note that $\mathbb{B}_\delta^X(x_0)$ is an open set in X containing x_0 and hence, setting $U := \mathbb{B}_\delta^X(x_0)$, we have proven the existence of an open set U

in X containing x_0 such that $f(U) \subseteq V$. Since V was an arbitrary open set in Y containing $f(x_0)$, it follows that property $(**)$ holds.

Suppose now that property $(**)$ holds. Since, for any $\varepsilon > 0$, $\mathbb{B}_\varepsilon^Y(f(x_0))$ is an open set in Y containing $f(x_0)$, we may apply $(**)$ with $\mathbb{B}_\varepsilon^Y(f(x_0))$ in the role of V to obtain the existence of an open set U in X containing x_0 such that $f(U) \subseteq V$. Since U is open and contains x_0, there exists $\delta > 0$ such that $\mathbb{B}_\delta^X(x_0) \subseteq U$. It then follows that

$$f(\mathbb{B}_\delta^X(x_0)) \subseteq f(U) \subseteq V = \mathbb{B}_\varepsilon^Y(f(x_0)).$$

Since $\varepsilon > 0$ was arbitrary, it follows that f is continuous at x_0. $\qquad\square$

Solution to Exercise 2.16. (i) The continuity of g at x_0 shows that there exists $\delta_0 > 0$ such that

$$\forall x \in X : d(x, x_0) < \delta_0 \implies |g(x) - g(x_0)| < |g(x_0)|/2.$$

Therefore, for all $x \in \mathbb{B}_{\delta_0}(x_0)$ we have that

$$|g(x)| \geq |g(x_0)| - |g(x) - g(x_0)| > |g(x_0)| - |g(x_0)|/2 = |g(x_0)|/2 > 0,$$

which implies in particular that $g(x) \neq 0$.

(ii) Fix $x_0 \in X$, arbitrary, and let δ_0 be as in part (i). For all $x \in \mathbb{B}_{\delta_0}(x_0)$ we now obtain, using the preceding inequality, that

$$
\begin{aligned}
\left| \frac{f(x)}{g(x)} - \frac{f(x_0)}{g(x_0)} \right| &= \left| \frac{(f(x) - f(x_0))g(x_0) + f(x_0)(g(x_0) - g(x))}{g(x)g(x_0)} \right| \\
&\leq \frac{|f(x) - f(x_0)||g(x_0)|}{|g(x)||g(x_0)|} + \frac{|f(x_0)||g(x) - g(x_0)|}{|g(x)||g(x_0)|} \\
&\leq \frac{2|f(x) - f(x_0)|}{|g(x_0)|} + \frac{2|f(x_0)||g(x) - g(x_0)|}{|g(x_0)|^2}
\end{aligned}
$$

Fix $\varepsilon > 0$, arbitrary. Since f is continuous at x_0, there exists $\delta_1 > 0$ such that

$$\forall x \in X : x \in \mathbb{B}_{\delta_1}(x_0) \implies |f(x) - f(x_0)| < \frac{|g(x_0)|\varepsilon}{4}.$$

Also, since g is continuous at x_0, there exists $\delta_2 > 0$ such that

$$\forall x \in X : x \in \mathbb{B}_{\delta_2}(x_0) \implies |g(x) - g(x_0)| < \frac{|g(x_0)|^2\varepsilon}{4(1 + |f(x_0)|)}.$$

Set $\delta := \min\{\delta_0, \delta_1, \delta_2\}$. Combining the above inequalities, we obtain that

$$\forall x \in X : x \in \mathbb{B}_\delta(x_0) \implies \left| \frac{f(x)}{g(x)} - \frac{f(x_0)}{g(x_0)} \right| < \varepsilon.$$

Since $\varepsilon > 0$ was arbitrary, this shows that f/g is continuous at x_0. Since $x_0 \in X$ was arbitrary, it follows that f/g is continuous on X. $\qquad\square$

Solution to Exercise 2.17. It is of course possible to prove that the sets in question are open using the definition of an open set. However, it is easier to use the hint.

(i) By Exercise 2.6, the function $f : X \longrightarrow \mathbb{R}$ given by $f(x) = d(x, a)$ for all $x \in X$ is continuous from (X, d) to $\mathbb{R}$ with the standard metric. Since $\{x \in X : d(x, a) > r\} =$

$f^{-1}((r,\infty))$ and (r,∞) is an open set in $\mathbb{R}$, it follows that $\{x \in X : d(x,a) > r\}$ is open in (X,d).

(ii) Similarly, by Exercise 2.7, the function $g : X \longrightarrow \mathbb{R}$ given by

$$g(x) = d(x,A) \quad \text{for all } x \in X$$

is continuous from (X,d) to $\mathbb{R}$ with the standard metric. Since $\{x \in X : d(x,A) < r\} = g^{-1}((-\infty,r))$ and $\{x \in X : d(x,A) > r\} = g^{-1}((r,\infty))$, and both $(-\infty,r)$ and (r,∞) are open in $\mathbb{R}$, it follows that $\{x \in X : d(x,A) < r\}$ and $\{x \in X : d(x,A) > r\}$ are open in (X,d). $\qquad \square$

Solution to Exercise 2.18. It is obvious that X is a vector subspace of the vector space

$$B([0,1],\mathbb{R}) = \{f : [0,1] \longrightarrow \mathbb{R} : f \text{ is bounded on } [0,1]\}.$$

It is straightforward to check that $\|\cdot\|_\infty$ satisfies $(N1)$ and $(N2)$. To check $(N3)$, note that, for any $p,q \in X$ we have, for every $y \in [0,1]$,

$$|p(y) + q(y)| \le |p(y)| + |q(y)| \le \|p\|_\infty + \|q\|_\infty,$$

from where $(N3)$ follows.

We now prove that $\|\cdot\|_1$ is a norm on X. By using well known properties of the Riemann integral for continuous functions on $[0,1]$, we easily obtain $(N1)$ and $(N2)$. Also, for any $p,q \in X$, we have that

$$\begin{aligned}
\|p+q\|_1 &= \int_0^1 |p(x) + q(x)| \mathrm{d}x \\
&\le \int_0^1 |p(x)| + |q(x)| \mathrm{d}x \\
&= \int_0^1 |p(x)| \mathrm{d}x + \int_0^1 |q(x)| \mathrm{d}x \\
&= \|p\|_1 + \|q\|_1,
\end{aligned}$$

which proves that $(N3)$ also holds for $\|\cdot\|_1$.

Now note that, for every $p \in X$,

$$\|p\|_1 = \int_0^1 |p(x)| \mathrm{d}x \le \int_0^1 \|p\|_\infty \mathrm{d}x = \|p\|_\infty.$$

Suppose for a contradiction that $\|\cdot\|_\infty$ and $\|\cdot\|_1$ are equivalent. Then there exists $M > 0$ such that

$$\|p\|_\infty \le M\|p\|_1 \quad \text{for all } p \in X.$$

In particular, the above should be satisfied by p_n, for each $n \in \mathbb{N}$, where $p_n(x) = x^n$ for all $x \in [0,1]$. Note however that $\|p_n\|_\infty = 1$ for each $n \in \mathbb{N}$, while

$$\|p_n\|_1 = 1/(n+1) \quad n \in \mathbb{N}.$$

The above inequality would imply that $n + 1 \le M$ for all $n \in \mathbb{N}$, which is clearly impossible. This contradiction shows that the two norms are not equivalent. $\qquad \square$

Solution to Exercise 2.19. (i) It is obvious that ρ satisfies $(M1)$ and $(M2)$, since d does. To prove $(M3)$ for ρ, we need to show that, for all $x, y, z \in X$,

$$\min\{1, d(x, z)\} \le \min\{1, d(x, y)\} + \min\{1, d(y, z)\}.$$

This is equivalent to proving that

$$1 \le \min\{1, d(x, y)\} + \min\{1, d(y, z)\} \quad \text{or} \quad d(x, z) \le \min\{1, d(x, y)\} + \min\{1, d(y, z)\}.$$

We prove this by considering various cases, as follows.

Firstly, if either $d(x, y) \ge 1$ or $d(y, z) \ge 1$, then clearly

$$\min\{1, d(x, y)\} + \min\{1, d(y, z)\} \ge 1,$$

since one of the two terms in the left-hand side equals 1 and the other is non-negative.

Secondly, if $d(x, y) < 1$ and $d(y, z) < 1$ then, using $(M3)$ for d, we obtain

$$\min\{1, d(x, y)\} + \min\{1, d(y, z)\} = d(x, y) + d(y, z) \ge d(x, z).$$

We have thus proved the required inequality in all possible cases.

(ii) Observe first that, for any two points $x, y \in X$, if $d(x, y) < 1$ then $\rho(x, y) = d(x, y)$, whilst if $\rho(x, y) < 1$ then $d(x, y) = \rho(x, y)$. This implies that $\mathbb{B}_s^d(x) = \mathbb{B}_s^\rho(x)$ for any $x \in X$ and any $s \in (0, 1)$. Note now that a subset A of X is open in (X, d) if and only if for every $x \in A$ there exists $s \in (0, 1)$ such that $\mathbb{B}_s^d(x) \subseteq A$. (While the definition of an open set is that A is open if and only if there exists $r > 0$ such that $\mathbb{B}_r^d(x) \subseteq A$, observe that if such an r exists and $r \ge 1$, then any $s \in (0, 1)$ satisfies $\mathbb{B}_s^d(x) \subseteq \mathbb{B}_r^d(x) \subseteq A$.) A similar property characterises then open sets in (X, ρ). Since $\mathbb{B}_s^d(x) = \mathbb{B}_s^\rho(x)$ for any $x \in X$ and any $s \in (0, 1)$, it follows that a subset A of X is open in (X, d) if and only if it is open in (X, ρ). This shows that the metrics d and ρ are topologically equivalent.

(iii) It is obvious that $\rho(x, y) \le d(x, y)$ for all $x, y \in X$. Suppose for a contradiction that d and ρ are Lipschitz equivalent, where d is the standard metric on $\mathbb{R}$. Then there exists a constant $M > 0$ such that $|x - y| \le M \min\{1, |x - y|\}$ for all $x, y \in \mathbb{R}$. Since $\min\{1, |x - y|\} \le 1$ for all $x, y \in \mathbb{R}$, the preceding inequality implies that $|x - y| \le M$ for all $x, y \in \mathbb{R}$, which is clearly impossible. This contradiction shows that the metrics are not Lipschitz equivalent. $\qquad\square$

Solution to Exercise 2.20. (i) It is obvious that ρ satisfies $(M1)$ and $(M2)$, since d does. To prove $(M3)$ for ρ, we need to show that, for all $x, y, z \in X$,

$$\frac{d(x, z)}{1 + d(x, z)} \le \frac{d(x, y)}{1 + d(x, y)} + \frac{d(y, z)}{1 + d(y, z)}. \tag{$*$}$$

Fix $x, y, z \in X$, arbitrary. Since the function $t \mapsto \frac{t}{1+t}$ is nondecreasing on $[0, \infty)$, whilst $(M3)$ for d means that $d(x, z) \le d(x, y) + d(y, z)$, we obtain that

$$\frac{d(x, z)}{1 + d(x, z)} \le \frac{d(x, y) + d(y, z)}{1 + d(x, y) + d(y, z)} = \frac{d(x, y)}{1 + d(x, y) + d(y, z)} + \frac{d(y, z)}{1 + d(x, y) + d(y, z)}.$$

The required inequality $(*)$ now follows from the above by using the inequalities

$$\frac{d(x, y)}{1 + d(x, y) + d(y, z)} \le \frac{d(x, y)}{1 + d(x, y)}, \qquad \frac{d(y, z)}{1 + d(x, y) + d(y, z)} \le \frac{d(y, z)}{1 + d(y, z)}.$$

(ii) Note that, for any $r \in (0, \infty)$, two points $x, y \in X$ satisfy $d(x,y) < r$ if and only if $\rho(x,y) < r/(1+r)$. Also, for any $s \in (0,1)$, two points $x, y \in X$ satisfy $\rho(x,y) < s$ if and only if $d(x,y) < s/(1-s)$. This shows that

$$\mathbb{B}_r^d(x) = \mathbb{B}_{r/(1+r)}^\rho(x) \quad \text{for any } x \in X \text{ and } r > 0,$$

whilst $\mathbb{B}_s^\rho(x) = \mathbb{B}_{s/(1-s)}^d(x)$ for any $x \in X$ and $s \in (0,1)$. Let A be an arbitrary open set in (X, d). Then for any $x \in A$ there exists $r > 0$ such that $\mathbb{B}_r^d(x) \subseteq A$, and hence $\mathbb{B}_{r/(1+r)}^\rho(x) \subseteq A$. Therefore, A is open in (X, ρ) also. On the other hand, let A be an arbitrary open set in (X, ρ). Then for any $x \in A$ there exists $s \in (0,1)$ such that $\mathbb{B}_s^\rho(x) \subseteq A$, and hence $\mathbb{B}_{s/(1-s)}^d(x) \subseteq A$. Therefore, A is open in (X, d) also. In conclusion, the metrics d and ρ are topologically equivalent.

(iii) It is obvious that $\rho(x,y) \le d(x,y)$ for all $x, y \in X$. Suppose for a contradiction that d and e are Lipschitz equivalent, where d is the standard metric on $\mathbb{R}$. Then there exists a constant $m > 0$ so that

$$m|x - y| \le \frac{|x - y|}{1 + |x - y|} \quad \text{for all } x, y \in \mathbb{R} \text{ with } x \ne y.$$

This implies that

$$1 + |x - y| \le \frac{1}{m} \quad \text{for all } x, y \in \mathbb{R} \text{ with } x \ne y$$

which is clearly impossible, since one can choose for example $x := 1/m$ and $y := 0$, for which the above inequality fails. The contradiction so obtained shows that the metrics are not Lipschitz equivalent. $\qquad \square$

Solution to Exercise 2.21. (i) Firstly, the convergence of the series in the definition of d follows from the inequality

$$\frac{1}{2^k} \frac{|x_k - y_k|}{1 + |x_k - y_k|} \le \frac{1}{2^k}$$

for all $k \in \mathbb{N}$ and any $x = (x_1, x_2, \ldots)$, $y = (y_1, y_2, \ldots) \in \mathbf{s}$, combined with the fact that

$$\sum_{k=1}^\infty \frac{1}{2^k} < +\infty.$$

It is obvious that d satisfies $(M1)$ and $(M2)$. We now check $(M3)$ for d. Let $x = (x_1, x_2, \ldots)$, $y = (y_1, y_2, \ldots)$, $z = (z_1, \ldots, z_k, \ldots) \in \mathbf{s}$, arbitrary. It is a consequence of Exercise 2.20 that

$$\rho(a, b) := \frac{|a - b|}{1 + |a - b|} \quad \text{for all } a, b \in \mathbb{R}$$

defines a metric on $\mathbb{R}$, a fact which implies that, for any $k \in \mathbb{N}$,

$$\frac{|x_k - z_k|}{1 + |x_k - z_k|} \le \frac{|x_k - y_k|}{1 + |x_k - y_k|} + \frac{|y_k - z_k|}{1 + |y_k - z_k|}.$$

Since the above inequality holds for all $k \in \mathbb{N}$, it follows easily that d satisfies $(M3)$. Therefore, d is a metric on $\mathbf{s}$.

(ii) Suppose first that $x^{(n)} \longrightarrow x^{(0)}$ in $(\mathbf{s}, d)$. Then, for any fixed $k \in \mathbb{N}$, arbitrary, we have that

$$0 \le \frac{1}{2^k} \frac{|x_k^{(n)} - x_k^{(0)}|}{1 + |x_k^{(n)} - x_k^{(0)}|} \le d(x^{(n)}, x^{(0)}) \longrightarrow 0 \quad \text{as } n \to \infty,$$

so that

$$\frac{|x_k^{(n)} - x_k^{(0)}|}{1 + |x_k^{(n)} - x_k^{(0)}|} \longrightarrow 0 \quad \text{as } n \to \infty.$$

Since, as proved in Exercise 2.20, the metric ρ above is topologically equivalent to the usual metric in $\mathbb{R}$, it follows that $|x_k^{(n)} - x_k^{(0)}| \longrightarrow 0$ as $n \to \infty$, so that $x_k^{(n)} \longrightarrow x_k^{(0)}$ as $n \to \infty$, as required.

Suppose now that $x_k^{(n)} \longrightarrow x_k^{(0)}$ as $n \to \infty$, for each $k \in \mathbb{N}$. We want to show that $x^{(n)} \longrightarrow x^{(0)}$ in $(\mathbf{s}, d)$. Fix $\varepsilon > 0$, arbitrary. Let $K \in \mathbb{N}$ be such that $1/2^K < \varepsilon/2$. Then, for any $n \in \mathbb{N}$, we have

$$\sum_{k=K+1}^{\infty} \frac{1}{2^k} \frac{|x_k^{(n)} - x_k^{(0)}|}{1 + |x_k^{(n)} - x_k^{(0)}|} \le \sum_{k=K+1}^{\infty} \frac{1}{2^k} = \frac{1}{2^K} < \frac{\varepsilon}{2}. \tag{$*$}$$

On the other hand, since for any $k \in \{1, \ldots, K\}$ we have that $x_k^{(n)} \longrightarrow x_k^{(0)}$, it follows that there exists $N \in \mathbb{N}$ such that, for all $n \ge N$,

$$|x_k^{(n)} - x_k^{(0)}| < \varepsilon/2 \quad \text{for all } k \in \{1, \ldots, K\}.$$

(The existence of such an N depends crucially on the fact that we require the above inequalities to hold simultaneously for only *finitely* many values of k, namely $k \in \{1, \ldots, K\}$, and *not* for all $k \in \mathbb{N}$.) This implies that, for any $n \ge N$, we have

$$\sum_{k=1}^{K} \frac{1}{2^k} \frac{|x_k^{(n)} - x_k^{(0)}|}{1 + |x_k^{(n)} - x_k^{(0)}|} \le \sum_{k=1}^{K} \frac{1}{2^k} |x_k^{(n)} - x_k^{(0)}| \le \sum_{k=1}^{K} \frac{1}{2^k} \frac{\varepsilon}{2} < \frac{\varepsilon}{2}. \tag{$\dagger$}$$

Combining $(*)$ and $(\dagger)$, we obtain that $d(x^{(n)}, x^{(0)}) < \varepsilon$ for all $n \ge N$. Since $\varepsilon > 0$ was arbitrary, it follows that $x^{(n)} \longrightarrow x^{(0)}$ in $(\mathbf{s}, d)$. $\qquad\square$

Solution to Exercise 2.22. (i) Since Y is a subspace of $\mathbb{R}$ with the standard metric, one has, for every $y_0 \in Y$ and $r > 0$, that

$$\mathbb{B}_r^Y(y_0) = Y \bigcap \mathbb{B}_r^{\mathbb{R}}(y_0) = Y \bigcap (y_0 - r, y_0 + r).$$

Using this fact, we obtain

$$\mathbb{B}_{1/2}(0) = [0, 1/2); \quad \mathbb{B}_2(2) = (0, 1) \bigcup [2, 3]; \quad \mathbb{B}_1(4) = \{4\};$$

$$\mathbb{B}_5(4) = [0, 1) \bigcup [2, 3] \bigcup \{4\} \bigcup [5, 9); \quad \mathbb{B}_3(6) = \{4\} \bigcup [5, 9).$$

(ii) We know that the open sets of Y are exactly those of the form $Y \bigcap V$, where V is an open set in $\mathbb{R}$. We also know that every interval (a, b) is open in $\mathbb{R}$, and every union of open sets is open (in any topological space). Since $[0, 1) = Y \bigcap (-1, 1)$, $(2, 3) = Y \bigcap (2, 3)$, $(0, 1) \bigcup \{4\} = Y \bigcap ((0, 1) \bigcup (7/2, 9/2))$, and $[5, 6) = Y \bigcap (4, 6)$, it follows that $[0, 1)$, $(2, 3)$, $[0, 1) \bigcup \{4\}$, and $[5, 6)$ are open in Y. On the other hand, since $6 \in [5,6]$, but for any $r > 0$, $\mathbb{B}_r^Y(6) = Y \bigcap (6 - r, 6 + r)$ is not contained in $[5,6]$, it follows that $[5,6]$ is not open in Y. $\quad\square$

Solution to Exercise 2.23. Let $x \in \overline{A}$, arbitrary. Then there exists $(x_n)_1^\infty \subseteq A$ such that $d(x_n, x) \longrightarrow 0$. In particular, for every $\varepsilon > 0$ there exists $N \in \mathbb{N}$ such that $d(x_N, x) < \varepsilon$. Recalling that $x_N \in A$, we have that ε is not a lower bound for $\{d(x,a) : a \in A\}$. Since this is true for every $\varepsilon > 0$, and since 0 is clearly a lower bound for $\{d(x,a) : a \in A\}$, it follows that indeed $d(x, A) = 0$.

Let now $x \notin \overline{A}$, arbitrary. Then, by the definition of $\overline{A}$, there exists $r > 0$ such that $\mathbb{B}_r(x) \bigcap A = \emptyset$. This means that $d(x,a) \geq r$ for every $a \in A$. This implies that $0 < r \leq d(x, A)$, and therefore $d(x, A) \neq 0$. $\quad\square$

Solution to Exercise 2.24. By the definition of a supremum, we have that $x \leq \beta$ for all $x \in A$, and that for every $\varepsilon > 0$ there exists $x_\varepsilon \in A$ such that $x_\varepsilon > \beta - \varepsilon$. In particular, $x_\varepsilon \in A \bigcap (\beta - \varepsilon, \beta + \varepsilon)$, and hence $A \bigcap (\beta - \varepsilon, \beta + \varepsilon) \neq \emptyset$. Since this is true for all $\varepsilon > 0$, and in $\mathbb{R}$ with the usual metric we have that $\mathbb{B}_\varepsilon(\beta) = (\beta - \varepsilon, \beta + \varepsilon)$, it follows that $\beta \in \overline{A}$. A similar argument shows that $\alpha \in \overline{A}$. $\quad\square$

Solution to Exercise 2.25. Let $z_0 \in \overline{A} \bigcap \overline{B}$. Then there exists sequences $(a_n)_1^\infty \subseteq A$ and $(\mathbb{B}_n)_1^\infty \subseteq B$ such that $a_n \longrightarrow z_0$ and $b_n \longrightarrow z_0$ in (X,d) as $n \to \infty$. It follows that $d(a_n, z_0) \longrightarrow 0$ and $d(b_n, z_0) \longrightarrow 0$ as $n \to \infty$, and therefore

$$0 \leq d(a_n, b_n) \leq d(a_n, z_0) + d(b_n, z_0) \longrightarrow 0 + 0 = 0 \quad \text{as } n \to \infty,$$

so that, by the Squeeze Theorem, $d(a_n, b_n) \longrightarrow 0$ as $n \to \infty$. Since $a_n \in A$ and $b_n \in B$ for each $n \in \mathbb{N}$, the above implies that the set of nonnegative numbers $\{d(a,b) : a \in A, b \in B\}$ contains arbitrarily small numbers, which means that its infimum is 0, so that $d(A, B) = 0$ indeed. $\quad\square$

Solution to Exercise 2.26. (i) Since A is bounded, there exist $x_0 \in X$ and $M > 0$ such that
$$d(z, x_0) \leq M \quad \text{for all } z \in A.$$
Consider now an arbitrary $x \in \overline{A}$. Then there exists $(x_n)_1^\infty \subseteq A$ such that $x_n \longrightarrow x$. Since $d(x_n, x_0) \leq M$ for all $n \in \mathbb{N}$, by passing to the limit as $n \to \infty$ we obtain that $d(x, x_0) \leq M$ as well. Thus we have proved that
$$d(x, x_0) \leq M \quad \text{for all } x \in \overline{A},$$
which means that $\overline{A}$ is bounded.

(ii) The fact that $A \subseteq \overline{A}$ implies that $\operatorname{diam} A \leq \operatorname{diam} \overline{A}$. On the other hand, let $\varepsilon > 0$ arbitrary, and let $x, y \in \overline{A}$, arbitrary. The definition of $\overline{A}$ shows that there exist $x_\varepsilon \in \mathbb{B}_\varepsilon(x) \bigcap A$ and $y_\varepsilon \in \mathbb{B}_\varepsilon(y) \bigcap A$. Therefore, we deduce that
$$d(x, y) \leq d(x, x_\varepsilon) + d(x_\varepsilon, y_\varepsilon) + d(y_\varepsilon, y) \leq \operatorname{diam} A + 2\varepsilon.$$
Since $x, y \in \overline{A}$ were arbitrary, it follows that
$$\operatorname{diam} \overline{A} \leq \operatorname{diam} A + 2\varepsilon,$$
and since $\varepsilon > 0$ was arbitrary, it follows that
$$\operatorname{diam} \overline{A} \leq \operatorname{diam} A.$$

In conclusion, diam $\overline{A} = $ diam A. $\qquad\qquad\qquad\qquad\qquad\qquad\qquad\square$

Solution to Exercise 2.27. One can easily check, using the definitions, that, for any $p \in [1, \infty)$,

$$\|x\|_\infty \leq \|x\|_p \leq m^{1/p}\|x\|_\infty \qquad \text{for all } x \in \mathbb{F}^m.$$

This shows that, for any $p \in [1, \infty)$, $\|\cdot\|_p$ is equivalent to $\|\cdot\|_\infty$. On the other hand, if $r \in [1, \infty)$, we also have

$$\|x\|_\infty \leq \|x\|_r \leq m^{1/r}\|x\|_\infty \qquad \text{for all } x \in \mathbb{F}^m.$$

Combining the two relations above, we deduce that, for any $p, r \in [1, \infty)$,

$$m^{-1/r}\|x\|_r \leq \|x\|_p \leq m^{1/p}\|x\|_r \qquad \text{for all } x \in \mathbb{F}^m.$$

This shows that, for any $p, r \in [1, \infty)$, $\|\cdot\|_p$ and $\|\cdot\|_r$ are also equivalent. $\quad\square$

Solution to Exercise 2.28. Let $q \in (1, \infty)$ be the conjugate exponent of $\beta/\alpha \in (1, \infty)$. Then, for any $f \in C([a, b], \mathbb{F})$, by applying Hölder's inequality to the functions $x \mapsto 1$ and $x \mapsto |f(x)|^\alpha$ on the interval $[a, b]$, with the pair of conjugate exponents q and β/α, we obtain:

$$\int_a^b 1 \cdot |f(x)|^\alpha \, dx \leq \left(\int_a^b 1^q \, dx \right)^{1/q} \left(\int_a^b (|f(x)|^\alpha)^{\beta/\alpha} \, dx \right)^{\alpha/\beta}$$

$$= (b-a)^{(\beta-\alpha)/\beta} \left(\int_a^b |f(x)|^\beta \, dx \right)^{\alpha/\beta}.$$

Rearranging this yields

$$\left(\int_a^b |f(x)|^\alpha \, dx \right)^{1/\alpha} \leq (b-a)^{(\beta-\alpha)/\alpha\beta} \left(\int_a^b |f(x)|^\beta \, dx \right)^{1/\beta},$$

which proves the required inequality with $M := (b-a)^{(\beta-\alpha)/\alpha\beta}$. $\qquad\square$

Solution to Exercise 2.29. Let A be a countable dense subset of X. The fact that A is dense means that $\mathbb{B}_r(x) \bigcap A \neq 0$ for every $x \in X$ and $r > 0$. Consider now the set $C = \{(a, n) \in A \times \mathbb{N} : \mathbb{B}_{1/n}(a) \bigcap Y \neq \emptyset\}$. Then C is a subset of $A \times \mathbb{N}$, which is a countable set (due to the countability of both A and $\mathbb{N}$), and therefore C is itself countable. For every $(a, n) \in C$, let us pick some element $y_{a,n} \in \mathbb{B}_{1/n}(a) \bigcap Y$, and let $P := \{y_{a,n} : (a, n) \in C\}$. Then P is a subset of Y, and is a countable set, because it is indexed by the countable set C. We claim that P is also dense in Y. Let $z \in Y$ and $r > 0$, arbitrary. We need to show that $\mathbb{B}_r(z)$ contains an element of P. Let $n \in \mathbb{N}$ be such that $2/n < r$. Since A is dense in X, there exists $a \in \mathbb{B}_{1/n}(z) \bigcap A$. Then $z \in \mathbb{B}_{1/n}(a) \bigcap Y$, which implies that $(a, n) \in C$. Consider now the corresponding element $y_{a,n} \in P$, where $y_{a,n} \in \mathbb{B}_{1/n}(a) \bigcap Y$. Then

$$d(z, y_{a,n}) \leq d(a, z) + d(a, y_{a,n}) < 1/n + 1/n = 2/n < r,$$

so that $y_{a,n} \in \mathbb{B}_r(z) \bigcap P$. Since $z \in Y$ and $r > 0$ were arbitrary, this proves that P is dense in Y. Since P was also countable, it follows that (Y, d) is separable. $\quad\square$

15.3   Solutions to the exercises in Chapter 3

Solution to Exercise 3.1. Let $(x_n)_1^\infty$ be an arbitrary Cauchy sequence in (X, d_0). Taking $\varepsilon_0 := 1/2$ in the definition of a Cauchy sequence, we obtain that there exists $N \in \mathbb{N}$ such $d_0(x_n, x_m) < 1/2$ for all $n, m \geq N$. The definition of a discrete metric d_0 shows that $x_n = x_m$ for all $n, m \geq N$, and hence that $x_n = x_N$ for all $n \geq N$. Setting $x_0 := x_N$, it is straightforward to check that $x_n \longrightarrow x_0$ in (X, d_0) as $n \to \infty$. This shows that (X, d_0) is complete. □

Solution to Exercise 3.2. Observe that $(\mathbb{Z}, d)$ is actually a metric subspace of $\mathbb{R}$ with the usual metric (that is: the metric d is exactly the restriction to $\mathbb{Z}$ of the usual metric on $\mathbb{R}$). Since $\mathbb{R}$ is complete with respect to the usual metric, for the completeness $(\mathbb{Z}, d)$ is complete it is both sufficient and necessary that $\mathbb{Z}$ closed in $\mathbb{R}$. But

$$\mathbb{R} \setminus \mathbb{Z} = \bigcup_{n \in \mathbb{Z}} (n, n+1),$$

so that $\mathbb{R} \setminus \mathbb{Z}$ is open (since any open interval is an open set, and any union of open sets is open), and hence $\mathbb{Z}$ is closed in $\mathbb{R}$. Therefore, $(\mathbb{Z}, d)$ is complete. □

Solution to Exercise 3.3. Note that A and B are subsets of $\mathbb{R}$, whilst C is a subset of $\mathbb{R}^2$, and since $\mathbb{R}$ and $\mathbb{R}^2$ are complete with respect with the usual metric, each of the sets A, B and C is complete if and only if it is closed.

Now,

$$\mathbb{R} \setminus A = (-\infty, 0) \bigcup (1, \infty) \bigcup \left(\bigcup_{n=1}^{\infty} \left(\frac{1}{n+1}, \frac{1}{n} \right) \right),$$

which is obviously open, so that A is closed in $\mathbb{R}$, and therefore complete.

It is well known that $\mathbb{Q}$ is dense in $\mathbb{R}$ (meaning that every element of $\mathbb{R}$ is a limit of a sequence of rationals). This easily implies that

$$\overline{B} = [0, 1] \neq B.$$

This shows that B is not closed in $\mathbb{R}$, and therefore not complete.

We claim that C is closed in $\mathbb{R}^2$. Let $(x_0, y_0) \in \overline{C}$, arbitrary. Then there exists a sequence $((x_n, y_n))_1^\infty \subseteq C$ such that $(x_n, y_n) \longrightarrow (x_0, y_0)$ in $\mathbb{R}^2$. This means that $x_n \longrightarrow x_0$, $y_n \longrightarrow y_0$ in $\mathbb{R}$. Since $(x_n, y_n) \in C$, it follows that $x_n > 0$ and $x_n y_n \geq 1$ for all $n \in \mathbb{N}$. Passing to the limit as $n \to \infty$, we obtain $x_0 \geq 0$ and $x_0 y_0 \geq 1$. The second inequality implies that $x_0 \neq 0$ which, together with the first inequality, implies $x_0 > 0$. Then the second inequality may be rewritten as $y_0 \geq 1/x_0$. It follows that $(x_0, y_0) \in C$. We have therefore proved that $\overline{C} \subseteq C$, which means that C is closed, and therefore complete. □

Solution to Exercise 3.4. (i) Axioms (M1) and (M2) are straightforward. To check (M3), note that, for any $x, y, z \in \mathbb{R}$,

$$\rho(x, z) = |\arctan x - \arctan z|$$
$$= |(\arctan x - \arctan y) + (\arctan y - \arctan z)|$$

and hence

$$\rho(x, z) \le |\arctan x - \arctan y| + |\arctan y - \arctan z|$$
$$= \rho(x, y) + \rho(y, z),$$

where we have used the triangle inequality for $|\cdot|$. Thus (M3) holds, and hence ρ is a metric on $\mathbb{R}$.

(ii) We construct an example of a Cauchy sequence in $(\mathbb{R}, \rho)$ which is not convergent in $(\mathbb{R}, \rho)$. Let us define $x_n = n$ for all $n \in \mathbb{N}$. Then, since $\arctan n \longrightarrow \pi/2$ as $n \to \infty$, we have that for any $\varepsilon > 0$ there exists $N \in \mathbb{N}$ such that

$$|\arctan n - \pi/2| < \varepsilon/2 \quad \text{for all } n \ge N.$$

But then, for any $n, m \ge N$, we have

$$\rho(x_n, x_m) = |\arctan n - \arctan m| \le |\arctan n - \pi/2| + |\arctan m - \pi/2| < \varepsilon.$$

This shows that $(x_n)_1^\infty$ is a Cauchy sequence in $(\mathbb{R}, \rho)$. On the other hand, suppose for a contradiction that $x_n \longrightarrow x_0$ in $(\mathbb{R}, \rho)$, for some $x_0 \in \mathbb{R}$. Then $\rho(x_n, x_0) \longrightarrow 0$, so that

$$|\arctan n - \arctan x_0| \longrightarrow 0 \quad \text{as } n \to \infty.$$

Since $\arctan n \longrightarrow \pi/2$ as $n \to \infty$, it follows that $\arctan x_0 = \pi/2$, which is impossible. This shows that $(x_n)_1^\infty$ is not convergent in $(\mathbb{R}, \rho)$. In conclusion, $(\mathbb{R}, \rho)$ is not complete. $\qquad \square$

Solution to Exercise 3.5. (i) Axioms (M1) and (M2) are straightforward, while (M3) is an immediate consequence of the triangle inequality for $|\cdot|$ on $\mathbb{R}$.

(ii) We construct an example of a Cauchy sequence in $(\mathbb{N}, \rho)$ that is not convergent in $(\mathbb{N}, \rho)$. Indeed, let us define $x_n = n$ for all $n \in \mathbb{N}$. Fix $\varepsilon > 0$, arbitrary, and let $N \in \mathbb{N}$ be such that $2/N < \varepsilon$. Then, for any $n, m \ge N$, we have

$$\rho(x_n, x_m) = \left| \frac{1}{n} - \frac{1}{m} \right| \le \frac{2}{N} < \varepsilon.$$

This shows that $(x_n)_1^\infty$ is a Cauchy sequence in $(\mathbb{N}, \rho)$. We now claim $(x_n)_1^\infty$ is not convergent in $(\mathbb{N}, \rho)$. Suppose for a contradiction that $(x_n)_1^\infty$ converges, and let $x_0 \in \mathbb{N}$ be such that $x_0 = \lim_{n \to \infty} x_n$ in $(\mathbb{N}, \rho)$. This means that

$$\rho(x_n, x_m) = \left| \frac{1}{n} - \frac{1}{x_0} \right| \to 0 \quad \text{as } n \to \infty,$$

which implies that $1/x_0$, a fact which is obviously impossible. This shows that $(x_n)_1^\infty$ is not convergent in $(\mathbb{N}, \rho)$. In conclusion, $(\mathbb{N}, \rho)$ is not complete.

$\qquad \square$

Solution to Exercise 3.6. Let $(x_n)_1^\infty$ be an arbitrary Cauchy sequence in $Y \bigcup Z$. Since

$$\mathbb{N} = \{n \in \mathbb{N} : x_n \in Y\} \bigcup \{n \in \mathbb{N} : x_n \in Z\},$$

and the former set is infinite, it follows that at least one of the two latter sets is infinite. It follows that there exists a subsequence of $(x_n)_1^\infty$ that is either fully contained in Y, or fully contained in Z. Since both Y and Z are complete, we obtain that this subsequence converges, to a limit that belongs either to Y or to Z, and is therefore in $Y \bigcup Z$. However, any Cauchy sequence with a convergent subsequence is necessarily convergent itself, and therefore $(x_n)_1^\infty$ has a limit in $Y \bigcup Z$. This shows that $Y \bigcup Z$ is complete. □

Solution to Exercise 3.7. Let $m > 0$ and $M > 0$ be such that

$$md(x,y) \le \rho(x,y) \le Md(x,y) \quad \text{for all } x, y \in X.$$

Since the Lipschitz equivalence of two metrics implies their topological equivalence, and since the convergence of a sequence depends only on the topology (and not on the metric), it follows that the metric spaces (X, d) and (X, ρ) have exactly the same convergent sequences.

We now prove that the metric spaces (X, d) and (X, ρ) also have exactly the same Cauchy sequences. Let $(x_n)_1^\infty$ be an arbitrary Cauchy sequence in (X, d). Let $\varepsilon > 0$, arbitrary. Since $(x_n)_1^\infty$ is a Cauchy sequence in (X, d), there exists $N \in \mathbb{N}$ such that $d(x_n, x_p) < \varepsilon/M$ for all $n, p \ge N$. But then, for any $n, p \ge N$, we have that

$$\rho(x_n, x_p) \le Md(x_n, x_p) < M(\varepsilon/M) = \varepsilon.$$

Since $\varepsilon > 0$ was arbitrary, it follows that $(x_n)_1^\infty$ is a Cauchy sequence in (X, ρ). In a similar way one can prove that any Cauchy sequence in (X, ρ) is also a Cauchy sequence in (X, d).

Suppose now that (X, d) is a complete metric space. Let $(x_n)_1^\infty$ be an arbitrary Cauchy sequence in (X, ρ). Then the preceding considerations show that $(x_n)_1^\infty$ is a Cauchy sequence in (X, d) also. Since (X, d) is complete, it follows that $x_n \longrightarrow x_0$ in (X, d), for some $x_0 \in X$. Then, because of the topological equivalence of the metrics, we have that $x_n \longrightarrow x_0$ in (X, ρ) also. Therefore, (X, ρ) is a complete metric space. In a similar way one can prove that if (X, ρ) is a complete metric space then (X, d) is a complete metric space. □

Solution to Exercise 3.8. We first show that Y is a vector subspace of $C([-1, 1], \mathbb{R})$. Let $f, g \in Y$, arbitrary. Then $f(-1) = f(1)$ and $g(-1) = g(1)$. This implies that $(f + g)(-1) = (f + g)(1)$, which shows that $f + g \in Y$. Let $f \in Y$ and $\lambda \in \mathbb{R}$, arbitrary. Then $f(-1) = f(1)$, and hence $(\lambda f)(-1) = (\lambda f)(1)$, which shows that $\lambda f \in Y$. In conclusion, Y is a vector subspace of $C([-1, 1], \mathbb{R})$.

We know that $(C([-1, 1], \mathbb{R}), \|\cdot\|_\infty)$ is a Banach space. We now show that Y is closed in $(C([-1, 1], \mathbb{R}), \|\cdot\|_\infty)$, which would imply that $(Y, \|\cdot\|_\infty)$ is itself a Banach space. For this, we consider an arbitrary sequence $(f_n)_1^\infty \subseteq Y$ which converges in $(C([-1, 1], \mathbb{R}), \|\cdot\|_\infty)$ to some $f \in C([-1, 1], \mathbb{R})$. We need to show that $f \in Y$. As we know, convergence in $(C([-1, 1], \mathbb{R}), \|\cdot\|_\infty)$ means exactly uniform convergence, which implies pointwise convergence. Thus,

$$f_n \longrightarrow f \text{ in } (C([-1, 1], \mathbb{R}), \|\cdot\|_\infty) \text{ as } n \to \infty$$

implies in particular that $f_n(-1) \longrightarrow f(-1)$, and $f_n(1) \longrightarrow f(1)$. Thus, we may pass to the limit as $n \to \infty$ in the relation $f_n(-1) = f(1)$ for all $n \in \mathbb{N}$, to obtain $f(-1) = f(1)$. This shows that $f \in Y$, as required. In conclusion, Y is closed in $(C([-1, 1], \mathbb{R}), \|\cdot\|_\infty)$, and therefore $(Y, \|\cdot\|_\infty)$ is a Banach space. □

Solution to Exercise 3.9. We first show that $\mathbf{c}_0$ is a vector space, by checking that it is a vector subspace of ℓ^∞. Let us consider arbitrary $x = (x_1, \ldots, x_k, \ldots) \in \mathbf{c}_0$ and $y = (y_1, \ldots, y_k, \ldots) \in \mathbf{c}_0$. Then $x_k \longrightarrow 0$ and $y_k \longrightarrow 0$, so that $(x_k + y_k) \longrightarrow 0$ also as $k \to \infty$, which shows that $(x + y) \in \mathbf{c}_0$. Similarly, if $x = (x_1, \ldots, x_k, \ldots) \in \mathbf{c}_0$ and $\lambda \in \mathbb{F}$ are arbitrary, then $x_k \longrightarrow 0$ implies $\lambda x_k \longrightarrow 0$ also, so that $\lambda x \in \mathbf{c}_0$.

Since $(\ell^\infty, \|\cdot\|_\infty)$ is a Banach space, we now prove that $(\mathbf{c}_0, \|\cdot\|_\infty)$ is a Banach space by checking that $\mathbf{c}_0$ is a closed subset of ℓ^∞. Let $x^{(0)} \in \overline{\mathbf{c}_0}$, arbitrary, where $\overline{\mathbf{c}_0}$ denotes the closure of $\mathbf{c}_0$ in $(\ell^\infty, \|\cdot\|_\infty)$. Then there exists a sequence $(x^{(n)})_1^\infty \subseteq \mathbf{c}_0$ such that $x^{(n)} \longrightarrow x^{(0)}$ in $(\ell^\infty, \|\cdot\|_\infty)$. We now show that $x^{(0)} \in \mathbf{c}_0$. Fix $\varepsilon > 0$, arbitrary. Since $x^{(n)} \longrightarrow x^{(0)}$ in $(\ell^\infty, \|\cdot\|_\infty)$ as $n \to \infty$, there exists $N \in \mathbb{N}$ such that

$$\|x^{(N)} - x^{(0)}\|_\infty < \varepsilon/2.$$

Since $x^{(N)} \in \mathbf{c}_0$, it follows that there exists $K \in \mathbb{N}$ such that $|x_k^{(N)}| < \varepsilon/2$ for all $k \geq K$. Then, for any $k \geq K$, we have

$$\begin{aligned}
|x_k^{(0)}| &\leq |x_k^{(0)} - x_k^{(N)}| + |x_k^{(N)}| \\
&\leq \|x^{(N)} - x^{(0)}\|_\infty + |x_k^{(N)}| \\
&< \varepsilon/2 + \varepsilon/2 \\
&= \varepsilon.
\end{aligned}$$

Since $\varepsilon > 0$ was arbitrary, this shows that $x^{(0)} \in \mathbf{c}_0$. We have therefore proved that $\overline{\mathbf{c}_0} \subseteq \mathbf{c}_0$, and hence $\mathbf{c}_0$ is a closed subset of ℓ^∞. In conclusion, $(\mathbf{c}_0, \|\cdot\|_\infty)$ is a Banach space. $\qquad\square$

Solution to Exercise 3.10. (i) We show that $\mathbf{c}$ is a vector space by showing that it is a vector subspace of ℓ^∞. For this, we need to check that, for any $x, y \in \mathbf{c}$, we have that $x + y \in \mathbf{c}$, and that, for any $x \in \mathbf{c}$ and any $\lambda \in \mathbb{F}$, we have that $\lambda x \in \mathbf{c}$. This is indeed so because, if

$$x = (x_1, \ldots, x_k, \ldots) \quad \text{and} \quad y = (y_1, \ldots, y_k, \ldots)$$

are such that $\lim_{k \to \infty} x_k$ and $\lim_{k \to \infty} y_k$ exist, then clearly $\lim_{k \to \infty}(x_k + y_k)$ exists, whilst if $\lim_{k \to \infty} x_k$ exists, then $\lim_{k \to \infty} \lambda x_k$ exists too.

(ii) Since $(\ell^\infty, \|\cdot\|_\infty)$ is a Banach space, it suffices to show that $\mathbf{c}$ is a closed subspace of $(\ell^\infty, \|\cdot\|_\infty)$. Let us therefore consider an arbitrary element $x^{(0)} \in \overline{\mathbf{c}}$, where $\overline{\mathbf{c}}$ denotes the closure of $\mathbf{c}$ in $(\ell^\infty, \|\cdot\|_\infty)$. Then there exists a sequence $(x^{(n)})_1^\infty \subseteq \mathbf{c}$ which converges to $x^{(0)}$ in $(\ell^\infty, \|\cdot\|_\infty)$ as $n \to \infty$. We need to show that $x^{(0)} \in \mathbf{c}$, and for this it suffices, since $\mathbb{F}$ is complete, to show that $(x_k^{(0)})_1^\infty$ is a Cauchy sequence.

Fix $\varepsilon > 0$, arbitrary. Since $x^{(n)} \longrightarrow x^{(0)}$ in $(\ell^\infty, \|\cdot\|_\infty)$ as $n \to \infty$, there exists $N \in \mathbb{N}$ such that

$$\|x^{(N)} - x^{(0)}\|_\infty < \varepsilon/3.$$

Since $x^{(N)} \in \mathbf{c}$, it follows that $(x_k^{(N)})_1^\infty$ is a Cauchy sequence, and therefore there exists $K \in \mathbb{N}$ such that

$$|x_i^{(N)} - x_j^{(N)}| < \varepsilon/3 \quad \text{for all } i, j \geq K.$$

Then, for any $i, j \geq K$, we have

$$\begin{aligned}
|x_i^{(0)} - x_j^{(0)}| &\leq |x_i^{(0)} - x_i^{(N)}| + |x_i^{(N)} - x_j^{(N)}| + |x_j^{(N)} - x_j^{(0)}| \\
&\leq 2\|x^{(N)} - x^{(0)}\|_\infty + |x_i^{(N)} - x_j^{(N)}| \\
&< 2\varepsilon/3 + \varepsilon/3 \\
&= \varepsilon.
\end{aligned}$$

This shows that $(x_k^{(0)})_1^\infty$ is a Cauchy sequence, and hence $x^{(0)} \in \mathbf{c}$, as required. Thus $(\mathbf{c}, \|\cdot\|_\infty)$ is a Banach space. $\qquad\square$

Solution to Exercise 3.11. Clearly Y is a vector subspace of the normed space $(\ell^\infty, \|\cdot\|_\infty)$, and therefore is itself a normed space. To show that $(Y, \|\cdot\|_\infty)$ is not complete, it suffices to show that it is not closed in $(\ell^\infty, \|\cdot\|_\infty)$. To this aim, let

$$x^{(0)} := (1, 1/2, 1/3, \ldots, 1/k, \ldots) \in \ell^\infty \setminus Y.$$

Let also, for any $n \in \mathbb{N}$,

$$x^{(n)} := (1, 1/2, \ldots, 1/n, 0, 0, \ldots) \in Y.$$

Since

$$\|x^{(n)} - x^{(0)}\|_\infty = \frac{1}{n+1} \longrightarrow 0 \text{ as } n \to \infty,$$

it follows that $x^{(n)} \longrightarrow x^{(0)}$ in $(\ell^\infty, \|\cdot\|_\infty)$ as $n \to \infty$. This shows that $x^{(0)} \in \overline{Y} \setminus Y$, and hence Y is not closed in $(\ell^\infty, \|\cdot\|_\infty)$, and therefore is not a Banach space. $\qquad\square$

Solution to Exercise 3.12. Consider, for each $n \in \mathbb{N}$, an element $x^{(n)} \in Y$ given by

$$x^{(n)} = (0, \ldots, 0, 1/n^2, 0, \ldots)$$

(that is, where the nth entry of $x^{(n)}$ is $1/n^2$ and all the other entries are 0) and consider the series $\sum_{n=1}^\infty x^{(n)}$ in $(Y, \|\cdot\|)$. Then

$$\sum_{n=1}^\infty \|x^{(n)}\|_\infty = \sum_{n=1}^\infty \frac{1}{n^2} < +\infty,$$

so that the series is absolutely convergent. On the other hand, we claim that there does not exist any $s \in Y$ such that $\sum_{n=1}^\infty x^{(n)} = s$ in $(Y, \|\cdot\|_\infty)$. Indeed, suppose for a contradiction that such an $s = (s_1, \ldots, s_k, \ldots) \in Y$ exists, and let $K \in \mathbb{N}$ be such that $s_k = 0$ for all $k > K$. Consider now the sequence $(s^{(m)})_{m \geq 1}$ of partial sums of $\sum_{n=1}^\infty x^{(n)}$, that is,

$$s^{(m)} = x^{(1)} + \ldots + x^{(m)} \quad \text{for all } m \in \mathbb{N}.$$

Then clearly, for any $m \in \mathbb{N}$, we have that $s^{(m)} = (1, 1/2^2, \ldots, 1/m^2, 0, 0, 0, \ldots)$. But then, for every $m > K$, we have that

$$\|s^{(m)} - s\|_\infty \geq |s_{K+1}^{(m)} - s_{K+1}| = \frac{1}{(K+1)^2},$$

and this contradicts the fact that $s^{(m)} \longrightarrow s$ in $(Y, \|\cdot\|_\infty)$ as $m \to \infty$. (Underlying the above argument is the fact that $\sum_{n=1}^{\infty} x^{(n)}$ is convergent in the larger space $(\ell^{\infty}, \|\cdot\|_{\infty})$ to the limit $(1, 1/2^2, 1/3^2, \ldots, 1/n^2, \ldots)$, which however does not belong to Y, so that by the uniqueness of limits in ℓ^{∞}, the series cannot converge to any limit in Y.) □

Solution to Exercise 3.13. Let $(x_n)_1^{\infty}$ be an arbitrary Cauchy sequence in $(X, \|\cdot\|)$. We need to show that $(x_n)_1^{\infty}$ converges in $(X, \|\cdot\|)$. First, the Cauchy condition ensures that there exists $n_1 \in \mathbb{N}$ such that

$$\|x_n - x_{n_1}\| \leq \frac{1}{2} \quad \text{for all } n \geq n_1.$$

Then, the Cauchy condition ensures that there exists $n_2 \in \mathbb{N}$ with $n_2 > n_1$, such that

$$\|x_n - x_{n_2}\| \leq \frac{1}{2^2} \quad \text{for all } n \geq n_2.$$

Suppose now that, for some $p \in \mathbb{N}$ with $p \geq 2$, we have constructed $n_1 < n_2 < \ldots < n_{p-1}$. Then, using again the Cauchy condition, we obtain the existence of $n_p \in \mathbb{N}$ with $n_p > n_{p-1}$, such that

$$\|x_n - x_{n_p}\| \leq \frac{1}{2^p} \quad \text{for all } n \geq n_p.$$

In this way we construct a subsequence $(x_{n_p})_{p \geq 1}$ of $(x_n)_1^{\infty}$, which satisfies, in particular,

$$\|x_{n_{p+1}} - x_{n_p}\| \leq \frac{1}{2^p} \quad \text{for all } p \in \mathbb{N}. \tag{$*$}$$

Consider now the series $x_{n_1} + \sum_{p=2}^{\infty} (x_{n_p} - x_{n_{p-1}})$. The inequality $(*)$ ensures that

$$\|x_{n_1}\| + \sum_{p=2}^{\infty} \|x_{n_p} - x_{n_{p-1}}\| \quad \text{converges,}$$

which shows that the above series is absolutely convergent. It follows therefore, using the assumption, that the series converges in $(X, \|\cdot\|)$. But the sequence of partial sums of the series is exactly $(x_{n_p})_{p \geq 1}$, so that this sequence converges in $(X, \|\cdot\|)$. However, it is always the case that if a Cauchy sequence has a convergent subsequence, then the entire sequence converges. We therefore obtain that $(x_n)_1^{\infty}$ converges. This shows that $(X, \|\cdot\|)$ is a Banach space. □

Solution to Exercise 3.14. We prove that $(d(x_n, y_n))_1^{\infty}$ is a Cauchy sequence of real numbers. Fix $\varepsilon > 0$, arbitrary. Since $(x_n)_1^{\infty}$ and $(y_n)_1^{\infty}$ are Cauchy sequences, the exists $N \in \mathbb{N}$ such that $d(x_n, x_m) < \varepsilon/2$ and $d(y_n, y_m) < \varepsilon/2$ for all $n \geq N$. It follows, using Exercise 2.2(i), that for all $n, m \geq N$, we have:

$$|d(x_n, y_n) - d(x_m, y_m)| \leq d(x_n, x_m) + d(y_n, y_m) < \varepsilon/2 + \varepsilon/2 = \varepsilon.$$

This shows that $(d(x_n, y_n))_1^{\infty}$ is a Cauchy sequence. Since $(\mathbb{R}, |\cdot|)$, it follows that the sequence $(d(x_n, y_n))_1^{\infty}$ converges. □

Solution to Exercise 3.15. Let $(x_n)_1^{\infty}$ be an arbitrary Cauchy sequence in (X, d). We want to show that $(x_n)_1^{\infty}$ converges. Since Y is dense in X, there exists, for each

$n \in \mathbb{N}$, a point $y_n \in Y$ such that $d(x_n, y_n) < \frac{1}{n}$. We claim that $(y_n)_1^\infty$ is also a Cauchy sequence. Fix $\varepsilon > 0$, arbitrary. Since $(x_n)_1^\infty$ is Cauchy, there exists $N_1 \in \mathbb{N}$ such that $d(x_n, x_m) < \varepsilon/2$ for all $n, m \geq N_1$. Let $N_2 \in \mathbb{N}$ be such that $\frac{2}{N_2} < \varepsilon/2$. Let $N := \max\{N_1, N_2\}$. Then, for every $n, m \geq N$, we have

$$d(y_n, y_m) \leq d(y_n, x_n) + d(x_n, x_m) + d(x_m, y_m)$$
$$< \frac{1}{n} + \frac{1}{m} + d(x_n, x_m)$$
$$\leq \frac{2}{N} + d(x_n, x_m)$$
$$< \varepsilon.$$

Thus $(y_n)_1^\infty$ is indeed a Cauchy sequence. Since $(y_n)_1^\infty \subseteq Y$, by assumption there exists $y_0 \in X$ such that $y_n \longrightarrow y_0$ in (X, d), which means that $d(y_n, y_0) \longrightarrow 0$. But then

$$d(x_n, y_0) \leq d(x_n, y_n) + d(y_n, y_0) < \frac{1}{n} + d(y_n, y_0) \longrightarrow 0 + 0 = 0.$$

By the Squeeze Theorem it follows that $d(x_n, y_0) \longrightarrow 0$, which means that $x_n \longrightarrow y_0$ in (X, d). In conclusion, (X, d) is complete. $\qquad\square$

Solution to Exercise 3.16. The metric space $[a, b]$ with the usual metric is complete, being a closed subspace of the complete metric space $\mathbb{R}$. We now prove that, for every $x, y \in [a, b]$, we have

$$|f(x) - f(y)| \leq \alpha|x - y|. \qquad (*)$$

Since $\alpha \in (0, 1)$, this would mean that f is a contraction on $[a, b]$, and therefore the required result that f has a unique fixed point in $[a, b]$ would follow from the Contraction Mapping Principle. Consider therefore $x, y \in [a, b]$, arbitrary. It suffices to consider the case $x \neq y$, and then we may assume with no loss of generality that $x < y$. Since f is continuous on $[x, y]$ and differentiable on (x, y), the Mean Value Theorem (from Real Analysis) shows that there exists $z \in (x, y)$ such that

$$f(y) - f(x) = f'(z)(y - x).$$

Taking absolute values and using the assumption that $|f'(z)| \leq \alpha$, we obtain $(*)$. This completes the proof. $\qquad\square$

15.4   Solutions to the exercises in Chapter 4

Solution to Exercise 4.1. By axiom $(T3)$, any finite intersection of open sets is open, and by property $(\widetilde{T2})$, any intersection of closed sets is closed.

Since $U \setminus F = U \cap (X \setminus F)$, and both U and $X \setminus F$ are open sets, it follows that $U \setminus F$ is open.

Since $F \setminus U = F \cap (X \setminus U)$, and both F and $X \setminus U$ are closed sets, it follows that $F \setminus U$ is closed. $\qquad\square$

Solution to Exercise 4.2. Axiom $(T1)$ obviously holds. Let $\{U_i\}_{i \in I}$ be an arbitrary family of sets in $\mathcal{T}$. If at least one of these sets is $\mathbb{N}$, then the union is $\mathbb{N}$, which belongs to $\mathcal{T}$. If some of the sets are empty, then they don't contribute anything to the union, and may be discarded. Suppose therefore that, for each $i \in I$, we have that $U_i = A_{k_i}$, for some $k_i \in \mathbb{N}$. If the set $\{k_i : i \in I\}$ is unbounded, then clearly $\bigcup_{i \in I} U_i = \mathbb{N}$, so that $\bigcup_{i \in I} U_i \in \mathcal{T}$. If the set $\{n_i : i \in I\}$ is bounded, then it has a maximal element k^*, and then

$$\bigcup_{i \in I} U_i = \{1, \ldots, k^*\} = A_{k^*},$$

so that $\bigcup_{i \in I} U_i \in \mathcal{T}$ in this case too. In conclusion, axiom $(T2)$ holds. Suppose now that $U_1, \ldots, U_n$ are in $\mathcal{T}$, for some $n \in \mathbb{N}$. If at least one of these sets is empty, then their intersection is empty, so it belongs to $\mathcal{T}$. If some of the sets in this finite family are $\mathbb{N}$, then their presence does not affect the intersection, and may be discarded. Suppose therefore that, for each $i \in \{1, \ldots, n\}$, we have that $U_i = A_{k_i}$ for some $k_i \in \mathbb{N}$. Let k_* be the minimal element of the finite set $\{k_1, \ldots, k_n\}$. Then clearly

$$U_1 \bigcap \cdots \bigcap U_n = \{1, \ldots, k_*\} = A_{k_*},$$

so that

$$U_1 \bigcap \cdots \bigcap U_n \in \mathcal{T}.$$

In conclusion, axiom $(T3)$ holds too. Therefore, $\mathcal{T}$ is a topology on $\mathbb{N}$. $\qquad \square$

Solution to Exercise 4.3. We are given that $\emptyset \in \mathcal{T}$. On the other hand, $X \setminus X = \emptyset$, a finite set, hence $X \in \mathcal{T}$. Therefore, $\mathcal{T}$ satisfies axiom $(T1)$. Let $\{U_i\}_{i \in I}$ be any family of sets in $\mathcal{T}$, for some index set I. Then

$$X \setminus \left(\bigcup_{i \in I} U_i \right) = \bigcap_{i \in I} (X \setminus U_i).$$

If there exists $i_0 \in I$ such that $U_{i_0} \neq \emptyset$, then $X \setminus U_{i_0}$ is finite, and since

$$\bigcap_{i \in I} (X \setminus U_i) \subseteq X \setminus U_{i_0},$$

it follows that $X \setminus (\bigcup_{i \in I} U_i)$ is a finite set, thus proving that $\bigcup_{i \in I} U_i \in \mathcal{T}$. On the other hand, if $U_i = \emptyset$ for all $i \in I$, then

$$\bigcup_{i \in I} U_i = \emptyset \in \mathcal{T}.$$

Therefore, $\mathcal{T}$ satisfies axiom $(T2)$. Let $U_1, \ldots, U_n$ be sets in $\mathcal{T}$, for some $n \in \mathbb{N}$. If $U_{i_0} = \emptyset$ for some $i_0 \in \{1, \ldots, n\}$, then $\bigcap_{i=1}^{n} U_i = \emptyset \in \mathcal{T}$. If $U_i \neq \emptyset$ for all $i \in \{1, \ldots, n\}$, then $X \setminus U_i$ is finite for all $i \in \{1, \ldots, n\}$, and hence

$$X \setminus \left(\bigcap_{i=1}^{n} U_i \right) = \bigcup_{i=1}^{n} (X \setminus U_i)$$

is a finite union of finite sets, and therefore a finite set itself, thus proving that

$$\bigcap_{i=1}^{n} U_i \in \mathcal{T}.$$

Therefore, $\mathcal{T}$ satisfies axiom $(T3)$. In conclusion, $\mathcal{T}$ is a topology on X. $\square$

Solution to Exercise 4.4. Fix $x_0 \in X$, arbitrary. Let V be an arbitrary open set containing x_0. Since V is non-empty, it follows that $X \setminus V$ is a finite set, and hence it is of the form $X \setminus V = \{a_1, \ldots, a_p\}$ for some $p \in \mathbb{N}$, where $a_1, \ldots, a_p \in X$ with $a_i \neq a_j$ for all $i, j \in \{1, \ldots, p\}$ with $i \neq j$. Let

$$K = \left\{ k \in \{1, \ldots, p\} : \text{ there exists } n \in \mathbb{N} \text{ such that } x_n = a_k \right\}.$$

Then K is a finite set and, using the assumption that $x_i \neq x_j$ for all $i, j \in \mathbb{N}$ with $i \neq j$, it follows that for each $k \in K$ there exists a unique $n_k \in \mathbb{N}$ such that $x_{n_k} = a_k$. Set:

$$N := 1 + \max \left\{ n_k : k \in K \right\}.$$

Then for all $n \geq N$ we have that $x_n \notin (X \setminus V)$, and hence $x_n \in V$. Since $V \in \mathcal{T}$ with $V \ni x_0$ was arbitrary, it follows that indeed $x_n \longrightarrow x_0$ in $(X, \mathcal{T})$ as $n \to \infty$. $\square$

Solution to Exercise 4.5. (i) The fact that U and V are non-empty open sets in $(X, \mathcal{T})$ means that $X \setminus U$ and $X \setminus V$ are finite. This implies that

$$X \setminus (U \textstyle\bigcap V) = (X \setminus U) \textstyle\bigcup (X \setminus V)$$

is also a finite set. Since X itself is infinite, it follows that $U \bigcap V$ cannot be empty. (ii) Suppose for a contradiction that f is not constant. Then there exists $x, y \in X$ such that $f(x) \neq f(y)$. Let $\alpha := f(x)$ and $\beta := f(y)$, with $\alpha \neq \beta$. Then there exists $\varepsilon > 0$ sufficiently small, such that

$$(\alpha - \varepsilon, \alpha + \varepsilon) \textstyle\bigcap (\beta - \varepsilon, \beta + \varepsilon) = \emptyset.$$

Set

$$U := f^{-1}((\alpha - \varepsilon, \alpha + \varepsilon)), \quad V := f^{-1}((\beta - \varepsilon, \beta + \varepsilon)).$$

Since f is continuous, we have that $U, V \in \mathcal{T}$, and obviously U, V are not empty. By part (i), $U \bigcap V$ is non-empty. Let $z \in U \bigcap V$. Then

$$f(z) \in (\alpha - \varepsilon, \alpha + \varepsilon) \textstyle\bigcap (\beta - \varepsilon, \beta + \varepsilon) = \emptyset,$$

a contradiction. In conclusion, f must be constant on X. $\square$

Solution to Exercise 4.6. Let $x, y \in X$, arbitrary. Let $U, V \in \mathcal{T}$ be such that $U \ni x$, $V \ni y$, arbitrary. Then U and V are not empty, and hence $X \setminus U$ and $X \setminus V$ are finite sets. Then

$$X \setminus (U \textstyle\bigcap V) = (X \setminus U) \textstyle\bigcup (X \setminus V)$$

is a finite set too. Since X itself is infinite, it follows that $U \bigcap V$ cannot be empty. This shows that $(X, \mathcal{T})$ is not a Hausdorff space. $\square$

Solution to Exercise 4.7. Since $x_i \neq x_j$ for each $i, j \in \{1, 2, 3\}$ with $i \neq j$, the fact that $(X, \mathcal{T})$ is a Hausdorff space implies that there exist $U_{i,j}, U_{j,i} \in \mathcal{T}$ such that $U_{i,j} \ni x_i$, $U_{j,i} \ni x_j$, and $U_{i,j} \bigcap U_{j,i} = \emptyset$. Let us define

$$U_1 := U_{1,2} \textstyle\bigcap U_{1,3}, \quad U_2 := U_{2,1} \textstyle\bigcap U_{2,3}, \quad U_3 := U_{3,1} \textstyle\bigcap U_{3,2}.$$

It is then easy to check that $U_k \in \mathcal{T}$ and $U_k \ni x_k$ for each $k \in \{1, 2, 3\}$, whilst for each $i, j \in \{1, 2, 3\}$ with $i \neq j$ we have that

$$U_i \bigcap U_j \subseteq U_{i,j} \bigcap U_{j,i} = \emptyset,$$

and hence $U_i \bigcap U_j = \emptyset$. □

Solution to Exercise 4.8. Since any finite union of closed sets in a topological space is closed, it suffices to prove that any subset of X having exactly one element is closed. Let $x \in X$, arbitrary. To prove that $\{x\}$ is closed, we shall prove the equivalent statement that $X \setminus \{x\}$ is open. Let $y \in X \setminus \{x\}$, arbitrary. Since $(X, \mathcal{T})$ is a Hausdorff space, there exists $U_y \in \mathcal{T}$, $V_y \in \mathcal{T}$, with $x \in U_y$, $y \in V_y$, such that $U_y \bigcap V_y = \emptyset$. In particular, $y \in V_y \subseteq X \setminus \{x\}$. This implies that

$$X \setminus \{x\} = \bigcup_{y \in X \setminus \{x\}} V_y,$$

and since each V_y is an open set, it follows that $X \setminus \{x\}$ is an open set, as required. □

Solution to Exercise 4.9. Let $x_0 \in X$, arbitrary. Since $g(x_0) \neq 0$, the continuity of g at x_0 shows that there exists $U_0 \in \mathcal{T}$ with $U_0 \ni x_0$ such that

$$\forall x \in X : x \in U_0 \implies |g(x) - g(x_0)| < |g(x_0)|/2.$$

Therefore, for all $x \in U_0$ we have that

$$|g(x)| \geq |g(x_0)| - |g(x) - g(x_0)| > |g(x_0)| - |g(x_0)|/2 = |g(x_0)|/2 > 0.$$

For all $x \in U_0$ we now obtain, using the preceding inequality, that

$$\left| \frac{f(x)}{g(x)} - \frac{f(x_0)}{g(x_0)} \right| = \left| \frac{(f(x) - f(x_0))g(x_0) + f(x_0)(g(x_0) - g(x))}{g(x)g(x_0)} \right|$$
$$\leq \frac{|f(x) - f(x_0)| |g(x_0)|}{|g(x)| |g(x_0)|} + \frac{|f(x_0)| |g(x) - g(x_0)|}{|g(x)| |g(x_0)|}$$
$$\leq \frac{2|f(x) - f(x_0)|}{|g(x_0)|} + \frac{2|f(x_0)| |g(x) - g(x_0)|}{|g(x_0)|^2}$$

Let V be an open set in $\mathbb{R}$ with $V \ni f(x_0)$. Then there exists $\varepsilon > 0$ such that

$$\left(\frac{f(x_0)}{g(x_0)} - \varepsilon, \frac{f(x_0)}{g(x_0)} + \varepsilon \right) \subseteq V.$$

Since f is continuous at x_0, there exists $U_1 \in \mathcal{T}$ with $U_1 \ni x_0$ such that

$$\forall x \in X : x \in U_1 \implies |f(x) - f(x_0)| < \frac{|g(x_0)| \varepsilon}{4}.$$

Also, since g is continuous at x_0, there exists $U_2 \in \mathcal{T}$ with $U_2 \ni x_0$ such that

$$\forall x \in X : x \in U_2 \implies |g(x) - g(x_0)| < \frac{|g(x_0)|^2 \varepsilon}{4 \max\{|f(x_0)|, 1\}}.$$

Set $U := U_0 \bigcap U_1 \bigcap U_2$. Then $U \in \mathcal{T}$ and $U \ni x_0$. Combining the above inequalities, we obtain that

$$x \in U \implies \left| \frac{f(x)}{g(x)} - \frac{f(x_0)}{g(x_0)} \right| < \varepsilon.$$

Thus we have shown that there exists $U \in \mathcal{T}$ with $U \ni x_0$ such that

$$\forall x \in X : x \in U \implies \frac{f(x)}{g(x)} \in V.$$

This shows that f/g is continuous at x_0. Since $x_0 \in X$ was arbitrary, it follows that f/g is continuous on X. $\qquad\square$

Solution to Exercise 4.10. (i) Let $x \in A^\circ$, arbitrary. Then there exists $U \in \mathcal{T}$ with $U \ni x$ such that $U \subseteq A$. Since $A \subseteq B$, it follows that $U \subseteq B$, and hence $x \in B^\circ$. The arbitrariness of x shows that $A^\circ \subseteq B^\circ$. Let now $x \in \overline{A}$, arbitrary. Then $U \bigcap A \neq \emptyset$ for every $U \in \mathcal{T}$ with $U \ni x$. Since $A \subseteq B$, it follows that $U \bigcap B \neq \emptyset$ for every such U. The arbitrariness of x shows that $\overline{A} \subseteq \overline{B}$.
(ii) Since $A \bigcap B \subseteq A$ and $A \bigcap B \subseteq B$, part (i) implies that

$$(A \bigcap B)^\circ \subseteq A^\circ \text{ and } (A \bigcap B)^\circ \subseteq B^\circ,$$

so that

$$(A \bigcap B)^\circ \subseteq A^\circ \bigcap B^\circ.$$

On the other hand, let $x \in A^\circ \bigcap B^\circ$, arbitrary. Then $x \in A^\circ$ and $x \in B^\circ$, so that there exist $U, V \in \mathcal{T}$ with $U \ni x$, $V \ni x$, such that $U \subseteq A$, $V \subseteq B$. It follows that $U \bigcap V \in \mathcal{T}$, $U \bigcap V \ni x$, and $U \bigcap V \subseteq A \bigcap B$. This implies that $x \in (A \bigcap B)^\circ$. The arbitrariness of x shows that

$$A^\circ \bigcap B^\circ \subseteq \left(A \bigcap B \right)^\circ.$$

The validity of the two inclusions shows that $(A \bigcap B)^\circ = A^\circ \bigcap B^\circ$.

The property just proved is valid also with $X \setminus A$ and $X \setminus B$ in the roles of A and B, namely

$$\left((X \setminus A) \bigcap (X \setminus B) \right)^\circ = (X \setminus A)^\circ \bigcap (X \setminus B)^\circ.$$

Using De Morgan's Laws and a known property stating that $(X \setminus C)^\circ = \overline{X \setminus C}$ for every subset C of X, we obtain that

$$\left(X \setminus (A \bigcup B) \right)^\circ = (X \setminus \overline{A}) \bigcap (X \setminus \overline{B})$$

and, furthermore, that

$$X \setminus \overline{A \bigcup B} = X \setminus (\overline{A} \bigcup \overline{B}),$$

so that $\overline{A \bigcup B} = \overline{A} \bigcup \overline{B}$, as required.
(iii) It is known that $C^\circ = C$ for every open set C. Applying that property for A°, which is an open set, we obtain that $(A^\circ)^\circ = A^\circ$. It is also known that $\overline{C} = C$ for every closed set C. Applying that property for $\overline{A}$, which is a closed set, we obtain that $\overline{\overline{A}} = \overline{A}$. $\qquad\square$

Solution to Exercise 4.11. Let $x \in A^\circ$, arbitrary. Then there exists $U \in \mathcal{T}$ with $U \ni x$ such that $U \subseteq A$. It follows that $U \subseteq A \bigcup B$, and hence $x \in (A \bigcup B)^\circ$. This shows that $A^\circ \subseteq (A \bigcup B)^\circ$. Similarly, $B^\circ \subseteq (A \bigcup B)^\circ$. Therefore,

$$A^\circ \bigcup B^\circ \subseteq \left(A \bigcup B\right)^\circ.$$

Let now $x \in \overline{A \bigcap B}$, arbitrary. Then for every $U \in \mathcal{T}$ with $U \ni x$ we have that

$$U \bigcap (A \bigcap B) \neq \emptyset.$$

It follows that $U \bigcap A \neq \emptyset$ for any such U, and hence $x \in \overline{A}$. This shows that $\overline{A \bigcap B} \subseteq \overline{A}$. Similarly, $\overline{A \bigcap B} \subseteq \overline{B}$. Therefore,

$$\overline{A \bigcap B} \subseteq \overline{A} \bigcap \overline{B}.$$

We now construct the required examples when $X = \mathbb{R}$ with the usual topology. Take $A := (-1, 0) \bigcup (0, 1)$ and $B := \{0\}$. Then A is open, and thus $A^\circ = A$, whilst clearly $B^\circ = \emptyset$. On the other hand, $A \bigcup B = (-1, 1)$ which is open, and thus

$$(A \bigcup B)^\circ = (-1, 1) \neq (-1, 0) \bigcup (0, 1) = A^\circ \bigcup B^\circ.$$

Take $A := (-1, 0)$ and $B := (0, 1)$. Then $A \bigcap B = \emptyset$, so $\overline{A \bigcap B} = \emptyset$. On the other hand, $\overline{A} = [-1, 0]$ and $\overline{B} = [0, 1]$, and thus

$$\overline{A} \bigcap \overline{B} = \{0\} \neq \emptyset = \overline{A \bigcap B}. \quad \square$$

Solution to Exercise 4.12. By definition, $\partial A = \overline{A} \setminus A^\circ$.
(i) If A is closed in X, then $\overline{A} = A$, so that

$$\partial A = A \setminus A^\circ \subseteq A.$$

If, on the other hand, $\partial A \subseteq A$, this means that $\overline{A} \setminus A^\circ \subseteq A$. It follows that $\overline{A} \subseteq A \bigcup A^\circ$, and therefore, since $A^\circ \subseteq A$, that $\overline{A} \subseteq A$. Since it is also true that $A \subseteq \overline{A}$, it follows that $A = \overline{A}$, which implies that A is closed in X.
(ii) If $\partial A = \emptyset$, then $\overline{A} \subseteq A^\circ$. Since in general $A^\circ \subseteq A \subseteq \overline{A}$, it follows that $A^\circ = A = \overline{A}$. The fact that $A = A^\circ$ implies that A is open, whilst the fact that $A = \overline{A}$ implies that A is closed. If, on the other hand, A is both open and closed, then $A^\circ = A$ and $\overline{A} = A$, so that $\partial A = \emptyset$. $\quad \square$

Solution to Exercise 4.13. We are going to use the result that f is continuous on X if and only if

$$f^{-1}(V) \in \mathcal{T} \quad \text{for any } V \in \mathcal{S}. \tag{$*$}$$

Suppose first that f is continuous on X. Then $(*)$ holds. Let A be an arbitrary subset of Y. Then $A^\circ \in \mathcal{S}$, and therefore, by $(*)$, $f^{-1}(A^\circ) \in \mathcal{T}$. On the other hand, since $A^\circ \subseteq A$, it follows that

$$f^{-1}(A^\circ) \subseteq f^{-1}(A).$$

Since $f^{-1}(A^\circ)$ is an open set contained in $f^{-1}(A)$ and $(f^{-1}(A))^\circ$ is the largest open set contained in $f^{-1}(A)$, it follows that

$$f^{-1}(A^\circ) \subseteq (f^{-1}(A))^\circ,$$

as required.

Suppose now that $f^{-1}(A^\circ) \subseteq (f^{-1}(A))^\circ$ for any subset A of Y. We check that f is continuous on X by verifying $(*)$. Let $V \in \mathcal{S}$, arbitrary. Then $V^\circ = V$. By taking $A := V$, we therefore obtain, from the assumed property of f, that $f^{-1}(V) \subseteq (f^{-1}(V))^\circ$. Since the reverse inclusion is also true (because $C^\circ \subseteq C$ for any set C), it follows that $f^{-1}(V) = (f^{-1}(V))^\circ$. This implies that $f^{-1}(V) \in \mathcal{T}$ (because C° is an open set for any set C). We have therefore verified $(*)$, which shows that f is continuous on X. $\qquad\square$

Solution to Exercise 4.14. Suppose first that f is continuous on X. Let A be any subset of X. Since $\overline{f(A)}$ is a closed subset of Y, the continuity of f implies that $f^{-1}(\overline{f(A)})$ is a closed subset of X. Obviously $A \subseteq f^{-1}(\overline{f(A)})$. Since $\overline{A}$ is the smallest closed set containing A, it follows that $\overline{A} \subseteq f^{-1}(\overline{f(A)})$, which is equivalent to $f(\overline{A}) \subseteq \overline{f(A)}$.

Suppose now that $f(\overline{A}) \subseteq \overline{f(A)}$ for every subset A of X. Let F be any closed subset of Y. Define $A := f^{-1}(F)$. Then, in particular, $f(A) \subseteq F$. By assumption, we know that

$$f(\overline{A}) \subseteq \overline{f(A)} \subseteq \overline{F} = F,$$

where we have used the fact that F is closed. But this implies that $\overline{A} \subseteq f^{-1}(F) = A$. It follows that $\overline{A} = A$ and hence A is a closed set in X. Recalling the definition of A, we have thus proved that $f^{-1}(F)$ is closed in X for every set F closed in Y. This shows that f is continuous on X. $\qquad\square$

Solution to Exercise 4.15. (i) We prove the equivalent statement that the set $C := \{x \in X : f(x) \neq g(x)\}$ is open in X. Let $x \in C$, arbitrary. Then $f(x) \neq g(x)$. Since $(Y, \mathcal{S})$ is a Hausdorff space, there exist $V_1 \in \mathcal{S}$, $V_2 \in \mathcal{S}$, with $V_1 \ni f(x)$, $V_2 \ni g(x)$, such that $V_1 \cap V_2 = \emptyset$. Let

$$U_1 := f^{-1}(V_1) \quad \text{and} \quad U_2 := g^{-1}(V_2).$$

Since f and g are continuous on X, it follows that $U_1, U_2 \in \mathcal{T}$. Also, it is clear that $x \in U_1 \cap U_2$. Note now that, for any $y \in U_1 \cap U_2$ we have that $f(y) \in V_1$ and $g(y) \in V_2$, and therefore, since $V_1 \cap V_2 = \emptyset$, that $f(y) \neq g(y)$. Denoting $U := U_1 \cap U_2$, it follows that $U \in \mathcal{T}$, $U \ni x$ and $U \subseteq C$. Since $x \in C$ was arbitrary, the property just proved implies that C is open.
(ii) Let

$$F := \{x \in X : f(x) = g(x)\},$$

so that $A \subseteq F$, where A is a dense subset of X. From Part (i) we know that F is a closed set, and hence $\overline{F} = F$. Taking closures in the relation $A \subseteq F$, we therefore obtain $X = \overline{A} \subseteq \overline{F} = F$, which shows that $F = X$, as required. $\qquad\square$

15.5　Solutions to the exercises in Chapter 5

Solution to Exercise 5.1. (i) A is not sequentially compact, since it is not closed: the sequence $((n+1)/n)_1^\infty$ is contained in A but its limit, 1, is not contained in A. B is not sequentially compact, since it is obviously unbounded. C is not sequentially compact since it is not closed: since $\mathbb{Q}$ is dense in $\mathbb{R}$, any real number in $[0, 1]$ is the limit of a sequence in $\mathbb{Q} \cap [0, 1]$, so that $\overline{C} = [0, 1] \neq C$.

(ii) D is sequentially compact since it is both closed and bounded: D is closed since $D = f^{-1}(\{4\})$, where

$$f : \mathbb{R}^2 \longrightarrow \mathbb{R}, \quad f(x, y) := (x - 1)^2 + (y + 2)^2$$

is a continuous function on $\mathbb{R}^2$ and $\{4\}$ is a closed subset of $\mathbb{R}$, and D is bounded since $D \subseteq \mathbb{D}_2((1, -2))$. E is sequentially compact: an easy way to see this is to observe that $E = g(\mathbb{S}_1(0))$, where

$$\mathbb{S}_1(0) = \{(x, y) : x^2 + y^2 = 1\}$$

is clearly a sequentially compact subset of $\mathbb{R}^2$, and

$$g : \mathbb{R}^2 \longrightarrow \mathbb{R}^2 \quad g(x, y) := (x^4 + y^5, x^5 + y^4) \text{ for all } (x, y) \in \mathbb{R}^2$$

is a continuous function on $\mathbb{R}^2$. Then F is not sequentially compact, since it is not bounded: F contains the unbounded sequence $((0, n\pi))_1^\infty$. □

Solution to Exercise 5.2. Consider, for each $n \in \mathbb{N}$, a point $x_n \in F_n$. Since (X, d) is sequentially compact, there exist $x_0 \in X$ and a subsequence $(x_{n_k})_1^\infty$ of $(x_n)_1^\infty$ such $x_{n_k} \longrightarrow x_0$ as $k \to \infty$. We claim that $x_0 \in \bigcap_{m \in \mathbb{N}} F_m$. Fix $m \in \mathbb{N}$, arbitrary. We now prove that $x_0 \in F_m$. Since

$$n_1 < n_2 < n_3 < \cdots n_k < \cdots,$$

there exists $K \in \mathbb{N}$ such that $n_k \geq m$ for all $k \geq K$. It follows that, for all $k \geq K$, $x_{n_k} \in F_{n_k} \subseteq F_m$, so that $(x_{n_k})_{k \geq K} \subseteq F_m$. Since F_m is closed and $x_{n_k} \longrightarrow x_0$ as $k \to \infty$, it follows that $x_0 \in F_m$. Since $m \in \mathbb{N}$ was arbitrary, we have therefore proved that $x_0 \in \bigcap_{m \in \mathbb{N}} F_m$, so that $\bigcap_{m \in \mathbb{N}} F_m \neq \emptyset$. □

Solution to Exercise 5.3. (i) Suppose for a contradiction that $d(A, B) = 0$. Using the definition of $d(A, B)$ as an infimum, we obtain that for every $n \in \mathbb{N}$ there exist $a_n \in A$, $b_n \in B$ such that $d(a_n, b_n) < 1/n$. This implies that $d(a_n, b_n) \longrightarrow 0$ in $\mathbb{R}$ as $n \to \infty$. Since A is sequentially compact, there exists $a_0 \in A$ and a subsequence $(a_{n_k})_1^\infty$ of $(a_n)_1^\infty$ such that $a_{n_k} \longrightarrow a_0$ in (X, d). This means that $d(a_{n_k}, a_0) \longrightarrow 0$ as $k \to \infty$. Since

$$0 \leq d(b_{n_k}, a_0) \leq d(b_{n_k}, a_{n_k}) + d(a_{n_k}, a_0) \longrightarrow 0 + 0 = 0 \quad \text{as } k \to \infty,$$

it follows by the Squeeze Theorem that $b_{n_k} \longrightarrow a_0$ in (X, d). Since $(b_{n_k})_1^\infty \subseteq B$ and B is closed, it follows that $a_0 \in B$. But this contradicts the fact that $A \cap B = \emptyset$. Hence indeed $d(A, B) > 0$.

(ii) The definition of $d(A, B)$ as an infimum shows that for every $n \in \mathbb{N}$ there exist $a_n \in A$, $b_n \in B$ such that

$$d(A, B) \le d(a_n, b_n) < d(A, B) + 1/n.$$

Since $(a_n)_1^\infty \subseteq A$ and A is sequentially compact, there exists $a_0 \in A$ and a subsequence $(a_{n_k})_1^\infty$ of $(a_n)_1^\infty$ such that $a_{n_k} \longrightarrow a_0$ as $k \to \infty$. Since $(b_{n_k})_1^\infty \subseteq B$ and B is sequentially compact, there exists $b_0 \in B$ and a subsequence $(b_{n_{k_\ell}})_1^\infty$ of $(b_{n_k})_1^\infty$ such that $b_{n_{k_\ell}} \longrightarrow b_0$ as $\ell \to \infty$. Since

$$a_{n_{k_\ell}} \longrightarrow a_0 \quad \text{and} \quad b_{n_{k_\ell}} \longrightarrow b_0 \quad \text{as} \ n \to \infty,$$

it follows by a familiar argument that

$$d(a_{n_{k_\ell}}, b_{n_{k_\ell}}) \longrightarrow d(a_0, b_0) \quad \text{as} \ \ell \to \infty.$$

Note also that

$$d(A, B) \le d(a_{n_{k_\ell}}, b_{n_{k_\ell}}) < d(A, B) + \frac{1}{n_{k_\ell}} \quad \text{for all} \ \ell \in \mathbb{N}.$$

We deduce from the Squeeze Theorem that $d(A, B) = d(a_0, b_0)$. Recalling also that $a_0 \in A$, $b_0 \in B$, this proves the required result.

(iii) The fact that A and B are closed is a consequence of the fact that their complements are (countable) unions of open intervals, and therefore open sets. The fact that $A \cap B = \emptyset$ is true since the equation $m = n - 1/n$ has no solution with $m \in \mathbb{N}$, $n \in \mathbb{N}$. Indeed, $1/n \notin \mathbb{Z}$ when $n \in \mathbb{N}$ except when $n = 1$, but in that case $n - 1/n = 0 \notin \mathbb{N}$. Note that

$$1/n \in \{d(a, b) : a \in A, b \in B\},$$

as one can see by taking $a := n$, $b := n - 1/n$. Thus $0 \le d(A, B) \le 1/n$ for all $n \in \mathbb{N}$, which implies that indeed $d(A, B) = 0$. $\qquad\square$

Solution to Exercise 5.4. Let $(y_n)_1^\infty \subseteq f(K)$ be an arbitrary sequence in $f(K)$. By the definition of $f(K)$, there exists $(x_n)_1^\infty \subseteq K$ such that $y_n = f(x_n)$ for all $n \in \mathbb{N}$. Since K is sequentially compact, there exists a subsequence $(x_{n_k})_1^\infty$ of $(x_n)_1^\infty$ such that $x_{n_k} \longrightarrow x_0$ for some $x_0 \in K$ as $k \to \infty$. The continuity of f at x_0 shows that $f(x_{n_k}) \longrightarrow f(x_0)$ as $k \to \infty$, and hence

$$y_{n_k} \longrightarrow f(x_0) \in f(K) \quad \text{as} \ k \to \infty.$$

This proves that $f(K)$ is sequentially compact. $\qquad\square$

Solution to Exercise 5.5. We prove the uniqueness claim first. Let x_0 and $\tilde{x}_0$ be such that $f(x_0) = x_0$, $f(\tilde{x}_0) = \tilde{x}_0$. If $x_0 \ne \tilde{x}_0$, then

$$0 < d(x_0, \tilde{x}_0) = d(f(x_0), f(\tilde{x}_0)) < d(x_0, \tilde{x}_0),$$

a contradiction. Hence there exists at most one fixed point of f. We now prove the existence of a fixed point of f. Let

$$\alpha := \inf \{d(x, f(x)) : x \in X\}.$$

By the definition of an infimum, there exists a sequence $(x_n)_1^\infty$ such that $\alpha \le$

$d(x_n, f(x_n))$ for all $n \in \mathbb{N}$, and $d(x_n, f(x_n)) \longrightarrow \alpha$ as $n \to \infty$. Since (X, d) is sequentially compact, there exist $x_0 \in X$ and a subsequence $(x_{n_k})_1^\infty$ of $(x_n)_1^\infty$ such that $x_{n_k} \longrightarrow x_0$ in (X, d) as $k \to \infty$. Note that the assumption on f implies that f is continuous on X (namely, $\delta := \varepsilon$ works in the definition of continuity). Hence $f(x_{n_k}) \longrightarrow f(x_0)$ as $k \to \infty$. It now follows that

$$d(x_{n_k}, f(x_{n_k})) \longrightarrow d(x_0, f(x_0)) \quad \text{as} \quad k \to \infty,$$

and hence $d(x_0, f(x_0)) = \alpha$. We now show that $\alpha = 0$, which would imply that $x_0 = f(x_0)$. Suppose for a contradiction that $\alpha > 0$, so that $x_0 \neq f(x_0)$. Let $x_{00} = f(x_0)$. Then, by assumption,

$$d(x_{00}, f(x_{00})) = d(f(x_0), f(x_{00})) < d(x_0, x_{00}) = d(x_0, f(x_0)) = \alpha,$$

contradicting the definition of α as an infimum. Hence indeed $\alpha = 0$, which implies that $x_0 = f(x_0)$, as required. $\square$

Solution to Exercise 5.6. Using the Hausdorff condition for x and for each $y \in K$ we obtain that there exist open sets U_y, V_y such that $x \in U_y$, $y \in V_y$, and $U_y \cap V_y = \emptyset$. Since $K \subseteq \bigcup_{y \in K} V_y$, where $V_y \in \mathcal{T}$ for all $y \in K$, and K is a compact set, there exist $n \in \mathbb{N}$ and $y_1, \ldots, y_n \in K$, such that $K \subseteq \bigcup_{k=1}^n V_{y_k}$. By setting

$$V := \bigcup_{k=1}^n V_{y_k}, \quad U := \bigcap_{k=1}^n U_{y_k},$$

one can easily check that U and V are open sets, $x \in U$, $K \subseteq V$, and $U \cap V = \emptyset$. $\square$

Solution to Exercise 5.7. (i) To show that Y is closed, we prove the equivalent statement that $X \setminus Y$ is open. Since Y is compact then, according to the preceding question, for any $x \in X \setminus Y$ there exists open sets U and V such that $x \in U$, $Y \subseteq V$, and $U \cap V = \emptyset$. This implies, in particular, that for any $x \in X \setminus Y$, we have that $x \in U \subseteq X \setminus Y$, where $U \in \mathcal{T}$, a property which implies that $X \setminus Y$ is open.
(ii) Let $\{U_i\}_{i \in I} \subseteq \mathcal{T}$, arbitrary, be such that $Y \subseteq \bigcup_{i \in I} U_i$. This implies that

$$X = \left(\bigcup_{i \in I} U_i \right) \bigcup (X \setminus Y).$$

Observe that $(X \setminus Y) \in \mathcal{T}$, and hence the family of sets consisting of $\{U_i\}_{i \in I}$ and $X \setminus Y$, constitutes an open cover of X. Using the compactness of X, we deduce that there exist $n \in \mathbb{N}$ and $i_1, \ldots, i_n \in I$ such that

$$X = \left(\bigcup_{k=1}^n U_{i_k} \right) \bigcup (X \setminus Y).$$

This implies, however, that $Y \subseteq \bigcup_{k=1}^n U_{i_k}$. The argument given proves that Y is indeed compact. $\square$

Solution to Exercise 5.8. Let $\{U_i\}_{i \in I} \subseteq \mathcal{T}$ be an arbitrary open cover of X, so that $X = \bigcup_{i \in I} U_i$. Clearly, there exists $i_0 \in I$ such that $U_{i_0} \neq \emptyset$. Hence one must have that $X \setminus U_{i_0}$ is a finite set. Denote the elements of $X \setminus U_{i_0}$ by $x_1, \ldots, x_n$, for some $n \in \mathbb{N}$. Since $X = \bigcup_{i \in I} U_i$, it follows that for each $k \in \{1, \ldots, n\}$ there exists $i_k \in I$ so that $x_k \in U_{i_k}$. It therefore follows that $X = \bigcup_{k=0}^{n} U_{i_k}$, and thus we have shown that $\{U_i\}_{i \in I}$ has the finite subcover $\{U_{i_k}\}_{k \in \{0, \ldots, n\}}$. Therefore, $(X, \mathcal{T})$ is compact. $\qquad\square$

Solution to Exercise 5.9. Let us denote the above property by $(*)$. Suppose first that $(X, \mathcal{T})$ is compact. Let $\{F_i\}_{i \in I}$ be an arbitrary family of closed sets such that $\bigcap_{i \in I} F_i = \emptyset$. Taking complements and using De Morgan's Laws we get

$$\bigcup_{i \in I} (X \setminus F_i) = X.$$

Since each F_i is closed, it follows that $X \setminus F_i$ is an open set for all $i \in I$. The compactness of X shows that there exists a finite subset J of I such that $\bigcup_{i \in J} (X \setminus F_i) = X$. Taking again complements, we get $\bigcap_{i \in J} F_i = \emptyset$. This proves $(*)$. Now suppose that $(*)$ holds. Consider an arbitrary open cover $\{U_i\}_{i \in I}$ of X. Taking complements and using De Morgan's Laws in the relation $X = \bigcup_{i \in I} U_i$ we get that

$$\bigcap_{i \in I} (X \setminus U_i) = \emptyset.$$

Since each U_i is open, it follows that $X \setminus U_i$ is a closed set for all $i \in I$. By applying $(*)$ we obtain that there exists a finite subset J of I such that

$$\bigcap_{i \in J} (X \setminus U_i) = \emptyset.$$

Taking again complements, we get $\bigcup_{i \in J} U_i = X$. Therefore $(X, \mathcal{T})$ is compact. $\quad\square$

Solution to Exercise 5.10. Since K is compact, $K \subseteq \bigcup_{i=1}^{\infty} U_i$, and $\{U_i\}_{i \in \mathbb{N}} \subseteq \mathcal{T}$, it follows that there exists a finite subset J of $\mathbb{N}$ such that $K \subseteq \bigcup_{i \in J} U_i$. Let $i_0 := \max\{i : i \in J\}$. The assumption $(*)$ ensures that $K \subseteq U_{i_0}$, as required. $\qquad\square$

Solution to Exercise 5.11. Let $\{U_i\}_{i \in I}$ be an arbitrary open cover of $Y \bigcup Z$. Then

$$Y \subseteq \bigcup_{i \in I} U_i, \quad Z \subseteq \bigcup_{i \in I} U_i.$$

Since both Y and Z are compact, it follows that there exist finite sets $J_Y, J_Z \subseteq I$ such that $Y \subseteq \bigcup_{i \in J_Y} U_i$, and $Z \subseteq \bigcup_{i \in J_Z} U_i$. But then we obviously have

$$Y \bigcup Z \subseteq \bigcup_{i \in (J_Y \cup J_Z)} U_i,$$

and thus the given open cover has a finite subcover. It follows that $Y \bigcup Z$ is indeed compact. $\qquad\square$

Solution to Exercise 5.12. Let, for each $n \in \mathbb{N}$, $e^{(n)} = (0, \ldots 0, 1, 0, \ldots)$, where the n-th entry is 1, and the rests of the entries are 0. Then $e^{(n)} \in \ell^{\infty}$ for all $n \in \mathbb{N}$, $e^{(n)}$ belongs to the closed unit ball $D_1(0)$ in $(\ell^{\infty}, \|\cdot\|_{\infty})$, and

$$\|e^{(i)} - e^{(j)}\|_{\infty} = 1 \text{ for all } i, j \in \mathbb{N} \text{ with } i \neq j.$$

This shows that no subsequence of $(e^{(n)})_1^\infty$ is a Cauchy sequence in $(\ell^\infty, \|\cdot\|_\infty)$, and thus no subsequence converges.   □

Solution to Exercise 5.13. Let $x_0 \in X \setminus Y$, arbitrary, and let

$$d(x_0, Y) := \inf\{\|x_0 - y\| : y \in Y\}.$$

We now prove that the above infimum is attained, that is, there exists $w \in Y$ such that $d(x_0, Y) = \|x_0 - w\|$. Indeed, let $\gamma := d(x_0, Y)$. It follows that, for every $n \in \mathbb{N}$, there exists $y_n \in Y$ such that

$$\gamma \le \|x_0 - y_n\| < \gamma + 1/n.$$

This implies that, for all $n \in \mathbb{N}$,

$$\|y_n\| \le \|y_n - x_0\| + \|x_0\| < \|x_0\| + \gamma + 1,$$

so that the sequence $(y_n)_1^\infty$ is bounded in $(Y, \|\cdot\|)$. Since $(Y, \|\cdot\|)$ is finite-dimensional, any bounded sequence in it has a convergent subsequence. Thus there exists a subsequence $(y_{n_k})_1^\infty$ of $(y_n)_1^\infty$, and $w \in Y$, such that $y_{n_k} \longrightarrow w$ in $(Y, \|\cdot\|)$, and hence in $(X, \|\cdot\|)$ too. Passing to the limit as $k \to \infty$ in the inequality

$$\gamma \le \|x_0 - y_{n_k}\| < \gamma + 1/n_k \quad \text{for all } k \in \mathbb{N},$$

we obtain that $\|x_0 - w\| = \gamma$, as required. Since $x_0 \notin Y$, this also shows that $\gamma > 0$. (An alternative proof of the fact that the infimum in the definition of $d(x_0, Y)$ is attained may be given by applying the Weierstrass Theorem to the continuous function $y \mapsto \|x_0 - y\|$ on a suitably chosen closed ball in Y, which is a (sequentially) compact set because Y is finite-dimensional. Note that Y itself is not (sequentially) compact, because it is not bounded. Some care is required in the choice of the above ball, which is why the more direct proof given above seems preferable.)

With w chosen as above, we now define

$$z = \frac{x_0 - w}{\|x_0 - w\|}.$$

Then $\|z\| = 1$ and, for any $y \in Y$, we have

$$\begin{aligned}
\|z - y\| &= \left\| \frac{x_0 - w}{\|x_0 - w\|} - y \right\| \\
&= \frac{\|x_0 - (w + \|x_0 - w\|y)\|}{\|x_0 - w\|} \\
&= \frac{\|x_0 - (w + \|x_0 - w\|y)\|}{\gamma} \\
&\ge 1,
\end{aligned}$$

since $w + \|x_0 - w\|y \in Y$. This completes the proof.   □

15.6   Solutions to the exercises in Chapter 6

Solution to Exercise 6.1. By using that E and F are elementary sets, we may find boxes B_k and C_k such that $E = \bigcup_{k=1}^{n} B_k$ and $F = \bigcup_{k=1}^{m} C_k$. Then

$$E \cup F = \left(\bigcup_{k=1}^{n} B_k \right) \cup \left(\bigcup_{k=1}^{m} C_k \right).$$

By a known result in the main text, we can find disjoint boxes D_j and index sets J_k and J_k' such that $B_k = \bigcup_{j \in J_k} D_k$ and $C_k = \bigcup_{j \in J_k'} D_k$. But then

$$E \setminus F = \left(\bigcup_{k=1}^{n} \bigcup_{j \in J_k} D_k \right) \setminus \left(\bigcup_{k'=1}^{m} \bigcup_{j' \in J_k'} D_{j'} \right) = \bigcup_{k=1}^{n} \bigcup_{j \in J_k} \left(D_k \setminus \left(\bigcup_{k'=1}^{m} \bigcup_{j' \in J_k'} D_{j'} \right) \right).$$

As the D_k's are disjoint, we have that

$$D_k \setminus \left(\bigcup_{k'=1}^{m} \bigcup_{j' \in J_k'} D_{j'} \right)$$

is either D_k or $\emptyset$ in any case it is a box. Hence we see that $E \setminus F$ can be written as a union of boxes. $\qquad \square$

Solution to Exercise 6.2. We proceed in the same way as in the solution of Exercise 6.1.

1. We write $E = \bigcup_{k=1}^{n} B_k$ and $F = \bigcup_{l=1}^{m} C_l$ where B_k and C_l are boxes. Then

$$E \cup F = \left(\bigcup_{k=1}^{n} B_k \right) \cup \left(\bigcup_{k=1}^{m} C_k \right).$$

2. By Exercise 6.1, we know that $E \setminus F$ and $F \setminus E$ are also elementary sets. We have shown that the union of elementary sets is again elementary and so $E \triangle F$ is also elementary.

3. Using the special form of E, we have

$$x + E = \left\{ x + y \;\middle|\; y \in \bigcup_{j=1}^{n} B_j \right\} = \{ x + y \mid \exists j : y \in B_j \} = \bigcup_{j=1}^{n} (x + B_j)$$

and hence the translation of a box is a box. $\qquad \square$

Solution to Exercise 6.3. We need to check the axioms of a ring. One needs to identify which of the two operations plays the role of the addition and which plays the role of the multiplication. By using that

$$(E \cup F) \setminus G = (E \setminus G) \cup (F \setminus G), \qquad E \setminus (F \setminus G) = (E \setminus F) \cup (E \cap F \cap G),$$

we see that both operations are commutative and also

$$
\begin{aligned}
(E \Delta F) \Delta G &= ((E \setminus F) \cup (F \setminus E)) \, \Delta G \\
&= [((E \setminus F) \cup (F \setminus E)) \setminus G] \cup [G \setminus ((E \setminus F) \cup (F \setminus E))] \\
&= (E \setminus (F \cup G)) \cup (F \setminus (E \cup G)) \cup (G \setminus (E \cup F)) \cup (E \cap F \cap G).
\end{aligned}
$$

The expression we derived for $(E \Delta F) \Delta G$ is symmetric in E, F, G and hence it has to be equal to $E \Delta (F \Delta G)$. Hence the operation Δ is associative. The neutral element for the disjoint union is the empty set, as

$$
E \Delta \emptyset = (E \setminus \emptyset) \cup (\emptyset \setminus E) = E
$$

and the inverse element is E itself, because

$$
E \Delta E = (E \setminus E) \cup (E \setminus E) = \emptyset
$$

Thus, the operation Δ plays the role of the addition. Therefore, we have to check the validity of the distribution law. Indeed, we have

$$
\begin{aligned}
E \cap (F \Delta G) &= E \cap ((F \setminus G) \cup (G \setminus F)) \\
&= (E \cap (F \setminus G)) \cup (E \cap (G \setminus F)) \\
&= ((E \cap F) \setminus (E \cap G)) \cup ((E \cap G) \setminus (E \cap F)) \\
&= (E \cap F) \Delta (E \cap G).
\end{aligned}
$$

The conclusion follows. $\qquad\square$

Solution to Exercise 6.4. Let $(B_n)_1^\infty$ be a cover of F by non-degenerate boxes. Then, this is also a cover of E because $E \subseteq F \subseteq \bigcup_1^\infty B_n$. Hence, E has a larger set of admissible coverings than F. Since the infimum is taken over a larger set, thus it is smaller or equal:

$$
\mathcal{L}^{d*}(E) = \inf_{(B_n)_1^\infty : E \subseteq \bigcup_n B_n} \sum_{n=1}^\infty \mathcal{V}^d(B_n) \leq \inf_{(B_n)_1^\infty : F \subseteq \bigcup_n B_n} \sum_{n=1}^\infty \mathcal{V}^d(B_n) = \mathcal{L}^{d*}(F).
$$

Thus, $\mathcal{L}^{d*}(E) \leq \mathcal{L}^{d*}(F)$. $\qquad\square$

Solution to Exercise 6.5. $E \in \mathcal{L}(\mathbb{R}^d)$ means that for all $A \subseteq \mathbb{R}$ holds that

$$
\mathcal{L}^{d*}(A) \geq \mathcal{L}^{d*}(A \cap E) + \mathcal{L}^{d*}(A \cap E^c).
$$

As $(E^c)^c = E$, the definition is invariant under swapping E with E^c. $\qquad\square$

Solution to Exercise 6.6. We just have to check the definition. Indeed,

$$
\mathcal{L}^{d*}(A \cap \mathbb{R}^d) + \mathcal{L}^{d*}(A \cap (\mathbb{R}^d)^c) = \mathcal{L}^{d*}(A) + \mathcal{L}^{d*}(A \cap \emptyset) = \mathcal{L}^{d*}(A) + 0
$$

and hence $\mathbb{R}^d \in \mathcal{L}(\mathbb{R}^d)$. $\qquad\square$

Solution to Exercise 6.7. First note that by the definitions it is very easy to prove that for any box $B \subseteq \mathbb{R}^d$, tB is also a box and $\mathcal{V}^d(tB) = t^d \mathcal{V}^d(B)$. Fix any $E \in \mathcal{L}(\mathbb{R}^d)$ and $\lambda > 0$. Then,

$$\mathcal{L}^d(\lambda E) = \inf \left\{ \sum_1^\infty \mathcal{V}^d(B_i) \,\Big|\, B_1, B_2, \ldots \text{ non-degenerate boxes}, \lambda E \subseteq \bigcup_1^\infty B_i \right\}.$$

Then, $\lambda E \subseteq \bigcup_1^\infty B_i$ if and only if $E \subseteq \bigcup_1^\infty \frac{1}{\lambda} B_i$. Further,

$$\mathcal{L}^d(E) = \inf \left\{ \sum_1^\infty \mathcal{V}^d(D_i) \,\Big|\, D_1, D_2, \ldots \text{ non-degenerate boxes}, E \subseteq \bigcup_1^\infty D_i \right\}.$$

Let $(B_i)_1^\infty$ be a cover of λE by boxes. Then, $\left(\frac{1}{\lambda} B_i\right)_1^\infty$ is a cover of E by boxes. Hence,

$$\mathcal{L}^d(E) \leq \sum_1^\infty \mathcal{V}^d \left(\frac{1}{\lambda} B_i\right) = \frac{1}{\lambda^d} \sum_1^\infty \mathcal{V}^d(B_i)$$

and by taking "inf" over all covers of λE by boxes, we get

$$\mathcal{L}^d(E) \leq \frac{1}{\lambda^d} \mathcal{L}^d(\lambda E).$$

Finally, for the given $A \in \mathcal{L}(\mathbb{R}^d)$ and $t > 0$, we apply this first to $E := A$ and $\lambda := t$ and then to $E := \frac{1}{t} A$ and to $\lambda := \frac{1}{t}$.

The conclusion regarding the invariance under translations is a consequence of Exercise 6.2 and the definition of the volume of boxes. $\qquad\square$

Solution to Exercise 6.8. Fix $A \subseteq \mathbb{R}^d$. Since every cover of A by cubes $(Q_i)_1^\infty$ is itself a cover by boxes $(B_i)_1^\infty$ of special form, we have that

$$\mathcal{Q}^{d*}(A) := \inf \left\{ \sum_{i=1}^\infty \mathcal{V}^d(Q_i) \,\Big|\, (Q_i)_1^\infty \text{ non-degenerate cubes}, A \subseteq \bigcup_{i=1}^\infty Q_i \right\}$$

$$\geq \inf \left\{ \sum_{i=1}^\infty \mathcal{V}^d(B_i) \,\Big|\, (B_i)_1^\infty \text{ non-degenerate boxes}, A \subseteq \bigcup_{i=1}^\infty B_i \right\}$$

$$= \mathcal{L}^{d*}(A).$$

Conversely, let $\varepsilon > 0$. Then, there exist boxes $(B_i)_1^\infty$ such that $A \subseteq \bigcup_{i=1}^\infty B_i$ and

$$\sum_{i=1}^\infty \mathcal{V}^d(B_i) \leq \mathcal{L}^{d*}(A) + \varepsilon.$$

Since the volume of each box B_i is the same as the volume of its closure $\overline{B}_i \supseteq B_i$, we may assume all the boxes $(B_i)_1^\infty$ are closed. Then, every box B_i can be filled up with countably many disjoint cubes $(Q_{ij})_{j=1}^\infty$ such that

$$B_i = \bigcup_{j=1}^\infty Q_{ij}, \quad \forall\, i \in \mathbb{N}.$$

This can be done recursively as follows (we give the details for simplicity for $d = 2$ only): given a fixed box $B \in (B_i)_{i=1}^\infty$ with side lengths a, b, if $a = b$, we are done and set $Q_1 := B$, $Q_k := \emptyset$, $k \geq 2$. If $a \neq b$, suppose $b > a$ (or else swap names of the sides). Suppose further by a change of coordinates that $B = [0, a] \times [0, b]$ Then, B can be written as

$$B = \Big([0, a] \times [0, a]\Big) \bigcup \Big([0, a] \times (a, b]\Big).$$

Set $Q_1 := [0, a] \times [0, a]$. If $a = b - a$, then set $Q_2 := [0, a] \times (a, b]$ and $Q_k := \emptyset$, $k \geq 3$. If not, then either $a < b - a$ or $a > b - a$. In either case, we can fill up $[0, a] \times (a, b]$ with the union of a cube and a box. It can be easily shown that this recursive construction exhausts B and one thus obtained the desired sequence. Hence, since the volume of each cube is the same as its Lebesgue measure and the latter is σ-additive, we have

$$\mathcal{V}^d(B_i) = \mathcal{L}^d(B_i) = \mathcal{L}^d\left(\bigcup_{j=1}^{\infty} Q_{ij}\right) = \sum_{j=1}^{\infty} \mathcal{L}^d(Q_{ij}) = \sum_{j=1}^{\infty} \mathcal{V}^d(Q_{ij}).$$

Since the double sequence $(Q_{ij})_{i,j=1}^{\infty}$ covers A as well, i.e.

$$A \subseteq \bigcup_{i=1}^{\infty} \bigcup_{j=1}^{\infty} Q_{ij},$$

we have

$$\mathcal{Q}^{d*}(A) \leq \sum_{i=1}^{\infty} \sum_{j=1}^{\infty} \mathcal{V}^d(Q_{ij}) = \sum_{i=1}^{\infty} \mathcal{V}^d(B_i) \leq \mathcal{L}^{d*}(A) + \varepsilon.$$

Since ε is arbitrary, we obtain $\mathcal{Q}^{d*}(A) \leq \mathcal{L}^{d*}(A)$. □

Solution to Exercise 6.9. Let Π^{d-1} be any hyperplane in $\mathbb{R}^d$, which by the invariance of the Lebesgue measure under isometries we may assume that is $\mathbb{R}^{d-1} \times \{0\} \subseteq \mathbb{R}^d$. Consider the parallelogram

$$P_{\ell, \varepsilon} = [-\ell, +\ell]^{d-1} \times [-\varepsilon, \varepsilon] \subseteq \mathbb{R}^d.$$

Then, $\mathcal{L}^d(P_{\ell, \varepsilon}) = (2\ell)^{d-1} 2\varepsilon$. Hence, for all $\varepsilon > 0$, If $Q_\ell := [-\ell, \ell]^{d-1} \times \{0\}$, we have that

$$\mathcal{L}^d\big(\Pi^{d-1} \cap Q_\ell\big) \leq \mathcal{L}^d(P_{\ell, \varepsilon}) = 2^n \ell \varepsilon \longrightarrow 0, \quad \text{as } \varepsilon \to 0.$$

Hence, since

$$\Pi^{d-1} = \bigcup_{l=1}^{\infty} [-\ell, +\ell]^{d-1},$$

we have

$$\mathcal{L}^d(\Pi^{d-1}) = \lim_{m \to \infty} \sum_{\ell=1}^{m} \mathcal{L}^d(Q_{\ell+1}) = \lim_{m \to \infty} 0 + \dots + 0 = 0. \quad □$$

Solution to Exercise 6.10. We have

$$(A \cup B) \setminus (A \cap B) = (A \setminus B) \cup (B \setminus A)$$

and

$$A = (A \setminus B) \cup (A \cap B), \qquad B = (B \setminus A) \cup (A \cap B).$$

Hence, by using that $A \setminus B \subseteq (A \cup B) \setminus (A \cap B)$, we have

$$
\begin{aligned}
\mathcal{L}^d(A) &= \mathcal{L}^d(A \setminus B) + \mathcal{L}^d(A \cap B) \\
&\leq \mathcal{L}^d((A \cup B) \setminus (A \cap B)) + \mathcal{L}^d(A \cap B) \\
&= \mathcal{L}^d(A \cap B) \\
&\leq \mathcal{L}^d(B).
\end{aligned}
$$

Since the given relation is symmetric with respect to A and B, by replacing A with B and repeating the proof above we get that $\mathcal{L}^d(B) \leq \mathcal{L}^d(A)$. As an outer measure is also sub-additive and monotone, the same result holds for the outer measure. □

Solution to Exercise 6.11. This is a direct consequence of Carathéodory's Theorem (item (viii)) and Lemma 6.14 (item (v)). □

Solution to Exercise 6.12. For each m and all n holds as the supremum is an upper bound that $a_{n,m} \leq \sup_{k \in \mathbb{N}} a_{k,m}$. Hence

$$
\sum_{m=1}^{M} a_{n,m} \leq \sum_{m=1}^{M} \sup_{k \in \mathbb{N}} a_{k,m}.
$$

Both sides are monotone in M and hence the limit exists and we get that

$$
\sum_{m=1}^{\infty} a_{n,m} \leq \sum_{m=1}^{\infty} \sup_{k \in \mathbb{N}} a_{k,m}.
$$

As this holds for all n the right hand side is an upper bound and hence the least upper bound of the left hand side has to be smaller, that is

$$
\sup_{n \in \mathbb{N}} \sum_{m=1}^{\infty} a_{n,m} \leq \sum_{m=1}^{\infty} \sup_{k \in \mathbb{N}} a_{k,m}. \quad \square
$$

Solution to Exercise 6.13. On the one hand,

$$
\sum_{n=1}^{\infty} \left(\frac{m}{2^n} - \frac{2m}{3^n} \right) = m \sum_{n=1}^{\infty} \frac{1}{2^n} - 2m \sum_{n=1}^{\infty} \frac{2}{3^n}
$$

recall that for $|x| < 1$ holds

$$
\sum_{n=1}^{\infty} x^n = x \sum_{n=0}^{\infty} x^n = \frac{x}{1-x}
$$

and hence

$$
\sum_{n=1}^{\infty} \left(\frac{m}{2^n} - \frac{2m}{3^n} \right) = \frac{m/2}{1-1/2} - 2\frac{m/3}{1-1/3} = m - m = 0
$$

and thus we get that

$$
\sum_{m=1}^{\infty} \sum_{n=1}^{\infty} \left(\frac{1}{2^n} - \frac{2}{3^n} \right) = 0.
$$

On the other hand,

$$
\sum_{m=1}^{\infty} \left(\frac{1}{2^n} - \frac{2}{3^n} \right) = \begin{cases} \infty, & \text{for } n \geq 2, \\ -\infty, & \text{for } n = 1. \end{cases}
$$

□

15.7   Solutions to the exercises in Chapter 7

Solution to Exercise 7.1.

1. For the first two sets, we have

$$C_1 = \left[0, \frac{1}{3}\right] \cup \left[\frac{2}{3}, 1\right]$$

and

$$C_2 = \left[0, \frac{1}{9}\right] \cup \left[\frac{2}{9}, \frac{3}{9}\right] \cup \left[\frac{6}{9}, \frac{7}{9}\right] \cup \left[\frac{8}{9}, 1\right].$$

By induction, all sets C_n are closed and since intersection of closed sets is closed, we infer that $\bigcap_{n=1}^{\infty} C_n$ is also closed. Since $C_n \subseteq [0, 1]$ for all n, it is also bounded and hence compact.

2. There exists an one to one correspondence between C and $[0, 1]$ by mapping the triadic expansion to the binary expansion, that is $\sum_k \frac{a_k}{3^k} \mapsto \sum_k \frac{a_k/2}{2^k}$, because $a_k \neq 1$ for all k. Hence C has the same cardinality as $\mathbb{R}$.

3. As C_{n-1} is the union of 2^n disjoint intervals of length $1/3^n$, it is an elementary set and also $\mathcal{L}^1(C_n) = \frac{2^n}{3^n}$.

4. Since $\mathcal{L}^1(C_{n-1}) = (2/3)^n$ and $\mathcal{L}^1(C_1) < \infty$, by the lower continuity of the measure we have

$$\mathcal{L}^1(C) = \mathcal{L}^1\left(\bigcap_{n=1}^{\infty} C_n\right) = \lim_{n \to \infty} \mathcal{L}^1(C_n) = \lim_{n \to \infty} \left(\frac{2}{3}\right)^n = 0. \quad \square$$

Solution to Exercise 7.2.

• Let $(A_i)_1^{\infty} \subseteq \mathfrak{M}$. Recall that $(\bigcap_{n=1}^{\infty} A_n)^c = \bigcup_{n=1}^{\infty} (A_n)^c$. Also, $A_n \in \mathfrak{M}$ and $(A_n)^c \in \mathfrak{M}$, whence $\bigcup_{n=1}^{\infty} (A_n)^c \in \mathfrak{M}$. This in turn implies

$$\bigcap_{n=1}^{\infty} (A_n) = \left(\bigcup_{n=1}^{\infty} (A_n)^c\right)^c \in \mathfrak{M}.$$

• If $A, B \in \mathfrak{M}$, then $B \setminus A = B \cap (A^c) \in \mathfrak{M}$ since as we have just shown that $\mathfrak{M}$ is closed under intersections too. $\quad \square$

Solution to Exercise 7.3.

• We have $B = A \cup (B \cap A^c)$ and the union is disjoint. Hence $\mu(B) = \mu(A) + \mu(B \cap A^c)$. As $\mu(B \cap A^c) \geq 0$ we get $\mu(B) \geq \mu(A)$.

• Since $B = A \cup (B \setminus A)$ and the union is disjoint, we have $\mu(B) = \mu(A) + \mu(B \setminus A)$. Since $0 \leq \mu(A) < \infty$, we can subtract it from both sides. $\quad \square$

Solution to Exercise 7.4. We recall that

$$\mathcal{H}_\delta^{s*}(A) = \inf\left\{\sum_{k=1}^{\infty} \alpha(s) \left(\frac{\text{diam}(C_k)}{2}\right)^s : (C_k)_1^{\infty} \subseteq \mathbb{R}^d, \ A \subseteq \bigcup_{k=1}^{\infty} C_k, \ \text{diam}(C_k) \leq \delta\right\},$$

and

$$\mathcal{H}^{s*}(A) = \lim_{\delta \to 0} \mathcal{H}^{s*}_\delta(A) = \sup_{\delta > 0} \mathcal{H}^{s*}_\delta(A).$$

Note that we readily have $\mathcal{H}^{s*}(\emptyset) = 0$.

Let $A \subseteq B$. Then any sequence $(C_k)_1^\infty \subseteq \mathbb{R}^d$ with $B \subseteq \bigcup_{k=1}^\infty C_k$ and $\mathrm{diam}(C_k) \le \delta$ satisfies also $(C_k)_1^\infty \subseteq \mathbb{R}^d$ with $A \subseteq \bigcup_{k=1}^\infty C_k$ and $\mathrm{diam}(C_k) \le \delta$. Hence any $(C_k)_1^\infty$ that appears in the infimum in the definition of $\mathcal{H}^{s*}_\delta(B)$ also appears in the infimum in the definition of $\mathcal{H}^{s*}_\delta(A)$, but there can appear also further $C_1, C_2, \ldots$ and hence $0 \le \mathcal{H}^{s*}_\delta(A) \le \mathcal{H}^{s*}_\delta(B)$. By taking the limit $\delta \downarrow 0$, we obtain $\mathcal{H}^{s*}(A) \le \mathcal{H}^{s*}(B)$.

Consider $A_1, A_2, \ldots$ and $\varepsilon > 0$. By the definition of the infimum for each n there exists $C_1^{(n)}, C_2^{(n)}, \ldots \subseteq \mathbb{R}^d$ and $A_n \subseteq \bigcup_{k=1}^\infty C_k^{(n)}$ and $\mathrm{diam}(C_k^{(n)}) \le \delta$ such that

$$\sum_{k=1}^\infty \alpha(s) \left(\frac{\mathrm{diam}(C_k^{(n)})}{2} \right)^s \le \mathcal{H}^{s*}_\delta(A_n) + \varepsilon 2^{-n}.$$

Then,

$$C_1^{(1)}, C_2^{(1)}, C_1^{(2)}, C_1^{(3)}, C_2^{(2)} \cdots \subseteq \mathbb{R}^d$$

is a cover of $\bigcup_{n=1}^\infty A_n$ with the required properties such that it appears in the infimum in the definition of $\mathcal{H}^{s*}_\delta \left(\bigcup_{n=1}^\infty A_n \right)$ and hence

$$\mathcal{H}^{s*}_\delta \left(\bigcup_{n=1}^\infty A_n \right) \le \sum_{j=1}^\infty \sum_{l=1}^j \alpha(s) \left(\frac{\mathrm{diam}(C_l^{(j-l)})}{2} \right)^s$$

$$= \sum_{n=1}^\infty \sum_{k=1}^\infty \alpha(s) \left(\frac{\mathrm{diam}(C_k^{(n)})}{2} \right)^s$$

$$\le \sum_{n=1}^\infty \mathcal{H}^{s*}_\delta(A_n) + \sum_{n=1}^\infty \varepsilon 2^{-n}$$

$$= \sum_{n=1}^\infty \mathcal{H}^{s*}_\delta(A_n) + \varepsilon$$

By taking the limit $\delta \to 0$ we obtain that $\mathcal{H}^{s*}$ is σ-subadditive. $\square$

Solution to Exercise 7.5. Since $f - g$ is $\mathfrak{M}$-measurable, we have that $\{f \le g\} = \{f - g \le 0\} = (f - g)^{-1}((-\infty, a]) \in \mathfrak{M}$. $\square$

Solution to Exercise 7.6. Clearly, for all $B \in \mathcal{P}(X)$ holds $\delta_x(B) \ge 0$. Furthermore, $\delta(\emptyset) = 0$ because $x \notin \emptyset$. Finally, let $(B_n)_1^\infty$ a sequence of mutually disjoint sets. If $x \notin \bigcup_{n=1}^\infty B_n$, then $x \notin B_n$ for all $n \in \mathbb{N}$. Hence $\delta_x(B_n) = 0$ for all $n \in \mathbb{N}$ and

$$\delta_x \left(\bigcup_{n=1}^\infty B_n \right) = 0, \qquad \sum_{n=1}^\infty \delta_x(B_n) = 0.$$

If $x \in \bigcup_{n=1}^\infty B_n$, then there exists exactly one k such that $x \in B_k$ and $x \notin B_n$ for all $n \ne k$. Hence $\delta_x(B_n) = 0$ for $n \ne k$ and $\delta_x(B_k) = 1$. Hence

$$\delta_x \left(\bigcup_{n1}^\infty B_n \right) = 1, \qquad \sum_{n=1}^\infty \delta_x(B_n) = 1. \quad \square$$

Solution to Exercise 7.7. As f and g are non-negative, $f \cdot g$ and αf is non-negative. By the approximation result of measurable by simple functions, there exist monotone increasing sequences of simple function $(\varphi_n)_1^\infty$ such that $\lim_{n \to \infty} \varphi_n = f$ and $(\psi_n)_1^\infty$ such that $\lim_{n \to \infty} \psi_n = g$. As the product operation is a continuous function $\mathbb{R} \times \mathbb{R} \longrightarrow \mathbb{R}$, we obtain

$$\lim_{n \to \infty} (\varphi_n(x)\psi_n(x)) = \left(\lim_{n \to \infty} \varphi_n(x) \right) \left(\lim_{n \to \infty} \psi_n(x) \right) = f(x)g(x)$$

Since for each $n \in \mathbb{N}$, φ_n and ψ_n are simple functions, we can represent them as

$$\varphi_n = \sum_{k=1}^{N^{(n)}} a_k^{(n)} \chi_{A_k^{(n)}}, \quad \psi_n = \sum_{j=1}^{M^{(n)}} b_j^{(n)} \chi_{B_j^{(n)}}$$

and hence the product is a simple function as well, since

$$\varphi_n(x)\psi_n(x) = \left(\sum_{k=1}^{N^{(n)}} a_k^{(n)} \chi_{A_k^{(n)}} \right) \left(\sum_{j=1}^{M^{(n)}} b_j^{(n)} \chi_{B_j^{(n)}} \right) = \sum_{k=1}^{N^{(n)}} \sum_{j=1}^{M^{(n)}} b_j^{(n)} a_k^{(n)} \chi_{A_k^{(n)} \cap B_j^{(n)}}.$$

Further, for $\alpha \geq 0$ we see that $\alpha \varphi_n$ converges to αf and as αf is a simple function we see that αf is measurable. $\quad\square$

Solution to Exercise 7.8. The characterisation of measurability for non-negative functions f, g implies that there exist monotone increasing sequences simple function $(\varphi_n)_1^\infty$ such that $\lim_{n \to \infty} \varphi_n = f$ and $(\psi_n)_1^\infty$ such that $\lim_{n \to \infty} \psi_n = g$. By the limit rule for the addition one gets that

$$\lim_{n \to \infty} \left(\varphi_n(x) + \psi_n(x) \right) = \left(\lim_{n \to \infty} \varphi_n(x) \right) + \left(\lim_{n \to \infty} \psi_n(x) \right) = f(x) + g(x)$$

as the sum of two simple functions is a simple function we get that $f + g$ is measurable.

Let now $\varphi = \sum_{k=1}^{n} a_k \chi_{A_k}$ and $\psi = \sum_{j=1}^{m} b_j \chi_{B_j}$ be simple functions. We can rewrite them as

$$\varphi = \sum_{k=1}^{n} \sum_{j=1}^{m} a_k \chi_{A_k \cap B_j}, \quad \psi = \sum_{j=1}^{m} \sum_{k=1}^{n} b_j \chi_{A_k \cap B_j}$$

as for k, j such that $A_k \cap B_j \neq \emptyset$ we get for $x \in A_k \cap A_j$ that

$$\max \left\{ \varphi(x), \psi(x) \right\} = \max \left\{ a_k \chi_{A_k \cap B_j}(x), b_j \chi_{A_k \cap B_j(x)} \right\}$$
$$= \max \left\{ a_k, b_j \right\} \chi_{A_k \cap B_j}(x)$$

and as the $A_k \cap B_j$ are disjoint we obtain that

$$= \sum_{j'=1}^{m} \sum_{k'=1}^{n} \max \left\{ a_{k'}, b_{j'} \right\} \chi_{A_{k'} \cap B_{j'}}(x)$$

which is a simple function.

Let now f and g be measurable functions. Then, it is sufficient to show that

$\max\{\varphi_n, \psi_n\} \to \max\{f, g\}$. Indeed, let x be such that $f(x) > g(x)$, then for n large enough $\varphi_n(x) > g(x)$ and hence

$$\max\{\varphi_n(x), \psi_n(x)\} = \varphi_n(x) \longrightarrow f(x) = \max\{f(x), g(x)\},$$

as $n \to \infty$. □

Solution to Exercise 7.9.
1. As f, g are measurable, there exist monotone increasing sequences of simple functions φ_n and ψ_n with $\lim_{n \to \infty} \varphi_n = f$ and $\lim_{n \to \infty} \psi_n g$. The function $\varphi_n + \psi_n$ is simple and as $\varphi_n \le \varphi_{n+1}$ and $\psi_n \le \psi_{n+1}$ hence also

$$\varphi_n + \psi_n \le \varphi_{n+1} + \psi_{n+1}$$

and thus

$$\lim_{n \to \infty} (\varphi_n + \psi_n) = f + g.$$

Then, by the definition of the integral we have

$$\int_X (f + g)\, d\mu = \lim_{n \to \infty} \int_X (\varphi_n + \psi_n)\, d\mu.$$

Using that the limit of a sum of sequences is the sum of the limit we get that

$$\int_X (f + g)\, d\mu = \lim_{n \to \infty} \int_X \varphi_n\, d\mu + \lim_{n \to \infty} \int_X \psi_n\, d\mu = \int_X f\, d\mu + \int_X g\, d\mu.$$

2. If $\alpha = 0$, it is obvious. If $\alpha > 0$, we have

$$\int_X (\alpha f)\, d\mu = \sup\left\{ \int_X \phi\, d\mu \,\middle|\, \phi \text{ simple}, \phi \le \alpha f \right\}$$

$$= \sup\left\{ \int_X \alpha \frac{1}{\alpha} \phi\, d\mu \,\middle|\, \phi \text{ simple}, \frac{1}{\alpha}\phi \le f \right\}$$

$$= \alpha \sup\left\{ \int_X \frac{1}{\alpha} \phi\, d\mu \,\middle|\, \phi \text{ simple}, \frac{1}{\alpha}\phi \le f \right\}$$

$$= \alpha \sup\left\{ \int_X \psi\, d\mu \,\middle|\, \phi \text{ simple}, \psi \le f \right\}$$

$$= \alpha \int_X f\, d\mu. \quad \square$$

Solution to Exercise 7.10. As φ is simple it can be written as $\varphi = \sum_{k=1}^n a_k \chi_{A_k}$ for $A_1, \ldots, A_n \in \mathfrak{M}$. Then, $\alpha\varphi = \sum_{k=1}^n \alpha \cdot a_k \chi_{A_k}$ which is also a simple function.

1. Follows by the definition of integral for simple functions $\varphi = \chi_A$.

2. As $\alpha\varphi = \sum_{k=1}^n \alpha \cdot a_k \chi_{A_k}$ we have by the definition of the integral for simple functions that

$$\int_X \varphi\, d\mu = \sum_{k=1}^n \alpha a_k \mu(A_k) = \alpha \sum_{k=1}^n a_k \mu(A_k) = \alpha \int_X \varphi\, d\mu.$$

3. As ψ is simple, it can be written in the standard form $\psi = \sum_{j=1}^{m} b_j \chi_{B_j}$ with $B_1, \ldots, B_m \in \mathfrak{M}$ and $B_j \cap B_{j'} = \emptyset$ for $j \neq j'$. By using that $\bigcup_{j=1}^{m} B_j = X$, we have

$$\varphi = \sum_{k=1}^{n} a_k \chi_{A_k \cap (\bigcup_{j=1}^{m} B_j)} = \sum_{k=1}^{n} \sum_{j=1}^{m} a_k \chi_{A_k \cap B_j}$$

and analogously

$$\psi = \sum_{j=1}^{m} \sum_{k=1}^{n} b_j \chi_{A_k \cap B_j}.$$

Note that

$$(A_k \cap B_j) \bigcap (A_{k'} \cap B_{j'}) = (A_k \cap A_{k'}) \bigcap (B_j \cap B_{j'})$$

if $k \neq k'$ or $j \neq j'$. Hence $\psi \leq \varphi$ implies that for each part k, j with $A_k \cap B_j \neq \emptyset$ holds that

$$b_j = b_j \chi_{A_k \cap B_j}(x) = \psi(x) \leq \varphi(x) = a_k \chi_{A_k \cap B_j} = a_k$$

where $x \in A_k \cap B_j$. This implies that

$$\int_X \psi \, d\mu = \sum_{j=1}^{m} b_j \mu(B_j) = \sum_{j=1}^{m} b_j \mu \left(B_j \cap \left(\bigcup_{k=1}^{n} A_k \right) \right) = \sum_{j=1}^{m} \sum_{k=1}^{n} b_j \mu(B_j \cap A_k)$$

the latter summands are non-zero only if $A_k \cap B_j \neq \emptyset$ and hence it is

$$\int_X \psi \, d\mu \leq \sum_{j=1}^{m} \sum_{k=1}^{n} a_k \mu(B_j \cap A_k) = \int_X \varphi \, d\mu. \quad \square$$

Solution to Exercise 7.11. For $m \in \mathbb{N}$ fixed holds that

$$\psi_m \leq \lim_{n \to \infty} \psi_n = \sup_{n \in \mathbb{N}} \phi_n = \lim_{n \to \infty} \varphi_n$$

and hence

$$\int_X \psi_m \, d\mu \leq \int_X \lim_{n \to \infty} \varphi_n \, d\mu = \lim_{n \to \infty} \int_X \varphi_n \, d\mu,$$

for all $m \in \mathbb{N}$, where the equality above is a consequence of the monotone convergence theorem because $\varphi_m, \psi_m \geq 0$. As the left hand side is monotone increasing in m, we get that

$$\lim_{m \to \infty} \int_X \psi_m \, d\mu \leq \lim_{n \to \infty} \int_X \varphi_n \, d\mu.$$

The other inequality follows the same way. $\quad \square$

Solution to Exercise 7.12. Let $(f_n)_1^\infty$ be a sequence of $\mathfrak{M}$-measurable functions on $(X, \mathfrak{M})$.

1. We have that

$$\left(\inf_{n \in \mathbb{N}} f_n \right)^{-1} (]a, \infty) = \left\{ x \in X \ \Big| \ \inf_{n \in \mathbb{N}} f_n(x) > a \right\}$$

$$= \left\{ x \in X \ \Big| \ \text{exists } n \in \mathbb{N} : \ f_n(x) > a \right\}$$

$$= \bigcup_{n=1}^{\infty} f_n^{-1}((a, \infty]).$$

Since all f_n are $\mathfrak{M}$-measurable, we get that $f_n^{-1}((a, \infty]) \in \mathfrak{M}$ and therefore the same is true for their union.

2. By the definition, we have $\limsup\limits_{n \to \infty} f_n(x) = \inf\limits_{n \in \mathbb{N}} \sup\limits_{k \geq n} f_k(x)$. Hence

$$\inf_{n \in \mathbb{N}} \sup_{k \geq n} f_k(x) > a$$

if and only if for all n holds that $\sup_{k \geq n} f_k(x) > a$. The latter means that there exists some $k \geq n$ such that $f_k(x) > a$. Hence

$$\left(\limsup_{n \to \infty} f_n \right)^{-1} ((a, \infty]) = \bigcup_{n=1}^{\infty} \bigcap_{n=k}^{\infty} f_k^{-1}((a, \infty]). \quad \square$$

Solution to Exercise 7.13.
1. We have

$$A = \left\{ x \in X \ : \ \limsup_{n \to \infty} f_n(x) - \liminf_{n \to \infty} f_n(x) = 0 \right\}.$$

By a known result both $\limsup_{n \to \infty} f_n$ and $\liminf_{n \to \infty} f_n$ are measurable and hence the set A is measurable.

2. Note that

$$\left(\chi_A \lim_{n \to \infty} f_n \right)(x) = \begin{cases} \lim\limits_{n \to \infty} f_n, & \text{if } x \in A \\ 0, & \text{if } x \notin A \end{cases}$$

$$= \begin{cases} \liminf\limits_{n \to \infty} f_n, & \text{if } x \in A \\ 0, & \text{if } x \notin A \end{cases}$$

$$= \left(\chi_A \liminf_{n \to \infty} f_n \right)(x)$$

As both χ_A and $\liminf\limits_{n \to \infty} f_n$ are measurable, their product is also measurable. $\quad \square$

Solution to Exercise 7.14. We set $A := \{ x \in X \,|\, f(x) > a \} = f^{-1}((a, \infty)) \in \mathfrak{M}$. Then

$$f^a = a \chi_A + f \chi_{A^c}.$$

As A and A^c are in $\mathfrak{M}$ we get that the four functions a, χ_A, f, and χ_{A^c} are measurable. Since the sums and products of measurable functions are measurable, the result ensues. $\quad \square$

Solution to Exercise 7.15. Let $f : \mathbb{R} \longrightarrow \mathbb{R}$ be a monotone function. We may assume that f is monotone increasing (i.e., non-decreasing), or else consider $-f$ and note that f is measurable if and only if $-f$ is. Then, $\{ f \geq a \}$ is of the form $[b(a), \infty)$ where

$$b(a) := \inf \{ x \in \mathbb{R} \,|\, f(x) \geq a \}.$$

Hence, f is $\mathcal{L}^1$-measurable since $[b(a), \infty)$ is a Borel set (and hence an element of $\mathcal{L}(\mathbb{R})$). $\quad \square$

Solution to Exercise 7.16. Let $f : [a, b] \longrightarrow \mathbb{R}$ be monotone increasing (i.e., non-decreasing) and non-negative. In Exercise 7.15 we showed that it is $\mathcal{L}([a,b])$-measurable, hence the Lebesgue integral

$$\int_{[a,b]} f(x)\,\mathrm{d}\mathcal{L}^1(x)$$

exists, although it might be $+\infty$. Hence, $f \in L^+([a,b], \mathcal{L}(\mathbb{R}))$. $\square$

Solution to Exercise 7.17. By a result in the main text, f is measurable if and only if for all $B \in \mathcal{B}(\mathbb{R})$ holds that $f^{-1}(B) \in \mathfrak{M}$. Let I be an open interval. Then, $g^{-1}(I) \in \mathcal{B}(\mathbb{R})$ as g is Borel measurable. By the aforementioned lemma, also $f^{-1}(g^{-1}(I)) \in \mathfrak{M}$. Hence it it enough to note that

$$
\begin{aligned}
f^{-1}(g^{-1}(I)) &= \left\{ x \in \mathbb{R} \,\middle|\, f(x) \in g^{-1}(I) \right\} \\
&= \left\{ x \in \mathbb{R} \,\middle|\, f(x) \in \{ y \in \mathbb{R} \,|\, g(y) \in I \} \right\} \\
&= \left\{ x \in \mathbb{R} \,\middle|\, g(f(x)) \in I \right\} \\
&= (g \circ f)^{-1}(I). \quad \square
\end{aligned}
$$

Solution to Exercise 7.18. Singleton sets are in $\mathcal{B}(\mathbb{R})$, since for any $x \in \mathbb{R}$ we have $\{x\} = \bigcap_1^\infty (x - 1/n, x + 1/n)$. As all the open intervals are in $\mathcal{B}(\mathbb{R})$ and a σ-algebra is closed under countable intersections, we get that $\{x\} \in \mathcal{B}(\mathbb{R})$. Given that $\mathbb{Q} = \bigcup_{x \in \mathbb{Q}} \{x\}$ and $\mathbb{Q}$ is countable, we get that $\mathbb{Q} \in \mathcal{B}(\mathbb{R})$, because σ-algebras are closed under countable unions. $\square$

Solution to Exercise 7.19. The Borel σ-algebra is by definition the smallest σ-algebra that contains all open sets.

• Let $\mathcal{G}$ be the smallest σ-algebra that contains all closed sets. A set is open if and only if its complement is closed and a σ-algebra is closed under taking complements. Hence $\mathcal{G}$ contains all open sets as well. As $\mathcal{B}(\mathbb{R})$ is the smallest σ-algebra with this property we get that $\mathcal{B}(\mathbb{R}) \subseteq \mathcal{G}$. Conversely, $\mathcal{B}(\mathbb{R})$ contains with the same argument all closed sets. As $\mathcal{G}$ is the smallest σ-algebra which contains all closed sets we get $\mathcal{G} \subseteq \mathcal{B}(\mathbb{R})$.

• Let $\mathcal{G}$ be the smallest σ-algebra which contains all open boxes. As open boxes are particular open sets, $\mathcal{B}(\mathbb{R})$ contains all open boxes. As $\mathcal{G}$ is the smallest σ-algebra with this property we obtain $\mathcal{G} \subseteq \mathcal{B}(\mathbb{R})$. We showed in a theorem that all open sets can be written as a countable union of open boxes. Given that σ-algebras are closed under countable unions, we obtain that any open set is in $\mathcal{G}$. Since $\mathcal{B}(\mathbb{R})$ is the smallest σ-algebra which contains all open sets, we obtain that $\mathcal{B}(\mathbb{R}) \subseteq \mathcal{G}$.

• Let $\mathcal{G}$ denote the smallest σ-algebra which contains all boxes. As any interval can be written as the countable union and intersection of open intervals because

$$[a,b) = \bigcap_{n=1}^\infty \left(a - \frac{1}{n}, b \right), \quad (a,b] = \bigcap_{n=1}^\infty \left(a, b + \frac{1}{n} \right), \quad [a,b] = \bigcap_{n=1}^\infty \left(a - \frac{1}{n}, b + \frac{1}{n} \right)$$

it follows that any box can be written as the intersection of open boxes. However, the countable intersection of open boxes is also in $\mathcal{B}(\mathbb{R})$ and hence $\mathcal{B}(\mathbb{R})$ contains all boxes. On the other hand, since $\mathcal{G}$ is the smallest σ-algebra with this property, we get $\mathcal{G} \subseteq \mathcal{B}(\mathbb{R})$. Conversely, as $\mathcal{G}$ contains all open boxes and by the previous $\mathcal{B}(\mathbb{R})$ is also the smallest σ-algebra with this property we get that $\mathcal{B}(\mathbb{R}) \subseteq \mathcal{G}$. $\qquad\square$

Solution to Exercise 7.20. (1) $\implies$ (2): It suffices to note that for any $a \in \mathbb{R}$, $(-\infty, a]$ is a Borel set. So, if f by definition inverts Borel sets to sets lying in $\mathfrak{M}$, then $f^{-1}((-\infty, a]) = \{f \le a\} \in \mathfrak{M}$.

(2) $\implies$ (3): For any $a < b$ in $\mathbb{R}$, we have

$$(a, b) = (-\infty, b) \setminus (-\infty, a]$$

$$= \left(\bigcup_{k=1}^{\infty} \left(-\infty, b + \frac{1}{k} \right] \right) \setminus (-\infty, a]$$

and hence, since f^{-1} commutes with complements and unions-intersections, we have

$$f^{-1}((a, b)) = f^{-1}\left((-\infty, b) \setminus (-\infty, a] \right)$$

$$= \left(\bigcup_{k=1}^{\infty} f^{-1} \left(\left(-\infty, b + \frac{1}{k} \right] \right) \right) \setminus f^{-1}((-\infty, a]).$$

Since for any $t \in \mathbb{R}$ we have $f^{-1}((-\infty, t]) \in \mathfrak{M}$, the above shows that $f^{-1}((a, b)) \in \mathfrak{M}$.

(3) $\implies$ (1): By a known lemma, in order for f to be measurable, it suffices to invert open sets $U \subseteq \mathbb{R}$ to sets in $\mathfrak{M}$. By assumption, we have $f^{-1}((a, b)) \in \mathfrak{M}$, i.e. the inverse image of open intervals is in $\mathfrak{M}$. Open subsets of $\mathbb{R}$ are not just intervals, but it is known that every open set $U \subseteq \mathbb{R}$ can be written a countable union of disjoint open intervals:

$$U = \bigcup_{k=1}^{\infty} (a_k, b_k).$$

Hence, we have

$$f^{-1}(U) = \bigcup_{k=1}^{\infty} f^{-1}((a_k, b_k))$$

which shows that $f^{-1}(U) \in \mathfrak{M}$. $\qquad\square$

Solution to Exercise 7.21. Let $A \subseteq X$ and $t \in \mathbb{R}$. Then, we have

$$(\chi_A)^{-1}((-\infty, t]) = \begin{cases} X, & t \ge 1, \\ X \setminus A, & 0 \le t < 1, \\ \emptyset, & t < 0. \end{cases}$$

Hence, $A \in \mathfrak{M}$ if and only if χ_A is $\mathfrak{M}$-measurable. $\qquad\square$

Solution to Exercise 7.22. We have

$$\mu(A \cup B) = \mu(A \setminus B) + \mu(B \setminus A) + \mu(A \cap B)$$

$$= \left(\mu(A) - \mu(A \cap B) \right) + \left(\mu(B) - \mu(A \cap B) \right) + \mu(A \cap B)$$

$$= \mu(A) + \mu(B) - \mu(A \cap B).$$

The above calculation makes sense since $\mu(A \cap B) < \infty$ (otherwise we would have $+\infty - \infty$).   □

Solution to Exercise 7.23.

1. We have $(a, \infty) \in \mathcal{G}$ if and only if $f^{-1}((a, \infty)) \in \mathfrak{M}$. Since f is by assumption $\mathfrak{M}$-measurable, the conclusion follows.

2. This follows from the fact that the inverse image map $f^{-1} : \mathcal{P}(\mathbb{R}) \longrightarrow \mathcal{P}(X)$ commutes with complements, unions and intersections. We need to prove that

(a) $\emptyset, \mathbb{R} \in \mathcal{G}$.

(b) If $A \in \mathcal{G}$, then $\mathbb{R} \setminus A \in \mathcal{G}$.

(c) If $(A_k)_1^\infty \subseteq \mathcal{G}$, then $\bigcup_1^\infty A_k \in \mathcal{G}$.

(a): We have that $f^{-1}(\emptyset) = \emptyset \in \mathfrak{M}$ and $f^{-1}(\mathbb{R}) = X \in \mathfrak{M}$.
(b): If $A \in \mathcal{G}$, then $f^{-1}(A) \in \mathfrak{M}$ and hence $f^{-1}(\mathbb{R} \setminus A) = X \setminus f^{-1}(A) \in \mathfrak{M}$.
(c): We argue similarly.

3. Follows directly from a result in the main text.   □

15.8   Solutions to the exercises in Chapter 8

Solution to Exercise 8.1. The function φ is a simple function because it can be written as $\varphi = 1\chi_{(1,2]} + 2\chi_{\{2\}}$ and hence the integral is

$$\int_\mathbb{R} \varphi \, d\mathcal{L}^1 = \mathcal{L}^1((1,2]) + 2\mathcal{L}^1(\{2\}) = 1 + 2 \cdot 0 = 1. \quad \square$$

Solution to Exercise 8.2. The intervals we exclude in the k-th step are given by $C_k \setminus C_{k+1}$, where C_k are the closed sets of the k-th step and $C_0 := [0, 1]$. This means that

$$f = \sum_{k=0}^\infty k \, \chi_{C_k \setminus C_{k+1}}$$

So we can write $f := \sup_{n \in \mathbb{N}} f_n$ for the simple functions

$$f_n = \sum_{k=0}^n k \, \chi_{C_k \setminus C_{k+1}}.$$

Note that the sequence $(f_n)_1^\infty$ is increasing because all summands are non-negative. Thus, by the Monotone Convergence Theorem we have

$$\int_X f \, d\mathcal{L}^1 = \lim_{n \to \infty} \int_X f_n \, d\mathcal{L}^1 = \lim_{n \to \infty} \sum_{k=0}^n k \, \mathcal{L}(C_k \setminus C_{k+1}) =$$

$$= \lim_{n \to \infty} \sum_{k=0}^n k \, \frac{2^k}{3^{k+1}} = \sum_{k=0}^\infty k \, \frac{2^k}{3^{k+1}}$$

and hence

$$\int_X f \, d\mathcal{L}^1 = \frac{2}{9} \sum_{k=1}^{\infty} k \frac{2^{k-1}}{3^{k-1}} = \frac{2}{9} \frac{d}{d\alpha}\bigg|_{\alpha=2/3} \sum_{k=0}^{\infty} \alpha^k =$$
$$= \frac{2}{9} \frac{2/3}{(1-2/3)^2} = \frac{4}{3}. \quad \square$$

Solution to Exercise 8.3. By using the definition of positive and negative part of functions we have the identity

$$(f + g)^+ + f^- + g^- = (f + g)^- + f^+ + g^+.$$

Hence,

$$\int_X (f + g)^+ \, d\mu + \int_X f^- \, d\mu + \int_X g^- \, d\mu = \int_X \left((f + g)^+ + f^- + g^- \right) d\mu$$
$$= \int_X \left((f + g)^- + f^+ + g^+ \right) d\mu$$
$$= \int_X (f + g)^- \, d\mu + \int_X f^+ \, d\mu + \int_X g^+ \, d\mu.$$

By noting that

$$f(x) + g(x) \le \max\{f(x), 0\} + \max\{g(x), 0\},$$

we infer

$$(f + g)^+ \le f^+ + g^+$$

and similarly we see that $(f + g)^- \le f^- + g^-$. Hence, either all three integrals

$$\int_X f^+ \, d\mu, \quad \int_X g^+ \, d\mu, \quad \int_X (f + g)^+ \, d\mu$$

are finite, or all three integrals

$$\int_X f^- \, d\mu, \quad \int_X g^- \, d\mu, \quad \int_X (f + g)^- \, d\mu$$

are finite. In the latter case, we get the result by reordering the terms. In the former case, if either of $\int_X f^- \, d\mu$ or $\int_X g^- \, d\mu$ is infinite, it follows that $\int_X (f+g)^- \, d\mu$ has to be infinite as well. We then obtain the required results with both sides being equal to $-\infty$. $\quad \square$

Solution to Exercise 8.4. For each fixed x the function $x \mapsto 1 + \frac{x^{2017}}{n}$ is decreasing and positive and hence $x \mapsto \frac{1}{1+\frac{x^{2017}}{n}}$ is increasing and positive. Hence by the monotone convergence theorem

$$\lim_{n\to\infty} \int_{-1}^{1} \left(\frac{1}{1 + \frac{x^{2017}}{n}} \right) dx = \int_{-1}^{1} \left(\lim_{n\to\infty} \frac{1}{1 + \frac{x^{2017}}{n}} \right) dx.$$

The latter is easy to be computed, because

$$\lim_{n\to\infty} \left(1 + \frac{x^{2017}}{n} \right) = 1$$

and hence we get as result

$$\lim_{n\to\infty} \int_{-1}^{1} \left(\frac{1}{1 + \frac{x^{2017}}{n}} \right) dx = \int_{-1}^{1} dx = 2. \quad \square$$

Solution to Exercise 8.5. For $\alpha \geq 0$, the function is continuous on $[0, \infty)$ and hence Borel measurable. For $\alpha < 0$, the function it is not defined at zero. From the general theory of Lebesgue integration follows that it does not matter for the integral which value we chooses at the 1 point $\{0\}$ since it has measure zero. It suffices to have the function being $\mathcal{L}^1$-a.e. defined.

1. Let $R > 0$ and consider first $\alpha > 0$. On $[0, R]$ the function f_α is bounded and continuous. Hence it is Riemann-integrable and the by a known theorem the Riemann-integral coincides with the Lebesgue integral. Hence,

$$\int_{[0,R]} f_\alpha(x) \, d\mathcal{L}^1(x) = \frac{R^{\alpha+1}}{\alpha+1}.$$

Let us consider the case $\alpha \leq 0$. Since $\{0\}$ has Lebesgue measure zero, it is enough to consider $f_\alpha \chi_{(0,R]}$, since

$$\int_{(0,R]} f_\alpha \, d\mathcal{L} = \int_{[0,R]} f_\alpha \, d\mathcal{L}^1.$$

We want to use the Riemann integral to compute the integral of this function. The function f_α is not bounded which is a necessary property in order to compare the Lebesgue with the Riemann integral. For each $\varepsilon > 0$ and each $R > \varepsilon$ the function $f_\alpha \chi_{[\varepsilon,R]}$ is bounded and converges monotonically to $f_\alpha \chi_{(0,R]}$. Hence by the monotone convergence theorem

$$\lim_{n\to\infty} \int_{[1/n,R]} f_\alpha \, d\mathcal{L}^1 = \int_{(0,R]} f_\alpha \, d\mathcal{L}^1.$$

Using the relation between Lebesgue and Riemann integral again we get that

$$\int_{(0,R]} f_\alpha \, d\mathcal{L}^1 = \begin{cases} \lim_{n\to\infty} \dfrac{R^{\alpha+1} - \frac{1}{n^{\alpha+1}}}{\alpha+1}, & \alpha \neq -1, \\[2mm] \lim_{n\to\infty} \left(\ln(R) - \ln(1/n) \right), & \alpha = -1, \end{cases}$$

$$= \begin{cases} \dfrac{R^{\alpha+1}}{\alpha+1}, & -1 < \alpha, \\[2mm] +\infty, & \alpha \leq -1. \end{cases}$$

Therefore, the function $x \mapsto x^\alpha$ is integrable on $[0, R]$ if and only if $\alpha > -1$.

2. We note that $\chi_{[0,R]} f_\alpha$ is non-negative and converges monotonically to f_α, hence

$$\lim_{R\to\infty} \int_{(0,R]} f_\alpha \, d\mathcal{L}^1 = \int_{(0,\infty)} f_\alpha \, d\mathcal{L}^1.$$

Using the previous result we get $\int_{(0,\infty)} f_\alpha(x) \, d\mathcal{L}^1(x) = \infty$ for all α. Hence f_α is not integrable on $(0, \infty)$ for any α.

3. It suffices to note that we can split the integral over $[-R, R]$ to integrals over $[-R, 0]$ and $[0, R]$. We have already calculated the integral over $[0, R]$, whilst the integral over $[-R, 0]$ vanishes. □

Solution to Exercise 8.6. For each $a \in \bar{\mathbb{R}}$ we have that $f^{-1}([a, \infty]) = \mathbb{Q}$ and hence f is Borel measurable. As the function f differs from the function constant equal to zero only on countably many points, the integral is zero ($\chi_{\mathbb{Q}}$ is $\mathcal{L}^1$-a.e. equal to zero). Hence $\int_{\mathbb{R}} \chi_{\mathbb{Q}} \, d\mathcal{L}^1 = 0$ because of the convention $\infty \cdot 0 = 0$.

We can also show this directly in the following way. Let φ be a simple function such that $0 \leq \varphi \leq f$. As φ is simple it is of the form $\sum_{k=1}^n a_k \chi_{A_k}$. As $\varphi \leq f$ we have that $A_k \subseteq \mathbb{Q}$. Hence $\mathcal{L}^1(A_k) \leq \mathcal{L}^1(\mathbb{Q}) = 0$. Thus by the definition of integral for simple functions we get

$$\int_X \varphi \, d\mathcal{L}^1 = \sum_{k=1}^n a_k \mathcal{L}^1(A_k) = 0.$$

By the definition of integral as supremum of the integrals of simple functions we get that

$$\int_X f \, d\mathcal{L}^1 = 0. \quad \square$$

Solution to Exercise 8.7.
1.We first consider the case of scalar $f : X \longrightarrow \mathbb{R}$. Since f is $\mathfrak{M}$-integrable, that is $\int_X f^\pm \, d\mu < \infty$. Consider first $c > 0$. Then, $(cf)_\pm = cf_\pm$ and hence

$$\int_X (cf) \, d\mu = \int_X (cf)^+ \, d\mu - \int_X (cf)^- \, d\mu$$
$$= \int_X cf^+ \, d\mu - \int_X cf^- \, d\mu$$
$$= c \left(\int_X f^+ \, d\mu - \int_X f^- \, d\mu \right)$$
$$= c \int_X f \, d\mu$$

For $c < 0$ holds $(cf)^+ = |c|f^-$ and $(cf)^- = |c|f^+$, thus

$$\int_X (cf) \, d\mu = \int_X (cf)^+ \, d\mu - \int_X (cf)^- \, d\mu$$
$$= \int_X |c|f^- \, d\mu - \int_X |c|f^+ \, d\mu$$
$$= |c| \left(\int_X f^- \, d\mu - \int_X f^+ \, d\mu \right)$$
$$= |c| \left(- \int_X f \, d\mu \right)$$
$$= c \int_X f \, d\mu.$$

The case $c = 0$ is trivial, whilst the case of vector-valued $f : X \longrightarrow \mathbb{R}^d$ follows from the previous, since then $f = (f_1, \ldots, f_d)$ and the previous holds for each f_i.

2. Let first $f : X \longrightarrow \mathbb{R}$ be scalar and μ-integrable. Then, f^+ and f^- have finite integral. Since $|f| = f^+ + f^-$ and both f^+ and f^- are non-negative and we showed in the main text that the integral for non-negative functions is additive we get that

$$\int_X |f| \, d\mathcal{L}^1 = \int_X f^+ \, d\mathcal{L}^1 + \int_X f^- \, d\mathcal{L}^1 < \infty.$$

Conversely, assume that $|f|$ has a finite integral. As $0 \leq f^{\pm} \leq |f|$ also both f^+ and f^- have a finite integral. The case of vector-valued $f : X \longrightarrow \mathbb{R}^d$ follows again from the previous since $|f| = \sqrt{\sum_1^d (f_i)^2}$. Hence, $|f|$ has finite integral if and only if each of the $|f_i|$ has finite integral and the latter happens if and only if all the $2d$-many non-negative functions $f_i^{\pm}$ have finite integral.

3. Let $f : X \longrightarrow \mathbb{R}^d$ satisfy $\int_X |f| \, d\mu < \infty$. By the Chebyshev inequality we have that

$$\mu \left(\{ |f| \geq n \} \right) \leq \frac{1}{n} \int_X |f| \, d\mu < \infty.$$

Hence f is finite μ-a.e. on X because by the lower continuity of the measure

$$\mu \left(\{ |f| = \infty \} \right) = \mu \left(\bigcap_{n=1}^{\infty} \{ |f| \geq n \} \right)$$

$$= \lim_{n \to \infty} \mu \left(\{ |f| \geq n \} \right)$$

$$\leq \lim_{n \to \infty} \frac{1}{n} \int_X |f| \, d\mu = 0. \quad \square$$

Solution to Exercise 8.8. As $\lim_{y \to \infty} y e^{-y} = 0$, for each fixed $x > 0$ it follows that

$$\lim_{n \to \infty} n^2 x^2 e^{-n^2 x^2} = 0.$$

Furthermore, for each $\varepsilon > 0$ there exists an $R_\varepsilon > 0$ such that for $y \geq R_\varepsilon$ holds that $0 \leq y e^{-y} \leq \varepsilon$. On the other hand, for $0 \leq y \leq R_\varepsilon$ holds that $y e^{-y} \leq y \leq R_\varepsilon$. Hence we get that

$$0 \leq y e^{-y} \leq \begin{cases} R_\varepsilon, & \text{for } 0 \leq y \leq R_\varepsilon, \\ \varepsilon, & \text{for } y > R_\varepsilon, \end{cases}$$

which is clearly bounded by $R_\varepsilon + \varepsilon$. Hence

$$0 \leq n^2 x^2 e^{-n^2 x^2} \leq R_\varepsilon + \varepsilon,$$

which gives

$$0 \leq \frac{n^2 x e^{-n^2 x^2}}{1 + x^2} \leq \frac{R_\varepsilon + \varepsilon}{x + x^3} \leq \frac{R_\varepsilon + \varepsilon}{x^3}$$

because for $x > 0$ we have

$$\frac{1}{x + x^3} \leq \frac{1}{x^3}.$$

We define

$$g(x) := \frac{R_\varepsilon + \varepsilon}{x^3}$$

and we note that $\int_0^\infty g \, d\mathcal{L} < \infty$, because

$$\int_a^\infty \frac{1}{x^3} \, d\mathcal{L}^1(x) = \frac{1}{2a^2}.$$

Hence by the Dominated Convergence Theorem, for $a > 0$ we have

$$\lim_{n\to\infty} \int_a^\infty \frac{n^2 x e^{-n^2 x^2}}{1 + x^2} \, d\mathcal{L}^1(x) = \int_a^\infty \lim_{n\to\infty} \frac{n^2 x e^{-n^2 x^2}}{1 + x^2} \, d\mathcal{L}^1(x) = 0.$$

We now consider the case $a = 0$. Arguing as before we see that

$$\lim_{n\to\infty} \int_1^\infty \frac{n^2 x e^{-n^2 x^2}}{1 + x^2} \, d\mathcal{L}^1(x) = 0.$$

Hence it remains to consider

$$\lim_{n\to\infty} \int_0^1 \frac{n^2 x e^{-n^2 x^2}}{1 + x^2} \, d\mathcal{L}^1(x).$$

First, let us check that the integral is finite. Indeed,

$$0 \le \frac{n^2 x e^{-n^2 x^2}}{1 + x^2} \le \frac{n^2 x}{1 + x^2} \le n^2 x$$

and the right hand side is bounded and continuous on $[0, 1]$. Thus the Lebesgue and the Riemann integral coincide. We change variables by taking $y = n^2 x^2$, or equivalently $x = \sqrt{y}/n$. Thus, $dy = n^2 2x \, dx = 2n\sqrt{y} \, dx$ and

$$\int_0^{n^2} \frac{y e^{-y}}{\sqrt{y}/n + y^{3/2}/n^3} \frac{1}{2\sqrt{y}n} \, dy = \frac{1}{2} \int_0^{n^2} \frac{e^{-y}}{1 + y/n} \, dy$$

$$= \frac{1}{2} \int_0^\infty \chi_{[0,n^2]}(y) \frac{e^{-y}}{1 + y/n} \, dy.$$

Clearly

$$0 \le \chi_{[0,n^2]}(y) \frac{e^{-y}}{1 + y/n} \le e^{-y} =: g(y)$$

and g has a finite integral on $[0, 1]$. Hence

$$\lim_{n\to\infty} \int_{[0,1]} \frac{n^2 x e^{-n^2 x^2}}{1 + x^2} \, d\mathcal{L}^1(x) = \frac{1}{2} \int_0^\infty \lim_{n\to\infty} \chi_{[0,n^2]}(y) \frac{y e^{-y}}{1 + y/n} \, dy$$

$$= \frac{1}{2} \int_0^\infty y e^{-y} \, dy.$$

As the function $y \mapsto y e^{-y}$ is continuous the Lebesgue integral is equal to the Riemann integral and hence we can compute it in the usual way using integration by parts we get

$$\frac{1}{2} \int_0^\infty y e^{-y} \, dy = \frac{1}{2} \int_0^\infty e^{-y} \, dy = \frac{1}{2}.$$

Summarising,

$$\lim_{n\to\infty} \int_0^\infty \frac{n^2 x e^{-n^2 x^2}}{1 + x^2} \, dy = \frac{1}{2}. \quad \square$$

Solution to Exercise 8.9. The assertion means that $f\chi_{N^c} \le g\chi_{N^c}$ on X for a set $N \in \mathfrak{M}$ with $\mu(N) = 0$. By integrating, it follows that

$$\int_X f\chi_{N^c} \, d\mu \le \int_X g\chi_{N^c} \, d\mu.$$

So it remains to be shown that $\int_X f \, d\mu = \int_X f\chi_{N^c} \, d\mu$ and for g analogously. As $N^c \subseteq X$ and $f \geq 0$, it follows that

$$\int_X f \, d\mu \geq \int_X f\chi_{N^c} \, d\mu.$$

Let φ a simple function with $\varphi \leq f$. Then, $\varphi\chi_{N^c} \leq f\chi_{N^c}$. Furthermore, using that φ is simple, that is $\varphi = \sum_{k=1}^n a_k\chi_{A_k}$ with $A_1, \ldots, A_n$ measurable and disjoint, we get that

$$\int_X \varphi \, d\mu = \sum_{k=1}^n a_k\mu(A_k) = \sum_{k=1}^n a_k\mu(A_k \cap N^c) + \sum_{k=1}^n a_k\mu(A_k \cap N).$$

By using that $\mu(A_k \cap N) \leq \mu(N) = 0$, we get

$$\int_X \varphi \, d\mu \leq \int_X \chi_{N^c}\varphi \, d\mu \leq \int_X f\chi_{N^c} \, d\mu.$$

Taking the supremum over all simple functions $\varphi \leq f$, we deduce that

$$\int_X f \, d\mu \leq \int_X f\chi_{N^c} \, d\mu. \quad \square$$

Solution to Exercise 8.10. First note that for each fixed $x \geq 1$ holds that

$$\lim_{n \to \infty} \frac{\sqrt{x}}{1 + nx^2} = 0.$$

Second, note that

$$0 \leq \frac{1}{1 + nx^2} \leq \frac{1}{1 + x^2}$$

hence it remains to prove that the function $g(x) := \frac{1}{1+x^2}$ is integrable on $[1, \infty)$. Also, for all $x > 1$ we have

$$\frac{\sqrt{x}}{1 + x^2} \leq \frac{\sqrt{x}}{x^2} = x^{-3/2}$$

and hence

$$\int_1^\infty g(x) \, dx \leq \int_1^\infty x^{-3/2} \, dx = \frac{1}{2}.$$

Then by the Dominated Convergence Theorem follows that

$$\lim_{n \to \infty} \int_1^\infty \frac{\sqrt{x}}{1 + nx^2} \, dx = \int_1^\infty \lim_{n \to \infty} \frac{\sqrt{x}}{1 + nx^2} \, dx = 0. \quad \square$$

Solution to Exercise 8.11. Since the interval $[-1, 1]$ is bounded and closed (hence compact) the continuous function $x \mapsto x\sqrt[n]{1 + x^2e^x}$ will have a maximum and a minimum on this interval. In order to apply the Dominated Convergence theorem we have to show that this bound is independent of n. The term $1 + x^2e^x$ satisfies

$$1 \leq 1 + x^2e^x \leq 1 + e.$$

Hence

$$1 \leq \sqrt[n]{1 + x^2e^x} \leq \sqrt[n]{1 + e} \leq 1 + e,$$

since for any $y \geq 1$ holds that $\sqrt[n]{y} \leq y$. So all together we get that

$$\left| x \sqrt[n]{1 + x^2 e^x} \right| \leq 1 + e.$$

Define now $g(x) := 1 + e$ which is non-negative and

$$\int_{-1}^{1} g(x)\, \mathrm{d}x = 1 + e$$

because for continuous functions the Lebesgue and the Riemann integral coincides. It remains to check that the pointwise limit exists, but for $y > 0$ holds that *smash* $\lim_{n\to\infty} x \sqrt[n]{y}$ and thus

$$\lim_{n\to\infty} x \sqrt[n]{1 + x^2 e^x} = x$$

by using that $1 \leq 1 + x^2 e^x$. In conclusion, by Lebesgue's dominate convergence we get

$$\lim_{n\to\infty} \int_{-1}^{1} x \sqrt[n]{1 + x^2 e^x}\, \mathrm{d}x = \int_{-1}^{1} x\, \mathrm{d}x.$$

By using that for continuous functions the Lebesgue and the Riemann integral coincide, we infer that the limit vanishes. $\qquad\square$

Solution to Exercise 8.12. By assumption, for any $\varepsilon > 0$, there exists and $N = N(\varepsilon)$ such that for all $n \geq N$, holds

$$\sup_{x \in [a,b]} \left| f_n(x) - f(x) \right| < \varepsilon. \tag{$*$}$$

Hence, if $U(f, P)$ and $L(f, P)$ denote respectively the upper and lower Riemann sums of f with respect a partition P of $[a, b]$, we easily see that

$$\left| U(f, P) - U(f_n, P) \right| \leq 2\varepsilon(b - a),$$
$$\left| L(f, P) - L(f_n, P) \right| \leq 2\varepsilon(b - a).$$

Let us give more details for the first case. We have

$$\left| U(f, P) - U(f_n, P) \right| \leq \sum_{k=0}^{n-1} \left| \sup_{x \in [x_k, x_{k+1}]} f(x) - \sup_{x \in [x_k, x_{k+1}]} f_n(x) \right| (x_{k+1} - x_k).$$

Hence it is sufficient to show that

$$\left| \sup_{x \in [x_k, x_{k+1}]} f(x) - \sup_{x \in [x_k, x_{k+1}]} f_n(x) \right| \leq 2\varepsilon.$$

Note that for any $\varepsilon > 0$, there exist x_ε and $x_\varepsilon^{(n)}$ in $[x_k, x_{k+1}]$ such that

$$\sup_{x \in [x_k, x_{k+1}]} f(x) = f(x_\varepsilon), \qquad \sup_{x \in [x_k, x_{k+1}]} f_n(x) = f(x_\varepsilon^{(n)}).$$

By $(*)$, it holds that

$$\left| f_n(x_\varepsilon) - f(x_\varepsilon) \right| \leq \varepsilon, \qquad \left| f_n(x_\varepsilon^{(n)}) - f(x_\varepsilon^{(n)}) \right| \leq \varepsilon$$

and hence

$$\left| f(x_\varepsilon) - f_n\big(x_\varepsilon^{(n)}\big) \right| \le 2\varepsilon,$$

as needed. Next note that

$$\sup_P L(f_n, P) \le \sup_P L(f, P) + 2\varepsilon(b - a)$$
$$\le \inf_P U(f, P) + 2\varepsilon(b - a)$$
$$\le \inf_P U(f_n, P) + 4\varepsilon(b - a).$$

As f_n is Riemann-integrable, we get

$$\int_a^b f_n(x)\,dx \le \sup_P L(f, P) + 2\varepsilon(b - a)$$
$$\le \inf_P U(f, P) + 2\varepsilon(b - a)$$
$$\le \int_a^b f_n(x)\,dx + 4\varepsilon(b - a)$$

and hence

$$\lim_{n\to\infty} \left| \sup_P L(f, P) - \int_a^b f_n(x)\,dx \right| = \lim_{n\to\infty} \left| \inf_P L(f, P) - \int_a^b f_n(x)\,dx \right| = 0,$$

as required.

15.9 Solutions to the exercises in Chapter 9

Solution to Exercise 9.1. The fact that the two norms are not equivalent on ℓ^p follows from the fact that the set

$$\left\{ \frac{\|x\|_{\ell^p}}{\|x\|_{\ell^r}} : x \in \ell^p \setminus \{0\} \right\}$$

is unbounded in $[0, \infty)$. This is indeed so because, by taking for any $n \in \mathbb{N}$,

$$x^{(n)} := \left(1, \frac{1}{2^{1/p}}, \dots, \frac{1}{n^{1/p}}, 0, 0, \dots \right),$$

we see that $x^{(n)} \in \ell^p$ and, if $r \in (p, \infty)$,

$$\frac{\|x^{(n)}\|_{\ell^p}}{\|x^{(n)}\|_{\ell^r}} = \frac{\left(\sum_{k=1}^n \frac{1}{k}\right)^{1/p}}{\left(\sum_{k=1}^n \frac{1}{k^{r/p}}\right)^{1/r}} \longrightarrow \infty \quad \text{as } n \to \infty,$$

because

$$\sum_{k=1}^\infty \frac{1}{k} = +\infty, \qquad \sum_{k=1}^\infty \frac{1}{k^{r/p}} < +\infty.$$

On the other hand, if $r = \infty$, then

$$\frac{\|x^{(n)}\|_p}{\|x^{(n)}\|_{\ell^r}} = \left(\sum_{k=1}^{n} \frac{1}{k} \right)^{1/p} \longrightarrow \infty \quad \text{as } n \to \infty.$$

Thus, the two norms are not equivalent. □

Solution to Exercise 9.2. We may use the fact that, for $\alpha \geq 0$,

$$\sum_{k=1}^{\infty} \frac{1}{k^\alpha} = +\infty \quad \text{if } 0 \leq \alpha \leq 1 \qquad \text{and} \qquad \sum_{k=1}^{\infty} \frac{1}{k^\alpha} < +\infty \quad \text{if } \alpha > 1.$$

This shows that, by defining

$$x := \left(1, \frac{1}{2^{1/p}}, \ldots, \frac{1}{k^{1/p}}, \ldots \right),$$

we have that $x \notin \ell^p$ and $x \in \ell^r$ for all $r \in (p, \infty)$, and we also obviously have $x \in \ell^\infty$. This establishes the required result in (ii), of which the particular case $p = 1$, $r = 2$ establishes (i). □

Solution to Exercise 9.3. It is easy to check that Y is a vector subspace of ℓ^2. We now claim that $\overline{Y} = \ell^2$. Let $x = (x_1, x_2, \ldots) \in \ell^2$, arbitrary. Let, for each $n \in \mathbb{N}$, $x^{(n)} = (x_1, x_2, \ldots, x_n, 0, 0, \ldots)$. Then $(x^{(n)})_1^\infty \subseteq Y$ and

$$\|x^{(n)} - x\|_2^2 = \sum_{k=n+1}^{\infty} |x_k|^2 \longrightarrow 0 \quad \text{as } n \to \infty.$$

This shows that $x \in \overline{Y}$. In conclusion, $\overline{Y} = \ell^2$, as claimed. □

Solution to Exercise 9.4. By using Hölder inequality, namely that

$$\left(\frac{1}{|\Omega|} \int_\Omega |f(x)|^p dx \right)^{1/p} \leq \|f\|_{L^\infty(\Omega)},$$

we obtain

$$\limsup_{p \to \infty} \|f\|_{L^p(\Omega)} \leq \|f\|_{L^\infty(\Omega)}.$$

For the opposite inequality, we may use Chebyshev inequality applied to $|f|^p \in L^1(\Omega)$ with

$$t := (\|f\|_{L^\infty(\Omega)} - \varepsilon)^p$$

for some $\varepsilon > 0$ small enough fixed, to get

$$\left(\frac{1}{|\Omega|} \int_\Omega |f(x)|^p dx \right)^{1/p} \geq (\|f\|_{L^\infty(\Omega)} - \varepsilon) \left| \{ |f| > \|f\|_{L^\infty(\Omega)} - \varepsilon \} \right|^{1/p}.$$

Note now that

$$\left| \{ |f| > \|f\|_{L^\infty(\Omega)} - \varepsilon \} \right| > 0$$

directly from the definition of the essential supremum. Then, let *first* $p \to \infty$ and *subsequently* $\varepsilon \to 0$ to deduce that

$$\liminf_{p \to \infty} \|f\|_{L^p(\Omega)} \geq \|f\|_{L^\infty(\Omega)}.$$

Can you explain why you need the ε in the first place? $\square$

Solution to Exercise 9.5. It suffices to show the equality when $f \neq 0$, namely when there is E measurable with $|E| > 0$ and $f > 0$ a.e. on E. Observe that since $f \in L^\infty(\Omega)$, you can divide both sides of the desired equality by

$$c := \|f\|_{L^\infty(\Omega)} > 0$$

and hence is suffices to prove it when $|f| \leq 1$ a.e. on Ω, because $|f/c| \leq 1$ a.e. on Ω. Consequently, in this case, for $p \geq 1$ we have

$$|f|^p \leq |f| \quad \text{a.e. on } \Omega,$$

and for a.e. $x \in \Omega$, $|f(x)|^p \longrightarrow |f(x)|$ as $p \to 1^+$. Since $|\Omega| < \infty$, the function f is in $L^1(\Omega)$ as a result of Hölder's Inequality. By applying the Dominated Convergence Theorem, we infer that

$$\alpha_p := \int_\Omega |f(x)|^p \mathrm{d}x \longrightarrow \int_\Omega |f(x)| \mathrm{d}x,$$

as $p \to 1^+$. Finally, by using that

$$(\alpha_p)^{\frac{1}{p}} = e^{\frac{1}{p} \ln \alpha_p}$$

we conclude with $(\alpha_p)^{\frac{1}{p}} \longrightarrow \|f\|_{L^1(\Omega)}$ as $p \to 1^+$. $\square$

Solution to Exercise 9.6. Apply Chebyshev's inequality to $|f_m - f|^p$ and use that $\|f_m - f\|_{L^p(\Omega)}^p \longrightarrow 0$ as $m \to \infty$. $\square$

Solution to Exercise 9.7. It can be readily seen that we may reduce the proof of completeness of $L^p(\Omega, w)$ to the case of $w \equiv 1$, by proving convergence of the Cauchy sequences as follows:

$$\|f_m - f_k\|_{L^p(\Omega,w)} \longrightarrow 0 \quad \Longleftrightarrow \quad \left\| w^{\frac{1}{p}} f_m - w^{\frac{1}{p}} f_k \right\|_{L^p(\Omega)} \longrightarrow 0,$$

as $m, k \to \infty$. Then, by using that $w \geq c > 0$ on Ω, we can construct a limit function of the form $w^{\frac{1}{p}} f$:

$$w^{\frac{1}{p}} f_m \longrightarrow g = \left(g w^{-\frac{1}{p}}\right) w^{\frac{1}{p}} =: w^{\frac{1}{p}} f. \quad \square$$

Solution to Exercise 9.8.
(a) We may use the Mean Value theorem to find that

$$\int_m^{m^2} f(t)\, \mathrm{d}t = f(t_m)(m^2 - m),$$

for some $t_m \in [m, m^2]$, $m \in \mathbb{N}$. Hence, since $\|f\|_{L^1(\mathbb{R})} < \infty$, it can be easily seen that $t_m \longrightarrow \infty$ and $f(t_m) \longrightarrow 0$ as $m \to \infty$. For

(b) We need to construct an example which says that if $g(t_m) \to 0$ along a sequence, then it is **not true** that
$$\lim_{t \to \infty} g(t) = 0.$$
Such a function g can be constructed as a piecewise affine continuous function which is zero except for a sequence of triangular tall positive "spikes" with small area. For example, for each $k \in \mathbb{N}$ consider the triangle E_k with base $[k - 2^{-k}, k + 2^{-k}]$ and height 1 and take as g the function whose graph equals the sides of the triangles on $[k - 2^{-k}, k + 2^{-k}]$, $k \in \mathbb{N}$ and is zero otherwise, namely on all intermediate intervals
$$\left[k + 2^{-k}, (k+1) - 2^{-(k+1)} \right], \quad k \in \mathbb{N}.$$

Then, this function is by construction continuous, on each vertex (at k's) we have $g(k) = 1$ and since each triangle E_k has area 2^{-k}, by the geometric series we have
$$\int_0^\infty f(t)\,dt \;=\; \sum_{k=1}^\infty |E_k| \;=\; \sum_{k=1}^\infty 2^{-k}.$$

The conclusion follows. $\qquad\qquad\square$

Solution to Exercise 9.9. Modify each triangle E_k of Exercise 9.8(b) in order to have height k and make the base smaller, in order to have again that each triangle has area 2^{-k} as before and hence the resulting h satisfies $h(k) = k \to \infty$ but still $\|h\|_{L^1(0,\infty)} < \infty$. This construction says that it may happen that along a sequence the function goes to zero but along an other may go to infinity and yet its integral on $(0, \infty)$ may be finite. $\qquad\qquad\square$

Solution to Exercise 9.10.

(1) This is a direct consequence of the fact that $\mathcal{P}(X)$ contains all subsets of X.

(2) Clearly $\sharp(A) \geq 0$ as the number of points is always a non-negative integer. Also, $\sharp(\emptyset) = 0$. Let $A_1, A_2, \ldots$ be a sequence of mutually disjoint sets. If $\bigcup_1^\infty A_n$ has only finitely many elements, then at most finitely many of the A_i are non-empty. Hence there exist an $N \in \mathbb{N}$ such that all $A_n = \emptyset$ for $n \geq N + 1$. Thus,
$$\sharp\left(\bigcup_{n=1}^\infty A_n \right) = \sharp\left(\bigcup_{n=1}^N A_n \right) = \left\{ \text{number of elements of } \bigcup_{n=1}^N A_n \right\}$$
which by mutual disjointness is just
$$= \sum_{n=1}^N \left\{ \text{number of elements of } A_n \right\} = \sum_{n=1}^N = \sharp(A_i).$$

If $\bigcup_1^\infty A_n$ has infinitely many elements we have to consider two cases:
Case 1: one of the A_k contains infinitely many elements. In this case
$$\infty = \sharp(A_k) \leq \sum_{n=1}^\infty \sharp(A_n) \leq \infty$$

which is equal to $\sharp\left(\bigcup_{n=1}^{\infty} A_n\right) = \infty$.

Case 2: Infinite many A_k are non empty. In this case

$$\sum_{n=1}^{\infty} \mu(A_n) \geq \left\{\text{number of } A_k \text{ which are non-empty}\right\} = \infty$$

which is again equal to $\sharp\left(\bigcup_{1}^{\infty} A_n\right) = \infty$.

(3) As a function f is measurable if for each $c \in \mathbb{R}$ the set $\{n \in \mathbb{N} \mid f(n) \geq c\}$ is measurable. This is obviously true as all subsets of $\mathbb{N}$ are $\sharp$-measurable.

(4) We show this by starting from simple functions. Let f be a simple function, that is f can be written as $f = \sum_i c_i \chi_{A_i}$ for sets $A_i \subseteq \mathbb{N}$, constants c_i and the sum is only over finitely many summands. For such a function by definition

$$\int_{\mathbb{N}} f(x) \, \mathrm{d}\sharp(x) = \sum_i c_i \sharp(A_i).$$

On the other hand,

$$\sum_{n=1}^{\infty} f(n) = \sum_{n=1}^{\infty} \sum_i c_i \chi_{A_i}(n)$$

and as the sum in i contains only finite many summands, we have

$$\sum_{n=1}^{\infty} f(n) = \sum_i c_i \sum_{n=1}^{\infty} \chi_{A_i}(n) = \sum_i c_i \left\{\text{number of elements in } A_i\right\} = \int_{\mathbb{N}} f(x) \, \mathrm{d}\sharp(x).$$

Thus the formula holds for f simple.

Let now f be a non-negative measurable function. By the characterisation of measurable functions, it follows that it can be written as the monotone limit of measurable simple non-negative functions $f_k : \mathbb{N} \longrightarrow [0, \infty)$. Hence, the left hand side of the desired equality by the monotone convergence theorem satisfies

$$\int_{\mathbb{N}} f(n) \, \mathrm{d}\sharp(n) = \lim_{k \to \infty} \int_{\mathbb{N}} f_k(n) \, \mathrm{d}\sharp(n).$$

Let us now consider the right hand side. Obviously, we have

$$\sum_{n=1}^{\infty} f_k(n) \leq \sum_{n=1}^{\infty} f(n).$$

If we have $\sum_1^{\infty} f(n) = \infty$, then for any $B > 0$ there exists an N such that $\sum_{n=1}^{N} f(n) \geq B$. As for each n holds $\lim_{k \to \infty} f_k(n) = f(n)$, by the definition of f and as there are only finite many summands we obtain that for each $\varepsilon > 0$ there exits an K such that for all $k \geq K$ and all $n \leq N$ holds $f(n) - f_k(n) \geq \varepsilon$. Hence for all $k \geq K$ holds

$$\sum_{n=1}^{N} f_k(n) \geq B - N\varepsilon$$

and thus in the limit

$$\lim_{k \to \infty} \sum_{n=1}^{N} f_k(n) \geq B - N\varepsilon.$$

As this holds for all ε we get

$$\lim_{k \to \infty} \sum_{n=1}^{N} f_k(n) \geq B.$$

Clearly this implies

$$\lim_{k \to \infty} \sum_{n=1}^{\infty} f_k(n) \geq \lim_{k \to \infty} \sum_{n=1}^{N} f_k(n) \geq B.$$

As this holds for all B, we get that also

$$\lim_{k \to \infty} \sum_{n=1}^{\infty} f_k(n) = \infty.$$

If $\sum_{n \in \mathbb{N}} f(n) < \infty$, then for each $\varepsilon > 0$ there exists an N with

$$\sum_{n=1}^{\infty} f(n) \leq \sum_{n=1}^{N} f(n) + \varepsilon.$$

There exists a K such that for all $k \geq K$ and all $n \leq N$ holds $f(n) - f_k(n) \leq \varepsilon$. Hence, for all $k \geq K$

$$\sum_{n=1}^{\infty} f(n) \leq \sum_{n=1}^{N} f_k(n) + (N + 1)\varepsilon.$$

Thus it holds also in the limit

$$\sum_{n=1}^{\infty} f(n) \leq \lim_{k \to \infty} \sum_{n=1}^{N} f_k(n) + (N + 1)\varepsilon$$

and in particular

$$\sum_{n=1}^{\infty} f(n) \leq \lim_{k \to \infty} \sum_{k=1}^{\infty} f_k(n) + (N + 1)\varepsilon.$$

As this holds for all ε we finally obtain

$$\sum_{n=1}^{\infty} f(n) \leq \lim_{k \to \infty} \sum_{n=1}^{\infty} f_k(n).$$

(5) Assume f is integrable. This is equivalent to that $|f|$ is integrable. Hence by the previous result and the definition of integral for real-valued functions (via $f = f^+ - f^-$) this is equivalent to the fact that $\sum_{1}^{\infty} |f(n)|$ converges.

(6) This is just the Dominated Convergence theorem theorem for $f(n) := a_n$ and $g(n) := b_n$.

(7) By to the definition and the additivity of the integral, by defining $h_k(n) := a_{n,k}$ we have

$$\sum_{k=1}^{N} \sum_{n=1}^{\infty} a_{k,n} = \sum_{k=1}^{N} \int_X h_k(n) \, d\sharp(n) = \int_X \sum_{k=1}^{N} h_k(n) \, d\sharp(n).$$

By defining $f_N(n) := \sum_{k=1}^{N} h_k(n)$, we have

$$|f_N(n)| \le \sum_{k=1}^{N} |h_k(n)| = \sum_{k=1}^{N} |a_{k,n}|.$$

Finally, we define $g(n) := \sum_{k=1}^{\infty} |a_{k,n}|$. Then $|f_N| \le g$ and

$$\int_X g(n)\, d\sharp(n) = \sum_{n=1}^{\infty} \sum_{k=1}^{\infty} |a_{k,n}| < \infty.$$

Hence by part (6), the result follows. $\qquad\square$

Solution to Exercise 9.11. It suffices to note that the given hypothesis can be restated as

$$a\,\chi_A(x) \le f(x) \le b\,\chi_A(x), \qquad x \in X,$$

which by integration gives the desired result, since

$$\mu(A) = \int_X \chi_A\, d\mu. \quad \square$$

Solution to Exercise 9.12. We may take $[a,b] := [-1,1]$ and $f(x) := -\chi_{[-1,0)} + \chi_{[0,1]}$. Then, we have

$$\int_1^1 f(x)\, dx = 0$$

but $f(\xi) \ne 0$ for any $\xi \in [-1,1]$. Hence, the second form of the mean value theorem cannot be true for discontinuous functions. $\qquad\square$

Solution to Exercise 9.13. Suppose that $f(x) \le g(x)$ for a.e. $x \in \mathbb{R}^n$, where $f, g \in L^1_{\mathrm{loc}}(\mathbb{R}^n)$. Fix $\varepsilon > 0$ and consider the mollifier $(\eta^\varepsilon)_{\varepsilon>0}$ as in the main text. Then, since $\eta^\varepsilon \ge 0$, we have

$$f(y)\,\eta^\varepsilon(y - x) \le g(y)\,\eta^\varepsilon(y - x)$$

for a.e. $x, y \in \mathbb{R}^n$. By integrating the above with respect to $y \in \mathbb{R}^n$, we obtain $f^\varepsilon(x) \le g^\varepsilon(x)$ for a.e. $x \in \mathbb{R}^n$ and hence this holds for all $x \in \mathbb{R}^n$ by the continuity of the mollified functions. To see the second part of the claim, note that by the definition of the L^∞ norm, we have the inequality

$$-\|f\|_{L^\infty(\Omega)} \le f(x) \le +\|f\|_{L^\infty(\Omega)}$$

for a.e. $x \in \mathbb{R}^n$. We applying the first part to the above inequality, we conclude. $\quad\square$

15.10 Solutions to the exercises in Chapter 10

Solution to Exercise 10.1. Let $(x_n)_1^\infty$ be the given sequence of elements of X and $x_0 \in X$, where $(X, \langle \cdot, \cdot \rangle)$ is an inner product space. Then, we have

$$
\begin{aligned}
\left\| x_n - x_0 \right\|^2 &= \langle x_n - x_0, x_n - x_0 \rangle \\
&= \| x_n \|^2 + \| x_0 \|^2 - \overline{\langle x_n, x_0 \rangle} - \langle x_n, x_0 \rangle \\
&\longrightarrow \| x_0 \|^2 + \| x_0 \|^2 - \overline{\langle x_0, x_0 \rangle} - \langle x_0, x_0 \rangle \\
&= 0,
\end{aligned}
$$

an $n \to \infty$. Hence, $x_n \longrightarrow x_0$ as $n \to \infty$ in X. $\square$

Solution to Exercise 10.2. It can be easily verified by the orthogonality relations that the orthogonal complement of the set

$$
Y := \left\{ x = (x_1, x_2, \ldots) \in \ell^2 : x_{3k} = 0 \text{ for } k \in \mathbb{N} \right\}.
$$

is given by

$$
Y^\perp = \left\{ x = (x_1, x_2, \ldots) \in \ell^2 : x_{3k-1} = x_{3k-2} = 0 \text{ for } k \in \mathbb{N} \right\}.
$$

Indeed, it suffice to note that if $\langle a, y \rangle$ for all $y \in Y$, we may choose $y = t e_i$ for $t \in \mathbb{F}$ to find that $t \neq 0$ only when $i = 3k - 2, 3k - 1$, $k \in \mathbb{N}$. $\square$

Solution to Exercise 10.3. By the Hölder and the Parseval inequalities, we have

$$
\left(\sum_{k=1}^{\infty} |\langle x, e_k \rangle \langle y, e_k \rangle| \right)^2 \leq \left(\sum_{k=1}^{\infty} |\langle x, e_k \rangle|^2 \right) \left(\sum_{k=1}^{\infty} |\langle y, e_k \rangle|^2 \right) \leq \| x \|^2 \| y \|^2,
$$

for all $x, y \in X$. $\square$

Solution to Exercise 10.4. Suppose that such a measure μ existed. Then, it would have to be σ-additive on $\mathcal{B}(H)$ and hence

$$
\mu \left(\bigcup_{i=1}^{\infty} U_i \right) = \sum_{i=1}^{\infty} \mu(U_i)
$$

for any sequence $(U_i)_1^\infty$ of disjoint Borel subsets of H. By property (3), the measure is translation invariant and any open set contains an open ball. By properties (1)-(2), it follows that there exists an $r_0 > 0$ such that for any ball of radius $r \in (0, r_0)$ we have $0 < \mu(\mathbb{B}_r(x)) = a_r < \infty$ for all $x \in H$ and $r \in (0, r_0)$. By the Riesz Lemma (Theorem 5.26) and the proof of Theorem 5.25, the open ball of radius equal to one contains a countable disjoint union of open balls of the same radii $\alpha \in (0, 1)$. For $\alpha = r_0/2$, this leads to a contradiction to the σ-additivity of the measure. $\square$

Solution to Exercise 10.5. Let $\{e_i : i \in \mathbb{N}\}$ be an orthonormal basis of the Hilbert space $(H, \langle \cdot, \cdot \rangle)$, so that every $x \in H$ has the expansion with respect to the basis as

$$x = \sum_{i=1}^{\infty} \langle x, e_i \rangle \, e_i.$$

We define the function

$$||| \cdot ||| \; : \; H \longrightarrow \mathbb{R}, \quad |||x||| := \sum_{i=1}^{\infty} 2^{-i} |\langle x, e_i \rangle|.$$

Then, $||| \cdot |||$ is a norm not equivalent to the original norm $\| \cdot \|$ induced by the inner product, namely $\|x\|^2 = \langle x, x \rangle$. Indeed, we have $\|e_j\| = 1$, whilst

$$|||e_j||| = \sum_{i=1}^{\infty} 2^{-i} |\langle e_j, e_i \rangle| = \sum_{i=1}^{\infty} 2^{-i} \delta_{ij} = 2^{-j}.$$

Hence, there is no $C > 0$ such that $|||x|||/C \leq \|x\| \leq C \|x\|$ for all $x \in H$. $\qquad \square$

Solution to Exercise 10.6. For $n = 2$, this is just the standard parallelogram law:

$$\|x_1 + x_2\|^2 + \|x_1 - x_2\|^2 = 2\|x_1\|^2 + \|x_2\|^2.$$

The general case follows easily by induction on $n \in \mathbb{N}$. More precisely, assume it holds for $x_1, \ldots, x_k \in H$ where $k \in \mathbb{N}$ and show it holds for $x_1, \ldots, x_k, x_{k+1} \in H$, by applying the above to $x_1 + \cdots + x_k$ and x_{k+1}. $\qquad \square$

Solution to Exercise 10.7. The proof is essentially a calculation in order to check that the given function is indeed an inner product on X. Obviously, $\langle \cdot, \cdot \rangle$ is continuous in each variable separately and also symmetric, that is $\langle x, y \rangle = \langle y, x \rangle$. By using the parallelogram identity, we have that it is additive in each variable separately, since

$$\langle x + y, z \rangle = \langle x, z \rangle + \langle y, z \rangle.$$

By induction, we can show that $N\langle x, y \rangle = \langle Nx, y \rangle$ for all $x, y \in H$ and all $N \in \mathbb{N}$. Hence, for any two integers $n, m \in \mathbb{N}$, we have

$$\left\langle \frac{n}{m} x, y \right\rangle = n \left\langle \frac{1}{m} x, y \right\rangle = m \frac{n}{m} \left\langle \frac{1}{m} x, y \right\rangle = \frac{n}{m} \left\langle \frac{m}{m} x, y \right\rangle = \frac{n}{m} \langle x, y \rangle.$$

Since $\mathbb{Q}$ is dense in $\mathbb{R}$ and every $q \in \mathbb{Q}$ can be represented as $p = \pm n/m$ for $n\, m \in \mathbb{N}$, the conclusion follows by an approximation argument and the continuity of the function $\langle \cdot, \cdot \rangle$ in each variable separately. The complex case is completely analogous, the only difference being that the function $\langle \cdot, \cdot \rangle$ is not symmetric but conjugate symmetric, that is $\langle x, y \rangle = \overline{\langle y, x \rangle}$. $\qquad \square$

Solution to Exercise 10.8. Consider $H = \ell^2$ and as $Y \subseteq H$ take

$$Y := \left\{ x = (x_1, x_2, \ldots) \in \ell^2 \; \middle| \; \text{exists } k = k(x) \in \mathbb{N} : x_k = 0 \text{ for } k > k(x) \right\}.$$

Then, we have that the orthogonal complement $Y^\perp$ of Y in ℓ^2 vanishes, that is

$Y^\perp = \{0\}$. This happens because for any point $x \in H \setminus \{0\}$, we have that $\langle x, e_k \rangle = 0$ for $k = 1, \ldots, k(x)$. $\qquad\square$

Solution to Exercise 10.9. This is an immediate consequence of the Bessel inequality. Since $\{a_k : k \in \mathbb{N}\}$ is an orthonormal sequence, for any $x \in X$, the series $\sum_{k=1}^\infty |\langle x, a_k \rangle|^2$ converges, and

$$\sum_{k=1}^\infty |\langle x, a_k \rangle|^2 \leq \|x\|^2.$$

For $x := e_i$, this gives

$$\sum_{k=1}^\infty |\langle e_i, a_k \rangle|^2 \leq \|e_i\|^2 = 1.$$

It suffices to notice that $e_{ik} = \langle e_i, a_k \rangle$ and that to recall from real Analysis that if a series convergence the corresponding sequence converges to zero. $\qquad\square$

Solution to Exercise 10.10. Let S be the set of the definition on the statement. Then, we have

$$\|x\|^2 = \sum_{i=1}^\infty |\langle x, e_i \rangle|^2 < \sum_{i=1}^\infty \left(1 + \frac{1}{i}\right) |\langle x, e_i \rangle|^2 \leq 1.$$

On the other hand, the sequence $(x_k)_1^\infty$ given by

$$x_k := \frac{e_k}{\sqrt{1 + \frac{1}{k}}}$$

satisfies that $(x_k)_1^\infty \subseteq S$ because

$$\sum_{i=1}^\infty \left(1 + \frac{1}{i}\right) |\langle x_k, e_i \rangle|^2 = \sum_{i=1}^\infty \left(1 + \frac{1}{i}\right) \left|\left\langle \frac{e_k}{\sqrt{1 + \frac{1}{k}}}, e_i \right\rangle\right|^2 \leq 1.$$

However, by the Parseval identity we have

$$\|x_k\|^2 = \sum_{i=1}^\infty \left|\left\langle \frac{e_k}{\sqrt{1 + \frac{1}{k}}}, e_i \right\rangle\right|^2 = \frac{1}{1 + \frac{1}{k}} \longrightarrow 1, \qquad \text{as } k \to \infty.$$

Therefore,

$$\sup_{s \in S} \|s\| \geq \lim_{k \to \infty} \|x_k\| = 1$$

and hence no point $s \in S$ can realise the supremum. $\qquad\square$

Solution to Exercise 10.11. Suppose by contradiction that the infinite-dimensional Hilbert space H admitted a countable Hamel (algebraic) basis $A = \{a_i : i \in \mathbb{N}\}$. For each $n \in \mathbb{N}$, we consider the vector subspace

$$F_n := \text{span}[\{a_1, \ldots, a_n\}] \subseteq H.$$

Then, each F_n is a closed subspace and since

$$H = \bigcup_{n=1}^{\infty} F_n,$$

at least one of the sets F_n, say F_{n_0}, has non-empty interior, as a consequence of the Baire's Theorem 3.15. This is a contradiction since F_{n_0} is a finite dimensional subspace and cannot contain an open ball of the Hilbert space H. Hence, the space H can not have a countable Hamel basis. $\qquad\square$

15.11 Solutions to the exercises in Chapter 11

Solution to Exercise 11.1. (1) Let x, y be arbitrary elements of $\overline{Y}$. Then there exists sequences $(x_n)_1^{\infty}$ and $(y_n)_1^{\infty}$ contained in Y such that $x_n \longrightarrow x$ and $y_n \longrightarrow y$ in $(X, \| \cdot \|)$ as $n \to \infty$. Since Y is a vector subspace, we have that $(x_n + y_n) \in Y$ for all $n \in \mathbb{N}$. Since

$$x_n + y_n \longrightarrow x + y \quad \text{as } n \to \infty.$$

it follows that $(x + y) \in \overline{Y}$. In a similar way we can show that, for any $x \in \overline{Y}$ and $\lambda \in \mathbb{F}$, we have that $\lambda x \in \overline{Y}$. In conclusion, $\overline{Y}$ is also a vector subspace of X.
(2) Suppose that $Y^{\circ} \neq \emptyset$, and let $x_0 \in X$ and $r > 0$ be such that $\mathbb{B}_r(x_0) \subseteq Y$. We first claim that $\mathbb{B}_r(0) \subseteq Y$. Indeed, since any $x \in \mathbb{B}_r(0)$ can be written as

$$x = (x_0 + x) - x_0,$$

where $(x_0 + x) \in \mathbb{B}_r(x_0) \subseteq Y$ and $x_0 \in \mathbb{B}_r(x_0) \subseteq Y$, and since Y is a vector subspace, it follows that $x \in Y$. Therefore, $\mathbb{B}_r(0) \subseteq Y$.

We now deduce that $Y = X$. Indeed, consider any $x \in X$ with $x \neq 0$. Then we can write

$$x = \frac{2\|x\|}{r} \left(\frac{r}{2} \frac{x}{\|x\|} \right),$$

and, since $\frac{r}{2} \frac{x}{\|x\|} \subseteq \mathbb{B}_r(0) \subseteq Y$, and since Y is a vector subspace, it follows that $x \in Y$. Therefore $(X \setminus \{0\}) \subseteq Y$, and hence $Y = X$, as required. $\qquad\square$

Solution to Exercise 11.2. (1) Let $f_0 \in C([a, b], \mathbb{R})$, arbitrary. The linearity of T is obvious. Note that, for any $f \in C([a, b], \mathbb{R})$, the following holds:

$$\begin{aligned}
\left| T(f) - T(f_0) \right| &= \left| f(a) + f(b) - f_0(a) - f_0(b) \right| \\
&\leq \left| f(a) - f_0(a) \right| + \left| f(b) - f_0(b) \right| \\
&\leq 2\|f - f_0\|_{\infty}.
\end{aligned}$$

This implies that, for any $\varepsilon > 0$, choosing $\delta := \varepsilon/2$ we have that

$$\forall f : \quad \|f - f_0\|_{\infty} < \delta \quad \Longrightarrow \quad |T(f) - T(f_0)| < \varepsilon.$$

This shows that T is continuous at f_0. The arbitrariness of f_0 shows that T is continuous as a function $C([a, b], \mathbb{R}) \longrightarrow \mathbb{R}$.

(2) Note that $F = T^{-1}(\{0\}) = \{T = 0\}$. Since $\{0\}$ is a closed subset of $\mathbb{R}$, the continuity of T implies, by a known result in the main text, that $T^{-1}(\{0\})$ is a closed subset in $C([a, b], \mathbb{R})$, which is the required result. $\qquad \square$

Solution to Exercise 11.3. (1) Checking that $\| \cdot \|_{C^k}$ is a norm on the normed vector space $C^k([0, 1])$ is a triviality. It suffices to show that the space is complete with respect to the metric topology generated by this norm. With a simple finite induction argument on the derivatives of the functions, it suffices to show that $C^1([0, 1])$ is complete with respect to the C^1 convergence on $[0, 1]$. Suppose that $(f_n)_1^\infty$ is a Cauchy sequence in $C^1([0, 1])$ with respect to $\| \cdot \|_{C^1}$. Then,

$$\max_{t \in [0,1]} |f_n(t) - f_m(t)| + \max_{t \in [0,1]} |f_n'(t) - f_m'(t)| \longrightarrow 0,$$

as $n, m \to \infty$. Since $C^0([0, 1])$ is complete, there exist $f, F \in C^0([0, 1])$ such that $f_n \longrightarrow f$ and $f_n' \longrightarrow F$, both modes of convergence being uniform on $[0, 1]$ as $n \to \infty$. In order to conclude, we need to show that f' exists on $[0, 1]$ (as 1-sided derivative at the endpoints), is continuous and $F \equiv f'$. Indeed, fix $t \neq 0$. If $x = 0$ in the inequality below we choose $t > 0$ and if $x = 1$ we choose $t < 0$, so that $x + t \in [0, 1]$ when $|t|$ is small enough and hence $f(x + t)$ is well defined. By utilising the Fundamental Theorem of Calculus, we estimate

$$\left| F(x) - \frac{f(x + t) - f(x)}{t} \right| \leq \left| F(x) - \frac{f_m(x + t) - f_m(x)}{t} \right|$$
$$+ \left| \frac{f_m(x + t) - f_m(x)}{t} - \frac{f(x + t) - f(x)}{t} \right|$$
$$\leq \left| F(x) - \frac{1}{t} \int_x^{x+t} f_m'(z) \, dz \right|$$
$$+ \left| \frac{f_m(x + t) - f(x + t)}{t} - \frac{f_m(x) - f(x)}{t} \right|$$
$$\leq \left| \frac{1}{t} \int_x^{x+t} [F(x) - f_m'(z)] \, dz \right| + \frac{2}{|t|} \| f_m - f \|_{C^0}.$$

By letting $m \to \infty$ for t fixed, we get

$$\left| F(x) - \frac{f(x + t) - f(x)}{t} \right| \leq \left| \frac{1}{t} \int_x^{x+t} [F(x) - F(z)] \, dz \right|$$
$$\leq \frac{|(x + t) - x|}{|t|} \sup_{z \in [x-|t|, x+|t|] \cap [0,1]} \left| F(x) - F(z) \right|.$$

By letting $t \to 0$, by the continuity of F we conclude that f' exists and equals f, as desired.

(2) This is immediate from the definition of the C^k-norm:

$$\| Tf \|_{C^{k-1}} = \max_{p \in \{0, \dots, k-1\}} \max_{t \in [0,1]} \left| (Tf)^{(p)}(t) \right|$$
$$= \max_{p \in \{0, \dots, k-1\}} \max_{t \in [0,1]} \left| f^{(p+1)}(t) \right|$$
$$\leq \max_{p \in \{0, \dots, k\}} \max_{t \in [0,1]} \left| f^{(p)}(t) \right|$$
$$= \| f \|_{C^k}. \qquad \square$$

Solution to Exercise 11.4. The method of the solution is essentially identical to that of Exercise 11.3, the only difference being the integrals will be in the Lebesgue sense and it is higher dimensional. Therefore, we refrain from providing any further details. □

Solution to Exercise 11.5. Consider the sequence of functions in the hint

$$f_n := (n(n+1))^{1/p} \chi_{(\frac{1}{n+1}, \frac{1}{n})}, \quad n \in \mathbb{N}.$$

Then, for each $n \in \mathbb{N}$ we have $\|f_n\|_{L^p(0,1)} = 1$. In addition, the sequence $(f_n)_1^\infty$ is linearly independent since for $n \neq m$, the characteristic functions $\chi_{(\frac{1}{n+1}, \frac{1}{n})}$ and $\chi_{(\frac{1}{m+1}, \frac{1}{m})}$ are non-vanshing on disjoint sets. Consider the linear operator

$$T : \quad \ell^p \longrightarrow L^p(0,1)$$

defined for any $x = (x_1, x_2, \dots) \in \ell^p$ as

$$Tx := \sum_{m=1}^\infty x_m f_m.$$

Then, T is an isometry, since

$$\|Tx\|_{L^p}^p = \int_0^1 \left| \sum_{m=1}^\infty x_m f_m \right|^p \, d\mathcal{L}^1$$

$$= \sum_{n=1}^\infty \int_{1/(n+1)}^{1/n} \left| \sum_{m=1}^\infty x_m f_m \right|^p \, d\mathcal{L}^1$$

$$= \sum_{n=1}^\infty |x_n|^p \int_{1/(n+1)}^{1/n} |f_n|^p \, d\mathcal{L}^1$$

$$= \sum_{n=1}^\infty |x_n|^p \int_{1/(n+1)}^{1/n} \left| (n(n+1))^{1/p} \chi_{(\frac{1}{n+1}, \frac{1}{n})} \right|^p \, d\mathcal{L}^1$$

$$= \sum_{n=1}^\infty |x_n|^p$$

$$= \|x\|_{\ell^p}.$$

It follows that the space ℓ^p can be identified with the image $T\ell^p = \text{span}[\{f_n : n \in \mathbb{N}\}] \subseteq L^p(0,1)$. □

Solution to Exercise 11.6. The operator is indeed continuous. Suppose for the sake of contradiction that this were not the case. Then, there exists some sequence $(x_n)_1^\infty \subseteq X$ such that $x_n \longrightarrow 0$ as $n \to \infty$, but $Tx_n \nrightarrow 0$ as $n \to \infty$. By passing to a subsequence, we may assume that $\|Tx_{n_k}\| \geq \varepsilon_0$ for some $\varepsilon_0 > 0$ and all $k \in \mathbb{N}$. Consider the sequence of X given by

$$y_k := \frac{x_{n_k}}{\sqrt{\|x_{n_k}\|}}, \quad k \in \mathbb{N}.$$

Then, we have that $\|y_k\| \longrightarrow 0$, but

$$\|Ty_k\| = \frac{\|Tx_{n_k}\|}{\sqrt{\|x_{n_k}\|}} \geq \frac{\varepsilon_0}{\sqrt{\|x_{n_k}\|}} \longrightarrow \infty, \quad \text{as } k \to \infty,$$

leading to the desired contradiction. $\qquad\square$

Solution to Exercise 11.7. Let $T : X \longrightarrow Y$ be an one to one linear operator between normed spaces. Obviously, (1) is equivalent to (2) by homogeneity and a rescaling argument (dividing by the norms). Further, (1) implies (3). Conversely, suppose that (3) holds and for the sake of contradiction that (2) fails. Then, there exists a unit vector $x \in X$ with $\|x\| = 1$ such that $\|Tx\| = t < 1$. It follows that $\|x\|/t > 1$ and hence $\|T(x/t)\| = 1$. This implies that there exists $z \in \mathbb{B}^X$ with $Tz = T(x/t)$, a contradiction to the injectivity of T. Hence, (2) follows. $\qquad\square$

Solution to Exercise 11.8. Let $(x_n)_1^\infty \subseteq X$ be a sequence for which $(Tx_n)_1^\infty \subseteq T(X) \subseteq Y$ satisfies $Tx_n \longrightarrow y$ as $n \to \infty$ for some $y \in Y$. We shall show that $y \in T(X)$, i.e. that there exists $x \in X$ such that $y = Tx$. Since $Tx_n \longrightarrow y$ as $n \to \infty$, it follows that $(Tx_n)_1^\infty$ is Cauchy in Y. By our hypothesis, $(x_n)_1^\infty$ is Cauchy in X:

$$\|x_n - x_m\| \leq \frac{\|Tx_n - Tx_m\|}{\delta}, \quad n, m \in N.$$

By completeness, there exists $x \in X$ such that $x_n \longrightarrow x$ as $n \to \infty$. Since T is bounded linear, it follows that $Tx_n \longrightarrow Tx$ as $n \to \infty$ and by the uniqueness of limits it follows that $Tx = y$, as claimed. The fact that T is invertible follows from the hypothesis of boundedness from below. Finally, for $x := T^{-1}y$, $y \in Y$, we have

$$\|T^{-1}y\| \leq \frac{\|T(T^{-1}y)\|}{\delta} = \frac{1}{\delta}\|y\|$$

and hence T^{-1} is bounded too. $\qquad\square$

Solution to Exercise 11.9. The idea is to use induction. Fix a point $x \in X \setminus \{0\}$. Let $x_1 \in M$ be a point selected so that

$$\|x - x_1\| \leq \frac{\|x\|}{2}.$$

This is possible by the density of M. Supposing $x_1, \ldots, x_k$ have been selected, we select a point $x_{k+1} \in M$ such that

$$\left\|x - (x_1 + \cdots + x_k) - x_{k+1}\right\| \leq \frac{\|x\|}{2^k}.$$

This is again possible by the density of M. Then, the above estimates which is valid for any $k \in \mathbb{N}$ implies that the series $\sum_1^\infty x_k$ converges and also

$$x = \lim_{N\to\infty} \sum_1^N x_k = \sum_1^\infty x_k.$$

Further, we have

$$\|x_k\| = \left\|\left(x - \sum_{n=1}^{k-1} x_n\right) - \left(x - \sum_{n=1}^{k} x_n\right)\right\| \leq \frac{\|x\|}{2^{k-1}} + \frac{\|x\|}{2^k} = \frac{3\|x\|}{2^k},$$

as claimed. □

Solution to Exercise 11.10. Let $(X, \|\cdot\|)$ be a separable Banach space and $D \subseteq X$ a dense subset of X. Then, we define the subset D_1 of the sphere $\mathbb{S}^X$ given by

$$D_1 := \left\{ \frac{x}{\|x\|} \; : \; x \in D \right\} \subseteq \mathbb{S}^X.$$

Then, D_1 is dense in $\mathbb{S}^X$. Indeed, fix $z \in \mathbb{S}^X$ and $\varepsilon \in (0, 1/2)$. Since D is dense in X, there exists $x \in D$ such that $\|z - x\| < \varepsilon$. Then, by the triangle inequality it follows that $|\|x\| - 1| \leq \|z - x\| < \varepsilon$. Hence

$$\left\| z - \frac{x}{\|x\|} \right\| = \frac{|\|x\|z - x\||}{\|x\|} \leq \frac{\|z - x\| + \|z\|\,|\|x\| - 1|}{1 - \varepsilon} \leq \frac{\varepsilon + \varepsilon}{1 - \varepsilon} \leq 4\varepsilon$$

and as a result D_1 is indeed dense in $\mathbb{S}^X$.

Conversely, suppose that $S_1 = \{z_n : n \in \mathbb{N}\} \subseteq \mathbb{S}^X$ is a dense sequence of points in the sphere. Let $Q := \{q_m : m \in \mathbb{N}\} \subseteq \mathbb{F}$ be a dense sequence in the field $\mathbb{F}$. We consider the subset S of X given by

$$S := \left\{ q_m z_n \; : \; n, m \in \mathbb{N} \right\} \subseteq X.$$

Then, by the results of Chapter 1 it follows that S is countable. Further, we claim that S is dense in X. To see this, fix $x \in X$ and $\varepsilon > 0$. By the density of S_1, there exists $z \in S_1$ such that $\|z - (x/\|x\|)\| < \varepsilon$. Further, there exists $q \in Q$ such that $|\|x\| - q| < \varepsilon$. Then, $zq \in S_1$ and also

$$\|x - qz\| = \left\| \frac{x}{\|x\|}\|x\| - qz \right\| \leq \left\| \frac{x}{\|x\|} \right\| \,|\|x\| - q| + |q| \left\| \frac{x}{\|x\|} - z \right\| \leq \varepsilon + |q|\varepsilon$$

and since $|q| < \|x\| + \varepsilon$, the conclusion follows. □

Solution to Exercise 11.11. Consider the space $L^1(\Omega)$ for some open domain $\Omega \subseteq \mathbb{R}^n$ with the Lebesgue measure. Consider the spaces of simple and of smooth compactly supported functions:

$$\Sigma(\Omega) := \left\{ f = \sum_1^m a_i \chi_{A_i} \; : \; (a_i)_1^m \subseteq \mathbb{F}, \; (A_i)_1^m \subseteq \Omega \text{ measurable disjoint} \right\}$$

$$C_c^\infty(\Omega) := \left\{ f \in C^\infty(\Omega) \; : \; \text{supp}(f) \text{ is a compact subset of } \Omega \right\}.$$

Then, by the results of Chapters 8-9 it follows that $\Sigma(\Omega)$ and $C_c^\infty(\Omega)$ are both dense in $L^1(\Omega)$, but their only common element is the zero function. □

Solution to Exercise 11.12. In the Hilbert space $X = L^2(0, 1)$, we consider the closed convex sets

$$A := \left\{ h \in L^2(0, 1) : h \geq 1 \text{ a.e. on } (0, 1) \right\}, \quad B := \left\{ t \, \text{id} : t \in \mathbb{R}, \, \text{id}(x) = x \right\}.$$

Then, there exists no hyperplane separating A from B. Indeed, suppose for the sake of contradiction that there exists a functional $\phi \in L^2(0,1)^*$ such that $\phi(f) \geq \phi(g)$ for all $f \in A$ and $g \in B$. Then, ϕ is non-negative on the set

$$A - B = \left\{ h \in L^2(0,1) \, \middle| \, \exists \, t \in \mathbb{R} \, : \, h(x) = f(x) \geq 1 - tx, \text{ a.e. } x \in (0,1) \right\}.$$

Fix $f \in L^2(0,1)$ and $n \in \mathbb{N}$. Let us choose $t_n \in \mathbb{R}$ large enough such that

$$\left\| \max\left\{ 1 - t_n \mathrm{id} + |f|, 0 \right\} \right\|_{L^2(0,1)} \leq \frac{1}{4^n}.$$

This is possible by the Dominate convergence theorem because we have

$$0 \leq \max\left\{ 1 - t\mathrm{id} + |f|, 0 \right\} \longrightarrow 0 \text{ as } t \to \infty.$$

We define the L^2 function

$$g := \sum_{n=1}^{\infty} 2^n \max\left\{ 1 - t_n \mathrm{id} + |f|, 0 \right\}.$$

Then, we obtain

$$\pm f(x) + 2^{-n} g(x) \geq 1 - t_n x, \quad \text{a.e. } x \in (0,1).$$

This yields that $\pm\phi(f) + 2^{-n}\phi(g) \geq 0$ and as a result $\phi \equiv 0$, which is a contradiction. Hence, A, B cannot be separated. $\qquad\square$

Solution to Exercise 11.13. (1) By definition,

$$\|f - g\|_\infty = \sup_{x \in [-\pi, \pi]} |f(x) - g(x)| = \sup_{x \in [-\pi, \pi]} |\sin x - \cos x|.$$

Consider the function $h : [-\pi, \pi] \longrightarrow \mathbb{R}$ given by

$$h(x) = \sin x - \cos x \quad \text{for all } x \in [-\pi, \pi].$$

The function h is continuous on $[-\pi, \pi]$ and differentiable on $(-\pi, \pi)$, and so by standard results in Real Analysis, is has a maximum and minimum value on $[-\pi, \pi]$, and each of them is attained either at an endpoint of the interval or at a critical point. Note that $h(-\pi) = 1$, $h(\pi) = 1$ and

$$h'(x) = \cos x + \sin x \quad \text{for all } x \in (-\pi, \pi).$$

The critical points of h (the points where $h'(x) = 0$) are $x_1 = -\pi/4$ and $x_2 = 3\pi/4$. Note that

$$h(x_1) = -\frac{\sqrt{2}}{2} - \frac{\sqrt{2}}{2} = -\sqrt{2},$$

$$h(x_2) = \frac{\sqrt{2}}{2} + \frac{\sqrt{2}}{2} = \sqrt{2}.$$

It follows that

$$|h(x_1)| = |h(x_2)| = \sqrt{2} \quad \text{and} \quad |h(x)| \leq \sqrt{2} \quad \text{for all } x \in [-\pi, \pi].$$

The above implies that $\|f - g\|_\infty = \sqrt{2}$.

(2) The set $\overline{\mathbb{B}}_r(f)$ is closed (as a closed ball) and the set $\{g\}$ is compact, since it is a singleton set. By the geometric Hahn-Banach theorem, for any $r < \sqrt{2}$, there exists an affine hyperplane of the form $\{\phi = a\}$ for sone $\phi \in \big(B([-\pi, \pi], \mathbb{R})\big)^*$ and some $a \in \mathbb{R}$ that separates the sets strictly. $\qquad\square$

Solution to Exercise 11.14 Consider the spaces ℓ^∞, c as given and the shift operator $S : \ell^\infty \longrightarrow \ell^\infty$. Consider the subspace

$$(S - I)\ell^\infty := \Big\{ Sx - x \, : \, x \in \ell^\infty \Big\}.$$

Then, for any $(x_2 - x_1, \ldots, x_{n+1} - x_n, \ldots) \in (S-I)\ell^\infty$, and any $c \in \mathbb{R}$, we have that

$$\big(x_2 - x_1, \ldots, x_{n+1} - x_n, \ldots\big) = (c, \ldots, c, \ldots)$$

if and only $c = 0$, because the above equality implies $(x_1, x_2, \ldots) = (c, 2c, \ldots)$ and $(c, 2c, \ldots)$ must be in ℓ^∞. Hence, the vector spaces $((S-I)\ell^\infty)$ and $\mathrm{span}[\{(1, 1, \ldots)\}]$ intersect only at $\{0\} \subseteq \ell^\infty$. We define the linear functional

$$\phi \; : \quad X := \big((S - I)\ell^\infty\big) \bigoplus \mathrm{span}[\{(1, 1, \ldots)\}] \longrightarrow \mathbb{R},$$

by taking

$$\phi\Big((x_2, x_2, \ldots) + (c, c, \ldots)\Big) := c,$$

Then, it can be easily checked that $\|\phi\|_{X^*} = 1$. By the Hahn-Banach Theorem, there exists a linear functional $\Phi(:= \mathrm{Lim}) \in (\ell^\infty)^*$ such that $\Phi|_X = \phi$ and $\|\Phi\|_{X^*} = 1$, thus showing (1). By definition, $\phi \equiv 0$ on $(S - I)\ell^\infty$ and $\phi \equiv I$ on $\mathrm{span}[\{(1, 1, \ldots)\}]$. Hence

$$\Phi(Sx) - \Phi x = \Phi(Sx - x) = \phi(Sx - x) = 0,$$

for all $x \in \ell^\infty$, therefore showing (2) and also

$$\Phi(c, c, \ldots) = c,$$

showing (3) as well. Let now $x = (x_1, x_2, \ldots) \in \ell^\infty$ be such that $x_n \geq 0$ for all $n \in \mathbb{N}$. Set also $t := \|x\|_{\ell^\infty}$ and note that $0 \leq x_n \leq t$ for all $n \in \mathbb{N}$. Then, we have

$$t - \Phi(x) = \Phi\Big((t, t, \ldots) - (x_1, x_2, \ldots)\Big) \leq \Big\|(t, t, \ldots) - (x_1, x_2, \ldots)\Big\| \leq t$$

which shows that $\Phi(x) \geq 0$, establishing (4). Finally, fix $x = (x_1, x_2, \ldots) \in \ell^\infty$ and consider $s < t$ in $\mathbb{R}$ such that

$$s < \liminf_{n \to \infty} x_n, \quad \limsup_{n \to \infty} x_n < t.$$

Then, there exists $N \in \mathbb{N}$ large enough such that $s < x_n < t$, for all $n \geq N$. Hence, by symbolising the N-th iterate $S \circ \cdots \circ S$ of the shift S as S^N, we have that the sequences

$$y := S^N x - (s, s, \ldots) = (x_N - s, x_{N+1} - s, \ldots),$$
$$z := (t, t, \ldots) - S^N x = (t - x_N, t - x_{N+1}, \ldots),$$

have non-negative terms. Hence, by part (4) we have $\Phi(y), \Phi(z) \geq 0$, which implies

$$0 \leq \Phi(y) = \Phi(S^N x) - \Phi(s(1, 1, \ldots)) = \Phi(x) - s,$$
$$0 \leq \Phi(z) = \Phi(t(1, 1, \ldots)) - \Phi(S^N x) = t - \Phi(x).$$

Hence, $s < \Phi(x) < t$. This establishes both (5) and (6). $\qquad\square$

15.12 Solutions to the exercises in Chapter 12

Solution to Exercise 12.1. Yes, F is weakly LSC. To see this, note that each monomial $p_k(t) := a_{2k}t^{2k}$ is a convex, continuous and strictly increasing function on $[0, \infty)$, for any $k = 0, 1, \ldots, N$. It can be easily verified that the composition $p_k \circ \| \cdot \|$ is also convex and continuous on X, since $\| \cdot \|$ itself is convex and continuous: since

$$\|tx + (1 - t)y\| \le t\|x\| + (1 - t)\|y\|$$

for any $x, y \in X$ and $t \in [0, 1]$, we have

$$a_{2k}\Big(\|tx + (1 - t)y\|\Big)^{2k} \le a_{2k}\Big(t\|x\| + (1 - t)\|y\|\Big)^{2k}$$
$$\le t\Big(a_{2k}\|x\|^{2k}\Big) + (1 - t)\Big(a_{2k}\|t\|^{2k}\Big).$$

Further, the sum of of finitely many convex continuous functions is also convex and continuous, as a direct consequence of the definitions. By a theorem in the main text, the function F is weakly LSC, since it is convex and strongly continuous. □

Solution to Exercise 12.2. Given a sequence $\{x_m\}_1^\infty$ in a Banach space $(X, \| \cdot \|)$, we need to construct a *separable* Banach space Y such that $\{x_m\}_1^\infty \subseteq Y \subseteq X$. We set

$$Y := \overline{\mathrm{span}\,[\{x_1, x_2, \ldots\}]}.$$

i.e. Y is the norm closure of the vector space whose elements are (finite) linear combinations of the form $\sum_{a \in A} \lambda_a x_{i_a}$, $\lambda_a \in \mathbb{R}$, $x_{i_a} \in (x_k)_1^\infty$ and $\sharp(A) < \infty$ ($\sharp$ is the counting measure, so this condition says the set A is finite). We note that Y contains a countable dense set, i.e. the set of linear combinations with rational coefficients:

$$Y_0 := \mathrm{span}_{\mathbb{Q}}[\{x_1, x_2, \ldots\}] = \left\{\sum_{a \in A} \lambda_a x_{i_a} \,\Big|\, \lambda_a \in \mathbb{Q},\ x_{i_a} \in (x_k)_1^\infty,\ \sharp(A) < \infty\right\}.$$

The set Y_0 is countable because it equals a countable union of countable sets:

$$Y_0 = \bigcup_{k \in \mathbb{N}} Y_0^k, \quad Y_0^k := \mathrm{span}_{\mathbb{Q}}[\{x_1, \ldots, x_k\}] \cong \mathbb{Q}^k. \quad □$$

Solution to Exercise 12.3. Let $x, y \in X$ be two linearly independent vectors of a Banach space $(X, \| \cdot \|)$ with $\|x\| = \|y\| = 1$. The given vector space

$$Y := \mathrm{span}[\{x, y\}] = \{z \in X : z = ax + by,\ a, b \in \mathbb{R}\}.$$

is isomorphic to the plane $\mathbb{R}^2$, via the map which sends the basis elements $\{x, y\} \subseteq Y$ to the standard basis of $\mathbb{R}^2$ (and extended by linearity), that is $x \mapsto (1, 0)$ and $y \mapsto (0, 1)$. Hence, since the space Y is two-dimensional, its weak topology coincides with its norm topology. □

Solution to Exercise 12.4. Let $(X, \mathcal{T})$ be a compact topological space and consider the Banach space $C(X) = C(X, \mathbb{R})$, endowed with the sup-norm

$$\|f\|_\infty = \sup_{x \in X} |f(x)|.$$

If $\Phi : [0, \infty) \longrightarrow [0, \infty)$ is a strictly increasing continuous function, then Φ is invertible with increasing continuous inverse $\Phi^{-1} : [0, \infty) \longrightarrow [0, \infty)$. The central point of the argument below is the Real Analysis fact that Φ commutes with suprema and infima:

$$\sup_{x \in X} \Phi(I(x)) = \Phi\left(\sup_{x \in X} I(x) \right),$$

$$\inf_{x \in X} \Phi(I(x)) = \Phi\left(\inf_{x \in X} I(x) \right),$$

for **any** set $X \neq \emptyset$ and any function $I : X \longrightarrow [0, \infty)$. (This can be verified directly from the definition of suprema/infima as being the least upper bounds/greatest lower bounds, together with the observation that

$$\sup_{x \in X} I(x) = \sup I(X). \ )$$

Hence, for the function

$$E \ : \ C(X) \longrightarrow \mathbb{R}, \quad E(f) = \sup_{x \in X} \Phi(|f(x)|).$$

we have

$$E(f) = \sup_{x \in X} \Phi(|f(x)|) = \Phi\left(\sup_{x \in X} |f(x)| \right).$$

The desired conclusion, which is

$$E(f) \leq \liminf_{m \to \infty} E(f_m),$$

when $f_m \longrightarrow f$ as $m \to \infty$ in $C(K)$, can be written equivalently as

$$\Phi\left(\sup_{x \in X} |f(x)| \right) \leq \Phi\left(\liminf_{m \to \infty} \sup_{x \in X} |f_m(x)| \right),$$

and by applying the strictly increasing function Φ^{-1}, the above is equivalent to

$$\sup_{x \in X} |f(x)| \leq \liminf_{m \to \infty} \sup_{x \in X} |f_m(x)|.$$

The latter is merely the statement of sequential weak LSC of the norm of the space $C(K)$, which is true. $\qquad\Box$

Solution to Exercise 12.5. The solution is identical to that of Exercise 12.4, the only difference being that one has to replace the supremum norm

$$\|f\|_\infty = \sup_{x \in X} |f(x)|$$

by its measure-theoretic sibling, the essential supremum norm

$$\|f\|_{L^\infty(X, \mu, \mathbb{F})} = \operatorname*{ess\,sup}_{x \in X} |f(x)| = \inf \left\{ a \geq 0 : \mu(\{|f| > a\}) = 0 \right\}. \quad \Box$$

Solution to Exercise 12.6. In general, the function E_p is not even defined as an $\mathbb{R}$-valued (strongly continuous) function on $L^p(X, \mu, \mathbb{F})$ for $p < \infty$. Indeed, it may well happen that $|f|^p \in L^1(X, \mu, \mathbb{F})$, but $\Phi(|f|) \notin L^p(X, \mu, \mathbb{F})$. This is always true for infinitely-many functions f, unless Φ has linear growth at infinity, that is if it satisfies some estimate of the type

$$\Phi(t) \leq C_1 t + C_2,$$

for some $C_1, C_2 > 0$ and all $t > 0$. $\qquad\square$

Solution to Exercise 12.7. The sufficiency direction follows from the results in the main text. Conversely, fix $\varepsilon > 0$ and approximate any fixed $g \in X^*$ by f in the given set in a way that $\|f - g\|_* < \varepsilon$. Then, since $g(x_n) \longrightarrow g(x)$ as $n \to \infty$, we have

$$\left| f(x_n) - f(x) \right| \leq \left| f(x_n) - g(x_n) \right| + \left| g(x_n) - g(x) \right| + \left| g(x) - f(x) \right|$$

which gives

$$\limsup_{n \to \infty} \left| f(x_n) - f(x) \right| \leq \varepsilon \sup_{n \in \mathbb{N}} \|x_n\| + 0 + \varepsilon \|x\|$$

and the conclusion follows by letting $\varepsilon \to 0$. $\qquad\square$

Solution to Exercise 12.8. Since $(x_n)_1^\infty$ is Cauchy in X, there exists $x \in X$ such that $x_n \longrightarrow x$ as $n \to \infty$. Suppose for the sake of contradiction that $x \neq 0$. Then, since strong convergence implies weak convergence, it follows that $x_n \longrightarrow 0$ as $n \to \infty$. Since by assumption we have $x_n \longrightarrow 0$, it follows that $x = 0$, which is a contradiction. $\qquad\square$

Solution to Exercise 12.9. Consider a weakly open neighbourhood subbasis element of the origin $0 \in X$ of the form

$$\mathcal{U} = \left\{ x \in X \: : \: |f_k(x)| < \varepsilon, \ k = 1, \ldots, n \right\}$$

for some $\varepsilon > 0$ and $f_1, \ldots, f_n \in X^*$. Since each of the inverse images $f_k^{-1}(\{0\})$ are hyperplanes (i.e. have codimension 1), it follows that the intersection $\bigcap_{k=1}^n f_k^{-1}(\{0\})$ has codimension equal to n. Since X is an infinite-dimensional normed space, it follows that necessarily there exists a point $x \in \mathbb{S}^X \bigcap \mathcal{U} \neq \emptyset$. $\qquad\square$

Solution to Exercise 12.10. The set

$$S := \left\{ e_i \: : \: i \in \mathbb{N} \right\} \bigcup \{0\} \subseteq \ell^\infty$$

where $e_i = (0, \ldots, 0, 1, 0, \ldots)$ with 1 in the ith position, is not norm compact. Indeed, $\|e_i\|_{\ell^\infty} = 1$ for all $i \in \mathbb{N}$, whilst $\|0\|_{\ell^\infty} = 0$. Hence, the sequence $(e_i)_1^\infty \subseteq S$ has no norm convergent subsequence because, for $i < j$,

$$\|e_i - e_j\|_{\ell^\infty} = \sup_{k \in \mathbb{N}} \left| e_{ik} - e_{jk} \right| = \sup_{k \in \mathbb{N}} \left| (\ldots, 0, \underbrace{1}_{i\text{th place}}, 0, \ldots) - (\ldots, 0, \underbrace{1}_{j\text{th place}}, 0, \ldots) \right|$$

and hence $\|e_i - e_j\|_{\ell^\infty} = 1$. However, we have $e_i \longrightarrow 0$ as $i \to \infty$. Indeed, for any $u \in \ell^1$, we have that

$$\sum_{k=1}^\infty u_k \, e_{ik} = u_i \longrightarrow 0 \quad \text{as } i \to \infty,$$

because $\|u\|_{\ell^1} = \sum_1^\infty |u_i| < \infty$. Hence, the set S is weakly sequentially compact because any sequence has a weakly converging subsequence. $\qquad\square$

Solution to Exercise 12.11. Let $\mathcal{U}$ be the weakly open neighbourhood subbasis element of the origin generated by vectors $x^1, \ldots, x^n \in \ell^2$ and $\varepsilon > 0$. We define $z = (z_1, z_2, \ldots) \in \ell^2$, whose ith component is given by

$$z_i := \sum_{k=1}^n |x_i^k| = \sum_{k=1}^n \left|\langle e_i, x^k \rangle_{\ell^2}\right|, \quad i \in \mathbb{N},$$

where for each $k \in \{1, \ldots, n\}$, we have $x^k = (x_1^k, x_2^k, \ldots) \in \ell^2$. Then, for an infinite number of indices i we have $|z_i|^2 < \varepsilon/i$, or else $z \notin \ell^2$. Hence, for an infinite number of indices i we have

$$\left|\langle \sqrt{i}e_i, x^k \rangle_{\ell^2}\right|^2 < \varepsilon \quad \text{for } k = 1, \ldots, n.$$

It follows that

$$\mathcal{U} \cap \left\{ \sqrt{n}e_n : n \in \mathbb{N} \right\} \neq \emptyset.$$

However, no subsequence satisfies $\sqrt{i}e_i \longrightarrow 0$ along the sequence of indices $(i_k)_1^\infty$, because $\|\sqrt{i}e_i\|_{\ell^2} = \sqrt{i} \longrightarrow \infty$, as $i \to \infty$, and we know that weakly convergent sequences have to be strongly bounded. In conclusion, the weak topology is not metrisable since this would contradict the sequential characterisation of the closure in metric spaces. $\qquad\square$

Solution to Exercise 12.12. By replacing x_n by $x_n - x$, we may assume that $x_n \longrightarrow 0$ as $n \to \infty$. By the weak convergence assumption, we have $\|x_n\| \leq M$ for some $M > 0$ and all $n \in \mathbb{N}$. Select $n_1 := 1$ and, assuming $\{x_{n_1}, \ldots, x_{n_k}\}$ have been selected, we select $x_{n_{k+1}}$ such that

$$\langle x_{n_j}, x_{n_{k+1}} \rangle_{\ell^2} \leq \frac{1}{k} \quad \text{for all } j \leq k.$$

Then, we have

$$\left\| \frac{x_{n_1} + \cdots + x_{n_{k+1}}}{k+1} \right\|^2 \leq \frac{(k+1)M^2 + 2 \cdot 1 + 2 \cdot \frac{2}{2} + \cdots + \frac{2k}{k}}{(k+1)^2} \longrightarrow 0,$$

as $k \to \infty$. The conclusion ensues. $\qquad\square$

Solution to Exercise 12.13. Suppose for the sake of contradiction that $(\mathcal{U}_n)_1^\infty$ were a countable neighbourhood basis of the origin of X for the weak topology. Without loss of generality, we may assume that the terms of the sequence $(\mathcal{U}_n)_1^\infty$ have the form of subbasis elements, that is for each fixed $n \in \mathbb{N}$ there exists $N_n \in \mathbb{N}$ such that

$$\mathcal{U}_n = \bigcap_{i=1}^{N_n} \left\{ x \in X : \left| f_{i,n}(x) \right| < \varepsilon_n \right\}$$

for a certain sequence of number $(\varepsilon_n)_1^\infty \subseteq (0, \infty)$ and certain functionals

$$\left\{ f_{i,n} : i = 1, \ldots, N_n \right\} \subseteq X^*.$$

By the hint, there exist $f \in X^*$ which does not lie in the linear span of the sequence

$$\bigcup_{n=1}^{\infty} \{f_{i,n} : i = 1, \ldots, N_n\} = \{f_{i,n} : i = 1, \ldots, N_n, n \in \mathbb{N}\}.$$

Then, the weakly open neighbourhood of the origin given by $\mathcal{U} := f^{-1}((-1, 1))$ does not contain any of the $\mathcal{U}_n$'s. Indeed, the latter would mean that f is a linear combination of elements of the sequence. $\qquad\square$

15.13 Solutions to the exercises in Chapter 13

Solution to Exercise 13.1. Consider a Hilbert space $(X, \langle \cdot, \cdot \rangle)$ and let $x, y \in X$ with $\|x\|, \|y\| \leq 1$ and $0 < \varepsilon < \sqrt{2}$. By the Parallelogram Identity, we have

$$
\begin{aligned}
\left\| \frac{x+y}{2} \right\|^2 &= \frac{\|x\|^2 + \|y\|^2}{4} - \left\| \frac{x-y}{2} \right\|^2 \\
&\leq \frac{1}{2} - \frac{\varepsilon^2}{4} \\
&= 1 - \frac{\varepsilon^2 + 2}{4} \\
&=: 1 - \delta(\varepsilon).
\end{aligned}
$$

Hence, the norm which is induced by the inner product is uniformly convex. Suppose now that $x_n \longrightarrow x$ and $\|x_n\| \longrightarrow \|x\|$ as $n \to \infty$. The desired conclusion that $x_n \longrightarrow x$ as $n \to \infty$ follows from the result which allows to strengthen weak to strong convergence. $\qquad\square$

Solution to Exercise 13.2.
(ii) $\implies$ (i): The weak* topology on X makes all functionals $\hat{x} \in \hat{X} \subseteq X^{**}$ continuous. Hence $f_m \longrightarrow f$ with respect to $\mathcal{T}^{w^*}$ implies $f_m(x) = \hat{x}(f_m) \longrightarrow \hat{x}(f) = f(x)$, as $m \to \infty$.

(i) $\implies$ (ii): Let $\mathcal{U}_x \in \mathcal{T}^{w^*}$ be a weakly* open neighbourhood of $f \in X^*$. Then, $\mathcal{U}_0 := \mathcal{U}_f - f \in \mathcal{T}^{w^*}$ is a weakly open neighbourhood of the origin $0 \in X^*$. Hence, there exists an $\varepsilon > 0$ and $x_1, \ldots, x_N \in X$ such that

$$\mathcal{U}_0 \supseteq \bigcap_{k=1}^{N} \hat{x}_k^{-1}((-\varepsilon, \varepsilon)).$$

Since $f_m \overset{*}{\longrightarrow} f$, we have $(f_m - f)(x) \longrightarrow 0$ as $m \to \infty$ for any $x \in X$. Since $x_1, \ldots, x_N$ are finitely many, we have

$$\max_{k=1,\ldots,N} \left| (f_m - f)(x_k) \right| \longrightarrow 0$$

as $m \to \infty$. By choosing m_0 large, we have, for $m \geq m_0$,

$$\left| (f_m - f)(x_k) \right| \leq \varepsilon, \quad k = 1, \ldots, N$$

and hence we obtain $f_m - f \in \mathcal{U}_0$, namely $f_m \in \mathcal{U}_f$, as desired. □

Solution to Exercise 13.3.
(i) Suppose that $f_m \xrightarrow{*} f$ in X^*, as $m \to \infty$. By the Banach-Steinhaus theorem (Uniform Boundedness principle), it suffices to show that for each $x \in X$, the set

$$\{f_m(x) : m \in \mathbb{N}\} \subseteq \mathbb{R}$$

is bounded. This however follows directly from the fact that $f_m(x) \longrightarrow f(x)$ as $m \to \infty$ because convergent sequences are bounded.

(ii) Suppose again that $f_m \xrightarrow{*} f$ in X^*, as $m \to \infty$. We have

$$|f(x)| \leq |(f_m - f)(x)| + |f_m(x)|$$
$$\leq |(f_m - f)(x)| + \|f_m\|_* \|x\|$$

and hence, in the limit

$$|f(x)| \leq \liminf_{m \to \infty} \left(|(f_m - f)(x)| + \|f_m\|_* \|x\| \right) \leq \|x\| \liminf_{m \to \infty} \|f_m\|_*.$$

Hence, by the definition of the dual norm,

$$\|f\|_* = \sup_{x \in \mathbb{B}^X} |f(x)| \leq \sup_{f \in \mathbb{B}^{X^*}} \left(\|x\| \liminf_{m \to \infty} \|f_m\|_* \right) \leq \liminf_{m \to \infty} \|f_m\|_*.$$

Hence, the norm of X^* is indeed sequentially weakly* LSC.

(iii) Suppose that $x_m \longrightarrow x$ strongly in X and $f_m \xrightarrow{*} f$ weakly* in X^*, both as $m \to \infty$. By the triangle inequality, we have

$$\left| f_m(x_m) - f(x) \right| \leq \|f_m\|_* \|x_m - x\| + \left| (f_m - f)(x) \right|.$$

Note that $\left| (f_m - f)(x) \right| \longrightarrow 0$ as $m \to \infty$ and by item (i) we have

$$\sup_{m \in \mathbb{N}} \|f_m\|_* < \infty.$$

Also, we have $\|x_m - x\| \longrightarrow 0$, as $m \to \infty$. In conclusion, we see that

$$\limsup_{m \to \infty} \left| f_m(x_m) - f(x) \right| \leq \left(\sup_{m \in \mathbb{N}} \|f_m\|_* \right) \limsup_{m \to \infty} \|x_m - x\|$$
$$+ \limsup_{m \to \infty} \left| (f_m - f)(x) \right|$$
$$= 0.$$

Thus, $f_m(x_m) \longrightarrow f(x)$, as $m \to \infty$. □

Solution to Exercise 13.4. Obviously, (1) implies (2). Conversely, assume (2) and fix $g \in X^*$ and $\varepsilon > 0$. By the density of D in X^*, there exists $f \in D$ such that $\|f - g\|_* < \varepsilon$. Further, by the convergence assumption, for the selected $f \in D$ we have $|f(x_m) - f(x)| < \varepsilon$ for all $m \in \mathbb{N}$ greater or equal to some $m(\varepsilon) \in \mathbb{N}$. Hence, we have

$$|g(x_m) - g(x)| \leq |g(x_m) - f(x_m)| + |f(x_m) - f(x)| + |g(x) - f(x)|$$
$$\leq \|g - f\|_* \|x_m\| + |f(x_m) - f(x)| + \|g - f\|_* \|x\|$$
$$\leq \|g - f\|_* M + \varepsilon + \|g - f\|_* M$$
$$\leq \varepsilon(2M + 1),$$

for all $m \geq m(\varepsilon)$. If follows that $x_m \longrightarrow x$ as $m \to \infty$, in the space X. $\qquad\square$

Solution to Exercise 13.5. The solution is nearly identical to that of Exercise 13.4, but we give the details anyway. Obviously, (1) implies (2). Conversely, assume (2) and fix $x \in X$ and $\varepsilon > 0$. By the density of D in X, there exists $z \in D$ such that $\|x - z\| < \varepsilon$. Further, by the convergence assumption, for the selected $z \in D$ we have $|f_m(z) - f(z)| < \varepsilon$ for all $m \in \mathbb{N}$ greater or equal to some $m(\varepsilon) \in \mathbb{N}$. Hence,

$$
\begin{aligned}
\left| f_m(x) - f(x) \right| &\leq \left| f_m(x) - f_m(z) \right| + \left| f_m(z) - f(z) \right| + \left| f(z) - f(x) \right| \\
&\leq \|f_m\|_* \|x - z\| + \left| f_m(z) - f(z) \right| + \|f\|_* \|x - z\| \\
&\leq M\|x - z\| + \varepsilon + M\|x - z\| \\
&\leq \varepsilon(2M + 1),
\end{aligned}
$$

for all $m \geq m(\varepsilon)$. If follows that $f_m \overset{*}{\longrightarrow} f$ as $m \to \infty$, in the space X^*. $\qquad\square$

Solution to Exercise 13.6. The required example is given by the space $c_0 \subseteq \ell^\infty$ of eventually vanishing sequences (endowed with the ℓ^∞ norm) and by considering the sequence $(e_n)_1^\infty \subseteq \ell^1 \subseteq (\ell^\infty)^*$, which is the standard sequence by $e_n = (0, \dots, 0, 1, 0, \dots)$ with 1 in the nth position. We leave the verification of the claim to the reader. $\qquad\square$

Solution to Exercise 13.7. Suppose for the sake of contradiction that $(\mathcal{U}_n)_1^\infty$ were a countable neighbourhood basis of the origin of X^* for the weak* topology. Without loss of generality, we may assume that the terms of the sequence $(\mathcal{U}_n)_1^\infty$ have the form of subbasis elements, that is for any fixed $n \in \mathbb{N}$ there exists $N_n \in \mathbb{N}$ such that

$$
\mathcal{U}_n = \left\{ f \in X^* : \left| f(x_{i,n}) \right| < \varepsilon_n, \ i = 1, \dots, N_n \right\}
$$

for a certain sequence $(\varepsilon_n)_1^\infty \subseteq (0, \infty)$ and certain points

$$
\left\{ x_{i,n} : i = 1, \dots, N_n \right\} \subseteq X.
$$

It follows that there exist $x \in X$ which does not lie in the linear span of the sequence

$$
\left\{ x_{i,n} : i = 1, \dots, N_n, \ n \in \mathbb{N} \right\}.
$$

Then, the weakly* open neighbourhood of the origin given by $\mathcal{U} := (\hat{x})^{-1}((-1, 1))$ does not contain any of the $\mathcal{U}_n$'s. Indeed, the latter would mean that x is a linear combination of elements of the sequence. $\qquad\square$

Solution to Exercise 13.8. Recall that $(c_0)^* = \ell^1$ and also that $(c_0)^{**} = \ell^\infty$. For any $x \in c_0$, we have

$$
\sum_{i=1}^\infty e_{k_i} x_i = x_k \longrightarrow 0 \quad \text{as } k \to \infty
$$

and hence $e_i \overset{*}{\longrightarrow} 0$ as $i \to \infty$. However, $e_i \not\to 0$ as $i \to \infty$ because for $x := (1, 1, \dots) \in \ell^\infty$, we have

$$
\sum_{i=1}^\infty e_{k_i} x_i = x_k = 1 \not\to 0 \quad \text{as } k \to \infty. \quad \square
$$

Solution to Exercise 13.9. Let us recall that $(c_0)^* = \ell^1$ and $(c_0)^{**} = \ell^\infty$. It is obvious that f is norm continuous since

$$|f(x)| = \left| \sum_{i=1}^\infty x_i \right| \leq \|x\|.$$

Consider the standard basis $(e_i)_1^\infty$ with $e_i = (0, \ldots, 1, 0, \ldots)$ (having 1 in the ith position). Then, by Exercise 13.8 we have that $e_i \overset{*}{\longrightarrow} 0$ as $i \to \infty$. However, $f(e_i) = 1 \nrightarrow 0$ as $i \to \infty$. $\qquad\square$

Solution to Exercise 13.10. Fix $x_0 \in X$ with $\|x_0\| \leq 1$ and consider the subbasis element of the weak topology of the form

$$\mathcal{U} = \Big\{ x \in X \; : \; |f_i(x - x_0)| < \varepsilon, \; i = 1, \ldots, n \Big\}$$

for some $\varepsilon > 0$ and $f_1, \ldots, f_n \in X^*$. Select a point $x \in X \setminus \{0\}$ such that

$$x \in \bigcap_{i=1}^n f_i^{-1}(\{0\})$$

and note that the continuous function $\|\cdot\| : X \longrightarrow \mathbb{R}$ restricted to the affine line

$$L := \Big\{ x_0 + tx \; : \; t \in \mathbb{R} \Big\} \subseteq X$$

necessarily attains the value 1 somewhere on the line L (why?). This establishes that $\mathcal{U} \cap \mathbb{B}^X \neq \emptyset$, showing the desired weak*-density. To see that the sphere is a weakly* G_δ set, we set

$$\mathcal{U}_n := \Big\{ x \in \mathbb{B}^X \; : \; \|x\| > 1 - \frac{1}{n} \Big\}.$$

Then, each $\mathcal{U}_n$ is weakly* open and also

$$\mathbb{S}^X = \bigcap_{n=1}^\infty \mathcal{U}_n. \quad\square$$

15.14   Solutions to the exercises in Chapter 14

Solution to Exercise 14.1. First note that, in the finite measure space $(X, \mathfrak{M}, \mu)$, the functional

$$E \; : \; L^\infty(X, \mu, \mathbb{R}^d) \longrightarrow \mathbb{R}, \qquad E(f) := \int_X \Phi(f(x)) \, d\mu(x),$$

is well defined when $\Phi : \mathbb{R}^d \longrightarrow \mathbb{R}$ a convex continuously differentiable function, $d \in \mathbb{N}$. Indeed, if $f \in L^\infty(X, \mu, \mathbb{R}^d)$, then the composition $\Phi(f) = \Phi \circ f$ is also μ-measurable and also it is essentially bounded, since one can easily verify from the definition of the essential supremum that

$$|\Phi(f(x))| \leq \mu\text{-ess}\sup_X |\Phi(f)| \leq \sup_{\overline{\mathbb{B}}_R(0)} |\Phi|$$

for μ-a.e. $x \in X$, where

$$R := \mu\text{-ess} \sup_X |f|.$$

Hence, by Hölder's inequality and the finiteness of $\mu(X)$, it follows that $\Phi(f) \in L^1(X, \mu, \mathbb{R})$. Next, since Φ is convex, for any $t \in (0,1]$ and any $p, q \in \mathbb{R}^d$, we have

$$\Phi(tq + (1-t)p) \leq t\Phi(q) + (1-t)\Phi(p).$$

Equivalently,

$$\frac{\Phi(p + t(p-q)) - \Phi(p)}{t} \leq \Phi(q) - \Phi(p)$$

and by letting $t \to 0^+$, since Φ is differentiable the left hand side converges to the directional derivative of Φ at p is the direction of $q - p$:

$$D\Phi(p) \cdot (q - p) \leq \Phi(q) - \Phi(p).$$

Suppose now that $f_m \xrightarrow{*} f$ in $L^\infty(X, \mu, \mathbb{R}^d)$ as $m \to \infty$. By the above inequality, we have

$$D\Phi(f) \cdot (f_m - f) \leq \Phi(f_m) - \Phi(f),$$

μ-a.e. on X. Note that the vector-valued function $D\Phi(f) : X \longrightarrow \mathbb{R}^d$ is actually in $L^1(X, \mu, \mathbb{R}^d)$, as a consequence of Hölder's inequality, the finiteness of $\mu(X)$ and the continuity of $D\Phi$, by arguing as in the beginning of the solution. By integrating the last inequality above, we infer that

$$\int_X D\Phi(f) \cdot (f_m - f) \, d\mu \leq \int_X \Phi(f_m) \, d\mu - \int_X \Phi(f) \, d\mu.$$

Since $f_m \xrightarrow{*} f$ in $L^\infty(X, \mu, \mathbb{R}^d)$ as $m \to \infty$, it follows that

$$\int_X D\Phi(f) \cdot (f_m - f) \, d\mu \longrightarrow 0$$

as $m \to \infty$, because $D\Phi(f)$ is in $L^1(X, \mu, \mathbb{R}^d)$ and of course it is independent of $m \in \mathbb{N}$. Hence, by passing to the limit, we infer that

$$\int_X \Phi(f) \, d\mu \leq \liminf_{m \to \infty} \int_X \Phi(f_m) \, d\mu.$$

Thus, we have obtain

$$E(f) \leq \liminf_{m \to \infty} E(f_m)$$

and as a result E is sequentially weakly* LSC on $L^\infty(X, \mu, \mathbb{R}^d)$. $\qquad \square$

Solution to Exercise 14.2. This exercise is mostly computational, therefore we omit the details. The central point is to use integration by parts and that if $\phi \in C_c^\infty(0, 2\pi)$, then $\phi(0) = \phi(2\pi) = 0$. Recall also the identities

$$\cos(mt) = \left(\frac{\sin(mt)}{m}\right)', \quad \cos^2(mt) - \frac{1}{2} = \frac{\cos(2mt)}{2}.$$

which are needed in order the calculate the integrals. $\qquad \square$

Solution to Exercise 14.3. One direction is obvious, since any characteristic function $\phi := \chi_E$ is in $L^{p'}(\Omega)$ and we know from a theorem is the main text that weakly convergent sequences are strongly bounded. Conversely, suppose that the norm bound holds true and that the integrals over E converge, for any measurable set E of finite measure. By linearity of the integral, for any simple function with representation

$$\phi = \sum_{j=1}^{m} a_j \chi_{E_j}, \quad (a_j)_1^m \subseteq \mathbb{R}, \ (E_j)_1^m \subseteq \Omega \text{ measurable, disjoint, with } |E_j| < \infty,$$

we have the convergence

$$\int_{\Omega} f_m \phi \longrightarrow \int_{\Omega} f\phi$$

as $m \to \infty$. Since simple functions are dense in $L^{p'}(\Omega)$, we may use Exercise 13.4 and the bound on $\|f_m\|_{L^p(\Omega)}$ to infer that

$$\int_{\Omega} f_m \psi \longrightarrow \int_{\Omega} f\psi$$

as $m \to \infty$, for any fixed $\psi \in L^{p'}(\Omega)$. $\qquad\square$

Solution to Exercise 14.4. It is nearly identical to Exercise 14.3 and therefore we omit the details, the only essential difference being that one has to quote Exercise 13.5 instead of Exercise 13.4. $\qquad\square$

Solution to Exercise 14.5. Fix $f \in L^\infty(\mathbb{R}^n)$ and consider for each $\varepsilon > 0$ its mollifier

$$f^\varepsilon := f * \eta^\varepsilon \in C^\infty(\mathbb{R}^n).$$

Then, $f^\varepsilon \in L^\infty(\mathbb{R}^n)$, since for any $x \in \mathbb{R}^n$ we have

$$|f^\varepsilon(x)| = \left| \int_{\mathbb{R}^n} f(x-y)\, \eta^\varepsilon(y)\, dy \right| \le \int_{\mathbb{R}^n} |f(y)|\, \eta^\varepsilon(x-y)\, dy \le \|f\|_{L^\infty(\mathbb{R}^n)},$$

We also know already that

$$f^\varepsilon(x) \longrightarrow f(x) \quad \text{as} \quad \varepsilon \to 0 \quad \text{for a.e. } x \in \mathbb{R}^n.$$

By the Dominated Convergence Theorem and Hölder's inequality, the last estimate implies that for any Lebesgue measurable set $A \subseteq \mathbb{R}^n$ with finite measure, we have

$$f^\varepsilon \longrightarrow f, \quad \text{as } \varepsilon \to 0 \text{ in } L^p(A),$$

for any $1 \le p < \infty$, because the function $\|f\|_{L^\infty(\mathbb{R}^n)}\chi_A$ is in $L^p(A)$ and all f^ε's are dominated by it. We conclude by applying Exercise 13.5. $\qquad\square$

Solution to Exercise 14.6. The first part of the solution is nearly identical to that of Exercise 14.5. To conclude, it suffice to show that we can modify the construction in order to achieve a compactly supported approximation. The trick is to consider an appropriate cut-off of the mollifier. Namely, in the setting of the solution to Exercise 14.5, we set

$$f_\varepsilon := \zeta^\varepsilon \left(f * \eta^\varepsilon \right) \in C_c^\infty(\Omega),$$

where $\zeta^{\varepsilon} \in C_c^{\infty}(\Omega)$ is a compactly supported function such that

$$\zeta^{\varepsilon} \equiv 1, \quad \text{on } \Omega \cap \left\{ x \in \Omega : \text{dist}(x\partial\Omega) > 2\delta(\varepsilon) \right\},$$

$$\zeta^{\varepsilon} \equiv 0, \quad \text{on } \mathbb{R}^n \setminus \left(\Omega \cap \left\{ x \in \Omega : \text{dist}(x\partial\Omega) > \delta(\varepsilon) \right\} \right),$$

$$\|D\zeta^{\varepsilon}\|_{L^{\infty}(\Omega)} \leq \frac{1}{\delta(\varepsilon)},$$

for some $\delta(\varepsilon) > 0$ small, chosen such that the measure of

$$\Omega \cap \left\{ x \in \Omega : \text{dist}(x\partial\Omega) < 2\delta(\varepsilon) \right\}$$

is less than ε. The remaining of the argument follows the exact same lines of the proof of Theorem 9.44. $\square$

Solution to Exercise 14.7. It suffices to notice that for $1 < p < \infty$ the space L^p is reflexive and therefore uniformly convex. We conclude by applying Proposition 13.32 about "strengthening weak to strong convergence". Alternatively, one can prove it directly by following the steps of the proof of Proposition 13.32. $\square$

Solution to Exercise 14.8. Fix $u \in C_c^1(\Omega)$, $x \in \Omega$ and $y \in \partial\Omega$. Since $u = 0$ on $\partial\Omega$, we have that $u(y) = 0$. By the Calculus' fundamental theorem applied to the function $t \mapsto u(tx + (1-t)y)$, the chain rule allows us to infer that

$$\begin{aligned} |u(y) - u(x)| &= \left| \int_0^1 \frac{d}{dt} \left(u(tx + (1-t)y) \right) dt \right| \\ &= \left| \int_0^1 Du(tx + (1-t)y) \cdot (y - x) \, dt \right| \\ &\leq \int_0^1 \left| Du(tx + (1-t)y) \right| |y - x| \, dt \\ &\leq |y - x| \max_{t \in [0,1]} \left| Du(tx + (1-t)y) \right| \\ &\leq \text{diam}(\Omega) \|Du\|_{L^{\infty}(\Omega)} \end{aligned}$$

where in the last step we have used that the straight line segment $[x, y]$ lies in $\overline{\Omega}$, by the convexity of Ω. Since $u(y) = 0$, we conclude by taking the supremum with respect to all $x \in \Omega$. $\square$

Solution to Exercise 14.9. Let $p \in [1, \infty)$ and suppose that $u_m \rightharpoonup u$ in $L^p(\mathbb{R}^n)$ as $m \to \infty$. Fix $\varepsilon > 0$ and consider the standard mollifiers $\{\eta^{\varepsilon} : \varepsilon > 0\} \subseteq C_c^{\infty}(\mathbb{R}^n)$. Since each η^{ε} is compactly supported and smooth, it follows that $\eta^{\varepsilon} \in L^{p'}(\mathbb{R}^n)$, where $p' = p/(p - 1) \in (1, \infty]$ is the conjugate exponent. Fix $x \in \mathbb{R}^n$ and consider the translated function $\eta(\cdot - x)$. Since $\eta(\cdot - x)$ is also in $C_c^{\infty}(\mathbb{R}^n)$ and hence in $L^{p'}(\mathbb{R}^n)$, by the definition of weak convergence and the Riesz Representation Theorem, we have

$$\int_{\mathbb{R}^n} u_m(y) \, \eta^{\varepsilon}(y - x) \, dy \longrightarrow \int_{\mathbb{R}^n} u(y) \, \eta^{\varepsilon}(y - x) \, dy$$

as $m \to \infty$, for any fixed point $x \in \mathbb{R}^n$. Since the integrals above coincide with the convolutions $u_m * \eta^\varepsilon$ and $u * \eta^\varepsilon$ respectively, we indeed see that $u_m * \eta^\varepsilon \longrightarrow u * \eta^\varepsilon$ a.e. on $\mathbb{R}^n$ as $m \to \infty$. If $p = \infty$ and $u_m \overset{*}{\rightharpoonup} u$ in $L^\infty(\mathbb{R}^n)$ as $m \to \infty$, the result is still true by the definition of weak* convergence and the Riesz Representation Theorem. □

Solution to Exercise 14.10. Let $\mu : \mathcal{L}(\mathbb{R}) \longrightarrow [0, \infty]$ be defined by

$$\mu(A) := \lim_{n \to \infty} \frac{\mathcal{L}^1(A \cap (0, n))}{n}$$

where the limit is understood as the Banach limit "Lim" defined in Exercise 11.14.

(1) If $A \in \mathcal{L}(\mathbb{R})$ is such that $\mathcal{L}^1(A) = 0$, then $\mathcal{L}^1(A \cap (0, n)) = 0$ and hence $\mu(A) = 0$. Thus, $\mu << \mathcal{L}^1$.

(2) This is consequence of the linearity of the Banach limit. Consider two sets $A, B \in \mathcal{L}(\mathbb{R})$ with $A \cap B = \emptyset$. Then,

$$\mathcal{L}^1\Big((A \cup B) \bigcap (0, n)\Big) = \mathcal{L}^1(A \cap (0, n)) + \mathcal{L}^1(B \cap (0, n))$$

and hence

$$\mu(A \cup B) = \mu(A) + \mu(B).$$

(3) By setting $A_k := (k - 1, k]$ for each $k \in \mathbb{N}$, we have that

$$\mu(A_k) = \lim_{n \to \infty} \frac{\mathcal{L}^1((k - 1, k) \cap (0, n))}{n} = 0$$

whilst

$$\mu\left(\bigcup_{k=1}^{\infty} A_k\right) = \mu((0, \infty)) = \lim_{n \to \infty} \frac{\mathcal{L}^1((0, n))}{n} = 1 \neq 0 = \sum_{k=1}^{\infty} \mu(A_k).$$

The conclusion follows. □

Bibliography

[1] L. Ambrosio, G. Da Prato, A.C.G. Mennucci, *Introduction to Measure Theory and Integration*, Lecture Notes 10 (Scuola Normale Superiore di Pisa), Birkhaüser, 2011.

[2] R. Bartle, *The Elements of Integration*, John Wiley & Sons Inc., 1966.

[3] H. Bauer, *Probability theory and elements of measure theory*, Probability and Mathematical Statistics, Academic Press, 2nd edition, 1982.

[4] H. Brezis, *Functional Analysis, Sobolev Spaces and Partial Differential Equations*, Springer Universitext, 2011.

[5] M. Capinski, E. Kopp, *Measure, Integral and Probability*, Springer Undergraduate Text, 2005.

[6] V. A. Churkin, *A continuous version of the Hausdorff - Banach - Tarski paradox*, Algebra and Logic 49 (1), 91 - 98 (2010).

[7] J. B. Conway, *A Course in Functional Analysis*, Springer, 1985.

[8] C. Costara, D. Popa, *Exercises in Functional Analysis*, Springer, 2003.

[9] B. Dacorogna, *Direct Methods in the Calculus of Variations*, 2nd Edition, Volume 78, Applied Mathematical Sciences, Springer, 2008.

[10] J. Dieudonné, *Foundations of Modern Analysis*, Academic Press, 1960.

[11] N. Dunford, J. T. Schwartz, *Linear Operators* volumes I, II, III, Wiley Interscience, 1972.

[12] R. E. Edwards, *Functional Analysis: Theory and Applications*, Dover Books on Mathematics, 2003.

[13] L. C. Evans, *Partial Differential Equations*, AMS, Graduate Studies in Mathematics Vol. 19.1, Second Edition, 2010.

[14] L. C. Evans, R. Gariepy, *Measure theory and fine properties of functions*, Studies in advanced mathematics, CRC press, Taylor & Francis Group, Revised edition, 2015.

[15] M. Fabian, P. Habala, P. Hajek, V. Montesinos Santalucia, J. Pelant, V. Zizler, *Functional Analysis and Infinite-Dimensional Geometry*, CMS Books in Mathematics, Springer, 2001.

[16] G. B. Folland, *Real Analysis: Modern techniques and their applications*, 2nd edition, Wiley-Interscience, 1999.

[17] L. C. Florescu, C. Godet-Thobie, *Young measures and compactness in metric spaces*, De Gruyter, 2012.

[18] I. Fonseca, G. Leoni, *Modern methods in the Calculus of Variations: L^p spaces*, Springer Monographs in Mathematics, 2007.

[19] D. Gilbarg, N. Trudinger, *Elliptic Partial Differential Equations of Second Order*, Classics in Mathematics, reprint of the 1998 edition, Springer.

[20] P. Halmos, *Measure theory*, Graduate Texts in Mathematics 18, Springer, 2nd edition, 1978.

[21] R. Holmes, *Geometric Functional Analysis and Its Applications*, Springer, 1975.

[22] E. Kasner, J. R. Newman, *Mathematics and the Imagination*, Dover, 2001 (reprint of the 1940 edition).

[23] N. Katzourakis, *An Introduction to viscosity Solutions for Fully Nonlinear PDE with Applications to Calculus of Variations in L^∞*, Springer Briefs in Mathematics, 2015.

[24] B. Kirchheim, *Rigidity and geometry of microstructures*, Issue 16 of Lecture notes, Max-Planck-Institut fr Mathematik in den Naturwissenschaften Leipzig, 2003, 116 pages.

[25] A. N. Kolmogorov, S. V. Fomin, *Introductory Real Analysis*, Dover, revised English edition, 1970.

[26] E. Kreyszig, *Introductory Functional Analysis with Applications*, John Wiley & Sons, revised edition, 1989.

[27] P.D. Lax, *Linear Algebra and Its Applications*, Wiley-Interscience, 2nd edition, 2007.

[28] R. E. Megginson, *An introduction to Banach space theory*, Graduate Texts in Mathematics 183, New York: Springer-Verlag, 1998.

[29] G. H. Moore, *The axiomatisation of linear algebra: 18751940*, Historia Mathematica 22 (3), 262 - 303 (1995).

[30] J. R. Munkres, *Topology*, Pearson's, 2nd edition, 1999.

[31] P. Pedregal, *Parametrized Measures and Variational Principles*, Birkhäuser, 1997.

[32] W. Rudin, *Real and complex analysis*, McGraw-Hill, 3rd edition, 1987.

[33] W. Rudin, *Functional Analysis*, International series in pure and aapplied mathematics, 2nd edition, McGraw Hill, 1991.

[34] B. Rynne, M. A. Youngson, *Linear Functional Analysis*, Springer-Verlag, 2nd edition, 2008.

[35] H. H. Schaefer, *Topological vector spaces*, New York: The MacMillan Company, 1966.

[36] E. Schechter, *Handbook of Analysis and its Foundations*, Academic Press, 1997.

[37] E. M. Stein, R. Shakarchi, *Real analysis: measure theory, integration, and Hilbert spaces*, Princeton lectures in Analysis, 2005.

[38] K. Stromberg, *The Banach - Tarski paradox*, The American Mathematical Monthly 86 (3), 151 - 161 (1979).

[39] W. A. Sutherland, *Introduction to metric and topological spaces*, Oxford University Press, 2nd edition, 2008.

[40] L. Steen, J. A. Seebach Jr., *Counterexamples in Topology*, Springer-Verlag, 2nd edition, 1978.

[41] T. Tao, *Introduction to Measure Theory*, American Mathematical Society, 2011.

[42] S. Wagon, *The Banach-Tarski Paradox*, Cambridge University Press, 1994.

[43] K. Yosida, *Functional Analysis*, Springer, 1965.

Index